Lecture Notes in Computer Science 9313

Commenced Publication in 1973
Founding and Former Series Editors:
Gerhard Goos, Juris Hartmanis, and Jan van Leeuwen

More information about this series at http://www.springer.com/series/7409

Reynold Cheng · Bin Cui
Zhenjie Zhang · Ruichu Cai
Jia Xu (Eds.)

Web Technologies
and Applications

17th Asia-Pacific Web Conference, APWeb 2015
Guangzhou, China, September 18–20, 2015
Proceedings

 Springer

Editors
Reynold Cheng
University of Hong Kong
Hong Kong
China

Ruichu Cai
University of Technology
Guangzhou
China

Bin Cui
Computer Science
Peking University
Beijing
China

Jia Xu
Guangxi University
Guangxi
China

Zhenjie Zhang
Advanced Digital Sciences Center (ADSC)
Singapore
Singapore

ISSN 0302-9743   ISSN 1611-3349   (electronic)
Lecture Notes in Computer Science
ISBN 978-3-319-25254-4   ISBN 978-3-319-25255-1   (eBook)
DOI 10.1007/978-3-319-25255-1

Library of Congress Control Number: 2015950917

LNCS Sublibrary: SL3 – Informations Systems and Applications, incl. Internet/Web, and HCI

Printed on acid-free paper

Springer International Publishing AG Switzerland is part of Springer Science+Business Media
(www.springer.com)

Preface

Welcome to the 17th Asia-Pacific Web Conference (APWeb 2015). APWeb has become a leading conference hosting top researchers as well as industrial players to exchange experiences with edge-cutting technologies for Web-based applications and Web data processing. The prosperity of big data has also propelled the technical advances of computer science on Web technologies, which is reflected by the passionate interest shown by the attendants and speakers at APWeb conferences. For the first time in the history of APWeb, China Telecom, one of the largest telecommunication companies in China, participated as a main sponsor and organizer of this academic conference, and actively contributed to the first data challenge competition in APWeb. Driven by the huge interest from industry, research investigations on web technologies are moving toward practical technologies to support businesses in the real world. Text analytics and recommendation systems were the most popular topics discussed during the research track of the conference, which contained both research and practical experience with the emergence of the Internet economy around the world.

The APWeb conference provides a unique venue for researchers and practitioners from all over the world. The conference successfully attracts submissions from Canada, China, Hong Kong, India, Japan, Luxembourg, and other countries. The Program Committee accepted 66 submissions in the research track, three submissions in the industrial track, and 7 submissions in the demonstration track. All these high-quality works cover a wide spectrum of Web-related data management problems, and provide a thorough overview of the rapid advances in technical solutions.

We would like to thank all authors for their contributions. We would like to express our gratitude to the Program Committee members for their huge effort in reviewing and discussing the submissions. Special thanks to our keynote speakers, Prof. Elisa Bertino, Prof. Wenfei Fan, and Prof. Xindong Wu. We also would like to express our gratitude to the local organizers and sponsors.

We hope the success of APWeb 2015 enhances the research on Web-based data management technologies, consolidates the connections between academia and industry, and creates an exciting exchange of ideas between all participants.

September 2015

Reynold Cheng
Bin Cui
Zhenjie Zhang

Organization

General Co-chairs

Kang Cai — China Telecom Co. Ltd., China
Zhifeng Hao — Guangdong University of Technology, China
Xuemin Lin — University of New South Wales, Australia

Program Committee Co-chairs

Reynold Cheng — The University of Hong Kong, Hong Kong, SAR China
Bin Cui — Peking University, China
Zhenjie Zhang — Advanced Digital Sciences Center, Singapore

Workshop Co-chairs

Xiaokui Xiao — Nanyang Technological University, Singapore
Jian Yin — Sun Yat-Sen University, China

Tutorial Co-chairs

Sai Wu — Zhejiang University, China
Ying Zhang — University of Technology Sydney, Australia

Panel Co-chairs

Zexiong Ma — China Telecom Co. Ltd., China
Jeffrey Yu — The Chinese University of Hong Kong, Hong Kong, SAR China
Zhiming Zhong — China Telecom Co. Ltd., China

Industrial Co-chairs

Hongyang Chao — Sun Yat-Sen University, China
Junjun Hu — China Telecom Co. Ltd., China
Ying Yan — Microsoft Research Asia, China

Demo Co-chairs

Xiaoli Li   Institute for Infocomm Research, Singapore
Xiaojun Xiao   Guangzhou Useease Information Technology
     Co. Ltd., China
Rong Zhang   East China Normal University, China

Big Data Challenge Co-chairs

Kang Chen   China Telecom Co. Ltd., China
Baozhong Wang   China Telecom Co. Ltd., China

Distinguished Lecture Series Chair

Aoying Zhou   East China Normal University, China

Research Students Symposium Co-chairs

Xiaohui Yu   Shandong University, China
Wenjie Zhang   University of New South Wales, Australia

Publication Chairs

Ruichu Cai   Guangdong University of Technology, China
Jia Xu   National University of Defense Technology,
     China

Publicity Co-chairs

Ilaria Bartolini   University of Bologna, Italy
Yin Yang   Hamad bin Khalifa University, Qatar
Bin Zhou   University of Maryland, Baltimore County,
     USA

Local Organization Co-chairs

Xiaohui Guo   China Telecom Co. Ltd., China
Jie Ling   Guangdong University of Technology, China
Lijuan Wang   Guangdong University of Technology, China
Wen Wen   Guangdong University of Technology, China

Steering Committee

Jeffrey Xu Yu (Chair)	The Chinese University of Hong Kong, China
Masaru Kitsuregawa	University of Tokyo, Japan
Jianzhong Li	Harbin Institute of Technology, China
Xuemin Lin	The University of New South Wales, Australia
Kyu-Young Whang	Korea Advanced Institute of Science and Technology, Korea
Ge Yu	Northeastern University, China
Yanchun Zhang	Victoria University, Australia
Aoying Zhou	East China Normal University, China
Xiaofang Zhou	The University of Queensland, Australia

Program Committee

Toshiyuki Amagasa	University of Tsukuba, Japan
Zhifeng Bao	University of Tasmania, Australia
Arnab Bhattacharya	Indian Institute of Technology, Kanpur, India
Sourav Bhowmick	Nanyang Technological University, Singapore
Tru Hoang Cao	Ho Chi Minh City University of Technology, Vietnam
Kang Chen	China Telecom Co. Ltd., China
Rui Chen	Hong Kong Baptist University, Hong Kong, SAR China
Wenliang Chen	Soochow University, China
Yueguo Chen	Renmin University, China
James Cheng	The Chinese University of Hong Kong, Hong Kong, SAR China
Zhihong Chong	Southeast University, China
Bing Tian Dai	Singapore Management University, Singapore
Markus Endres	Univresity of Augsburg, Germany
Yuan Fang	Institute for Infocomm Research, Singapore
Tom Z.J. Fu	Advanced Digital Sciences Center, Singapore
Gabriel Ghinita	University of Massachusetts at Boston, USA
João Bártolo Gomes	Institute for Infocomm Research, Singapore
Jun Gao	Peking University, China
Shenghua Gao	Shanghait Tech University, China
Yu Gu	Northeast University, China
Lifang He	University of Illinois at Chicago, USA
Bao-Quoc Ho	VNU-HCMUS, Vietnam
Liang Hong	Wuhan University, China
Haibo Hu	Hong Kong Baptist University, Hong Kong, SAR China
Zhuolin Jiang	Noah's Ark Lab, Huawei Technologies, Hong Kong, SAR China

Mizuho Iwaihara	Waseda University, Japan
Yoshiharu Ishikawa	Nagoya University, Japan
Yiping Ke	Nanyang Technological University, Singapore
Hady Lauw	Singapore Management University, Singapore
Sangkeun Lee	Oak Ridge National Laboratory, USA
Wookey Lee	Inha University, Korea
Feng Li	Microsoft Research, USA
Guoliang Li	Tsinghua University, China
Chen Lin	Xiamen University, China
Zhiyong Lin	Guangdong Polytechnic Normal University, China
Bo Liu	Guangdong University of Technology, China
Hua Lu	Aalborg University, Denmark
Mihai Lupu	Vienna University of Technology, Austria
Shuai Ma	Beihang University, China
Zakaria Maamar	Zayed University, United Arab Emirates
Alexander Markowetz	Bonn University, Germany
Anirban Mondal	Xerox Research Center India, India
Yang-Sae Moon	Kangwon National University, Korea
Yasuhiko Morimoto	Hiroshima University, Japan
Shinsuke Nakajima	Kyoto Sangyo University, Japan
Hiroaki Ohshima	Kyoto University, Japan
Sanghyun Park	Yonsei University, Korea
Dhaval Patel	IIT-R, India
Stavros Papadopoulos	Intel Labs, USA
Chenxi Qi	China Telecom Co. Ltd., China
Jianzhong Qi	University of Melbourne, Australia
Weining Qian	East China Normal University, China
Krishna Reddy	IIIT Hyderabad, India
Chaofeng Sha	Fudan University, China
Nan Tang	Qatar Computing Research Institute, Qatar
Quan Thanh Tho	Ho Chi Minh City University of Technology, Vietnam
James Thom	RMIT, Australia
Quan Thanh Tho	Ho Chi Minh City University of Technology, Vietnam
Alex Thomo	University of Victoria, Australia
Leong Hou U.	University of Macau, Macau
Taketoshi Ushiama	Kyushu University, Japan
Quang-Hieu V.U.	Khalifa University, United Arab Emirates
Baozhong Wang	China Telecom Co. Ltd., China
Hongzhi Wang	Harbin Institue of Technology, China
Junhu Wang	Griffith University, Australia

Raymond Wong	Hong Kong University of Science and Technology, Hong Kong, SAR China
Xiaochun Yang	Northeast University, China
Xiaowei Yang	South China University of Technology, China
Yunjie Yao	East China Normal University, China
Hongzhi Yin	University of Queensland, Australia
Man Lung Yiu	Hong Kong Polytechnical University, Hong Kong, SAR China
Shanshan Ying	National University of Singapore, Singapore
Haruo Yokota	Tokyo Institute of Technology, Japan
Ting Yu	Qatar Computing Research Institute, Qatar
Ganzhao Yuan	South China University of Technology, China
Dongxiang Zhang	National University of Singapore, Singapore
Meihui Zhang	Singapore University of Technology and Design, Singapore
Rui Zhang	University of Melbourne, Australia
Wen Zhang	Wuhan University, China
Vicent Zheng	Advanced Digital Sciences Center, Singapore
Jia Zhu	South China Normal University, China
Feida Zhu	Singapore Management University, Singapore
Lei Zou	Peking University, China

Industry Committee

Yu Cao	EMC Corporation, China
Rui Li	Tencent, China
Yuan Ni	IBM Research China, China
Shouke Qin	Baidu Research, China
Bin Shao	Microsoft Research Asia, China
WeeHyong	Tok Microsoft SQL Server, USA
Xintian Yang	Google, USA
Jianfeng Yan	Dazhong Dianping, China

Demo Committee

Yang-Sae Moon	Kangwon National University, South Korea
Xiaofeng He	East China Normal University, China
Wei Pan	Northwestern Polytechnical University, China
Sanjay Kumar Madria	Missouri University of Science and Technology, USA
Ming Gao	East China Normal University, China
Kyoung-Sook Kim	National Institute of Advanced Industrial Science and Technology, Japan
Kai (Kevin) Zheng	University of Queensland, Australia
Bin Yang	Aalborg University, Denmark

Contents

Research Track

On the Marriage of SPARQL and Keywords 3
Peng Peng, Lei Zou, and Dongyan Zhao

An Online Inference Algorithm for Labeled Latent Dirichlet
Allocation 17
Qiang Zhou, Heyan Huang, and Xian-Ling Mao

Efficient Buffer Management for PCM-Enhanced Hybrid Memory
Architecture 29
Kaimeng Chen, Peiquan Jin, and Lihua Yue

Efficient Location-Dependent Skyline Queries in Wireless Broadcast
Environments 41
*Yingyuan Xiao, Pengqiang Ai, Hongya Wang, Ching-Hsien Hsu,
and Wenxiang Cui*

Distance and Friendship: A Distance-Based Model for Link Prediction
in Social Networks 55
Yang Zhang and Jun Pang

Multi-Label Emotion Tagging for Online News by Supervised
Topic Model 67
Ying Zhang, Lili Su, Zhifan Yang, Xue Zhao, and Xiaojie Yuan

Distinguishing Specific and Daily Topics 80
Tao Ge, Wenzhe Pei, Baobao Chang, and Zhifang Sui

Matching Reviews to Object Based on 2-Stage CRF 92
*Zhang Yongxin, Li Qingzhong, Wang Dequan, Ding Yanhui,
Liu Congli, and Yan Zhongmin*

Discovering Restricted Regular Expressions with Interleaving 104
Feifei Peng and Haiming Chen

Efficient Algorithms for Distance-Based Representative Skyline
Computation in 2D Space 116
Taotao Cai, Rong-Hua Li, Jeffrey Xu Yu, Rui Mao, and Yadi Cai

Trustworthy Collaborative Filtering through Downweighting Noise and
Redundancy 129
Qiuxiang Dong, Zhi Guan, and Zhong Chen

A Co-ranking Framework to Select Optimal Seed Set for Influence
Maximization in Heterogeneous Network 141
  Yashen Wang, Heyan Huang, Chong Feng, and Xianxiang Yang

Hashtag Sense Induction Based on Co-occurrence Graphs 154
  Mengmeng Wang and Mizuho Iwaihara

Hybrid-LSH for Spatio-Textual Similarity Queries 166
  Mingdong Zhu, Derong Shen, Ling Liu, and Ge Yu

Sleep Quality Evaluation of Active Microblog Users 178
  Kai Wu, Jun Ma, Zhumin Chen, and Pengjie Ren

An Ensemble Matchers Based Rank Aggregation Method for Taxonomy
Matching 190
  *Hailun Lin, Yuanzhuo Wang, Yantao Jia, Jinhua Xiong,
  Peng Zhang, and Xueqi Cheng*

Distributed XML Twig Query Processing Using MapReduce 203
  *Xin Bi, Guoren Wang, Xiangguo Zhao, Zhen Zhang,
  and Shuang Chen*

Sentiment Word Identification with Sentiment Contextual Factors 215
  Jiguang Liang, Xiaofei Zhou, Yue Hu, Li Guo, and Shuo Bai

Large-Scale Graph Classification Based on Evolutionary Computation
with MapReduce.. 227
  Zhanghui Wang, Yuhai Zhao, Guoren Wang, and Yurong Cheng

Multiple Attribute Aware Personalized Ranking 244
  Weiyu Guo, Shu Wu, Liang Wang, and Tieniu Tan

Knowledge Base Completion Using Matrix Factorization 256
  Wenqiang He, Yansong Feng, Lei Zou, and Dongyan Zhao

MATAR: Keywords Enhanced Multi-label Learning for Tag
Recommendation 268
  Licheng Li, Yuan Yao, Feng Xu, and Jian Lu

Reverse Direction-Based Surrounder Queries 280
  Xi Guo, Yoshiharu Ishikawa, Aziguli Wulamu, and Yonghong Xie

Learning to Hash for Recommendation with Tensor Data 292
  Qiyue Yin, Shu Wu, and Liang Wang

Spare Part Demand Prediction Based on Context-Aware Matrix
Factorization 304
  Jianwei Ding, Yingbo Liu, Yuan Cao, Li Zhang, and Jianmin Wang

Location Sensitive Friend Recommendation in Social Network 316
    Xueqin Sui, Zhumin Chen, and Jun Ma

A Compression-Based Filtering Mechanism in Content-Based
Publish/Subscribe System... 328
    Qin Liu, Yiwen Zheng, and Kaile Wang

Sentiment Classification for Chinese Product Reviews Based on
Semantic Relevance of Phrase 340
    Heng Chen, Hai Jin, Pingpeng Yuan, Lei Zhu, and Hang Zhu

Overlapping Schema Summarization Based on Multi-label
Propagation.. 352
    *Man Yu, Chao Wang, Xiangrui Cai, Ying Zhang, Yanlong Wen,
    and Xiaojie Yuan*

Analysis of Subjective City Happiness Index Based on Large Scale
Microblog Data 365
    Kai Wu, Jun Ma, Zhumin Chen, and Pengjie Ren

Tree-Based Metric Learning for Distance Computation
in Data Mining 377
    Ming Yan, Yan Zhang, and Hongzhi Wang

GSCS – Graph Stream Classification with Side Information 389
    *Amit Mandal, Mehedi Hasan, Anna Fariha,
    and Chowdhury Farhan Ahmed*

A Supervised Parameter Estimation Method of LDA 401
    Liu Zhenyan, Meng Dan, Wang Weiping, and Zhang Chunxia

Learning Similarity Functions for Urban Events Detection by Mining
Hotline Phone Records 411
    Pengjie Ren, Peng Liu, Zhumin Chen, Jun Ma, and Xiaomeng Song

Answering Spatial Approximate Keyword Queries in Disks 424
    *Jinbao Wang, Donghua Yang, Yuhong Wei, Hong Gao,
    Jianzhong Li, and Ye Yuan*

Hashing Multi-Instance Data from Bag and Instance Level 437
    *Yao Yang, Xin-Shun Xu, Xiaolin Wang, Shanqing Guo,
    and Lizhen Cui*

A Multi-news Timeline Summarization Algorithm Based
on Aging Theory.. 449
    Jie Chen, Zhendong Niu, and Hongping Fu

Extended Strategies for Document Clustering with
Word Co-occurrences 461
    Yang Wei, Jinmao Wei, and Zhenglu Yang

A Lightweight Evaluation Framework for Table Layouts in MapReduce
Based Query Systems 473
    Feng Zhu, Jie Liu, Lijie Xu, Dan Ye, Jun Wei, and Tao Huang

An Extended Graph-Based Label Propagation Method for Readability
Assessment... 485
    Zhiwei Jiang, Gang Sun, Qing Gu, Lixia Yu, and Daoxu Chen

Simple is Beautiful: An Online Collaborative Filtering Recommendation
Solution with Higher Accuracy 497
    Feng Zhang, Ti Gong, Victor E. Lee, Gansen Zhao, and Guangzhi Qu

Random-Based Algorithm for Efficient Entity Matching............... 509
    *Pingfu Chao, Zhu Gao, Yuming Li, Junhua Fang, Rong Zhang,
    and Aoying Zhou*

A New Similarity Measure Between Semantic Trajectories Based on
Road Networks 522
    *Xia Wu, Yuanyuan Zhu, Shengchao Xiong, Yuwei Peng,
    and Zhiyong Peng*

Effective Citation Recommendation by Unbiased Reference Priority
Recognition 536
    Wen-Yang Lu, Yu-Bin Yang, Xiao-Jiao Mao, and Qi-Hai Zhu

UserGreedy: Exploiting the Activation Set to Solve Influence
Maximization Problem 548
    Wenxin Liang, Chengguang Shen, and Xianchao Zhang

AILabel: A Fast Interval Labeling Approach for Reachability Query on
Very Large Graphs.. 560
    *Feng Shuo, Xie Ning, Shen de-Rong, Li Nuo, Kou Yue,
    and Yu Ge*

Graph-Based Hybrid Recommendation Using Random Walk and Topic
Modeling 573
    Hai-Tao Zheng, Yang-Hui Yan, and Ying-Min Zhou

RDQS: A Relevant and Diverse Query Suggestion Generation
Framework... 586
    Hai-Tao Zheng and Yi-Chi Zhang

Minimizing the Cost to Win Competition in Social Network 598
    *Ziyan Liu, Xiaoguang Hong, Zhaohui Peng, Zhiyong Chen,
    Weibo Wang, and Tianhang Song*

Ad Dissemination Game in Ephemeral Networks 610
    *Lihua Yin, Yunchuan Guo, Yanwei Sun, Junyan Qian,
    and Athanasios Vasilakos*

User Generated Content Oriented Chinese Taxonomy Construction   623
    Jinyang Li, Chengyu Wang, Xiaofeng He, Rong Zhang,
    and Ming Gao

Mining Weighted Frequent Itemsets with the Recency Constraint   635
    Jerry Chun-Wei Lin, Wensheng Gan, Philippe Fournier-Viger,
    and Tzung-Pei Hong

Boosting Explicit Semantic Analysis by Clustering Paragraph Vectors
of Wikipedia Articles   647
    Hai-Tao Zheng and Wenzhen Wu

Research on Semantic Disambiguation in Treebank   658
    Lin Miao, Xueqiang Lv, Yunfang Wu, and Yue Wang

PDMA: A Probabilistic Framework for Diversifying Recommendation
Lists   670
    Yang-Hui Yan, Ying-Min Zhou, and Hai-Tao Zheng

User Behavioral Context-Aware Service Recommendation for
Personalized Mashups in Pervasive Environments   683
    Wei He, Guozhen Ren, Lizhen Cui, and Hui Li

On Coherent Indented Tree Visualization of RDF Graphs   695
    Qingxia Liu, Gong Cheng, and Yuzhong Qu

Online Feature Selection Based on Passive-Aggressive Algorithm with
Retaining Features   707
    Hai-Tao Zheng and Haiyang Zhang

Online Personalized Recommendation Based on Streaming Implicit
User Feedback   720
    Zhisheng Wang, Qi Li, Ye Liu, Wei Liu, and Jian Yin

A Self-learning Rule-Based Approach for Sci-tech Compound Phrase
Entity Recognition   732
    Tingwen Liu, Yang Zhang, Yang Yan, Jinqiao Shi, and Li Guo

Batch Mode Active Learning for Geographical Image Classification   744
    Zengmao Wang, Bo Du, Lefei Zhang, Wenbin Hu, Dacheng Tao,
    and Liangpei Zhang

A Multi-view Retweeting Behaviors Prediction in Social Networks   756
    Bo Jiang, Ying Sha, and Lihong Wang

Probabilistic Frequent Pattern Mining by PUH-Mine   768
    Wenzhu Tong, Carson K. Leung, Dacheng Liu, and Jialiang Yu

A Secure and Efficient Framework for Privacy Preserving Social
Recommendation   781
    Shushu Liu, An Liu, Guanfeng Liu, Zhixu Li, Jiajie Xu,
    Pengpeng Zhao, and Lei Zhao

DistDL: A Distributed Deep Learning Service Schema with GPU
Accelerating   793
    Jianzong Wang and Lianglun Cheng

A Semi-supervised Solution for Cold Start Issue
on Recommender Systems   805
    Zhifeng Hao, Yingchao Cheng, Ruichu Cai, Wen Wen,
    and Lijuan Wang

Industry Track

Hybrid Cloud Deployment of an ERP-Based Student Administration
System   821
    Simon K.S. Cheung

A Benchmark Evaluation of Enterprise Cloud Infrastructure   832
    Yong Chen, Xuanzhong Xie, and Jiayin Wu

A Fast Data Ingestion and Indexing Scheme for Real-Time
Log Analytics   841
    Haoqiong Bian, Yueguo Chen, Xiongpai Qin, and Xiaoyong Du

Demonstration Track

A Fair Data Market System with Data Quality Evaluation and
Repairing Recommendation   855
    Xiaoou Ding, Hongzhi Wang, Dan Zhang, Jianzhong Li,
    and Hong Gao

HouseIn: A Housing Rental Platform with Non-redundant Information
Integrated from Multiple Sources   859
    Jian Zhou, Zhixu Li, Qiang Yang, Jun Jiang, Jia Zhu, An Liu,
    Guanfeng Liu, and Lei Zhao

ONCAPS: An Ontology-Based Car Purchase Guiding System   863
    Jianfeng Du, Jun Zhao, Jiayi Cheng, Qingchao Su,
    and Jiacheng Liang

A Multiple Trust Paths Selection Tool in Contextual Online Social
Networks   867
    Linlin Ma, Guanfeng Liu, Guohao Sun, Lei Li, Zhixu Li, An Liu,
    and Lei Zhao

EPEMS: An Entity Matching System for E-Commerce Products   871
    Lei Gao, Pengpeng Zhao, Victor S. Sheng, Zhixu Li, An Liu,
    Jian Wu, and Zhiming Cui

PPS-POI-Rec: A Privacy Preserving Social Point-of-Interest
Recommender System   875
    Xiao Liu, An Liu, Guanfeng Liu, Zhixu Li, Jiajie Xu,
    Pengpeng Zhao, and Lei Zhao

Incorporating Contextual Information into a Mobile Advertisement
Recommender System   879
    Ke Zhu, Yingyuan Xiao, Pengqiang Ai, Hongya Wang,
    and Ching-Hsien Hsu

Author Index   883

Research Track

On the Marriage of SPARQL and Keywords

Peng Peng, Lei Zou*, and Dongyan Zhao

Peking University, China
{pku09pp,zoulei,zhaodongyang}@pku.edu.cn

Abstract. To maximize the advantages of both SPARQL and keyword search, we introduce a novel paradigm that combines both of them and propose a hybrid query (called an SK query) that integrates SPARQL and keyword search in this paper. In order to answer SK queries efficiently, a structural index is devised and a novel integrated query algorithm is proposed. We evaluate our method in large real RDF graphs and experiments demonstrate both effectiveness and efficiency of our method.

1 Introduction

As more and more knowledge bases become available, the question of how end users to access this body of knowledge becomes of crucial importance. As the de facto standard of a knowledge base, RDF (Resource Description Framework) repository is a collection of triples, denoted as ⟨subject, predicate, object⟩. An RDF dataset can be represented as a graph, where subjects and objects are vertices connected by labeled edges (i.e., predicates). Fig. 1 shows an example RDF graph, which is a part of a well-known knowledge base Yago [11]. The numbers besides the vertices are IDs.

SPARQL query language is a standard way to access RDF data and is based on the subgraph (homomorphism) match semantic [9]. More specifically, SPARQL can be represented by a query graph Q. Answering SPARQLs equals to finding subgraph matches of query graph Q over RDF graph G. Hence, the query intension should be modeled as a query graph Q to enable SPARQL query processing. In contrast, keyword search provides an intuitive way of specifying fuzzy information needs. Generally speaking, most existing keyword search techniques over RDF graphs take the following assumption: *A small-size substructure containing all keywords is an informative result.*

In fact, we observe in some real applications, the user's query intent cannot be well modeled using one query type. To better illustrate this, consider the following example.

Example 1. Which actors/actresses played in Philadelphia are mostly related to Academy Award and Golden Globe Award in Yago?

In this example, "actors/actresses played in Philadelphia" is a precise query requirement. It is easy to write down a SPARQL statement. However, the possible relationship between actors/actresses with Academy Award and Golden Globe Award is arbitrary. We cannot use SPARQL to represent the query intension. Thus, we resort to keyword search to discover potential relationships.

---

* Corresponding author.

© Springer International Publishing Switzerland 2015
R. Cheng et al. (Eds.): APWeb 2015, LNCS 9313, pp. 3–16, 2015.
DOI: 10.1007/978-3-319-25255-1_1

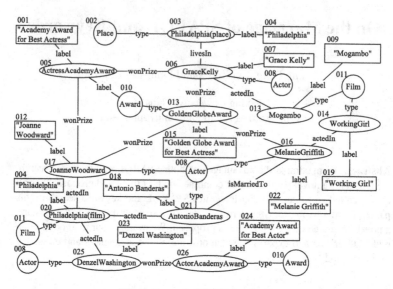

Fig. 1. Example RDF Data Graph

Motivated by the above example, we propose a hybrid query that integrates SPARQL and keyword (called an SK query) over large RDF graphs. The whole query requirement includes two parts. The first is "actors/actresses played in Philadelphia" which is a precise query criterion and we represent it as a SPARQL query[1]. The second part is "related to the Academy Award and Golden Globe Award", which is a fuzzy query requirement. Thus, it is better modeled as a "keyword search" to discover possible relationships between an actor and an award. "Shortest path distance" is a widely used measure to evaluate the relationship strength [15]. The shorter the path between entities is, the stronger their relationship is.

Let us recall Example 1 again. We issue the SK query $\langle Q, q \rangle$ as in Fig. 2(a). Fig. 2(b) shows three different results. First, there are three different subgraph matches of query Q, i.e., M_1, M_2 and M_3. Then, the keywords are matched in different literal vertices, i.e., 001, 015 and 026. The distance between a subgraph match M and a keyword in q is the shortest distance between M and one vertex containing keywords. We find that M_1 is the closest to the two keywords. It says "Joanne Woodward starring in Philadelphia film won both Academy Award and Golden Globe Award". Obviously, this is an informative answer to the query in Example 1.

In summary, we made the following contributions in this paper.

1. We propose a new query paradigm over RDF data combining keywords and SPARQL(called an SK query), and design a novel solution for this problem.
2. We design an index to speed up SK query processing. We propose a frequent star pattern-based index to reduce the search space.

---

[1] How to translate a natural language query into a SPARQL query automatically is out of the scope of this paper. In this work, we assume that users can specify SPARQL for the precise query criteria and use keywords properly to express his/her fuzzy query intentions.

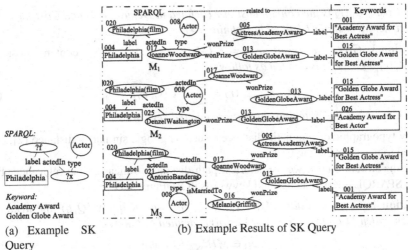

Fig. 2. Example SK Query and Its Results

3. We evaluate the effectiveness and efficiency of our method in real large RDF graphs and conclude that our methods are better than comparative models in both effectiveness and query response time.

The remainder of this paper is organized as follows: Section 2 defines the preliminary concepts. Section 3 gives an overview of our approach. We introduce a structural index to efficiently find the candidates of variables in Section 4. We discuss how to compute the results of SK queries in Section 5. Experimental results are presented in Section 6. Related works and the final conclusion are drawn 8.

2 Preliminaries

In this section, we introduce the fundamental definitions used in this paper. An RDF dataset consists of a number of triples, which is corresponding to an RDF graph. An RDF data graph G is denoted as $\langle V(G), E(G), L \rangle$, where $V(G)$ is the set of vertices corresponding to all subjects and objects in RDF data; $E(G)$ is the set of edges in G; and L is a finite set of edge labels, i.e. predicates.

The SK query is to find k SPARQL matches that are top-k closest to all keywords.

Definition 1. *An SK (SPARQL & Keyword) query is a pair $\langle Q, q \rangle$, where Q is a SPARQL query graph and q is a set of keywords $\{w_1, w_2, ..., w_n\}$.*

Given an SK query $\langle Q, q \rangle$, the result of $\langle Q, q \rangle$ in a G is a pair $\langle M, \{v_1, v_2, ..., v_n\} \rangle$, where M is a match of Q and v_i $(i = 1, ..., n)$ is a literal vertex containing keyword w_i.

Given an SK query $\langle Q, q \rangle$, the cost of a result $r = \langle M, \{v_1, v_2, ..., v_n\} \rangle$ contains two parts. The first part is the content cost and the second part is the structure cost.

Definition 2. *Given a result $r = \langle M, \{v_1, v_2, ..., v_n\} \rangle$, the cost of r is defined as follows:*

$$Cost(r) = Cost_{content}(r) + Cost_{structure}(r) \qquad (1)$$

where $Cost_{content}(r)$ is the content cost of r (defined in Definition 3) and $Cost_{structure}(r)$ is the structure cost of r (defined in Definition 4).

Definition 3. *Given a result* $r = \langle M, \{v_1, v_2, ..., v_n\}\rangle$, *the content cost of* $r = \langle M, \{v_1, v_2, ..., v_n\}\rangle$ *is defined as follows:*

$$Cost_{content}(r) = \sum_{i=1}^{i=n} C(v_i, w_i) \tag{2}$$

where $C(v_i, w_i)$ *is the matching cost between* v_i *and keyword* w_i.

Any typographic or linguistic distances, such as google similarity distance [2], can be used to measure $C(v_i, w_i)$.

For the structure cost, we use shortest path distance to measure the relevances between SPARQL query matches and keywords in RDF graph G.

Definition 4. *Given a result* $\langle M, \{v_1, v_2, ..., v_n\}\rangle$ *for an SK query* $\langle Q, q\rangle$, *the distance between match M and vertex* v_i $(i = 1, ..., n)$ *is defined as follows.*

$$d(M, v_i) = MIN_{v \in M}\{d(v, v_i)\} \tag{3}$$

where v *is a vertex in M and* $d(v, v_i)$ *is the shortest path distance between* v *and* v_i *in RDF graph G.*

Then, the structure cost *of a result* $r = \langle M, \{v_1, v_2, ..., v_n\}\rangle$ *is defined as follows.*

$$Cost_{structure}(r) = \sum_{i=1}^{i=n} d(M, v_i) \tag{4}$$

(Problem Definition.) *Given an SK query* $\langle Q, q\rangle$ *and a parameter k, our problem is to find k results (Definition 1), which have the k-smallest costs.*

3   Overview

In this section, we give an overview of the different steps involved in our process of SK query, which is depicted in Fig. 3. In this paper, we are concerned with the challenge of efficiently finding the results of SK queries. We propose an approach in which the best results of the SK query are computed using the graph exploration. We detail the different steps of the approach below.

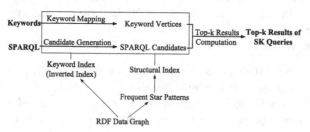

Fig. 3. Overview of Our Approach

Keyword Mapping. In the offline phase, we create an inverted index storing a map from keywords to its locations in the RDF graph. In the online phase, we map keywords to vertices based on the inverted index. We call the vertices mapping keywords *keyword vertices*. In this paper, because our primary focus is finding matches of SPARQL query and their relations to keywords, we use an existing IR engine to analyze given keywords, perform an imprecise matching, and return a list of graph elements having labels that are syntactically or semantically similar.

Candidate Generation. When we find a vertex reachable to elements of all keywords, we need to run subgraph homomorphism to check whether there exist some matches (of Q) containing v. As we know, subgraph homomorphism is not efficient due to its high complexity [4]. Thus, we propose a filter-and-refine strategy to reduce the number of subgraph homomorphism operations. The basic idea is to build a frequent star pattern-based structural index to filter out some vertices that are not in any subgraph match of Q. We call them "*dummy*" vertices and do not perform subgraph homomorphism algorithm for them. We will detail how to build the structural index in Section 4.1 and how to use the index to reduce the candidates of all variables in Section 4.2.

Top-k Results Computation. Based on the keyword vertices and variables' candidates, we propose a solution based on graph exploration to compute out the top-k results of SK queries. Our approach starts graph exploration from all keyword vertices, and explores to their neighboring vertices recursively until the distances between a vertex and keyword vertices have been computed out. The exploration terminates when the top-k results have been computed. We also propose some early stop strategies for top-k computation to reach early termination after obtaining the top-k results, instead of searching the data graph for all results. We discuss the detail of top-k results computation in Section 5.

4   Candidate Generation Based on Structural Index

In this section, we first introduce a structural index based on a certain kind of patterns in Section 4.1. Then, we discuss how to generate the candidate lists of variables based on our structural index in Section 4.2.

4.1   Structural Index

In this section, we propose a frequent star pattern-based index. We mine some frequent star patterns in G. For each frequent star S, we build an inverted list $L(S)$ that includes all vertices (in RDF graph G) contained by at least one match of S. A reason for selecting stars as index elements is that SPARQL queries tend to contain star-shaped subqueries, for combining several attribute-like properties of the same entity [6].

We propose a sequential pattern mining-based method to find frequent star patterns in RDF graphs. For each entity vertex in an RDF graph, we sort all its adjacent edges in lexicographic order of edge labels (i.e. predicates). These sorted edges can form a sequence. For example, vertex "Philadelphia(film)" has five adjacent edges, that are $\langle actedIn, actedIn, actedIn, name, type \rangle$. Then, we employ the existing sequential pattern mining algorithms, such as PrefixSpan[7], to find frequent sequential patterns, where each sequential pattern corresponds to a star pattern in RDF graphs. For example, assume that the minimal support count $s = 2$, $\langle actedIn, type \rangle$ and $\langle actedIn, type, $

wonPrize⟩ are two frequent sequential patterns for the RDF graph in Fig. 1. It is easy to know that a sequential pattern always corresponds to one star pattern. For ease of presentation, we use the terms "sequential patterns" and "star patterns" interchangeably in the following discussion.

For each frequent star pattern S, we maintain an inverted list $L(S) = \{v|S$ occurs in v's adjacent edge sequence}. Note that, to ensure the completeness of the indexing, we always choose the (absolute) support to be 1 for size-1 stars (star with only one edge). This method can guarantee no-false-negative, since all vertices (in G) are indexed in at least one inverted list. Then, we can find out vertices that cannot be contained by any matches as discussed in the following theorem.

Theorem 1. *Given a SPARQL query Q, a vertex v in graph G can be pruned (there exists no subgraph match of Q containing v) if the following equation holds.*

$$v \notin \bigcup_{S \in Q} L(S) \tag{5}$$

where $S \in Q$ means that S is included in Q.

Proof. Please refer to the full version [8] of our paper at arXiv.

4.2 Candidate Generation

Given an SK query, we first tag the vertices that can be pruned by Theorem 1. For each variable in SPARQL, we locate its candidates in RDF graph. Each variable can map to a predicate sequence according to the SPARQL statement. For example, variable "?a" of the SPARQL query in Fig. 2(a) has the predicate sequence ⟨*actedIn, type*⟩. Then, for each variable x, we look up our structural index and find the maximum pattern contained by x's predicate sequence. We load the vertex list of the maximum pattern as x's candidates. A vertex in at least one vertex lists of variables is not dummy. We define these pruned vertices as *dummy vertices* as follows.

Definition 5. *Dummy Vertex. Given a SPARQL query Q, a vertex v in graph G is called as* dummy vertex *if the following equation holds.*

$$v \notin \bigcup_{S \in Q} L(S) \tag{6}$$

where $S \in Q$ is a star pattern included in Q.

When the graph exploration meets a dummy vertex v during top-k results computation, we do not need to perform subgraph isomorphism algorithm beginning with v.

5 Top-k Results Computation

In this section, we introduce our approach for SK queries, which is based on the "graph exploration" strategy [12]. Our algorithm for searching top-k results of SK queries is shown in Algorithm 1. This algorithm consists of three parts: 1) graph exploration to find vertices connecting the keyword vertices, 2) generation of SPARQL matches from the vertices connecting the keyword vertices and 3) top-k computation. In the following, we will elaborate on these three tasks.

5.1 Graph Exploration

The objective of the exploration is to find vertices in the graph that connect these keyword vertices and compute their distances to these keyword vertices. Let V_i denote all literal vertices (in RDF graph G) containing keyword w_i.

To explore the RDF graph, we maintain a priority queue PQ_i for each keyword w_i. Each element in PQ_i is represented as $(v, p, |p|)$, where v is a vertex id, p is a path from a vertex in V_i to v and $|p|$ denotes the path distance. All elements in PQ_i are sorted in the non-descending order of $|p|$. Each keyword w_i is also associated with a result set RS_i. In order to keep track of information related to each vertex v, we associate v with a vector $d[v]$. If a vertex v is in RS_i ($i = 1, ..., n$), the shortest path distance is known. In this case, we set $d[v][i] = d(v, w_i)$; otherwise, we set $d[v][i] = null$.

Initially, the exploration starts with a set of vertices containing keywords. For each vertex v containing keyword w_i, an element $(v, \emptyset, 0)$ is created and placed into the queue PQ_i (Line 3 in Algorithm 1). During the exploration, at each step, we pick a queue PQ_i ($i = 1, ..., n$) to expand in a round-robin manner (Line 5 in Algorithm 1). We assume that we pop the queue head $(v, p, |p|)$ from PQ_i. When a queue head $(v, p, |p|)$ is popped from queue PQ_i, we insert it into result set RS_i and set $d[v][w_i] = |p|$ (Line 6 in Algorithm 1). We can prove that the following theorem holds.

Algorithm 1. Graph Exploration for Top-k Results of SK Queries

---

Input: RDF data graph G, SK query $\langle Q, q \rangle$, $\{V_1, ..., V_n\}$ where V_i is the set of vertices containing keyword w_i, priority queues $\{PQ_1, ..., PQ_n\}$.

Output: Top-k results R of $\langle Q, q \rangle$.

1 **for** *each vertices set V_i* **do**
2 $\quad$ **for** *each vertex v in V_i* **do**
3 $\quad\quad$ Insert $(v, \emptyset, 0)$ into PQ_i;
4 **while** $\exists i\ 1 \le i \le n \wedge PQ_i \neq \emptyset$ **do**
5 $\quad$ **for** $i = 1, ..., n$ **do**
6 $\quad\quad$ Pop the PQ_i's head $(v, p, |p_i|)$, set $d[v][i] = |p_i|$ and insert it into RS_i;
7 $\quad\quad$ **for** *each adjacent edge $\overline{vv'}$ to v* **do**
8 $\quad\quad\quad$ **if** *there exists another element $(v', p', |p'|)$ in PQ_i* **then**
9 $\quad\quad\quad\quad$ **if** $|p'| > |p| + 1$ **then**
10 $\quad\quad\quad\quad\quad$ Replace $(v', p', |p'|)$ with $(v', p \cup \overline{vv'}, |p_i| + 1)$ in PQ_i;
11 $\quad\quad\quad$ **else**
12 $\quad\quad\quad\quad$ Insert $(v', p \cup \overline{vv'}, |p| + 1)$ in PQ_i;
13 $\quad\quad$ **if** *v is a fully-seen vertex and v is a candidate of vertex u of Q* **then**
14 $\quad\quad\quad$ Initialize a function f with (u, v) and call Algorithm 2 to find matches from f;
15 $\quad\quad$ **for** *each match M containing vertex v* **do**
16 $\quad\quad\quad$ **if** *all vertices in M are fully-seen vertices* **then**
17 $\quad\quad\quad\quad$ Use M to update R and the upper bound δ of top-k results
18 $\quad\quad$ Update the cost of all partially-seen matches, the lower bound cost θ_1 of partially-seen matches and the lower bound cost θ_2 of un-seen vertices;
19 $\quad$ **if** $\delta < \theta_1 \wedge \delta < \theta_2$ **then**
20 $\quad\quad$ Break;
21 Return R.

Theorem 2. *When a queue head* $(v, p, |p|)$ *is popped from queue* PQ_i, *the following equation holds.*

$$d(v, w_i) = d[v][i] = |p| \tag{7}$$

Proof. Please refer to the full version [8] of our paper at arXiv.

When a queue head $(v, p, |p|)$ is popped from queue PQ_i, the distance between v and keyword w_i has been computed out. We also say that v is *seen* by keyword w_i.

Definition 6. *Seen by Keyword.* *When a queue head* $(v, p, |p|)$ *is popped from queue* PQ_i, *we say vertex* v *is* seen *by keyword* w_i.

Assume that $(v, p, |p|)$ is popped from queue PQ_i. For each incident edge $\overline{vv'}$ to v, we obtain a new element $(v', p \cup \overline{vv'}, |p| + 1)$, where $p \cup \overline{vv'}$ means appending an edge to p. Then, we check whether there exists another element $(v', p', |p'|)$ that has the identical vertex v' with the new element $(v', p \cup \overline{vv'}, |p| + 1)$, where $|p'| > |p| + 1$. If so, we replace $(v', p', |p'|)$ with $(v', p \cup \overline{vv'}, |p| + 1)$ in PQ_i (Line 10 in Algorithm 1). If there exists no element for v', we insert $(v', p \cup \overline{vv'}, |p| + 1)$ into PQ_i directly(Line 12 in Algorithm 1).

Definition 7. *Fully-seen Vertex, Partially-seen Vertex and Un-seen Vertex.* *Given a vertex* v, *if* v *is seen by all keywords* w_i ($i = 1, ..., n$), v *is called a* fully-seen *vertex; if* v *is not a fully-seen vertex but it has been seen by at least one keyword,* v *is called a partially-seen vertex; if* v *is not seen by any keyword,* v *is called an un-seen vertex.*

At each step, we check whether the vertex v just popped from the queue has been seen by all keywords. If all dimensions of v's vector $d[v]$ are not null, it means that all keywords have seen vertex v, i.e., v is a *fully-seen* vertex. When we meet a fully-seen vertex v, we will employ a subgraph homomorphism algorithm to find matches containing v (Line 13-14 in Algorithm 1).

5.2   Generation of SPARQL Matches

When we explore a fully-seen vertex v, it means that we have known the distance between v and each keyword w_i. The next step is to compute SPARQL matches containing vertex v if any. Here, we perform subgraph homomorphism algorithm to find matches (of query Q) containing v. The matching process consists of determining a function f that associates vertices of Q with vertices of G. The matches are expressed as a set of pairs (u, v) ($u \in Q$ and $u \in G$). A pair (u, v) represents the matching of a vertex u of query Q with a vertex v of RDF graph G. The set of vertex pairs (u, v) constitutes a SPARQL match.

A high-level description of finding matches from vertex v is outlined in Algorithm 2. Initially, we introduce a candidate vertex pair (u, v) to expand the current function f. Assume that we introduce a new candidate vertex pair (v', u') into the current function f to form another function f'.

5.3   Top-k Computation

The native solution for computing top-k results of an SK query is to explore the graph until that all vertices (in RDF graph G) are fully-seen vertices. Then, according to the

---

Algorithm 2. SPARQL Matching Algorithm

---

1 **if** *all vertices of query Q have been matched in the function f* **then**
2 | Return;
3 Select an unmatched u' adjacent to a matched vertex u in the function f
4 **for** *each neighbor v' of $f(u)$* **do**
5 | **if** *v' is a candidate of u' and v' is not a fully-seen vertex* **then**
6 | | $f' \leftarrow f \cup (u', v')$ and call Algorithm 2 with f'

---

results' cost, we can find the top-k results. Obviously, this is an inefficient solution especially when G is very large. In this subsection, we design an early-stop strategy.

Let us consider a snapshot of some iteration step in Algorithm 1. All subgraph matches of SPARQL query Q can be divided into three categories: fully-seen matches, partially-seen matches and un-seen matches.

Definition 8. *Fully-seen Match, Partially-seen Match and Un-seen Match. Given a match M of SPARQL query Q, if all vertices in M are fully-seen vertices, M is called a fully-seen match; if M is not a fully-seen match and M contains at least one fully-seen vertex, it is called a partially-seen match; otherwise, it is called an un-seen match.*

The idea of our early-stop strategy is as follows. We can compute the cost of fully-seen matches according to Definition 4. Then, we use the fully-seen matches to find a threshold δ, which is the k-th smallest cost so far. We also compute the lower bounds θ_1 and θ_2 for partially-seen matches and un-seen matches. The algorithm can early stop if $\delta < \theta_1 \wedge \delta < \theta_2$. Note that, if there are less than k fully-seen matches so far, δ is ∞.

Given a partially-seen match, we compute the lower bound of its cost as follows.

Theorem 3. *Given a partially-seen match M of SPARQL query Q, v is a partially-seen vertex or an un-seen vertex in the match. The following equation holds.*

$$Cost(M) = \sum_{1 \leq i \leq n} d(v, w_i) \geq \sum_{d[v][w_i] \neq null \wedge 1 \leq i \leq n} d[v][w_i] + \sum_{d[v][w_i] = null \wedge 1 \leq i \leq n} |p_i|$$

where $d[v][w_i]$ is the i-th dimension of v's vector corresponding to keyword w_i, and $|p_i|$ corresponds to the current queue head $(v, p_i, |p_i|)$ in queue PQ_i.

Proof. Please refer to the full version [8] of our paper at arXiv.

According to Theorem 3, the lower bound θ_1 of all partially-seen matches can be defined as follows.

Definition 9. *Lower Bound θ_1 of Partially-seen Matches. The lower bound θ_1 of all partially-seen matches is as follows.*

$$\theta_1 = MIN_{M \in PS}(MIN_{v \in M}(\sum_{d[v][w_i] \neq null \wedge 1 \leq i \leq n} d[v][w_i] + \sum_{d[v][w_i] = null \wedge 1 \leq i \leq n} |p_i|)) \quad (8)$$

where PS denotes all partially-seen matches.

On the other hand, let us consider the lower bound of an un-seen match M. There are two kinds of vertices in M, i.e., partially-seen vertices and un-seen vertices.

Theorem 4. *For an un-seen vertex v, if threshold $\delta \neq \infty$, the following equation holds.*

$$\delta \leq \sum_{1 \leq i \leq n} d(v, w_i) \tag{9}$$

Proof. Please refer to the full version [8] of our paper at arXiv.

According to Theorem 4, it is not necessary to consider un-seen vertices to define the lower bound for un-seen matches. Therefore, we define the lower bound for all un-seen matches as follows.

Definition 10. *Lower Bound θ_2 of Un-seen Matches. The lower bound θ_2 for all un-seen matches is as follows.*

$$\theta_2 = \underset{v \in PSet}{MIN}(\sum_{d[v][w_i] \neq null \wedge 1 \leq i \leq n} d[v][w_i] + \sum_{d[v][w_i] = null \wedge 1 \leq i \leq n} |p_i|) \tag{10}$$

where PSet contains all partially-seen vertices so far, $d[v][w_i]$ is the i-th dimension of the v's vector corresponding to keyword w_i and $|p_i|$ corresponds to the current queue head $(v, p_i, |p_i|)$ in queue PQ_i.

With the increasing of the iteration steps, some partially-seen matches become fully-seen matches. Then, the threshold δ, θ_1 and θ_2 are updated accordingly.

6 Experiments

6.1 Datasets and Setup

We use two real-world RDF graphs Yago and DBPedia in our experiments. Here, all sample queries are shown in Appendix of the full version [8] at arXiv. Our experiments are conduct on a machine with 2 Ghz Core 2 Duo processor, 16G RAM memory and running Windows Server 2008. All experiments are implemented in Java. We use MySQL to store the indices. The details about the two datasets are as follows.

Yago. Yago[2] extracts facts from Wikipedia and integrates them with the WordNet thesaurus. The RDF graph has $19,012,849$ edges and $12,811,222$ vertices. We define 8 sample SK queries for Yago.

DBPedia and QALD. DBPedia [3] is an RDF dataset extracted from Wikipedia. The DB-Pedia contains $73,766,900$ edges and $13,100,739$ vertices. QALD [4] is an evaluation campaign on question answering over linked data. In this campaign, the committee provides some questions. Each question is annotated with some recommended keywords and the answers that these queries retrieve. Note that, many questions in QALD are so simple to be split into a SPARQL query and some keywords. Thus, we only select 10 complex queries from QALD for evaluation.

---

[2] http://www.mpi-inf.mpg.de/yago-naga/yago/

[3] http://downloads.dbpedia.org/3.7/en/

[4] http://greententacle.techfak.uni-bielefeld.de/~cunger/qald/index.php?x =challenge&q=2

6.2 Effectiveness Study

In this section, we compare our method with a classical keyword search algorithm BANKS [1] over Yago and DBPedia to show the effectiveness of our method. Furthermore, since each resource in DBPedia is annotated by Wikipedia documents, we design a stronger baseline named as "Annotated SPARQL" for DBPedia, which first finds out all matches of the SPARQL query, then ranks these matches by how closely the corresponding Wikipedia documents match the keywords.

NDCG@k over Yago. In order to quantify the effectiveness of SK query, we evaluate the NDCG (Normalized Discounted Cumulative Gain [5]) of both SK query and the keyword search. Since there are no golden standards, we invite 10 volunteers to judge the result quality. Specifically, we ask each volunteer to rate the goodness of the results returned by SK query and the keyword search method. The score is between 1 and 5. Higher the score, better the result.

Table 1. Average NDCG Values over Yago

	NDCG@3	NDCG@5	NDCG@10
BANKS	0.3455	0.39	0.4643
SK query	0.815	0.868	0.872

Table 1 reports NDCG@k values by varying k from 3 to 10 in Yago. SK query outperforms the traditional keyword search by 0%-50%. The reason is that Yago has a complex schema and keywords may result in lots of ambiguity in Yago. It means that the superiority of SK query is more pronounced in semantic-rich data.

MAP over DBPedia. Since QALD provides the standard answers of each queries, we evaluate the MAP (Mean Average Precision [13]) to compare the SK query with BANKS and "Annotated SPARQL". Table 2 reports MAP values of our ten QALD queries.

Table 2. MAP Value over DBPedia & QALD

	BANKS	Annotated SPARQL	SK query
MAP	0.012	0.192	0.205

Both SK query and annotated SPARQL outperforms the traditional keyword search by a order of magnitude. The MAP value of the "annotated SPARQL" is smaller than the SK query. This is because that the "annotated SPARQL" can do well when the documents associated with the matches contains the keywords. In other words, the "annotated SPARQL" can work, only when the relation between the matches and the keywords is explicit. However, in pracitce, the relation between the matches and the keywords is often implicit. Then, the SK query do better.

6.3 Efficiency Study

In this section, we evaluate the efficiency of SK query in large real graphs. Here, the default number of returned results is set to be 10. Because there is no existing method for SK queries, we evaluate our approach with two baselines, i.e., the *exhaustive computing* and the *naive backward search* for efficiency study. "Exhaustive computing" works as follows: we first find all subgraph matches of Q (in RDF graph G) by existing techniques; then, we compute the shortest path distances between these subgraph matches and the vertices containing keywords on the fly; finally, the matches with shortest distances to keywords are returned as answers. The *naive backward search* is to run the backward search algorithm until that all vertices have been fully-seen by keywords. Then, according to the results' cost, we find the top-k results.

Offline Performance. We report the index size and index construction time in Table 3. Since our structural index is based on the efficient sequential pattern mining, we can finish the structural index construction in several minutes.

Table 3. Index Size and Index Construction time

	Index Construction Time(s)	Index Size(MB)
Yago	176.67	844.066
DBPedia	600.263	283.559

Online Performance. In this section, we evaluate the efficiency of our method. Fig. 4 shows the time cost of the three methods. The performance of our method is always not worse than the comparative methods. Especially for Q_3 on Yago, our method only takes one fifth of the exhaustive-computing. This is because that the matches of these SPARQLs are close to vertices containing keywords. Thus, the query processing can terminate soon.

Note that, because our inverted index for keywords are stored in disk, keyword mapping processing costs much time and takes up a large part of the total time. Hence, it is difficult for our method to improve the efficiency by orders of magnitudes.

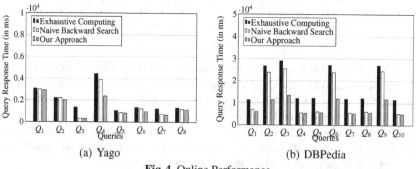

(a) Yago   (b) DBPedia

Fig. 4. Online Performance

7   Related Work

To the best of our knowledge, although there exist a few previous works [3,14] for the hybrid query combined SPARQL and keyords, there has been no existing work on SK query defined as the above. Elbassuoni et al. [3] assumes that each RDF triple may have associated documents. Then, Elbassuoni et al. extend the triple patterns in SPARQL with keyword conditions. CE^2 [14] assumes that each resource associate with a document. Then, CE^2 extend the variables in SPARQL with keyword conditions. Nonetheless, many current RDF datasets do not provide documents to annotate triples or resources. Hence, these methods cannot handle our example queries.

In addition, [10] defines a new query language that blends keyword search with structured query processing. [16] translates natural language questions into SPARQL queries.

8   Conclusions

In this paper, we first propose a new kind of query (SK query) that integrates SPARQL and keywords. To handle this kind of query, we build up a frequent star pattern-based index and propose a method based on graph exploration. With two real RDF graphs, we show that our method can outperform the baseline both in effectiveness and efficiency.

Acknowledgments. This paper was supported in part by National High-tech R&D Program (863 Program) under Grant 2015AA015402, National Science Foundation of China (NSFC) under Grant 61370055. This work was also supported by Tencent.

References

1. Aditya, B., Bhalotia, G., Chakrabarti, S., Hulgeri, A., Nakhe, C., Parag, Sudarshan, S.: BANKS: Browsing and Keyword Searching in Relational Databases. In: VLDB (2002)
2. Cilibrasi, R., Vitányi, P.M.B.: The google similarity distance. IEEE Trans. Knowl. Data Eng. 19(3), 370–383 (2007)
3. Elbassuoni, S., Ramanath, M., Schenkel, R., Weikum, G.: Searching RDF graphs with sparql and keywords. IEEE Data Eng. Bull. 33(1), 16–24 (2010)
4. Garey, M.R., Johnson, D.S.: Computers and Intractability: A Guide to the Theory of NP-Completeness. W. H. Freeman and Company, San Francisco (1979)
5. Järvelin, K., Kekäläinen, J.: IR evaluation methods for retrieving highly relevant documents. In: SIGIR, pp. 41–48 (2000)
6. Neumann, T., Weikum, G.: RDF-3X: A RISC-style Engine for RDF. PVLDB 1(1), 647–659 (2008)
7. Pei, J., Han, J., Mortazavi-Asl, B., Pinto, H., Chen, Q., Dayal, U., Hsu, M.: Prefixspan: Mining sequential patterns by prefix-projected growth. In: ICDE, pp. 215–224 (2001)
8. Peng, P., Zou, L., Zhao, D.: On the marriage of SPARQL and keywords. CoRR, abs/1411.6335 (2014)
9. Pérez, J., Arenas, M., Gutierrez, C.: Semantics and complexity of SPARQL. ACM Trans. Database Syst. 34(3) (2009)

10. Pound, J., Ilyas, I.F., Weddell, G.E.: Expressive and flexible access to web-extracted data: a keyword-based structured query language. In: SIGMOD, pp. 423–434 (2010)
11. Suchanek, F.M., Kasneci, G., Weikum, G.: Yago: a core of semantic knowledge. In: WWW, pp. 697–706 (2007)
12. Tran, T., Wang, H., Rudolph, S., Cimiano, P.: Top-k Exploration of Query Candidates for Efficient Keyword Search on Graph-Shaped (RDF) Data. In: ICDE, pp. 405–416 (2009)
13. Turpin, A., Scholer, F.: User performance versus precision measures for simple search tasks. In: SIGIR, pp. 11–18 (2006)
14. Wang, H., Tran, T., Liu, C., Fu, L.: Lightweight integration of IR and DB for scalable hybrid search with integrated ranking support. J. Web Sem. 9(4), 490–503 (2011)
15. Zou, L., Chen, L., Özsu, M.T.: Distancejoin: Pattern match query in a large graph database. PVLDB 2(1), 886–897 (2009)
16. Zou, L., Huang, R., Wang, H., Yu, J.X., He, W., Zhao, D.: Natural language question answering over RDF: a graph data driven approach. In: SIGMOD, pp. 313–324 (2014)

An Online Inference Algorithm for Labeled Latent Dirichlet Allocation

Qiang Zhou, Heyan Huang, and Xian-Ling Mao

Beijing Engineering Research Center of High Volume Language Information
Processing and Cloud Computing Applications,
Department of Computer Science and Technology,
Beijing Institute of Technology, Beijing 100081, China
{qzhou,hhy63,maoxl}@bit.edu.cn

Abstract. Using topic models to analyze documents is a popular method in text mining. Labeled Latent Dirichlet Allocation(Labeled LDA) is one of them that is widely used to model tagged documents and to solve relevant problems, such as tagged document visualization, snippet extraction and so on. However, traditional batch inference for Labeled LDA, which runs over entire document collection, is computationally expensive and not suitable for large scale corpora and text streams. In this paper, we develop an efficient online algorithm for Labeled LDA, called *online Labeled LDA*(online-LLDA). It is based on particle filter, a Sequential Monte Carlo approximation technique. Our experiments show that online-LLDA significantly outperforms batch algorithm(batch-LLDA) in time, while preserving equivalent quality.

Keywords: Online Inference, Online Labeled LDA, Particle Filter.

1 Introduction

In recent years, topic modeling has emerged as a relatively new approach to analyze text data. Lots of topic models, such as Probabilistic Latent Semantic Indexing(PLSI)[10], Latent Dirichlet Allocation(LDA)[3] and Hierarchical Dirichlet Processes(HDP) [18] have been proposed to model documents without tags.

Meanwhile, more and more pages in many websites, especially social media websites, are born with tags, or labels. These tags are usually assigned by users, and carry a lot of information about the pages. For example, a twitter user posts a tweet, and associates it with some hashtags, which produces a tagged document. Also, generally, category names of news pages can also be regarded as tags of pages.

Therefore, a number of topic models have been proposed to model labeled documents. Labeled LDA is one of them proposed by Ramage et al. in 2009, and has many extensions including PLDA[15], hLLDA[11] and so on. In Labeled LDA, a topic is defined as a word distribution, and each topic corresponds to a label. Thus, topics of each labeled document are restricted within its labels.

© Springer International Publishing Switzerland 2015
R. Cheng et al. (Eds.): APWeb 2015, LNCS 9313, pp. 17–28, 2015.
DOI: 10.1007/978-3-319-25255-1_2

Also, through selecting the topic with highest probability for each word, Labeled LDA assigns the most appropriate label to each individual word. With this ability, it can solve many text analysis problems, such as credit attribution, tagged document visualization, snippet extraction and multi-labeled text classification[12][16]. For example, in tagged document visualization, annotating important words with their labels provides users a short semantic description of documents; In snippet extraction, by finding out snippet that contains most specific label, appropriate snippet can be extracted.

Traditional inference algorithms like Gibbs sampling and variational inference require multiple passes through the whole corpus. These algorithms are often referred to as *off-line* or *batch* algorithm. Obviously, it's difficult to apply batch-LLDA for large scale data. Considering the task of clustering all tagged web pages in *del.icio.us*, a popular social bookmarking website, it can be accomplished by using the topic distributions learned by Labeled LDA. Since it contains massive amounts of pages, and we have to run through all pages for many times, it will be time consuming and infeasible. Also, when documents arrive in a stream, batch algorithms do not work. For example, twitter users update status whenever they like to, so it will generate a flow of tweets. If we use batch algorithm to analyze these tweets, once a new tweet is available, we have to run the algorithm again over all tweets including previously seen ones to obtain the newest model parameters. That will be slower and slower as the number of tweets increases, which is unbearable in the real world. Thus, it's necessary to develop an efficient algorithm to solve the problem.

To the best of our knowledge, there is no existing work to obtain parameters of Labeled LDA by using an online style algorithm. Aiming at solving this problem, we propose an online algorithm, called online-LLDA, which incrementally updates model parameters when a document arrives. In this case, the algorithm do not need to visit the previous documents that have been processed. Consequently, these documents can be abandoned to save memory. Differently from batch Gibbs sampling, we use another sampling-based framework, particle filter[5], to make the algorithm work in online mode. We evaluate our method on two datasets, and show that online-LLDA performs better than batch-LLDA in time, while achieving equivalent effect.

The rest of the paper is organized as follows. We conclude the related work in Section 2. Our approach is introduced in Section 3. We demonstrate effectiveness and efficiency of our algorithm in section 4. The final conclusion and future work will be discussed in section 5.

2   Related Work

Most existing researches on LDA-like model utilize various inference algorithms, such as variational Bayesian[3][17], Gibbs sampling[7][8], expectation propagation [13] and belief propagation[21] to obtain the parameters. Unfortunately, most of them are batch algorithms.

Thus, a host of online learning methods have been developed for topic models. Some of them focus on modeling large amount of documents efficiently based on LDA. Typical researches include TM-LDA[19], On-Line LDA[1] and so on. By minimizing the error between predicted topic distribution and the real distribution generated by LDA, TM-LDA captures the latent topic transitions in temporal documents. On-Line LDA uses topics learned from previous documents by LDA as the prior of following topics. It was designed by Alsumait et al.(2008) to detect and track topics in an online fashion.

Some other work try to improve inference algorithms themselves. Banerjee et al.(2007) presented online variants of vMF, EDCM and LDA. Their experiments illustrated faster speed and better performance on real-world streaming text[2]. Hoffman et al.(2010) developed an online variational Bayesian algorithm[9] for LDA based on online stochastic optimization. In their approach, they thought of LDA as a probabilistic factorization of the matrix of word counts into a matrix of topic weights and a dictionary of topics. Thus, they used online matrix factorization techniques to obtain an online fashion schema of LDA. Canini et al.(2009) also proposed an online version algorithm[4] using Gibbs sampling. Yao et al.(2009) compared several batch methods mentioned above, and introduced a new algorithm, SparseLDA[20] to accelerate learning process and consequently achieved nearly 20 times faster speed than traditional LDA.

To conclude, a large number of prior work have made great efforts on designing appropriate online algorithms for LDA. However, less work focus on improving the efficiency for Labeled LDA. In this paper, we propose a novel online learning method for Labeled LDA, called online-LLDA.

3 Method

In this section, we first review the Labeled LDA model along with its original batch algorithm. Then, we analyze the feasibility of applying particle filter on Labeled LDA in the following part. Finally, we present online-LLDA in detail.

3.1 Preliminary

Labeled LDA is a generative model for labeled documents that extends LDA. With the assumption that topics of labeled document are relevant with its labels, it adds a constraint on topic generation. Formally, for each document d in the collection of documents D, it can only pick up topics associated with its labels. As the graphical model shows in Figure 1, topics θ are determined by both topic prior and labels. Let K be the number of unique labels in all current observed documents, which equals the number of topics. Each document d has a label set $L_d = \{l_1^{(d)}, l_2^{(d)}, ..., l_{K_d}^{(d)}\}$ where K_d is the number of labels. For the description convenience, $\Lambda_d = \{\lambda_1^{(d)}, \lambda_2^{(d)}, ..., \lambda_{K_d}^{(d)}\}$ is defined to represent the label index set in document d. For each element $\lambda_j^{(d)}$ in Λ_d, we have $\lambda_j^{(d)} = \sum_{k=1}^{K} k \cdot I\{\lambda_j^{(d)} = l_k\}$, where I is the indicator function.

The original inference approach of Labeled LDA is a batch algorithm(batch-LLDA)[14], that iteratively samples topics of all words using the conditional probability of latent variable z. For a word i, it is sampled according to the following equation:

$$P(z_i = k | \mathbf{z}_{\neg i}, \mathbf{w}) \propto I\{k \in \Lambda_d\} \cdot \left(\frac{n_{k,\neg i}^{(w_i)} + \beta}{n_{k,\neg i}^{(\cdot)} + V\beta} \cdot \frac{n_{d,\neg i}^{(k)} + \alpha}{n_{d,\neg i}^{(\cdot)} + K\alpha} \right) \tag{1}$$

where w_i is a word in document d, V is the size of vocabulary, K is the number of topics. $n_{k,\neg i}^{(w_i)}$ is the number of times word w_i is assigned to topic k, excluding w_i from all documents. $n_{k,\neg i}^{(\cdot)}$ is the corresponding summation over words. $n_{d,\neg i}^{(k)}$ is the count of words in document d with topic assignment k, excluding w_i. And $n_{d,\neg i}^{(\cdot)}$ is the corresponding summation over topics. α and β are prior parameters.

With Equation (1), batch Gibbs sampler can constantly sample topic for each word. After the burn-in period, topic assignments $\mathbf{z}$ are available for estimation. To ensure the convergence, the number of iterations is usually set as big as enough. Hence, it will take too much time for batch-LLDA to reach the burn-in point, which leads to inefficiency.

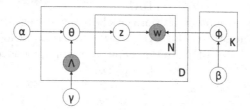

Fig. 1. Graphical model of Labeled LDA.

3.2 Feasibility Analysis

Labeled LDA can be viewed as a state space model, if we regard the latent variable topic z as a state variable, and the word w as an observation variable. Particle filter is a kind of efficient Markov-Chain Monte Carlo(MCMC) sampling method for estimating parameters of state space model. Differently from Gibbs sampling, it's an online algorithm.

Particle filter can be applied to solve the inference problem of Labeled LDA. (1) Suppose x is the hidden state variable and y is the observation variable in particle filter. The objective of particle filter is to estimate the values of hidden state variables given observations, that is $P(\mathbf{x}|\mathbf{y})$. Coincidently, what we want to acquire in Labeled LDA is the joint distribution $P(\mathbf{z}|\mathbf{w})$, which is also the probability of state variables given observations. (2) Particle filter assumes that observations are conditionally independent, and observation y_k is only determined by x_k. Obviously, based on the bag-of-words assumption, Labeled LDA fulfills this requirement.

In this paper, we use Rao-Blackwellised particle filter(RBPF), an enhanced version of particle filter as our approximation method. RBPF integrates out the latent variables, and thus makes the solution simpler. In our algorithm, we have

P particles, and each particle p represents an answer that we desire, namely the posterior distributions of topics. In our implementation, a particle p stores all topic assignments of words, together with an importance weight $\omega^{(p)}$ indicating the importance of particle. What's more, a reassignment process is added to enhance the quality of samples.

3.3   Online Labeled LDA

In this section, we introduce an online learning method by using particle filter. As demonstrated in Algorithm 1, the overall algorithm consists of two phases: **initialization phase** and **online phase**. Initialization phase accomplishes the task of launching the online phase, while online phase continually processes every word w_i in a newly arrived document d and generates new parameter set Φ_d after the entire document has been processed.

In *initialization phase* (line 1 to line 4), for each particle, we apply batch-LLDA on an initial corpus E that contains a small fraction of documents. After running over, we get initial topic assignments of all initial words, along with sufficient statistics. These values are stored into each particle, which are useful in the online phase.

In *online phase* (line 5 to line 17), we first initialize particle weights with equal values and then process documents in a text stream one after another. Here, N_d represents the length of document d. Model parameters will be updated every time a document is processed. A new sampling equation is used as shown in Equation (2). In this equation, $\neg i$ has different meaning from Equation (1). In Equation (1), it represents all words or topics except i, but here it aims at excluding i from currently observed words. Thus, $\mathbf{i}\neg i$ represents first $i-1$ words, $\mathbf{i}\neg j$ represents first i words except j. This difference is essential between batch-LLDA and online-LLDA. Also notice that when $i = 1$, $\mathbf{i}\neg i$ represents nothing, that is another reason why we need an initialization phase.

$$P(z_i^{(p)} = k | \mathbf{z}_{\mathbf{i}\neg i}^{(p)}, \mathbf{w}_i) \propto I\{k \in \Lambda_d\} \cdot \left(\frac{e_{k,w_i}^{(p)} + n_{k,\mathbf{i}\neg i}^{(w_i)(p)} + \beta}{e_{k,\cdot}^{(p)} + n_{k,\mathbf{i}\neg i}^{(\cdot)(p)} + V\beta} \cdot \frac{n_{d,\mathbf{i}\neg i}^{(k)(p)} + \alpha}{n_{d,\mathbf{i}\neg i}^{(\cdot)(p)} + T_d\alpha} \right) \quad (2)$$

Since we use all currently observed words including words in initial corpus, the word count in Equation (2) includes two parts. The first part is the word count of initial collection of documents, and the second part is the word count of documents coming in a stream in the online phase. The superscript p in all notations indicates the particle index. $e_{k,w_i}^{(p)}$ is the number of times word w_i is assigned to topic k in initial corpus, and $e_{k,\cdot}^{(p)}$ is the summation. n is similar with those in Equation (1). T_d is the number of unique labels when document d is available, namely the current topic number. All sampled topics should be restricted within its document label set.

In this algorithm, $P(z_i^{(p)} | \mathbf{z}_{\mathbf{i}\neg i}^{(p)}, \mathbf{w}_i)$ in Equation (2) is selected as the proposal distribution, so the importance weights are updated as $\hat{\omega}^{(p)} = \omega^{(p)} P(w_i | \mathbf{z}_{\mathbf{i}\neg i}^{(p)}, \mathbf{w}_i)$, and then normalized to sum to 1. $P(w_i | \mathbf{z}_{\mathbf{i}\neg i}^{(p)}, \mathbf{w}_i)$ is the probability of word i

---

Algorithm 1. Online Labeled LDA

Initialization Phase:

1   **for** $p = 1, ..., P$ **do**
2   **while** *burn-in point is not reached* **do**
3   draw topic for each word w_i in initial corpus E, using Equation (1)

4   calculate sufficient statistics for each particle

Online Phase:

5   initialize importance weights $\omega^{(p)} = 1/P$ for any $p \in \{1, ..., P\}$
6   **for** $d = 1, ..., D$ **do**
7   **for** $i = 1, ..., N_d$ **do**
8   **for** $p = 1, ..., P$ **do**
9   draw sample $z_i^{(p)}$ using Equation (2)
10            $\hat{\omega}^{(p)} = \omega^{(p)} P(w_i | \mathbf{z}_{i \neg i}^{(p)}, \mathbf{w}_i)$
11   normalize $\omega^{(p)}$ for any $p \in \{1, ..., P\}$, to sum to 1
12   calculate $\hat{N}_{eff}$ using Equation (3)
13   **if** $\hat{N}_{eff} \leq N_{thresh}$ **then**
14   Sampling Importance Resample process
15   draw sample $z_r^{(p)}$ for any $r \in R(i)$, $p \in \{1, ..., P\}$ using Equation (2)
16   set $\omega^{(p)} = 1/P$ for any $p \in \{1, ..., P\}$

17   generate new parameter set $\boldsymbol{\Phi}_d$

---

in topic $z_i^{(p)}$, which is sampled in line 9. It is calculated using all currently sampled topic assignments, as the equation tells. Next, effective sample size $\hat{N}_{eff}$ is calculated as:

$$\hat{N}_{eff} = 1/\sum_{j=1}^{P}(\omega^{(j)})^2 \tag{3}$$

It is an estimation of N_{eff}, that measures the efficiency of the method and controls the algorithm to avoid degeneracy[6]. A sampling importance resample procedure (line 14) will be run if $\hat{N}_{eff}$ is no more than threshold N_{thresh}. P particles are resampled with replacement according to the importance weights. Then old particles are replaced with the new ones. After this process, particles with small weight have high possibility to be eliminated. This process reflects the "survival of the fittest" law, "excellent" solutions should be inherited. Intuitively, N_{thresh} decides the frequency of resample, and thus influences the effectiveness and speed of the algorithm.

In addition, we add a reassignment period (line 15) after resample to improve the quality of samples. Since words coming in online phase are only sampled once, the result might be inaccurate. We solve this problem by picking up some words randomly, and reassigning topics to them. $R(i)$ is an index set, containing a fixed number of indexes that no more than i. These indexes are randomly selected from $\{1, ..., i\}$, and represent the words reached earlier than word i including word i itself. For each element r in $R(i)$, we sample a new topic according to Equation (2). Obviously, when $|R(i)|$ is big enough, online-LLDA will degenerate to a batch algorithm, since previous words will be reassigned constantly.

Generally, our final objective, the joint distribution $P(\mathbf{z}|\mathbf{w})$ as we mentioned in section 3.2 is approximated as below:

$$P(\mathbf{z}|\mathbf{w}) \approx \sum_{p=1}^{P} \omega^{(p)} I(\mathbf{z}, \mathbf{z}^{(p)}) \qquad (4)$$

where $I(\mathbf{z}, \mathbf{z}^{(p)})$ is an indicator function,

$$I(\mathbf{z}, \mathbf{z}^{(p)}) = \begin{cases} 1 & \text{if } \mathbf{z} \text{ equals } \mathbf{z}^{(p)} \\ 0 & \text{otherwise} \end{cases} \qquad (5)$$

In other words, particles with same vector $\mathbf{z}$, have same $P(\mathbf{z}|\mathbf{w})$. And it equals the sum of weights of these particles. Then, topic assignments $\mathbf{z}^*$ for parameter estimation are calculated as follows:

$$\mathbf{z}^* = \arg\max_{\mathbf{z}^{(p)}} P(\mathbf{z}^{(p)}|\mathbf{w}) \qquad (6)$$

However, in reality, we found that it cost too much time to check two particles whether they share the same $\mathbf{z}$, since $\mathbf{z}$ is a vector whose length is the size of all words. We also found that there were seldom same particles. Therefore, we modify this procedure to choose the particle with biggest weight for estimation. The modified $\mathbf{z}^*$ is:

$$\mathbf{z}^* = \mathbf{z}^{(p^*)}, \quad p^* = \arg\max_{p} \omega^{(p)} \qquad (7)$$

With $\mathbf{z}^*$, we can compute word count and estimate parameters we need. After document d is processed we can get $\mathbf{\Phi}_d$. It contains word distributions, or topics for all labels.

4   Experiment

We evaluated both efficiency and effectiveness of our method on two real datasets through three experiments. The first one compares the interpretability of two algorithms. The second one shows the effectiveness of batch-LLDA and online-LLDA, by using perplexity as metric. And the last one demonstrates the power of our online-LLDA in fast processing document.

4.1   Experiment Setting

The first dataset is a collection of papers crawled from the ACM website[1], which consists of 1712 full papers of four conferences (CIKM,SIGIR,SIGKDD

---

[1] http://dl.acm.org/

and WWW) from the year 2011 to the year 2013. We refer to this corpus as **Conf** in the rest of this paper. We used keywords in the papers as labels. The second dataset, called **Twitter**, is a corpus of tweets downloaded from Twitter, which contains about 2 million tweets. We used hashtags as labels in this corpus. The detailed information of datasets is listed in Table 1. All experiments were run on a server with an Intel Xeon E3-1230 3.3 GHz CPU and 32GB memory.

Table 1. Dataset Description

Dataset	Conf	Twitter
Document Size	1,712	2,180,548
Number of Labels	3,537	5,615
Vocabulary Size	17,697	14,977

In all of our experiments, α and β are fixed at 0.1, and burn-in iteration time is 1000 for batch-LLDA. Since $|R(i)|$, N_{thresh} and P influences the effectiveness and runtime of online-LLDA, we should make a tradeoff. In our experiments, $|R(i)|$ is set as 30 for the smaller dataset Conf and 10 for Twitter, N_{thresh} is 1.5, and P is 10.

4.2   Topic Visualization

In this experiment, we ran both batch-LLDA and online-LLDA over two entire corpora. Then, we picked up top ten words of each topic learned from each algorithm to represent a topic. Each topic is associated with its corresponding label. Table 2 compares the results of two algorithms on **Conf** corpus, while Table 3 shows the **Twitter**'s.

Most topics generated by our online-LLDA are as interpretable as batch-LLDA. And these topics are highly relevant with corresponding labels. For example, in Conf, the top words "topic","topics" and "modeling" explain the label "topic modeling" well; "user", "item" and "recommendation" are frequently used to describe "recommender systems"; In Twitter, when talking about "politics", people are most likely to discuss "found", "obama" and "health".

4.3   Document Modeling

In computational linguistics, perplexity is a widely used metric that measures how well a language model predicts words in the test corpus. Lower perplexity indicates higher likelihood of the test corpus, or better generation performance of the model. Generally, it is computed as Equation (8). Since topics in Labeled LDA are relevant with labels, when a label in the test corpus do not appear in the training corpus, the algorithm will not assign any topics that relates with this label on words. Thus, only prior parameter α can influence the word distribution θ in this case. It means that, in this topic, all words share the same probability to

Table 2. Top ten words from five topics with labels learned in the **Conf** corpus.

Label	Top Ten Words	
	batch-LLDA	online-LLDA
topic modeling	topic, topics, model, document, word, words, number, segment, lda, distribution	topic, topics, modeling, entity, dirichlet, mining, lda, distributions, name, text
deep learning	semantic, gram, embedding, deep, training, latent, vocabulary, layer, vector, hashing	deep, hashing, letter, layer, neural, gram, unsupervised, speech, layers, dimensionality
recommender systems	user, users, item, recommendation, model, based, data, items, recommender, systems	user, item, recommendation, items, data, model, ratings, recommender, system, rating
sentiment analysis	sentiment, set, based, analysis, opinion, words, approach, data, results, pages	sentiment, opinion, based, analysis, classification, pages, performance, data, using, approach
svm	classification, sampling, data, methods, results, sentiment, class, learning, svm, tweets	methods, svm, time, performance, optimization, problem, table, accuracy, words, training

Table 3. Top ten words from five topics with labels learned in the **Twitter** corpus.

Label	Top Ten Words	
	batch-LLDA	online-LLDA
apple	apple, iphone, ipod, new, itunes, event, touch, via, snow, nano	apple, ipod, iphone, new, itunes, event, snow, touch, via, leopard
dogs	dogs, dog, new, please, korean, million, animals, pls, news, too	dogs, dog, pets, update, pet, animals, new, cats, please, pls
gift	gift, super, shop, album, handmade, rock, art, music, cover, recycled	gift, perfect, lover, consider, friend, love, store, jacket, any, friends
photography	photography, photo, photos, new, via, digital, camera, how, photographer, get	photography, photo, new, photos, via, digital, how, camera, get, blog
politics	politics, found, obama, news, health, speech, us, obamas, care, glenn	politics, found, obama, health, speech, care, obamas, glenn, beck, us

be generated, which leads to low ability of predicting words in unseen documents and high perplexity.

$$
perplexity(D_{test}) = \exp \left\{ - \frac{\sum\limits_{d=1}^{M} \log p(\mathbf{w}_d)}{\sum\limits_{d=1}^{M} N_d} \right\} \tag{8}
$$

In this experiment, we simulated the situation that documents are coming in a stream. In our online-LLDA, model parameters are updated every time a new document arrives. We use 200 documents in Conf and 5% tweets in Twitter for initialization. For the sake of fairness, we excluded topics that were sampled in initialization phase when generating the model. We computed perplexity of the held-out test dataset at some points using the current model learned by online-LLDA. As for batch-LLDA, we computed perplexity at the same points, however each run has to use all of the documents previously seen.

Figure (2) shows that, in both corpora, perplexity is much higher at the start, and then declines as the seen documents number increases. It is reasonable since the seen documents used for training are not enough at first, and lack most of labels in test corpus. As the training dataset grows, the learned model fits the test corpus better, and perplexity converges. In Conf, we achieved better effect than batch-LLDA. In Twitter, the two curves are very close, and online-LLDA's is lower than batch-LLDA's as we can see in the zoomed fragment. It's not surprise since online-LLDA incorporates a resample process and a reassignment period. Also, better particles will remain according to the algorithm, which leads to better effectiveness. Since the number of tweets is large, and a single tweet is short, the result in Twitter is satisfactory. Notice that the document numbers in Twitter are in logarithmic scale.

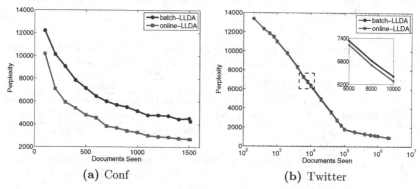

(a) Conf   (b) Twitter

Fig. 2. Perplexity On Two Datasets

4.4   Efficiency

In this experiment, we evaluated the efficiency for each algorithm, which is the main purpose of online-LLDA. We used the same points as we described in section 4.3, and recorded the training time using current observed documents.

When new documents arrive, batch-LLDA has to run over all documents for many iterations again. Since the time for each iteration grows with the number of documents, the total time for batch-LLDA grows fast. However, online-LLDA does not need to do this, it should only process the new coming document and update parameters to get a new model. As Figure (3) shows, batch-LLDA costs much more time to get the new parameters compared with online-LLDA, especially when the number of observed documents is large. When running over the Conf corpus, the training time of online-LLDA is less than 400 seconds, while batch-LLDA takes more than 3,000 seconds, which is about 8 times longer. In twitter, batch-LLDA takes more than 10,000 seconds, while oline-LLDA takes less than 6,000 seconds.

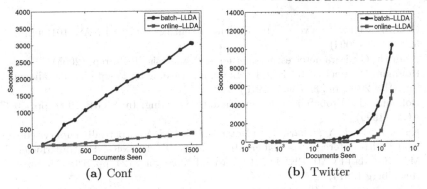

Fig. 3. Training Time On Two Datasets

5   Conclusion and Future Work

In this paper, we proposed an online method for Labeled LDA inference, called online-LLDA. We analyzed the feasibility of applying particle filter, a MCMC method on Labeled LDA and demonstrated that it's a feasible choice for Labeled LDA. We presented our algorithm in detail and conducted several experiments on two real datasets. Our experiment results clearly show that online-LLDA performs as good as batch-LLDA, and costs much less time.

In the future, we will explore the following directions. (1) We will speed up this algorithm through optimizing the data structure of particle and word count. (2) We will try other types of method, including online version of variational Bayesian, expectation propagation, belief propagation, and so on.

Acknowledgment. The work was supported by National Natural Science Foundation of China (Grant No. 61402036 and No. 3070021501109), and 863 Program of China (Grant No. 2015AA015404).

References

1. AlSumait, L., Barbará, D., Domeniconi, C.: On-line lda: Adaptive topic models for mining text streams with applications to topic detection and tracking. In: ICDM 2008, pp. 3–12. IEEE (2008)
2. Banerjee, A., Basu, S.: Topic models over text streams: A study of batch and online unsupervised learning. In: SDM, vol. 7, pp. 437–442. SIAM (2007)
3. Blei, D.M., Ng, A.Y., Jordan, M.I.: Latent dirichlet allocation. The Journal of Machine Learning Research 3, 993–1022 (2003)
4. Canini, K.R., Shi, L., Griffiths, T.L.: Online inference of topics with latent dirichlet allocation. In: International Conference on Artificial Intelligence and Statistics, pp. 65–72 (2009)
5. Doucet, A., De Freitas, N., Murphy, K., Russell, S.: Rao-blackwellised particle filtering for dynamic bayesian networks. In: UAI 2000, pp. 176–183. Morgan Kaufmann Publishers Inc. (2000)
6. Doucet, A., Godsill, S., Andrieu, C.: On sequential monte carlo sampling methods for bayesian filtering. Statistics and Computing 10(3), 197–208 (2000)

7. Griffiths, T.L., Steyvers, M.: Finding scientific topics. PNAS 101(suppl. 1), 5228–5235 (2004)
8. Heinrich, G.: Parameter estimation for text analysis. Tech. rep. (2005)
9. Hoffman, M., Bach, F.R., Blei, D.M.: Online learning for latent dirichlet allocation. In: NIPS 2010, pp. 856–864 (2010)
10. Hofmann, T.: Probabilistic latent semantic indexing. In: SIGIR 1999, pp. 50–57. ACM (1999)
11. Kang, D., Park, Y., Chari, S.N.: Hetero-labeled LDA: A partially supervised topic model with heterogeneous labels. In: Calders, T., Esposito, F., Hüllermeier, E., Meo, R. (eds.) ECML PKDD 2014, Part I. LNCS, vol. 8724, pp. 640–655. Springer, Heidelberg (2014)
12. Li, X., Ouyang, J., Zhou, X.: Supervised topic models for multi-label classification. Neurocomputing (2014)
13. Minka, T., Lafferty, J.: Expectation-propagation for the generative aspect model. In: UAI 2002, pp. 352–359. Morgan Kaufmann Publishers Inc. (2002)
14. Ramage, D., Hall, D., Nallapati, R., Manning, C.D.: Labeled lda: A supervised topic model for credit attribution in multi-labeled corpora. In: EMNLP 2009, pp. 248–256. Association for Computational Linguistics (2009)
15. Ramage, D., Manning, C.D., Dumais, S.: Partially labeled topic models for interpretable text mining. In: KDD 2011, pp. 457–465. ACM (2011)
16. Rubin, T.N., Chambers, A., Smyth, P., Steyvers, M.: Statistical topic models for multi-label document classification. Machine Learning 88(1-2), 157–208 (2012)
17. Teh, Y.W., Newman, D., Welling, M.: A collapsed variational bayesian inference algorithm for latent dirichlet allocation. In: NIPS 2006, pp. 1353–1360 (2006)
18. Teh, Y.W., Jordan, M.I., Beal, M.J., Blei, D.M.: Hierarchical dirichlet processes. Journal of the American Statistical Association 101(476) (2006)
19. Wang, Y., Agichtein, E., Benzi, M.: Tm-lda: efficient online modeling of latent topic transitions in social media. In: KDD 2012, pp. 123–131. ACM (2012)
20. Yao, L., Mimno, D., McCallum, A.: Efficient methods for topic model inference on streaming document collections. In: KDD 2009, pp. 937–946. ACM (2009)
21. Zeng, J., Cheung, W.K., Liu, J.: Learning topic models by belief propagation. IEEE Transactions on Pattern Analysis and Machine Intelligence 35(5), 1121–1134 (2013)

Efficient Buffer Management for PCM-Enhanced Hybrid Memory Architecture

Kaimeng Chen[1], Peiquan Jin[1,2], and Lihua Yue [1,2]

[1]School of Computer Science and Technology,
University of Science and Technology of China, Hefei 230027, China
[2]Key Laboratory of Electromagnetic Space Information, Chinese Academy of Sciences,
Hefei 230027, China
jpq@ustc.edu.cn

Abstract. Recently, *phase change memory* (PCM) has been considered in memory architecture to serve as an extension to DRAM, due to its special properties of byte-addressibility, low energy consumption, and high read performance. However, PCM has a lower write speed than DRAM. Besides, it has a limited write endurance. Therefore, the co-existence of PCM and DRAM in main memory urges a careful buffer-management policy to avoid frequent writes to PCM. To address this problem, we present the first approach that reduces PCM writes by efficient page exchanges and page replacements. Specially, we propose two clock data structures to maintain DRAM and PCM pages, and devise a page exchange method to make recently-updated pages reside in DRAM. In addition, differing from previous studies that do not consider the influence of page replacements on PCM writes, we present a new page replacement algorithm to reduce page replacement on PCM. With this mechanism, we can reduce PCM writes efficiently while keeping a high hit ratio. We conduct trace-driven experiments on both synthetic and real traces. The experimental results suggest that our proposal can greatly reduce PCM writes and maintain a high hit ratio for PCM/DRAM-based hybrid memory architecture.

Keywords: Phase change memory, Hybrid memory, Buffer management.

1 Introduction

Phase change memory (PCM) is one type of random-access memories. Compared with DRAM, PCM has the advantage of non-volatility, high density, and low energy consumption. On the other hand, it is superior to other non-volatile storage such as flash memory in that it supports byte addressability and has higher access speeds. Thus, PCM is regarded as an alternative of future random-access memories and receives much attention in recent years [1-3]. However, PCM has some limitations, e.g., high write latency, limited lifecycle, high price, and slower access speed than DRAM. Therefore, it is not feasible to completely replace DRAM with PCM in current memory architectures. Instead, a more exciting idea is to use PCM to enhance DRAM and construct a hybrid buffer for operating systems or database systems, as shown in Fig. 1. Thus, we can utilize the advantages of both PCM and DRAM [4-6].

© Springer International Publishing Switzerland 2015
R. Cheng et al. (Eds.): APWeb 2015, LNCS 9313, pp. 29–40, 2015.
DOI: 10.1007/978-3-319-25255-1_3

One challenge for PCM/DRAM-based hybrid memory architecture is that we have to cope with heterogeneous memories. Traditional buffer management schemes are mostly designed on the basis of DRAM-only main memory and do not consider the special properties of PCM. Simply employing an existing buffer management policy for PCM/DRAM-based hybrid memory systems will lead to too many writes to PCM, which will worsen overall performance and shorten the lifecycle of PCM. An intuitive approach to reduce PCM writes is to make write requests focusing in DRAM [4-6]. However, a read request will also introduce a write to PCM, if it triggers a page replacement on PCM.

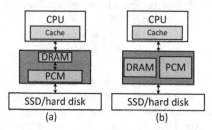

Fig. 1. PCM-enhanced hybrid memory architecture

In this paper, we present an efficient buffer management scheme for PCM/DRAM-based hybrid memory systems shown in Fig. 1(b). Differing from previous works, we not only aim to keep write-intensive pages in DRAM and thus make most writes focus on DRAM, but also try to avoid frequent page-replacements on PCM (because they will lead to extra PCM writes). In summary, we make the following contributions in this paper:

(1) We present a new buffer management policy named D-CLOCK for PCM/DRAM-based hybrid memory architecture. D-CLOCK uses two clock data structures to maintain DRAM and PCM pages, and employs a page exchange method to make write-intensive pages reside in DRAM. In addition, we present a new page replacement algorithm to reduce page replacements on PCM. To the best of our knowledge, our work is the first one considering optimizing page replacement algorithms to reduce PCM writes in hybrid memory. (**Section 3**)

(2) We conduct trace-driven experiments in a simulated PCM/DRAM-based hybrid main memory environment under six synthetic traces and one real OLTP trace, and compare our proposal with several baseline methods. The experimental results suggest that our proposal can greatly reduce PCM writes and maintain a high hit ratio for PCM/DRAM-based hybrid memories. (**Section 4**)

2   Related Work

Due to the different access latencies and write endurance of PCM and DRAM in hybrid memory architecture, conventional page replacement algorithms designed for DRAM-based memory are not suitable for managing pages in hybrid memory. For conventional page replacement algorithms, hit ratio is a key metric to evaluate the performance of page replacement algorithms, but in PCM-enhanced hybrid memory architecture, PCM write count becomes another critical metric.

The LRU algorithm is one conventional page replacement algorithm that has been widely used in database systems based on disks or new storage devices such as flash memory [7, 8]. LRU sorts pages in memory according to their recent reference time. When a page fault occurs and no free frames in memory, the least recently used page is selected as victim for page replacement. LRU shows good performance for temporal locality workloads [9], but it has the overhead of page movement in the list. The CLOCK algorithm was proposed as an approximation to LRU [10]. It organizes pages in memory as a circular list and sets a reference bit for each page. When a page hits in memory, its reference bit is set to 1. When a page fault occurs, CLOCK starts from the page next to the last evicted page and sequentially scans pages in the circular list to find a page with a reference bit 0. When a page with reference bit 1 is checked, its reference bit is set to 0.

Since the write latency of PCM is larger than DRAM and it has limit endurance, page replacement algorithms for hybrid memory architecture have to consider reducing PCM writes to improve the overall performance. So far, based on conventional algorithms, some page replacement algorithms aiming to reduce PCM writes in hybrid memory have been presented. CLOCK-DWF is a CLOCK-based algorithm for hybrid memory [4]. It manages DRAM and PCM separately, places pages associated with read requests in PCM and pages for write requests on DRAM. When a page in PCM is hit by a write request, the page is moved to DRAM. To get a free frame in DRAM, CLOCK-DWF moves a write-cold page in DRAM to PCM. To get a free frame in PCM, it uses the CLOCK algorithm in PCM to free a page. APP-LRU is an LRU-based buffer management scheme for PCM-based hybrid memory [5]. It records the access history of each page using a history table to identify the read and write intensity of pages. As a consequence, read-intensive pages are stored in PCM and write-intensive pages are saved in DRAM. When a page fault occurs, APP-LRU frees the page in the LRU position, and identifies the requested page as a DRAM page or a PCM page according to its historical access pattern and request type. MHR-LRU [6] is also an LRU-based algorithm for hybrid memory. It performs page migration between PCM and DRAM to reduce PCM writes and maintain a high hit ratio.

CLOCK-DWF, APP-LRU, and MHR-LRU are all designed to keep write-intensive pages in DRAM to absorb future write requests, but they do not consider the influence of page replacements in PCM writes. On the other hand, a page replacement in PCM will introduce a write operation to PCM, and thus it is better to devise a new page replacement scheme for PCM-enhanced hybrid memory so that we can reduce the PCM writes incurred by inappropriate page replacements.

3   D-Clock

3.1   Main Idea

The motivation of D-CLOCK is twofold. First, we need to let most write-intensive pages reside in DRAM so as to reduce writes to PCM. Second, we aim to avoid frequent page-replacements in PCM because each page replacement will introduce a write operation in PCM.

Therefore, we propose two solutions in D-CLOCK to reduce writes to PCM. Particularly, we propose a page exchange method to put write-intensive pages in DRAM and an improved clock-based page replacement algorithm to further reduce PCM writes.

Our design on page exchange and page replacement is based on a dual-clock (D-CLOCK) buffer structure, which is shown in Figure 2. The D-CLOCK structure includes a *hybrid-clock* structure and an additional *DRAM-clock* structure. The *hybrid-clock* structure maintains all the frames in PCM and DRAM, while the additional *DRAM-clock* structure only maintains DRAM frames. The DRAM-clock is used to select a page from DRAM, which is needed in our page-exchange algorithm and page-replacement algorithm. Similar with the traditional clock approach, each page in the hybrid-clock and DRAM-clock has a reference bit. In addition, each DRAM page has a write-count field that records the write count of the DRAM page since it is used for a disk page.

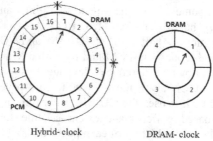

Fig. 2. The dual-clock (D-CLOCK) buffer structure

3.2 Page Exchange

We present a page exchange method to keep write-intensive pages in DRAM, thus we can reduce PCM writes. Specially, we consider two situations where a page exchange is needed:

(1) When a page in PCM is hit by a write request, we select a page in DRAM and exchange it with a PCM page. According to the temporal locality of page requests, we can expect that a recently-written page is much likely to be updated by next requests. Therefore, it is better to put recently-updated pages in DRAM and thus future updates can be concentrated on DRAM rather than in PCM. Fig. 3(a) shows an example of this kind of page exchanges.

(2) When a page replacement is triggered by a write request and the victim is in PCM, we exchange the victim page in PCM with a DRAM page. Thus, the requested page will be finally put in DRAM. The objective of this kind of page exchanges is similar to that of (1), i.e., to put recently-updated pages in DRAM. Fig. 3(b) shows an example of this kind of page exchanges.

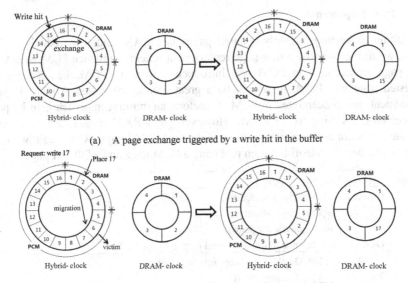

(a)   A page exchange triggered by a write hit in the buffer

(b)   A page exchange triggered by a page replacement in PCM

Fig. 3. Page-exchange examples of D-CLOCK

Algorithm 1 presents the page exchange process. The most critical part in the algorithm is selecting an appropriate DRAM page for exchange. As we aim to put write-intensive pages in DRAM, it is reasonable to select a read-intensive page from DRAM. According to this idea, we compare the write count of a DRAM page with the average write count of all DRAM pages, i.e., T_w. If the write count of the page is less than T_w, the page is selected as the victim to be exchanged with a PCM page. Otherwise, we decrease the write count of the page by an aging factor α and continue to check remaining pages in DRAM-clock.

Algorithm 1: *page_exchange*

Input: *D: DRAM-clock* structure
       p: the page in PCM to be exchanged

Output: a reference to the exchanged page *p* in DRAM

1:   *s* ← the page pointed by the clock-hand of *DRAM-clock*
2:   **while** (true) **do**
3:      **if** *s.write_count* $< T_w$ **then** // T_w *is the average write count of DRAM pages*
4:         exchange the location between *p* and *s*;
5:         **return** a reference to *p* in DRAM;
5:      **else**
6:         *s.write_count=s.write_count−α*;
7:         *s* = the next frame to *s* in DRAM-clock;

3.3   Page Replacement

In addition to keeping recently-updated pages in DRAM and thus reducing PCM writes, we further propose a new page replacement scheme to reduce PCM writes. As each page replacement in PCM will introduce a write to PCM, i.e., writing the requested page into PCM, there will be a great number of writes to PCM if page replacements are concentrated in PCM. Therefore, an intuitive approach is to let page replacements focusing on DRAM. However, if DRAM pages are frequently requested, it is not appropriate to always replace a DRAM page. As a result, we need to consider a better tradeoff between replacing a PCM page and a DRAM page.

Algorithm 2: *page_request*

Input: D: DRAM-clock structure

      H: Hybrid-clock structure

      a page request q for page p

Output: the reference to the requested page p in the buffer

1:   **if** p is in DRAM **then**   /*page hits in DRAM*/

2:     **if** q is a read request **then**

3:        $p.reference\_bit = 1$;

4:     **else**

5:        $p.write\_count++$; $p.reference\_bit = 1$;

6:     **return** p;

7:   **if** p is in PCM **then**   /*page hits in PCM*/

8:     **if** q is a read request **then**

9:        $p.reference\_bit = 1$;

10:    **else**

11:       $p \leftarrow page\_exchange(D, p)$;

12:       $p.write\_count++$; $p.reference\_bit = 1$;

13:    **return** p;

     /*page miss*/

14: **if** $W_{bal} < (PCM\ size\ /\ DRAM\ size)$ **then**

15:    get a free page v in *Hybrid-Clock* by the clock algorithm;

16:    **if** v is a PCM frame **then**

17:      $W_{bal} = W_{bal} + 1$;

18:    **elseif** v is a DRAM frame and $W_{bal}\ != 0$ **then**

19:      $W_{bal} = W_{bal} - 1$;

20:    **if** q is a write request and v is a PCM frame **then**

21:      $s \leftarrow page\_exchange(D, v)$;

22:      insert p into s;

23:    **else**

24:      insert p into v;

25: **else** /* limit the number of replacements in PCM*/

26:    get free page v in *DRAM-Clock* by the clock algorithm;

27:    insert p into v;

28:    $W_{bal} = 0$; /*reset W_{bal}*/

29: **return** p;

In this paper, we introduce a write-balance factor W_{bal} to control the tradeoff between replacing PCM and DRAM. When a PCM page is selected as victim for page replacement, we increase W_{bal} increase by 1. When a DRAM page is selected as victim and we decrease W_{bal} by 1 (unless $W_{bal} = 0$). Thus, a higher value of W_{bal} indicates that there have been many page-replacements occurring in PCM.

Algorithm 2 shows the detailed page-request algorithm of D-CLOCK. When a DRAM page is hit by a request, we set its reference bit to 1, and increase its write count by 1 if the request is a write request (Line 1~6). When a PCM page is hit by a request, we set its reference bit to 1 if the request is a read request; otherwise we exchange the PCM page with a DRAM page. Then, the reference bit of the requested page (now in DRAM) is set to 1 and its write count is increased by 1 (Line 7~14). If a request does not hit on both PCM and DRAM, we need to perform a page replacement operation. In order to determine whether a replacement should be performed in PCM or DRAM, we check the write-balance factor W_{bal}. If $W_{bal} < \frac{PCM\ Size}{DRAM\ Size}$, we get a free page from the hybrid-clock. If $W_{bal} \geq \frac{PCM\ Size}{DRAM\ Size}$, we get a free page from the DRAM-clock.

Here, we set the ratio between PCM size and DRAM size as the threshold. This threshold is based on the experimental result shown in Fig. 4, where we implement the traditional clock algorithm to find out the change of the number of page replacements in PCM with respect to different ratios between the PCM size and the DRAM size. The result indicates that a higher ratio between PCM and DRAM will incur more page-replacements in PCM. Therefore, in this paper we set the threshold as the ratio between PCM and DRAM.

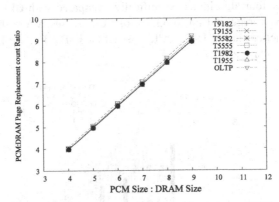

Fig. 4. Page replacements in PCM w.r.t. different ratios between PCM and DRAM

4   Performance Evaluation

4.1   Experimental Settings

We implement a hybrid-memory simulator to evaluate the performance of our proposal and four previous methods, including CLOCK, CLOCK-DWF [4], APP-LRU [5], and MHR-LRU [6]. APP-LRU and MHR-LRU are both based on LRU, and the other two are based on a clock buffer structure. The default ratio between DRAM and PCM is set to 1:4. The parameters used in CLOCK-DWF and APP-LRU are the same as that in the original papers [4, 5].

We conduct trace-driven experiments. Specially, we use DISKSIM [13] to generate six types of traces with different access patterns, which are shown in Table 1. In addition, we use a real OLTP trace from a bank system.

Table 1. Synthetic traces and real OLTP trace

Type	Total References	Different Pages Accessed	Read/Write Ratio	Locality
T9182	300,000	10,000	90% / 10%	80% / 20%
T9155	300,000	10,000	90% / 10%	50% / 50%
T5582	300,000	10,000	50% / 50%	80% / 20%
T5555	300,000	10,000	50% / 50%	50% / 50%
T1982	300,000	10,000	10% / 90%	80% / 20%
T1955	300,000	10,000	10% / 90%	50% / 50%
OLTP	607,390	51,880	77% / 23%	

4.2   Write Count on PCM

The PCM write counts of D-CLOCK and other algorithms are shown in Fig. 5 and Fig. 6. For the synthetic traces, D-CLOCK performs better on reducing PCM writes than all the other four algorithms. Specifically, compared with CLOCK, D-CLOCK reduces 27% on average and up to 43.7% of PCM writes, while the average PCM write-count reduction rate of D-CLOCK over CLOCK-DWF, APP-LRU, and MHR-LRU is 7.09%, 6.7%, and 13.7%.

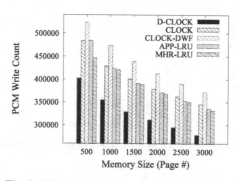

Fig. 5. PCM write count on OLTP workloads

As Fig. 5 shows, D-CLOCK also has the least PCM write-count than its competitors. In particular, it reduces 14.8% on average of PCM writes over the CLOCK algorithm.

Specially, under the read-intensive workloads, as shown in Fig. 6(a) and (b), D-CLOCK can still efficiently reduce PCM writes. Moreover, under low-hit-ratio situations, such as Fig. 6(b), (d), and (f), or high-locality workloads with small memory size, D-CLOCK can greatly reduce PCM writes than CLOCK-DWF, APP-LRU, and MHR-LRU. These results show that the page-replacement scheme of D-CLOCK is efficient on reducing PCM writes.

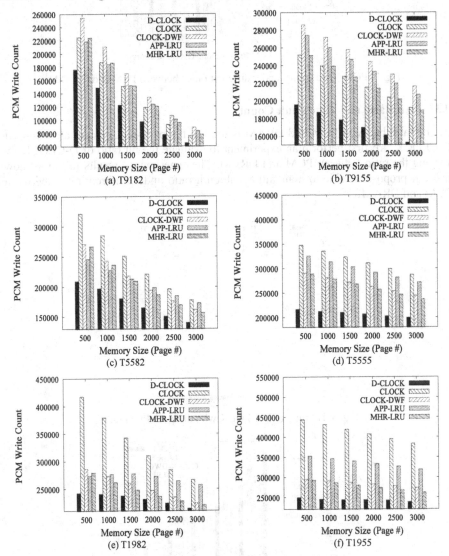

Fig. 6. PCM write counts on synthetic workloads

We also measure the PCM write counts of our proposal under different ratios between DRAM and PCM. In this experiment, we use the OLTP trace and Fig. 7 shows the results. Fig. 7 shows that D-CLOCK can adapt to the change of ratio between DRAM and PCM.

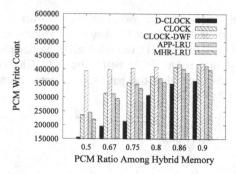

Fig. 7. PCM write counts under different ratios between PCM and DRAM

4.3   Hit Ratio and Page Fault Count

As the ratio between PCM and DRAM is used in the page-replacement algorithm of D-CLOCK, we first perform an experiment to evaluate the hit ratio of D-CLOCK by varying the ratio between PCM and DRAM. The experimental results in Fig. 8 show that our proposal is able to maintain a stable hit ratio under different ratios between PCM and DRAM.

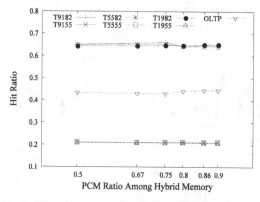

Fig. 8. Hit ratio of D-CLOCK with different PCM ratio

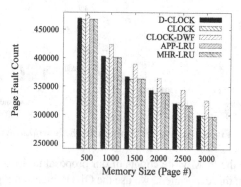

Fig. 9. Page fault count on the OLTP trace

Next, we use the number of page faults to measure hit ratio and compare the hit ratio of D-CLOCK with other algorithms by varying memory size. The experimental results are shown in Fig. 9 and Fig. 10. Compared with CLOCK, D-CLOCK has almost the same number of page faults under different traces and different memory sizes. Specially, for the traces with a high locality, D-CLOCK has fewer page faults than CLOCK and CLOCK-DWF, as shown in Fig. 10(a), (c), and (e). Compared with APP-LRU and MHR-LRU that have the same hit ratio as LRU, D-CLOCK also shows better performance under workloads with a high locality.

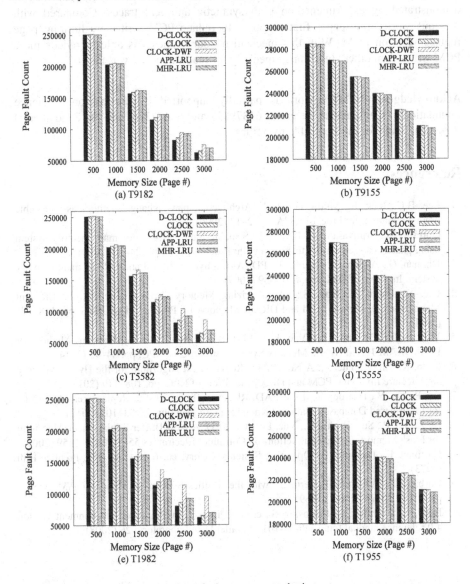

Fig. 10. Page fault count on synthetic traces

5   Conclusion

PCM has emerged as one of the most promising memories to be used in main memory hierarchy. Many studies propose to construct hybrid memory architecture involving PCM and DRAM to utilize the advantages of both media. In this paper, we focus on PCM-enhanced hybrid memory architecture and study the buffer management scheme for such hybrid architecture. Aiming to reduce PCM writes, we propose a page-exchange mechanism and a new page-replacement algorithm whose efficiency is demonstrated by experiments on both synthetic and real traces. Compared with previous works, this paper first proposes to reduce PCM writes by limiting page replacements on PCM. With this mechanism, our proposal is able to reduce more PCM writes than other algorithms especially for read-intensive workloads.

Acknowledgements. This work is partially supported by the National Science Foundation of China under the grant (61379037 and 61472376) and the Fundamental Research Funds for the Central Universities.

References

1. Lee, B.C., Ipek, E., Mutlu, O., et al.: Architecting phase change memory as a scalable dram alternative. In: Proc. of ISCA, pp. 2–13 (2009)
2. Qureshi, M., Srinivasan, V., Rivers, J.: Scalable high performance main memory system using phase-change memory technology. In: Proc. of ISCA, pp. 24–33 (2009)
3. Dhiman, G., Ayoub, R., Rosing, T.: PDRAM: a hybrid PRAM and DRAM main memory system. In: Proc. of DAC, pp. 664–669 (2009)
4. Lee, S., Bahn, H., Noh, S.: Characterizing Memory Write References for Efficient Management of Hybrid PCM and DRAM Memory. In: Proc. of MASCOTS, pp. 168–175 (2011)
5. Wu, Z., Jin, P., Yang, C., Yue, L.: APP-LRU: A New Page Replacement Method for PCM/DRAM-Based Hybrid Memory Systems. In: Proc. of NPC, pp. 84–95 (2014)
6. Chen, K., Jin, P., Yue, L.: A Novel Page Replacement Algorithm for the Hybrid Memory Architecture Involving PCM and DRAM. In: Proc. of NPC, pp. 108–119 (2014)
7. Jin, P., Ou, Y., Haerder, T., Li, Z.: ADLRU: An Efficient Buffer Replacement Algorithm for Flash-based Databases. Data and Knowledge Engineering 72, 83–102 (2012)
8. Li, Z., Jin, P., Su, X., Cui, K., Yue, L.: CCF-LRU: A New Buffer Replacement Algorithm for Flash Memory. IEEE Transactions on Consumer Electronics 55(3), 1351–1359 (2009)
9. Coffman, E., Denning, P.: Operating Systems Theory, ch. 6, pp. 241–283. Prentice Hall (1973)
10. Corbato, F.: A Paging Experiment with the Multics System. In Honor of P.M. Morse, pp. 217–228. MIT Press (1969)
11. Bucy, J., Schindler, J., Schlosser, S., et al.: The disksim simulation environment version 4.0 reference manual (cmu-pdl-08-101). Parallel Data Laboratory: 26 (2008)

Efficient Location-Dependent Skyline Queries in Wireless Broadcast Environments

Yingyuan Xiao[1,2], Pengqiang Ai[1], Hongya Wang[3],
Ching-Hsien Hsu[4], and Wenxiang Cui[1]

[1] Tianjin University of Technology, 300384, Tianjin, China
[2] Tianjin Key Lab of Intelligence Computing and Novel Software Tech., 300384, China
[3] Donghua University, 201620, Shanghai, China
[4] Chung Hua University, 30012, Taiwan

Abstract. We study the problem of location-dependent skyline query (*LDSQ*) processing in wireless broadcast environments in this paper. Compared with answering the skyline queries in a conventional setting, two new issues arise while processing location-dependent skyline in wireless broadcast environments. First, the result of an *LDSQ* is closely related to the query point's location; secondly, query processing strategies must take the linear property of wireless broadcast media and limited battery life of mobile devices into consideration. To address these new issues, this paper proposes an efficient solution for *LDSQ* processing in wireless broadcast environments. In particular, data objects to be disseminated are first divided into two parts via pre-computation in the broadcast server, and then a novel air data organization scheme is designed in the broadcast disk. At the mobile client end, an energy-efficient *LDSQ* processing algorithm is presented. To demonstrate the efficiency of our solution, extensive experiments are conducted along with detailed performance analysis.

Keywords: *LDSQ*, data broadcast, air data organization scheme.

1 Introduction

The ever growing popularity of mobile devices and rapid advent of wireless technology have given rise to a new class of mobile services, i.e., location-based services (LBS). LBS enable users of mobile devices equipped with GPS to search for facilities such as car-parks, shops and restaurants close to their route. In general, mobile clients send location-dependent queries to an LBS server from where the corresponding location-related information is returned as query results. Conventional location-dependent queries (e.g., range query and *k* nearest neighbor query) purely focus on the proximity of objects and hence may not be sufficient for some complex decision-making applications that involve both spatial proximity and non-spatial attributes of objects. Motivated by this observation, a new type of location-dependent queries called *location-dependent skyline query* (*LDSQ*) was recently proposed [1, 2]. *LDSQ* synthesizes the properties of both location-dependent query and skyline query. Specifically, an *LDSQ*

© Springer International Publishing Switzerland 2015
R. Cheng et al. (Eds.): APWeb 2015, LNCS 9313, pp. 41–54, 2015.
DOI: 10.1007/978-3-319-25255-1_4

involves a mobile client q and a set of spatial objects S, with each object in S having a set of non-spatial attributes P. *LDSQ*s issued by mobile clients are to retrieve those objects that are not *location-dependently dominated* by others. Here, an object t is said to location-dependently dominate another object u, if t is closer to the query point q than u and at the same time t dominates u with respect to P.

Different from conventional location-dependent queries, *LDSQ* is able to provide a skyline-like spatial search service in that both the proximity and non-spatial attributes of objects are considered in query processing. However, unlike the regular skyline queries whose result is only determined by the actual query itself and irrelevant to the location of query point, the result of an *LDSQ* is closely relevant to the location of the mobile client, i.e., the location of the mobile client is an important parameter of the *LDSQ*. To answer *LDSQ*s, a typical client/server scheme is employed in [1, 2], i.e., each mobile client sends the *LDSQ* request together with its current location to the LBS server through a GSM/3G/Wi-Fi service provider and the LBS server is responsible for processing each *LDSQ*. While conceptually correct, the client/server scheme may suffer serious scalability problems caused by the ever growing number of mobile devices and the sharply increasing query loads [3]. A promising solution to the above problem is the wireless broadcast model [3, 4]. In this model, the LBS server, acting as a broadcast server, periodically "pushes" data to mobile clients over the wireless broadcast channel, and the mobile clients tune in the broadcast channel and process their queries locally. The main advantage of the broadcast model is that it can support an arbitrary number of users/queries, since no query processing takes place at the LBS server and the network capacity is a constant factor and not affected by the number of mobile clients.

Motivated by scalability challenges faced in the conventional client/server scheme, we study the problem of *LDSQ* processing in wireless broadcast environments in this paper. To the best of our knowledge, it is the first effort to process *LDSQ* in the wireless broadcast model. The linear property of wireless broadcast media and limited battery life of mobile clients make this problem particularly interesting and challenging. Our contributions can be summarized as follows: 1) we propose a novel air data organization scheme for answering *LDSQ*s in wireless broadcast environments; 2) we develop an energy-efficient search algorithm for *LDSQ*s based on the proposed data organization scheme; and 3) we evaluate the performance of the proposed method. The rest of this paper is organized as follows. Section 2 reviews related work. Section 3 defines the problem studied in this work. Section 4 presents the processing techniques for *LDSQ*s in wireless broadcast environments. Section 5 evaluates our method through extensive experiments, and Section 6 concludes this paper.

2   Related Work

In this section, we first review existing work related to skyline query processing and then discuss techniques for location-dependent query processing in wireless broadcast environments.

The skyline operator was introduced into the database community by Borzsonyi et al. [5]. Since then a large number of algorithms have been proposed for conventional skyline queries [6-10]. As an extension of the conventional skyline query, location-dependent skyline query (*LDSQ*) was first proposed in [1]. *LDSQ* integrates skyline query with location-dependent query to provide a skyline-like spatial search service. In the literature [1], the concept of valid scope was formulated to address the validation issue of query results, and the *δ-scanning* algorithm was presented for efficient *LDSQ* processing in a client/server scheme. Furthermore, Xiao et al. [2] proposed a novel *LDSQ* processing method based on peer-to-peer sharing, which leverages query results cached in the neighboring peers to improve the query performance. The main difference between [1, 2] and our work is they employ a client/server scheme while this paper processes *LDSQ* in the wireless broadcast model.

Location-dependent queries in wireless broadcast environments have been investigated in recent years [11-14]. The related work mainly focuses on range queries and *k*NN queries. Literatures [11] construct a linear index structure based on Hilbert Curve for range query processing. Bugra et al. [12] exploit *k*NN query processing in sequential-access R-trees. They evaluate the effect of different broadcast organizations on the tuning time, and propose the use of histograms to enhance the pruning capabilities of the search algorithms. The distributed spatial index (DSI) [13] is proposed for supporting both range and *k*NN queries. The basic idea of DSI is to divide the whole data objects into multiple frames and associate with each frame an index table. The index table maintains information regarding to the Hilbert-Curve values of data objects to be broadcast with specific waiting interval from the current frame. Kyriakos et al. [14] present the Broadcast Grid Index (BGI) method, which is suitable for both snapshot and continuous spatial queries. BGI indexes the data with a regular grid, which partitions data space into square cells of equal size. The abovementioned methods purely focus on the proximity of objects and hence are not suitable for *LDSQ* in which both spatial and non-spatial attributes of objects must be considered simultaneously as search criteria.

3   Preliminary

Let S be the set of spatial objects stored at the LBS server and every object $t \in S$ has both spatial location attribute denoted by $L(t)$ and a set of non-spatial preference attributes denoted by $P = \{p_1, p_2, \ldots, p_m\}$. We use $t[p_i]$ to denote the i-th non-spatial attribute value of t and $d(L(t), L(q))$ to represent the distance between t and query point q where $d(.)$ denotes a distance metric obeying the triangle inequality. As a mobile client usually acts as a query issuer, we will use *mobile client* and *query point* interchangeably in the following sections. Given $t, u \in S$, let $t \prec u$ denote t dominates u with respect to P, i.e., $t \prec u \leftrightarrow (\forall p_i \in P, t[p_i] \leq u[p_i]) \wedge (\exists p_k \in P, t[p_k] < u[p_k])$. We denote a skyline result over S by $sk(S)$, and formally, $sk(S) = \{o \in S | \nexists k \in S \ s.t. \ k \prec o\}$. For convenience, we use $\textbf{ds}(t)$ to denote the set $\{u \in S | u \prec t\}$, and $\textbf{sk}(S)^C$ to denote the set of those objects not belonging to $sk(S)$, i.e., $sk(S)^C = S - sk(S)$. In the following, we formally define the *LDSQ* and related concepts.

Definition 1 (*Location-Dependent Dominance* $\prec_q$): Given two spatial objects t, u and a query point q, we say $t\prec_q u$ with respect to q, if $t\prec u$ with respect to non-spatial preference attributes and $d(L(t), L(q))<d(L(u), L(q))$, formally, $t \prec_q u \leftrightarrow t\prec u \wedge d(L(t), L(q))<d(L(u), L(q))$.

Definition 2 (*Location-dependent Skyline*): Given a set of spatial objects S and a query point q, we define $lsk(S, q)=\{o \in S \mid \nexists k \in S \text{ s.t. } k \prec_q o\}$ as the location-dependent skyline of q.

Definition 3 (*LDSQ*): An *LDSQ* is to find the location-dependent skyline over a given set of spatial objects with respect to a given query point.

Based on the above definitions, we know that $lsk(S, q)$ depends not only on the data set S but is closely related to the location of q. It has been proved in [1] that for any query point q, $sk(S) \subseteq lsk(S, q)$. Let $\Phi(sk(S)^C, q)$ denote the set of objects belonging to $sk(S)^C$ and *not location-dependently dominated* by any other objects of S with respect to q, i.e., $\Phi(sk(S)^C, q) = \{u \mid u \in sk(S)^C \wedge u \in lsk(S, q)\}$. We can easily prove the following theorem.

Theorem 1. For any query point q, $lsk(S, q) = sk(S) + \Phi(sk(S)^C, q)$ holds.

Theorem 1 is self-evident by the above definitions, so its proof is omitted. By Theorem 1, answering an *LDSQ* with respect to a query point q can be transformed into computing the union of $sk(S)$ and $\Phi(sk(S)^C, q)$. As $sk(S)$ is fixed for any query point unless the updates to S happen, we assume $sk(S)$ is known via pre-processing. (i.e., $sk(S)$ can be obtained in advance by executing any regular skyline algorithm over S at the LBS server). Consequently, the crucial issue of answering an *LDSQ* is to how to compute $\Phi(sk(S)^C, q)$ efficiently. Further, we can prove the following theorem.

Theorem 2. For a given $t \in sk(S)^C$, if $\forall k \in \{u \in S \mid d(L(u), L(q)) < d(L(t), L(q))\}$, $k \notin ds(t)$ then $t \in lsk(S, q)$.

Proof: Assume $t \notin lsk(S, q)$, i.e., $\exists k \in S \text{ s.t. } k\prec t \wedge d(L(k), L(q)) < d(L(t), L(q))$. Since $d(L(k), L(q))<d(L(t), L(q))$, $k \in \{u \in S \mid d(L(u), L(q)) < d(L(t), L(q))\}$. And since $k\prec t$, $k \in ds(t)$. Consequently, $\exists k \in \{u \in S \mid d(L(u), L(q)) < d(L(t), L(q))\}$ and $k \in ds(t)$ which leads to a contradiction with the known condition that $\forall k \in \{u \in S \mid d(L(u), L(q)) < d(L(t), L(q))\}$, $k \notin ds(t)$. As a result, the assumption is wrong, i.e., $t \in lsk(S, q)$ is proved.

Theorem 2 provides an insight into how we judge whether an object in $sk(S)^C$ belongs to $lsk(S, q)$ for a given query point q.

4   *LDSQ* Processing in Wireless Broadcast Environments

The wireless broadcast model leverages the computational capabilities of mobile clients, and pushes the query processing tasks to the client side while the broadcast server is only responsible for organizing and broadcasting data. To process *LDSQ* in wireless broadcast environments, the following two issues should be addressed: 1) how to organize the broadcast cycle at the LBS server end, and 2) how to design the efficient *LDSQ* processing algorithm for mobile clients. In this section, we first propose a novel air data organization scheme in Section 4.1. Then, in Section 4.2 we present an efficient algorithm for processing *LDSQ*s at the client side.

4.1 Air Data Organization Scheme

In this subsection, we propose a novel air data organization scheme, called *air scheme based on classification and Z-order curve* (*ASCZ*). Considering the fact that $sk(S)$ is fixed unless the updates to S happens, *ASCZ* first divides S into $sk(S)$ and $sk(S)^C$ using a regular skyline algorithm in the LBS server. Then, Z-order curve is leveraged to order the objects of $sk(S)^C$ based on the spatial proximity. Z-order curve orders the objects of $sk(S)^C$ and clusters them in blocks to facilitate efficient dominance tests and space pruning [10]. Further, these ordered objects are divided into $m \times k$ distinct segments in order, denoted by $sk_1(S)^C$, $sk_2(S)^C$, ..., $sk_k(S)^C$, $sk_{k+1}(S)^C$, $sk_{k+2}(S)^C$, ..., $sk_{2k}(S)^C$, ..., $sk_{mk-k+1}(S)^C$, $sk_{mk-k+2}(S)^C$, ..., $sk_{mk}(S)^C$. Finally, $m \times k$ distinct segments and $sk(S)$ are organized as a broadcast cycle based on (k, m) interleaving scheme, i.e., start with $sk_1(S)^C$, every k consecutive $sk_i(S)^C$ are preceded by a complete $sk(S)$, and $sk(S)$ is repeated m times in a broadcast cycle.

The basic logical unit on a broadcast channel is called a *bucket*. Each bucket is a unit of information that is sent on the broadcast channel. A bucket is made up of a fixed number of *packet*—the smallest unit of message transfer in network, similar to the page concept in conventional storage systems. All buckets are typically assumed to be of the same size for convenience and uniformity. In *ASCZ*, a set of contiguous buckets storing a complete $sk(S)$ is called a *determinable segment* while a set of contiguous buckets storing an $sk_i(S)^C$ is called a *pending segment*. In addition to data objects, each segment contains some extra index information to help locate the designated data segment and assist the space pruning in answering *LDSQ*s. Fig. 1 shows the detailed structures of a determinable segment and a pending segment. The first packet of a data segment provides general control information and is called *header* packet.

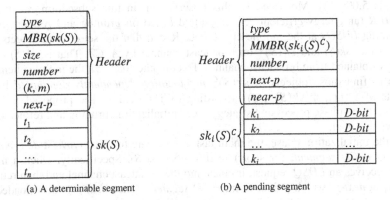

(a) A determinable segment   (b) A pending segment

Fig. 1. The structure of data segments in *ASCZ*

As shown in Fig. 1(a), the header packet of a determinable segment contains (i) *type* (indentifying data segment type, i.e., determinable segment or pending segment), (ii) $MBR(sk(S))$, i.e., the minimum bounding rectangle of $sk(S)$, (iii) *size*, i.e., the object size in bytes, (iv) *number*, i.e., the object number of $sk(S)$, (v) (k, m), i.e., every k consecutive pending segments preceded by a complete determinable segment and the number m of determinable segments per broadcast cycle, and (vi) the offset *next-p* to the beginning of the next determinable segment. Similarly, the header packet of a pending segment, shown in Fig. 1(b), contains (i) *type* (indentifying data segment type, i.e., determinable segment or pending segment), (ii) $MBR(sk_i(S)^C)$, i.e., the minimum bounding rectangle of $sk_i(S)^C$, (iii) *number*, i.e., the object number of $sk_i(S)^C$, (iv) the offset *next-p* to the beginning of the next determinable segment, (v) the offset *near-p* to the beginning of the next pending segment. In addition, a bit (D-bit) is set for each object in pending segments. For an object t in a pending segment, if there is at least one object u at some pending segment satisfying $u \prec t$, the D-bit of t is set as 1; otherwise, its D-bit is set as 0. The organization structure of $ASCZ$ is beneficial to answering $LDSQs$ in broadcast environments because it facilitates efficient dominance tests and the space pruning, as we will illustrate in detail in the sequel.

4.2   The Proposed Method

One naïve solution for processing $LDSQs$ in the wireless broadcast model is that each mobile client listens to the broadcast channel continuously until all data objects are downloaded, and then computes the requested $LDSQ$ locally by means of an existing $LDSQ$ processing algorithm based on client/server scheme, such as δ-scanning algorithm [1]. Clearly, the naïve solution leads to a long tuning time and very high memory requirements at mobile clients, which are prohibitively expensive. Note that mobile devices often have much smaller RAM memory than desktop or laptop computers. More importantly, only a fraction of this memory is available to applications. To remedy the deficiency of the naïve solution, mobile clients should be able to tune selectively based on the characteristics of the aforementioned $ASCZ$ and only buffer those objects which are likely to be location-dependent skyline objects.

To provide energy conservation, a smart mobile device is designed to support two operation modes: *active mode* (i.e., listen mode) and *doze mode* (i.e., the power conserving mode). The ratio of power consumption in the active mode to that in the doze mode is 5,000 [3]. Motivated by this observation, in this subsection, we propose *EESMPR* (an energy-efficient search method based on pruning and refinement) for computing $LDSQs$ at the mobile client side. Recall that the set of data objects S are divided into $sk(S)$ and $sk(S)^C$ on broadcast channel by $ASCZ$. That is, $sk(S)$ can be directly obtained from broadcast channel. To compute $lsk(S, q)$, the mobile client only needs to find those objects in $sk(S)^C$ *not location-dependently dominated* by any objects of S (i.e., $\Phi(sk(S)^C, q)$) according to Theorem 1. The proposed *EESMPR* adopts a three-phase processing strategy, i.e., initialization, pruning and refinement, to achieve this goal.

In the initialization phase, the main task is to locate the *next earliest determinable segment* (*NED segment*, for short) to download $sk(S)$. Specifically, when a mobile client receives an $LDSQ$ request, it tunes into the broadcast channel and then retrieves the offset *next-p*, which points to the *NED segment*. Once *next-p* is downloaded, the mobile client immediately switches to the *doze* mode for saving power until the *NED segment* arrives. When the *NED segment* arrives, the mobile client downloads the whole *NED segment* from the broadcast channel into its own memory space, and then all downloaded data objects (i.e., $sk(S)$) are used to construct a *temple-list*, which will be described in detail. For ease of presentation, we call these pending segments appearing behind the *NED* segment but before the reappearance of the *NED* segment in the next broadcast cycle *waiting pending segments*. For each *waiting pending segment*, the pruning and refining operations are executed at the pruning and refinement phases, which are critical parts of *EESMPR*. Specifically, for each *waiting pending segment*, the pruning phase is first employed to check if the segment can be pruned directly. For those waiting pending segments that are not be pruned directly, the refinement phase is responsible for location-dependent dominance tests to find those *not location-dependently dominated* objects. Clearly, efficient space pruning and location-dependent dominance tests are vital for *EESMPR*.

In order to efficiently judge whether a waiting pending segment can be directly pruned, we have the following theorem, which leverages two distance metrics, i.e., *Mindist(.)* and *Maxdist(.)*. Given a query point q and a minimum bounding rectangle *MBR*, *Mindist(q, MBR)* is defined by 0 when q is inside *MBR* or on the perimeter of *MBR*; otherwise *Mindist(q, MBR)* is the minimum distance between q and *MBR*. Similarly, we simply define *Maxdist(q, MBR)* as the maximum distance between q and *MBR* in this paper.

Theorem 3. Given a query point q and a pending segment $sk_i(S)^C$, if *Mindist(q, $MBR(sk_i(S)^C)$)>Maxdist(q, MBR(sk(S)))*, then $\forall t \in sk_i(S)^C$, $t \notin lsk(S, q)$.

Proof: Assume $\exists k \in sk_i(S)^C$, $k \in lsk(S, q)$. Since $k \in sk_i(S)^C$, there is at least one object $u \in sk(S)$ with $u \prec k$. By the definitions of *Mindist(.)* and *Maxdist(.)*, $d(L(q), L(k)) \geq$ *Mindist(q, $MBR(sk_i(S)^C)$)* and *Maxdist(q, MBR(sk(S))) $\geq d(L(q), L(u))$*. As *Mindist(q, $MBR(sk_i(S)^C)$)) > Maxdist(q, MBR(sk(S)))*, $d(L(q), L(u)) < d(L(q), L(k))$. Consequently, $u \prec_q k$, which contradicts our assumption $k \in lsk(S, q)$. The proof is completed.

Theorem 3 shows if the minimum distance between q and $MBR(sk_i(S)^C)$ is greater than the maximum distance between q and *MBR(sk(S))*, then the pending segment $sk_i(S)^C$ cannot contain any location-dependent skyline object, i.e., $sk_i(S)^C$ can be directly pruned. Compared with the conventional method based on expensive location-dependent dominance tests, the space pruning method based on Theorem 3 only needs to simply compute two distances.

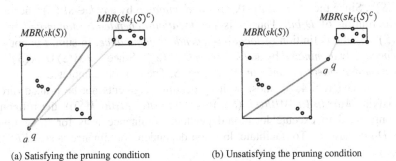

(a) Satisfying the pruning condition   (b) Unsatisfying the pruning condition

Fig. 2. An illustration of Theorem 3

Fig. 2 depicts the two different cases satisfying and unsatisfying the pruning condition in Theorem 3. The case satisfying the pruning condition of Theorem 3 is shown in Fig. 2(a) where the yellow line segment denotes $Mindist(q, MBR(sk_i(S)^C))$, the blue line segment indicates $Maxdist(q, MBR(sk(S))$ and $Mindist(q, MBR(sk_i(S)^C))>Maxdist(q, MBR(sk(S)))$. For this case, the pending segment $sk_i(S)^C$ can be directly pruned according to Theorem 3. Fig. 2(b) shows the other case where the minimum distance between query point q and $MBR(sk_i(S)^C)$ (i.e., the yellow line segment) is less than the maximum distance between q and $MBR(sk(S))$ (i.e., the blue line segment). That is, the pruning condition of Theorem 3 is not satisfied in Fig. 2(b). Therefore, the pending segment $sk_i(S)^C$ cannot be directly pruned in the case and the refinement processing is required. The refinement processing is responsible for location-dependent dominance tests to find all location-dependent skyline objects from those waiting pending segments, which cannot be pruned directly. Because location-dependent dominance test is an expensive operation, it is important for the refinement phase to efficiently reduce the number of executing location-dependent dominance test. Let $sk_i(S)^C$ unsatisfy the pruning condition of Theorem 3 (i.e., not be pruned directly). We have the following lemma through analysis.

Lemma 1. Given a query point q, if $Maxdist(q, MBR(sk_i(S)^C)) \leq Mindist(q, MBR(sk(S)))$, then $\forall t \in sk_i(S)^C$, t is not *location-dependently dominated* by any $k \in sk(S)$.

Proof: By the definitions of $Maxdist(.)$ and $Mindist(.)$, $Maxdist(q, MBR(sk_i(S)^C)) \geq d(L(q), L(t))$ for any $t \in sk_i(S)^C$ and $d(L(q), L(k)) \geq Mindist(q, MBR(sk(S)))$ for any $k \in sk(S)$. Since $Maxdist(q, MBR(sk_i(S)^C)) \leq Mindist(q, MBR(sk(S)))$, $d(L(q), L(t)) \leq d(L(q), L(k))$ for any $t \in sk_i(S)^C$ and any $k \in sk(S)$, which means that $\forall t \in sk_i(S)^C$, t is not *location-dependently dominated* by any $k \in sk(S)$. The proof is completed.

By Lemma 1, we can easily prove the following Theorem 4.

Theorem 4. Given $t \in sk_i(S)^C$ and a query point q, if $Maxdist(q, MBR(sk_i(S)^C)) \leq Mindist(q, MBR(sk(S)))$ and the D-bit of t is 0, then $t \in lsk(S, q)$.

Proof: In order to prove $t \in lsk(S, q)$, we only need to prove t is not *location-dependently dominated* by any $k \in S$ according to the definition of $lsk(S, q)$. First, it is self-evident by Lemma 1 that t is not *location-dependently dominated* by any $k \in sk(S)$. Then, we need to prove t is not *location-dependently dominated* by any $k \in sk(S)^C$. Since the D-bit of t is 0, t is not *dominated* by any $k \in sk(S)^C$ according to the definition of D-bit. Thus, t is not *location-dependently dominated* by any $k \in sk(S)^C$ by the definition of *location-dependent dominance*. In sum, t is not *location-dependently dominated* by any $k \in sk(S) \cup sk(S)^C$. Since $S = sk(S) \cup sk(S)^C$, t is not *location-dependently dominated* by any $k \in S$. The proof is completed.

Based on Theorem 4, for those waiting pending segments not be pruned directly but satisfying $Maxdist(q, MBR(sk_i(S)^C)) \leq Mindist(q, MBR(sk(S)))$, the refinement phase only need to execute location-dependent dominance tests for those objects whose D-bits are 1. To facilitate location-dependent dominance tests, *EESMPR* constructs a *temple-list* in memory space at the mobile client side. At the beginning the *temple-list* contains all objects of $sk(S)$ downloaded from the broadcast channel, together with the distances between these objects and the mobile client (i.e., query point q). That is, for each object t_i of $sk(S)$, the *temple-list* stores an entry of the form $<t_i, d(L(q), L(t_i))>$, and all these entries are sorted in ascending order of their distance values in the *temple-list*.

For each $k \in sk_i(S)^C$ violating the conditions in Theorems 3 and 4, location-dependent dominance test is done based on the *temple-list* according to the following two cases: 1) if k is *location-dependently dominated* by any object in the *temple-list*, it is discarded directly, 2) if k is not *location-dependently dominated* by any object in the *temple-list*, it is inserted in ascending order of distance values, and all objects in the *temple-list location-dependently dominated* by k, if exist, are dropped, as we depict formally in Algorithm 1.

As shown in Algorithm 1, these objects satisfying the conditions listed in Theorem 4 (lines 2-4 in Algorithm 1) are processed according to the abovementioned case 2. For those objects unsatisfying the conditions in Theorem 4, location-dependent dominance test is done by Theorem 2 (lines 7-11 and lines 14-18 in Algorithm 1). The function *MaintenanceList* () is responsible for inserting an object into the *temple-list* in ascending order of distance values and deleting those objects in the *temple-list location-dependently dominated* by the object. The formalized description of *MaintenanceList* () is given in Algorithm 2.

Algorithm 1. *LDominanceTest* $(sk_i(S)^C)$

Input: $sk_i(S)^C$ $(1 \leq i \leq m \times k)$
Output: updated *temple-list*
1: **if** $Maxdist(q, MBR(sk_i(S)^C)) \leq Mindist(q, MBR(sk(S)))$
2: | **for** each $k \in sk_i(S)^C$ **do**
3: | | $flag \leftarrow 1$;
4: | | **if** $k.D\text{-}bit = 0$ **then**   // $k.D\text{-}bit$ denotes the $D\text{-}bit$ of object k
5: | | $MaintenanceList$ $(k, temple\text{-}list)$;
6: | | **else**
7: | | | **for** each $<u, d(L(q), L(u))>$ in the *temple-list* satisfying $d(L(q), L(u)) < d(L(q), L(k))$
8: | | | | **if** $u \prec k$ **then**
9: | | | | $flag \leftarrow 0$; break;
10: | | | **if** $flag = 1$ **then**
11: | | | $MaintenanceList$ $(k, temple\text{-}list)$;
12: **else**
13: | **for** each $k \in sk_i(S)^C$ **do**
14: | | $flag \leftarrow 1$;
15: | | **for** each $<u, d(L(q), L(u))>$ in the *temple-list* satisfying $d(L(q), L(u)) < d(L(q), L(k))$
16: | | | **if** $u \prec k$ **then**
17: | | | $flag \leftarrow 0$; break;
18: | | **if** $flag = 1$ **then**
19: | | | $MaintenanceList$ $(k, temple\text{-}list)$;

Algorithm 2 *MaintenanceList* (*k, temple-list*)

Input: object *k* and the *temple-list*
Output: updated *temple-list*

1: Insert entry $<k, d(L(q), L(k))>$ into the *temple-list* in ascending order of distance values;
2: **for** each $<u, d(L(q), L(u))>$ in the *temple-list* satisfying $d(L(q), L(u) > d(L(q), L(k))$ **do**
3:    **if** $k<u$ **then**
4:        Delete entry $<u, d(L(q), L(u))>$ from the *temple-list*;

Assume that a mobile client (*MC*) receives an *LDSQ* request. Now, we formalize *EESMPR* in Algorithm 3.

Algorithm 3 *EESMPR* ()

Input: the current location of *MC* (i.e., $L(q)$)
Output: *LDSQ* result (i.e., $lsk(S, q)$)

1: *counter* ← 1; *temple-list* ← ∅;
2: *MC* tunes into the broadcast channel for getting the offset *next-p* to the *NED segment*;
3: Once *next-p* being downloaded, *MC* immediately switches to the *doze mode* until the *NED segment* arrives;
4: When the *NED segment* arriving, *MC* downloads the whole *NED segment* and all objects of $sk(S)$ together with the distances from them to $L(q)$ are inserted into *temple-list* in ascending order of distance values;
5: **for** each sequent $sk_i(S)^C$ until *counter* > $m×k$ **do**
6:    *counter* ← *counter*+1;
7:    download the header of the pending segment $sk_i(S)^C$;
8:    **if** $Mindist(q, MBR(sk_i(S)^C)) > Maxdist(q, MBR(sk(S)))$
9:        switch to the *doze mode* until the next pending segment arriving;
10:   **else**
11:       *LDominanceTest* $(sk_i(S)^C)$;
12: return all objects of *temple-list*;

---

5   Performance Evaluation

In this section, we evaluate the performance of the proposed method (*EESMPR*) through extensive experiments. Similar to other works in the literature, the performance metrics of concern in this paper are the *tuning time* and the *access latency*. In what follows, we will first describe the experimental settings and then present the simulation results.

5.1   Experiment Settings

Following the empirical study in [1], we use two real datasets, namely *School* and *NBA*, to evaluate the proposed method. The spatial attributes of *School* represent the 2D geographical coordinates of schools in United States. Each school object has six non-spatial attributes, which are generated randomly. *NBA* contains 13 non-spatial dimensionalities, which corresponds to the statistics of one NBA player's performance in 13aspects. The spatial 2D attributes of *NBA* are generated uniformly. In terms of non-spatial attributes, *School* is an independent data set while *NBA* follows anti-correlated distribution.

A mobile client initiates *LDSQ*s continuously. The *tuning time* and the *access latency* for a single execution of *LDSQ* are measured independently and the location of the mobile client is updated randomly every time a new *LDSQ* is invoked. The experimental results are the average *tuning time* and *access latency* over the total runs of all *LDSQ*s. Due to the lack of comparable baseline methods, our simulation experiments mainly focus on evaluating the impact of different parameter settings over the performance of the proposed method. Table 1 summarizes the main parameters and their settings. In each experiment we vary a single parameter, while setting the remaining ones to fixed values.

Table 1. Main simulation parameters

Parameter	Description	Range / Default
m	The frequency of the determinable segment $sk(S)$ appearing in a broadcast cycle	1~10 / 5
k	The number of consecutive $sk_i(S)^c$ behind a determinable segment $sk(S)$	1~12 / 6
l	The number of the smaller minimum bounding rectangles which are employed to replace $MBR(sk(S))$ in modified $ASCZ$	2~10 / 5
No_1	The number of data objects in *School* dataset	49152
No_2	The number of data objects in *NBA* dataset	17408
bb	Broadcast bandwidth	144kbps

5.2   Simulation Results

First, we study the effect of m and k on the access latency of the proposed method. Fig. 3 depicts the access latency as a function of k under m=1, 2, 4, 6, 8, 10. According to *EESMFR*, the access latency T_{access} of a single execution of *LDSQ* mainly consists of the duration T_{probe} from tuning into the broadcast channel to the *NEA determinable segment* arriving and the time T_{bcast} required by broadcasting all data in a complete broadcast cycle, i.e., $T_{access} \approx T_{probe} + T_{bcast}$. The average T_{probe} is equal to ($\frac{T_{determinable} + k \times T_{pending}}{2}$), where $T_{determinable}$ denotes the needed time of broadcasting a determinable segment and $T_{pending}$ represents the needed time of broadcasting a pending segment. It is easy to see $T_{bcast} = m \times T_{determinable} + m \times k \times T_{pending}$. Note that the size of a pending segment is affected by m and k when the size of data set S is fixed. Thus, $T_{pending}$ varies with m and k. We can see from Fig. 3 that the access latency on *School* dataset is greater than that on *NBA* dataset. This is because the much larger number of objects in *School* dataset causes the longer T_{bcast}. We can also see from Fig. 3 that the access latency increases with k for fixed m, and the rising rate of the access latency become much higher when m is greater. This is not surprising because the increment of k causes the increase of the number of pending segments in a broadcast cycle, and then the increase of the number of pending segments leads to the corresponding growth in header information. In a nutshell, the growth of k causes the increase of T_{bcast} and the increase of m will accelerate the increase of the number of pending segments. Fig. 3(a) shows that the maximum access latency is reached at $m=1$ for *School* dataset. This is because T_{probe} is equal to $\frac{T_{bcast}}{2}$ when $m=1$. Another observation is T_{access} gradually decreases when m grows from 1 to 4 and become the smallest for $m=4$. Then T_{access} begins to increase with m (i.e. $m=4$ to 10). Fig. 3(b) shows the maximum and minimum access latency are obtained when $m=10$ and $m=2$ for *NBA* dataset respectively.

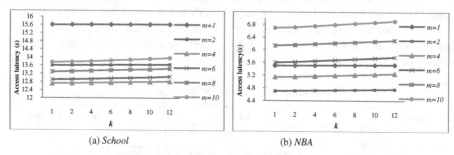

(a) *School*   (b) *NBA*

Fig. 3. Access latency vs. k at $m=1, 2, 4, 6, 8, 10$

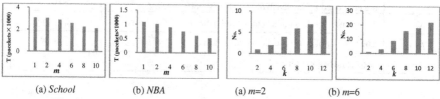

(a) *School*   (b) *NBA*   (a) $m=2$ (b) $m=6$

Fig. 4. Tuning time vs. m **Fig. 5.** Number of the pruned segments vs. k

Then, we study the effect of m and k on the tuning time of the proposed method. We use the number of packets monitored by the mobile client to measure the tuning time because the tuning time is in direct proportion to the number of monitored packets. Fig. 4 plots the tuning time as a function of m when the other parameters are set to their default values. As we expect, the tuning time decreases with the increase of m. This is because the increase of m causes the larger possibility of a pending segment being pruned. Fig. 5 depicts the number of the pruned pending segments as a function of k for $m=2$ and $m=6$, respectively, using *School* dataset. We can see from Fig. 5 that the number of the pruned pending segments increases with k. The reason is the same as the one for Fig. 4. Finally, we study the effect of l on the access latency and tuning time. Fig. 6 illustrates the access latency as a function of l. The results show that the access latency increases slightly with l varying from 2 to 10. The reason is that the increase in l leads to the increase of the length of the broadcast cycle. However, the impact on the length of the broadcast cycle is very limited compared with the size of the dataset itself, and thus the effect of l on the access latency is almost negligible. Fig. 7 shows the tuning time as a function of l. As we expect, the tuning time decreases with l. In addition, we can see from Fig. 7 that the decreasing trend of the tuning time becomes slow with the increase of l. This is not surprising because the fine-grained clustering on $sk(S)$ has a very limited contribution over relaxing the pruning condition when l increases to a certain point.

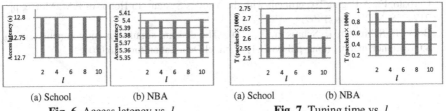

(a) School   (b) NBA   (a) School   (b) NBA

Fig. 6. Access latency vs. l **Fig. 7.** Tuning time vs. l

6   Conclusion

A distinguishing feature of the wireless broadcast model is "one-transmission, infinite-share", which results in almost unlimited scalability. Motivated by the ever growing number of mobile devices and the sharply increasing query loads faced in mobile environments, in this paper we study the problem of *LDSQ* processing in wireless broadcast environments. We propose a novel air data organization scheme based on classification and Z-order curves. Then, an energy-efficient search method based on filter and refinement is present. We conduct extensive experiments and comprehensive performance analysis, which demonstrate the effectiveness and efficiency of our method.

Acknowledgment. This work is supported by the NSF of China (No. 61170174, 61370205) and Tianjin Training plan of University Innovation Team (No.TD12-5016) .

References

1. Lee, B.C., Ipek, E., Mutlu, O., et al.: Architecting phase change memory as a scalable dram alternative. In: Proc. of ISCA, pp. 2–13 (2009)
2. Qureshi, M., Srinivasan, V., Rivers, J.: Scalable high performance main memory system using phase-change memory technology. In: Proc. of ISCA, pp. 24–33 (2009)
3. Dhiman, G., Ayoub, R., Rosing, T.: PDRAM: a hybrid PRAM and DRAM main memory system. In: Proc. of DAC, pp. 664–669 (2009)
4. Lee, S., Bahn, H., Noh, S.: Characterizing Memory Write References for Efficient Management of Hybrid PCM and DRAM Memory. In: Proc. of MASCOTS, pp. 168–175 (2011)
5. Wu, Z., Jin, P., Yang, C., Yue, L.: APP-LRU: A New Page Replacement Method for PCM/DRAM-Based Hybrid Memory Systems. In: Proc. of NPC, pp. 84–95 (2014)
6. Chen, K., Jin, P., Yue, L.: A Novel Page Replacement Algorithm for the Hybrid Memory Architecture Involving PCM and DRAM. In: Proc. of NPC, pp. 108–119 (2014)
7. Jin, P., Ou, Y., Haerder, T., Li, Z.: ADLRU: An Efficient Buffer Replacement Algorithm for Flash-based Databases. Data and Knowledge Engineering 72, 83–102 (2012)

8. Li, Z., Jin, P., Su, X., Cui, K., Yue, L.: CCF-LRU: A New Buffer Replacement Algorithm for Flash Memory. IEEE Transactions on Consumer Electronics 55(3), 1351–1359 (2009)
9. Coffman, E., Denning, P.: Operating Systems Theory, ch. 6, pp. 241–283. Prentice Hall (1973)
10. Corbato, F.: A Paging Experiment with the Multics System. In Honor of P.M. Morse, pp. 217–228. MIT Press (1969)
11. Bucy, J., Schindler, J., Schlosser, S., et al.: The disksim simulation environment version 4.0 reference manual (cmu-pdl-08-101). Parallel Data Laboratory: 26 (2008)

Distance and Friendship: A Distance-Based Model for Link Prediction in Social Networks

Yang Zhang[1] and Jun Pang[1,2]

[1] Faculty of Science, Technology and Communication
[2] Interdisciplinary Centre for Security, Reliability and Trust
University of Luxembourg, Luxembourg
{yang.zhang,jun.pang}@uni.lu

Abstract. With the emerging of location-based social networks, study on the relationship between human mobility and social relationships becomes quantitatively achievable. Understanding it correctly could result in appealing applications, such as targeted advertising and friends recommendation. In this paper, we focus on mining users' relationship based on their mobility information. More specifically, we propose to use distance between two users to predict whether they are friends. We first demonstrate that distance is a useful metric to separate friends and strangers. By considering location popularity together with distance, the difference between friends and strangers gets even larger. Next, we show that distance can be used to perform an effective link prediction. In addition, we discover that certain periods of the day are more social than others. In the end, we use a machine learning classifier to further improve the prediction performance. Extensive experiments on a Twitter dataset collected by ourselves show that our model outperforms the state-of-the-art solution by 30%.

1 Introduction

Online social networks have gained a huge success during the past decade and play an important role in our daily life. For example, people publish statuses on Facebook, read news through Twitter and share photos on Instagram. During the past five years, with the development and deployment of mobile devices, such as smart phones and tablets, people begin to use social network services more often on mobile devices. For example, Facebook has 703 million daily active mobile users, and 30% of all the Facebook users only use mobile devices for Facebook services.[1] One interesting and important service related to mobile social network applications is location sharing which can be achieved through a number of localization and positioning techniques, including the GPS satellite navigation system, Wi-Fi-based positioning system, and mobile phone tracking. Nowadays, it is very common to see that users publish photos or statuses that are labeled with the corresponding locations. Moreover, a new type of social network services, namely location-based social networks (LBSNs), has emerged, e.g., Foursquare and Yelp. In LBSNs, users may just share their location to participate in some kinds of social games or get coupons and discounts from restaurants and shops.

---

[1] http://newsroom.fb.com/

© Springer International Publishing Switzerland 2015
R. Cheng et al. (Eds.): APWeb 2015, LNCS 9313, pp. 55–66, 2015.
DOI: 10.1007/978-3-319-25255-1_5

Human movement or mobility have been studied for a long time. With more and more people's location information becoming available through social networks, quantitative study on human mobility becomes achievable (e.g., see [1,2,3,4,5,6,7]). Understanding human mobility can result in appealing applications such as targeted advertising and friends recommendation. In a broader context, it also helps us to tackle the challenges that we are facing at the moment, e.g., urban planning, disease spread, pollution control, etc.

In this paper, we focus on mining social relationship between users based on their mobility information (see related work in Section 5). The main idea of our method is to measure the geographical distance between two users and use this distance to predict whether they are friends.

Contributions. Our contributions in this work are summarized as follows.

- We profile each user's mobility using his frequent movement areas and propose several metrics to quantify the distance between two users' frequent movement areas. Through data analysis, we discover that distance is an effective metric to separate friends and strangers.
- We exploit an important property of location, namely location popularity, to further adjust the distance between users. The adjusted distance achieves a better performance in differentiating friends and strangers.
- We directly exploit distance between two users to predict their friendship, and the prediction performance is promising.
- Moreover, we discover that certain periods of the day are more social than others w.r.t. friendship prediction accuracy. We combine all the distances under these social time periods into a machine learning classifier to further improve the performance of our friendship predictor.

Through extensive experiments on a Twitter dataset collected by ourselves, we have demonstrated that our predictor outperforms the state-of-the-art solution by 30%. Note that the experiment code as well as the dataset are available upon request.

Organization. After the introduction, we present the notations as well as the dataset used in the paper in Section 2. The relationship between distance and user friendship in LBSNs is analyzed in Section 3. We present our friendship predictor as well as its experimental evaluation in Section 4. Related works are summarized in Section 5. Section 6 concludes the paper with future work.

2 Preliminaries

We first introduce notations we use in the paper, and then describe the dataset we collect to conduct the experiments.

Notations. Each user is denoted by u and set U contains all the users. If two users are friends, then there exists an edge between these two users. The social network thus forms a undirected graph $G = (U, E)$, where E contains all the edges between users. We use ℓ to represent a location and it corresponds to a point denoted by (lat, lon). For two locations ℓ and ℓ', $d(\ell, \ell')$ represents their Euclidean distance. A user visiting

Fig. 1. Check-ins in New York

Table 1. Dataset summary

Starting date	30/11/2014
Ending date	17/02/2015
# of check-ins	2,543,776
# of users	175,566
# of edges	9,490,426
# of active users	4,673
# of edges among active users	37,382

a location, namely check-in, is represented as a tuple $\langle u, t, \ell \rangle$, where t is the time when the check-in happens. Without ambiguity, we use location and check-in interchangeably in the rest of the paper.

Dataset. We exploit Twitter's streaming API[2] to collect geo-tagged tweets in New York city from November 30th, 2014 to February 17th, 2015. When a user shares a geo-tagged tweet at a location, we say that he checks in at that location. In total, we have collected 2,543,776 check-ins from 175,566 users. Each check-in is represented as

$$\langle userID, time, latitude, longitude \rangle.$$

Figure 1 depicts a sample geographical distribution of the check-ins in our dataset. To obtain the social network data, we exploit Twitter's REST API[3] to query each user's followers and followees. Two users are considered friends if they follow each other, this totally gives us a social network with 9,490,426 edges. In the rest of the paper, we only concentrate on users who have at least 50 and no more than 500 check-ins.[4] In the dataset there are totally 4,673 users of this kind, and we refer them as *active users*. The detailed description of the dataset is listed in Table 1.

3 Distance and Friendship

In this section, we first summarize each user's check-ins as his frequent movement areas. Then, based on two users' frequent movement areas, we propose several metrics to quantify the distance between them. In the end, we study the relationship between distance and friendship.

3.1 Frequent Movement Areas

A user does not visit places randomly in a city. Instead, studies have shown that a user's mobility is centered around several points such as home and office [8].

---

[2] https://dev.twitter.com/streaming/overview
[3] https://dev.twitter.com/rest/public
[4] Users with more than 500 check-ins within the time period (2.5 months) are normally public accounts. For instance, @NewYorkCP (New York Press) publishes almost 9,000 tweets at the exact same location.

In Figure 2, the user u mainly visits places around up-east side and Empire state building, while another user u' visits Midtown and Brooklyn often. To measure the geographical distance between two users based on their check-ins, we first need to summarize the places each of them has been, then define the distance based on this summarization. Clustering is the natural solution for this task. In this work, we exploit the centroid-based hierarchical clustering to profile each user's mobility. The cut-off distance of the clustering is set to 500 meters which is a meaningful human movement range. In Section 4, we will further study the sensitivity of this linkage distance.

All the places that a user has visited are then grouped into several clusters, we use the central point of a cluster to represent this cluster. Each central point is named as a *frequent movement area* of the user. A user u's mobility is then summarized by all his frequent movement areas denoted by $m(u)$. The frequent movement areas of the two users u and u' are depicted in Figure 2 as well.

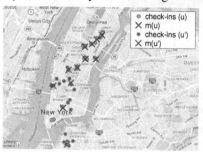

Fig. 2. Frequent movement areas of two users

3.2   Distance Between Users

After profiling each user's check-ins, the distance between two users u and u' can be measured by the distance between their frequent movement areas, i.e., between the two sets $m(u)$ and $m(u')$. Our goal is to measure the distance between these two sets. There are several metrics proposed to quantify the distance between sets. In this work we adopt the following three ones.

Minimal Distance. The minimal distance between u and u' is defined as

$$mind(u, u') = \min(pd(u, u'))$$

where min gives the smallest number in a set of values, and $pd(u, u')$ is the *pairwise distance* between the frequent movement areas of u and u' formally defined as

$$pd(u, u') = \{d(\ell, \ell') \mid \forall \, (\ell, \ell') \in m(u) \times m(u')\}.$$

Here, the minimal distance is simply the distance between two frequent movement areas, one from each user, that are closest to each other.

Average Distance. Besides minimal distance, we also exploit the average distance between u and u' as another metric, it is defined as

$$avgd(u, u') = \frac{\sum_{d(\ell, \ell') \in pd(u, u')} d(\ell, \ell')}{\mid pd(u, u') \mid}.$$

Hausdorff Distance. Hausdorff distance is another classic metric for quantifying the distance between two sets. Here, we define the Hausdorff distance between u and u' as

$$hausd(u, u') = \max\{\sup_{\ell \in m(u)} \inf_{\ell' \in m(u')} d(\ell, \ell'), \sup_{\ell' \in m(u')} \inf_{\ell \in m(u)} d(\ell, \ell')\},$$

where sup represents the supremum and inf the infimum.

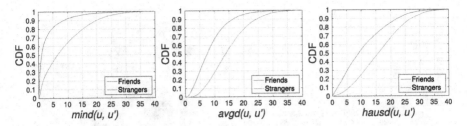

Fig. 3. CDF of distances between friends and strangers

3.3 Distance and Friendship

Previous works [9,8,10] have shown that the distance between two users' home locations is related to their relationship. Concretely, they have discovered that friends tend to live closer than strangers. Generally speaking, we would like to check if friends' frequent movement areas are closer than the ones of strangers.

Figure 3 depicts the cumulative distribution functions (CDF) of distances between friends as well as strangers under the three distance metrics. As we can see, the friends and strangers are easily separable by all the three metrics. In particular, more than 80% of friends' minimal distances are less than 5km while only 40% of strangers' minimal distances have the same value. This indicates that the minimal distance between two users is an effective metric to separate friends and strangers. The same happens with the average distance. On the other hand, the difference driven by the Hausdorff distance is relatively smaller compared with the other two metrics. We further study the relationship of social strength (quantified by embeddedness) and the minimal distance between users. As shown in Figure 4, with the increase of $mind(u, u')$, the embeddedness between them drops. This indicates the minimal distance betwen users is correlated with their social strength as well.

3.4 Location Popularity

The popularity of locations can potentially affect the distance between friends and strangers. For two users who both have frequent movement areas near a popular place, such as a metro station, the chance that they know each other is low since they may just happen to take the metro everyday. On the other hand, as we show in the previous section the short distance between these two frequent movement areas is a strong indicator that these two users are friends. Therefore, to find a meaningful distance metric to separate friends and strangers, it is necessary to take location popularity into account.

In [11], the authors propose a metric named *location entropy* to quantify a location's popularity. It is defined as follows

$$locent(\ell) = - \sum \frac{|ci(\ell, u)|}{|ci(\ell)|} \log \frac{|ci(\ell, u)|}{|ci(\ell)|},$$

where $ci(\ell)$ represents all the check-ins at location ℓ and $ci(\ell, u)$ contains the check-ins of u at ℓ. Popular places have higher location entropies than unpopular ones. Figure 5

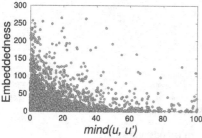
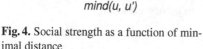
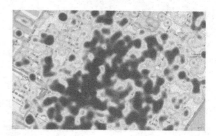

Fig. 4. Social strength as a function of minimal distance

Fig. 5. Heat map of Midtown w.r.t. location entropy

depicts a heatmap of Midtown w.r.t. location entropy. Since we focus on each user's frequent movement areas, we further define the location entropy of a frequent movement area as the average location entropy of all locations in that cluster represented by the frequent movement area. Similarly, more popular a frequent movement area is, its location entropy is also getting higher. Moreover, instead of considering location entropy directly, we use *location diversity* defined as

$$locdiv(\ell) = \exp(locent(\ell))$$

to represent a location's popularity.

After having the distance between two frequent movement areas, we multiply the distance by the maximal location diversity of these two frequent movement areas. The adjusted pairwise distance between u and u', namely $ldpd(u, u')$ is

$$ldpd(u, u') = \{d(\ell, \ell') \cdot \max(locdiv(\ell), locdiv(\ell')) \mid \forall (\ell, \ell') \in m(u) \times m(u')\}.$$

With this adjustment, the distance measure between popular frequent movement areas is increased, while the long ones between unpopular places are reduced. Based on this new pairwise distance, we redefine the minimal and average distances between users accordingly (denoted by $ldmind(u, u')$ and $ldavgd(u, u')$). In addition, the adjusted Hausdorff distance w.r.t. location diversity is defined as

$$ldhausd(u, u') = \max\{ld\_di\_hausd(u, u'), ld\_di\_hau\_d(u', u)\},$$

where $ld\_di\_hausd(u, u')$ is defined as

$$\sup_{\ell \in m(u)} \inf_{\ell' \in m(u')} (d(\ell, \ell') \cdot \max(locdiv(\ell), locdiv(\ell'))).$$

Figure 6 depicts the CDF of the adjusted distances between friends and strangers. As we can see, $ldmind(u, u')$ can better differentiate friends and strangers compared with the other two metrics. Almost 70% of friends' $ldmind(u, u')$ are less than 10 while the value is less than 20% for strangers. Moreover, we notice that this difference is even larger than $mind(u, u')$ depicted in Figure 3. In Section 4, we will demonstrate the effectiveness of this adjusted minimal distance on friendship prediction. On the other hand, the separations driven by $ldavgd(u, u')$ and $ldhausd(u, u')$ get less clear when compared with $avgd(u, u')$ and $hausd(u, u')$.

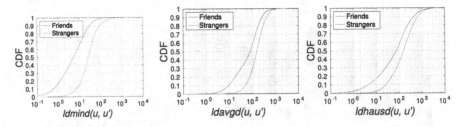

Fig. 6. CDF of adjusted distances between friends and strangers

4   Link Prediction

Link prediction plays an essential part in the context of social networks. For example, recommending friends to a newly joined user is crucial for attracting the user to stay with the social network service. In Section 3, we have shown that friendship is related to distance, i.e., friends' frequent movement areas are closer than strangers. In this section, we aim to predict friendship between users based on their distances.

4.1   Experiment Setup and Metrics

All our experiments are conducted on a machine with 2.6GHz Intel Core i7 processor and 8Gb memory. We extract all the friends pairs from the active users in New York and randomly sample the same number of stranger pairs to construct a balanced dataset.

We exploit ROC curve and three other standard metrics including *AUC* (area under the ROC curve), *Accuracy* and *F1score* to evaluate our friendship predictor. Let *TP*, *FP*, *TN* and *FN* denote true positive, false positive, true negative and false negative, respectively. Accuracy and F1score are defined as the following:

$$Accuracy = \frac{|TP| + |TN|}{|TP| + |FP| + |FN| + |TN|};$$

$$F1score = 2 \cdot \frac{Precision \cdot Recall}{Precision + Recall}, \text{ with}$$

$$Precision = \frac{|TP|}{|TP| + |FP|}, \quad Recall = \frac{|TP|}{|TP| + |FN|}.$$

4.2   Prediction Based on Distances

In Section 3, we have proposed 6 different metrics to quantify the distances between two users including $mind(u, u')$, $avgd(u, u')$, $hausd(u, u')$, $ldmind(u, u')$, $ldavgd(u, u')$ and $ldhausd(u, u')$. We start by directly using these distances to predict whether u and u' are friends or not. More precisely, we tune a threshold τ and predict pairs of users whose distances are less than τ to be friends. Figure 7 shows the AUC value of all the different distances for predicting friendships. Among all the six distances, $ldmind(u, u')$ achieves the best performance (AUC = 0.81) followed by $mind(u, u')$. On the other hand, $ldhausd(u, u')$ has the worst performance, with AUC equals to 0.65.

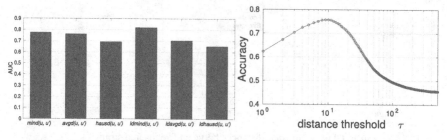

Fig. 7. AUC under different distances   **Fig. 8.** Accuracy vs. threshold τ

This result is consistent with the friends and strangers separation results in Figure 3 and Figure 6, where $ldmind(u, u')$ achieves the best result on differentiating friends and strangers while $ldhausd(u, u')$ performs the worst. We also notice that adding location diversity to distances only improves the perform of the minimal distance. Meanwhile, $ldavgd(u, u')$'s performance gets even worse than $avgd(u, u')$. This indicates that the adjusted average distance cannot capture the distance between users very well. The same happens with the adjusted Hausdorff distance.

Since the results of all the distance-based link predictors are obtained by tuning the threshold τ, we proceed by studying which is the optimal threshold for predicting friendships. Here, we focus on the best performance distance, i.e., $ldmind(u, u')$ and find τ w.r.t. prediction accuracy. As shown in Figure 8, when the threshold τ falls into the range [8, 11], the accuracy achieves the highest value (0.75). In the following experiments, we set τ to 10 when computing accuracy and F1score for $ldmind(u, u')$.

4.3 Parameter Sensitivity

In our whole settings, there is only one parameter to adjust, which is the cut-off distance used in the hierarchical clustering algorithm for finding each user's frequent movement areas (see Section 3.1). For the above evaluation as well as the evaluation in Section 3, we have set it to 500m. Next, we focus on the sensitivity of this cut-off distance.

We have performed friendship predictions through $ldmind(u, u')$ on multiple cut-off distances. As we can see from Figure 9, our predictor achieves the best performance when the cut-off distance falls between [400m, 700m]. Moreover, we also notice that F1score drops when the distance becomes longer (such as 5km and 10km). This is expected considering the human movement range. For example, if the cut-off distance is set to 10km, then a user probably will only have one frequent movement area that covers the whole city. This is too coarse-grained and cannot properly reflect the user's mobility. Based on this analysis, we set the cut-off distance as 500m in our experiments.

4.4 Time and Friendship

So far, our distance-based friendship predictor only considers mobility information from users (i.e., frequent movement areas) and locations (i.e., location popularity). On the other hand, time also plays an essential role on users' mobility. For example, a user may check in at places close his office during the working hours while visiting bars and cinemas at his spare time. Intuitively, if two users are close at a social time, the

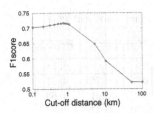
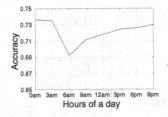
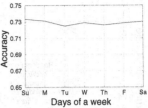

Fig. 9. Parameter vs. F1score   **Fig. 10.** Hour vs. Accuracy   **Fig. 11.** Day vs. Accuracy

probability for them being friends is high. For example, if u and u' are close (appearing at a same metro station) at 8am in the morning, then we are less confident that they are friends, compared with the case when their frequent movement areas are close during the midnight. Next, we verify whether this hypothesis holds in general.

We consider time on the daily scale as well as the weekly scale, respectively. For the daily scale, we divide a day by eight with each part being a three hour range staring from 0am. We extract each user's check-ins under different time periods and construct the corresponding frequent movement areas. The social level of each time periods is evaluated by the accuracy on predicting friendship through $ldmind(u, u')$. For the weekly scale, we consider users' frequent movement areas on each day (from Sunday to Saturday) and perform an evaluation similar to the daily scale.

As shown in Figure 10, on the daily scale, 0-3am is the most social time of the day followed by 3-6am and 9-12pm. This indicates that if two users' frequent movement areas are close at these hours, the chance that they are friends is high compared with other hours. There is a sudden drop when the period is from 6am to 9am, this means that 6am to 9am is the least social hours of the day, most probably because 6-9am is the commuting hours of the day. On the other hand, the prediction accuracy stays stable on a weekly scale with Sunday and Saturday slightly higher than the weekdays (Figure 11).

4.5 Combining Features with Machine Learning

We have shown that different distance measures between users together with location and time information can capture different aspects of friendship. As a consequence, it is natural to ask whether combining all the information together can further improve the results of link prediction.

Feature Description. For two users u and u', we adopt $mind(u, u')$, $avgd(u, u')$ and $hausd(u, u')$ as three features for the classifier. Besides, we take the maximal value and standard deviation of the pairwise distance, i.e., $pd(u, u')$, as another two features as well. We also consider all the above distances' adjusted versions, i.e., taking into account location popularity. This totally gives us 10 features. As some hours are more social than the others on the daily scale, besides the general distance between two users (10 features), we further take their distances at 0-3am and 3-6am into account. Each time period provides 10 features. In total, we collect 30 features for each pair of users.

Learning Techniques. Regarding the machine learning classifier, we have tried logistic regression, support vector machine, gradient boosting machine and random forest. It turns out that random forest outperforms the other algorithms. Thus we adopt random

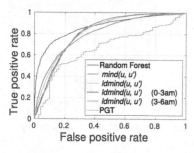

Fig. 12. ROC of Random Forest

Table 2. Performance comparison with PGT

	AUC	Accuracy	F1 score
Random Forest	0.87	0.79	0.78
$ldmind(u, u')$	0.82	0.75	0.71
$mind(u, u')$	0.77	0.61	0.68
$ldmind(u, u')$ (0-3am)	0.82	0.73	0.65
$ldmind(u, u')$ (3-6am)	0.81	0.74	0.68
PGT	0.66	0.62	0.63

forest as our classifier. We put 70% of user pairs into our training set and the rest 30% are used for testing. In all sets, we perform 10-fold cross validation.

Comparison with the State-of-the-Art Model. For baseline models, except for some of the distance-based predictors presented before, we also choose a state-of-the-art friendship predictor that exploits the location information, namely the PGT model, proposed in [12]. For each pair of users, PGT first extracts their meeting events. Then, the model considers personal, global and temporal factors based on the meeting events to discover friendship. For the personal factor, PGT proposes a density-based function for each user. If the place of a meeting event is less visited by the user, PGT will value more on this event. For the global factor, PGT adopts the location entropy to adjust the meeting location popularity. In the end, PGT introduces the temporal factor to penalize the meeting event that happens closely with the following events. According to the experiment results in [12], PGT outperforms other location information based link prediction models including [13,11,14]. In our experiments, we follow the parameter settings as specified in [12] for the PGT model including the distance and time for discovering a meeting event and two parameters used in the personal and temporal factors.

Results. The experiment results including the ROC curve and three evaluation metrics are listed in Figure 12 and Table 2. As we can see, our random forest achieves a good prediction result and it outperforms all the other distance-based predictors. Besides, all our models outperform the PGT model. In particular, the machine learning classifier achieves a 30% improvement. We believe this is because PGT only focuses on meeting events between two users, which is too strict since friends can hang out at the same place but they do not have to check in together. In a broader view, according to the homophily theory [15], friends have similar interests. In the context of mobility, this means that friends tend to visit similar places such as same kinds of shops or bars, and this does not have to happen at the same time. Our model considers two users' mobility at a macro level (through their frequent movement areas) which naturally captures the concept of homophily, while PGT and other solutions such as EBM [14] fail to do so.

5 Related Work

With the development of location-based social networks, research on analyzing the social relationship and mobility has attracted a lot of attention. The research can be roughly partitioned into two groups. One is to use friendship to understand mobility, such as location prediction [8,16] and recommendation [17]; the other is to use mobility information to infer friendship. Our work belongs to the second group.

Authors of [13] propose a probabilistic model to infer friendships from location data shared on Flickr. Their model considers both temporal and spatial information. However, they make a strong assumption that each user only has one friend which is not the case in the real-life applications. Cranshaw et al. [11] propose to use a machine learning classifier to infer the friendship between two users. The features they consider include the ones related to locations as well as the social network structure. Besides, they also propose location entropy to characterize the popularity of a location. The effectiveness of location entropy has been demonstrated in [14,12], we use it to measure the distance between users in this work as well. In [14], the authors propose an entropy-based model, namely EBM. The model first extracts a vector of meeting events between two users, then EBM builds two components based on these meeting events. The first component of EBM is named diversity which is a Rényi entropy formalization on the meeting events vector. More locations two users visit together, better chance they will be friends. The second component is the weighted frequency which exploits location entropy to penalize meeting events at popular locations. Then diversity and weighted frequency are fitted by a linear regression to two users' social strength quantified by Katz score. With the learned model, by tuning a threshold on Katz score, friendship prediction is achieved. More recently, Wang et al. [12] propose the PGT model that we use in this work as the baseline model for friendship prediction.

Advantages of Our Model. Besides prediction performance, our solution has the following advantages compared with the two recently developed models, i.e., EBM and PGT. First, our model is intuitive and easy to implement. There is only one parameter to adjust in our solution which is the cut-off distance used in the clustering algorithm. On the other hand, both EBM and PGT have several parameters to adjust. Second, both EBM and PGT compute their model components based on the locations two users have visited together, i.e., their meeting events. However, meeting events largely depend on the parameter settings such as the time range and location distance. In PGT, the authors consider two check-ins to be a meeting event if their locations' distance is less than 30m and they happen within one hour. How to set these parameters to find a meeting event is rather unclear. Moreover, as mentioned in Section 4, meeting events cannot capture important social behaviors such as homophily. On the other hand, our model studies two users' distance at a macro level which can naturally capture the concept of homophily.

6 Conclusion

In this work, we have proposed to exploit the distance between two users to predict their friendship. We further integrated location popularity and time information into the prediction. With experiments, we have demonstrated the effectiveness of our approach.

There are several directions we would like to pursue in the future. First, we plan to integrate location and time semantics into the process of summarizing users' frequent movement areas. Based on the new summarized frequent movement areas, we aim to further analyze the relation between friendship and distance. Second, our experiments only focus on the check-in data in New York, we also want to check whether our discoveries in this paper generally hold in other cities or countries.

References

1. González, M., Hidalgo, C., Barabási, A.L.: Understanding individual human mobility patterns. Nature **453** (2008) 779–782
2. Song, C., Koren, T., Wang, P., Barabási, A.L.: Modelling the scaling properties of human mobility. Nature Physics **6**(10) (2010) 818–823
3. Simini, F., González, M., Maritan, A., Barabási, A.L.: A universal model for mobility and migration patterns. Nature **484** (2012) 96–100
4. Yan, Z., Chakraborty, D., Parent, C., Spaccapietra, S., Aberer, K.: Semantic trajectories: Mobility data computation and annotation. ACM Transactions on Intelligent Systems and Technology **4**(3) (2013) 49
5. Ying, J.J.C., Lee, W.C., Tseng, V.S.: Mining geographic-temporal-semantic patterns in trajectories for location prediction. ACM Transactions on Intelligent Systems and Technology **5**(1) (2013) 2
6. Chen, X., Pang, J., Xue, R.: Constructing and comparing user mobility profiles for location-based services. In: Proc. 28th ACM Symposium on Applied Computing (SAC), ACM (2013) 261–266
7. Chen, X., Pang, J., Xue, R.: Constructing and comparing user mobility profiles. ACM Transactions on the Web **8**(4) (2014) 21
8. Cho, E., Myers, S.A., Leskovec, J.: Friendship and mobility: user movement in location-based social networks. In: Proc. 17th ACM Conference on Knowledge Discovery and Data Mining (KDD), ACM (2011) 1082–1090
9. Backstrom, L., Sun, E., Marlow, C.: Find me if you can: Improving geographical prediction with social and spatial proximity. In: Proc. 19th International Conference on World Wide Web (WWW), ACM (2010) 61–70
10. McGee, J., Caverlee, J., Cheng, Z.: Location prediction in social media based on tie strength. In: Proc. 22nd ACM International Conference on Information & Knowledge Management (CIKM), ACM (2013) 459–468
11. Cranshaw, J., Toch, E., Hone, J., Kittur, A., Sadeh, N.: Bridging the gap between physical location and online social networks. In: Proc. 12th ACM International Conference on Ubiquitous Computing (UbiComp), ACM (2010) 119–128
12. Wang, H., Li, Z., Lee, W.C.: PGT: Measuring mobility relationship using personal, global and temporal factors. In: Proc. 14th IEEE International Conference on Data Mining (ICDM), IEEE (2014) 570–579
13. Crandalla, D.J., Backstrom, L., Cosley, D., Suri, S., Huttenlocher, D., Kleinberg, J.: Inferring social ties from geographic coincidences. Proceedings of the National Academy of Sciences **107**(52) (2010) 22436–22441
14. Pham, H., Shahabi, C., Liu, Y.: EBM: an entropy-based model to infer social strength from spatiotemporal data. In: Proc. 2013 ACM International Conference on Management of Data (SIGMOD), ACM (2013) 265–276
15. Tang, J., Chang, Y., Liu, H.: Mining social media with social theories: a survey. ACM SIGKDD Explorations Newsletter **15**(2) (2014) 20–29
16. Pang, J., Zhang, Y.: Exploring communities for effective location prediction. In: Proc. 24th World Wide Web Conference (Companion Volume) (WWW), ACM (2015) 87–88
17. Gao, H., Tang, J., Hu, X., Liu, H.: Content-aware point of interest recommendation on location-based social networks. In: Proc. 29th AAAI Conference on Artificial Intelligence (AAAI), The AAAI Press (2015)

Multi-Label Emotion Tagging for Online News by Supervised Topic Model

Ying Zhang[1,2], Lili Su[1], Zhifan Yang[1], Xue Zhao[1], and Xiaojie Yuan[1,2,*]

[1] College of Computer and Control Engineering,
Nankai University, China
[2] College of Software, Nankai University, China
{zhangying,sulili,yangzhifan,zhaoxue,yuanxiaojie}@dbis.nankai.edu.cn

Abstract. An enormous online news services provide users with interactive platforms where users can freely share their subjective emotions, such as sadness, surprise, and anger, towards the news articles. Such emotions can not only help understand the preferences and perspectives of individual users, but also benefit a number of online applications to provide users with more relevant services. While most of previous approaches are intended for recognizing a single emotion of the author, it has been observed that different emotions of the readers are more representative of the news articles. Therefore, this paper focuses on predicting readers' multiple emotions evoked by online news. To the best of our knowledge, this is the first research work for addressing the task. This paper proposes a novel supervised topic model which introduces an additional emotion layer to associate latent topics with evoked multiple emotions of readers. In particular, the model generates a set of latent topics from emotions, followed by generating words from each topic. The experiments on the real dataset from online news service demonstrate the effectiveness of the proposed approach in multi-label emotion tagging for online news.

Keywords: Emotion Tagging, Multi-Label Classification, Online News, Supervised Topic Model, Reader's Perspectives.

1 Introduction

With the explosion of social media over the past decade, more and more user-generated data is available on the Web for expressing users' opinions and emotions. Among various types of social media, online news is an important form that attracts billions of users to read, and actively respond. Opinions and emotions of individual users often have huge impacts on other users and the community. Users often have subjective emotions such as sadness, surprise and anger after reading the online news articles. Grasping such emotions can help understand the perspectives and preferences of individual users, and thus may facilitate online publishers to provide more personalized services or statistically study readers'

---

* Corresponding author.

© Springer International Publishing Switzerland 2015
R. Cheng et al. (Eds.): APWeb 2015, LNCS 9313, pp. 67–79, 2015.
DOI: 10.1007/978-3-319-25255-1_6

attitudes toward social events. Emotion tagging for online news is an application of the research area of opinion mining and sentiment analysis, which has attracted much attention in information retrieval and natural language processing communities.

Conventional emotion tagging research primarily focuses on determining the emotions of the authors who created the articles, while few focus on the reader's perspective. Reader's emotion prediction study aims to explore the reader's emotions aroused by the affective text. Evidences have shown that reader's emotions are not necessarily consistent with writer's [16]. Moreover, the existing works on emotion prediction from reader's perspectives [6] are mostly based on single-label classification techniques which are concerned with learning from a set of online news that are each associated with a single emotion label. It is conflict to the observation that some news articles could belong to multiple emotional states. An increasing number of social news websites provide a novel service via which users can share their emotions after reading news articles by voting for a set of predefined emotion tags. Fig. 1 shows an example from a popular Chinese news website (i.e., Sina News[1]). We can observe that the news article evokes more than one major emotion, including sad, angry and sympathetic. Therefore, it is desirable that automatic emotion tagging be regarded as a multi-label classification task which can predict one or more than one emotions. To the best of our knowledge, the existing literature contains no study on multi-label reader's emotion tagging for online news, which is to be discussed in this paper.

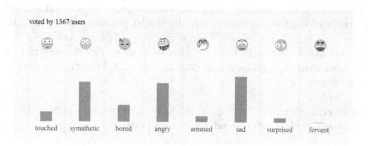

Fig. 1. An example of emotion tags voted by 1,367 users for a Sina news article.

Most existing multi-label classification methods are word-level models. However, the underlying assumption that "word" is the only essential feature in emotion tagging has many flaws. one same word always has different emotions in different subjects or contexts. Besides, some background words will bring noise because each word is regarded independently in the word-level emotion lexicon, many relevant words associated with certain emotions are usually mixed with background noisy words which do not convey affective meaning [1]. Therefore, it is more sensible to associate emotions with a specific emotional topic, which

---

[1] http://news.sina.com.cn/society/

means a real-world event, object or abstract entity that indicates the subject or context of the emotion, instead of only a single word.

In this paper, we propose a multi-label supervised emotion-topic model (ML-sETM) for tagging the emotions of the document effectively. In more details, the model follows a three-step generation process, which first generates an emotion from a document-specific emotion distributions with an emotion prior, then generates a latent topic conditioned on the emotion and finally generates document terms based on the latent topic as the Latent Dirichlet Allocation (LDA) [2]. The ML-sETM model is supervised by the observed emotion label set and associates each topic with reader emotions jointly as well as distinguishes between different emotions of the same word since the expression of emotion is characterized by its topic rather than an individual word.

We evaluate the proposed model on an online news corpus collected from the Sina news website. Experimental results show that the proposed model concerns multi-label classification for news articles and can classify the documents into emotion categories more effectively compared to the state-of-art models.

To the best of our knowledge, this is the first piece of research that focuses on modeling multi-label emotion tagging for online news from reader's perspectives. Our proposed model can be also generalized and applied to the other multi-label applications.

The rest of the paper is organized as follows. Section 2 reviews some related work. Section 3 formalizes the multi-label emotion tagging problem and describes our proposed ML-sETM model. Datasets and experimental results are discussed in Section 4. The last section provides conclusions and possible future work.

2   Related Work

Sentiment analysis [7] has become an important subfield of information management. Past researches on sentiments delivered by documents focused on detecting the sentiment that the authors of the documents were expressing. Pang *et al.* [8] exploited the positive or negative sentiment contained in movie reviews by applying support vector machines (SVM) with word unigram features. Quan and Ren [9] built the emotion lexicons for eight basic emotions and computed the similarities between sentences and the eight emotion lexicons to recognize the emotions presented in a sentence. Mishne [4] classified blog text according to the emotion reported by its author during the writing using SVM.

While the author might insert certain emotion and opinions towards the facts by omitting or stressing upon some aspect of text, the interpretation of the text largely depends on the personal background knowledge. Due to this emotion bias, only emotions of readers can accurately represent the emotion categories of online news. Therefore, researches begin to focus on analyzing the emotion expressed explicitly by readers. Lin *et al.* [6] firstly studied the emotions of readers and classified documents into reader-emotion categories. They also studied the approaches to ranking reader emotions of documents by either minimizing pairwise ranking errors or utilizing regression to model emotion distributions [5]. Zhang *et al.* [19,20] proposed a meta-classification model and corresponding domain adaption approaches for predicting emotions for the comments of online news from reader's perspective.

Since news always evokes multiple emotions of readers, predicting emotions of news can be formulated as a multi-label classification task to more accurately describe the emotions of readers. Tsoumakas and Katakis [13] reviewed the multi-label classification methods and grouped the existing methods into two main categories: problem transformation methods and algorithm adaption methods [18]. The first category is to transform the multi-label classification problem either into one or more single-label classification or regression problems, such as Binary Relevance (BR), Classifier Chains (CC) [12], Calibrated Label Ranking (CLR) [3] and RAndom k-labELsets (RAkEL) [15]. The second category is to extend specific learning algorithms to handle multi-label data directly. MLkNN is considered as one of the widely-used algorithm adaption methods [18].

However, most multi-label classification methods mainly focus on exploiting the sentiments of individual words. The same word in different subjects or contexts may express different emotions. As a result, it is more sensible to associate emotions with a specific emotional topic instead of only a single word. Bao *et al.* [1] proposed an emotion-topic model (ETM) to explore the sentiments of particular topics. ETM is a joint emotion-topic model that introduces an intermediate layer into Latent Dirichlet Allocation (LDA) [5] to bridge the gap between emotions and topics. Unfortunately, ETM is criticized for only annotating emotions from the writer's perspective. Therefore, Rao *et al.* [10,11] came up with the sentiment topic models and affective topic model to associate latent topics with evoked emotions of readers. Nonetheless, few of topic-based sentiment analysis methods can be applied to solve the multi-label emotion classification problems.

In light of these considerations, we here develop a supervised topic model for multi-label emotion classification for online news from reader's perspective.

3   Multi-Label Emotion Tagging

This section introduces our proposed supervised topic model for multi-label emotion tagging for online news, namely Multi-label supervised Emotion-Topic Model (ML-sETM). Before providing the details of this model, we first briefly introduce the problem definition and notations that will be used to describe the ML-sETM model.

3.1   Problem Definition

In this paper, we cast predicting emotions for online news into a multi-label classification problem that classifies news articles into multiple emotion tags. The problem setting is as follows. Given a corpus of online news $D = \{d_1, d_2, \ldots, d_{|D|}\}$ with a vocabulary $\mathcal{V}$, an emotion category set is defined as $\mathcal{E} = \{e_1, e_2, \ldots, e_M\}$. Multi-label emotion tagging problem is to tag online news with a corresponding subset of predefined emotion labels $E \subseteq \mathcal{E}$ utilizing a set of labeled news articles $\mathcal{D}$. A news article is described as a feature vector $w = \{w_i\}(i = 1, 2, \ldots, V)$ where V is the number of distinct words in vocabulary $\mathcal{V}$. For convenience, we associate each news article with a target indicator variable $\mu = \{\mu_k\}(k = 1, 2, \ldots, M)$ denotes which emotions evoked by the news article. More precisely, μ_k equals to 1 if the news article d is tagged with emotion e_k or 0 otherwise. Thus the labeled news article set $\mathcal{D}$ can be denoted as $\{(w^{(1)}, \mu^{(1)}), (w^{(2)}, \mu^{(2)}), \ldots (w^{(n)}, \mu^{(n)})\}$, where n is the number of labeled news articles. The set of all possible emotion label combinations can be written as $\mathcal{U} = \{0, 1\}^M$, thus the multi-label classifier can be expressed as a decision function $f : \mathcal{D} \to \mathcal{U}$.

Traditional classification models dealing with multi-labeled data simply treat words individually without considering the contextual information. While it is more sensible to utilize topic model to discover the topics from document, it fails to bridge the connection between reader's emotions and effective words. So our objective is to accurately model the connection between reader's emotions and specific topics and utilize the emotion label set to guide the topic generation.

3.2   Multi-Label Supervised Emotion-Topic Model

In this paper, we propose a supervised topic model for multi-label emotion tagging for online news called multi-label supervised emotion-topic model (ML-sETM). The Latent Dirichlet Allocation (LDA) model, also referred to as topic model, is an unsupervised learning technique for extracting topics from a corpus. However, as an unsupervised model it can not directly applicable to document classification problem and fails to bridge the connection between reader's emotions and words. The proposed ML-sETM model incorporates supervision of observed emotion labels to guide the document generation process and accounts for emotion labels by introducing an additional emotion layer to traditional LDA.

The proposed multi-label supervised emotion-topic model (ML-sETM) is represented as a probabilistic graphical model in Fig. 2. Formally, let M be the number of emotion labels and K be the number of latent topics in the corpus. Each topic is modeled as a multinomial distribution φ over observed words and each emotion label is modeled as a multinomial distribution θ over topics. ε and z are corresponding emotion and topic assignment for word w. For document d, both the words as well as the emotion labels are observed variables. The random vector δ is sampled from a symmetric Dirichlet distribution with hyperparameter γ and supervised by its emotion labels μ. Given the distribution over emotions δ, the ML-sETM model generates latent topics and words similar to LDA model. The generative process for ML-sETM is shown below.

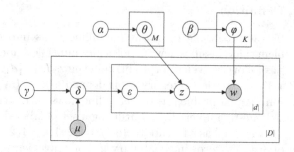

Fig. 2. The graphical model of Multi-Label Supervised Emotion-Topic Model

1. For each emotion choose $\theta \sim Dirichlet(\cdot|\alpha)$
2. For each topic choose $\varphi \sim Dirichlet(\cdot|\beta)$
3. Choose $\delta_d \sim Dirichlet(\cdot|\gamma)$
   a) Generate $\delta^{(d)} = \mu^{(d)} \times \delta_d$
   b) For each word w_i in document d:
   - Choose an emotion $\varepsilon_i \sim Multinomial(\delta^{(d)})$
   - Choose a topic $z_i \sim Multinomial(\theta_{\varepsilon_i})$
   - Choose a word $w_i \sim Multinomial(\varphi_{z_i})$

$\delta^{(d)}$ is restricted to only over the emotions corresponding to its emotion labels $\mu^{(d)}$ by setting the probability of irrelevant emotions to zero in order to ensure that all the emotion assignments are limited to the emotion labels of online news.

3.3   Model Learning and Inference

Since it is intractable to perform an exact inference for the multi-labeled emotion-topic model, we use an approximate inference method based on Gibbs sampling. According to the above generative process, the joint probability of all the random variables for a document collection is

$$P(\varepsilon, z, w, \delta, \theta, \varphi; \alpha, \beta, \gamma) = P(\theta|\alpha)P(\varphi|\beta)P(\delta|\gamma)P(\varepsilon|\delta)P(z|\varepsilon, \theta)P(w|z, \varphi) \quad (1)$$

For each word, we estimate the posterior distribution on emotion e and topic z based on the following conditional probabilities, which can be derived by marginalizing the above joint probabilities in Equation 1.

$$P(\varepsilon_i = e|\varepsilon_{-i}, z, w; \alpha, \beta, \gamma) \propto \frac{\alpha + n_{e,z,\cdot}^{-i}}{K\alpha + \sum_{z'} n_{e,z',\cdot}^{-i}} \times \frac{\gamma + N_{d_i,e}^{-i}}{M\gamma + \sum_{e'} N_{d_i,e'}^{-i}} \times \mu^{(d_i)} \quad (2)$$

$$P(z_i = z|z_{-i}, \varepsilon, w; \alpha, \beta) \propto \frac{\alpha + n_{\varepsilon,z,\cdot}^{-i}}{K\alpha + \sum_{z'} n_{\varepsilon,z',\cdot}^{-i}} \times \frac{\beta + n_{\cdot,z,w_i}^{-i}}{V\beta + \sum_w n_{\cdot,z,w}^{-i}} \quad (3)$$

Where e and z are the candidate emotion and topic for sampling, respectively. $d_i \in D$ indicates the document from which the current word w_i is sampled. $n_{e,z,\cdot}^{-i}$

is the number of times topic z has been assigned to emotion e, $N_{d_i,e}^{-i}$ means the number of words in document d_i has been assigned to emotion e. Similarly, $n_{\cdot,z,w_i}^{-i}$ is the number of times a word w has been assigned to topic z. The superscript $-i$ means the count that does not include the current assignment of emotion and topic for word w_i, respectively.

In more details, we start the algorithm by randomly assigning all the words to emotions and topics. Then we repeat Gibbs sampling on each word in the document collection by applying the Equation 2 and the Equation 3 sequentially. This sampling process is repeated for I iterations when the stop condition is met. With the sampled topics and emotions available, it is easy to estimate the distribution of δ, θ and φ as follows:

$$\delta_{d,e} = \frac{\gamma + N_{d,e}}{M\gamma + \sum_e N_{d,e}} \tag{4}$$

$$\theta_{e,z} = \frac{\alpha + n_{e,z,\cdot}}{K\alpha + \sum_k n_{e,k,\cdot}} \tag{5}$$

$$\varphi_{z,w} = \frac{\beta + n_{\cdot,z,w}}{V\beta + \sum_{w'} n_{\cdot,z,w'}} \tag{6}$$

Once the emotion-topic distribution $\theta_{e,z}$ and topic-word distribution $\varphi_{z,w}$ are learned from the training set, during testing time (document labels are unknown), our model is free to sample from any of M emotion labels, and any K topics in ML-sETM. For no ambiguity, given the optimal $\theta_{e,z}$ and $\varphi_{z,w}$, we simply reword a full cycle update process of the inference method here as follows:

1. Update the assignment of the word tokens to one of M emotions by

$$P(\varepsilon_i = e | \varepsilon_{-i}, z, w; \alpha, \beta, \gamma) \propto \theta_{e,z} \times \frac{\gamma + N_{d_i,e}^{-i}}{M\gamma + \sum_{e'} N_{d_i,e'}^{-i}} \tag{7}$$

2. Update the assignment of the word tokens to one of K topics by

$$P(z_i = z | z_{-i}, \varepsilon, w; \alpha, \beta) \propto \theta_{e,z} \times \varphi_{z,w} \tag{8}$$

3. After I iterations, we predict a testing document d via the estimation of posterior over emotions $\delta_{d,e}$ by the Equation 4. A threshold method [14] defined as the Equation 9 can decide whether an emotion is included in the final output for a given testing document.

$$\frac{g_j}{max(g_1, \ldots, g_M)} \geq p \tag{9}$$

where p is the threshold and g_j, $(j = 1, \ldots, M)$ is the predicted score of emotion label e_j for a given test document.

4   Experiments

In this section, we first introduce the experimental setups including datasets, comparison baselines and evaluation metrics. We then report an extensive set of experiment results of proposed model and baseline methods and finally explore the influence of parameters in our model. Analysis and discussions are presented based on the results.

4.1   Experimental Setups

Dataset. We collected most-viewed online news between January and June in 2011 from the society channel of Sina News. The online users voted these news articles with one of the eight social emotions, including touched, sympathetic, bored, angry, amused, sad, surprised and fervent. We can only utilize this Chinese dataset since we have not found similar emotion votes services in English yet, but the proposed model is language independent. We extracted the title and main body as the content of each news article, then segmented all news articles to terms with a Chinese word segmentation software ICTCLAS [17] and at last removed the most general terms and the most sparse terms. A threshold method defined as Equation 9 assigned multiple emotion labels for the news articles by selecting the representative emotion labels voted by online users. The applied dataset after preprocessing and cleaning contained 4,383 distinct terms in 4,654 news articles. The statistics of top ranked emotion label sets are listed in Table 1. In the whole dataset, 25.01% have two emotion labels, 10.06% have three emotion labels and about 3.5% have four or more emotion labels. As previously discussed, a large amount of news articles evoke more than one emotion. It is therefore necessary to classify emotions with multiple labels rather than only a single label.

Table 1. The statistics of top ranked emotion label sets.

	Emotion Labels	Number	Portion		Emotion Labels	Number	Portion
1	{Angry}	1,828	39.28%	6	{Sympathetic, Sad}	152	3.27%
2	{Touched}	430	9.24%	7	{Angry, Sad}	139	2.99%
3	{Amused}	359	7.71%	8	{Bored, Angry, Amused}	123	2.64%
4	{Angry, Amused}	326	7.00%	9	{Sad}	107	2.30%
5	{Bored, Amused}	167	3.59%	10	{Sympathetic, Angry, Sad}	70	1.50%

Comparison Baselines. We have selected several well-known modern high-performing methods from the literature for our benchmark comparison. For easy reference, Table 2 lists all the algorithms compared in the experiments, their corresponding abbreviations, some default parameters, and citations. We adapt Multi-class Logistic Regression (MLR) to the multi-label emotion tagging problem using the same threshold method, due to MLR's effective performance in the single-label classification task [19]. The Support Vector Machine (SVM)

Table 2. The descriptions of comparing algorithms in the experiments.

Key	Algorithm	Parameters	Citation
MLR	Multi-class Logistic Regression		Zhang *et al.* [19]
BR	Binary Relevance		Tsoumakas and Katakis [13]
CC	Classifier Chains		Read *et al.* [12]
MLkNN	Multi-label k-nearest neighbor	$k=10$	Zhang and Zhou [18]
RAkEL	RAndom K labEL subsets	$m=5, k=3$	Tsoumakas and Vlahavas [15]
CLR	Calibrated Label Ranking		Fürnkranz *et al.* [3]

is chosen as base classifier for problem transformation methods including BR, CC, RAkEL and CLR, since SVM performs best among several base classifiers in the multi-label classification task [13].

Evaluation Metrics. The evaluation metrics are different from those used in single-labeled classification problems. Here we use an evaluation dataset of emotion multi-label examples $(x_i, Y_i)(i = 1, 2, \ldots n)$, where $Y_i \subseteq \mathcal{E}$ is the set of true labels and $\mathcal{E} = \{e_1, e_2, \ldots, e_M\}$ is the set of all labels. Given an instance x_i, the set of labels predicted by a multi-label classifier is denoted as Z_i, where $Z_i \subseteq \mathcal{E}$. We use the following measures in our experimental evaluation, which are widely used in the literature and indicative of the performance of multi-label classification methods [13].

1. *Hamming Loss*
   $HLoss$ is one of the most frequently used criteria, which computes the symmetrical difference between Y_i and Z_i averaged over all test examples, the smaller the value of $HLoss$, the better the performance is. It is defined as:

$$HLoss = \frac{1}{n} \sum_{i=1}^{n} \frac{|Y_i \triangle Z_i|}{M}$$

   where $\triangle$ stands for the symmetric difference of two sets, which is the set-theoretic equivalent of the exclusive disjoint in boolean logic.

2. *subsetAcc*
   subsetAcc is a very strict evaluation measure as it requires the predicted set of labels to be an exact match of the true set of labels, which is defined as:

$$subsetAcc = \frac{1}{n} \sum_{i=1}^{n} I(Y_i = Z_i)$$

   where $I(true) = 1$ and $I(false) = 0$.

3. *F − measure*
   The $F − measure$ is the harmonic mean between precision and recall, common to information retrieval, which is defined as:

$$F - measure = \frac{1}{n} \sum_{i=1}^{n} \frac{2|Y_i \cap Z_i|}{|Y_i| + |Z_i|}$$

4.2 The Performance of Multi-Label Emotion Tagging Models

Experiments in this section investigate the effectiveness of our proposed ML-sETM model and baseline methods with 10-fold cross-validation in the multi-label emotion tagging task.

The hyperparameters α and β of ML-sETM are set to symmetric Dirichlet priors with values of $\alpha = 50/K$, $\beta = 0.1$, similar to previous work [2]. The hyperparameter γ is set as $50/K$. For the topic number, we find ML-sETM performs best when $K = 650$ by empirical study. The choice of topic number and γ will be discussed in Section 4.3 in detail.

Table 3 summarizes the evaluation results of the proposed ML-sETM model and baseline methods. Obviously, the proposed ML-sETM model leads the performance in most cases. Based on these results, we can make the following observations: 1) MLR performs worst in terms of all evaluation measures, which validates the necessity of multi-label classification methods of emotion tagging for online news. 2) MLkNN performs worse according to $HammingLoss$ and $subsetAcc$ and better according to $F - measure$. This means that besides the emotion labels that are originally relevant, MLkNN predicts non-relevant emotion labels as relevant. 3) Among the four kinds of Problem Transformation methods, BR performs worst since BR tackles the multi-label classification problem by simply splitting the original multi-labeled dataset into M single-labeled datasets, which will easily lead to data skew problem. CC and RAkEL perform better than BR and MLkNN by considering the correlations between the different emotion labels of each instance. This reflects that the emotion label correlations may provide helpful extra information. 4) ML-sETM consistently outperforms all the existing multi-label classification algorithms above. This indicates that it is more effective to associate topic information of online news with emotion labels. In addition, a paired t-test shows that all improvements of ML-sETM over BR, MLkNN, CC, RAkEL and CLR are statistically significant with p-value < 0.001.

Table 3. Experimental results on $HammingLoss$, $subsetAcc$ and $F - Measure$.

Algorithm	$HammingLoss$ ↓	$subsetAcc$ ↑	$F - Measure$ ↑
MLR	0.2258±0.0174	0.2353±0.0302	0.7331±0.0297
MLkNN	0.1641±0.0076	0.3318±0.0255	0.8014±0.0117
BR	0.1627±0.0081	0.3378±0.0207	0.7760±0.0113
CC	0.1617±0.0082	0.3636±0.0192	0.7894±0.0093
CLR	0.1612±0.0074	0.3363±0.0216	0.7684±0.0110
RAkEL	0.1611±0.0091	0.3580±0.0199	0.7952±0.0097
ML-sETM	**0.1572±0.0100**	**0.3758±0.0326**	**0.8021±0.0156**

4.3 The Influence of Parameters to ML-sETM Model

The proposed ML-sETM model is conditioned on four parameters: the Dirichlet hyperparameters α, β, γ and the number of latent topics K. Experiments in this section were conducted to explore the influence of different values of latent topics and hyperparameter γ to our proposed model.

Influence of Latent Topic Number. We first evaluate the influence of latent topic number K by fixing α, β and γ and explore the consequence of varying K, ranging from 50 to 800. In particular, we set $\alpha = \gamma = 50/K$ and $\beta = 0.1$ with 500 iterations and 10-fold cross-validation. Fig. 3 shows the performance on ML-sETM with a growing number of topic numbers. We omitted *subsetAcc* due to it's consistency with $F - Measure$ and the limited space. It can be seen that the performance of ML-sETM model converges when K is larger than 400 and reaches a peak at $K = 650$. This indicates that when the topic number equals to 650, our ML-sETM is being rich enough to fit the information available in the data, yet not as complex as being fitting noise. Therefore we can conclude that the topic number can be treated as the discriminate granularity of the ML-sETM model and it operates a tradeoff between the generality and specificity.

Influence of Hyperparameters. There are three hyperparameters in ML-sETM model α, β and γ , the choice of which has important implications for the results produced by ML-sETM model. Similar to previous work [1][2], α and β are set to symmetric Dirichlet prior with the value of $50/K$ and 0.1, respectively. The hyperparameter γ represents the prior observation counts for the number of times emotions sampled from topic before any emotion label is observed, which can reflect how different the documents are in terms of emotion labels. We fix topic number $K = 400$ and vary the value of γ from 0.01 to 1.

The experimental results with different values of γ is exhibited in Fig 4. We can make the following observations: 1) The ML-sETM model gets best performance when the value of γ equals to $50/K$. 2) The variance of HammingLoss and F-Measure are 0.0093 and 0.0151 respectively, which are all smaller than the variance of different topic numbers. This indicates that the performance of ML-sETM model with hyperparameter γ is more stable than parameter K.

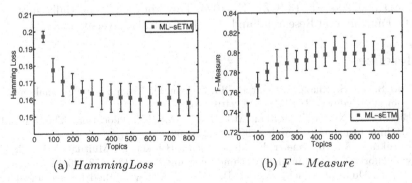

(a) $HammingLoss$ (b) $F - Measure$

Fig. 3. $HammingLoss$ and $F - Measure$ with different topic numbers.

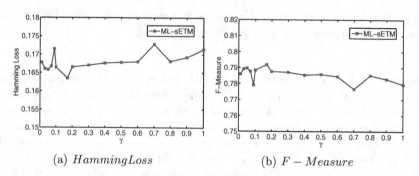

(a) $HammingLoss$ (b) $F - Measure$

Fig. 4. $HammingLoss$ and $F - Measure$ with different values of hyperparameter γ.

5 Conclusions and Future Work

This paper proposes a novel framework to address the task of predicting multiple reader's emotion for online news by supervised topic model. Specifically, we incorporate supervision of observed emotion labels to guide the topic generation and introduce an additional emotion layer into LDA to associate each topic with reader's emotions jointly. As far as we know, this is the first piece of research on multi-label emotion tagging for online news from reader's perspectives. The empirical results on real dataset have shown the effectiveness of our approach.

This work is an initial step towards a promising research direction. In the future work, we will propose more sophisticated topic models by taking account of dependencies among emotions additionally. Furthermore, we will also conduct a more comprehensive set of experiments to evaluate the proposed model on more large-scale testbeds.

Acknowledgements. This work is partially supported by National Natural Science Foundation of China under Grant No. 61170184, 61402242, and 61402243, National 863 Program of China under Grant No. 2015AA015401, as well as Tianjin Municipal Science and Technology Commission under Grant No. 13ZCZDGX02200, No. 13ZCZDGX01098. This work is also partially supported by the Fundamental Research Funds for the Central Universities of China.

References

1. Bao, S., Xu, S., Zhang, L., Yan, R., Su, Z., Han, D., Yu, Y.: Mining social emotions from affective text. IEEE TKDE 24(9), 1658–1670 (2012)
2. Blei, D.M., Ng, A.Y., Jordan, M.I.: Latent dirichlet allocation. The Journal of Machine Learning Research 3, 993–1022 (2003)
3. Fürnkranz, J., Hüllermeier, E., Mencía, E.L., Brinker, K.: Multilabel classification via calibrated label ranking. Machine Learning 73(2), 133–153 (2008)
4. Hu, Y., Duan, J., Chen, X., Pei, B., Lu, R.: A new method for sentiment classification in text retrieval. In: Dale, R., Wong, K.-F., Su, J., Kwong, O.Y. (eds.) IJCNLP 2005. LNCS (LNAI), vol. 3651, pp. 1–9. Springer, Heidelberg (2005)

5. Lin, K.H.Y., Chen, H.H.: Ranking reader emotions using pairwise loss minimization and emotional distribution regression. In: Proceedings of the Conference on Empirical Methods in Natural Language Processing, pp. 136–144. ACL (2008)

6. Lin, K.H.Y., Yang, C., Chen, H.H.: Emotion classification of online news articles from the reader's perspective. In: IEEE/WIC/ACM Conference on Web Intelligence and Intelligent Agent Technolog, vol. 01, pp. 220–226. IEEE (2008)

7. Pang, B., Lee, L.: Opinion mining and sentiment analysis. Foundations and Trends in Information Retrieval 2(1-2), 1–135 (2008)

8. Pang, B., Lee, L., Vaithyanathan, S.: Thumbs up?: sentiment classification using machine learning techniques. In: Proceedings of the ACL Conference on EMNLP, vol. 10, pp. 79–86. ACL (2002)

9. Quan, C., Ren, F.: Sentence emotion analysis and recognition based on emotion words using ren-cecps. International Journal of Advanced Intelligence 2(1), 105–117 (2010)

10. Rao, Y., Li, Q., Mao, X., Wenyin, L.: Sentiment topic models for social emotion mining. Information Sciences 266, 90–100 (2014)

11. Rao, Y., Li, Q., Wenyin, L., Wu, Q., Quan, X.: Affective topic model for social emotion detection. Neural Networks 58, 29–37 (2014)

12. Read, J., Pfahringer, B., Holmes, G., Frank, E.: Classifier chains for multi-label classification. Machine Learning 85(3), 333–359 (2011)

13. Tsoumakas, G., Katakis, I.: Multi-label classification: An overview. Dept. of Informatics, Aristotle University of Thessaloniki, Greece (2006)

14. Tsoumakas, G., Papadopoulos, A., Qian, W., Vologiannidis, S., D'yakonov, A., Puurula, A., Read, J., Švec, J., Semenov, S.: WISE 2014 challenge: Multi-label classification of print media articles to topics. In: Benatallah, B., Bestavros, A., Manolopoulos, Y., Vakali, A., Zhang, Y. (eds.) WISE 2014, Part II. LNCS, vol. 8787, pp. 541–548. Springer, Heidelberg (2014)

15. Tsoumakas, G., Vlahavas, I.P.: Random k-labelsets: An ensemble method for multilabel classification. In: Kok, J.N., Koronacki, J., Lopez de Mantaras, R., Matwin, S., Mladenič, D., Skowron, A. (eds.) ECML 2007. LNCS (LNAI), vol. 4701, pp. 406–417. Springer, Heidelberg (2007)

16. Yang, C., Lin, K.H., Chen, H.H.: Emotion classification using web blog corpora. In: IEEE/WIC/ACM Conference on Web Intelligence, pp. 275–278. IEEE (2007)

17. Zhang, H., Yu, H., Xiong, D., Liu, Q.: HMM-based Chinese lexical analyzer ICT-CLAS. In: SIGHAN Workshop on Chinese Language Processing (2003)

18. Zhang, M.L., Zhou, Z.H.: Ml-knn: A lazy learning approach to multi-label learning. Pattern Recognition 40(7), 2038–2048 (2007)

19. Zhang, Y., Fang, Y., Quan, X., Dai, L., Si, L., Yuan, X.: Emotion tagging for comments of online news by meta classification with heterogeneous information sources. In: Proceedings of SIGIR Conference, pp. 1059–1060. ACM (2012)

20. Zhang, Y., Zhang, N., Si, L., Lu, Y., Wang, Q., Yuan, X.: Cross-domain and cross-category emotion tagging for comments of online news. In: Proceedings of SIGIR Conference, pp. 627–636. ACM (2014)

Distinguishing Specific and Daily Topics

Tao Ge, Wenzhe Pei, Baobao Chang, and Zhifang Sui

MOE Key Laboratory of Computational Linguistics,
School of EECS, Peking University, Beijing, 100871, China
Collaborative Innovation Center for Language Ability, Xuzhou 221009, China
{getao,peiwenzhe,chbb,szf}@pku.edu.cn

Abstract. The task of distinguishing specific and daily topics is useful in many applications such as event chronicle and timeline generation, and cross-document event coreference resolution. In this paper, we investigate several numeric features that describe useful statistical information for this task, and propose a novel Bayesian model for distinguishing specific and daily topics from a collection of documents based on documents' content. The proposed Bayesian model exploits mixture of Poisson distributions for modeling probability distributions of the numeric features. The experimental results show that our approach is promising to solve this problem.

Keywords: specific and daily topics, numeric features, Bayesian model, mixture of Poisson distribution.

1 Introduction

Among techniques attempting to alleviate the impact of information overload, topic models can serve to organize information to users according to topics, helping them easily acquire knowledge. For people who want to learn about important events in the past, they can just glance over the topics discovered from a news archive without reading every document in the collection. However, it is common that some of the topics found are time-general; in other words, they concern daily and trivial topics which are considered having little retrospective value. A concrete example of daily topics is the topic document $D1$ talks about.

($D1$) The Shenzhen composite index rose slightly 0.2 points to 615.39 on Friday.
($D2$) An Alaska Airlines jet with at least 60 people on board crashed late Monday afternoon.

As we can see, $D1$ talks about a topic involving ups and downs of stock index, which is reported every day or every week regularly. From a retrospective viewpoint, the topic about temporary stock index has little value for general audience. In contrast, $D2$ concerns a specific topic about an important event – air crash. If one wants to learn about important events in some year (e.g. 2000), what he is interested in should be the specific topics like $D2$ discusses.

© Springer International Publishing Switzerland 2015
R. Cheng et al. (Eds.): APWeb 2015, LNCS 9313, pp. 80–91, 2015.
DOI: 10.1007/978-3-319-25255-1_7

Distinguishing specific topics (time-specific) and daily topics (sometimes called time-general, routine or recursive topics) is a useful task for many applications in natural languages processing and data mining. One typical example is event chronicle generation [6] and timeline generation [8,7]. As we know, chronicles and timelines have been adopted as a way to organize information by many websites such as Wikipedia and Facebook due to their readability and briefness. For generating a chronicle/timeline, it is necessary to distinguish specific topics from daily topics since only specific topics should be included.

Also, distinguishing specific and daily topics can help improve cross-document event coreference resolution systems which aim to find coreferential events in different documents. As Ge et al. [5] considered, the performance of cross-document event coreference resolution systems is likely to be affected by daily topics shown as follows:

(*D3*) The Germany's DAX index was down by 0.2 percent yesterday.
(*D4*) The Germany's DAX index was down by 0.2 percent on Wednesday.

Note that $D3$ and $D4$ were written in 2001 and 2004 respectively. If a cross-document event coreference resolution system can distinguish specific and daily topics, it will not consider these two events coreferential.

In this paper, we focus on distinguishing daily and specific topics without depending on documents' timestamps . The reasons why we do not use documents' timestamps are: First, the timestamp of documents, especially on the web, is unavailable (missing) or incredible due to arbitrary copy-paste behaviors such as retweet and reprint on the web, as [1,3,5] reported; Second, analyzing time distribution of topic is not feasible or reliable unless whole text stream during a time period is available; in other words, if the database used is just a sample of some text stream, the time distribution of a topic does not make sense. We investigate several numeric features for this task. For modeling the numeric features, we propose a novel Bayesian model with mixture of Poisson distributions. The experiments evaluate our model in terms of temporal distribution and semantic coherence and show that our model is more effective to discover and distinguish daily and specific topics from a collection of documents than conventional Bayesian topic models in an unsupervised manner.

The main contribution of this paper is: (**1**) we propose a Bayesian model for effectively discovering and distinguishing specific and daily topics; (**2**) we propose a general framework for incorporating numeric features into Bayesian models; (**3**) we give some measures for evaluating models for this task.

2   Methodology

2.1   Features for Distinguishing Specific and Daily Topics

As discussed in section 1, we attempt to distinguish between daily and specific topics using only textual features. For representing textual information, categorical features (also called nominal features) such as n-grams are usually adopted in NLP tasks. However, categorical features alone are not enough for describing statistical information such as the count of numerals in a document, which is important for this task. We introduce numeric features describing useful statistical information of documents for this task.

#PERSON Named Entities: Named entities are often discussed during specific time period. Intuitively, if a topic involves a number of PERSON names then this topic is less likely to be a daily topic. Hence, **the count of PERSON named entities** in a document is selected as a feature.

#Numerals: It is easy to understand that the documents concerning daily topics usually involve something that frequently changes such as temporary stock price, which should be updated in time. The most salient features for describing such variations and fluctuations are numerals. If a document contains too many numerals, it is very likely that the article may talk about a daily topic. Based on the intuition, **the count of numerals** in a document is selected as a feature.

#IDF: In addition to document-based features, we also investigate an important corpus-based feature – inverse document frequency (IDF) which is a measure of whether a term is common or rare across a collection of documents. Intuitively, if IDF of salient words of a topic tend to be high, this topic will be less likely to be frequently mentioned all the time. According to our analysis of corpus of Gigaword, we find that the terms whose idf is greater than 60% of MAXIDF is distinguished for determining a topic to some extent. Thus, we refer terms in set H which is defined as follows to high-idf terms.

$$H = \{w | idf(w) > \text{MAXIDF} \times 0.6\}$$

Given that **the counts of high-idf terms** in a document can indicate whether the document concerns a daily topic or a specific topic to some extent, we select it as a feature for the task.

2.2   A Bayesian Model with Mixtures of Poisson Distributions

For distinguishing specific and daily topics, we adopt a Bayesian model which is proved to be suitable for tasks concerning topics due to its ability of discovering topics hidden in a text collection and its flexibility that it can introduce various features with their dependencies.

In section 2.1, a variety of integer numeric features used for distinguishing specific and daily topics are discussed. Effective to reflect some statistical information as the features seem, it is somewhat difficult to directly incorporate these features such as into a Bayesian model. Although numeric features can be discretized into several categorical bins in advance which can be modeled by a multinomial distribution, that appears not very applicable. On one hand, if the numeric variables are discretized into too many bins, data sparsity problem would arise, which may have an adverse effect on the performance. On the other hand, if the number of bins is too small, the discretization might be unreasonable.

To deal with this problem, we assume that the integer numeric features follow Poisson distributions which are proved to be very suitable for expressing the

Table 1. Notations used in our model

Symbols	Descriptions
M, N, K	The number of documents in a corpus, tokens in a document and topics respectively
F	The number of features used. If we use all the three features discussed in section 2.1, then $F = 3$.
w	A word in a document
z	The topic assigned to a document
p	Topic type label indicating whether a topic is daily. $p \in \{0, 1\}$
f	Features discussed in section 2.1, which could be the count of PERSON named entity, numerals or high-idf tokens in a document
θ	The topic distribution
π	The Bernoulli distribution for p
Φ	The distribution for words given a topic
λ	The parameter of poisson distributions.
α, β	The hyperparameters for Dirichlet prior of θ and Φ respectively
γ	The hyperparameter for Beta prior of π

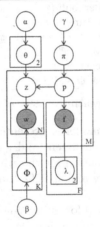

Fig. 1. Graphical illustration of our model.

probability distribution of the integer numeric variables such as the count of a term in a document, and use the poisson distributions in our Bayesian model. The plate diagram of our Bayesian model is given in figure 1 and notations of our model are summarized in table 1.

The left part of the plate diagram of our model shown in figure 1 is similar to general latent dirichlet allocation (LDA) [2]. But unlike the general LDA model in which a document is a mixture of topics, our model assumes that a document corresponds to only one topic. Moreover, our model introduces another latent variable p which indicates the type of a topic (daily or specific), and poisson distributions which aim to express the probability distribution of the count of PERSON named entities, numerals and high-idf terms in a document respectively. As shown in figure 1, the numeric features can be considered as being generated by a component of the mixture of Poisson distributions.

In addition, our model assumes that the topic distributions under different topic types are different, which is reasonable. While constructing topics, not only does our model consider the word distribution, but also takes into consideration the numeric features. In this way, we can even distinguish topics through the difference of the numeric features even if they have the similar word distribution. For example, a document about temporary stock price and another document talking about a specific stock market crash might be assigned to two different topics in our model due to differences of the numeric features while they might be in the same topic cluster in conventional Bayesian topic models because their word distributions are similar. As a result, specific topics found by our model can be more time-specific and daily topics found by our model could be more time-general than those found by conventional Bayesian topic models.

Draw $\theta \sim$Dirichlet(α) for each topic type p
Draw $\Phi \sim$Dirichlet(β) for each topic k
Draw $\pi \sim$Beta(γ)
For each document m:
    Draw $p \sim$Bernoulli(π)
    For each feature f_i:
        Draw $f_i \sim$Poisson(λ_i^p)
    Draw $z \sim$Multi(θ^p)
    For each token w in m:
        Draw $w \sim$Multi(Φ^z)

Fig. 2. The generative story of our model

The generative story of our model is presented in figure 2. Note that f_i in figure 2 is the ith feature of a document and $i \in \{1, 2, ..., F\}$, λ_i^p denotes the parameter of the Poisson distribution for the ith feature of a document whose topic type is p. In this way, daily and specific topics can be distinguished while constructing topics.

Model inference and the method for estimation of parameters λ of the Poisson distributions are to be discussed in detail in section 2.3.

2.3 Model Inference and Parameter Estimation

For model inference, we use Gibbs sampling approach to sample latent variables p and z. Specifically, for a given document m, the conditional probabilities of its latent variable p and z are shown in (1) and (2) respectively:

$$P(p_m | \boldsymbol{p}_{\neg m}, \boldsymbol{z}, \boldsymbol{f}(m); \gamma, \alpha)$$

$$= \frac{c_p + \gamma}{\sum_p (c_p + \gamma)} \times \frac{c_{z,p} + \alpha}{\sum_z (c_{z,p} + \alpha)} \times \prod_{i=1}^{F} P(f_i(m) | \lambda_i^p) \tag{1}$$

$$P(z_m | \boldsymbol{z}_{\neg m}, \boldsymbol{p}, \boldsymbol{w}; \alpha, \beta)$$

$$= \frac{c_{z,p} + \alpha}{\sum_z (c_{z,p} + \alpha)} \times \prod_{w \in W_m} \frac{c_{z,w} + \beta}{\sum_w (c_{z,w} + \beta)} \tag{2}$$

where W_m denotes tokens in document m, $c_{z,p}$ is the number of documents whose topic and topic type are z and p respectively, $c_{z,w}$ is the number of words w in documents whose topic is z, c_p is the number of documents whose topic type is p and $\boldsymbol{f}(m) = [f_1(m), f_2(m), ..., f_F(m)]$ in which $f_i(m)$ is the ith feature of document m. $P(f_i(m)|\lambda_i^p)$ can be easily computed using the probability mass function of Poisson distributions.

Now, the problem for our model is how to estimate the parameters λ of the Poisson distributions. As mentioned in section 2.2, the features can be regarded as being generated by their corresponding mixture of poisson distributions; thus, we use an EM-based method to estimate the parameters of the Poisson distributions. As is known, the only parameter λ of a Poisson distribution is the expectation of the distribution. Therefore, whenever we finish sampling the latent variable p for document m, we re-estimate λ of the Poisson distribution λ_i^p using maximum likelihood estimation (MLE) as shown in (3).

$$\lambda_i^p = \frac{\sum_{j:p_j=p} f_i(j)}{\sum_{j:p_j=p} 1} \tag{3}$$

where $f_i(j)$ denotes the ith feature of document j.

In our method of parameter estimation, (1) and (3) can be considered as E-step and M-step respectively. Different from the general EM algorithm, the latent variable p sampled in the E-step is a hard estimation instead of a soft one for efficiency. In this way, the parameter λ_i^p of the Poisson distributions can be estimated during sampling.

3 Experiments and Evaluations

3.1 Experimental Setting

Since there is no standard benchmark dataset for this task, we use all the news articles written in the year 2000 of the Xinhua News Agency in Gigaword English Corpus to evaluate our model. This dataset contains 99,538 news articles involving a variety of subjects.

As preprocessing, we use Stanford CoreNLP toolkit [10] to do lemmatization, named entity extraction and POS tagging. In this way, we can obtain PERSON named entities and numerals whose count is used as features in our Bayesian model as mentioned in section 2. Also, we compute idf of all terms in the vocabulary except numerals whose idf is set to 0. We normalized the count of PERSON named entities, numerals and high-idf terms in document m as follows:

$$f_{norm}(m) = \lfloor \frac{f(m)}{length(m)} \times 100 \rfloor$$

where $f(m)$ denotes a feature of document m which could be the count of PERSON named entities, numerals or high-idf tokens in document m and $length(m)$ is document m's length.

Table 2. Estimated parameters λ of poisson distributions of different feature combinations. λ_1, λ_2 and λ_3 are parameters of Poisson distributions for the count of named entities, numerals and high-idf terms respectively

Feature	p	λ_1	λ_2	λ_3	topic type
per	0	8.09	-	-	specific
	1	0.79	-	-	daily
num	0	-	3.59	-	specific
	1	-	25.07	-	daily
idf	0	-	-	20.34	specific
	1	-	-	7.25	daily
per+num	0	3.43	3.51	-	speccfic
	1	2.26	24.65	-	daily
per+idf	0	7.65	-	20.72	specific
	1	1.12	-	8.94	daily
num+idf	0	-	3.70	13.47	specific
	1	-	25.36	9.82	daily
All	0	1.12	25.48	7.67	daily
	1	3.72	4.14	14.09	specific

We empirically set hyperparameters $\alpha = 0.05$, $\beta = 0.01$, $\gamma = 0.5$. The number of topics K is set to 100. We simply select the topic with the largest probability for a document as the document's topic. Formally, for document m,
$$topic(m) = argmax_z P_m(z)$$
where $P_m(z)$ denote the probability of a topic for document m and it can be estimated as follows:

$$P_m(z = k) = \frac{\sum_{i=1}^{S} \delta(z^{(i)} = k)}{S}$$

where $z^{(i)}$ is the topic sampled for document m at the ith iter after burn-in, $\delta(.)$ is an indicator function and S is the number of iterations of after burn-in.

In order to verify the effectiveness of features we use, we try different combinations of features. Parameters λ of poisson distributions after the model converges is shown in table 2. It should be noted that even though the topic type of a given topic can be indicated by the topic type label p ($p \in \{0, 1\}$), the meaning of p's value is unknown since our Bayesian model is an unsupervised approach. In other words, we do not know the topic type label p of a daily topic should be 0 or 1. Therefore, we must make clear what the exact meaning of p's values are. Based on the intuition discussed in section 2.1, we can use the estimated parameter λ of the mixture of Poisson distributions of the integer numeric features under different topic type p to help understand p's value's meaning. Intuitively, as for the daily topic type, its average count of PERSON named entities and high-idf terms per document should be less while its average count of numerals per document should be much more than the counterpart of the specific topic type. After making clear the meaning of p's values (as shown in table 2), we set a high confidence threshold (0.9) for determining the daily topics. Specifically, if the probability of

Table 3. Temporal perplexity of daily topics and specific topics under different feature combinations. Intuitively, for daily topics, the larger temporal perplexity, the better; for the specific topics, the smaller, the better. As for Δ_{avg}, the larger, the better.

Feature	topic type	#topics	max	min	avg	Δ_{avg}
per	daily	36	**362.04**	5.78	192.67	65.55
	specific	64	**308.69**	5.66	127.12	
num	daily	15	**362.04**	51.63	183.55	32.38
	specific	85	324.03	6.06	151.17	
idf	daily	18	**362.04**	145.01	250.73	131.3
	specific	82	317.37	5.82	**119.43**	
per+num	daily	18	357.05	18.77	181.02	29.85
	specific	82	319.57	11.96	151.17	
per+idf	daily	27	**362.04**	37.79	218.27	93.77
	specific	73	319.57	8.11	124.5	
num+idf	daily	13	**362.04**	**174.85**	**257.78**	**132.41**
	specific	87	319.57	**4.63**	125.37	
All	daily	16	**362.04**	140.07	232.32	85.29
	specific	84	319.57	7.06	147.03	

a topic z to be a daily topic is larger than 0.9 (i.e. $P(p = 0|z) > 0.9$, assuming that $p = 0$ indicates that the a topic is daily), the topic z will be considered as a daily topic; otherwise, the topic is considered specific. The probability $P(p|z)$ can be estimated as follows:

$$P(p|z) = \frac{(c_{z,p} + \alpha)}{\sum_p (c_{z,p} + \alpha)}$$

3.2   Experimental Results

In this section, we evaluate and compare our method with other Bayesian models in terms of temporal perplexity and log-likelihood per document.

Temporal Perplexity. Since there is no golden standard for evaluating whether a topic or a document concerns a daily topic or not, we alternatively use an indirect way to evaluate our model – using temporal distribution to measure a topic's distribution over time. If the temporal distribution of a topic is almost uniform, then the topic might be a daily topic. In contrast, if the temporal distribution of a topic fluctuates significantly over time or the number of articles involving the topic surge during a short period of time, the topic is more likely to be a specific topic. Therefore, it is possible to use temporal distribution to help evaluate our model. Inspired by the perplexity measure in information theory, we define *temporal perplexity* (*TP* for short) for measuring the temporal distribution of a topic, as shown in (4).

$$TP(z) = 2^{-\sum_t \frac{c_{z,t}}{c_z} \times \log_2 \frac{c_{z,t}}{c_z}} \tag{4}$$

where t denotes a time epoch whose granularity can be either a day, a week or a month, $c_{z,t}$ denotes the number of documents involving topic z at t and c_z is

Table 4. Comparison between NB, LDA and our model in terms of temporal perplexity

Model	topic type	#topics	max	min	avg	Δ_{avg}
LDA	daily	13	315.04	132.27	248.26	21.86
	specific	87	362.04	33.20	226.40	
NB	daily	13	303.40	30.41	198.91	70.75
	specific	87	362.04	9.81	128.16	
ours	daily	13	**362.04**	**174.85**	**257.78**	**132.41**
	specific	87	**319.57**	**4.63**	**125.37**	

the number of documents involving z across the collection. According to (4), the more uniform the temporal distribution of a topic is, the larger the TP of the topic will be. Therefore, the temporal perplexity of daily topics should be larger than that of specific topics.

Table 3 reports the temporal perplexity of the identified daily topics and the specific topics under different combinations of features. Note that the temporal granularity is day. In table 3, max, min and avg for a topic type denote the maximal, minimal and average temporal perplexity of topics of the topic type (daily or specific). Δ_{avg} denotes the difference between the average temporal perplexity of the identified daily topics and that of the specific topics:

$$\Delta_{avg} = avg_{daily} - avg_{specific}$$

As shown in table 3, the features we used (i.e. count of PERSON named entities, numerals and high-idf terms in one document) are all capable of distinguishing daily and specific topics, which is reflected by a positive Δ_{avg} value. Among these features, the count of high-idf terms seems to be the most effective, which achieves the largest Δ_{avg} since IDF is a corpus-based feature, just as temporal perplexity which is also a measure based on a corpus. Hence, compared with the other document-based features, the count of high-idf terms appears to be more correlated with temporal perplexity.

As for the combination of features, it is not difficult to find that the *num+idf* combination performs best. This combination achieves the highest temporal perplexity for the identified daily topics in terms of max, min, avg as well as Δ_{avg}. As discussed in section 2.1, the count of numerals in a document is an important feature for distinguishing between specific topics and daily topics. Although the feature alone may not result in a large Δ_{avg}, it is very helpful in improving the performance when it is combined with *idf*. However, the combination of all the three features does not achieve a better result than *num+idf* and *per+idf* combinations. The possible reason is that the combination of *per* and *num* may affect the Bayesian model. As mentioned above, the task is to distinguish between the documents about daily topics which tend to contain few PERSON named entities and many numerals, and the documents about specific topics which tend to contain many PERSON named entities and few numerals. Nevertheless, it is not uncommon that a news article contains both few PERSON named entities and few numerals, or contains both many PERSON named entities and many numerals (e.g. list of top Premier League goal scorers). When *per* and *num* are simultaneously selected as features, the two features may interfere each other, which perhaps leads to a poor performance. In contrast, other feature combinations seem less likely to suffer from such a problem.

In addition, we compare our model with two typical Bayesian topic model – Naive Bayes (NB) and Latent Dirichlet Allocation (LDA). For NB and LDA, we first detect topics and then use idf+num feature to distinguish specific and daily topics. Formally, we define a score for topic z as follows:

$$score_z = \#numeral_z - \#highidf_z$$

where features (e.g. $\#numeral_z$) are average of their counterparts of documents whose topic is z.

According to the intuition in section 2.1, the higher score, the more likely the topic is to be daily. Since our model identifies 13 daily topics with idf+num features, we identify the 13 topics with the highest score as daily topics for NB and LDA model. The performance of NB and LDA is given in table 4. It can be easily seen that our model performs much better than LDA and NB in terms of temporal perplexity. One main reason is that our model considers the numeric information while constructing topic clusters, as discussed in section 2.2. When a document contains many high-idf terms and few numerals, it would be more likely to be assigned to a specific topic cluster. In contrast, NB and LDA consider only word distribution when constructing topic clusters. Hence, in NB and LDA, a specific topic (e.g. stock market crash) and a daily topic (e.g. temporary stock price) might be in the same topic cluster owing to similar word distribution while they are less likely to be the same cluster in our model due to difference of numeric features. Therefore, specific topics found by our models tend to be more time-specific and daily topics found by our models tend to be more time-general. Figure 3 also verifies the claim. It is shown that our model can find more extremely daily(TE>300) and specific(TE<50) topics than those found by NB and LDA.

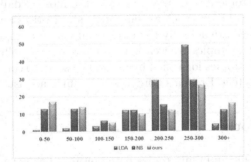

Fig. 3. The number of topics in intervals of temporal perplexity

Max Log-likelihood Per Document. In addition to temporal perplexity which evaluates the temporal distribution of topics found by our model, we also use Max log-likelihood per document to compare our model with NB and LDA in terms of semantic coherence which is an important measure for evaluating topic models.

As is known, topic models assume a collection of documents are generated by mixture of language models. Thus, we can use the idea for evaluating language models for evaluations. Log-likelihood per document is one for such measures for topic model evaluation [4]. Since our work assumes that a document is associated with only one topic, therefore, we use max log-likelihood per document instead of log-likelihood for evaluations. Formally, max log-likelihood per document ($MLLPD$) for a test set T is computed as (5):

$$MLLPD(T) = \frac{\sum_{d \in T} \sum_{w_i \in d} log P(w_i|z_d^*)}{|T|} \tag{5}$$

where $z_d^* = argmax_z \prod_{w_i \in d} P(w_i|z)$

Table 5. Max Log-likelihood per documents under different number of topics

#topics	20	50	100
LDA	-2779.36	-2870.44	-2967.11
NB	-2600.58	-2584.08	-2594.29
ours	-2597.41	-2587.22	-2594.90

We use the news articles written during June 2000 by Associated Press Worldstream as test set which contains 4,392 news articles. Table 5 shows the comparison of $MLLPD$ between NB, LDA and our model on this test set. It can be seen that our model performs almost the same with NB and better than LDA, which verifies that incorporating the numeric features while constructing topic clusters do not affect the semantic coherence of topics.

At last, we list the top 5 daily and specific topics identified by our model (using *num+idf* feature combination) as well as their top words in table 6 for comparing the two topic types. As we can see, the most of daily topics concern economic and financial issues such as temporary stock prices and fluctuations of exchange rates, which are actually the most common daily topic in news genre. In addition, some daily topics are weather forecasts and air pollution reports. By contrast, the specific topics identified by our model seem to correspond to one or more specific events (e.g. Eritrea's border issue), which are usually time-specific.

4   Related Work

Distinguishing specific and daily topics based on text has not been well studied so far. As far as we know, Ge et al. [5] is the most related work to ours, which identifies daily events based on text for avoiding arbitrarily considering them coreferential. Unlike our unsupervised approach, they used an extra collection of documents with timestamps to generate a training set (but not golden standard) and trained a maximum entropy classifier with several categorical features like unigrams for identifying daily events. Another similar work is done by Li and Cardie [7], who used a hierarchical dirichlet process (HDP) model [9] to recognize

Table 6. Examples of daily and specific topics identified by our approach (idf+num)

Daily					Specific				
1	2	3	4	5	1	2	3	4	5
dollar	index	degree	dollar	pollution	Iraq	Ethiopia	China	disease	drug
index	stock	breeze	exchange	air	oil	Eritrea	trade	EU	seize
rupee	fall	max	U.S	city	OPEC	UN	minister	ban	police
gold	rise	min	pound	level	sanction	peace	economic	U.S.	kilogram
turnover	NASDAQ	gentle	hongkong	report	minister	border	development	trade	heroin
silver	company	cloudy	British	pollution	Kuwait	Taliban	WTO	animal	myanmar

time-specific and time-general topics for generating timelines for individuals from Twitter data.

Different from the previous work, our model only uses textual information and thus can be applied to any collection of documents even if their timestamps are unreliable or unavailable.

5   Conclusion

In this paper, we investigate several numeric features and propose a novel Bayesian model with mixtures of Poisson distributions for distinguishing daily and specific topics. Our proposed model can be easily generalized to other tasks for incorporating numeric features in Bayesian models.

Acknowledgements. We thank the anonymous reviewers for their valuable comments. This paper is supported by National Key Basic Research Program of China 2014CB340504 and NSFC project 61375074. The contact author of this paper is Zhifang Sui.

References

1. Alonso, O., Gertz, M., Baeza-Yates, R.: Clustering and exploring search results using timeline constructions. In: CIKM (2009)
2. Blei, D.M., Ng, A.Y., Jordan, M.I.: Latent dirichlet allocation. The Journal of Machine Learning Research (2003)
3. Chambers, N.: Labeling documents with timestamps: Learning from their time expressions. In: ACL (2012)
4. Doyle, G., Elkan, C.: Accounting for burstiness in topic models. In: ICML (2009)
5. Ge, T., Chang, B., Li, S., Sui, Z.: Event-based time label propagation for automatic dating of news articles. In: EMNLP (2013)
6. Ge, T., Pei, W., Ji, H., Li, S., Chang, B., Sui, Z.: Bring you to the past: Automatic generation of topically relevant event chronicles. In: ACL (2015)
7. Li, J., Cardie, C.: Timeline generation: Tracking individuals on twitter. In: WWW (2014)
8. Swan, R., Allan, J.: Automatic generation of overview timelines. In: SIGIR (2000)
9. Teh, Y.W., Jordan, M.I., Beal, M.J., Blei, D.M.: Sharing clusters among related groups: Hierarchical dirichlet processes. In: NIPS (2004)
10. Toutanova, K., Klein, D., Manning, C., et al.: Stanford core nlp (2013)

Matching Reviews to Object Based on 2-Stage CRF

Zhang Yongxin[1,2,*], Li Qingzhong[3], Wang Dequan[4], Ding Yanhui[5],
Liu Congli[6], and Yan Zhongmin[3]

[1] School of Mathematical Sciences,
Shandong Normal University, Jinan 250014, China
[2] Shandong Provincial Key Laboratory of Software Engineering, Jinan, China
[3] School of Computer Science and Technology, Shandong University, Jinan 250101,
China
[4] School of Computer Science, Fudan University, Shanghai 200433, China
[5] School of Information Science and Engineering,
Shandong Normal University, Jinan 250014, China
[6] Department of Information Technology, QILU Bank, Jinan 250001, China
waterzyx@hotmail.com, {lqz,yzm}@sdu.edu.cn, dequanwang86@126.com,
dingyanhui@gmail.com, lcl@qlbchina.com

Abstract. With the development of web data integration, it poses a new
challenge how to match relevant reviews to integrated database objects
and provide users the more complete holistic views of entities. According to the features of web data integration and reviews from web, we
proposed a method based on 2-layer Conditional Random Fields(CRF)
to match reviews to database objects. On the one hand, our method
leverages the integrated structured entity and significantly reduces the
dependence on manually labeled training data. On the other hand, we
employ semi-Markov CRF to recognize the structured entities and exploit a variety of entity-level and pattern-level recognition clues available
in a database of entities and labeled reviews, thereby effectively resolving
the entity variety and improving the accuracy of the entity recognition.
Experiments in multiple domains show that our method can substantially superior to traditional tf-idf based methods as well as a recent
language model-based method for the review matching problem.

Keywords: Web Data Integration, Review Matching, Conditional Random Fields, Semi-Markov.

1 Introduction

Compared to traditional data integration, the entity type concerned in web data
integration is more various which includes the focus target entities as well as the
relevant entities, providing users the more complete views of entities. Among
the relevant entities, review is the very important information because reviews
can reflect the opinions of users to products and events and are the critical

---

* Corresponding author.

© Springer International Publishing Switzerland 2015
R. Cheng et al. (Eds.): APWeb 2015, LNCS 9313, pp. 92–103, 2015.
DOI: 10.1007/978-3-319-25255-1_8

information for the decision-making. Linking reviews with the relevant entities can not only provide users the holistic view of entities but also is the stepping stone to enabling more sophisticated applications such as opinion analysis and data mining. Therefore, it poses a new challenge to web data integration how to match reviews to integrated database objects.

With regard to reviews and structured entities from the same data source, we can easily match them through clues such as the link relationship of web pages and co-occurrence of them. However, with the autonomy of web data source and the ease of reviews publishing, a number of challenges need to be addressed in designing an effective solution to the task of matching reviews and structured entities. First, with the autonomy of web data source, each source describes the same entity with different manner and some varieties of entity exist in them, such as abbreviation, alias and typographical or orthographical errors, etc. These all pose difficulties for the task of matching reviews to integrated database entities.

Second, review web pages might mention other objects that are peripheral to the review and the information revealed about the object in a review might be partial or surrogate. Naturally, a review can refer to the object which it comments. But a review can also refer to other relevant entities besides the main entity. In addition, as reviews are published by users with a free form, an entity can be mentioned as a part of entity name or alias.

Fortunately, we also get some inspirations from the features of web data integration. First, the integrated entities can be viewed as knowledge base which implicit the pattern or semantic features of the entities and their attributes. Therefore, they can be used to guide for object recognition and matching as the external dictionaries. Moreover, reviews can also mention some entity attributes besides entity names, and it also provides useful clues for our matching task.

Our contributions. In this paper we proposed a method based on 2-layer CRF to match reviews to database objects. First, our method leverages the integrated structured entity and significantly reduces the dependence on manually labeled training data. Second, we employ semi-Markov CRF to recognize the entities in reviews and exploit a variety clues including entity-level dictionary features, thereby effectively resolving the entity variety and improving the accuracy of the entity recognition. Finally, our experiments and extensive analysis show that our method outperforms traditional tf-idf based methods and a recent language model-based method.

This paper is organized as follows. We briefly review some related research efforts in Section 2, and introduce the (semi-Markov) CRF model in Section 3. The overview and detail of the proposed approach is introduced in Section 4. Experimental evaluations are reported in Section 5, and in the last section we draw conclusions and point out some future directions.

2   Related Works

Little study had so far been done on matching reviews to database entities in data integration, and the work close to this is entity resolution [1,2] that is the problem of identifying duplicate references that refer to the same real world entity. The reference is an appearance of an entity in data source which can be a record of database or an entity name in a review. In the field of entity resolution, the variation closest to our problem is reference disambiguation whose goal is to identify the database entities which the references in documents refer to. The most representative work is conducted by the scholars of University of California, Irvine and they propose a research prototype named RelDC [3]. However, document in reference disambiguation is the text with latent structure such as bibliography or address data, so that there is often a correspondence between named entities and their attributes. Whereas the reviews in this problem are free texts without certain fixed formats, so we need to identify the named entities in reviews in order to matching reviews to database entities.

In addition, there are some researches which focus on associating of unstructured content with structured data. The most representative works were conducted by IBM India research lab which proposed two research prototypes: SCORE [4] and EROCS [5]. SCORE extracts keywords from query results on structured data and uses them to submit keyword queries that retrieve supplementary information, and vice versa. The EROCS system, which uses information extraction and entity matching, is closest in spirit to our problem; they, however, employ tf-idf to match, which has been shown to be sub-optimal [6].

Yahoo research lab proposed a generative language model [6,7] to match reviews to object without extraction of named entities. But this coarse-grained method can hardly guarantee the high accuracy of entity matching. So we need to establish the corresponding between entities in reviews and database attributes at first, and then conduct fine-grained entity matching.

A recent trend of named entity recognition is finding approximate matches in the text with respect to a large dictionary of known entities, as the domain knowledge encoded in the dictionary helps to improve the extraction performance [8,9]. The challenge of these methods is the dictionary structuring because there are seldom dictionaries available according to different application domain. Whereas large scale integrated entity information in data integration system is the natural dictionary that can be used to guide the entity recognition task.

CRF [10] is a discriminative probabilistic model which was proposed based on Maximum Entropy Model and Hidden Markov Model, and it is the best model for sequence data segment and labeling. CRFs have been applied under different scenario such as NLP and information extraction. As the extension of CRFs, semi-Markov CRF (semi-CRF) relaxes the constraint of CRF and is more applicable to entity identification task.

3   Semi-Markov CRF

3.1   Linear-Chain CRF

Conditional Random Fields (CRFs) [6] is a type of discriminative undirected probabilistic graphical model which is used to encode known relationships between observations and construct consistent interpretations. As it does not require the hypothesis of rigorous independence such as HMM (Hidden Markov Model) and overcomes the shortcoming of label bias just like MEMM (Maximum Entropy Markov Model) and other non-generative directed graphical models, CRF is the best statistical machine learning model of sequence labeling.

Definition 1. *For an undirected graph* $G = <V, E>$, $Y = \{Y_v | v \in V\}$ *is indexed by the vertices of G. Then* (X, Y) *is a conditional random field when the random variables* Y, *conditioned on* X, *obey the Markov property with respect to the graph:*
$Pr(Y_v | X, Y_w, w \neq v) = Pr(Y_v | X, Y_w, w \sim v)$, *where* $w \sim v$ *denotes that* w *and* v *are neighbors in G.*

According to the basic theory of CRF, the conditional distribution of labeling sequence y with a given observed sequence x is modeled as follow:

$$Pr(y|x) = \frac{exp(W \cdot F(x, y))}{Z(x)}$$

where $Z(x) = \sum_{y'} e^{W \cdot F(x, y')}$ is a normalization factor and $F(x, y) = \sum_{i=1}^{|x|} f(y_i, y_{i-1}, x, i)$, where $f(y_i, y_{i-1}, x, i)$ are characteristic functions and $W = W_1 \cdots W_K$ is a vector composed of weights which each characteristic function corresponds to.

During the process of model training, a CRF model can be learned for a given training sample and a predefined characteristic functions where the weight vector parameter W can be learned using MLE (maximum likelihood estimation), gradient descent algorithms, or Quasi-Newton methods. And for the process of inference, for a given observed sequence x, the corresponding labeling sequence can be inferred by the CRF model trained before. The most probable labeling sequence is:

$$argmax_y Pr(y|x) = argmax_y W \cdot F(x, y) = argmax_y W \cdot \sum_j f(y_j, y_{j-1}, x, j)$$

This inference can be implemented by a dynamic programming algorithm such as Viterbi.

3.2   Semi-Markov CRF

Semi-CRF extends CRF so that several tokens in a segment have the same tag. Especially in entity identification, semi-CRF is a natural and intuitive model since entities mostly consist of multiple words and higher accuracy is achieved if features could be defined over the entire proposed entity.

For notation, let $X = \{x_1, x_2, \cdots, x_n\}$ be a sequence of tokens and $S = \langle S_1 \cdots S_p \rangle$ is the corresponding segmentation of X. Each segment S_j contains a start position t_j, an end position u_j, and a tag y_j. The segments S complete cover X without overlapping each other, i. e. it always satisfies $t_1 = 1$, $u_p = |X|$, $1 \leq t_j \leq u_j \leq |X|$, $t_{j+1} = u_j + 1$. A Semi-CRF models the conditional probability distribution over segmentation S for a given input sequence X as follows:

$$Pr(s|x, W) = \frac{exp(W \cdot F(x, s))}{Z(x)}$$

where $Z(x) = \sum_{s'} exp(W \cdot F(x, s'))$ is a normalization factor, W is a weight vector for F, $F(x, s) = \sum_{i=1}^{|x|} f(j, x, s)$ and $f(j, x, s)$ is a feature function.

During the training of semi-CRF, we conduct maximum likelihood estimation for the weight vector W by gradient ascent just like [11]. For inference, we use a semi-Markov analog of the usual Viterbi algorithm [11]. The complexity of the algorithm is $O(KLN)$ where K is the number of tags, L is the upper bound of segments length and N is the number of tokens.

4   Matching Method

During the process of matching reviews to database entities, the input is integrated structured data and relevant reviews and the output is the associations of reviews and entities. Formally, given an integrated database which includes some relational tables, we only concern the target entity table denoted as E since our goal is to match reviews to the review objects. Let $A_1 \cdots A_m$ refer to the attributes of E. We set aside a few part of the reviews (R_L) to label named entities by attribute tags and the rest is the test set denoted as R_T. In addition, since reviews are collected according to the target entities, we assume that there is at least one corresponding object in E for each review.

In this paper, we proposed a method based on 2-layer CRF to match reviews to database objects and the flowchart of our method is illustrated in Figure 1.

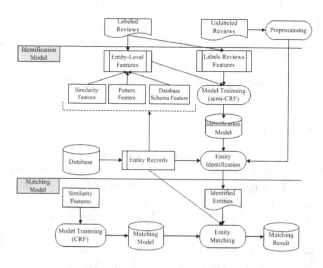

Fig. 1. The flowchart to the proposed approach

4.1   Preprocessing

As reviews in web are published by users, the form and language of reviews are free and at will. For our task that we expect to recognize the terms which can identify the object of reviews, so we need to preprocess the reviews so that filtering the irrelevant terms and noises.

At first, we conduct word segment and part-of-speech tagging for each review from R_T. Since the attribute values of the target entity table are content words, we retain nouns, numbers and times by filtering stop words and other terms. In addition, the reviews often refer to some attributes of the target entity table besides the entity names, so these terms can be views as identifying terms. We measure the symbolic capacity of terms utilizing the spirit of TFIDF as shown follow.

$$w(t) = tf(t) \cdot idf(t) = \frac{n(t)}{N_{(R_T)}} \cdot log(\frac{|R_T|}{|\{r : t \in r\}|})$$

where t is a term, $n(t)$ is the occurrences of t in all reviews, N_{R_T} is the occurrences of terms in all reviews after preprocessing, R_T denotes the amount of reviews and $|\{r : t \in r\}|$ represents the number of reviews which contain the term t.

After preprocessing, we get the term set $t_1 \cdots t_n$ for a review r of R_T.

4.2   Entity Identification

The entity identification is to establish the correspondence between terms of a review r and the attribute tags. Since reviews may contain irrelevant terms in reviews after preprocessing, we add an attributes A_0 in the attribute set which is used to label the irrelevant terms. So the attribute set is $A = A_0, A_1, \ldots, A_m$.

There are some efforts in the field of named entity recognition (NER), and CRF model is proved the best model in sequence data labeling. Current NER methods based on CRF operate by sequentially classifying words as to whether or not they participate in an entity attribute, while often multi-word segments of reviews can represent complete semantic meanings. Especially in our problem that we hope to identify named entities according to the correspondence of entities and attributes, it is more natural and reasonable manner to label segments instead of individual term. We choose Semi-CRF as the entity identification model since the model provides an high-accuracy and natural mechanism for extracting entities in the presence of databases.

During the process of entity identification based on Semi-CRF, it is the key of accuracy how to choose suitable features. As the integrated database gets large, in most cases the entity to be extracted from reviews will exist in the database. We therefore need models that can effectively exploit this valuable piece of information. In this paper, we summary four types of features that the first three types are derived from database named entity-level dictionary features and the fourth type is derived from labeled reviews.

(1) Similarity features

Since each review has at least one corresponding record in target entity set, often the segment identified from a review corresponds to an attribute value from target entity set in a slightly different form. So, we can exploit these to perform better identification by adding similarity features as proposed in [12]. For example, we can add a feature of maximum cosine similarity of a identified segment from and an attribute (person_name).

$$f(y_i, y_{i-1}, r, 3, 5) = max_{v \in person\_name \ in \ E} cosine(t_3 t_4 t_5, v) \cdot [y_i = person\_name]$$

(2) Patten features

Though the identified segments are not completely consistent with the attribute values of target entity set because of the presence of entity variety, different entity attributes have very distinctive patterns that can be generalized to identify these entities in reviews. We define some pattern features just like [13]. We first define some regular expression patterns of short and non-overlapping sub-pattern. Then for each attribute value in target entity set, we generalize each of the tokens to the pattern it matches. Finally, for values belong to a same attribute, we collect the corresponding regular expression patterns as the pattern sequence of the attribute.

For example, we can summary the regular expression pattern sequence of an attribute (person_name) as follow:

id	1	2	3	4
pattern	$[A - Z]$	$[A - Z]+$	$[A - Z]\{punct\}$	$[A - Z][a - z]+$

where $[A - Z]\{punct\}$ denotes a capital letter and a punctuation, $[A - Z][a - z]+$ denotes a capital letter and some lower-case letters. Then the text string "Alon Halevy" will get the pattern-id sequence: "4,4", whereas "W W. Cohen" is "1,3,4".

(3) Database schema features

The integrated structured data also contain a variety of others schema clues including attribute value length, data type and other data integrity constraints such as Check constraint. These clues can serve as the features for entity identification.

(4) Features from labeled reviews

Labeled reviews also provide available clues to identify entities besides features from database. First, labeled reviews also demonstrate pattern features of labeled entities. Second, the likely ordering of entity labels within a sequence is also useful. Since these features are common ones used in entity identification based on CRF, it unnecessary gives more details.

Especially, an important concern about similarity features and pattern features is efficiency. Since we need to conduct similarity computation and regular expression pattern matching for each attribute value, they can be relatively expensive to compute, particularly for a large scale integrated data. In the experiments below, we pre-process integrated by building an inverted index on each attribute and dynamically update the index with the process of data integration. During the process of off-line model training, we conduct top-K search to find the approximate string or matching pattern.

4.3 Entity Matching

After entity identification, for each review r we get the identified segments $s_1 \cdots s_p$ and the corresponding attribute tags $A_1 \cdots A_p$. The task of entity matching is to match a review r to an entity e in the target entity set E.

We identify if the labeled segments of a review r and the attribute values of e refer to one entity in order to determine if the main object of the review is e. At the core of this method is a binary classifier on various similarity measures between a pair of references. We can use any binary classifier like CRF, SVM, decision tree and logistic regression. And the features typically denote various kinds of attribute level similarity functions like Edit distance, Soundex, Jaccard and TFIDF. In this paper, we use a CRF with a normal version. Formally, given an review $r(s_1 \cdots s_p)$ and an entity $e(a_1 \cdots a_p)$, $a_1 \cdots a_p$ is the entity values of e which attribute labels correspond to $A_1 \cdots A_p$ where the attribute labels may be duplicated. For a review r and an entity e, CRF determines the prediction c where c is "1" if e is the main object of r and "0" otherwise.

$$Pr(c|r,e) = \frac{exp(W \cdot F(c,r,e))}{Z(r,e)}$$

where $Z(r,e)$ is a normalization factor, the feature vector $F(c,r,e)$ corresponds to various similarity measures between the text segments and W is a weight vector of F.

5 Experiment Result and Analysis

The real-world data from two domains are collected and used to evaluate the performance of our method. We used MySQL to store the structured database and Lucene to index the text data. We used roughly 5% of the reviews for training and the rest is used for testing. In our method, we use the CRF model with a Semi-Markov and normal version from an open source project named CRF. In order to examine the effectiveness of our method, we perform experiments in the following aspects: (1) The overall performance of our method; (2) The effects of changing the size of the training sample.

5.1 Dataset

(1)Movies

The movie dataset contains top 250 movie information from IMDB, www.imdb.com, which include six attributes such as name, year, type, director list, act list and runtime. And IMDB also provides reviews of a movie. We develop a program to crawl and extract the movie information as well as reviews from IMDB. Make sure that each of the 250 movies has the corresponding review(s). To add an extra degree of difficulty, we also collect additional 250 movies from IMDB for confusion. To sum up, the movies dataset contains 500 movies and 161,062 reviews.

Our task is to match a given review with its corresponding movie in the movie database. Since the reviews were extracted from the IMDB for the corresponding movies, we know the true match for each review and it serves as the gold standard to evaluate our method. In addition, we randomly select 5 percent reviews to annotate the attributes and the rest of reviews serve as the test set (similarly hereinafter).

(2) Restaurants

We obtain restaurant dataset just like [6]. First, we crawl and extract a set of reviews from the Yelp website, yelp.com, where we collect 1,875,630 reviews of 5,610 restaurants from San Francisco. Many of the reviews in Yelp are very short and do not contain any identifying information like good environment, good service. We process the dataset to retain only reviews contain identifying information so that a human can identify the restaurant corresponding to the review. Then we retain 1,588,072 reviews.

In the dataset from Yelp, there are also restaurants and reviews simultaneously. To validate the performance of our method in different data sources, we obtain a much more extensive list of restaurant objects from San Francisco in the Yahoo! Local database which contains 7,985 restaurants and consists of ten attributes including name, genre, telephone, address, price, etc. For each review, our task is to find the corresponding restaurant in the Yahoo! Local database. According to the setting of our problem, it is guaranteed that each review must have its corresponding restaurant. So we select reviews from 4,535 restaurants in Yelp so as to satisfy the condition. The restaurant dataset contains 7,985 restaurants from the Yahoo! Local database and 1,365,821 reviews in Yelp. In addition, we match restaurant information from Yahoo! Local and Yelp along with approximate address name matches. Thus we get the association between restaurants from Yahoo! Local and reviews from Yelp and serves as the gold standard.

5.2 Evaluation Criteria

To evaluate the performance comprehensively, several criteria that were widely used before are taken: Recall, Precision and F1-measure. A brief definition is given as follows.

Defining A as the total number of reviews to be matched, B as the number of reviews which are correctly matched, C as the number of reviews which are wrongly matched.

Then Recall, Precision and F1-measure are defined as:

$$Recall = \frac{B}{A}, \quad Precision = \frac{B}{B+C}, \quad F1 = \frac{2 \times Precision \times Recall}{Precision + Recall}$$

5.3   Experimental Results and Analysis

In order to examine the effectiveness of our method comprehensively, we perform experiments in the following aspects: (1) The performance on the review matching task; (2) The effects of changing the training sample size.

(1) Performance on the review matching task

The experimental results on two datasets are showed. A comparison with TFIDF [6] and the method based on language model (denoted as RLM) [6] is made. Table 2 list Recall, Precision and F1-measure on each dataset. From the reports in Table 2, it can be seen that our method achieves an obvious improvement compared with the other methods. Compared with RLM, the F1-measures on two different datasets are improved by 16.94% and 43.10% respectively. It substantiates that our fine-grained method based on 2-layer CRF has better performance that current coarse-grained methods.

In addition, the three methods archive better performance on the movie dataset than on the restaurant dataset. The main reason is that restaurant reviews are freer and the identifying information is less compared to movie reviews. The advantage of our method on the restaurant dataset is more prominent, and it also proves the adaptability of our method on review matching. Need of special note is that RLM label a half reviews for model training, it is not reasonable for large scale data. Whereas our method only need 5% training set, it is more scalable obviously.

Table 1. Performance on two datasets

Method	Movie Dataset			Restaurant Dataset		
	Recall	Precision	F1	Recall	Precision	F1
TFIDF	48.20%	82.04%	60.72%	9.28%	27.32%	13.85%
RLM	47.17%	92.32%	62.44%	18.06%	45.28%	25.82%
2Layer CRF	68.83%	93.74%	79.38%	59.34%	82.18%	68.92%

(2) Effects of changing the training sample size

To check the effect factors of our approach, we test the effectiveness of the size of training samples. We randomly select 1%, 3%, 5%, 7%, 9%, 11% reviews on two datasets to manually labeled, the rest is the test set. Figure 2 and Figure 3 show the F1-measures with increasing training sizes on the movie and restaurant data set respectively. We can see that a gradual improvement on F1-measures is obtained when increasing the training size in the two datasets. More interesting (after 5%), the slope of the two curves becomes flatter and flatter as increasing the training size. It shows that the bigger of training size, the better performance of our approach. But with the training size is bigger and bigger, its effectiveness will degrade gradually. For our approach, we only need a few training sample to get a superior performance. It proves that our approach less depends on manually labeled training data and is quiet qualified for the review matching task.

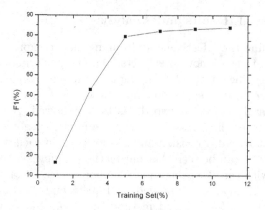

Fig. 2. Effects of changing the training sample size (In movie data set).

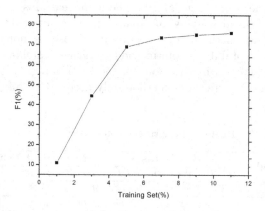

Fig. 3. The flowchart to the proposed approach (In restaurant data set).

6   Conclusions

It is a new challenge how to match relevant reviews to integrated database object in web data integration. We proposed a method based on 2-layer CRF to match reviews to database objects in accordance with the features of web data integration and reviews from web. First, our method leverages the integrated structured entity and significantly reduces the dependence on manually labeled training data; Second, we employ semi-Markov CRF to recognize the structured entities and exploit a variety of entity-level and pattern-level recognition clues available in a database of entities and labeled reviews, thereby effectively resolving the entity variety and improving the accuracy of the entity recognition. Finally, experiments in multiple domains show that our method can substantially superior to traditional tf-idf based methods as well as RLM based method.

Acknowledgment. This work was supported by the National Natural Science Foundation of China (No. 61303007, 61170038, 61472231, 61303005), Shandong Provincial Key Laboratory of Software Engineering (No. 2013SE02), the Outstanding Young Scientists Foundation Grant of Shandong Province (No. BS2013DX044), Postdoctoral Science Special Foundation of Shandong Province (No. 201303057), the Science and Technology Development Plan Project of Shandong Province (No. 2014GGX101019) and the Fundamental Research Funds of Shandong University (No. 2014JC025).

References

1. Whang, S.E., Menestrina, D., Koutrika, G., Theobald, M., Garcia-Molina, H.: Entity resolution with iterative blocking. In: SIGMOD, Rhode Island, USA, pp. 219–232 (2009)
2. Dong, X., Halevy, A.Y., Madhavan, J.: Reference reconciliation in complex information spaces. In: SIGMOD, Maryland, USA, pp. 85–96 (2005)
3. Kalashnikov, D.V., Mehrotra, S.: Domain-Independent Data Cleaning via Analysis of Entity-Relationship Graph. ACM Trans. Database Syst. 31(2), 716–767 (2006)
4. Roy, P., Mohania, M., Bamba, B., Raman, S.: Towards automatic association of relevant unstructured content with structured query results. In: CIKM, Bremen, Germany, pp. 405–412 (2005)
5. Chakaravarthy, V.T., Gupta, H., Roy, P., Mohania, M.: Efficiently linking text documents with relevant structured information. In: VLDB, Seoul, Korea, pp. 667–678 (2006)
6. Dalvi, N., Kumar, R., Pang, B., Tomkins, A.: Matching reviews to objects using a language model. In: EMNLP, Singapore, pp. 609–618 (2009)
7. Dalvi, N., Kumar, R., Pang, B., Tomkins, A.: A translation model for matching reviews to objects. In: CIKM, Hong Kong, China, pp. 167–176 (2009)
8. Wang, W., Xiao, C., Lin, X., Zhang, C.: Efficient approximate entity extraction with edit distance constraints. In: SIGMOD, Rhode Island, USA, pp. 759–770 (2009)
9. Weld, D.S., Hoffmann, R., Wu, F.: Using wikipedia to bootstrap open information extraction. ACM SIGMOD Record 37(4), 62–68 (2009)
10. Lafferty, J., McCallum, A., Pereira, F.: Conditional random fields: Probabilistic models for segmenting and labeling sequence data. In: ICML, MA, USA, pp. 282–289 (2001)
11. Sarawagi, S., Cohen, W.W.: Semi-markov conditional random fields for information extraction. In: NIPS, British Columbia, Canada, pp. 1185–1192 (2004)
12. Cohen, W.W., Sarawagi, S.: Exploiting dictionaries in named entity extraction: Combining semi-markov extraction processes and data integration methods. In: SIGKDD, Washington, USA, pp. 22–25 (2004)
13. Mansuri, I.R., Sarawagi, S.: Integrating unstructured data into relational databases. In: ICDE, Atlanta, GA, USA, p. 29 (2006)

Discovering Restricted Regular Expressions with Interleaving*

Feifei Peng[1,2] and Haiming Chen[1]

[1]State Key Laboratory of Computer Science,
Institute of Software, Chinese Academy of Sciences, Beijing 100190, China
[2] University of Chinese Academy of Sciences
{pengff,chm}@ios.ac.cn

Abstract. Discovering a concise schema from given XML documents is an important problem in XML applications. In this paper, we focus on the problem of learning an unordered schema from a given set of XML examples, which is actually a problem of learning a restricted regular expression with interleaving using positive example strings. Schemas with interleaving could present meaningful knowledge that cannot be disclosed by previous inference techniques. Moreover, inference of the *minimal* schema with interleaving is challenging. The problem of finding a *minimal* schema with interleaving is shown to be NP-hard. Therefore, we develop an approximation algorithm and a heuristic solution to tackle the problem using techniques different from known inference algorithms. We do experiments on real-world data sets to demonstrate the effectiveness of our approaches. Our heuristic algorithm is shown to produce results that are very close to optimal.

Keywords: schema inference, interleaving, partial orders, description.

1 Introduction

When XML is used for data-centric applications such as integration, there may be no order constraint among siblings [1]. Meanwhile, the relative order within siblings may be still important. For example, consider a ticket system with two ticket machines, where there are two bunches of tourists lining up waiting to buy tickets. Each group has two tourists. We can then define the unordered schema for the ticket system. The ordered groups preserve only the relative order of their members. This not only allows individual tourists to insert themselves within a group, but also lets two groups interleave their members. The exact XML Schema Definition (XSD) for the purchasing sequence can be essentially represented as $g1.m1^*g1.m2^*g2.m1^*g2.m2^*$ $|g2.m1^*g2.m2^*g1.m1^*g1.m2^*$ | $g1.m1^*g2.m1^*g1.m2^*g2.m2^*$ $|g1.m1^*g2.m1^*g2.m2^*g1.m2^*$ $|g2.m1^*g1.m1^*$ $g2.m2^*g1.m2^*$ $|g2.m1^*g1.m1^*g1.m2^*g2.m2^*$, where $gi.mj^*$ means the jth member in the ith group can buy zero or more tickets. The representation is

---

* Work supported by the National Natural Science Foundation of China under Grant Nos. 61472405, 61070038.

minimal by Definition 2. It shows the length of the exact regular expression can be exponential when compared to the number of members in sequences.

Actually, $(g1.m1|g1.m2|g2.m1|g2.m2)^*$ is used in practice [3] instead of the minimal ones, which may permit invalid XML documents (i.e., over-permissive). For example, it may permit the second member in the sequence of the first group to purchase tickets before the first member. There are many negative consequences of over-permissive [3]. Thus it is necessary to study how to infer an unordered minimal schema for this kind of XML documents.

Previous researches on XML Schema inference have been done mainly in the context of ordered XML, which can be reduced to learn regular expressions. Gold [9] showed the class of regular expressions is not identifiable in the limit. Therefore numerous papers (e.g.[2,5,6,12]) studied inference algorithms of restricted classes of regular expressions. Most of them were based on properties of automata. Bex et al. [2] proposed learning algorithms for single occurrence regular expressions (SOREs) and chain regular expressions (CHAREs). Freydenberger and Kötzing [12] gave more efficient algorithms learning a minimal generalization for the above classes. The approach is based on descriptive generalization [12] which is a natural extension of Gold-style learning.

However, there is no such kind of automata for regular expressions with interleaving since they do not preserve the total order among symbols. Thus we have to explore new techniques. While Ciucanu [13] proposed learning algorithms for two unordered schema formalisms: disjunctive multiplicity schemas (DMS) and its restriction, disjunction-free multiplicity schemas (MS), both of them disallow concatenation within siblings. Thus they are less expressive than ours. Moreover, the ordering information in our schema formalism can not be fully captured by the three characterizing triples used to construct a DMS or MS.

Inference algorithms in this paper use some similar techniques with algorithms mining global partial orders from sequence data [14,15,17]. However, the semantic concepts there are typically quite different from ours. Mannila et al. [15] tried to find mixture models of parallel partial orders. However, to learn unordered regular expressions, series parallel orders may not be sufficient since they can conflict with some data in the whole data set. Another restriction in the above method is that it can only be applied to strings where each symbol occurs at most once. Particularly, Gionis et al. [14] emphasised on recovering the underlying ordering of the attributes in high-dimensional collections of 0-1 data. An implicit assumption is that attribute can also occur at most once. For learning regular expressions with interleaving, symbols in strings can present any times and partial orders among siblings are independent with no violations. Hence many techniques from data mining are not directly applicable. Therefore, learning restricted regular expressions with interleaving remains a challenging problem.

In this paper, we address the problem of discovering a minimal regular expression with interleaving from positive examples. The main contributions of the paper are listed as follows:

- We propose a better and more suitable formalism to specify precise unordered XML: the subset of regular expressions with interleaving (SIREs). SIREs can express the content models succinctly and concisely. For example, the above example can be depicted as $(g1.m1^*g1.m2^*)\&(g2.m1^*g2.m2^*)$.
- We introduce the notion of SIRE-minimal in the terminology of [12] and some properties of SIRE-minimal.
- We prove the problem of finding a minimal SIRE is NP-hard and develop an approximation algorithm conMiner to find solutions with worst-case quality guarantees and a heuristic algorithm conDAG that mostly finds solutions of better quality as compared to the approximation algorithm conMiner.
- We conduct experiments comparing our methods with Trang [8] on real world data, incorporating small and large data sets. Our experiments show that conMiner and conDAG outperform existing systems on such data.

The rest of the paper is organized as follows. Section 2 contains basic definitions. In Section 3 we discuss properties of minimal-SIRE. In Section 4 an approximation algorithm conMiner and a heuristic algorithm conDAG are proposed. Section 5 gives the empirical results. Conclusions are drawn in Section 6.

2   Preliminaries

Let u and v be two arbitrary strings. By $u\&v$ we denote the set of strings that is obtained by interleaving of u and v in every possible way. That is, $u\&\varepsilon = \varepsilon\&u = u$, $v\&\varepsilon = \varepsilon\&v = v$. If both u and v are non-empty let $u = au', v = bv'$, a and b are single symbols, then $u\&v = a(u'\&v) \cup b(u\&v')$. The operator $\&$ can be then extended to regular languages as a binary operator in the canonical way. Let Σ be an alphabet of symbols. The regular expressions with interleaving over Σ are defined as: $\emptyset, \varepsilon$ or $a \in \Sigma$ is a regular expression, E_1^*, $E_1 E_2$, $E_1|E_2$, or $E_1\&E_2$ is a regular expression for regular expressions E_1 and E_2. They are denoted as RE($\&$). The language described by E is defined as follows: $L(\emptyset) = \emptyset$; $L(\varepsilon) = \varepsilon$; $L(a) = \{a\}$; $L(E_1^*) = L(E_1)^*$; $L(E_1 E_2) = L(E_1)L(E_2)$; $L(E_1|E_2) = L(E_1)\cup L(E_2)$; $L(E_1\&E_2) = L(E_1)\&L(E_2)$. E? and E^+ are used as abbreviations of $E + \epsilon$ and EE^*, respectively. We consider the subset of regular expressions with interleaving (SIREs) defined by the following grammar.

Definition 1. *The restricted class of regular expressions with interleaving (RREs) are RE($\&$) over Σ by the following grammar for any $a \in \Sigma$:*

$$S ::= T\&S|T$$
$$T ::= \varepsilon|a|a^*|TT$$

The subset of regular expressions with interleaving (SIREs) are those RREs in which every symbol can occur at most once. Since SIREs disallow repetitions of symbols, they are certainly deterministic and satisfy the UPA constraint required by the XML specification.

A partial order M for a string s is a binary relation that is reflexive, antisymmetric and transitive. We write $a \prec b$ if a is before b in the partial order. For string $s = x_1 \cdots x_l$, the *transitive closure* of s is denoted by $tr(s) = \{(x_i, x_j) | 1 \leq i < j \leq l\}$, where l is the length of s. For example $s = abcd$, $tr(s) = \{ab, ac, ad, bc, bd, cd\}$.

A partial-order set t is a set of symbols together with a partial ordering. We say $ab \in t$ if a precedes b in every string in a string collection. Consistent partial order set (CPOS) T is a set which contains all the disjoint partial-order sets t_i of the given examples. For example, consider $W = \{abcd, dabc\}$. Obviously, $a \prec b \prec c$, $T = \{abc, d\}$. The connection between CPOS and SIRE is directly. That is, given a CPOS, we can write it to the form of SIRE by combining all the elements in CPOS with &. For example, in this case the corresponding SIRE $s = abc\&d$. Therefore, the problem of finding a minimal SIRE can be reduced to the problem of finding a minimal CPOS.

3   Description

This section introduces the notion of minimal expressions. Roughly speaking minimal is the greatest lower bound of a language L within a class of expressions, which is conceptually similar with *infimum* in the terminology of mathematics.

Definition 2 ([12]). *Let $\mathcal{D}$ be a class of regular expressions over some alphabet Σ. A $\delta \in \mathcal{D}$ is called $\mathcal{D}$-minimal of non-empty language $S \subseteq \Sigma^*$, if $L(\delta) \supseteq S$ and there is no $\gamma \in \mathcal{D}$ such that $L(\delta) \supset L(\gamma) \supseteq S$.*

Proposition 1. *Let n be the number of alphabet symbols. The number of pairwise non-equivalent SIREs is $\mathcal{O}(\frac{4^n n!}{2(ln2)^{n+1}})$.*

Proof. Disregarding operators ?,+,*, the number of SIREs over a finite Σ is equivalent to the number of ordered partitioning $|\Sigma|$ symbols. The number of these partitions is given by the $|\Sigma|$th ordered Bell numbers [11]. For instance, if $\Sigma = \{a, b, c\}$, the 3th ordered Bell number $a(3) = 13$, and the ordered partitions of $\{a, b, c\}$ is $\{abc, acb, bac, bca, cab, cba, ab\&c, ba\&c, ac\&b, ca\&b, bc\&a, cb\&a, a\&b\&c\}$. They are also distinct partitions of SIREs over Σ. The ordered Bell number [10] can be approximated as $a(n) = \sum_{k=0}^{n} k! \binom{n}{k} \approx \frac{n!}{2(ln2)^{n+1}}$. Since every symbol a in Σ has four forms which can be represented as $a, a^?, a^+$ and a^*, the number of SIREs over Σ is $4^n a(n)$. Then $s(n) \approx \frac{4^n n!}{2(ln2)^{n+1}}$. $\qquad \square$

We can then prove the existence of minimal regular expressions for SIRE.

Proposition 2. *Let Σ be a finite alphabet. For every language $L \subseteq \Sigma^*$, there exists a SIRE-minimal SIRE δ_s.*

Proof. Assume there is a language L over Σ such that no expression $\alpha \in SIRE$ is SIRE-minimal. This implies that there is an infinite sequence $(\beta_i)_{i \geq 0}$ of expressions from SIRE with $\alpha = \beta_0$ and $L(\beta_i) \supset L(\beta_{i+1}) \supseteq L$ for all $i \geq 0$. This contradicts the fact that there are only a finite number of non-equivalent SIREs over Σ by Proposition 1. $\qquad \square$

Proposition 3. *For any example string set E over $\{a_1, \cdots, a_n\}$, let $S = s_1 \& \cdots \& s_l$ be a SIRE such that $E \subseteq L(S)$. S is a minimal SIRE if and only if:*
(1) the number of s_i is minimized and
(2) the size of each s_i is as large as possible.

The proof was omitted for space reasons.

In other words, a minimal SIRE is the most specific SIRE that consistent with the given example strings. For instance, all of $S_1 = a\&bc\&d$, $S_2 = abc\&d$ and $S_3 = ad\&bc$ can accept $E = \{abcd, adbc\}$. However, since $S_1 = (ad|da)\&bc = (ad\&bc)|(da\&bc) = S_3|(ad\&bc)$, we can get $L(S_1) \supset L(S_3)$ which means S_1 is not minimal. As for S_2 and S_3, since $L(S_2) = \{abcd, abdc, adbc, dabc\}$ and $L(S_3) = \{bcad, bacd, badc, abcd, abdc, adbc\}$, this means S_3 is not minimal. As we shall see, S_2 is a better approximation of E. In fact, S_2 can be verified to be a minimal by referring to Proposition 3.

4   Minimal SIREs

In this section, we first prove finding a minimal SIRE for a given set of strings is NP-hard by reducing from finding a maximum independent set of a graph, which is a well-known NP-hard graph problem [7]. Then we present learning algorithms that construct approximatively minimal SIREs.

4.1   Exact Identification

First, we introduce the notion of maximum independent set of a graph [7]. Consider an undirected graph $G(V, E)$, an independent set (IS) is a set that $\forall u, v \in IS$, $u, v \in V$ and $(u, v) \notin E$. The maximum independent set (MIS) problem consists in computing an IS of the largest size. Next, we define the problem `all_mis` which takes a graph G as input, finding a MIS S' of G by applying function `max_independent_set`, and repeating the step for subgraph $G[V - S']$ until there exists no vertex in the subgraph. In other words, `all_mis` is to divide V into disjoint subsets by `max_independent_set`. Clearly, problem `all_mis` is NP-hard. For example, consider $G(V, E)$ which $V = \{1, 2, 3, 4\}, E = \{(1, 2), (1, 3), (2, 3), (2, 4), (3, 4)\}$. The result of `max_independent_set(G)` is $\{1, 4\}$. The result of `max_independent_set(G[V-{1,4}])` is $\{2\}$. Thus the result of `all_mis(G)` is $\{\{1, 4\}, \{2\}, \{3\}\}$.

The main idea of finding a minimal SIRE is based on the observation that there are sets of conflicting siblings that cannot be divided into the same subset of CPOS. A pair xy is called forbid pair in a string database if both xy and yx exists in the transitive closure of strings. The set of forbid pairs is called a *constraint*. By Proposition 3, if we split the set of symbols in a *constraint* into several subsets $t_1, \cdots, t_n$ such that n is minimized and for each $i \in [1..n]$, t_i is the longest of its alternatives. Then the set of t_i where $i \in [1..n]$, is a minimal CPOS which can be transformed to a minimal SIRE.

Lemma 1. *Minimal SIRE finding problem is NP-hard.*

Proof. We demonstrate that `all_mis` can be reduced in polynomial time to minimal SIRE finding problem. Given an instance of `all_mis`, we can generate a corresponding instance of minimal SIRE finding as follows. For the graph G in `all_mis`, the reduction algorithm computes the *constraint* set by adding all edges in G to *constraint*, which is easily obtained in polynomial time. The output of the reduction algorithm is the instance set *constraint* of minimal SIRE finding problem. t_i in CPOS is the longest of its alternatives if and only if `all_mis` computes a maximum independent set at the ith step. Thus, minimal SIRE finding problem is equivalent to the original `all_mis`. Since `all_mis` is NP-hard, minimal SIRE finding problem is NP-hard.   □

4.2 Approximation Algorithm

The process of this approach is formalized in Algorithm 1. Algorithm 1 works in four steps and we illustrate them on the sample $E = \{abcd, aadbc, bdd\}$. The first step (lines 1-2) computes the non-constraint and constraint set using the function `tran_reduction`. The transitive closure of E is $tr = \{ab, ac, ad, bc, bd, cd, db, dc\}$. Add uv to *constraint* if $vu \in tr$. Add uv to *consist_tr* otherwise. We get $consist\_tr = \{ab, ac, ad, bc\}$ and $constraint = \{bd, cd, db, dc\}$. Construct an undirected graph G using element in *constraint* as edges. The second step (lines 3-7) is to select a MIS of G, add it to list *allmis* and delete the MIS and their related edges from G. The process is repeated until there exists no nodes in G. The problem of finding a maximum independent set is an NP-hard optimization problem. As such, it is unlikely that there exists an efficient algorithm for finding a maximum independent set of a graph. However, we can find a MIS in polynomial time with a approximation algorithm, e.g. the `clique_removal` algorithm proposed in [19] that finds the approximation of maximum independent set with performance guarantee $\mathcal{O}(n/(\log n)^2)$ by excluding subgraphs. For graph G, we obtain $allmis = \{\{b, c\}, d\}$. Next, we add the non-constraint symbols to the first MIS. Then we have $allmis = \{\{a, b, c\}, d\}$. The third step (lines 8-10) computes the topological sort for all subgraphs induced by subset of *consist_tr* and add the result to T. For the sample, it returns $T = \{abc, d\}$. Finally, the algorithm returns the SIRE whose corresponding counting operators $1, *, +, ?$ can be inferred using technique in algorithm `CRX` [4]. For the sample, it returns $a^*bc?\&d^+$.

When carefully implemented, *clique_removal* involves $(|V| + |E|)$ work [19], where $|E|$ is $\mathcal{O}(|V|^2)$. The total running time of conMiner is $(n^3 + m)$, where m is the sum of length of the input example strings, n the number of alphabet symbols.

4.3 Heuristic Algorithm

Although a number of approximation algorithms and heuristic algorithms have been developed for the maximum independent set problem, on any given instance, they may produce a SIRE that is very far from optimal. We introduce a heuristic directed acyclic graph construction algorithm directly computing a minimal SIRE. The main idea is to cluster the vertices of the existing directed graph

---

Algorithm 1. *conMiner(W)*

Input: Set of words $W = \{w_1, ..., w_n\}$
Output: a minimal SIRE T
1: $consist\_tr, constraint = tran\_reduction(W, T)$
2: $G = Graph(constraint)$
3: **while** $G.nodes()! = null$ **do**
4:    $v = clique\_removal(G)$
5:    $G = G - v$
6:    $allmis.append(v)$
7: $allmis[0] = allmis[0].union(alphabet(consist\_tr) - alphabet(constraint))$
8: **for** each $mis \in allmis$ **do**
9:    $H = Graph(mis, consist\_tr)$
10:    $T.append(topological\_sort(H))$
11: **return** $learner_{oper}(W, T)$

---

into several disconnected subgraphs. The graph is constructed incrementally to preserve CPOS within each vertex using a greedy approach. The pseudocode of algorithm conDAG is given in Algorithm 2.

The input to this algorithm is the same as the input of the conMiner. The algorithm maintains lists p, q as records to keep track of pairs violating the partial order constraint and lists s, t to record pairs violating the partial order constraint of the string under reading. Note that (a, b) violating the partial order constraint means there exist some $w_1, w_2 \in W$ such that $a \prec b$ in w_1 and $b \prec a$ in w_2.

Let ab be two adjacent symbols in a word w. The add_or_break function checks whether edge ab is added to the present graph G. If there exists no path from b to a, no path from a to b in G and edge ab will not make a connection between some $p[i]$ and $q[i]$, we add edge $a \rightarrow b$ in G. Self-loops such as $f \rightarrow f$ are always ignored since they have no influence on the partial order constraints. However, if there exist paths from b to a in G, $(a, b) \notin (p[i], q[i]), (q[i], p[i])$ and a, b are not in $p[i], q[i]$ at the same time for all $i < len(p)$, we should break all paths from b to a. The breakpoint can be found as below. Suppose there exists a path $u = b\alpha_1...a$, $\alpha_0 = b$ in G, and substring of w over $\{b, \alpha_1, ..., a\}$ is $\alpha_i...a$, then we delete edge $\alpha_{i-1} \rightarrow \alpha_i$, add edge $\beta \rightarrow \alpha_i$ for all nodes β that $\beta \rightarrow b$, and add edge $\alpha_{i-1} \rightarrow \gamma$ for all nodes γ that $a \rightarrow \gamma$. In the end, append string $b\alpha_1...\alpha_{i-1}$ to p,s and append string $\alpha_i...a$ to q,t. Now list p, s are $[b\alpha_1...\alpha_{i-1}]$ and q, t are $\alpha_i...a$.

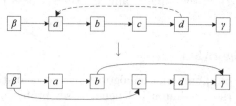

Fig. 1. This is an example to find the breakpoint

```
1:  function consistent(G, w, p, q)
2:      s, t := ∅, i := 1
3:      while i < |w| − 1 do
4:          if (w[i]! = w[i + 1]) ∧ ((w[i], w[i + 1]) ∉ (p, q), (q, p)) then
5:              add_or_break(G, w, w[i], w[i + 1], p, q, s, t)
6:          for j := 1 to |s| do
7:              if (w[i] ∈ s[j]) ∧ ((t[j][−1], w[i + 1]) ∉ (p, q)) then
8:                  add_or_break(G, w, t[j][−1], w[i + 1], p, q, s, t)
9:              if (w[i] ∈ t[j]) ∧ (s[j][−1], w[i + 1]) ∉ (p, q)) then
10:                 add_or_break(G, w, s[j][−1], w[i + 1], p, q, s, t)
11:         i + +
```

Example in Figure 1 shows how the function works. $W = \{\beta abcd\gamma, cda\}$, initialize empty list p,q,s,t and empty graph G. After reading w_1, list p,q,s,t are still empty. When reading $da \in w_2$, there already exists a path $abcd$ and $(d, a) \notin (p[i], q[i]), (q[i], p[i])$. We should break $abcd$. Since $substring(w_2, \{a, b, c, d\}) = cda$, breakpoint is c. Then we delete edge $b \to c$, and add edges $\beta \to c, b \to \gamma$. In the end, append string ab to list p,s and append string cd to list q,t.

The `consistent` function scans the whole string w by sequence to execute `add_or_break` function. Each time after reading two adjacent symbols ab, for all pairs $(\alpha_1 a\alpha_2, \alpha_3 c)$ or $(\alpha_3 c, \alpha_1 a\alpha_2) \in (s, t)$, handle cb likewise. Because $(\alpha_1 a\alpha_2, \alpha_3 c)$ or $(\alpha_3 c, \alpha_1 a\alpha_2) \in (s, t)$ declare $a \prec c$ and $c \prec a$ are in w, if $a \prec b$ in w, $c \prec b$ is also in w. Consider $acab$ as an example, c and a have been two parts after reading ca, a has been added to p and s and c added to q and t. After reading the next two symbols ab, add edge $a \to b$. Next we should consider cb since $a \in s[0], c \in t[0]$, thus add edge $c \to b$. The `topological_sort(g)` construct a topological ordering of DAG in linear time. The `learner_oper` is used to infer operators $?, +, *$ for each vertex.

The `conDAG` algorithm combines all the functions. The constructed graph is denoted by G and the corresponding set of partitions by C. In each iteration, it invokes `consistent` to update G using the ith string. Then it adds all the paths from the set of vertices of in-degree zero to the set of vertices of out-degree zero. To be able to calculate the largest independent partial-order plans, a preprocessing phase is implemented. First, we consider the elements of C in decreasing order of size. In each iteration, whenever we find two elements that the one contains elements of $p[i]$ and the other one contains elements of $q[i]$, we updates the shorter one by removing the common elements. Next, we merge all the lists in C that share common elements. The preprocess terminates when every symbol is included in one and only one list. The following steps of the algorithm are the same as the third and the forth step of the `conMiner`.

The time complexity analysis of this algorithm is straightforward. add_brea $k(G, w, a, b, p, q, s, t)$ can find all possible paths between two given nodes by modifying the DFS which needs $\mathcal{O}(|V|+|E|)$ steps. Breaking a circle requires $\mathcal{O}(|V|)$. Therefore, an overall time complexity for add_or_break is $\mathcal{O}(|V|+|E|)$. The length of w can at most be n and $|E|$ is $\mathcal{O}(|V|^2)$. So the total time of $consisitent$ is $\mathcal{O}(n^3)$.

Algorithm 2. $conDAG(W)$

Input: Set of unordered words $W = \{w_1, ..., w_t\}$
Output: a minimal SIRE
1: $consist\_tr, constraint = tran\_reduction(W, T)$
2: initialize graph G, $p, q := \emptyset$
3: **for** $i := 1$ to t **do**
4:      $consistent(G, w_i, p, q)$
5: $C = all\_paths(G, source, destination)$
6: remove the common elements from the shorter of $c_i, c_j \in C$ if $c_i[m] + c_j[n] \in$ constraint.
7: merge all lists that share common elements in C
8: **for** each mis in C **do**
9:      $H = Graph(mis, consist\_tr)$
10:      $T.append(topological\_sort(H))$
11: **return** $learner_{oper}(W, T)$

The $tran\_reduction$ computation requires $\mathcal{O}(n^2)$ time, where n is the number of distinct symbols. Each iteration requires $\mathcal{O}(n^3)$ time to maintain the graph. Computing all paths from source to destination can be done in $\mathcal{O}(n^2)$ time, and $topological\_sort(g)$ constructs a topological ordering of DAG in linear time, thus $\mathcal{O}(|V| + |E|)$ steps are sufficient. Inference of operators ?, +, * needs time $\mathcal{O}(m)$. Hence the time complexity of the algorithm is $\mathcal{O}(n^4 + m)$, where m is the sum of length of the input example strings, n the number of alphabet symbols.

To illustrate our algorithm, consider the example $E = \{abcd, aadbc, bdd\}$, $consist\_tr = \{ab, ac, ad, bc\}$, $constraint = \{bd, cd, db, dc\}$ in the above section. A directed graph which consists of vertex $V = \{a, b, c, d\}$ and edges $E = \{ab, bc, ad\}$ can be obtained. $p = \{bc\}$ and $q = \{d\}$. All paths from source to destination are $C = \{abc, ad\}$. Since $bd \in constraint$, $C[2]$ is updated by removing the common elements between $C[1]$ and $C[2]$. $C[2]$ is d. The final C is $\{abc, d\}$. The following steps are the same.

5   Experiments

In this section, we validate our approaches on real-life DTDs, and compare them with that of Trang [8]. All experiments were conducted on an IBM T400 laptop computer with a Intel Core 2 Duo CPU(2.4GHz) and 2G memory. All codes were written in python.

The number of corpora of XML documents with an interesting schema is rather limited. We obtained our real-life DTDs from the XML DATA repository maintained by Miklau [18]. Unfortunately, most of them are either not data-centric or not with a DTD. Specifically, We chose the DBLP Computer Science Bibliography corpus, a data-centric database of information on major computer science journals and proceedings. The code and data are available at http://lcs.ios.ac.cn/~pengff/projects.html. Table 1 lists the non-trivial element definitions in the above mentioned DTD together with the results derived by exact algorithm, heuristic algorithm conMiner, approximation algorithm

Table 1. Results of exact algorithm, conMiner, conDAG and Trang on DTDs

Element name Sample size	Original DTD Exact Minimal DTD Result of conMiner Result of conDAG Result of Trang
inproceedings 2122274 2122274 2122274 2122274	$(a_1\|a_2\|\cdots\|a_{22})^*$ $a_1{}^*a_{12}?a_5{}^*a_9?a_{18}?a_{15}{}^*\&a_3a_6a_{11}{}^*\&a_{19}{}^*\&a_{13}{}^*\&a_4\&a_{14}{}^*$ $a_5{}^*a_{18}?a_{15}{}^*\&a_{12}?a_9?a_{13}{}^*\&a_1{}^*a_{14}{}^*\&a_6a_{11}{}^*\&a_3\&a_4\&a_{19}{}^*$ $a_1{}^*a_4a_9?a_{11}{}^*a_{15}{}^*\&a_3a_{12}?a_5{}^*a_{18}?\&a_{13}{}^*\&a_6\&a_{14}{}^*\&a_{19}{}^*$ $(a_1\|a_3\|a_5\|a_6\|a_9\|a_{11}\|a_{12}\|a_{13}\|a_{14}\|a_{15}\|a_{18}\|a_{19})^+$
article 111608 111608 111608 111608	$(a_1\|a_2\|\cdots\|a_{22})^*$ $a_1{}^*a_{17}?a_5{}^*a_{12}?a_{15}{}^*\&a_3a_6a_{11}?\&a_{13}{}^*\&a_8\&a_{10}?\&a_{14}{}^*\&a_9?$ $a_{17}?a_{12}?a_9?a_{15}{}^*\&a_1{}^*a_6a_{11}?\&a_3\&a_5{}^*\&a_{13}{}^*\&a_8\&a_{10}?\&a_{14}{}^*$ $a_3{}^*a_{17}?a_6a_{11}?\&a_1{}^*a_8a_{12}?a_{15}{}^*\&a_{13}{}^*\&a_5{}^*\&a_{10}?\&a_{12}?\&a_9?$ $a_2?(a_1\|a_3\|a_5\|a_6\|a_8\|a_9\|a_{10}\|a_{11}\|a_{12}\|a_{13}\|a_{14}\|a_{15}\|a_{17})^+$
proceedings 3007 3007 3007 3007	$(a_1\|a_2\|\cdots\|a_{22})^*$ $a_2{}^*a_3{}^+a_{18}?a_{21}?a_8?a_{10}?a_{13}?a_{12}?a_{15}{}^*a_{19}?a_7?a_9?\&a_4?\&a_{17}?\&a_6\&a_{20}{}^*\&a_{11}?$ $a_2{}^*a_3{}^+a_{19}?a_{13}?a_{20}{}^*a_{15}{}^*a_{12}?\&a_4?\&a_7?a_8?a_9?\&a_{21}?a_{18}?a_{10}?\&a_6\&a_{17}?\&a_{11}?$ $a_2{}^*a_3{}^+a_8?a_{18}?a_{21}?a_{10}?a_9?a_{19}?a_{13}?a_7?a_{15}{}^*\&a_4?a_{12}?\&a_{17}?\&a_6\&a_{20}{}^*\&a_{11}?$ $a_2{}^*a_3{}^+(a_4\|a_6\|a_7\|a_8\|a_9\|a_{10}\|a_{11}\|a_{12}\|a_{13}\|a_{17}\|a_{18}\|a_{19}\|a_{20}\|a_{21})^+a_{15}{}^*$
incollection 1009 1009 1009 1009	$(a_1\|a_2\|\cdots\|a_{22})^*$ $a_1{}^*a_3a_4a_{17}?a_{20}?a_{16}?a_{11}?a_{15}{}^*a_{14}?\&a_{13}?a_{19}?\&a_5?\&a_6$ $a_1{}^*a_3a_{17}?a_6\&a_{15}{}^*a_{13}?a_{16}?a_{14}?\&a_4a_{11}?\&a_{20}?a_{19}?\&a_5?$ $a_1{}^*a_3a_4a_{17}?a_{11}?a_{15}{}^*a_{14}?\&a_6a_{20}?\&a_5?a_{16}?\&a_{13}?\&a_{19}?$ $(a_1\|a_3\|a_4\|a_5\|a_6\|a_{11}\|a_{13}\|a_{16}\|a_{17}\|a_{20})^+(a_{14}\|a_{15}{}^*)$
phdthesis 72 72 72 72	$(a_1\|a_2\|\cdots\|a_{22})^*$ $a_1a_3a_6a_{17}?a_{21}?a_{20}?a_9?a_{13}?a_{12}?\&a_{22}$ $a_1a_3a_6a_{12}?a_{21}?a_{22}a_{13}?a_{20}?\&a_{17}?a_9?$ $a_1a_3a_6a_{17}?a_{21}?a_{20}?a_{13}?a_9?a_{12}?\&a_{22}$ $a_1a_3a_6(a_{12}\|a_{21})?(a_9\|a_{17}\|a_{22})^+(a_{13}\|a_{20})?$

conDAG, and Trang. We implement the exact algorithm by replacing function `clique_removal` in conMiner with an exponential time algorithm proposed by S. Tsukiyama [20]. It can be verified that all expressions learned by exact algorithm, conDAG and conMiner are more strict than that of Trang and the original DTDs which indicates there exists much more over-permissive in both the original DTDs and the results of Trang. Moreover, although the results of conMiner and conDAG are not the ideal optimum, they are very close to that of exact algorithm thus are nearly as good.

There may exist many minimal expressions given a set of unordered strings. For instance, for `phdthesis`, the form of the result of conDAG is the same with the exact minimal expression. The orders among symbols of their first siblings, however, differ widely. This is due to the fact that a diagraph may have several different topological sorts. Therefore, we ignore the sequel in the symbols and only compare their simplified form which concerns only the length of each sibling and the number of interleavings. For example, the simplified result of exact algorithm for inproceedings is $6\&3\&1\&1\&1\&1$.

Table 2. Similarity of the results of conDAG and conMiner with exact algorithm, respectively.

Element name	Similarity	
	conDAG	conMiner
inproceedings	0.979610	0.729517
article	0.974022	0.838144
proceedings	0.995385	0.905593
incollection	0.732781	0.602769
phdthesis	1.000000	0.990992

We measure the similarity by formula $sim(vec(t_e), vec(t_c)) = sim_c(vec(t_e), vec(t_c)) * (num(vec(t_e))/num(vec(t_c)))$, where $vec(t_e)$ are the vector of the result of exact algorithm, $vec(t_c)$ the vector of the result of conDAG or conMiner, sim_c is the cosine similarity between two vector and num the number of interleavings. For the similarity between exact algorithm and con-Miner for inproceedings, $vec(t_e) = \{6, 3, 1, 1, 1, 1, 0\}$, $vec(t_c) = \{3, 3, 2, 2, 1, 1, 1\}$, we can get $sim(vec(t_e), vec(t_c)) = (33/(\sqrt[2]{49} * \sqrt[2]{29})) * (5/6) = 0.729517$. Table 2 shows clearly that conDAG yields concise super-approximations to the exact minimal expressions thus can find solutions of better quality as compared to the solutions found by the approximation algorithm.

6   Conclusion

This paper proposes a strategy for learning a class of regular expressions with interleaving: first, compute consistent partial order T, then equip each factor with counting operators. As future work, we will investigate several interesting problems inspired by this study. First, we would like to extend our algorithms for more expressive schemas, for example schemas allow disjunction "|" within siblings. Second, how to extend algorithms to mine all the independent frequent closed partial orders [17] is also an attractive topic.

Acknowledgement. We thank the users of Stack Overflow [21], for reminding us the maximum independent set problem.

References

1. Abiteboul, S., Bourhis, P., Vianu, V.: Highly expressive query languages for unordered data trees. In: Proceedings of the 15th International Conference on Database Theory, pp. 46–60 (2012)
2. Bex, G.J., Gelade, W., Martens, W., Neven, F.: Simplifying XML schema: effortless handling of nondeterministic regular expressions. In: Proceedings of the 2009 ACM SIGMOD International Conference on Management of Data, pp. 731–744 (2009)
3. Boneva, I., Ciucanu, R., Staworko, S.: Simple schemas for unordered XML. arXiv preprint arXiv:1303.4277 (2013)

4. Bex, G.J., Neven, F., Schwentick, T., Vansummeren, S.: Inference of concise DTDs from XML data. In: Proceedings of the 32nd International Conference on Very Large Data Bases, pp. 115–126. VLDB Endowment, September 2006

5. Bex, G.J., Neven, F., Vansummeren, S.: Inferring XML schema definitions from XML data. In: Proceedings of the 33rd international conference on Very large data bases, pp. 998–1009. VLDB Endowment, September 2007

6. Bex, G.J., Wouter, G., Neven, F., Vansummeren, S.: Learning deterministic regular expressions for the inference of schemas from XML data. ACM Transactions on the Web (TWEB) 4(4), 14 (2010)

7. Ignatiev, A., Morgado, A., Marques-Silva, J.: On reducing maximum independent set to minimum satisfiability. In: Sinz, C., Egly, U. (eds.) SAT 2014. LNCS, vol. 8561, pp. 103–120. Springer, Heidelberg (2014)

8. Clark, J.: Trang: Multi-format schema converter based on RELAX NG. http://www.thaiopensource.com/relaxng/trang.html

9. Gold, E.M.: Language identification in the limit. Information and Control 10(5), 447–474 (1967)

10. Bailey, R.W.: The number of weak orderings of a finite set. Social Choice and Welfare 15(4), 559–562 (1998)

11. de Koninck, J.M.: Those Fascinating Numbers. American Mathematical Soc. (2009)

12. Freydenberger, D.D., Kötzing, T.: Fast learning of restricted regular expressions and DTDs. In: Proceedings of the 16th International Conference on Database Theory, pp. 45–56 (2013)

13. Ciucanu, R., Staworko, S.: Learning schemas for unordered xml. arXiv preprint arXiv:1307.6348 (2013)

14. Gionis, A., Kujala, T., Mannila, H.: Fragments of order. In: Proceedings of the Ninth ACM SIGKDD International Conference on Knowledge Discovery and Data Mining, pp. 129–136 (2003)

15. Mannila, H., Meek, C.: Global partial orders from sequential data. In: Proceedings of the Sixth ACM SIGKDD International Conference on Knowledge Discovery and Data Mining, pp. 161–168 (2000)

16. Agrawal, R., Srikant, R.: Fast algorithms for mining association rules. In: Proc. 20th Int. Conf. Very Large Data Bases, vol. 1215, pp. 487–499. VLDB (1994)

17. Pei, J., Wang, H., Liu, J., Wang, K., et al.: Discovering frequent closed partial orders from strings. IEEE Transactions on Knowledge and Data Engineering 18(11), 1467–1481 (2006)

18. Miklau, G.: XMLData Repository, November 2002. http://www.cs.washington.edu/research/xmldatasets/

19. Boppana, R., Halldórsson, M.M.: Approximating maximum independent sets by excluding subgraphs. BIT Numerical Mathematics 32(2), 180–196 (1992)

20. Tsukiyama, S., Ide, M., Ariyoshi, H., Shirakawa, I.: A new algorithm for generating all the maximal independent sets. SIAM Journal on Computing 6(3), 505–517 (1977)

21. Algorithm to divide a set of symbols with constraints into minimun number of subsets. http://stackoverflow.com/q/29117747/4684328

Efficient Algorithms for Distance-Based Representative Skyline Computation in 2D Space

Taotao Cai[1], Rong-Hua Li[1], Jeffrey Xu Yu[2], Rui Mao[1], and Yadi Cai[1]

[1] Guangdong Province Key Laboratory of Popular High Performance Computers,
Shenzhen University, China
[2] The Chinese University of Hong Kong, Hong Kong
{taotaocai1992,iyadicai}@gmail.com, {rhli,mao}@szu.edu.cn,
yu@se.cuhk.edu.hk

Abstract. Representative skyline computation is a fundamental issue in database area, which has attracted much attention in recent years. A notable definition of representative skyline is the distance-based representative skyline (DBRS). Given an integer k, a DBRS includes k representative skyline points that aims at minimizing the maximal distance between a non-representative skyline point and its nearest representative. In the 2D space, the state-of-the-art algorithm to compute the DBRS is based on dynamic programming (DP) which takes $O(km^2)$ time complexity, where m is the number of skyline points. Clearly, such a DP-based algorithm cannot be used for handling large scale dataset due to the quadratic time cost. To overcome this problem, in this paper, we propose a new approximate algorithm called ARS, and a new exact algorithm named PSRS, based on a carefully-designed parametric search technique. We show that the ARS algorithm can guarantee a solution that is at most ϵ larger than the optimal solution. The proposed ARS and PSRS algorithms run in $O(k \log^2 m \log(T/\epsilon))$ and $O(k^2 \log^3 m)$ time respectively, where T is no more than the maximal distance between any two skyline points. We conduct extensive experimental studies over both synthetic and real-world datasets, and the results demonstrate the efficiency and effectiveness of the proposed algorithms.

1 Introduction

Skyline computation is a fundamental problem in database area [2], which has attracted considerable attention in database community. Given a set of d-dimensional points, the skyline is a subset of points that are not dominated by any other points. Here a point p dominating a point q $(p \neq q)$ means that q's coordinate is no larger than p's coordinate in all dimensions. The skyline operator is particularly useful in multi-criteria decision making related applications [2,10].

However, in many applications, the number of skyline points are typically very large, and even may be comparable to the size of the entire dataset [14]. To overcome this issue, a promising method is to select k representatives to represent the entire skyline [8,14]. In the literature, there are several definitions of representative skyline [8,14], including the notable domination-based representative skyline [8] and distance-based representative skyline [14], which will be surveyed in Section 6. As illustrated in [14], the defect of the domination-based representative skyline is that it may allocate many

R. Cheng et al. (Eds.): APWeb 2015, LNCS 9313, pp. 116–128, 2015.
DOI: 10.1007/978-3-319-25255-1_10

representatives to a dense region of the skyline points, while the distance-based representative skyline (DBRS) can overcome this drawback. Therefore, in this paper, we focus on the problem of computing the DBRS efficiently. Specifically, given an integer k, a DBRS consists of k representative skyline points that aims to minimize the maximal distance between a non-representative skyline point and its nearest representative. In the 2D space, there is a dynamic programming (DP) algorithm to compute the DBRS [14], which takes $O(km^2)$ time complexity, where m is the number of skyline points. Obviously, such a DP-based solution cannot be applied to large datasets due to the quadratic time overhead.

To speedup the DP-based algorithm, in this paper, we first propose an efficient approximate algorithm, called ARS, which runs in $O(k \log^2 m \log(T/\epsilon))$ time, where T is no more than the maximal distance between any two skyline points. We show that the ARS algorithm can guarantee a solution that is at most ϵ larger than the optimal solution. Second, we propose an exact algorithm, called PSRS, which runs in $O(k^2 \log^3 m)$ time. Unlike the DP-based algorithm, both the ARS and PSRS algorithms are based on a carefully-designed parametric search technique. Specifically, to compute the DBRS, we first solve a related decision problem described as follows. Given a distance τ, can we find k representatives such that the distance between a non-representative point and its nearest representative is no larger than τ? If that is the case, we refer to τ as a feasible solution, and otherwise it is non-feasible. Clearly, the minimal feasible solution of the decision problem is the optimal solution of the DBRS problem. To solve the decision problem, we propose a binary-search based greedy algorithm. The idea is that the algorithm greedily covers the skyline points using a circle with radius τ, and then checks whether all the skyline points can be covered by k circles. We prove that if the optimal solution of the DBRS problem is no smaller than τ, then such a greedy algorithm returns *true*, otherwise returns *false*. Based on this, the ARS algorithm makes use of a binary search technique to search the minimal feasible τ, which can obtain a near-optimal solution. However, for the PSRS algorithm, we first observe that there are k feasible solutions that can be the candidates for the optimal solution. Based on this key observation, the PSRS algorithm utilizes a binary search procedure to find all these k candidates, and then reports the minimal one. We conduct extensive experiments over both synthetic and real-world datsets, the results show that the ARS and PSRS algorithms are at least four orders of magnitude faster than the DP-based algorithm in medium-sized datasets. The main contributions of this paper are summarized as follows.

- We propose a near-optimal approximate algorithm ARS that runs in $O(k \log^2 m \log(T/\epsilon))$ time to solve the DBRS problem in 2D space.
- We also propose an efficient exact algorithm PSRS which takes $O(k^2 \log^3 m)$ time complexity to compute the DBRS in 2D space. Unlike the DP-based algorithm, both the ARS and PSRS algorithms are based on a new parametric search technique.
- We conduct comprehensive experimental studies to evaluate the proposed algorithms, and the results confirm our theoretical findings.

The rest of the paper is organized as follows. Section 2 describes the problem of distance-based representative skyline computation. In Sections 3 and 4, we propose the ARS and PSRS algorithms respectively. We report the experimental results in Section 5, and survey the related work in Section 6. Finally, we conclude this work and point out several future directions in Section 7.

2 Preliminaries

Let D be a set of 2D points, and S be a set of skyline points of D. Denote by n and m the cardinality of D and S respectively (i.e., $n = |D|$ and $m = |S|$). Further, we let K be the representative skyline that contains k points from S. Following [14], the definition of representative error of K, denoted by $E(K, S)$, is given by

$$E(K, S) = \max_{p \in S \backslash K} \{\min_{q \in K} ||p, q||\}, \tag{1}$$

where $||p, q||$ denotes the distance between points p and q. The representative error denotes the maximal distance between the non-representative point from $S \backslash K$ and its nearest representatives, which quantifies the representation quality of K. A good representative skyline should be with small $E(K, S)$. The goal of the distance-based representative skyline (DBRS) is to find a representative skyline K with minimal representative error $E(K, S)$. More formally, the DBRS problem is formulated as follows.

$$\min_{K} E(K, S)$$
$$s.t. \ |K| \leqslant k. \tag{2}$$

In [14], the authors proposed a dynamic programming algorithm to solve the above problem in 2D space. In the rest of this paper, we assume that the skyline R has already been computed from the dataset D and the skyline points are sorted in ascending order of their x-coordinate, which is a very common assumption in representative skyline literature [14,8]. Let $S = \{p_1, \cdots, p_m\}$ be the skyline points that is sorted in non-decreasing order of their x-coordinate. Denote by $S_j = \{p_1, \cdots, p_j\}$ a subset of S that includes the first j elements in S. Let $optE(j, t)$, for $0 \leq j \leq m, 0 \leq t \leq k$, be the representative error of the optimal size-t representative skyline of R_j. Let $r(i, j)$, for $1 \leq i \leq j \leq m$, be the radius of the smallest circle that covers all points in $\{p_i \cdots, p_j\}$ and centers at one of these points. Then, we have the following recursive formula to compute $optE(m, k)$ [14].

$$optE(j, t) = \min_{i=t}^{j-1} \{\max\{optE(i-1, t-1), r(i, j)\}\}. \tag{3}$$
$$optE(j, j) = 0$$

Armed with the above formula, we can easily devise a dynamic programming algorithm to compute $optE(m, k)$. By a careful implementation as shown in [14], the DP-based algorithm runs in $O(km^2)$ time. Clearly, such a DP-based solution cannot be used for large datasets due to the quadratic time cost. To overcome this issue, in the following sections, we shall present two efficient algorithms to compute the optimal representative skyline in 2D space. In the rest of this paper, we will focus on computing the optimal representative error, and it is straightforward to extend the proposed techniques to find the optimal representative points.

Algorithm 1. Compute the radius of the smallest circle $r(i, j)$

1: $i_0 \leftarrow i, j_0 \leftarrow j$;
2: **while** $i < j$ **do**
3:     $mid \leftarrow \lfloor (i + j)/2 \rfloor$, $id \leftarrow \|p_{i_0}, p_{mid}\|$, $jd \leftarrow \|p_{j_0}, p_{mid}\|$;
4:     **if** $id < jd$ **then**
5:         $i \leftarrow mid + 1$;
6:     **else if** $id > jd$ **then**
7:         $j \leftarrow mid$;
8:     **else**
9:         **return** id;
10: **return** $\max\{id, jd\}$;

---

3   The Approximate Algorithm

In this section, we propose a near-optimal approximate algorithm, called ARS, to compute the representative skyline in 2D space. Unlike the DP algorithm, the ARS algorithm is based on a dramatically different technique. The general idea of our technique is that to solve the minimization problem, we first solve a related decision problem which is described as follows. Given a distance τ, can we find a representative skyline with k points such that the distance between a non-representative point and its nearest representative is no larger than τ? If that is the case, we refer to τ as a feasible solution for the decision problem. It is easy to see that the representative error of the optimal representative skyline should be the minimal feasible solution of the decision problem. Once the decision problem can be efficiently solved, then we can use a binary search procedure (over all possible τ) to find the minimal feasible solution. Below, we first present an algorithm to solve the decision problem.

3.1   Solve the Decision Problem

Here we devise a greedy algorithm to solve the decision problem. The idea is that the algorithm greedily covers the skyline points from p_1 to p_m using a circle with radius τ, and then verifies whether all points can be covered by k circles. If so, then the distance τ is a feasible solution. Otherwise, τ is non-feasible which will be proved in Lemma 1. Before we proceed further, we first present an algorithm to compute the radius of the smallest circle that covers all points in $\{p_i, \cdots, p_j\}$ for $i \leq j$ and centers at one of these points, which will be frequently invoked in the ARS algorithm.

The general idea of our algorithm to compute the radius of the smallest circle is as follows. Let $d(i, t)$ (for $t \in \{i + 1, \cdots, j\}$) be the distance between the point i and point t. Then, one can easily derive that $d(i, t)$ is an increasing function with increasing t. Similarly, we can find that $d(j, t)$ is an decreasing function with increasing t. Since the center of the circle must be a point from $\{p_i, \cdots p_j\}$, the radius $r(i, j)$ of the smallest circle must be the distance $d(i, t)$ or $d(j, t)$ for a certain t (the one with minimal $|d(i, t) - d(j, t)|$). Based on the monotonic properties of the distance functions $d(i, t)$ and $d(j, t)$, we are able to devise a binary search algorithm to search the smallest radius. The detailed description of our algorithm is shown in Algorithm 1.

It is worth emphasizing that our binary-search based technique (Algorithm 1) is completely different from the technique developed in [14] to compute the smallest circle, in

---

Algorithm 2. TestDistance(τ, k)

1: $i \leftarrow 1$, $i_1 \leftarrow 1$;
2: **for** $l = 1$ to k **do**
3:     $j \leftarrow m$,$j_1 \leftarrow m$;
4:     **while** $i_1 < j_1$ **do**
5:         $mid \leftarrow \lfloor (i_1 + j_1)/2 \rfloor$;
6:         **if** $r(i, mid) \leq \tau$ **then**
7:             $i_1 \leftarrow mid + 1$;
8:         **else**
9:             $j_1 \leftarrow mid$;
10:    **if** $i_1 == m$ **then**
11:       **return** true;
12:    **if** $r(i, mid) \leq \tau$ **then**
13:       $i \leftarrow i_1 + 1, i_1 \leftarrow i$;
14: **return** false;

---

which the authors present a *collative pass* technique to compute the optimal radiuses for all pair (i, j) in $O(m^2)$ time. Instead of computing all radiuses, we focus on computing $r(i, j)$ on demand (i.e., we invoke Algorithm 1 only when $r(i, j)$ is requested.). It is easy to derive that the worst-case time complexity of Algorithm 1 is $O(\log m)$.

Equipped with Algorithm 1, we are ready to present the greedy algorithm to solve the decision problem. Specifically, we detail the greedy algorithm in Algorithm 2. In Algorithm 2, the algorithm sequentially find k circles with radius τ to cover the points from p_1 to p_m (lines 2-13). For the l-th circle, denoted by circle-l, the algorithm determines the rightmost skyline point covered by the circle by using a binary search procedure (lines 4-9). If the rightmost skyline point covered by circle-l is p_m and $l \leq k$, then we know that we can use at most k circles with radius τ to cover all the skyline points, and thus τ is a feasible solution (lines 10-11). Otherwise, τ is non-feasible, and thereby the algorithm returns false (line 14). In addition, it is worth mentioning that in line 6, the algorithm invokes Algorithm 1 to compute the radius $r(i, mid)$, which takes $O(\log m)$ time complexity. Putting it all together, we can easily derive that the worst-case time complexity of Algorithm 1 is $O(k \log^2 m)$, and the space complexity of Algorithm 2 is $O(m)$. The following lemma shows the correctness of the algorithm.

Lemma 1. *Let* $optE(m, k)$ *be the optimal representative error of the DBRS problem. If* $optE(m, k) > \tau$, *then TestDistance(τ, k) returns false, otherwise returns true.*

Proof. Since $optE(m, k)$ is the optimal solution, it is impossible to use k circles with radius smaller than $optE(m, k)$ to cover all the skyline points. Therefore, if $optE(m, k) > \tau$, TestDistance(τ, k) must return false. It remains to show when $optE(m, k) \leq \tau$, TestDistance(τ, k) returns true. For the algorithm TestDistance(τ, k), we assume that p_{s_i} be the rightmost point covered by the i-th circle with radius τ, where s_i is the index of this rightmost point. Since $optE(m, k) \leq \tau$, there is a way of covering all skyline points by using k circles with radius no more than τ. If the first circle of this covers includes the points $\{p_1, \cdots, p_t\}$, then $t \leq s_1$ (by the fact of the greedy procedure). As a result, there exists a way of covering points $\{p_{s_1+1}, \cdots, p_m\}$ by using $k - 1$ circles with radius no more than τ. This is because we can remove

Algorithm 3. ARS(S, m, ϵ, k)

1: $ub \leftarrow r(1, m)$, $lb \leftarrow 0$;
2: **while** $lb + \epsilon \geq ub$ **do**
3:     $\tau \leftarrow \lfloor (ub + lb)/2 \rfloor$;
4:     **if** TestDistance(τ, k) **then**
5:         $ub \leftarrow \tau$;
6:     **else**
7:         $lb \leftarrow \tau$;
8: **return** ub;

all the points in $\{p_{t+1}, \cdots, p_{s_1}\}$, and then use the optimal algorithm to covers all the points $\{p_{s_1+1}, \cdots, p_m\}$ by using $k - 1$ circles with radius no larger than τ. The same argument can be applied to the second circle, the third circle, and so forth. Hence, the lemma is established.

3.2 The ARS Algorithm

Recall that to solve the DBRS problem, we have to find the minimal feasible solution of the decision problem. To that end, we first find that the decision problem satisfies the monotonic property. That is to say, for any $\tau_1 > \tau_2$, if τ_2 is a feasible solution, then τ_1 is also a feasible solution. Then, based on the monotonic property of the decision problem, we can apply a binary search procedure to find the minimal feasible solution.

The ARS algorithm is detailed in Algorithm 3. Since τ is a real value, we may not obtain the minimal feasible solution exactly by using binary search. Therefore, in Algorithm 3, we use a parameter ϵ to balance the tradeoff between the accuracy and running time of the algorithm. In the experiments, we will show how ϵ affects the accuracy and the running time of the algorithm. Note that in Algorithm 3, one can easily derive that $ub = r(1, m)$ and $lb = 0$ are the upper and lower bounds of τ, thus we can use these bounds as two starting points for the binary search procedure. In addition, it is straightforward to show that the ARS algorithm can guarantee a solution that is at most ϵ larger than the optimal solution. The worst-case time complexity of the ARS algorithm is $O(k \log^2 m \log(T/\epsilon))$. Here $T = ub - lb$, which is no more than the maximal distance between any two skyline points. Additionally, it is easy to verify that the space complexity of ARS is $O(m)$.

4 The New Exact Algorithm

Here we present an exact algorithm to solve the DBRS problem. For a pair (i, j) with $1 \leq i \leq j \leq m$, let $c(i, j)$ be the circle that covers points $\{p_i, \cdots, p_j\}$ and centers at one of these points, and $r(i, j)$ be the radius of the circle $c(i, j)$. Then, we can easily verify that the optimal representative error for the DBRS problem must be equal to $r(i, j)$ for a certain pair (i, j). Therefore, to find the optimal solution, a naive algorithm is to enumerate all such pairs. However, the naive algorithm is not efficient, because there are $O(m^2)$ pairs needed to be enumerated in the worst case. The challenge is how to search the optimal solution over all $O(m^2)$ pairs efficiently.

Algorithm 4. PSRS (S, m, k)

```
1:  s_0 ← 1, s_k ← m;
2:  for i = 1 to k − 1 do
3:      left ← s_{i−1}, right ← m;
4:      while left < right do
5:          mid ← ⌊(left + right)/2⌋;
6:          τ ← r(s_{i−1}, mid);
7:          if TestDistance(τ, k) then
8:              right ← mid;
9:          else
10:             left ← mid + 1;
11:     s_i ← right;
12: return min_{i=1}^{k}{r(s_{i−1}, s_i)};
```

To tackle this challenge, we shall propose an efficient algorithm based on a deep analysis of the problem. Specifically, we find that to compute the optimal solution of the DBRS problem, we only need to check k candidates, which is significantly smaller than $O(m^2)$. Below, we first introduce some useful notations, and then detail our techniques to achieve this goal.

For $i = 0, \cdots, k$, we let $s_i \in \{1, \cdots, m\}$ be the indices of k skyline points with $s_0 = 1$, $s_k = m$, and $s_i < s_{i+1}$ for $i = 1, \cdots, k − 1$. We refer to p_{s_i} as the i-th breakpoint, because all the s_i for $i = 0, \cdots, k$ partition the entire skyline into k parts. Let $r(s_{i−1}, s_i)$ be the radius of the smallest circle denoted by $c(s_{i−1}, s_i)$ that covers all the points in $\{p_{s_{i−1}}, \cdots p_{s_i}\}$ and centers at one of these points. For completeness, we define $r(i, j) = −1$ whenever $j < i$. Then, we can compute the optimal representative error of the DBRS problem (i.e., $optE(m, k)$) based on the following lemma.

Lemma 2. *If the inequalities* $r(s_{i−1}, s_i − 1) < optE(m, k) \leq r(s_{i−1}, s_i)$ *for all* $i = 1, \cdots, k − 1$ *hold, then* $optE(m, k) = \min_{i=1}^{k}\{r(s_{i−1}, s_i)\}$.

Proof. First, we show that $r(s_{k−1}, s_k)$ is a feasible solution for the decision problem. To see this, we partition of the skyline into k parts which are $\{p_{s_0}, \cdots, p_{s_1} − 1\}, \cdots,$ $\{p_{s_{k−2}}, \cdots, p_{s_{k−1}} − 1\}$, and $\{p_{s_{k−1}}, \cdots, p_{s_k}\}$. Clearly, $\tau = \max\{\max_{i=1}^{k−1}\{r(s_{i−1}, s_i − 1), r(s_{k−1}, s_k)\}$ is a feasible solution for the decision problem, i.e., $\tau \geq optE(m, k)$. Since $r(s_{i−1}, s_i − 1) < optE(m, k)$ for all $i = 1, \cdots, k − 1$ hold, we have $\max_{i=1}^{k−1}\{r(s_{i−1}, s_i − 1) < r(s_{k−1}, s_k)$, and thereby $\tau = r(s_{k−1}, s_k)$ which is feasible. Second, we let $\tau_{\min} = \min_{i=1}^{k}\{r(s_{i−1}, s_i)\}$. Since $optE(m, k) \leq r(s_{i−1}, s_i)$ for all $i = 1, \cdots, k − 1$ hold, we have $optE(m, k) \leq \tau_{\min}$. To prove the lemma, it is sufficient to show that $optE(m, k) < \tau_{\min}$ is impossible. We assume to the contrary that $optE(m, k) < \tau_{\min}$. Suppose that $\tau_{\min} = r(s_{i−1}, s_i)$ for a certain i. It is easy to see that for the optimal solution, the best possible for the rightmost point covered by the first circle is $p_{s_1−1}$. Similarly, we can find that the best possible for the rightmost point covered by the i-th circle is $p_{s_i−1}$ for $i = 1, \cdots, k−1$. Following this logic, for the optimal solution, the points $\{p_{s_{k−1}}, \cdots, p_{s_k}\}$ must be covered by a circle, which contradicts to $optE(m, k) < \tau_{\min}$. This completes the proof.

Table 1. Datasets (m is the number of skyline points)

Dataset	m	Property
ID 1	14,735	Independent dataset
ID 2	1,087,893	Independent dataset
ACD 1	18,576	Anti-Correlated dataset
ACD 2	1,142,369	Anti-Correlated dataset
Urban	1,662	Real-world dataset

Based on Lemma 2, we only need to find all the s_i for $i = 1, \cdots, k-1$ that meet the condition $r(s_{i-1}, s_i - 1) < optE(m, k) \leq r(s_{i-1}, s_i)$. Once we determine all such s_i, then $r(s_{i-1}, s_i)$ for each $i = 1, \cdots, k-1$ is a candidate for the optimal solution, and we only need to take the minimal over all $r(s_{i-1}, s_i)$. We can determine all the s_i by sequentially invoking a binary search procedure based on the monotonic property of the decision problem. Specifically, if $r(s_{i-1}, s_i)$ is a feasible solution for the decision problem, then $r(s_{i-1}, s_i) \geq optE(m, k)$ and for all $s > s_i$ we have $r(s_{i-1}, s) \geq r(s_{i-1}, s_i)$. As a result, we are able to devise a binary search algorithm to determine the index of the first breakpoint s_1 such that $r(s_0, s_1 - 1) < optE(m, k) \leq r(s_0, s_1)$, and then given s_1, we can use the same binary search procedure to find s_2, and so on. The detailed description of our algorithm is outlined in Algorithm 4. We can easily derive that the time and space complexity of Algorithm 4 are $O(k^2 \log^3 m)$ and $O(m)$ respectively. The correctness of Algorithm 4 can be guaranteed by Lemma 2.

5   Experiments

In this section, we conduct comprehensive experiments to evaluate the proposed algorithms. To this end, we use the state-of-the-art DP based algorithm [14], called DPRS, as the baseline. We use the running time and the representative error as the metrics, and then we compare our algorithms ARS and PSRS with the baseline algorithm DPRS using these metrics. For the ARS algorithm, we set the parameter $\epsilon = 0.01$, unless otherwise specified. All algorithms are implemented in C++. All the experiments are conducted on a computer with 3.20GHz Intel(R) Core(TM) i5-3470 CPU and 6GB memory running Windows 8.1 operation system.

Datasets. In the experiments, we use five datasets. The first four datasets, named ID 1, ID 2, ACD 1 and ACD 2 respectively, are the synthetic datasets which are generated based on the method proposed in [2]. Specifically, for the ID 1 and ID 2 datasets, the (x, y)-coordinates of the skyline points are generated based on a uniform distribution. We generate 14,735 and 1,087,893 skyline points for the ID 1 and ID 2 datasets respectively. The ACD 1 and ACD 2 datasets are the anti-correlated dataset where the points are good in one dimension but they are bad in the other dimension. We use the method proposed in [2] to generate the ACD 1 and ACD 2 datasets which contain 18,576 and 1,142,369 skyline points respectively. The last dataset, called Urban, is a real-world dataset including 1,662 skyline points, which is previously used in urban computing community [20]. We download this dataset from http://research.microsoft.com/en-us/projects/urbancomputing/. In our experiments, we extract two dimensions of

Table 2. Running time of different algorithms in ID 1 dataset (in second)

k	5	10	15	20	25	30
DPRS	327.314	381.451	436.307	490.713	545.053	580.225
ARS	0.024	0.025	0.029	0.037	0.041	0.047
PSRS	0.023	0.024	0.024	0.027	0.028	0.030

Table 3. Running time of different algorithms in ID 2 dataset (in second)

k	5	10	15	20	25	30
DPRS	-	-	-	-	-	-
ARS	2.482	2.485	2.50	2.501	2.502	2.518
PSRS	2.50	2.528	2.531	2.602	2.634	2.66

each data point in the original dataset, where the first dimension represents the time, and the second dimension denotes the traffic flow, and then we compute the skyline. For a fair comparison, we normalize the (x, y)-coordinates of the points in all the datasets into the range $[0, 100]$. The detailed information of our datasets are shown in Table 1.

Efficiency. Here we report the running time of the DPRS, ARS, and PSRS algorithms. The results for varying k in all datasets are shown in Tables 2, 3, 4, 5, and 6 respectively. As can be seen, both the ARS and PSRS algorithms are significantly faster than the DPRS algorithm (the baseline algorithm), and the ARS algorithm is slightly more efficient than the PSRS algorithm. For example, in the ID 1 dataset (Table 2), we can see that the ARS and PSRS algorithms are 17,548 and 13,807 times faster than the DPRS algorithm on average. That is to say, the proposed ARS and PSRS algorithms are at least four orders of magnitude faster than the state-of-the-art DPRS algorithm in medium-sized datasets. Even in the smallest dataset Urban (Table 6), the ARS and PSRS algorithms are at least two orders of magnitude faster than the DPRS algorithm. In addition, in our two large datasets ID 2 and ACD 2 (Tables 3 and 5), the DPRS cannot get the solution in one day, while our ARS and PSRS algorithms can compute the solution in around 2.5 second. As desired, over all datasets, the running time of all the algorithms increase with increasing k. These results are consistent with the theoretical analysis presented in Sections 3 and 4.

Scalability. Here we test the scalability of the ARS and PSRS algorithms. To this end, we vary the number of skyline points from 10,000 to 1,000,000, and test the running time of our algorithms in ID 2 and ACD 2 datasets. We set $k = 10$, and similar results can also be observed for other k values. The results are depicted in Fig. 1. From Fig. 1, we can see that both the ARS and PSRS algorithms scale very well. When the datasets is large than 100,000, the increasing speed of the running time of the ARS and PSRS algorithms are lower than the increasing speed of the number of skyline points. The reason is that the time complexity of our algorithms are sublinear. The results further confirm that our algorithms can be applied to handle large datasets.

Effect of the Parameter ϵ. Here we study how the parameter ϵ affects the efficiency and effectiveness of the ARS algorithm. For this purpose, we vary ϵ from 0.001 to 0.1,

Table 4. Running time of different algorithms in ACD 1 dataset (in second)

k	5	10	15	20	25	30
DPRS	366.957	414.995	481.716	510.982	568.816	609.458
ARS	0.027	0.028	0.030	0.034	0.035	0.039
PSRS	0.028	0.030	0.031	0.046	0.052	0.063

Table 5. Running time of different algorithms in ACD 2 dataset (in second)

k	5	10	15	20	25	30
DPRS	-	-	-	-	-	-
ARS	2.727	2.750	2.751	2.753	2.768	2.794
PSRS	2.735	2.766	2.788	2.816	2.854	2.930

Table 6. Running time of different algorithms in Urban dataset (in second)

k	5	10	15	20	25	30
DPRS	2.371	3.546	4.672	7.768	8.881	10.053
ARS	0.015	0.015	0.015	0.015	0.016	0.016
PSRS	0.015	0.016	0.016	0.016	0.017	0.019

and use the running time and the representative error as two metrics to evaluate the efficiency and effectiveness of the ARS algorithm, respectively. Also, we set $k = 10$, and similar results can be observed for other k values. We test the ARS algorithm in ID 2 and ACD 2 datasets. The results of running time with varying ϵ are shown in Fig. 2. As can be seen, the running time decreases with increasing ϵ. When $\epsilon \leq 0.01$, the running time of the ARS algorithm dramatically decreasing with increasing ϵ. However, when $\epsilon > 0.01$, the curves are relatively smooth. The reason is that if ϵ is very small, then the binary search procedure in the ARS algorithm will take a relatively long time to converge to the near-optimal approximate solution. However, when ϵ is large, the binary search procedure converges quickly.

We report the results of the representative error of the ARS algorithm with different ϵ in Fig. 3. From Fig. 3, we can find that the representative error increases with increasing ϵ as desired. This is because ϵ controls the accuracy of the ARS algorithm. A large ϵ results in a low accuracy. Also, we can see that when $\epsilon = 0.01$, the representative error of the ARS algorithm is close to the exact representative error.

6 Related Work

Skyline Computation. The skyline operator was first introduced into database community by Borzsonyi et al. [2] in 2001, although the skyline computation problem can date back to 1975 [7]. Since then, a large number of skyline computation algorithms have been proposed. Previous results mainly cover three aspects: 1) centralized skyline query processing, including the BNL algorithm (block nested loop) [2], the SFS algorithm (sort filter skyline) [3], the divide and conquer algorithm [2], the bitmap algorithm [13],

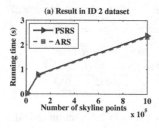
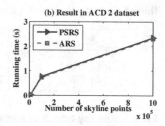

Fig. 1. Scalability of the PSRS and ARS algorithm

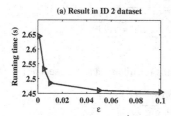
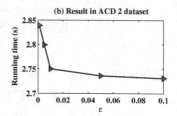

Fig. 2. Running time vs. ϵ (for the ARS algorithm)

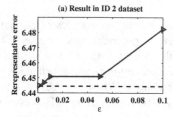
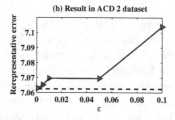

Fig. 3. Representative error vs. ϵ (dashed line denotes the exact representative error)

the index-based algorithm [13], the NN algorithms (Nearest Neighbor) [6], as well as the BBS (branch and bond skyline) algorithm [11]; 2) distributed skyline query processing, including skyline computing in the traditional distributed environment [1], skyline computing in the mobile distributed environment [5], and skyline computing in peer-to-peer networks [17,18], and so on; 3) Other skyline computing problems, including size estimation of the skyline [4], skyline query processing in any subspace [16], skyline query processing in all subspace [19,11], and skyline query processing in data flow [15], and so on.

Representative Skyline Computation. Since the traditional skylines has too many skyline points, in decision-making related applications, a better way is to find a small number of points that can best represent the entire skyline. Motivated by this, many new definitions of representative skyline have been proposed recently. Notable definitions include the domination-based representative skyline which is proposed by Lin et al. [8]. The goal of the domination-based skyline is to find k skyline points to maximize the number of points that can be dominated by any of the selected k skyline points.

Subsequently, Tao et al. [14] introduced a distance-based representative skyline definition, in which the goal is to minimize the distance between a non-representative skyline point and its closest neighbor. Sarma et al. [12] presented a representative skyline model based on the users' preference distributions, aiming at finding k representative skyline points such that the probability that a random user would click on one of them is maximized. More recently, Magnani et al. [9] proposed a new representative skyline definition which takes both significance and diversity of the data points into account. In this paper, we focus on developing efficient algorithms for computing the distance-based representative skyline in 2D space.

7 Conclusion

Representative skyline has a number of application in database community, especially in multi-objective decision making and real-time online service. A notable definition of representative skyline is the distance-based representative skyline (DBRS). However, in 2D space, the state-of-the-art algorithm for DBRS computation is based on dynamic programming (DP) which is costly for large scale datasets due to the quadratic time overhead. In this paper, we propose two much more efficient algorithms, called PSRS and ARS respectively, based on a carefully-designed parametric search technique. We show, in theory and experiments, that both PSRS and ARS algorithms are very efficient which are at least four orders of magnitude faster than the state-of-the-art algorithm in medium-sized datasets. There are several future directions needed to further investigate. First, it would be interesting to devise parallel and distributed counterparts for our algorithms. Second, the proposed techniques are tailored to two-dimensional data. An immediate question is how to extend our techniques to compute the DBRS in high dimensional data. Finally, it would also be interesting to further optimize our algorithms by developing non-trivial pruning techniques in the parametric search procedure.

Acknowledgements. The work was supported in part by (i) NSFC Grants (61402292, 61170076, U1301252, 61033009) and Natural Science Foundation of SZU (grant no. 201438); (ii) Research Grants Council of the Hong Kong SAR, China, 14209314 and 418512; (iii) China 863 (no. 2012AA010239) and Guangdong Key Laboratory Project (2012A061400024); (iv) National Key Technology Research and Development Program of the Ministry of Science and Technology of China2014BAH28F05. Dr. Rong-Hua Li is the corresponding author of this paper.

References

1. Balke, W.-T., Güntzer, U., Zheng, J.X.: Efficient distributed skylining for web information systems. In: Bertino, E., Christodoulakis, S., Plexousakis, D., Christophides, V., Koubarakis, M., Böhm, K. (eds.) EDBT 2004. LNCS, vol. 2992, pp. 256–273. Springer, Heidelberg (2004)
2. Börzsönyi, S., Kossmann, D., Stocker, K.: The skyline operator. In: ICDE, pp. 421–430 (2001)
3. Chomicki, J., Godfrey, P., Gryz, J., Liang, D.: Skyline with presorting. In: ICDE, pp. 717–719 (2003)

4. Godfrey, P.: Skyline cardinality for relational processing. In: Seipel, D., Turull-Torres, J.M.a. (eds.) FoIKS 2004. LNCS, vol. 2942, pp. 78–97. Springer, Heidelberg (2004)
5. Huang, Z., Jensen, C.S., Lu, H., Ooi, B.C.: Skyline queries against mobile lightweight devices in manets. In: ICDE, p. 66 (2006)
6. Kossmann, D., Ramsak, F., Rost, S.: Shooting stars in the sky: An online algorithm for skyline queries. In: VLDB, pp. 275–286 (2002)
7. Kung, H.T., Luccio, F., Preparata, F.P.: On finding the maxima of a set of vectors. J. ACM 22(4), 469–476 (1975)
8. Lin, X., Yuan, Y., Zhang, Q., Zhang, Y.: Selecting stars: The k most representative skyline operator. In: ICDE, pp. 86–95 (2007)
9. Magnani, M., Assent, I., Mortensen, M.L.: Taking the big picture: representative skylines based on significance and diversity. VLDB J. 23(5), 795–815 (2014)
10. Papadias, D., Tao, Y., Fu, G., Seeger, B.: An optimal and progressive algorithm for skyline queries. In: SIGMOD, pp. 467–478 (2003)
11. Pei, J., Jin, W., Ester, M., Tao, Y.: Catching the best views of skyline: A semantic approach based on decisive subspaces. In: VLDB, pp. 253–264 (2005)
12. Sarma, A.D., Lall, A., Nanongkai, D., Lipton, R.J., Xu, J.J.: Representative skylines using threshold-based preference distributions. In: ICDE, pp. 387–398 (2011)
13. Tan, K.-L., Eng, P.-K., Ooi, B.C.: Efficient progressive skyline computation. In: VLDB, pp. 301–310 (2001)
14. Tao, Y., Ding, L., Lin, X., Pei, J.: Distance-based representative skyline. In: ICDE, pp. 892–903 (2009)
15. Tao, Y., Papadias, D.: Maintaining sliding window skylines on data streams. IEEE Trans. Knowl. Data Eng. 18(2), 377–391 (2006)
16. Tao, Y., Xiao, X., Pei, J.: SUBSKY: efficient computation of skylines in subspaces. In: ICDE, p. 65 (2006)
17. Wang, S., Ooi, B.C., Tung, A.K.H., Xu, L.: Efficient skyline query processing on peer-to-peer networks. In: ICDE, pp. 1126–1135 (2007)
18. Wu, P., Zhang, C., Feng, Y., Zhao, B.Y., Agrawal, D.P., El Abbadi, A.: Parallelizing skyline queries for scalable distribution. In: Ioannidis, Y., et al. (eds.) EDBT 2006. LNCS, vol. 3896, pp. 112–130. Springer, Heidelberg (2006)
19. Yuan, Y., Lin, X., Liu, Q., Wang, W., Yu, J.X., Zhang, Q.: Efficient computation of the skyline cube. In: VLDB, pp. 241–252 (2005)
20. Zhang, F., Wilkie, D., Zheng, Y., Xie, X.: Sensing the pulse of urban refueling behavior. In: UbiComp, pp. 13–22 (2013)

Trustworthy Collaborative Filtering through Downweighting Noise and Redundancy

Qiuxiang Dong, Zhi Guan, and Zhong Chen

Institute of Software, School of EECS, Peking University, China
MoE Key Lab of High Confidence Software Technologies (PKU)
MoE Key Lab of Network and Software Security Assurance (PKU)
{dongqx,guanzhi,chen}@infosec.pku.edu.cn

Abstract. Proliferation of Electronic Commerce (EC) has revolutionized the way people purchase online. Web-based technologies enable people to more actively interact with merchants and service providers. Such purchasing logs and comments further lead to proliferation of recommender systems. Existing recommendation algorithms exploit either prior transactions or customer reviews to predict user interests towards certain items. Vast noise may be introduced into such information by fake raters, and information redundancy also makes recommender system entangled. In this work, we first examine user reviews and prior transactions to estimate user credibility and item importance to reduce effect from content polluters. Then we propose to alleviate the redundant information from homogeneous users based on link analysis. A unified framework is finally proposed to incorporate them in a mathematical formulation, which can be efficiently optimized. Experimental results on real world data reveal that our model can significantly outperform other baselines.

Keywords: Recommender System, Collaborative Filtering.

1 Introduction

The recent growth of online shopping has enabled an active interaction between customers and merchants. This increase is fueled not only by more products and service providers but also by the byproduct of online transactions: the huge amount of purchase history and customer reviews, which promote a better understanding of customer behaviors and preferences and lead to an effective recommender system.

Existing approaches try to exploit the user feedback from two perspectives. Collaborative Filtering (CF)-based methods recommend items through discovering purchasing patterns of users who purchase similar products and items which were purchased by similar group of users. Various techniques have been proposed to solve the problem. The underlying intuition is to reveal the "people who bought this item also bought" patterns hidden in prior transactions. Another stream of recommendation algorithms is to exploit the similarity between

© Springer International Publishing Switzerland 2015
R. Cheng et al. (Eds.): APWeb 2015, LNCS 9313, pp. 129–140, 2015.
DOI: 10.1007/978-3-319-25255-1_11

users and products. Their aim is to measure the distance in terms of content features, such as the profile of users and specifications of products. Due to the explosive increase of user reviews, there has been an increasing interest in leveraging the opinions of customer comments.

As the online market extends greater and greater, many problems of the traditional recommender systems have been exposed. A fundamental advantage of EC is its openness. It motivates users to contribute more contents and opinions to the platform. However, the openness makes such websites a main channel for fake reviews and spams, and it also makes the ratings highly biased.

As reported in [29], nearly one third of online reviews of some products are found to be fake and from spammers. Such reviews are for different malicious purposes and are definitely not helpful for recommendation. Various work has been done to investigate how to identify such spam in review. Thus, it is favorable if a recommender system can automatically reduce information brought by spams. Without considering the importance of ratings and reviews from different users, recommender systems will be entangled with such noise.

Since products in different departments are of different popularity, traditional recommendations suffer from a biased rating data. Ratings on products from popular departments are over emphasized, as many users' ratings are overlapped and thus are not that informative, while the useful ratings from less popular categories are ignored. It will be favorable if such information redundancy from the same group can be reduced.

In this paper, we aim to investigate how to solve the problem and propose a more trustworthy recommendation framework. Inspired by recent progress of review quality analysis, we first examine the reviews of customers to reveal their credibility. Through analyzing the links between customers and products, we further induce the importance of different products. The user credibility and product importance is combined into a unified framework. In order to incorporate both factors, several research challenges need to be addressed: 1) User credibility and product utility is an implicit index and is not available for most EC platforms; 2) Information redundancy widely exists between homogeneous customers.

To cope with the challenges mentioned above, we put forward a trustworthy framework which actively selects and upweights customers and items of higher utility simultaneously. Contributions of this work can be summarized as follows: 1)By studying the quality of user review as well as links between users and items, we calculate customer credibility and product utility. 2)A unified framework which actively considers the importance of customers and products is proposed. It is formulated in a decent mathematical form and can be optimized efficiently. 3)Various experiments are employed to test the effectiveness of our model and several competitive baselines. The data is obtained from real world websites and is publicly available.

The remaining sections of the paper are organized as follows: In Section 2, some related research results are discussed. The heuristics we employed to estimate user and product weight are discussed in Section 3. The proposed model is presented in Section 4. In Section 5 we introduce the experimental settings and results and the paper is finally concluded in Section 6.

2   Related Work

In this section, we will describe some related work on recommender systems [4]. There are various kinds of recommendation algorithms, here we briefly review two main genres.

Collaborative Filtering: Collaborative filtering (CF) may be the most successful tool in leveraging the user prior transactions to predict future interests. There are two types of collaborative filtering techniques, i.e., user-based and item-based CF. The user-based approach recommends an item to a user based on the opinions of other similar users who purchased it; Item-based CF approach recommends items to users based on the other items with high correlations. For both approaches, similarity measurement between users and items is of great significance. The widely used distance metrics include Pearson correlation coefficient, cosine-based similarity, vector space similarity and so on [1]. There are also several hybrid approaches [21]. In addition, associative retrieval techniques are applied in [12]. Authors of [11] try to propose algorithms for processing implicit feedbacks. An approach for incorporating externally specified aggregate ratings information into CF methods are introduced in [28].

Collaborative filtering has been frequently reduced into a latent factor problem in recent years. Models based on matrix factorization [17] have have proven to be useful and dominated leader board of recommendation challenges. Variants of matrix factorization using flexible regression priors [30] were also proven to be useful. However, it is difficult to explain predictions in latent factor models such as SVD. Supervised random walks for predicting and recommending links in social networks is adopted by Backstrom et al. [3]. A method called functional matrix factorization is proposed by Zhou et al. [31] to handle the cold-start problem. A collective probabilistic factor model is utilized [22] for Web site recommendation. On the basis of co-clustering, Leung et al. propose a collaborative location recommendation framework [19].

Our work differs from the existing CF methods from incorporating the weights of items and users. Most CF algorithms treat all instances as equal weight. However, the assumption does not hold in real applications as some experienced users' ratings are clearly more useful than others.

Content-based Recommender System: Another main category of recommendation algorithms are Content-based (CB) recommender systems. They try to recommend items similar to those a given user liked in the past. Specifically, the process of CB recommender system is performed in three steps [20]. In the content analyzer phase, the content of items is represented in a form suitable for the next processing steps, i.e., the items profiles. There are mainly three kinds of retrieval models for extracting items profiles from Web pages, news articles or product descriptions: keyword-based model (e.g., [2]), semantic analysis by using ontologies [5], semantic analysis by using encyclopedic knowledge sources (e.g., [18]). The profile learner module collects data representative of the user preferences and generalizes this data, so that the user profile can be constructed. The most used learning algorithms in generalization include naive Bayes classifier (e.g., [7]), relevance feedback [26] [2], and other methods, such as decision tree and nearest neighbor algorithms (e.g., [23],[25]). The final module—filtering component recommends relevant items to users based on the similarity [5] between the user profile and item profiles.

Recently, user tagging activity has been taken into account within content-based recommender systems. In [9], tag-based user profiles are exploited for producing music recommendations.

3   Customer and Product Modeling

In this section, we introduce our algorithms to predict the customer credibility and product utility.

3.1   Customer Credibility Modeling

In online customer review forum, users can often provide two kinds of information: a numerical rating, often a scalar from one to five/ten or "number of stars", together with a small paragraph describing their feedback. Our aim is to learn a credibility score from the user comments. Here, the credibility means how useful the users' ratings are for the recommendation.

Here we denote user-review matrix as $\mathbf{C} \in \mathbb{R}^{m \times n}$, user-rating matrix as $\mathbf{R} \in \mathbb{R}^{m \times p}$ and ground truth as $\mathbf{y} \in \mathbb{R}^m$, where m is the number of users and n is the number of features of user comment (normally the size of vocabulary of reviews) and p is the number of products. In our work, $\mathbf{y}$ is captured through measuring how close the user's rating is to the average rating. In order to avoid spamming, a product's rating will be used only if its rating number exceeds a predefined threshold. To better model the user reviews, we design different kinds of features based on previous work of review spam detection [14]. As related research has a long literature history and numerous research work has been proposed, different existing techniques can be used to extend our system. However, since our work focuses mainly on the learning framework, the optimal solution of evaluating utility of customer and product remains to be found in future work.

Although our solution of modeling customer credibility is not optimal, even near optimal solutions will also bring in some noises. The key issue is to reduce negative effect in the learning framework, which will be discussed later.

3.2   Product Utility Modeling

Since users will purchase products they are interested in, the transactions can build up links upon different customers and products. Such links can be regarded as an iterative reinforcement. In this work, in order to capture product utility, we adopt the HITS algorithm [15].

We regard each user as a hub node and product as an authority. The initial value of users is their credibility. By employing HITS on the user-product network, a utility score can be achieved for every product. Similarly, we refer utility of product as how useful the product's ratings are for the recommendation algorithm.

To avoid bringing noise into the model, we remove two kinds of links: If a user's rating is below average rating of the product, the links are removed; Links from low credible users are removed. The removal of edges is similar to the heuristics for detecting spammers in social media websites[10].

4   Trustworthy Collaborative Filtering

In this section, we will introduce the proposed Trustworthy Collaborative Filtering. The proposed framework enables the recommendation to be more trustworthy by downweighting nodes with lower credibility and products with less utility. Exclusive group regularizer is further introduced to reduce redundancy from homogeneous users.

4.1   Dual Weighted Dimension Reduction

Existing recommender systems often regard people equally weighted. In real applications, a user weight $\mathbf{T} \in \mathbb{R}^{1 \times m}$ and a product weight $\mathbf{Q} \in \mathbb{R}^{1 \times p}$ are often available, which are always ignored during building up recommendation models.

$$\min_{\mathbf{W}, \mathbf{H}} \sum_{i=1}^{m} \sum_{j=1}^{p} (R_{ij} - \sum_{k=1}^{K} W_{ik} H_{jk})^2. \tag{1}$$

Eq. 1 is a general matrix factorization form of collaborative filtering, which aims to build up two low rank factors through reducing the reconstruction loss. K is the number of latent dimensions, which is normally smaller than p. Clearly, each user vector $\mathbf{u}_i$ is regarded as equally weighted. In order to incorporate $\mathbf{T}$ and $\mathbf{Q}$, Eq.1 can be reformulated as follows:

$$\min_{\mathbf{W}, \mathbf{H}} \sum_{i=1}^{m} \sum_{j=1}^{p} T_i Q_j (R_{ij} - \sum_{k=1}^{K} W_{ik} H_{jk})^2 \tag{2}$$
$$s.t. \ \mathbf{W} \geq 0, \mathbf{H} \geq 0,$$

where T_i is the weight of user i and Q_j is the utility score of product j. By assigning small weight to less credible users and less useful products, their negative effect will be reduced. Since a negative value is difficult to be explained, similar to previous research, here we enforce the two factors $\mathbf{W}$ and $\mathbf{H}$ to be nonnegative.

Since Eq. 2 is not jointly convex to $\mathbf{W}$ and $\mathbf{H}$, a local minimum can be achieved through alternatively updating them.

We first fix $\mathbf{H}$ and update $\mathbf{W}$. When $\mathbf{H}$ is fixed, the problem can be reduced to

$$\min_{\mathbf{W}} \sum_{i=1}^{m} (\mathbf{R}_{i*} - \mathbf{W}_{i*}\mathbf{H}^T)\mathbf{A}_i(\mathbf{R}_{i*} - \mathbf{W}_{i*}\mathbf{H}^T)^T + tr(\boldsymbol{\Phi}\mathbf{W}) \tag{3}$$
$$s.t. \ \mathbf{W} \geq 0,$$

where $tr(\cdot)$ calculates the trace of a matrix and $\mathbf{A}_i = diag(E_{i*})$. $\mathbf{E}$ is the weight matrix which records weight of each entry $E_{ij} = T_i \times Q_j$. In order to solve the problem, we introduce a Langrange multiplier $\boldsymbol{\Phi} = [\boldsymbol{\Phi}_{ik}]$ for the nonnegative constraint. The partial derivatives with respect to W_{ik} are:

$$-2(\mathbf{R}_{i*}\mathbf{A}_i\mathbf{H})_k + 2(\mathbf{W}_{i*}\mathbf{H}^T\mathbf{A}_i\mathbf{H})_k + \boldsymbol{\Phi}_{ik}. \tag{4}$$

Then the derivatives can be reduced to the following form through using the KKT condition $\boldsymbol{\Phi}_{ik}U_{ik} = 0$:

$$[-(\mathbf{R}_{i*}\mathbf{A}_i\mathbf{H})_k + (W_{ik}\mathbf{H}^T\mathbf{A}_i\mathbf{H})_k]W_{ik} = 0. \tag{5}$$

Then we can get the update rule of W:

$$W_{ik} = W_{ik}\frac{(\mathbf{E}\otimes\mathbf{RH})_{ik}}{(\mathbf{E}\otimes(\mathbf{WH}^T)\mathbf{H})_{ik}}, \tag{6}$$

where $\otimes$ is the element wise product between matrices.

Similarly, when $\mathbf{W}$ is fixed, the update rule of $\mathbf{H}$ can be reduced to the following formulation:

$$H_{jk} = H_{jk}\frac{((\mathbf{E}\otimes\mathbf{R})^T\mathbf{W})_{jk}}{((\mathbf{E}\otimes(\mathbf{WH}^T))^T\mathbf{W})_{jk}}. \tag{7}$$

Here, the update rules are similar to the weighted NMF [8] when we denote the weight of users and products as a dual weight matrix $\mathbf{E}$.

4.2   Redundancy Reduction

A problem of directly using the weight is, members from communities with higher popularity and activeness will dominate the calculation, which is not desirable. In order to upweight the nodes from less active communities, we introduce the intra-group competition in this section.

In order to find the community structure of online users, we leverage the relationships between different users. We adopt the influence maximization approach [27] based on modularity. In order to build links between users, we calculate the number of products on which they both commented. If the number exceeds the predefined threshold, a link is built between the two users.

Modularity is introduced to measure the community structure of complex networks [6] [24].Modularity measures to what extent the links between different nodes are deviating from a random graph. Given degree of two nodes d_i and d_j, the expected number of links is $\frac{d_i d_j}{2m}$. Here 2m is the normalizing factor which represents the number of total edges in a graph.

$$\Pi = \frac{1}{2m} \sum_{i,j} \delta(s_i, s_j)(J_{ij} - \frac{d_i d_j}{2m}). \tag{8}$$

Eq. 8 depicts the formulation of modularity, where s_i is the community membership of node i and $\delta(s_i, s_j) = 1$ if $s_i = s_j$ and 0 otherwise. J is the adjacency matrix of the underlying graph. Thus, the more active the intragroup interaction is, the larger modularity will be achieved. Through maximizing the modularity of a network, optimal group memberships can be induced from the links between nodes.

For simplicity, here we introduce $\mathbf{B}$:

$$\mathbf{B} = \mathbf{A} - \frac{\mathbf{dd}^T}{2m}. \tag{9}$$

Then the modularity can be reformulated in the matrix form:

$$\Pi = \frac{1}{2m} Tr(\mathbf{S}^T \mathbf{B} \mathbf{S}), \tag{10}$$

where $S_{i,j}$ denotes the probability of assigning node i to affiliation j. Finding the optimal group memberships maximizing the modularity can be induced into finding the top-r eigenvectors of $\mathbf{B}$.

Then the group membership detection problem is reduced as a soft clustering problem. We employ standard K-means algorithm on the top-r eigenvectors of $\mathbf{B}$ to get the group memberships. For simplicity, we also use $\mathbf{S} \in \mathbb{R}^{m \times k}$ to denote such group memberships. We denote the number of total groups as k.

In order to reduce information redundancy from similar users, we introduce the exclusive regularizer here:

$$\ell_{1,2}(\mathbf{T}) = \frac{1}{2} \sum_{j=1}^{k} (\sum_{i}^{m} |T_i \times S_{ij}|)^2. \tag{11}$$

$\ell_{1,2}$ norm first imposes an ℓ_1 norm towards intra-group weight, and then calculates the sum of squares of different groups. The minimization process is similar to group LASSO, which first calculates ℓ_2 norm and then ℓ_1. The distinct feature of $\ell_{1,2}$ norm is that it achieves sparsity on the intra-group level. Concretely, it enforces more entries in a group to be zero. In the scenario of recommendation, such property reduces overlapped information from users in the same group.

By the definition of $\ell_{1,2}$ norm, Eq. (11) can be formulated as:

$$\ell_{1,2}(\mathbf{T}) = \frac{1}{2} \sum_{i=1}^{k} ||\mathbf{T}^T \mathbf{S}_{*,i}||^2 \tag{12}$$

$$= \frac{1}{2} \sum_{i=1}^{k} \mathbf{T}^T \mathbf{S}_{*,i} \mathbf{S}_{*,i}^T \mathbf{T} \tag{13}$$

$$= \frac{1}{2} \mathbf{T}^T \mathbf{S} \mathbf{S}^T \mathbf{T}. \tag{14}$$

The objective in Eq. 2 can then be reformulated as:

$$\min_{\mathbf{T},\mathbf{W},\mathbf{H}} \sum_i^m T_i(\mathbf{R}_{i*} - \mathbf{W}_{i*}\mathbf{H}^T)^T(\mathbf{R}_{i*} - \mathbf{W}_{i*}\mathbf{H}^T) + \frac{1}{2}\mathbf{T}^T\mathbf{SS}^T\mathbf{T} \qquad (15)$$
$$s.t. \quad \mathbf{W} \geq 0, \mathbf{H} \geq 0.$$

For simplicity, we replace $\mathbf{SS}^T$ by $\mathbf{D}$. Each entry $D_{i,j}$ denotes the correlation between two users. If the correlation $D_{i,j}$ is very large, it avoids user i and j's weights T_i and T_j to be large simultaneously, which finally results in the intra-group sparsity.

4.3   Optimization

The optimization objective is not jointly convex to $\mathbf{W}$, $\mathbf{H}$ and $\mathbf{T}$. Again, we update them in an alternative manner.

When $\mathbf{W}$ and $\mathbf{H}$ are fixed, the objective function can be reformulated as:

$$\min_{\mathbf{T}} \qquad \sum_{i=1}^m T_i K_i + \frac{1}{2}\mathbf{T}^T\mathbf{DT} \qquad (16)$$
$$\text{subject to} \quad lb < T_i < ub,$$

where $K_i = (\mathbf{R}_{i*} - \mathbf{Q}\otimes(\mathbf{W}_{i*}\mathbf{H}^T))^T(\mathbf{R}_{i*} - \mathbf{Q}\otimes(\mathbf{W}_{i*}\mathbf{H}^T))$. Obviously, Eq. (16) is a standard quadratic programming (QP) problem and can be solved efficiently. Here we introduce two parameters, the upper bound ub and lower bound lb, to control the intra-group sparsity. A higher ub gives more freedom to those influencers and makes more entries to zero; A higher lower bound lb alleviates the sparsity and forces the weight distribution to tend to the uniform distribution.

When $\mathbf{T}$ is fixed, the optimization has the same form with Eq.(2). Thus, $\mathbf{W}$ and $\mathbf{H}$ can be updated iteratively as denoted in Eq.(6) and Eq.(7).

The updating process keeps iterating until convergence or reaches the maximal number of iterations.

5   Experiments

5.1   Dataset

To validate our algorithm on real world applications, we selected the data set from Douban[1]. It contains various entities including books, users and user ratings and reviews towards different items. There are over 10 thousand users and 1 million ratings in our data set. We extract a dataset which contains both rating and reviews from the data set.

---

[1] http://www.douban.com/

5.2   Evaluation

In order to measure how the algorithm makes mistakes during recommendation, first, we adopt Mean Absolute Error (MAE):

$$MAE = \frac{\sum_{i=1}^{m} \sum_{j=1}^{p} |R_{ij} - (\mathbf{WH}^T)_{i,j}|}{m \times p}. \tag{17}$$

Here, a smaller MAE score means the algorithm makes fewer mistakes and thus is a better model.

In real applications, top-k performance is one of the most important criteria, since users may not spend that much time looking at low ranked recommended items. So we also adopt nDCG@5[13] in our experiment.

5.3   Settings

The main aim of our experiment is to test whether the incorporated user and product weights are useful for recommendation. So we adopt three baselines to compare with.

1) *CF*: The Collaborative Filtering method which only uses the rating data between users and products. Since there are various variants, we selected Low-rank matrix factorization [16] for our work.

2) *PU*: The method that only uses the Product Utility (PU).

3) *UC*: The method that only uses the User Credibility (UC).

4) *TCF*: The proposed Trustworthy Collaborative Filtering framework.

More complex and advanced variants of collaborative filtering are not selected in our work, since our work is a generalized framework of exploiting user credibility and product utility and can easily be extended and embedded with other CF algorithms. For the experiment, ten fold cross validation is employed for getting the final results. Data is randomly split into ten folds and one fold is selected as testing data while the rest as training data in each round. After ten rounds, the average MAE and nDCG are reported.

5.4   Experimental Results

Figure 1 illustrates the Mean Absolute Error of different methods. The lowest error is achieved by our proposed framework, which outperforms the methods that ignores either user credibility or product utility or both. This proves that the incorporated components are effective in improving recommendations. Collaborative Filtering achieves worst among all methods. PU method which only considers the product utility is the second worst method. This is possibly due to the noises brought by the user and product modeling algorithms. The UC method which only takes user credibility into consideration is the runner-up, which is most closest to the proposed framework. It proves that the proposed exclusive regularizer is useful for denoising and improving the recommendation.

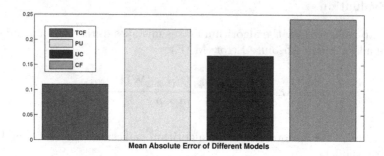

Fig. 1. The Mean Absolute Error of Different Models.

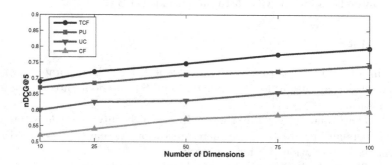

Fig. 2. The nDCG@5 Score of Different Models with Varying Dimensions.

Figure 2 illustrates the nDCG@5 score of different methods with varying number of latent dimensions (k). First, it further proves that our proposed framework is useful for recommending items. Second, it shows that the proposed framework increases steadily when the number of dimensions increases. The optimal dimensionality may be set based on different applications.

6   Conclusion and Future Work

In this work, we put forward a new recommendation framework, which can jointly take user and product utility into consideration during recommendation. The proposed framework is proven to be useful in reducing negative effect of noises and information redundancy. Since the incorporated user and product weights can be learnt separately, our model can easily be extended with different kinds of external knowledge, such as social network and knowledge base.

Since evaluating users and products is not our main focus, the optimal weighting methods are interesting problems and remain to be done in possible future work. We would like to investigate leveraging links and explicit group memberships in social networks in the future.

References

1. Adomavicius, G., Tuzhilin, A.: Toward the next generation of recommender systems: A survey of the state-of-the-art and possible extensions. IEEE Transactions on Knowledge and Data Engineering 17(6), 734–749 (2005)
2. Ahn, J.-W., Brusilovsky, P., Grady, J., He, D., Syn, S.Y.: Open user profiles for adaptive news systems: help or harm? In: Proceedings of the 16th International Conference on World Wide Web, pp. 11–20. ACM (2007)
3. Backstrom, L., Leskovec, J.: Supervised random walks: predicting and recommending links in social networks. In: Proceedings of the Fourth ACM International Conference on Web Search and Data Mining, pp. 635–644. ACM (2011)
4. Cai, Y., Leung, H.-F., Li, Q., Min, H., Tang, J., Li, J.: Typicality-based collaborative filtering recommendation. IEEE Transactions on Knowledge and Data Engineering 26(3), 766–779 (2014)
5. Cantador, I., Bellogín, A., Castells, P.: News@ hand: A semantic web approach to recommending news. In: Nejdl, W., Kay, J., Pu, P., Herder, E. (eds.) AH 2008. LNCS, vol. 5149, pp. 279–283. Springer, Heidelberg (2008)
6. Danon, L., Duch, J., Arenas, A., Díaz-guilera, A.: Comparing community structure identification. J. Stat. Mech. (2005)
7. Degemmis, M., Lops, P., Semeraro, G.: A content-collaborative recommender that exploits wordnet-based user profiles for neighborhood formation. User Modeling and User-Adapted Interaction 17(3), 217–255 (2007)
8. Du, L., Li, X., Shen, Y.-D.: Robust nonnegative matrix factorization via half-quadratic minimization
9. Firan, C.S., Nejdl, W., Paiu, R.: The benefit of using tag-based profiles. In: 2007 Latin American Web Conference, LA-WEB 2007, pp. 32–41. IEEE (2007)
10. Hu, X., Tang, J., Zhang, Y., Liu, H.: Social spammer detection in microblogging. In: IJCAI (2013)
11. Hu, Y., Koren, Y., Volinsky, C.: Collaborative filtering for implicit feedback datasets. In: ICDM 2008, pp. 263–272. IEEE (2008)
12. Huang, Z., Chen, H., Zeng, D.: Applying associative retrieval techniques to alleviate the sparsity problem in collaborative filtering. ACM Transactions on Information Systems (TOIS) 22(1), 116–142 (2004)
13. Järvelin, K., Kekäläinen, J.: Cumulated gain-based evaluation of ir techniques. ACM Transactions on Information Systems (TOIS) 20(4), 422–446 (2002)
14. Jindal, N., Liu, B.: Review spam detection. In: Proceedings of the 16th International Conference on World Wide Web, pp. 1189–1190. ACM (2007)
15. Kleinberg, J.M.: Authoritative sources in a hyperlinked environment. Journal of the ACM (JACM) 46(5), 604–632 (1999)
16. Koren, Y.: Factorization meets the neighborhood: a multifaceted collaborative filtering model. In: Proceedings of the 14th ACM SIGKDD International Conference on Knowledge Discovery and Data Mining, pp. 426–434. ACM (2008)
17. Koren, Y., Bell, R., Volinsky, C.: Matrix factorization techniques for recommender systems. Computer (8), 30–37 (2009)
18. Lees-Miller, J., Anderson, F., Hoehn, B., Greiner, R.: Does wikipedia information help netflix predictions? In: Seventh International Conference on Machine Learning and Applications, ICMLA 2008, pp. 337–343. IEEE (2008)
19. Leung, K.W.-T., Lee, D.L., Lee, W.-C.: Clr: a collaborative location recommendation framework based on co-clustering. In: Proceedings of the 34th International ACM SIGIR Conference on Research and Development In Information Retrieval, pp. 305–314. ACM (2011)

20. Lops, P., De Gemmis, M., Semeraro, G.: Content-based recommender systems: State of the art and trends. In: Recommender Systems Handbook, pp. 73–105. Springer (2011)
21. Ma, H., King, I., Lyu, M.R.: Effective missing data prediction for collaborative filtering. In: Proceedings of the 30th Annual International ACM SIGIR Conference on Research and Development In Information Retrieval, pp. 39–46. ACM (2007)
22. Ma, H., Liu, C., King, I., Lyu, M.R.: Probabilistic factor models for web site recommendation. In: Proceedings of the 34th International ACM SIGIR Conference on Research and Development in Information Retrieval, pp. 265–274. ACM (2011)
23. Middleton, S.E., Shadbolt, N.R., De Roure, D.C.: Ontological user profiling in recommender systems. ACM Transactions on Information Systems (TOIS) 22(1), 54–88 (2004)
24. Newman, M.: Modularity and community structure in networks. PNAS 103(23), 8577–8582 (2006)
25. Pazzani, M., Billsus, D.: Learning and revising user profiles: The identification of interesting web sites. Machine Learning 27(3), 313–331 (1997)
26. Rocchio, J.J.: Relevance feedback in information retrieval (1971)
27. Tang, L., Liu, H.: Relational learning via latent social dimensions. In: Proceedings of the 15th ACM SIGKDD International Conference on Knowledge Discovery and Data Mining. ACM (2009)
28. Umyarov, A., Tuzhilin, A.: Improving collaborative filtering recommendations using external data. In: Eighth IEEE International Conference on Data Mining, ICDM 2008, pp. 618–627. IEEE (2008)
29. Weiser, K.: A lie detector test for online reviewers. Business
30. Zhang, L., Agarwal, D., Chen, B.-C.: Generalizing matrix factorization through flexible regression priors. In: Proceedings of the Fifth ACM Conference on Recommender Systems, pp. 13–20. ACM (2011)
31. Zhou, K., Yang, S.-H., Zha, H.: Functional matrix factorizations for cold-start recommendation. In: Proceedings of the 34th International ACM SIGIR Conference on Research and Development in Information Retrieval, pp. 315–324. ACM (2011)

A Co-ranking Framework to Select Optimal Seed Set for Influence Maximization in Heterogeneous Network

Yashen Wang, Heyan Huang, Chong Feng, and Xianxiang Yang

Beijing Engineering Research Center of High Volume Language
Information Processing and Cloud Computing Applications,
School of Computer, Beijing Institute of Technology, Beijing, China
{yswang,hhy63,fengchong,yangxianxiang}@bit.edu.cn

Abstract. The rising popularity of social media presents new opportunities for one of the enterprise's most important needs—selecting most influential individuals in viral marketing, which has attracted increasing attention in both academia and industry. Most recent algorithms of influence maximization have demonstrated remarkable successes, however their applications are limited to homogeneous networks. In this paper, we formulate the problem of influence maximization in heterogeneous network, and propose a co-ranking framework to simultaneously select seed sets with different types. This framework is flexible and could adequately takes advantage of additional information implicit in the heterogeneous structure. We conduct extensive experiments using the data collected from ACM Digital Library, and the experimental results show that both the quality and the running time of the proposed algorithm rival the existing algorithms.

Keywords: Influence Maximization, Co-Ranking, Heterogeneous Network.

1   Introduction

As social network services such as Twitter connect individuals across the world, influence maximization has been actively researched with applications to viral marketing. Before applying viral marketing strategies on online social networks, some challenges need to be addressed firstly [1]: (i) how to determine the edge weights (influence strength) among different individuals; (ii) how to calculate the social influence given a seed set (a set of activated vertices at the beginning); (iii) how to select the optimal seed set, which has the maximum social influence. This paper mainly concentrates on solving the third challenge: *how to select optimal seed set in heterogeneous network*.

The idea of leveraging social influence maximization for marketing campaigns has been studied extensively [9, 10]. Most current heuristic algorithms are introduced and applied on *homogeneous networks* which contain only one type of vertex (individual) and edge/link (relationship). By contrast, in the real world, we have to consider multiple typed individual and relations simultaneously, e.g., academic network (Fig.1). Such sort of network is called *heterogeneous network* [8, 17]. However, the researches

© Springer International Publishing Switzerland 2015
R. Cheng et al. (Eds.): APWeb 2015, LNCS 9313, pp. 141–153, 2015.
DOI: 10.1007/978-3-319-25255-1_12

on social influence analysis have surprisingly largely overlooked this aspect. Although it is possible to come up with the problem of selecting seed set based solely on a homogeneous social network (e.g., collaboration network) [1, 2, 3, 14], those results are inherently limited, because the natural connection between researchers and their publications, and the social network among researchers are not fully leveraged.

To overcome the above drawbacks, we propose a novel co-ranking framework for selecting optimal seed sets in an iterative process on heterogeneous academic network (Fig.1). In [8, 15], authors and their publications (papers/tweets) are co-ranked under such heterogeneous network by simply coupling two random walks, which separately rank authors and their publications under PageRank [7]. We utilized different strategy in contrast with them: rather than simply coupling two random walks, we design an iterative procedure to achieve global ranking. By utilizing the implicit information in the heterogeneous networks, the rankings of authors and their documents depend on each other in a mutually reinforcing way. And as the results, individual's ranking-score reflects corresponding extent of influence. Individuals with high ranking-scores are selected into optimal seed set, and then propagate their influence (i.e., compute influence spread) under influence propagation model in corresponding homogeneous network.

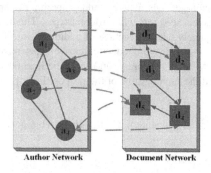

Author Network   **Document Network**

Fig. 1. The proposed co-ranking framework operates over an heterogeneous academic network including three sub-networks: (i) the social collaboration network connecting authors (undirected red links between authors indicate co-author relationship), (ii) the citation network connecting documents (the directed blue links between documents denotes citation relation), and (iii) the authorship network that ties the two previous networks together (the green edges are shown with dashed lines). Circles represent authors and rectangles represent documents.

To summarize, this work contributes on the following aspects:

- We formally formulate the problem of influence maximization in heterogeneous networks, and propose a co-ranking framework to select optimal seed set for this problem. Accordingly, the proposed framework could finish selecting seed sets for multiple typed sub-networks (homogeneous networks) simultaneously.
- Experimental results suggest that the proposed co-ranking framework is effective: (i) its influence spread is closed to even better than the best results produced by greedy algorithm [14], while it significantly reduces the running time; (ii) it is

scalable on networks with different types and sizes, even very large network with millions edges; (iii) it provides better performance than other co-ranking frameworks [8, 15] with less numbers of iterations and citation information.

- Our framework is flexible because it could be customized by redefining the affinity matrix (i.e., weights) and iterative procedure according to specific applications.

The outline of the paper is as follows. Section 2 surveys the related researches and Section 3 formally describes problem definition. Section 4 presents the co-ranking framework, and corresponding experimental results are shown in Section 5. Finally, we conclude the paper in Section 6.

2   Related Works

Influence Maximization. Domingos et al. were the first to study influence maximization as an algorithmic problem [5]. Kempe et al. [10] then formulated the problem as a discrete optimization problem by using Monte Carlo simulation to estimate the influence spread, and presented two basic influence models, the Independent Cascade (IC) model and the Linear Threshold (LT) model [9]. Then many heuristic algorithms were introduced for the IC model. Kim et al. proposed Independent Path Algorithm (IPA) [11], and Kimura and Saito proposed shortest-path based IC models [12]. Chen et al. proposed heuristic using degree discount [2] and MIA/PMIA model [3]. However, most of the aforementioned algorithms only focus on homogeneous networks.

Influence Analysis in Heterogeneous Network. Recently, influence analysis in heterogeneous network begins to attract researchers' attentions, and many works have tried to combine the sufficient information on heterogeneous networks to analyze influence diffusion [6] and to mine topic-level influence [17]. However, to our knowledge, rare attention has concentrated on influence maximization in heterogeneous network.

Co-ranking Framework. By efficiently taking advantage of the additional information implicit in the heterogeneous networks, the co-ranking framework has been initially developed for scientific impact measurement [8], tweet recommendation [15], and influence pattern discovery [17]. The main intuition behind co-ranking is that there is a mutually reinforcing relationship among multiple typed entities that could be reflected in the rankings. However, the proposed adaptation of co-ranking framework to the influence maximization task is novel to our knowledge.

3   Problem Definition

A heterogeneous network is a special type of information network with the underneath structure, which either contains multiple types of vertices or multiple types of edges.

DEFINITION 1. Heterogeneous Network. Define a network $G = (V, E)$. V is a set of vertices, which are classified into T types $V = \{V_t\}_{t=1}^{T}$ wherein V_t is a set of vertices with the t-th type. Edges in $E = \{(u, v) | u, v \in V\}$ represents the influence relationships among vertices. For $(u, v) \in E$, $p_{(u,v)}$ is the influence propagation probability from u to v which ranges over [0,1].

DEFINITION 2. Influence Maximization in Heterogeneous Network. For each V_t of V, let S_t be the subset of active vertices selected to initiate the influence propagation, which we call the *seed set*. Algorithm takes the graph G and a number k_t as input, and generates a seed set S_t of cardinality k_t. The final intention is that the expected number of finally activated vertices in V_t influenced by seed set S_t, which is usually called *influence spread* (denoted as $\sigma_t(S_t)$), is as large as possible under a certain influence propagation model. Note that, although seed sets with multiple typed vertices are selected in heterogeneous network simultaneously, the influence spread is calculated for each type of vertices (in corresponding homogeneous network) respectively.

4   A Co-ranking Framework to Select Optimal Seed Set

The proposed co-ranking framework operates over a heterogeneous network representing the documents, their authors, and the relationships between them. The conceptual scheme of the proposed co-ranking framework is illustrated in Fig. 2.

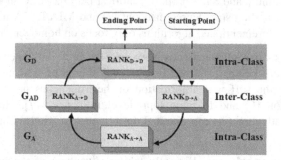

Fig. 2. The conceptual scheme of the proposed co-ranking framework to select seed set.

4.1   Data Model

We denote the entire heterogeneous graph as $G = (V, E)$, where $V = V_A \cup V_D$ is the set of vertices and $E = E_{AA} \cup E_{DD} \cup E_{AD}$ is the set of edges. G is composed of three sub-networks: (i) the *collaboration network* $G_A = (V_A, E_A)$ respecting authors, (ii) the *citation network* $G_D = (V_D, E_D)$ respecting documents, and (iii) the bipartite *authorship network* $G_{AD} = (V_{AD}, E_{AD})$ that ties authors (in G_A) and documents (in G_D) together.

The affinity matrix $\mathbf{M}$ of the heterogeneous network G is defined as follows,

$$\mathbf{M} = \begin{bmatrix} \mathbf{M_{AA}} & \mathbf{M_{AD}} \\ \mathbf{M_{DA}} & \mathbf{M_{DD}} \end{bmatrix}, \tag{1}$$

wherein $\mathbf{M_{AA}}$ represents the social relationship among authors, and $\mathbf{M_{DD}}$ represents the citation relationship among documents. $\mathbf{M_{AD}}$ and $\mathbf{M_{DA}}$ denote *bipartite* authorship (in G_{AD}), indicating the documents which a certain author has published and vice versa.

Collaboration Network (G_A). The collaboration network $G_A = (V_A, E_A)$, which represents the co-authorship among authors, is a weighted undirected graph. V_A is the author set with size of $n_A = |V_A|$, and E_A is the set of edges representing co-authorship ties. The individual author is denoted as $\{a_i | a_i \in V_A, i = 1, 2, ..., n_A\}$. We define its affinity matrix $\mathbf{M_{AA}}$ as follows: if two authors ever collaborated on a paper, there is an edge connecting them. Normally, the total number of edges is equal to the number of all the documents they collaborated on. So the element of $\mathbf{M_{AA}}$, is defined as follows:

$$M_{AA}[i][j] = \frac{connection(a_i, a_j)}{\sum_{k=1}^{n_A} connection(a_i, a_k)} \tag{2}$$

wherein $connection(a_i, a_j)$ represents number of edges between a_i and a_j. This definition follows the normal observation that, the less people an author collaborates with frequently, the closer their relation is. However, considering that paper with fewer authors strengthens the social ties among the corresponding authors, we define *social tie function* for each pair of a_i and a_j (given a document d_k) as follows.

$$\tau(a_i, a_j, d_k) = \frac{\mathbb{I}(M_{AD}[i][k] \neq 0, M_{AD}[j][k] \neq 0)}{|d_k|(|d_k|+1)/2}, \tag{3}$$

where $\mathbb{I}(M_{AD}[i][k] \neq 0, M_{AD}[j][k] \neq 0)$ is the *indicator function* of whether authors a_i and a_j collaborated on document d_k, and $|d_k|$ is the size of author list of d_k. So, adding up social ties from all documents, we obtain

$$M_{AA}[i][j] = \frac{connection(a_i, a_j) * \sum_{k=1}^{n_D} \tau(a_i, a_j, d_k)}{\sum_{k=1}^{n_A}(connection(a_i, a_k) * \sum_{l=1}^{n_D} \tau(a_i, a_k, d_l))}. \tag{4}$$

Citation Network (G_D). The citation network $G_D = (V_D, E_D)$ is an unweighted directed graph representing documents and their citation relationships, wherein V_D is the set of documents and links E_D among users are established by observing their citation behavior. The number of documents is $n_D = |V_D|$, and we denote the individual document as $\{d_i | d_i \in V_D, i = 1, 2, ..., n_D\}$. We denote $citation(d_i)$ as the total number of documents directly cited by d_i. Normally, if document d_i cites document d_j, there is a directed edge from d_i to d_j and $M_{DD}[i][j] = 1/citation(d_i)$ [7].

Authorship Network (G_{AD}). As shown in Fig. 1 and Fig. 2, the authorship network G_{AD} is a special graph leveraging the previous graphs: both of the previous two networks (G_A and G_D) are homogeneous networks, the authorship network G_{AD} makes the entire network heterogeneous. $G_{AD} = (V_{AD}, E_{AD})$ is the weighted bipartite graph that represents the authorship (i.e., the relation between a given document and its authors), wherein $V_{AD} = V_A \cup V_D$ and edges in E_{AD} connect each document with all of its authors.

In contrast with [8], we take the order of authors into account, so its edges are weighted as follows. $M_{AD}[i][j]$ and $M_{DA}[i][j]$ could not only indicates whether a document d_j is written by an author a_i, but also represents his contribution to this document because of considering the order of a_i in the author list of d_j. Thus, given author a_i and document d_j, we define the contribution of a_i to d_j as

$$order(a_i, d_j) = \frac{1}{position(a_i, d_j)}, \tag{5}$$

wherein $position(a_i, d_j)$ denotes the position of a_i in the author list of d_j. If a_i did not participate the publication of d_j, $order(a_i, d_j) = 0$. We assign $order(a_i, d_j)$ to $M_{AD}[i][j]$ straightway, rather than implementing following normalization:

$$M_{AD}[i][j] = \frac{order(a_i, d_j)}{\sum_{k=1}^{n_D} order(a_i, d_k)}, \tag{6}$$

Mainly because the normalization above decreases the author's influence in our co-ranking framework. However, normalization, which implies extent how a certain author contributes to a given document, is necessary for the definition of $M_{DA}[i][j]$:

$$M_{DA}[i][j] = \frac{order(a_j, d_i)}{\sum_{k=1}^{n_A} order(a_k, d_i)}. \tag{7}$$

So the construction of affinity matrix $\mathbf{M_{AD}}$ and $\mathbf{M_{AD}}$ is *asymmetric* (i.e., $\mathbf{M_{DA}}$ is not the transposed matrix of $\mathbf{M_{AD}}$) and reflects the asymmetric relationship between authors and their documents. Moreover, the constructions of G_A ($\mathbf{M_{AA}}$) and G_{AD} ($\mathbf{M_{AD}}$ and $\mathbf{M_{DA}}$) are strongly correlated, since G_{AD} implicitly encode co-authorship information.

4.2   Iterative Procedure

The proposed co-ranking framework operates following *iterative procedure* on G_A, G_D and G_{AD} mutually. Inspired by [8], which used parameters α and λ to regulate *intra-class* and *inter-class* random walks respectively, we also use a set of asymmetric parameters $\{\alpha_{AA}, \alpha_{AD}, \alpha_{DA}, \alpha_{DD}\}$ to determine the extent to which we want the rankings of documents and their authors depend on each other, with the following constraints:

$$\alpha_{AA} \in [0,1], \alpha_{AD} \in [0,1], \alpha_{DA} \in [0,1], \alpha_{DD} \in [0,1] \tag{8}$$

$$\alpha_{AA} + \alpha_{AD} + \alpha_{DA} + \alpha_{DD} = 1 \tag{9}$$

The iterative procedure is formulated as follows, wherein $\mathbf{RS_A^{(z)}}$ and $\mathbf{RS_D^{(z)}}$ are ranking-score vector for authors and documents in z-th iteration, and $RS_A^{(z)}(i)$ and $RS_D^{(z)}(j)$ are denoted as the ranking-scores of author a_i and document d_i, separately. To guarantee convergence, $\mathbf{RS_A^{(z)}}$ and $\mathbf{RS_D^{(z)}}$ are normalized after each iteration. The algorithm typically converges when the difference between the ranking-scores computed at two successive iterations, $diversity(z-1, z)$, falls below a threshold ε (10^{-4} in this study).

Step 1   Document ranks author ($RANK_{D \rightarrow A}$ in Fig. 2). The ranking-scores of documents are used to reinforce the scores of authors.

$$RS_A^{(z+1)}(i) = \alpha_{DA} \sum_{k=1}^{n_D} M_{DA}[k][i] * RS_D^{(z)}(k) \tag{10}$$

$$\mathbf{RS}_A^{(z+1)} = \mathbf{RS}_A^{(z+1)} / ||\mathbf{RS}_A^{(z+1)}|| \tag{11}$$

It means a highly authoritative document could increase ranking scores of its authors.

Step 2   Author ranks author ($RANK_{A \to A}$ in Fig. 2). The ranking-scores of authors are used to reinforce the scores of other authors during collaboration behavior.

$$RS_A^{(z+1)}(i) = \alpha_{AA}(1 - \beta + \beta * \sum_{k=1}^{n_A} M_{AA}[k][i] * RS_A^{(z)}(k)) \tag{12}$$

$$\mathbf{RS}_A^{(z+1)} = \mathbf{RS}_A^{(z+1)} / ||\mathbf{RS}_A^{(z+1)}|| \tag{13}$$

Wherein β is the *damping factor* as used in PageRank [7]. This formula is based on the normal observation that the influence of an author is enhanced if he (or she) co-authors with other highly ranked authors, and is somewhat similar to weighted PageRank [4]. As mentioned above, because of adding social tie function, $M_{AA}[i][j]$ could reflect the strength of collaboration between the two authors a_i and a_j.

Step 3   Author ranks document ($RANK_{A \to D}$ in Fig. 2). The ranking-scores of authors are used to reinforce the scores of documents.

$$RS_D^{(z+1)}(j) = \alpha_{AD} \sum_{k=1}^{n_A} M_{AD}[k][j] * RS_A^{(z)}(k) \tag{14}$$

$$\mathbf{RS}_D^{(z+1)} = \mathbf{RS}_D^{(z+1)} / ||\mathbf{RS}_D^{(z+1)}|| \tag{15}$$

This formula is based on the normal observation that more influential authors are more likely to publish more authoritative papers.

Step 4   Document ranks document ($RANK_{D \to D}$ in Fig. 2). The ranking-scores of documents are used to reinforce the scores of other documents during citation behavior.

$$RS_D^{(z+1)}(j) = \alpha_{DD}(1 - \beta + \beta * \sum_{k=1}^{n_D} M_{DD}[k][j] * RS_D^{(z)}(k)) \tag{16}$$

$$\mathbf{RS}_D^{(z+1)} = \mathbf{RS}_D^{(z+1)} / ||\mathbf{RS}_D^{(z+1)}|| \tag{17}$$

Wherein β is the damping factor. This formula means that the influence of a document is enhanced if it is cited by other highly ranked documents.

4.3   Co-ranking Framework

Based on the construction of sub-graphs (and affinity matrixes), and the design of the iterative procedure, let us sketch the co-ranking framework as shown in Algorithm 1. Wherein, $diversity(z - 1, z)$ denotes difference between the current iteration and the previous one. When the iteration procedure converges, we choose the top-k individuals (authors or documents) according to ranking-scores, and put them into corresponding seed set which will propagate their influence under the influence propagation models.

Algorithm 1. Co-ranking framework to select seed set in heterogeneous network

Input: $M_{AA}, M_{DD}, M_{AD}, M_{DA}, \alpha_{AA}, \alpha_{DD}, \alpha_{AD}, \alpha_{DA}, k_A, k_D$.

Initialization:

1. $RS_A(i) \leftarrow 0, \; i = 1, \ldots, n_A$.
2. $RS_D(j) \leftarrow$ ranking-scores based on PageRank, wherein $j = 1, \ldots, n_D$.
3. $z \leftarrow 1$.

Iterative Procedure:

1. **while** $(diversity(z - 1, z) > \varepsilon)$ **do**
2. $\quad z = z + 1$;
3. $\quad RANK_{D \to A}; \; RANK_{A \to A}; \; RANK_{A \to D}; \; RANK_{D \to D}$.
4. **end**

Output: select top-k_A(or top-k_D) individuals of RS_A(or RS_D), respectively.

---

5   Experiments and Results

To validate the performance of our co-ranking framework, we conduct experiments on both real-world and synthetic datasets, to compare the results of seed set selection generated from the proposed algorithm with those from various other heuristics.

5.1   Dataset

This paper employs the proposed algorithm and compares with other algorithms on academic networks firstly. We crawled information about (i) authors, (ii) their co-authorship, (iii) their publications, (iv) the citation relationships, and (v) the authorship in the domain of data mining from ACM Digital Library, and constructed two mutual real-world networks (details of dataset are shown in Table 1):

1. The first network, denoted as **Collaboration**, is an academic collaboration network, with vertices representing authors and edges representing co-authorship relations.
2. The second one, denoted as **Citation**, is an academic citation network, wherein vertices are documents published by the corresponding authors in **Collaboration**.

Table 1. Statistics of the academic network generated from ACM Digital Library

Collaboration Network	#author	56,327
	#co-authorship	264,055
	#co-authorship / #author	4.687
Citation Network	#paper	28,917
	#citation	197,422
	#citation / #paper	6.827
Authorship Network	#authorship	10.2098
	#authorship / #author	1.813
Time Interval		2007-2012

5.2   Alternative Algorithms

We compare the proposed method with several heuristics under IC/WC model:

CELFGreedy: The greedy algorithm with CELF optimization [14], The results of the original greedy algorithm are not reported here since its influence spread $\sigma_I(S)$ is the same as the CELF optimization while the latter runs too slowly (20,000 simulations).

DegreeDiscount: The degree discount heuristic developed for the uniform IC model with a propagation probability of 0.01 [2].

Distance: A simple heuristic that selects k vertices with the smallest average distance to other vertices in the network.

Degree: Another simple heuristic that selects k vertices with the largest degrees [2, 10].

PageRank: The popular algorithm used of ranking web page [7]. The transition probability of (u, v) indicates u's "vote" to v's ranking.

Co-Rank: The proposed co-ranking framework with iterative procedure in this paper.

Co-Rank-AD: The co-ranking framework proposed in [8] by simply coupling two random walks, which separately rank authors and their documents under **PageRank**.

Co-Rank-TR: [15] developed **Co-Rank-AD** by allowing the random walk on **Citation** to be time-variant to incorporate diversity.

HITS: Hyperlink Induced Topic Search [16] could cope with heterogeneous information, confirming ranking could be reinforced through interactions.

Random: As a baseline, the result of choosing vertices uniformly at random.

5.3 Experiment Setup

The seed set size k ranges from 1 to 50. For comparing time requirement, we report the running times for selecting $k = 50$ under IC model. All the experiments are run on a PC with a 2.93 GHz Intel Core 2 Duo Processor and 16GB memory.

Parameter Setting. We set the damping factor β in Eq.(12) and Eq.(16) as 0.85 following [8]. Large α_{AA} relies more heavily on the **Collaboration** network, whereas large α_{DD} places more emphasis on the **Citation** network. Especially, $\{\alpha_{AD}, \alpha_{DA}\}$ control the balance of the authorship graph and the large values take both graphs into account, and all our experiments assign them to be 0.24.

Generating the Propagation Probabilities ($p_{(u,v)}$). We use the following two models to generate those probabilities: (i) **IC model**: here we mainly report results on a small propagation probability of $p_{(u,v)} = 0.01$ [10]; (ii) **WC model**: In Weighted Cascade (WC) model [3], the propagation probability is reverse of the in-degree.

5.4 Influence Maximization on Collaboration Network

Fig. 3 (a, b) show the results of influence spreads on **Collaboration** under WC model and IC model. It is shown that **CELFGreedy** produces the best influence spread, however **Co-Rank** is very close to it. Comparing with others, **Co-Rank** also performs well: it matches the influence spread of **DegreeDiscount** while outperforms the rest heuristics which were also evaluated in [10]. Comparing to **HITS**, which also utilizes citation information from **Citation**, **Co-rank** is 9.14% (11.76%) better in WC (IC) model. **Random** as the baseline performs badly indicating that a careful seed section is important.

Table 1 shows the performance comparisons with other co-ranking frameworks, using the metric of coverage ratio ($\frac{\sigma_I(S)}{\#author}$ and $\frac{\sigma_I(S)}{\#paper}$). In **Collaboration** network, **Co-rank** is 5.08% and 2.75% better than **Co-Rank-AD** and **Co-Rank-TR** on average, especially in case of small seed set size. However, with size of seed set increasing, the performances of latter two become close to **Co-Rank**. This phenomenon could be explained as we introduce social tie function into the co-ranking framework, and the affinity matrix could better reflect the strength of collaboration. For running time, as shown in Fig. 3 (c), **CELFGreedy** is already quite slow (more than 15 hours), while **Co-rank** only takes 10.54 second and is comparable with **HITS** and **DegreeDiscount**.

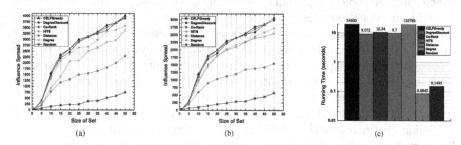

| | (a) | | (b) | | (c) |

Fig. 3. Experimental results on Collaboration network: (a) influence spreads under WC model, (b) influence spreads under IC model, and (c) running times under IC model.

Table 2. Comparison with other co-ranking frameworks using metric of coverage rate

#seed	Collaboration network			Citation network		
	Co-Rank	**Co-Rank-AD**	**Co-Rank-TR**	**Co-Rank**	**Co-Rank-AD**	**Co-Rank-TR**
10	**0.0256**	0.0226	0.0246	**0.0153**	0.0136	0.0149
20	**0.0461**	0.0419	0.0448	**0.0615**	0.0559	0.0604
30	**0.0553**	0.0527	0.0537	0.0804	0.0766	**0.0813**
40	**0.0623**	0.0605	0.0612	**0.0956**	0.0928	0.0954
50	**0.0692**	0.0685	0.0675	0.1035	0.1025	**0.1041**
Avg.	**0.0517**	0.04924	0.05036	**0.0713**	0.0683	0.0712

5.5   Influence Maximization on Citation Network

Fig. 4 (a, b) show the results on the **Citation** network. For the WC model, all the algorithms are separated into three groups: (i) **Co-Rank** almost match the **CELFGreedy** (even better in some points) and performs better than **DegreeDiscount** and **PageRank**. (ii) **Distance** performs slightly better than **Degree**, but **Co-Rank** has a large winning margin over them (almost 100% better them). (iii) As the baseline, **Random** performs badly enough as expected. What's more, on both **Collaboration** and **Citation**, **DegreeDiscount** performs worse under WC model than under IC model, because of being specially tuned for the uniform IC model. So, **Co-Rank** outperform **DegreeDiscount** in stability and suffers less limitations.

As shown in Table 2, **Co-Rank** still performs better than other co-ranking frameworks. However, **Co-Rank-TR** provides the similar performance mainly because of balancing popularity and diversity on **Citation** by using time-variant feature (it favors vertices with high popularity and there emerges a rich-get-richer effect as time goes by [15]). For running time under IC model, we see that **Co-Rank** only takes less than 6.27 seconds (similar to **PageRank**), while **CELFGreedy** costs about 10 hours.

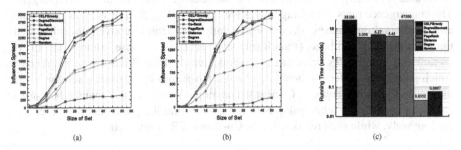

Fig. 4. Experimental results on Citation network: (a) influence spreads under WC model, (b) influence spreads under IC model, and (c) running times under IC model.

5.6   Experiments on Synthetic Networks

In order to test the scalability, we generate a family of networks with increasing sizes ($|V|$ ranges from 2K to 256K) using the LFR standard benchmark networks [13], which account for the heterogeneity of both degree and graph size. For each of the sizes, two synthetic networks are generated to simulate the **Collaboration** network and **Citation** network, and in every network, the degree distribution follows power-law distribution with exponent of 2.9, and the average degree is 2.5. We use the IC model and compute influence spread for 50 seed vertices in each case. Fig. 5(a) shows the running times of different algorithms on **Collaboration** networks.

Fig. 5(a) clearly shows that all algorithms are separated into two groups. Simple **Distance** and **CELFGreedy** have the worst slope and are certainly not scalable for large graphs. On the contrary, **Co-Rank** along with the rest heuristics, in other group in Fig. 5(a), could all scale up quite well and be easily scalable to large graphs.

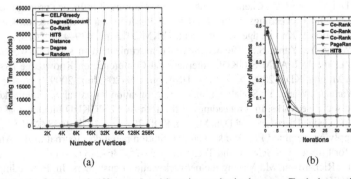

Fig. 5. (a) Scalability of different algorithms in synthetic datasets. Each data point is an average of 50 runs. (b) Convergence of different algorithms on Citation network under WC model.

5.7   Runtime and Ranking-Score Convergence

The computational complexity of **Co-Rank** is $O(t * |E|)$, while **Co-Rank-AD** costs more than $O(t * |E|^2)$, wherein t is the number of iterations before convergence. The convergence rates of ranking-scores are shown in Fig. 5(b). It is clear that, generally the co-ranking frameworks use less iterations to achieve convergence than traditional ranking algorithms (**PageRank** and **HITS**). Because almost all of the co-ranking frameworks implement several mutual **PageRank** for reinforcement, which guarantees better efficiency. Although **Co-Rank-TR** provides similar performance in **Citation** network, **Co-Rank** converges more quickly mainly because of capturing asymmetric relationship between authors and their documents in inter-class random walk, while such relationship in **Co-Rank-TR** is symmetric.

6   Conclusion

Unlike traditional influence maximization algorithms which solely focus on homogeneous network, we propose a co-ranking framework to select seed sets in heterogeneous network. By taking advantage of the additional information implicit in heterogeneous structure, the proposed algorithm simultaneously measures each type of individuals in a mutually reinforcing way. The experimental results on influence spread demonstrate its effectiveness and scalability: (i) it significantly reduces the running time of existing best greedy algorithms while maintaining influence spread guarantee; (ii) it provides better performance than other co-ranking frameworks with less iterations.

Acknowledgement. We would like to thank Zhaoyu Wang from UIUC for useful conversation. The work was supported by the National Basic Research Program of China (973 Program, Grant No. 2013CB329605, 2013CB329303) and National Natural Science Foundation of China (Grant No. 61201351).

References

1. Zhang, M., Dai, C., Ding, C., Chen, E.: Probabilistic solutions of influence propagation on social networks. In: Proceedings of the 22nd ACM International Conference on Conference on Information & Knowledge Management, CIKM 2013, pp. 429–438 (2013)
2. Chen, W., Wang, Y., Yang, S.: Efficient influence maximization in social networks. In: Proceedings of the 15th ACM SIGKDD International Conference on Knowledge Discovery and Data Mining, KDD 2009, New York, USA, pp. 199–208, June 2009
3. Chen, W., Wang, C., Wang, Y.: Scalable influence maximization for prevalent viral marketing in large-scale social networks. In: Proceedings of the 16th ACM SIGKDD International Conference on Knowledge Discovery and Data Mining, KDD 2010, pp. 1029–1038 (2010)
4. Ding, Y.: Applying weighted PageRank to author citation networks. Journal of the American Society for Information Science and Technology 62(2), 236–245 (2011)
5. Domingos, P., Richardson, M.: Mining the network value of customers. In: Proceedings of 7th ACM SIGKDD International Conference on Knowledge Discovery and data Mining, KDD 2001, New York, USA, pp. 57–66, August 2001

6. Galstyan, A., Musoyan, V., Cohen, P.: Maximizing influence propagation in networks with community structure. Physical Review E 79(5), 056102 (2009)
7. Brin, S., Page, L.: The anatomy of a large-scale hypertextual Web search engine. Computer Networks and ISDN Systems 30(1-7), 107–117 (1998)
8. Zhou, D., Orshanskiy, S.A., Zha, H., Giles, C.L.: Co-ranking authors and documents in a heterogeneous network. In: Proceedings of 7th IEEE International Conference on Data Mining, pp. 739–744, October 2007
9. Granovetter, M.: Threshold models of collective behavior. The American Journal of Sociology 83(1978), 1420–1443 (1978)
10. Kempe, D., Kleinberg, J., Tardos, É.: Maximizing the spread of influence through a social network. In: Proceedings of 9th ACM SIGKDD International Conference on Knowledge Discovery and Data Mining, KDD 2003, pp. 137–146 (2003)
11. Kim, J., Kim, S.-K., Yu, H.: Scalable and parallelizable processing of influence maximization for large-scale social networks. In: Proceedings of 2013 IEEE 29th International Conference on Data Engineering (ICDE), pp. 266–277, April 2013
12. Kimura, M., Saito, K.: Tractable models for information diffusion in social networks. In: Fürnkranz, J., Scheffer, T., Spiliopoulou, M. (eds.) PKDD 2006. LNCS (LNAI), vol. 4213, pp. 259–271. Springer, Heidelberg (2006)
13. Lancichinetti, A., Fortunato, S., Radicchi, F.: Benchmark graphs for testing community detection algorithms. Physical Review E 78(4), 046110 (2008)
14. Leskovec, J., Krause, A., Guestrin, C., Faloutsos, C., Van Briesen, J., Glance, N.: Cost-effective outbreak detection in networks. In: Proceedings of the 13th ACM SIGKDD International Conference on Knowledge Discovery and Data Mining, pp. 420–429 (2007)
15. Rui, Y., Lapata, M., Li, X.: Tweet recommendation with graph co-ranking. In: Proceedings of the 50th Annual Meeting of the Association for Computational Linguistics, ACL, Jeju, Republic of Korea, July 8-14, pp. 517–525 (2012)
16. Kleinberg, J.M.: Authoritative sources in a hyperlinked environment. Journal of the ACM (JACM) 46(5), 604–632 (1999)
17. Liu, L., Tang, J., Han, J., Jiang, M., Yang, S.: Mining topic-level influence in heterogeneous networks. In: Proceedings of the 19th ACM International Conference on Information and Knowledge Management, pp. 199–208. ACM (2010)

Hashtag Sense Induction
Based on Co-occurrence Graphs

Mengmeng Wang and Mizuho Iwaihara

The Graduate School of Information, Production and Systems, Waseda University
2-7 Hibikino, Wakamatsu-ku, Kitakyushu, Fukuoka 808-0135
emma@akane.waseda.jp,
iwaihara@waseda.jp

Abstract. Twitter hashtags are used to categorize tweets for improving search categorizing topic. But the fact that people can create and use hashtags freely leads to a situation such that one hashtag may have multiple senses. In this paper, we propose a method to induce senses of a hashtag in a particular time frame. Our assumption is that for a sense of a hashtag the context words around it are similar. Then we design a method that uses a co-occurrence graph and community detection algorithm. Both words and hashtags are nodes of the co-occurrence graph, and an edge represents the relation of two nodes co-occurring in the same tweet. A list of words with a high node degree representing a sense is extracted as a community of the graph. We take Wikipedia disambiguation list page as word sense inventory to refine the results by removing non-sense topics.

Keywords: Twitter, Hashtag, Sense Induction, Co-occurrence Graph, Wikipedia.

1 Introduction

Twitter is a popular online social network service where users can share information through tweets written within 140 characters each. Quite a large amount of data is created by users on Twitter every day. From the official Twitter statistic data, 500 million Tweets are sent per day[1]. Words in Tweets starting with a "#" symbol, used to mark keywords or topic in a Tweet are called hashtags[2]. People can use hashtags at their will, and there is no strict restriction on how many words to hashtags and what to hashtag. The loose rule of using hashtag and so much data is generated every day lead to arbitrary use of it. One of the results is that one hashtag may have multiple senses. For example: #SuperNatrual may refer to the phenomena beyond the physical theory, it also may refer to the US TV series.

Finding the senses for an ambiguous word is called word sense induction (WSI). Former researchers mainly focus on natural language word sense induction, and

---

[1] Twitter https://about.twitter.com/company
[2] Twitter https://support.twitter.com/entries/49309

© Springer International Publishing Switzerland 2015
R. Cheng et al. (Eds.): APWeb 2015, LNCS 9313, pp. 154–165, 2015.
DOI: 10.1007/978-3-319-25255-1_13

usually use unsupervised machine learning methods. All the methods are based on an assumption that the sense of a word could be explained by the context words around it [1]. There are three main types of methods [2]:

1. Cluster Approaches.

This type of method based on syntactic dependency statistics between words. A series of cluster algorithms like K-means, Bisecting K-means, Average-link are employed by computing the similarity of words.

2. Graph-based Approaches.

This type of approach is close to word clustering ones, but by the way of exploiting the graph. HyperLex is a successful algorithm inducing senses by identifying hubs. Other methods based on other graph patterns, namely Curvature Clustering, Square, Triangles, etc., aim to identify word sense using the local structure properties of the graph.

3. Translation-Oriented Approaches.

Translation-Oriented Approaches are for bilingual sense induction.

In this paper, we propose a method based on co-occurrence graph (see Figure 1) to induce the senses for hashtags. In a co-occurrence graph both hashtags and words are the nodes and an edge exists between two nodes when they appear in the same tweet. The graph is partitioned by a heuristic community detection algorithm based on modularity optimization [3].

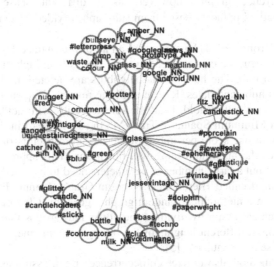

Fig. 1. Co-occurrence graph of hashtag #glass

We are facing two challenges:

1. Due to the characters of hashtags that they are irregular and that the use of them is easily influenced by popular events, some communities are not context words of senses; they are just a group of words highly related to the newly happened events.
2. To the best of our acknowledgment, we are the first to induce the sense of twitter hashtags. Also, because of the temporal character of hashtags, there is no sense inventory for all the hashtags, which means that we do not have a standard to evaluate our result.

In order to address these issues, first we consider filtering communities, so that the result can be more accurate; then we refine our result by mapping them to sense inventory. In our paper we utilize Wikipedia disambiguation list as sense inventory as there is no existing golden standard for hashtag.

The rest of the paper is organized as follows: In Section 2, we reviewed related works focus on word sense induction briefly. In Section 3, we introduce features of hashtags. The construction of our method is described in Section 4 and Section 5. In Section 6 we design experiments and discuss the result. Finally, a conclusion with future work is presented in Section 7.

2   Related Work

In this section we mainly highlight the works related to graph-based word sense induction. A co-occurrence graph follows the small world topology pattern that the nodes are highly clustered and the path length is short. All the works are based on word co-occurrence graphs, however, using different sense extraction algorithms. Our method induces senses from subgraphs divided by community detection algorithm.

V'eronis (V'eronis, 2004) has proposed an algorithm named Hyperlex [4]. The algorithm first constructs a co-occurrence graph. Nodes with highest relative frequency are selected as a hub. The hub and its neighbors are deleted, until there is no more eligible node as hub candidates. Repeat until the relative frequency of the candidates becomes lower than a threshold. Link the selected hubs with neighbors to the target word to build a Minimum Spanning Tree (MST). Each word in the MST is given a set of scores, with one score per hub. Add up the score vectors of all words, and the hub that receives the maximum score will be chosen. This approach could achieve a state of art result but a large number of parameters need to be tuned.

Another method called Hierarchical Random Graph (Ioannis P. Klapaftis et al.,), which is based on a hierarchical random graph algorithm (Clauset et al., 2008), constructs a hierarchical graph by dividing the nodes into groups that are further subdivided and so on. At different heights of the binary tree, this method can produce different levels of sense granularity [5].

Yet another method also utilizes co-occurrence graph, assuming that communities in the graph identify sense-specific contexts for each target word. It adopts the community detection algorithm called Link Clustering (Ahn et al., 2010), taking both multiple senses and sense granularity into consideration [6].

3   Hashtags

Hashtags are used to tag keywords of tweets. All the tweets with the same hashtag are categorized together, usually they are about the same topic but there is no strict constraint. Because of the rules of using hashtags and the limitation of tweet length, hashtags have many features that are different from natural words appearing on other media.

3.1   Rules to Use Hashtags

The start of a hashtag must begin with pound symbol "#". In the following characters, the first one must be a letter or an underscore; others can be a combination of letters, numbers and underscores. A hashtag must be no longer than 140 characters, and there is no other length limitation. Search by a hashtag in twitter is case-insensitive.

3.2   Features of Hashtag

Usage of Hashtags
There are four kinds of usage of hashtags.

1. Used as a phrase or part of the sentence, for example: #LeaveMeAlone, #PissedOff, #LateForWork.
2. Used as entities such as people, place, organization, object and so on, for example: #obama, #Japan, #AppleMacBook, #NBA.
3. Used to describe an event, for instance: #WordCup2014, #UTAvsCHA.
4. Used with functions, for instance: #AskTom (this hashtag is used to interact with tom).

Temporal Property
Hashtags have temporal features and are easily influenced by events. Usually a new hashtag is prompted by an event, for example: #HeForShe is a hashtag related to a campaign hosted by Emma Watson on 20 September, 2014. Around that day, many people talked about this issue, but when it finished the number of twitter users talking about it dropped. Another example is that, near Christmas, more and more people talking about it, so the hashtags related to Christmas are used extensively.

Type of Sense of Hashtag
Because a hashtag may have a temporal property, accordingly the sense of it also has temporal property. There are four types of senses due to our observation:

1. Eternal Sense. For instance: #rose, one sense of #rose is a kind of flower, this sense is inherited from natural language and it won't disappear.
2. Long-term Sense. For instance #rose, one of its senses is a fashion brand, another name of basketball player (Derrick Rose). This kind of senses will be used much when the brand or this player is popular. They will last for quite a long time, maybe for years or scores of years.

3. One-time sense. This kind of sense is event related. When an event ends, accordingly the use of the hashtag also comes to an end.
4. Periodical sense. Some events happen regularly. For instance, Olympic Games are held every four years, so senses of hashtags related to the games will be used in that frequency.

Algorithm 1.   Hashtag Sense Induction

Input: tweets T, Wikipedia disambiguation list articles W
Output: MappingMatrix
    //Preprocessing: Tokenization, lemmatization, stopwords filtering
  1.T' ←Preprocessing(T)
  2.W' {article$_i$} ←Preprocessing(W)
    //Community detection
  3. G(V,E) ←Co-occurrenceGraphConstruction(T')
  4.Comm{community$_j$} ←CommunityDetection(G(V,E))
  5. Cont{ContextWordGroup$_j$} ←Extract context words from Comm
    //Mapping to Wikipedia Disambiguation list
  6.Generate MappingMatrix={M$_{ij}$}
   7. **for** each m$_{ij}$ in MappingMatrix **do**
   8.   m$_{ij}$←Score(article$_i$, ContextWordGroup$_j$)
   9. **end for**
10. **repeat**
11.   **for** each column m$_j$ in MappingMatrix **do**
12.   **if** DispersionRatio(m$_j$) < threshold
13.   **then** delete m$_j$ from MappingMatrix
14.   Update (MappingMatrix)
15.   **end for**
16.   **for** any two columns m$_a$, m$_b$ in MappingMatrix **do**
17.   **if** Mapping(m$_a$)= =Mapping(m$_b$)
18.   **then** Merge(ContextWordGroup$_a$,ContextWordGroup$_b$)
19.   Update (MappingMatrix)
20.   **end for**
21. **until** Update (MappingMatrix) == false
22. **return** MappingMatrix

---

4   Hashtag Sense Induction System

4.1   The Outline of Our System

Our system accepts tweets as input, and executes preprocessing to generate a word list for each tweet. Each word including hashtags in the tweet is seen as nodes to build the co-occurrence graph. The edges are the relations that each pair of them appears in the same tweet. Communities in the co-occurrence graph are detected by the algorithm proposed by Vincent D. Blondel in 2008[3], and sensed are extracted from communities and represented in the form of a group of related context words.

4.2   Data set

We can collect tweets by the twitter search API and stream API. We collected 5000 tweets for each hashtag and a half of them are used to induce senses, and the other half are used to evaluate the result. As we have discussed that hashtags have temporal feature, it is unnecessary and impossible to induce all the senses for one hashtag.

4.3   Data Preprocessing

Tweets are different from natural language texts in online or offline media. A tweet has unique elements like a user name starting with symbol '@', URLs and hashtags. Also because of the length limitation of tweets, there exist quite a large number of twitter acronyms such as RT (retweet), fb (follow back) and idiosyncratic use of words such as cooooool (word 'cool' with strong emotion) in order to express more information and emotion in a short sentence.

Because retweets do not provide much semantic information, and a tweet can be retweeted quite a large number of times, we will filter out retweets and keep only one.

We apply the following steps to process each tweet:

The tweets are tokenized, and filtered according to a stop word list (including stop hashtags, punctuation), which has 5000 to 6000 tokens in total. Then all the tokens are transformed into their lowercase form and stemmed to their basic form. POS (part of speech) tagging for each word and hashtags is performed, and only the words with POS being noun are kept.

Fig. 2. Tweet

After preprocessing:
photo chief ebola survivor kenema sierraleone .

4.4   Construction of Hashtag Co-occurrence Graph

We define a hashtag co-occurrence graph $G = (V,E)$, such that the node set V is the union of hashtag set V_h and word set V_w The hashtag set and word set are not independent, but they are treated differently. Hashtags are used to categorize tweets so they are the keywords of this tweet. Compared to words, hashtags are keywords of tweets, so accordingly we should assign them a higher weight.

For each word or hashtag in a tweet, if it exists as a node in the graph, we increase the frequency of the node by 1; otherwise create a new node and set the initial frequency as 1. The weight of each word node is set as its frequency, and the weight of the hashtag node is calculated by multiplying its frequency with a coefficient. For any pair of nodes in the same tweet, there exists an edge between them; increment the frequency if the same edge appears in other tweets. The weight of an edge is set as its frequency.

Once the co-occurrence graph is constructed, we prune it by removing all the nodes with a frequency below 2. Nodes with low frequencies will not contribute much to sense induction. Furthermore, a pruned graph reduces the computation cost.

4.5   Community Detection and Sense Induction

For weighted graph G of N nodes, there are two phases of this algorithm. Fig. 3 is an example after community detection, where an identical color is assigned to each community.

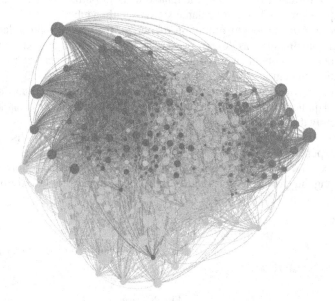

Fig. 3. Community detection for hashtag #train (visualized by Gephi)

First, each node was assigned to a different community. For each node v, v is moved to its neighbor's community which maximizes a gain. Repeat this process until all the gain for each node achieves to its best. Second, each community obtained from the first phase is treated as a node to generate a smaller graph. Repeat the two phases of the algorithm on this smaller network until there are no more changes and then a maximum of modularity is attained [3].

Before inducing the senses for hashtags, we should prune the graph by deleting the communities with a size below a threshold. The size of a community is measured by formula (1).

$$CommunitySize = \sum Ww + \sum Wh \tag{1}$$

The intuition is that, if a community has a very small size and all its nodes are not used very much, it cannot explain the sense of a hashtag very well.

In the next step, the nodes are ranked by their weights, and the top 40 nodes from each community are marked and extracted as context words for each promising sense.

5   Mapping to Wikipedia Disambiguation List

5.1   Wikipedia Disambiguation List as Sense Inventory

A Wikipedia disambiguation list is a page which lists all the articles for an ambiguous concept on Wikipedia. We call an entry in a disambiguation list a Wikipedia sense. We note that, since disambiguation lists are edited manually, not all the articles of the same name are covered. Also, not all the senses are registered in Wikipedia. By mapping context words of a sense to Wikipedia Disambiguation list, we can filter poor (infrequent) senses and merge highly similar context words into one group. The difference between Wikipedia data and Twitter data lies in two respects. First Twitter users are more responsive to newly happened and popular events; it means twitter data are more updated. However Wikipedia data are not that responsive due to that they are edited by different users and the content of which are quite long. Second Tweets are written by casual words to express feelings of users or to share interesting things, while Wikipedia articles are written in a formal way to explain the concept in its background, history, development, usages and so on.   Because of these differences Wikipedia disambiguation lists as word sense inventory has disadvantages. However it does improve the precision of the result.

Senses in a disambiguation list page and senses used in Twitter do not match perfectly. There is no match on Wikipedia for the sense newly appeared on Twitter. Also, most of the senses on Wikipedia disambiguation lists are not used in tweets, for the reason that most of the senses on Wikipedia are not newly appeared or are not used quite often.

Also because of the irregular feature of hashtag, not all of the hashtags can be found on Wikipedia. So in this paper we only use the hashtag in the form of normal words, such as #rose.

5.2   Mapping to Disambiguation List

Each Wikipedia sense on a disambiguation list is linked to an article. We discover a mapping between hashtag context words and Wikipedia senses, by calculating the similarity between the context words and the articles on the disambiguation list. Table 1 shows the mapping matrix. In this matrix the highest scores are selected as the winning sense.

Table 1. Mapping Matrix for #orange

	Sense1	Sense2
Orange_(UK)	0.5587	0.4834
Ornge	0.2930	0.4503
Orange	0.3409	0.4538,
Orange_Free_State	0.4288	0.5191
Orange_Music_Electronic_Company	0.2679	0.4875
Orange_(fruit)	0.3573	3.1651
Orange,_California	0.1508	0.3563,
Netherlands_national_football_team	0.5587	0.5943
Orange_S.A.	0.1794	0.5943
Syracuse_Orange	0.2032	0.5943
Agent_Orange	0.3493	0.2260
Orange_(colour)	3.7612	1.7510
Orangery	0.0975	0.2234
Orange,_Virginia	0.4637	0.2735
Orange,_New_South_Wales	0.3054	0.2248
Orange,_New_Jersey	0.3273	0.5943

We outline the process in Algorithm 1 from step 6 to step 21, and describe it below:

1. Delete the articles that are too short and preprocess all the articles on the disambig-uation list. Calculate TF-IDF for each word of all the articles. We obtain a mapping matrix by calculating similarity score of any pair of context words group and Wikipedia article. The similarity score is calculated by formula (2).

$$Score = \sum W * TFIDF \qquad (2)$$

Where W is the weight of each context word calculated when construct the co-occurrence graph.

2. For each group of context words, we obtain a score vector. In the score vector the highest score is regarded as the best match. We expect that in this score vector only one score is extremely high and others distribute in a limited lower range. So we filter unwanted senses by formula (3)

$$DispersionRatio = \frac{SD - sd}{sd} \qquad (3)$$

SD is the standard deviation of score vector; sd is standard deviation of score vector without the maximum score. Standard deviation is used to measure the dispersion of a group of numbers.

3. After filtering, if two groups of context words match to the same Wikipedia article, merge them into one group of context words.
4. Repeat 2 and 3 until there is no change.

In the sense matrix the highest scores are selected as the winning sense.

6   Experiments and Discussion

We chose 11 hashtags, and we collected 5000 tweets for each hashtag on Dec. 15th, 2014. We conducted experiments to induce senses for each hashtag separately. A portion of the results are showing in Table 2.

Table 2. Example of Induced Senses

hashtag		Sense
#orange	1	Orange (fruit)
		Juice make, fruit, fresh, carrot, recipe, cake, ginger, chocolate, apple, food, healthy, breakfast, banana, lemon, salad, smoothie, brown, tea, lunch
	2	Orange (colour)
		love, new, red, sunset, beautiful, black, blue, pink, color, sky, yellow, green, sun, photo, sunrise, flower, morning, design, colour, home, top
#season	1	Season (the portion of the year when games are played)
		ready, tax, wait, football, nfl, next, baseball, sat, greendot, team, leave, play, free, work, music, amazing, finish, training, book, patriot
	2	Season(the portion of the year)
		winter, snow, cold, nature, fall, weather, photo, holiday, autumn, bring, rain, rainy, landscape, friend, photography, grey, ski, stay, car, white
	3	Season(a cycle or set of episodes of a television program)
		episode, video, game, watch, dead, recap, first, trailer, walk, premiere, throne, tonight, amc, return, cut, abbey, downton, spoiler, blacklist, world

6.1   Discussions

We found the following differences of tweets from natural language texts in other media:

- Language usage preference on Twitter. Context words around hashtags are different from those around a word in other media. In twitter people share things that are interesting, funny, beautiful and novel. Because things with these features could catch more eyes, they are more likely to be retweeted and to induce discussions. Also by looking at the first sense of #orange in Table 1, quite a large number of words related to sky and sunset appear together with #orange. It is highly possible that people want to share when they see a sunset with orange sky.
- The use of hashtags is influenced by popular events and time. Since our data were collected before Christmas, more and more people were talking about Christmas and gifts. So in the senses we induced, there appear a number of words related to Christmas. Accordingly, #christmas or Christmas shows a very high weight in all the graphs. For example, the second sense of #cup is related to Christmas and Christmas gifts. Cup is a common choice for people when they buy gifts for their friends or families.

- Because Wikimedia disambiguation lists are not a golden standard, we could only induce elementary senses of a hashtag. Also because the lengths of articles on Wikipedia are not the same, the short ones are with a relatively low probability to be mapped to.

6.2   Evaluation

We evaluate our method in a way of information retrieval. For each sense, use the context words extracted from the co-occurrence graph as keywords to fetch tweets. Tweet is ranked by the score that calculated by summing up all the weights of keywords existing in it. Then let human to judge whether the sense of the hashtag in the top K tweets matches the sense we induced.

This evaluation method is biased to the measure of precision. The precision for each sense is polarized. If the mapping is good, the precision could be very high; if the mapping is poor, it could be very low. For hashtag #orange, the precision at 100 of the two senses is 100%, which means that the senses of #orange has been induced correctly. If the precision is too low, it means that either the mapping to Wikipedia disambiguation list or the extracting context words is wrong.

Resolution is a key parameter to decide the granularity of communities detected from co-occurrence graph. They have positive correlation.   We could get the highest average precision when resolution is 2. But compared to the result with resolution set 1, the induced senses decreased 31.58%.   We think the best result could be obtained when resolution is set 1. Fig. 4 shows the precision change at different resolution settings. For the case that, resolution is set 1(R=1), we can get a precision of 81.66% when we set K as 10. Then the precision drops slowly as the K increases. We can get a precision of 74.56% when K is 100.

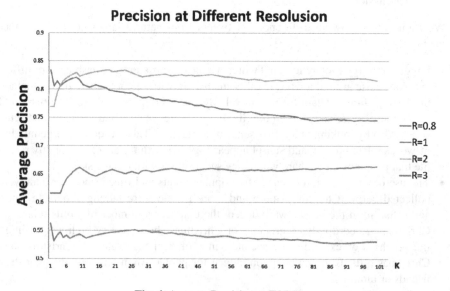

Fig. 4. Average Precision at TOP K

7   Conclusion and Future Work

This paper proposed a method to induce senses for hashtags through co-occurrence graphs. Senses of hashtags are induced from the communities detected from the co-occurrence graph. To the best of our knowledge we are the first to induce sense for hashtags and using tweets as data source. The most challenge we are facing is that there is no golden standard to evaluate the result. In order to solve this issue we designed a way of evaluation using information retrieval and human judgment.

In the future we will take sense granularity into consideration and induce hashtag senses at different granularity level. As we mentioned before, hashtags have temporal feature. In the future we try to utilize this feature and induce hashtag senses dynamically in real time. The newly emerging senses indicate that new events are happening or new things are coming into being.

References

1. Borovets, B., Van De Cruys, T.: Mining for meaning: The Extraction of Lexico-Semantic Knowledge from Text. Groningen Dissertations in Linguistics (2010)
2. Nasiruddin, M.: A state of the art of word sense induction: A way towards word sense disambiguation for under-resourced languages. In: TALN-RÉCITAL 2013, Les Sables d'Olonne, France, pp. 192–205 (2013)
3. Blondel, V.D., Guillaume, J.-L., Lambiotte, R., Lefebvre, E.: Fast unfolding of communities in large networks. J. Stat. Mech. Theory Exp. (10), 10008 (2008)
4. Véronis, J.: HyperLex: lexical cartography for information retrieval. Computer Speech & Language, pp. 223–252 (2004)
5. Klapaftis, I.P., Suresh, M.: Word sense induction & disambiguation using hierarchical random graphs. In: Proceedings of the 2010 Conference on EmpiricalMethods in Natural Language Processing, pp. 745–755 (2010)
6. Jurgens, D.: Word sense induction by community detection. In: Graph-based Methods for Natural Language Processing, pp. 24–28 (2011)

Hybrid-LSH for Spatio-Textual Similarity Queries

Mingdong Zhu[1], Derong Shen[1], Ling Liu[2], and Ge Yu[1]

[1] Northeastern University, China
mingdongzhu@hotmail.com, {shenderong,yuge}@ise.neu.edu.cn
[2] Georgia Institute of Technology, USA
lingliu@cc.gatech.edu

Abstract. Locality Sensitive Hashing (LSH) is a popular method for high dimensional indexing and search over large datasets. However, little efforts have put forward to utilizing LSH in mobile applications for processing spatio-textual similarity queries, such as find nearby shopping centers that have a top ranked hair salon. In this paper, we present hybrid-LSH, a new LSH method for indexing data objects according to both their spatial location and their keyword similarity. Our hybrid-LSH approach has two salient features: First our hybrid-LSH carefully combines the spatial location based LSH and textual similarity based LSH to ensure the correctness of the spatial and textual similarity based NN queries. Second, we present an adaptive query-processing model to address the fixed range problem of traditional LSH and to handle queries with varying ranges effectively. Extensive experiments conducted on both synthetic and real datasets validate the efficiency of our hybrid LSH method.

Keywords: similarity query, hybrid LSH, spatio-textual query.

1 Introduction

Location-based services have become more and more prevalent and have attracted significant attentions from both industry and academic community. Apple has an application to locate frequently used software; Yelp finds nearby restaurants of interest; and Facebook and FourSquare offers its members the capability to find his or her nearby (local) friends or points of interest.

One obvious way [1] to combine spatial distance and textual similarity is to use a spatial index to a spatial partition of the large dataset, which is most relevant to the spatial location of the query issuer, and for the given partition a string index is used to filter out those irrelevant objects by keywords matching and then rank the results using a hybrid ranking function that can combine spatial distance and textual similarity [2–4]. Spatio-textual data ordinarily are high-dimensional. For example, a typical microblog with location has 2 coordinates and some keywords, say 100, then the whole microblog, simultaneously considering location and text, contains $2\times100=200$ features(dimensions). Due to the curse of dimensionality,

© Springer International Publishing Switzerland 2015
R. Cheng et al. (Eds.): APWeb 2015, LNCS 9313, pp. 166–177, 2015.
DOI: 10.1007/978-3-319-25255-1_14

many traditional methods are not efficient. It's known that locality sensitive hashing(LSH) is a good method for similarity queries on high-dimensional data. However, none of previous work, to the best of our knowledge, has explored the feasibility of utilizing LSH for processing spatio-textual similarity queries. We argue that LSH is an attractive alternative method for processing spatio-textual similarity queries: First, LSH-based methods have no hierarchical structure, thus are easy to be maintained and scaled. Second, LSH can be directly used to hash spatially and textually similar objects to the same buckets, which can be obtained with less I/O cost.

In this paper, we design hybrid-LSH to process spatio-textual similarity queries, and the method treats spatial information and textual content of an object as a whole, rather than builds indices separately and combines two sets of query results when there is a query. The first challenge is how to design the hybrid-LSH such that hash buckets are conducted by considering spatial and textual similarity simultaneously. For each object, one hash value that reflects both spatial information and textual content considered as a whole, should be generated. The second challenge is how to make the hybrid LSH adaptive to spatio-textual similarity queries with different similarity ranges efficiently, because for LSH based methods their sensitive radii are fixed and it's difficult to answer queries with varying ranges.

To address these challenges, we propose a hybrid-LSH structure which considers both spatial and textual similarity, so that it is with high probability that spatially and textually similar objects are stored in the same bucket and can be found with one disk I/O. Then we present adaptive algorithms for queries with varying ranges. In addition, the hybrid-LSH's effectiveness and algorithms' accuracy are guaranteed by theoretical analysis. To summarize, we make the following contributions.

- By simultaneously considering spatial and textual similarity, we propose a hybrid-LSH and prove its theory features.
- To process query with varying ranges on the hybrid-LSH, we provide adaptive algorithms to process the query.
- We conduct extensive experiments on real and synthetic datasets in a distributed environment. Experimental results confirm the scalability and effectiveness of our approach.

2 Overview

Reference Model. Both the object model and the query model are defined with spatial location information and textual content consisting of keyword tokens. We assume that the spatial information and the textual content of objects are independent.

Formally, let P denote the universe of discourse, namely the set of spatial objects. Each object $p \in P$ is defined as a two-element tuple (loc, tok), where $p.loc$ is the spatial location information of object p and $p.tok$ is a set of tokens which represent the textual description of p. In order to compute spatio-textual similarity between two objects, say p_1 and p_2, we define a spatio-textual distance metric that combines spatial and textual similary through a weight parameter α, as shown in Eqn. 1.

$$\mathrm{DistST}(p_1, p_2) = \alpha * \mathrm{DistS}(p_1, p_2) + (1 - \alpha) * \mathrm{DistT}(p_1, p_2) \qquad (1)$$

$$\mathrm{DistS}(p_1, p_2) = \frac{\mathrm{dist}(p_1.loc, p_2.loc)}{dmax - dmin} \qquad (2)$$

$$\mathrm{DistT}(p_1, p_2) = 1 - \frac{(p_1.tok \cap p_2.tok)}{(p_1.tok \cup p_2.tok)} \qquad (3)$$

We use the normalized Euclidean distance of objects $p_1, p_2 \in P$, denoted as $\mathrm{dist}(p_1.loc, p_2.loc)$, to compute the spatial distance DistS, as shown in Eqn.2. $dmax$ and $dmin$ in Eqn.2 denote the maximum and minimum distance for pairs of objects in P. We use Jaccard distance [5] to measure the distance of textual similarity as shown in Eqn.3. Note that our hybrid LSH method is generic and independent of the specific distance functions used and thus can incorporate other spatial distance function and textual similarity distance function. For simplicity, B(q, D) denotes object set $\{o \in P | DistS(o.loc, q.loc) \leq D\}$, similarly, B($q, R$)=$\{o \in P | DistT(o.tok, q.tok) \leq R\}$, and B($q, D, R$)= $\{o \in P | \mathrm{DistS}(o, q) \leq D$ and DistT $(o, q) \leq R$ $\}$.

Given that spatial location based similarity is defined based on Euclidean distance and textual similarity is defined based on Jaccard distance, [6] and [7] describe the construction of an LSH family for Euclidean distance and the construction of an LSH family for Jaccard distance respectively.

Because the traditional (R, c)-NN problem [6,8] just adapts to objects with single data type. we extend the (R, c)-NN problem to (D, R, c)-NN for spatio-textual objects, which return an object $o \in$ B(q, cD, cR) if there exists an object $o^* \in$ B(q, D, R).

Related Work. There are many studies on spatial textual similarity query processing [2,3,9]. A good survey of techniques can be found in [10]. Generally they can be classified into two categories: tree-like style and grid style. Specifically for tree-like style, [2] proposes a new hybrid index structure Inverted File Quad-tree (IQ-tree) to manage a stream of Boolean range continuous(BRC) queries on a stream of incoming. [11] proposes a new indexing framework for processing the location-aware text retrieval query. [3] proposes a hybrid indexing structures called Intersection-Union-R tree (IUR-tree) and an efficient approach that take into account the fusion of location proximity and document similarity. For grid style, this category of indices combines a grid index with a text index (e.g., the inverted file). For example, [4] proposes a spatio-textual similarity search method (SEAL), which is a filter-and-verification framework.

3   Hybrid-LSH

In this section, we introduce the construction of hybrid-LSH. Concretely, each object $o \in P$ consists of spatial information s and textual content t. Assume that the following three parameters are provided: the spatial range d, the textual similarity range r, and the approximation factor $c > 1$. The hybrid LSH construction algorithm works as follows: For s we can find (d, cd, sp_1, sp_2)-sensitive LSH family, denoted by sH, for Euclidean distance (recall Section 2), while for t we can obtain (r, cr, tp_1, tp_2)-sensitive LSH family, denoted by tH, for Jaccard distance (recall Section 2). We combine the two hash families. In particular, let k_1 and k_2 denote the number of hash functions generated in sH and tH respectively. We define an LSH function family G= g: $S \to U^{(k_1+k_2)}$ with $k_1 + k_2$ hash functions such that g(o)={sh$_1$(o),$\cdots$,sh$_{k_1}$(o), th$_1$(o), $\cdots$, th$_{k_2}$(o)}, where th$_i$ $\in$tH and sh$_i$ $\in$sH. Let L be an integer, we choose L functions g$_1$,$\cdots$, g$_L$ from G independently and uniformly at random. During preprocessing, we store an indicator of object o in the bucket g$_i$(o) for i=1,$\cdots$, L. We define the spatial component as g$_s$(o), i.e., g$_s$(o)={sh$_1$(o),$\cdots$,sh$_{k_1}$(o)} and the textual component as g$_t$(o), i.e., g$_t$(o)={th$_1$(o),$\cdots$, th$_{k_2}$(o)}. Fig.1 is an illustration of hybrid-LSH.

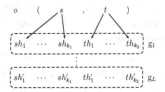

Fig. 1. An illustration of hybrid-LSH

Intuitively, hybrid-LSH is composed by k_1 hash values from (d, cd, sp_1, sp_2)-sensitive LSH and k_2 hash values from (r, cr, tp_1, tp_2)-sensitive LSH. Thus, hybrid-LSH is defined as $(d, r, c, \{sp_1, tp_1\}, \{sp_2, tp_2\})$-sensitive hybrid-LSH. Because sp_1, sp_2, tp_1 and tp_2 are determined by d, r and c, $(d, r, c, \{sp_1, tp_1\}, \{sp_2, tp_2\})$-sensitive is simplified as (d, r, c)-sensitive hybrid-LSH when no confusion occurs.

Base on hybrid-LSH, in order to process a (D, R, c)-NN query with query object q, simplified as query q, we first generate L hash value of q, i.e., g$_1$(q), $\cdots$, g$_L$(q), then search corresponding buckets of the hash values and randomly check $C * L$ objects in the buckets where C is a constant number. Let $o_1, \cdots, o_{C*L}$ be the checked objects. For any o_i, if DistS(o_i, q) $< cD$ and DistT(o_i, q) $< cR$, then we return YES and o_i, otherwise we return NO. Because there are probably lots of objects in all the buckets, it saves lots of computation to only check constant number of objects.

To ensure the correctness of the algorithm, the parameters k_1, k_2 and L are chosen so as to ensure the following properties hold with constant probability:

P_1 If there exists object o such that $\mathrm{DistS}(o,q) < D$ and $\mathrm{DistT}\ (o,q) < R$, denoted as $\mathrm{B}(q,D,R)$, then $g_i(o)=g_i(q)$ for some $i = 1,\cdots,L$.

P_2 The total number of collisions of q with the number of objects, which do not belong to $\mathrm{B}(q,cD,cR)$, is less than $3L$.

P_1 ensures objects who satisfy the query at least collide once for the L hash values. P_2 ensures if there is object who satisfies the query algorithm can find the object after checking $3L$ objects.

Theorem 1. *setting* $k_1 = \log_{sp_2}(1 - \sqrt{1 - \frac{1}{n}})$, $k_2 = \log_{tp_2}(1 - \sqrt{1 - \frac{1}{n}})$ *and*
$$L = (1 - \sqrt{1 - \frac{1}{n}})^{-(\log_{sp_2} sp_1 + \log_{tp_2} tp_1)}\ \text{guarantees that properties }P_1 \text{ and } P_2$$
hold with constant probability.

Proof. Let P_1 hold with probability p_1 and P_2 hold with p_2. Without loss of generality, assume that there is an object o^* which satisfies $\mathrm{DistS}(o^*.loc, q.loc) < D$ and $\mathrm{DistT}(o^*.tok, q.tok) < R$. Set $k_1 = \log_{sp_2}(1-\sqrt{1 - \frac{1}{n}})$ and $k_2 = \log_{tp_2}(1-\sqrt{1 - \frac{1}{n}})$. Use $\mathrm{P(A)}$ to denote the probability of event A. $\mathrm{P}(g(o')=g(q)$ & $o' \in (P- \mathrm{B}(q,cD) \cap \mathrm{B}(q,cR))$ (denoted as Pa) is not larger than $\mathrm{P}\ (g_s(o')=g_s(q)$ & $o' \in (P - \mathrm{B}(q,cD))\ |\ g_t(o') = g_t(q)$ & $o' \in (P - \mathrm{B}(q,cR))$ (denoted as Pb). $Pb = 1 - (1 - sp_2^{k_1})(1 - tp_2^{k_2}) = \frac{1}{n}$. so $Pa < \frac{1}{n}$, and then the expected number of objects allocated for g_i which don't satisfy the query condition is less than 1.5. The expected number of the objects for all g_i doesn't exceed $1.5L$, According to the Markov inequality, the probability that this number exceed $3L$ is less than $\frac{1}{2}$. so P_2 follows. The probability of $g(o^*)=g(q)$ is $sp_1{}^{k_1} * tp_1{}^{k_2}=(1 - \sqrt{1 - \frac{1}{n}})^{\log_{sp_2} sp_1 + \log_{tp_2} tp_1}$. By setting $L = (1 - \sqrt{1 - \frac{1}{n}})^{-(\log_{tp_2} tp_1 + \log_{sp_2} sp_1)}$, the probability that all $g_i(o^*) \neq g_i(q)$ is less than $\frac{1}{e}$. so the P_1 holds with the probability $1 - \frac{1}{e} > \frac{1}{2}$. □

4   Adaptive (D, R, c)-NN Query Processing with Hybrid-LSH

This section focuses on describing how to split a (D, R, c)-NN query into multiple subqueries with smaller query ranges such that each subquery can be processed directly using our hybrid LSH index. Recall Section 3, we show the structure of (d, r, c)-sensitive hybrid-LSH. For (D, R, c)-NN queries where $D \geq d$ and $R \geq r$, we need to decompose each (D, R, c)-NN query into multiple subqueries with spatial range d and textual range r.

4.1   Adaptive (D, R, c)-NN Query Processing

Let $d < D$ and $r < R$, and we build a hybrid-LSH with sufficiently small d and r so that ranges from query are bigger than them. The intent of the adaptive (D, R, c)-NN query method is to transform an original query into several queries with small query range which can be processed by the constructed hybrid-LSH. Then we give a formal definition of the transform as follows.

Definition 1. (*LSH query transform*) *LSH query transform is based on a single type* (r_1, r_2, p_1, p_2)-*sensitive LSH. If a query* (R, c)-*NN can be transformed into a query set S which contains several queries based on* (r_1, r_2, p_1, p_2)-*sensitive LSH, where* $r_1 < R$, $r_2 < cR$, *and S satisfies the following two properties, then S is a* (r_1, r_2, p_1, p_2)-*sensitive LSH query transform of q.*

P1 *the area or content which is covered by R in q is contained by the area or content which is covered by* r_1 *in S.*

P2 *the area or content which is covered by* r_2 *in S is contained by the area or content which is covered by cR in q.*

Then we show the effect of LSH query transform in Lemma 1, compared with the original query.

Lemma 1. *If there is an answer for a* (R, c)-*NN query, then a* (r_1, r_2, p_1, p_2)-*sensitive LSH query transform of the* (R, c)-*NN query can return a c-approximate answer with constant probability.*

Proof. Set QT be a LSH query transformation of a (R, c)-NN query q, where $QT = \{qt_i | 0 < i < |QT|\}$. If there is an object $o^* \in B(q, R)$, according to P1 in definition 1, then o^* locate in the query range of at least one query in QT, say qt_i. In addition, the objects which are outside of $B(q, cR)$, denoted as O, are outside of $B(qt_i, r)$. According to property 2 of LSH (R, c)-NN query in the [8], the number of collisions of st_i and O is less than $C * L$, So after checking $C * L$ objects st_i can return a c-approximate answer with constant probability. Then Lemma 1 follows. □

According to Theorem 1 we can use a LSH query transform to processing queries with varying ranges. However it is nontrivial to construct LSH query transform for spatial information or textual content. Now we present the LSH query transform methods for spatial information and textual content respectively.

For spatial information, it is similar with a disk covering problem to find a LSH query transform. To the best of our knowledge, optimal solutions of disk covering problems are only available for limited situations. We try to give a general algorithm and show it is a 3-approximate optimal method. Given a (d, cd, p_1, p_2)-sensitive LSH and a (D, c)-NN query with centre coordinate $p = (p_x, p_y)$, where $d < D$, we use several squares with sides $\sqrt{2}d$ to cover the area of $B(p, D)$, as shown in figure 2. The number of squares is $\lceil \frac{2D^2}{d^2} \rceil$. Then the (D, c)-NN queries from central points of the squares is a LSH query transformation, denoted as $ST = \{st_i\}$. Note that because $d < D$, ST is easy to satisfy the condition that the coverage of all queries in ST with radii cd is contained by the circle with radius cD of query q. According to the area formula, the low bound of optimal method is $\lceil \frac{D^2}{d^2} \rceil$. So it is easy to get an corollary in the following.

Corollary 1. *There is a constant c which makes ST a LSH query transformation of query q, and the LSH query transformation is a 3-approximate optimal method.*

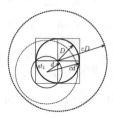

Fig. 2. LSH query transform for spatial data

For textual content, given a (r, cr, p_1, p_2)-sensitive min-hash and a textual (R, c)-NN query q, where $R > r$ and there are TL tokens in q, in order to find similar token sets by (r, cr, p_1, p_2)-sensitive LSH, we should consider two kinds of objects: objects in which the number of tokens are smaller than q and objects in which the number of tokens are bigger than q. For smaller objects, we should get rid of some tokens in q to match them. For an integer m specified later, we generate a query set which consists of all possible combinations of $(TL - m)$ tokens from tokens of q, which is denoted as DS. And for bigger objects, we should add some wildcard tokens in q to mach them. For an integer w specified later, we add w wildcard tokens to q. When q is hashed to a hash value, a wildcard token is hashed to all possible hash values. In this way we get another query set AS. Then by combining DS and AS we get a query set TT.

Let IN and UN be the size of intersection and union of two token sets, respectively. When parameters m and w satisfy the formulae 4, 5 and 6, it ensures that all similar objects of q are covered by the similar objects of query set TT, corresponding to P1 in definition 1. And when parameters m and w satisfy the formula 7, 8 and 9, it ensures that all c approximate similar objects of query set TT are covered by c approximate similar objects of q, corresponding to P2 in definition 1. By selecting the smallest m and w which satisfy the formula, corollary 2 follows.

$$\frac{IN}{UN} \geq 1 - R. \qquad (4)$$

$$\frac{IN}{UN - m} \geq 1 - r \qquad (5)$$

$$\frac{IN + w}{UN + w} \geq 1 - r \qquad (6)$$

$$\frac{IN}{UN} \leq 1 - cR \qquad (7)$$

$$\frac{IN}{UN - m} \leq 1 - cr \qquad (8)$$

$$\frac{IN + w}{UN + w} \leq 1 - cr \qquad (9)$$

Corollary 2. *when* $m = \lceil \frac{R-r}{(1-R)(1-r)}TL \rceil$ *and* $w = \lceil \frac{R-r}{(1-R)r}TL \rceil$, *where TL is number of tokens in the query q, query set TT is a LSH query transformation of query q.*

Proof. Due to the limited space, the proof is omitted.   □

Based on LSH query transform, the adaptive (D, R, c)-NN query algorithm is straightforward. The idea of the algorithm is to decompose a query for hybrid type data to queries for single type data, then generate LSH query transforms of the queries, join the LSH query transforms of two types, and lastly process the joined queries on hybrid-LSH. Specifically, for a (D, R, c)-NN query, it can be seen as a join of two single type queries (D, c)-NN and (R, c)-NN. First We find a (d, cd, sp_1, sp_2)-sensitive LSH query transform DT of (D, c)-NN and a (r, cr, tp_1, tp_2)-sensitive LSH query transform RT of (R, c)-NN query. By combining the set DT and RT in Cartesian product way, we generate a (d, r, c)-sensitive hybrid-LSH query transform, denoted as DRT. At last we process the DRT in the hybrid-LSH and return the query result.

Theorem 2. *If there is an answer for a (D, R, c)-NN query, the adaptive (D, R, c)-NN query algorithm can return a c-approximate answer with constant probability.*

Proof. The proof is straightfoward.   □

4.2 Multiple Adaptive Hybrid-LSHs

Shown in Theorem 1, each query in LSH query transformation at most checks $3L$ objects, so the number of buckets accessed by each query is directly proportional to the number of queries. Hence, the number of queries in hybrid-LSH query transformation is a direct indicator of query cost. The number of queries in hybrid-LSH query transformation is:

$$QN \leq (\frac{2D^2}{d^2} + 1)(C_{TL}^m + H^w) \tag{10}$$

According to Formula 10, when the query range is big, the query cost is very high, so we propose the multiple adaptive hybrid-LSHs.

The intent of the multiple adaptive hybrid-LSHs is to build many adaptive hybrid-LSHs to make the distance between queries and the closet adaptive hybrid-LSH small, which can significantly reduce the cost of query.

Shown in Formula 10, QN is in direct proportion to $\frac{D}{d}$ (spatially) and $(R-r)$ (textually). For (D, R, c)-NN queries where $MinD \leq D \leq MaxD$ and $MinR \leq R \leq MaxR$, we build $\{d_i, r_j, c\}$-sensitive adaptive hybrid-LSHs, where i, j are integers, $i \in (0, \log_b(\frac{MaxD}{MinD}))$, $j \in (0, \frac{MaxR-MinR}{t})$, $d_1 = MinD$, $\frac{d_{j+1}}{d_j} = b$, $r_1 = MinR$, $r_{i+1} - r_i = t$, and b, t are the common ratio and difference respectively.

174 M. Zhu et al.

Corollary 3. *Multiple adaptive hybrid-LSHs uses* $\log_b(\frac{MaxD}{MinD}) * \frac{MaxR-MinR}{t}$ *adaptive hybrid LSHs to process any* (D, R, c)-NN *query, where* $MinD < D < MaxD$ *and* $MinR < R <= MaxR$, *by hybrid-LSH query transformation with at most* $(2b^2+1)(C_{TL}^m + H^w)$, *where* $m = \frac{t}{(1-MaxR)^2}TL$, $w = \frac{t}{(1-MinR)(MinR-t)}TL$.

Proof. Set multiple adaptive hybrid-LSH $MAH = \{mah_{ij} \mid mah_{ij}$ is (d_i, r_j, c)-sensitive adaptive hybrid-LSH, $0 < i < \log_b \frac{MaxR}{MinR}$, $0 < j < \frac{MaxD-MinD}{t}$ $\}$. There is a query (qd, qr, c)-NN and $mah_{k,s}$ where $k = argmin_{(0<i<\log_b \frac{MaxR}{MinR}, qd>d_i)} \frac{qd}{d_i}$ and $s = argmin_{(0<i<\frac{MaxD-MinD}{t}, qr>r_j)}(qr - r_j)$. For the spatial part, $\frac{qd}{d_k} <= b$, then $(2\frac{D^2}{d^2} + 1) \leq 2b^2 + 1$. For the textual part, $qr - r_s \leq t$ and $qr > r_s$, so $\frac{qr-r_s}{(1-qr)(1-r_s)} < \frac{t}{(1-MaxR)^2}$, which is monotonically increasing function for qr, and $\frac{qr-r_s}{(1-qr)r_s} < \frac{t}{(1-MinR)(MinR-t)}$, which is monotonically decreasing function for qr. So $QN \leq (2b^2+1)*(C_{TL}^m + H^w)$ where $m = \frac{t}{(1-MaxR)^2}TL$, $w = \frac{t}{(1-MinR)(MinR-t)}TL$. The Corollary follows. $\square$

5 Experiments

Setup. In order to show scalability and maintainability, we built the adaptive hybrid-LSH (HLSH), multiple adaptive hybrid-LSH(MHLSH), and implemented the proposed (D, R, c)-NN in a distributed setting. For comparison, we also implemented state-of-the-art method SEAL [4] and LSHDSS [12]. SEAL uses hash based hybrid signature to process query, belongs to filter-and-verification framework and is an exact method, so for (D, R, c)-NN queries we stop the algorithm when one object which satisfies query condition is obtained. LSHDSS is a LSH based distributed similarity search algorithm, however it is designed for single data type, so we extended it to support Spatial-textual similarity query by executing separately and combining intermediate result. In this paper we mainly focus on high-dimensional spatio-textual data, so we compared with methods with little global information which can be easy maintained in a distributed setting when there are lots of updates, and due to expensive maintenance cost, methods with tree structures or hierarchical structures were not considered in our experiments.

Two datasets are used, The first one is a real dataset, which contains 0.5 million micro-blogs with location information collected from Sina microblog website, denoted as MicroBlog. According to the rule of microblog, each message must contain less than 140 characters and for MicroBlog, the longest meaningful spatial distance is 30km. After deleting the stop words, the average number of words of a message is 24, and in our experiments each word is taken as a token. The other dataset is a synthetic dataset, denoted as SynSet. SynSet has 1 million objects and objects' tokens are chosen from a word set, and each object's location is generated from a square area of which the side length is 100km.

All experiments are implemented in Cassandra v1.2.6. The Cassandra cluster consists of 10 machines with the same configuration: Intel(R) Core(TM) i7 Quad 870 @2.93GHz CPU and 8 GB RAM, and the same operating system: Ubuntu

10.04. To evaluate the performance of algorithms, we vary one parameter while keeping the others fixed at their default values. We run 20 times for each test, the average result of the query set is reported in the experiments.

Two performance measures are used: number of messages and accuracy. We count the number of messages from sending one query to obtaining result. The number of messages is an important indication of algorithm efficiency. Query accuracy [6] is used to measure the quality of the results. Specifically, o^* is the actual result and o are the returned results, query accuracy is $\frac{\text{DistST}_{(q,o)}}{\text{DistST}_{(q,o^*)}}$.

Effect of c. We first investigate the effect of the approximate ratio c on space consumption and query accuracy. Table 1 shows the performance of adaptive hybrid-LSH when $c = 2$ or $c = 3$. Because L is in direct proportion to dataset size, and index size is in direct proportion to L, the index size of MicroBlog is smaller than SynSet. As shown in table 1, the query accuracy of case $c = 3$ is a bit bigger than the case that of case $c = 2$, but it is still good, while the number of messages of case $c = 3$ is about $\frac{1}{3}$ of that of case $c = 2$. So it is worth trading a little accuracy for much higher query efficiency.

Table 1. Effect of c

HLSH	$c = 2$			$c = 3$		
	messages	accuracy	index size	messages	accuracy	index size
MicroBlog	608	1.52	1.18G	196	1.64	229M
SynSet	714	1.63	2.1G	216	1.72	575M

Performance of (D, R, c)-NN Query. We now show the results obtained from the (D, R, c)-NN query, Figures 3 show the number of messages and accuracy for $(1, 0.02)$-NN, $(1.5, 0.04)$-NN, $(2, 0.06)$-NN, $(2.5, 0.08)$-NN and $(3, 0.1)$-NN queries on MicroBlog and SynSet. Due to the limited space, in figures $(1, 0.02)$-NN, $(1.5, 0.04)$-NN, $(2, 0.06)$-NN, $(2.5, 0.08)$-NN and $(3, 0.1)$-NN are denoted as NN1, NN2, NN3, NN4 and NN5 respectively. As the distance and range increase, the number of messages cost of HLSH and SEAL increase notably at first and then decrease. When the distance and range increase for HLSH, some (D, R, c)-NN queries LSH-transforms are generated, which result in the increasing of the number of messages, and for SEAL, the efficiency of filtering of the algorithm degrade and it needs check more inverted lists and candidates. However, from $(2.5, 0.08)$ the objects which satisfy the queries increase, in turn the algorithms can terminate earlier and the number of disk accesses decreases. In contrast, the number of messages of MHLSH is stable. The number of messages cost of MHLSH is lower than that of SEAL, which is further lower than LSHDSS. From Figures 3(b) and 3(d), the average accuracy of HLSH is best, and the accuracy of MHLSH is better than that of SEAL.

Figures 4(a) and 4(b) illustrate the trends of number of messages and accuracy of the methods in terms of data size on SynSet. As the dataset size increases, the number of messages of SEAL increases fast, as well as that of LSHDSS. The

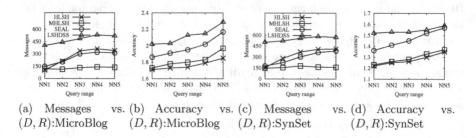

(a) Messages vs. (D, R):MicroBlog (b) Accuracy vs. (D, R):MicroBlog (c) Messages vs. (D, R):SynSet (d) Accuracy vs. (D, R):SynSet

Fig. 3. (D, R, c)-NN with varying (D, R)(Figures share the same key.)

reason is that, the number of objects encountered in an inverted list for SEAL is linearly proportional to the dataset size. As a result of two phases of similarity query for LSHDSS, the algorithm is sensitive to the data size. Due to the P_2 of Theorem 1, HLSH and MHLSH are relatively stable with the data size varying. The number of messages cost of MHLSH is still the lowest. Figure 4(b) shows that the accuracies of HLSH and MHLSH are better than those of SEAL and LSHDSS, and HLSH is slightly better than MHLSH at the cost of much more messages.

Figures 4(c) and 4(d) show the performance of (D, R, c)-NN query with different average number of tokens on SynSet. Obviously the number of tokens is similar to the dimensionality in traditional databases. Generally LSH methods are not sensitive to dimensionality, so it's easy to explain that MHLSH and LSHDSS are relatively not affected by the number of tokens. For HLSH, the number of LSH transform queries increases when the number of tokens increases, in result the number of messages of HLSH increases fast. For a fixed query distance and query range, SEAL is in direct proportion to the number of tokens. That's why the number of messages for SEAL increases as the number of tokens increases. Shown in Figure 4(d), the accuracy of HLSH is still the best and the accuracy of MHLSH is better than that of SEAL and that of LSHDSS.

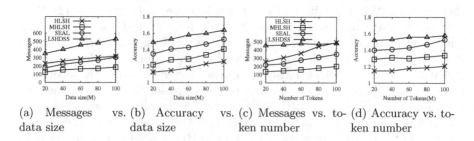

(a) Messages vs. data size (b) Accuracy vs. data size (c) Messages vs. token number (d) Accuracy vs. token number

Fig. 4. (D, R, c)-NN on SynSet(Figures share the same key.)

6   Conclusion

In this paper, we propose a hybrid-LSH scheme for the spatio-textual similarity query. We devise efficient adaptive (D, R, c)-NN algorithm and approximate k-NN method based on the hybrid-LSH scheme. Our theoretical studies show that the algorithms can have a guarantee on query quality. Results of empirical studies demonstrate that the paper's proposal offers scalability and efficiency.

Acknowledgments. This work is supported by the National Basic Research 973 Program of China under Grant (2012CB316201) and the National Natural Science Foundation of China under Grant (61472070). The first author is funded by China Scholarship Council. The third author is partially supported by the grants from NSF NetSE, SaTC, I/UCRC and Intel ICST on Cloud Computing.

References

1. Li, Z., Lee, K.C.K., Zheng, B., Lee, W.-C., Lee, D., Wang, X.: Ir-tree: An efficient index for geographic document search. IEEE Trans. on Knowl. and Data Eng. 23(4), 585–599 (2011)
2. Chen, L., Cong, G., Cao, X.: An efficient query indexing mechanism for filtering geo-textual data. In: Proceedings of the 2013 International Conference on Management of Data, pp. 749–760. ACM (2013)
3. Lu, J., Lu, Y., Cong, G.: Reverse spatial and textual k nearest neighbor search. In: Proceedings of the 2011 ACM SIGMOD International Conference on Management of Data, pp. 349–360. ACM (2011)
4. Fan, J., Li, G., Zhou, L., Chen, S., Hu, J.: Seal: Spatio-textual similarity search. Proceedings of the VLDB Endowment 5(9), 824–835 (2012)
5. Levandowsky, M., Winter, D.: Distance between sets. Nature 234(5323), 34–35 (1971)
6. Gan, J., Feng, J., Fang, Q., Ng, W.: Locality-sensitive hashing scheme based on dynamic collision counting. In: Proceedings of the 2012 ACM SIGMOD International Conference on Management of Data, pp. 541–552. ACM (2012)
7. Broder, A.Z., Glassman, S.C., Manasse, M.S., Zweig, G.: Syntactic clustering of the web. Computer Networks and ISDN Systems 29(8), 1157–1166 (1997)
8. Datar, M., Immorlica, N., Indyk, P., Mirrokni, V.S.: Locality-sensitive hashing scheme based on p-stable distributions. In: Proceedings of the 20th Annual Symposium on Computational geometry, pp. 253–262. ACM (2004)
9. Cao, X., Chen, L., Cong, G., Xiao, X.: Keyword-aware optimal route search. Proceedings of the VLDB Endowment 5(11), 1136–1147 (2012)
10. Chen, L., Cong, G., Jensen, C.S., Wu, D.: Spatial keyword query processing: an experimental evaluation. In: Proceedings of the 39th International Conference on VLDB, pp. 217–228. VLDB Endowment (2013)
11. Cong, G., Jensen, C.S., Wu, D.: Efficient retrieval of the top-k most relevant spatial web objects. Proceedings of the VLDB Endowment 2(1), 337–348 (2009)
12. Haghani, P., Michel, S., Aberer, K.: Distributed similarity search in high dimensions using locality sensitive hashing. In: Proceedings of the 12th International Conference on EDBT, pp. 744–755. ACM (2009)

Sleep Quality Evaluation
of Active Microblog Users

Kai Wu, Jun Ma, Zhumin Chen, and Pengjie Ren

School of Computer Science & Technology,
Shandong University, Jinan, 250101, China
kaiwusdu@gmail.com,
{majun,chenzhumin}@sdu.edu.cn,
andyren2012@hotmail.com

Abstract. In this paper, we propose a novel method to evaluate the sleep quality of Active Microblog Users(AMUs) based on Sina Microblog data, where Sina Microblog is the largest microblog platform with 500 million registered users in China. A microblog user is called AMU if s/he posts more than 100 microblogs during a year. Our study is meaningful because the amount of AMUs is huge in China and the results can reflect the lifestyle of these people. The primary works of this paper are as follows: First we successfully obtained 700 million microblogs from 0.55 million microblog users as our dataset. Then we detected the possible start and end sleep time of each AMU by a novel pattern and algorithm. Finally we designed an evaluation system to give the score of each AMU's sleep quality. In the experiment, we compared the sleep quality of AMUs in different cities of China and found the difference in topics between high and low score groups by LDA method.

Keywords: data mining, sleep quality, microblogging, social network.

1 Introduction

With the increasing popularity of social network sites, social network services have become information providers on a web scale, providing researchers a new opportunity on content mining. Microblog platforms are very popular in social networks, and recent progress of Web 2.0 applications has witnessed the rapid development of microblog in China (i.e., Sina Microblog[1]), which has already been one of the most important ways for people's online communications, especially on sharing information[2]. Sina Microblog provides similar services as other famous microblog platforms, such as Twitter[3]. Since its launch in August 2009, Sina Microblog has grown into the biggest Chinese microblog with 500 million registered users by the end of 2012[4]. Therefore, we choose Sina Microblog as our data resource. Although microblogging is increasingly popular, methods for organizing and providing access to microblog data are still in the early stages. In this paper, we propose a novel method of sleep quality evaluation by processing and mining large scale data on Sina Microblog. To the best

© Springer International Publishing Switzerland 2015
R. Cheng et al. (Eds.): APWeb 2015, LNCS 9313, pp. 178–189, 2015.
DOI: 10.1007/978-3-319-25255-1_15

of our knowledge, this is the first study that makes an effort to evaluate sleep quality totally based on large scale online data. Because of the characteristics of our method, only Active Microblog Users(e.g., the users who posts more than 100 microblogs per year) are useful in our research, and for simpleness, "AMU" means "Active Microblog User" in the rest of our paper.

It is obvious that surveying people's sleep quality is quite important in many interdisciplinary fields from academic interests to practical decision makings of government, enterprise and individual. However, in China, there are no related research about sleep quality until 2013. In 2013 and 2014, CHINESE MEDICAL DOCTOR ASSOCIATION, known as CMDA, released two reports[5,6] about sleep quality of Chinese people. This project was supported by a furniture company and was actualized by a Research Advisory Group. With many people involved and two months' hard work, they interviewed 8286 people face to face or online. Clearly, the number of samples is limited. In contrast with CMDA's method, our method by utilizing the large scale data of social network sites has many natural advantages. 1)The people whom we can survey are limitless in some degree. 2)Many features of people can be chosen to carry out our investigation, which includes living areas, occupations, sex, age and so on. 3)We can get most of the information on social network sites freely. However, limited to the usage of social network are more likely to be young people, the age structure tend to be younger in our survey. According to a survey about Sina Microblog, 90% of Sina Microblog users are people from age 18 to 35[7](about 80% in our survey). Therefore, the discovery in our research are more suitable to describe this group in China.

In our paper, we certainly could not get the accurate sleep time of a user, because people won't fall asleep immediately after posting the last microblog. However, according to the study of CMDA, 58.9% Chinese people have the habit of playing phone before going to sleep, and among them 60% people like to post microblogs using their mobile phones. That is to say, there is a great chance for AMUs to post microblogs before they go to sleep. Besides, normal AMUs usually post a lot of microblogs in their working or leisure time, but few microblogs in sleep time. Then by mining user microblog data, we can get the user's approximative sleep period and then evaluate it. Experiment will show that the sleep time we get is similar to CMDA's result.

The rest of this paper is organized as follows: In section 2, we introduce some related work on social network platform. In section 3, we explain the detail of our sleep quality evaluation method over Sina Microblog data. Section 4 illustrates our experimental results conducted with a real dataset of Sina Microblog. Section 5 concludes our work and makes a brief description of future work.

2   Related Work

Social network is becoming an important research topic in various fields. Early works by Java et al.[8], Zhao et al.[9] and Krishnamurthy et al.[10] examined the usage and role of Twitter in creating a social community on the basis of its basic functions. In these studies, Twitter was investigated for its social networking role, that is, how it would be used to send massive amounts of short messages about social activities.

Later, people gradually started to pay attention to the large amount of user data on social networks. Thanks to the advent of social networks, researchers can obtain crowd life logs and observe their lifestyles easily. Ren et al.[11] utilized the users historical data to discover meaningful tweets given a users interests. Fujisaka et al.[12] and Lee et al.[13,14] conducted detecting unusual social events such as festivals and disasters by turning up anomaly crowd behavior. Wakamiya et al.[15] proposed a method to find crowd viewing TV programs from people's tweets and rated the TV programs. Another of their work[16] used geolocation-based tweets to extract urban area characterization. Sue Jamison-Powell et al.[17] discussed insomnia and sleep disorders using microblogging social media service Twitter.

3   Evaluation of Sleep Quality over Sina Microblog

In this section, we introduce the detail of our evaluation method as follows: First we demonstrate how to model the users life patterns over their microblogs in section 3.1. Then we introduce how to extract the sleep pattern from user life pattern, and method of evaluating it in section 3.2.

3.1   Modeling User Life Pattern

In this paper, we proposed an explorative method to evaluate user sleep quality based on large scale online data. Specifically, China only has a single time zone, therefore time zone is not considered in our method. In addition, we assume that most people don't travel to other countries too much in a year and this assumption is true in most conditions for ordinary Chinese people. Our main idea is to detect users' sleep time from their microblogs. First we define the time period p_i to be the time period from i o'clock to $(i + 1)$ o'clock, $0 \leq i < 24$. Then we collect the total microblogs which a user posts in period i of every day in a long time(e.g., a year in this paper). After that, we define Microblog Time Distribution as follows:

$M = (m_0, m_1, m_2...m_{23})$

where m_i is the total number of microblogs which a user posted a year in period i. Then we define a user's life pattern as follows:

$L_p = ((p_0, r_0), (p_1, r_1), (p_2, r_2)...(p_{23}, r_{23}))$

where p_i is the percentage of microblogs which are published in period i, and r_i represents the change rate of current period compared with previous one. Their calculation can be got in formula (1) and (2).

$$p_i = \frac{m_i}{\sum_{i=0}^{23} m_i} \tag{1}$$

$$r_i = \frac{m_i - m_{ib}}{m_{ib} + P(x)} \tag{2}$$

where ib is the period before i (i.e.,$ib = (i+23)\%24$), $x = m_i - m_{ib}$. The function of $P(x)$ is defined as follows:

$$P(x) = \alpha \times e^{-\frac{x^2}{2 \times \beta^2}} + 1 \tag{3}$$

The parameter α, β in formula (3) are both empirically set 5 in this paper. $P(x)$ is a smooth function, we use it to avoid the situation that the denominator is zero. Also, it can help avoid the case that the change rate r is too big, when m_i, m_{ib} are both small and their difference x is close to the bigger one. For example, when "$m_i = 31, m_{ib} = 1$", the result of r_i is 3.48 with $P(x)$, compared with 30 without $P(x)$. In another condition of "$m_i = 130, m_{ib} = 100$", m_i, m_{ib} are larger but the difference between them is 30 the same as before. The result of r_i won't differ too much, 0.28 with $P(x)$, and 0.30 without $P(x)$.

Different users have different sleep habits, some users sleep early at night, others may go to bed rather late, and some users even sleep in the daytime. Although users won't always get up in the same time everyday, a user with regular sleep habit will get up in the same period in most days of a year. Therefore, there is some statistical regularity in most users' microblog data. Then we can find the start and end period of a user's sleep base on the vectors M and L_p. Table 1 shows an example of a Sina Microblog user data. The user's Microblog Time Distribution $M(m_i)$ can be obtained from our database. Then we get user life pattern $L_p(p_i, r_i)$ from M using the formulas above.

Table 1. Time Distribution M and Life Pattern L_p of a Microblog User.

Time Period(i)	0	1	2	3	4	5	6	7
$M(m_i)$	2	0	0	0	0	1	0	0
$L_p(p_i, r_i)$	0.009,-0.76	0,-0.26	0,0	0,0	0,0	0.004,0.17	0,-0.15	0,0
Time Period(i)	8	9	10	11	12	13	14	15
$M(m_i)$	5	10	18	22	22	17	13	4
$L_p(p_i, r_i)$	0.02,1.24	0.04,0.55	0.08,0.6	0.1,0.18	0.1,0	0.08,-0.19	0.06,-0.19	0.02,-0.6
Time Period(i)	16	17	18	19	20	21	22	23
$M(m_i)$	2	7	9	23	17	15	28	13
$L_p(p_i, r_i)$	0.009,-0.21	0.03,0.83	0.04,0.16	0.1,1.39	0.08,-0.23	0.07,-0.09	0.12,0.8	0.06,-0.52

3.2 Sleep Pattern Detection and Evaluation

After we get the user life pattern L_p, the next step is to detect the user sleep pattern S_p from it. The definition of S_p is as follows:

$S_p = (p_s, p_{(s+1)\%24}...p_e)$

where s, e represent the start and end period which are detected by algorithm 1. The main idea of algorithm 1 is based on the size of percentage p_i and change rate r_i. The detail introduction of algorithm 1 is as follows:

The input of algorithm 1 is User Life Pattern L_p which we defined in last section. The output is User Sleep Pattern S_p. Phase 1 from step 1 to 6 calculates

---

Algorithm 1. (User Sleep Pattern Detection)

Input:
    User Life Pattern : $L_p(p_i, r_i)$
Output:
    User Sleep Pattern : $S_p(p_s, p_{(s+1)\%24}...p_e)$
1: **for** $i = 0$ to 23 **do**
2:    **if** $(p_i <= 0.04$ *or* $r_i <= -1)$ and $p_{(i+23)\%24} >= 0.04$ and $p_{(i+25)\%24} <= 0.03$
3:    and $p_{(i+26)\%24} <= 0.03$ and $p_{(i+27)\%24} <= 0.03$ **then**
4:      s = i; break; //s is the start point of sleep time
5:    **end if**
6: **end for**
7: **for** $i = 0$ to 23 **do**
8:    j = (i + s)%24;
9:    **if** $p_j >= 0.03$ *or* $r_j >= 1$ **then**
10:      **if** $r_j >= 30$ **then**
11:        e = (j+23)%24; break; //e is the end point of sleep time
12:      **end if**
13:      **if** $p_{(j+25)\%24} >= 0.03$ or $p_{(j+26)\%24} >= 0.03$ **then**
14:        e = j; break;
15:      **end if**
16:    **end if**
17: **end for**
18: **return** $S_p(p_s, p_{(s+1)\%24}...p_e)$

---

the start period of sleep time, while phase 2 from step 7 to 17 gets the end period. In step 11, we assign end period e as the period before j if $r_j >= 30$, because there is a great probability that, on most days of a year, the user gets up in period $(j + 23)\%24$. On the other hand, if $r_j <= 30$, it means the user doesn't often got up in period $(j + 23)\%24$ of a year. Therefore, we will consider j as the end period of user sleep time if the condition in step 13 is satisfied.

Some threshold values of p_i and r_i in line 2,3,9,10 and 13 are set based on the experiment using training dataset which is not inclued in the experiment dataset below. For detail interpretation of line 2 and 3, if p_i is under 0.04 or r_i is under -1, we consider there is a probability that the user starts to sleep in period i in most days of a year. Then we will check whether the p values of next three periods are all under 0.03, and whether the p value of the period before is bigger than 0.04. If so, the start period of sleep time will be set i. The idea of setting end period is similar to start period. For the user data which is shown in Table 1, we can get start period and end period are 0 and 8 respectively by our algorithm, and the user's sleep time is 9 hours(including start period and end period).

Even if a user with a rather regular sleep habit, she won't go to sleep and get up in the same periods everyday. However, the start and end periods often keep the same in most days of a year, and the two periods of the rest days maybe a little earlier or later. Generally, our method can detect the two periods correctly if the user have a regular sleep habit. Results of experiment in next section can prove that algorithm 1 works well in our dataset. The concept of "sleep time" mentioned in the rest of our paper means the time between period s and period e (including s, e), i.e. $(e - s + 1 + 24)\%24$. We should notice that the sleep time detected by our method is more like an average number for a year rather than an accurate one for every day.

The performance of algorithm 1 will be discussed in detail in the experiment section. Assuming that we have gotten the sleep pattern S_p, the next work that we need to think over is how to evaluate it. Our evaluation method mainly considers two aspects: one is the length of S_p, and another is the quality of S_p. Generally, there is no unified or authoritative standard for the sleeping length of people. According to the instruction of CDC(U.S. Center for Disease Control), 7-9 hours' sleep is needed by an adult. Therefore, for length part, we choose a middle value of CDC's instruction - 8 hours as a boundary. User will get a score punishment if his sleep time is lower than 8 hours. In fact, the sleep time detected by our method will be a little longer than the actual time, for the reason that people won't fall asleep immediately after they stop posting microblogs, nor will they post microblogs immediately when they get up in the morning. Detail comparison will be shown in the experiment section. For quality part, considering that S_p represents a user's sleep situation of a whole year's time, p_i won't always be zero even if it is in the sleep pattern S_p. People may post microblogs to complaint if they are disturbed by phone calls or other noises when they are sleeping. Therefore, the larger the p_i is, the lower score a person will get. We can compute the score of a sleep pattern S_{score} as follows:

$$S_{score} = (\alpha \times L_{score} + (1 - \alpha) \times Q_{score}) \times 100 \tag{4}$$

$$L_{score} = 1/(1 + e^{-(l-\mu)}) \tag{5}$$

$$Q_{score} = \begin{cases} \frac{1}{8} \times \sum_{i=1}^{l} G(p_i) & l < 8 \\ \frac{1}{l} \times \sum_{i=1}^{l} G(p_i) & l >= 8. \end{cases} \tag{6}$$

$$G(x) = e^{-\frac{x^2}{2 \times \sigma^2}} \tag{7}$$

where L_{score}, Q_{score} represent the score of length part and quality part. l is the length of sleep pattern S_p. The right part of formula 5 is a sigmoid function to give a score of length part. The parameter μ is set 5 in this paper, which means a user will get $L_{score} = 0.5$ if he sleeps 5 hours, and 0.73, 0.88, 0.95 corresponding to 6, 7, 8 hours. The value of L_{score} will almost keep the same while sleep time continues to increase from 8 hours, and its upper limit is 1. Therefore, formula 5 is fit for our requirement.

The parameter α is set a small value 0.2 in formula 4, because Q_{score} also takes the length into account in some degree. The parameter σ of gauss function $G(x)$ is set 0.02. Formula 7, 9, 10 are all gauss function. The usage of them is similar to the sigmoid function of formula 5. On the basis of large experiments, we set parameters of gauss function empirically to make the score of sleep quality more reasonable. Notice that we use percentage instead of numbers in our score system to reduce the influence of the varying total amount of microblogs of different users. That is to say, a user who posts 1000 microblogs a year could also get a higher score than a user who posts 100 microblogs a year.

After we get the scores of users' sleep quality, we can evaluate the sleep quality of cities. The average score of city users is the first aspect which should be considered. Besides it, we define a user whose score is higher than 90 as a *healthy user* , user whose score is lower than 60 as an *unhealthy user*, and the rest of users are in a state of *subhealthy*. We will punish a city if the number of healthy users is small or the number of unhealthy users is large. The score of city sleep quality can be computed as follows:

$$C_{score} = \beta \frac{1}{|c|} \sum_{i \in c} S_{score}(i) + \frac{1-\beta}{2}(F(f_c) + O(o_c)) \times 100 \qquad (8)$$

$$F(x) = e^{-\frac{x^2}{2 \times \lambda^2}} \qquad (9)$$

$$O(x) = e^{-\frac{(x - x_{max})^2}{2 \times \gamma^2}} \qquad (10)$$

where $|c|$ is the number of users in city c. f_c, o_c are the percentage of unhealthy users and healthy users in city c. Three parameters λ, x_{max}, γ of gauss function (9) and (10) can be computed as follows: $\lambda = \frac{\sum_{c \in C} f_c}{|C|}$; $x_{max} = Max(o_c), c \in C$; $\gamma = \frac{\sum_{c \in C} o_c}{|C|}$. C is the set of all cities. The parameter β is set 0.7 in this paper.

4   Experiment

In this section, we describe our experiment to extract the sleep patterns of crowd which were mentioned in the section 3. First we computed the aforementioned life pattern respectively by cities and extracted the sleep patterns from them. Then we made a proper evaluation both for individual users and cities.

4.1   Experimental Dataset

In order to carry out our method of sleep quality evaluation, first we use our distributed crawling system to collect microblogs from Sina Microblog sites. The dataset covers 16 cities in China, containing 553,269 distinct Sina users, 304,809 active Sina Microblog users and 731 million microblogs in all. The information of a user's living city is stored in our database. Our method mentioned in section 3 is designed only for AMUs, for the reason that if a user posted too few microblogs a year, the user's data won't form a valid life pattern. The detail of our dataset is shown in Table 2.

Table 2. Statistics of our experiment dataset.

City	Users	Microblogs (Million)	Active Users	City	Users	Microblogs (Million)	Active Users
Beijing	51,114	80	33,504	Dalian	30,213	34	15,822
Shanghai	56,228	108	37,693	Xian	32,643	41	18,117
Guangzhou	35,437	58	23,835	Wuhan	29,320	37	15,913
Chengdu	33,936	44	19,774	Kunming	27,740	32	14,387
Jinan	26,252	32	15,397	Changsha	26,927	36	14,500
Qingdao	28,373	33	13,751	Xiamen	29,646	36	15,101
Harbin	31,969	36	16,852	Hangzhou	33,274	49	18,397
Changchun	28,167	29	13,016	Chongqing	52,030	46	18,750
Total	553,269	731	304,809				

4.2 Performance of User Sleep Quality Evaluation

For each AMU, we generate the user life pattern as we mentioned in section 3. Then we extract user sleep patterns from life patterns. Figure 1 shows some examples of user life and sleep pattern which are detected by algorithm 1. The ordinate of the figure represents the number of microblogs a user posts, while the abscissa represent the period. The area between two dotted lines is the sleep time of a user detected by our algorithm. We should pay attention that a period is one hour's time in this paper, and the sleep time includes both start and end period. Therefore, the sleep time in Figure 1 are 9, 6 and 9 respectively with scores of 94.6, 61.3 and 82.3, and the difference of the scores can be seen intuitively from the figures.

Our detection rate is about 74.2% for AMUs in the whole dataset. Considering that there exists some noise users in the dataset, for example, advertisement users who are microblogging all day long, users whose microblog time distribution can not form a regular life pattern, 74.2% is good enough for our research.

Next, we evaluate each city by the method we mentioned in section 3. Table 3 shows the users' average sleep time of each city. The overall average sleep time of China published by CMDA group is 7.5 hours in 2013, which is a little shorter than our analysis. One possible explanation which we have mentioned is that people may do some other things after they stop posting microblogs. The reason is the same in the morning.

From Table 3 we can see that the sleep time of year 2012 is a little shorter than that of 2011. But from 2012 to 2013, the sleep time of most cities even

Table 3. Average Sleep Time Of Cities.

City	2011	2012	2013	City	2011	2012	2013	City	2011	2012	2013
Beijing	8.226	8.064	8.085	Harbin	8.144	8.043	7.931	Changsha	8.144	8.028	8.069
Shanghai	8.254	8.122	8.154	Changchun	8.182	8.043	8.017	Xiamen	8.023	7.947	8.019
Guangzhou	8.121	8.089	8.180	Dalian	8.177	8.009	8.029	Hangzhou	8.221	8.133	8.165
Chengdu	8.209	8.109	8.123	Xian	8.101	7.972	8.012	Chongqing	8.202	8.123	8.153
Jinan	8.217	8.076	8.051	Wuhan	8.166	8.089	8.117				
Qingdao	8.196	8.072	8.085	Kunming	8.081	7.970	8.056				

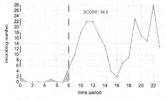
(a) A user with a high score of 94.6

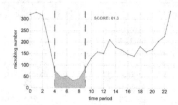
(b) A user with a low score of 61.3

(c) A daytime Sleeping user with a score of 82.3

Fig. 1. Three examples of different user life and sleep patterns.

become longer. We made a statistical test to show the difference between two adjacent years is significant by using spearman correlation coefficient.

Table 4. Spearman Correlation Coefficient.

Spearman-value	(2011,2012)	(2012,2013)
r-value	0.710	0.837
p-value	0.001	0.00001

Spearman correlation coefficient is commonly used in calculating the relevance of two distributions. It has two values: 1) r-value(from -1 to 1) shows the degree of correlation between two distributions and larger absolute value means more relevant. The plus or minus sign of r-value represents that whether the two distributions have positive or negative correlation. 2) p-value(from 0 to 1) shows the significant degree of r-value. Usually, we are confident to say r-value is significant when p-value is under 0.05, and very significant when p-value is under 0.01. The result of the statistical test is shown in Table 4 and from it we can conclude that the difference is statistically significant.

The average sleep time in China is stable in some degree, and almost every city in our experiment can reach 8 hours every year. However, in contrast with Figure 2, the situation of sleep quality is not optimistic - scores are generally low and most of them keep declining from year 2011 to 2013, merely the descend is smaller from 2012 to 2013 than that of 2011 to 2012. The reason for the descend

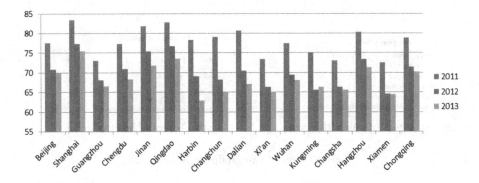

Fig. 2. Scores of some major cities in China.

from year 2012 to 2013 is that, although the sleep time in some cities increased, the frequency of microblogging behavior during sleep time increased too.

4.3 Reasoning by LDA Method

To reason the difference between high score users and low score users, we try to make a exploratory explanation by the popular method of Latent Dirichlet Allocation, known as LDA. In our experiment, we implemented the open source software jGibbLDA[19], a version of LDA model with Gibbs sampling. The original data comes from the database of city Shanghai, which contains 1120 unhealthy users and 3713 healthy users. We randomly selected 200 microblogs from each user to form the original text which LDA requires. Then after some necessary preprocessing steps(e.g., word segmentation by using ICTCLAS system, removing stop words and other noise words), we proceeded the LDA training of each group. To specify the number of topics in LDA, we tried several numbers from 10 to 50. The training results are not satisfied and varies a lot when the number is set under 20. When the topic number is larger than 20, the top topics which users involve most don't differ too much. Therefore, we choose a middle number 30 in our experiment. Other parameters are set default according to the instruction of jGibbLDA document.

We analyzed the result after the training process finished. We gave every latent topic a name based on the observation of the top frequent words in it. Figure 3 shows some topics with high percentage and from the figure we can see the two groups' topics mainly differ in two aspects. One is that the percentage of unhealthy group on topic "love, life and sentiment" is 11 points higher than that of healthy group. Another discovery is that healthy users have more relax topics, such as "movie, music", "food", "games, anime", "constellations". Maybe such difference in topics is one of the possible reasons which cause the gap in sleep quality scores between healthy users and unhealthy users.

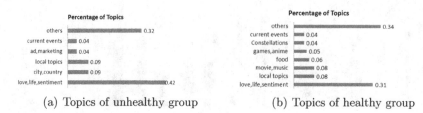

(a) Topics of unhealthy group   (b) Topics of healthy group

Fig. 3. Topics of two groups in Shanghai.

5   Conclusion and Future Work

In this paper, we proposed a novel method to evaluate sleep quality of AMUs based on crowd-sourced data of Sina Microblog. Meanwhile, we introduced a practical method of processing and mining of large scale data. In our work, we presented a complete set of system on how to evaluate people's sleep quality, which includes 1) collecting massive microblogs from Sina as the base data. 2) modeling user life pattern and sleep pattern from base data. 3) method of evaluating user sleep pattern and also the city sleep quality. 4) finding the difference between high score crowd and low score crowd in topics by LDA method. By conducting the experiment using the data we collected, we showed that our evaluating system is reasonable.

Our research shows that the sleep quality of AMUs is not optimistic in China, the sleep quality keeps declining from year 2011 to 2013. Therefore, both government, enterprises and individuals should be aware of such situation and work together to improve it. In the future work, first we will make a comparison of sleep quality between different group of AMUs who have the same feature such as interest, occupation. Another aspect is that we want to improve our evaluation method by adding sentiment analysis. For example, AMUs who post microblogs with happy sentiment before sleeping will have a score reward.

Acknowledgment. This work was supported by Natural Science Foundation of China (61272240, 60970047, 61103151), the Doctoral Fund of Ministry of Education of China (20110131110028), the Natural Science foundation of Shandong province
(ZR2012FM037), and Microsoft research fund (FY14-RES-THEME-25).

References

1. Sina Microblog. http://www.weibo.com/
2. Gao, Q., Abel, F., Houben, G.-J., Yu, Y.: A comparative study of users' microblogging behavior on sina weibo and twitter. In: Masthoff, J., Mobasher, B., Desmarais, M.C., Nkambou, R. (eds.) UMAP 2012. LNCS, vol. 7379, pp. 88–101. Springer, Heidelberg (2012)

3. Twitter. http://twitter.com/
4. Bao, P., Shen, H.W., Huang, J., Cheng, X.Q.: Popularity prediction in microblogging network: A case study on Sina Weibo. In: Proc. of WWW 2013, pp. 177–178 (2013)
5. CMDA Sleep Quality Report 1. http://www.cmda.gov.cn/xiehuixiangmu/jishupingjiatuiguangbu/bumendongtai/2013-03-20/11847.html
6. CMDA Sleep Quality Report 2. http://www.cmda.gov.cn/xiehuixiangmu/jishupingjiatuiguangbu/bumendongtai/2014-03-18/13018.html
7. Sohu report. http://it.sohu.com/20110519/n280623821.shtml
8. Java, A., Song, X., Finin, T., Tseng, B.: Why we twitter: understanding microblogging usage and communities. In: Proc. of the 9th WebKDD and 1st SNAKDD 2007 Workshop on Web Mining and Social Network Analysis, pp. 56–65 (2007)
9. Zhao, D., Rosson, M.B.: How and why people Twitter: the role that micro-blogging plays in informal communication at work. In: Proc. of the ACM 2009 International Conference on Supporting Group Work, pp. 243–252 (2009)
10. Krishnamurthy, B., Gill, P., Arlitt, M.: A few chirps about twitter. In: Proc. of WOSN 2008, pp. 19–24 (2008)
11. Ren, Z., Liang, S., Meij, E., de Rijke, M.: Personalized time-aware tweets summarization. In: Proc. of SIGIR 2013, pp. 513–522 (2013)
12. Fujisaka, T., Lee, R., Sumiya, K.: Discovery of user behavior patterns from geo-tagged micro-blogs. In: Proc. of ICUIMC 2010, pp. 246–255 (2010)
13. Lee, R., Sumiya, K.: Measuring geographical regularities of crowd behaviors for Twitter-based geo-social event detection. In: Proc. of LBSN 2010, pp. 1–10 (2010)
14. Lee, R., Wakamiya, S., Sumiya, K.: Discovery of unusual regional social activities using geo-tagged microblogs. In: Proc. of WWW Special Issue on Mobile Services on the Web, vol. 14(4), pp. 321–349 (2011)
15. Wakamiya, S., Lee, R., Sumiya, K.: Towards better tv viewing rates: exploiting crowd's media life logs over twitter for tv rating. In: Proc. of ICUIMC 2011, pp. 39:1–39:10 (2011)
16. Wakamiya, S., Lee, R., Sumiya, K.: Crowd-sourced urban life monitoring: urban area characterization based crowd behavioral patterns from twitter. In: Proc. of ICUIMC 2011, Article No 26 (2012)
17. Jamison-Powell, S., Linehan, C., Daley, L., Garbett, A., Lawson, S.W.: I can't get no sleep: discussing #insomnia on twitter. In: Proc. of SIGCHI 2012, pp. 1501–1510 (2012)
18. Blei, D.M., Ng, A.Y., Jordan, M.I.: Latent dirichlet allocation. The Journal of Machine Learning Research, 993–1022 (2003)
19. jGibbLDA. http://jgibblda.sourceforge.net/

An Ensemble Matchers Based Rank Aggregation Method for Taxonomy Matching

Hailun Lin[1], Yuanzhuo Wang[1], Yantao Jia[1], Jinhua Xiong[1], Peng Zhang[2], and Xueqi Cheng[1]

[1] Institute of Computing Technology, Chinese Academy of Sciences, Beijing, China
[2] Institute of Information Engineering, Chinese Academy of Sciences, Beijing, China
`linlunnian@software.ict.ac.cn`

Abstract. Taxonomy matching is an important operation of knowledge base merging. Several matchers for automating taxonomy matching have been proposed and evaluated in the knowledge base community. Studies reveal that there is no single taxonomy matcher suitable for any domain-specific taxonomy mapping, therefore an ensemble of taxonomy matchers is essential. In this paper, we propose taxonomy metamatching, a distributed computing framework for assembling taxonomy matchers and generating an optimal taxonomy mapping. And we introduce TRA, a Threshold Rank Aggregation algorithm for this problem. Experimental results show that TRA outperforms state-of-the-art approaches regardless of domains and scales of taxonomies, which demonstrates that TRA performs good adaptability to taxonomy matching.

Keywords: knowledge base merging, taxonomy matching, rank aggregation.

1 Introduction

As is known to all, a knowledge base is a formal collection of world knowledge including instances, relations and classes. And a taxonomy is a collection of classes, which are used to semantically classify or annotate knowledge items in a knowledge base [10]. In addition, different knowledge bases sometimes contain overlapping or complementing data, there has been growing interest in attempting to merge them by matching their common elements. As such, taxonomy matching is one of the most important operations in the process of knowledge base merging, which aims to match semantically equalled classes in different knowledge bases.

A number of taxonomy matchers have been established (e.g. [1,10,15,19]), either focusing on specific domains or aiming at providing a general technique in a wide variety of domains [17]. Most existing taxonomy matchers calculate similarities of classes between two taxonomies by utilizing various information, e.g., class names, structures, properties and instances. Specifically, these matchers usually define a similarity function by linearly combining the results of different similarities on different types of information.

© Springer International Publishing Switzerland 2015
R. Cheng et al. (Eds.): APWeb 2015, LNCS 9313, pp. 190–202, 2015.
DOI: 10.1007/978-3-319-25255-1_16

Despite the significant amount of matchers for taxonomy matching, most of them perform well only on fairly taxonomies from specific domains, and are unable to handle large-scale taxonomies as well as small-scale taxonomies [7]. This is because different knowledge bases generally use different terms or structures to represent their classes, and the space of possible matchings grows exponentially with the number of classes. Moreover, choosing a suitable one among these matchers for a given application is far from trivial. Firstly, the diversity of them complicates the choice of the most appropriate matcher. Secondly, there is no single dominant taxonomy matcher that performs the best, regardless of its application domain. In particular, due to the explosive emerging of Web big data, taxonomies become larger and more complex. Therefore, this paper aims to provide an efficient matching technique between different scales of taxonomies from various domains, on the premise of keeping high precision without greatly sacrificing recall.

To achieve the goal, we introduce a distributed computing framework *taxonomy metamatching*, which combines an ensemble of taxonomy matchers to generate an optimal aggregated taxonomy mapping between two taxonomies in parallel. It should be noted that a *taxonomy mapping* here refers to a set of semantically equalled class pairs between two taxonomies. The combination of taxonomy matchers can make full use of the advantages of each taxonomy matcher and potentially compensate for the weakness of each other. The core idea of our approach is based on the meta-search techniques developed in the area of Web search [5]. Furthermore, we introduce the threshold algorithm [6] to address the taxonomy metamatching problem. Experimental results show that our method outperforms state-of-the-art methods regardless of domains and scales of taxonomies, which demonstrates that our method performs good adaptability to taxonomy matching. In general, the main contributions of this paper are three-fold:

- We introduce *taxonomy metamatching*, a distributed computing framework which combines an ensemble of taxonomy matchers to generate an optimal aggregated taxonomy mapping between two taxonomies.
- We introduce TRA, a Threshold Rank Aggregation algorithm for taxonomy metamatching, with a self-tuning based strategy.
- We extensively evaluate our method on two benchmark datasets. Experimental results show that our method outperforms state-of-the-art methods, and performs good adaptability regardless of domains and scales of taxonomies.

The rest of this paper is organized as follows. In Section 2, we give a review of related work. Section 3 formulates the problem of taxonomy metamatching. Section 4 describes our method for taxonomy metamatching. Section 5 is devoted to the experimental results. Finally, the paper is concluded in Section 6.

2 Related Work

The problem of taxonomy matching has been extensively studied, particularly under the umbrella term of ontology matching. And there are a number of surveys that shed light on this problem. Choi et al. [3] provided a good overview of the problem in general. Shvaiko et al. [17] analyzed several state-of-the-art taxonomy matchers. These matchers mostly aim at calculating similarities between classes in terms of various information associated with the taxonomies.

Li et al. [13] presented a multistrategy matcher RiMOM for taxonomy matching, which dynamically combined string and structure strategies. But RiMOM had only been tested on small taxonomies. Jiménez-Ruiz et al. [9] presented LogMap matcher, which integrated string, semantic and structure information to find similar classes. Ba et al. [1] proposed ServOMap matcher to find similar classes based on an Ontology Server and exploited string and context information. Both LogMap and ServOMap are designed for biomedical ontologies. Suchanek et al. [19] presented PARIS, a probabilistic algorithm for taxonomy matching based on instances. Lee et al. [12] measured the similarity between two classes by linearly combining the instances and properties associated with classes. Lacoste et al. [10] proposed a simple greedy matching algorithm, named SiGMa, for aligning knowledge bases with millions of entities and facts. They used the string, properties and structures information to obtain equalled class pairs in a greedy local search manner. These methods exploiting instances would not perform well when instances associated with classes are sparse.

In addition, schema matching is a similar work to taxonomy matching. There are several surveys on schema matching (e.g. [16,18]). The goal of these work tries to construct a common schema from multiple data sources or to find the interscheme between different schemas. Although schema matching is similar to taxonomy matching, in comparison with schema matching, taxonomy matching has its own unique characteristics [13]. First, comparing with data schemas, taxonomies provide higher flexibility and more explicit semantics for defining data. Second, data schemas are usually defined for a specific database, whereas taxonomy is by nature reusable and sharable. Finally, in taxonomy, the number of knowledge representation primitives is much larger and more complex, e.g., cardinality constraints, disjoint classes and type-checking constraints. Therefore, the schema matching methods are no longer available to taxonomy matching.

3   Problem Formulation

In this section, we will study the problem of taxonomy metamatching. For this purpose, we will firstly describe the notation of taxonomy, then formalize taxonomy mapping and taxonomy mapping ordering for a taxonomy matcher, and finally formulate the taxonomy metamatching problem.

Gruber indicates that a taxonomy is an explicit specification of a conceptualization [8]. In this paper, we define taxonomy as the schemata of a knowledge base, which is a collection of classes to semantically classify or annotate information in a knowledge base. More formally, we have the following definition. A taxonomy is defined as a four-tuple: $T = (C, P, R, I)$, where C is the set of classes. P and I is the set of properties and instances related to C, respectively. R is the relationship function which gives a hierarchical relationship between classes, $R \subseteq C \times C$, $< c, c' > \in R$ denotes that class c is the subclass of c'. Under the notion of taxonomy, we define a taxonomy mapping between two taxonomies as follows:

Taxonomy Mapping. A taxonomy mapping is a set of semantically equalled class pairs between two taxonomies. More precisely, given two taxonomies $T = (C, P, R, I)$, $T' = (C', P', R', I')$, and a taxonomy matcher M, a taxonomy mapping ϕ generated by M can be denoted as:

$$\phi = \{(c_i, c_j, s_{ij}) | c_i \in C, c_j \in C'\}$$

where (c_i, c_j, s_{ij}) denotes an equalled class pair between T and T' with matching score s_{ij}. For any two class pairs (c_i, c_j, s_{ij}), $(c_k, c_l, s_{kl}) \in \phi$, subject to $i \neq k$ and $j \neq l$. We have $\cup\{c_i\} \subseteq C$ $(1 \leq i \leq |C|)$ and $\cup\{c_j\} \subseteq C'$ $(1 \leq j \leq |C'|)$.

In this paper, a taxonomy matcher can be any form of matcher constructed based on any information from taxonomies. Based on the above definition, we can define the ordering on taxonomy mappings as follows:

Taxonomy Mapping Ordering. Given two taxonomies $T = (C, P, R, I)$, $T' = (C', P', R', I')$, and a taxonomy matcher M. Let $\Omega = \{\phi = \{(c_i, c_j, s_{ij}) | c_i \in C, c_j \in C'\}\}$ is the set of all candidate taxonomy mappings generated by matcher M with an ordering $\succeq_M$. More precisely, for two taxonomy mappings $\phi, \phi' \in \Omega$, $\phi \succeq_M \phi'$ means that ϕ is better than ϕ'.

In the following, we elaborate on the process of a taxonomy matcher producing an ordered set of taxonomy mappings.

Assuming that $S : (C, C') \rightarrow \mathbf{R}$ is the similarity function used by matcher M, which computes the matching degree of a class pair between T and T'. Matcher M takes the following three steps to generate an ordered set of candidate taxonomy mappings between T and T':

1) Computing similarities of all class pairs between T and T' based on function S. Specifically, if $|C| = n, |C'| = n'$, the results produced in this step is an $n \times n'$ similarity matrix $A^{(M)}$, where $A_{i,j}^{(M)} = S(c_i, c_j)$ represents the matching score s_{ij} between the i-th class $c_i \in T$ and the j-th class $c_j \in T'$.
2) Generating all possible taxonomy mappings $\Omega = \{\phi = \{(c_i, c_j, s_{ij}) | c_i \in C, c_j \in C'\}\}$ for T and T', according to the similarity matrix $A^{(M)}$.
3) Ranking the taxonomy mappings in Ω. Specifically, using a scoring function f_M to quantify the quality of taxonomy mapping $\phi \in \Omega$. The function $f_M : \Omega \rightarrow \mathbf{R}$ is defined as the aggregation of the similarity associated with the individual equalled class pairs in ϕ. The quality of ϕ is computed as follows:

$$f_M(\phi, A^{(M)}) = f_M(A_{1,\phi(1)}^{(M)}, \ldots, A_{n,\phi(n)}^{(M)}) = \sum_{i=1}^{n} A_{i,\phi(i)}^{(M)} = \sum_{i=1}^{n} s_{i\phi(i)} \quad (1)$$

where $\phi(i)$ is the equalled class from T' with the i-th class from T.

Therefore, for two mappings $\phi, \phi' \in \Omega$, the ordering $\succeq_M$ on Ω is then computed as:

$$\phi \succeq_M \phi' \Leftrightarrow f_M(\phi, A^{(M)}) \geq f_M(\phi', A^{(M)})$$

After those three steps, we can obtain an ordered set of candidate taxonomy mappings between T and T' generated by matcher M. Now we can formulate the problem of taxonomy metamatching.

Taxonomy Metamatching. Given two taxonomies T and T', and m taxonomy matchers $M_1, \ldots, M_m$ with scoring function $f_{M_1}, \ldots, f_{M_m}$, respectively. Taxonomy metamatching is to generate an optimal taxonomy mapping between T and T' with respect to an ensemble of taxnomomy matchers $M_1, \ldots, M_m$.

Based on the definition, we will introduce how we solve the problem.

4   Taxonomy Metamatching

In this section, we will study the framework of taxonomy metamatching. Figure 1 summarizes the framework. In the framework, we firstly extract class features from taxonomies. Then, we develop a group of taxonomy matchers based on each of these features, respectively. And then, we apply those taxonomy matchers on two given taxonomies based on a distributed computing framework, such as MapReduce [4], to generate candidate taxonomy mappings in parallel. After that, we aggregate all taxonomy mappings generated by those taxonomy matchers. Finally, we obtain the optimal taxonomy mapping between taxonomies.

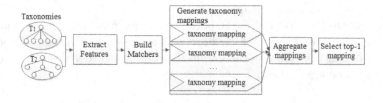

Fig. 1. Taxonomy metamatching framework overview

In what follows, we will firstly introduce the ensemble of taxonomy matchers that we used in this paper. Secondly, we will introduce how the taxonomy metamatching framework generates an optimal taxonomy mapping.

4.1   Taxonomy Matcher

We build four simplest taxonomy matchers as the source matchers for taxonomy metamatching, i.e., string-based, property-based, structure-based and semantic-based. In the following, we will introduce those matchers in details.

String-Based Taxonomy Matcher. This matcher computes the similarity between two classes in terms of class names. The similarity can be measured by Jaccard coefficient[1] or Edit distance[2]. In this paper, we adopt Edit distance.

---

[1] http://en.wikipedia.org/wiki/Jaccardindex
[2] http://en.wikipedia.org/wiki/Editdistance

Given two classes c_1 and c_2, we define the similarity between them as follows:

$$sim(c_1, c_2) = 1 - \frac{ed(c_1, c_2)}{\max(|c_1|, |c_2|)}$$

where $ed(c_1, c_2)$ is the edit distance between c_1 and c_2. $|\cdot|$ is the string length.

Property-Based Taxonomy Matcher. This matcher computes the similarity between two classes in terms of class properties. In this paper, we adopt Jaccard coefficient to calculate the property similarity. Given two classes c_1 and c_2, we define the property similarity between them as follows:

$$sim(c_1, c_2) = \frac{|c_1.P \cap c_2.P|}{|c_1.P \cup c_2.P|}$$

where $c_i.P$ is the set of properties of class c_i ($i = \{1, 2\}$); $c_1.P \cap c_2.P$ is the set of common properties between c_1 and c_2; $c_1.P \cup c_2.P$ is the union set of properties of c_1 and c_2; $|\cdot|$ is the set size.

Structure-Based Taxonomy Matcher. This matcher computes the similarity between two classes in terms of class neighbors (superclasses and subclasses). Like property-based matcher, we also use the Jaccard coefficient measure. Given two classes c_1 and c_2, we define the structure similarity between them as follows:

$$sim(c_1, c_2) = \frac{|N_{c_1} \cap N_{c_2}|}{|N_{c_1} \cup N_{c_2}|}$$

where $N_{c_i}(i = \{1, 2\})$ is the neighbor classes of c_i, which can be derived based on the relationship function R defined in taxonomy.

Semantic-Based Taxonomy Matcher. This matcher computes the similarity between two classes in terms of class semantics. There has been many studies that employ WordNet [14] to calculate the semantic relevance between two strings [2,11]. In this paper, we directly adopt the Lesk algorithm [2] to compute the semantic similarity between two classes. Given two classes c_1 and c_2, we define the semantic similarity between them as follows:

$$sim(c_1, c_2) = lesk\_sim(c_1, c_2)$$

where $lesk\_sim(c_1, c_2)$ [2] is the semantic similarity between class c_1 and c_2 based on the lexical database WordNet.

Based on those matchers, in the following, we will elaborate on the process of taxonomy metamatching to generate an optimal taxonomy mapping.

4.2 Taxonomy Mapping Generation

In this section, we will present how taxonomy metamatching framework generates the optimal taxonomy mapping. Given two taxonomies $T = (C, P, R, I)$, $T' = (C', P', R', I')$, and an ensemble of taxonomy matchers: string-based, property-based, structure-based and semantic-based. Let $m = 4$ denote the size of an ensemble of taxonomy matchers, $M_1, \ldots, M_m$ denotes these matchers, with score function $f_{M_1}, \ldots, f_{M_m}$, respectively. The framework takes the following two modules to generate an optimal taxonomy mapping between T and T'. In particular, we use a distributed computing framework to implement the parallelization in order to reduce the process time :

- **Taxonomy Mapping Ordering Generation.** Generating an ordered set of candidate taxonomy mappings Ω_{M_i} between T and T' for each individual matcher M_i. The generated ordering of each taxonomy matcher is denoted as $\succeq_{M_1}, \ldots, \succeq_{M_m}$, respectively. The ordering generation process of each matcher is according to the process described in Section 3.
- **The Optimal Taxonomy Mapping Selection.** Generating an aggregated mapping ordering $\succeq$ based on all the orderings $\succeq_{M_1}, \ldots, \succeq_{M_m}$. The ordering $\succeq$ is generated by aggregating the scores that each matcher M_i provides to the mappings in Ω. The quality of mapping ϕ in $\succeq$ is quantified using an ensemble aggregation function $F(\phi) : \Omega \to \mathbf{R}$, a function that aggregates the scores provided by $M_1, \ldots, M_m$:

$$F(\phi) = F(f_{M_1}(\phi, A^{(M_1)}), \ldots, f_{M_m}(\phi, A^{(M_m)})) = \sum_{i=1}^{m} \lambda_{M_i} f_{M_i}(\phi, A^{(M_i)}) \quad (2)$$

where λ_{M_i} is the weighting parameter. For two taxonomy mappings ϕ_1, ϕ_2, the aggregated ordering $\succeq$ on Ω is then defined as:

$$\phi_1 \succeq \phi_2 \Leftrightarrow F(\phi_1) \geq F(\phi_2)$$

The optimal taxonomy mapping selection is generalized to selecting the top-ranked mapping provided by the aggregated mapping ordering $\succeq$:

$$\phi^{(1)} = \arg\max\{F(\phi)|\phi \in \Omega\} \quad (3)$$

4.3   Threshold Rank Aggregation Algorithm

The problem of optimal aggregation of serval quantitatively ordered list has been studied in the context of Web search or database systems. The most efficient general algorithm for this problem is *Threshold* algorithm [6]. In this section, we present this algorithm from taxonomy metamatching perspective. In what follows, we describe the threshold rank aggregation algorithm in details.

Given two taxonomies $T = (C, P, R, I)$, $T' = (C', P', R', I')$, and an ensemble of matchers $\Psi = \{M_1, \ldots, M_m\}$ for T and T'. Here, we use λ_{M_i} to represent the weighting parameter for each matcher, $\lambda = \{\lambda_{M_1}, \ldots, \lambda_{M_m}\}$ is the parameters set. Let $\Phi \subseteq \Omega$ be the set of aggregated ordered taxonomy mappings, which is initialized as an empty set. The best taxonomy mapping $\phi^{(1)}$ can be generated by the threshold rank aggregation algorithm described in Algorithm 1.

Specifically, the algorithm firstly selects the top-ranked taxonomy mapping $\phi_{M_i}^{(1)}$ from each matcher M_i (see Lines 1-4). Then, it computes the aggregated score of selected mappings based on threshold selection (see Lines 5-14). Finally, we acquire the top-ranked mapping for taxonomies T, T' (see Lines 15). It is worth mentioning that the algorithm automatically updates the threshold τ_{ta} parameter (see Lines 7-10) based on existing selected taxonomy mapping.

Algorithm Analysis. Let $n = |C|$ and $n' = |C'|$ denote the number of classes in taxonomy T and T', respectively; m denote the taxonomy matchers size. Our method takes $n \times n'$ steps to compute the class similarities for each individual taxonomy matcher. Let $l = \max\{n, n'\}$, for each taxonomy matcher, the aggregated score function $F(\phi)$ runs in $O(l)$ time. Therefore, the algorithm runs in polynomial time $O(mnn' + ml)$.

Algorithm 1. The threshold rank aggregation algorithm

Input: $T = (C, P, R, I), T' = (C', P', R', I'), \Psi, \Phi, \lambda$;
Output: $\phi^{(1)}$;
1: **for all** $M_i \in \Psi$ **do**
2:    Compute similarity matrix $A^{(M_i)}$ between taxonomy classes C and C', and generate taxonomy mapping ordering $\succeq_{M_i}$;
3:    Select the top-ranked taxonomy mapping $\phi_{M_i}^{(1)}$ according to taxonomy mapping ordering $\succeq_{M_i}$ generated by matcher M_i;
4: **end for**
5: **for** $i = 1$ to m **do**
6:    Compute the aggregated score τ for $\phi_{M_i}^{(1)}$ based on the ensemble aggregation function $F(\phi_{M_i}^{(1)})$:

$$\tau = F(f_{M_1}(\phi_{M_i}^{(1)}, A^{(M_1)}), \ldots, f_{M_m}(\phi_{M_i}^{(1)}, A^{(M_m)}))$$

7:    **for** $j = 1$ to m **do**
8:       Let ϕ_j be the taxonomy mapping $\phi_{M_i}^{(1)}$ returned by the matcher M_j;
9:    **end for**
10:    Compute the aggregated score τ_{ta} as the threshold:

$$\tau_{ta} = F(f_{M_1}(\phi_1, A^{(M_1)}), \ldots, f_{M_m}(\phi_m, A^{(M_m)}))$$

11:    **if** $\tau \geq \tau_{ta}$ **then**
12:       $\Phi = \Phi \cup \{\phi_{M_i}^{(1)}\}$;
13:    **end if**
14: **end for**
15: **return** The top-ranked taxonomy mapping $\phi^{(1)}$ in Φ;

---

5   Experiments

In this section, we evaluate the effectiveness of our method TRA for taxonomy matching. We will: (1) compare the precision, recall and F1 measure of our method, to five state-of-the-art taxonomy matching methods. (2) analyze how the performance of our method changes with taxonomies of different size. (3) analyze the adaptability of our method across different domains.

5.1   Experimental Settings

All experiments were conducted on two servers running 64-bit Linux OS, with 16 core 2GHz AMD Opteron(tm) 6128 Processors and 32GB RAM. And we compared our method TRA with five methods described in Section 2, namely, SiGMa, YAM++, ServOMap, LogMap and IAMA. Specifically, we evaluated the performance using the standard metrics of precision, recall and F1 measure, based on the number of semantically equalled class pairs correctly matched. In the experiments, we used two benchmark datasets derived from ontology alignment evaluation initiative[3] (OAEI). The two datasets are available with corresponding ground truth data. We describe these datasets below.

The first dataset was obtained from the large biomedical test case (largebio), which aims at finding alignments between large and semantically rich biomedical ontologies. The test case has been split into two task involving "small" and "whole" fragments of the input ontologies, which aims to test the adaptability of matchers on taxonomies of different size. We used NCI, SNOMED ontologies

---

[3] http://oaei.ontologymatching.org/2013/

from this test case as test datasets and UMLS Metathesaurus as ground truth data.

The second one was obtained from the conference test case, whose goal is to find all correct correspondences within a collection of ontologies over the domain of organizing conferences. We evaluated the results of the approaches against the blind reference alignments (labeled as *ra*1 on the conference web-page), which includes all pairwise combinations between 7 different ontologies.

In the experiments, for the sake of simplicity, the weighting parameter λ_{M_i} used in TRA is set to be 1.0 for each taxonomy matcher. As SiGMa starts with an initial equalled class pair, which is declared as any initial matching pair assumed of good quality [10]. Therefore, in the experiments, we selected the root class pair between the benchmark datasets as the initial seed.

5.2   Experimental Results and Analysis

In this section, we will evaluate the adaptability of our method TRA and the baseline methods on different scales and different domains of taxonomies. For this purpose, we will firstly test these methods on the small and whole large-bio datasets, and then we will test them both on the largebio dataset and the conference dataset.

Evaluation on Datasets with Different Size. In this experiment, we tested TRA and the baseline methods on the small and whole fragments of largebio datasets. Table 1 presents the results. From the results, we can see that TRA obtains a relatively higher F1 measure than any of the baseline methods both on the small and whole tasks. Despite TRA's running time slightly higher than the greedy-based method SiGMa, TRA obtains an impressive F1 measure above 76% both on the small and whole largebio datasets, significantly improving over the SiGMa baseline.

Table 1. Results on the small and whole largebio dataset

Approach	small				whole			
	Precision	Recall	F1	Time(s)	Precision	Recall	F1	Time(s)
SiGMa	0.896	0.647	0.751	**57**	0.873	0.466	0.608	**169**
YAM++	0.967	0.611	0.749	391	0.881	0.601	0.714	713
LogMap	0.896	0.672	0.768	433	0.882	0.570	0.692	1,233
ServOMap	0.933	0.642	0.761	1,699	0.822	0.637	0.718	6,320
IAMA	0.965	0.439	0.604	100	0.917	0.439	0.593	207
TRA	**0.967**	**0.672**	**0.793**	**68**	**0.939**	**0.642**	**0.763**	**183**

In order to explicitly demonstrate the performance changes of these methods on taxonomies with different size, the results in Table 1 is graphically depicted in Figure 2. From Figure 2, it can be seen that all methods perform better on the small task than on the whole task. The most likely reason for this is that larger taxonomies also involve more possible candidate semantically equalled class pairs between two taxonomies, and it is harder to keep high precision without sacrificing recall, and vice versa. In spite of this, our method keeps on achieving the best precision, recall and F1 measure over the small and whole

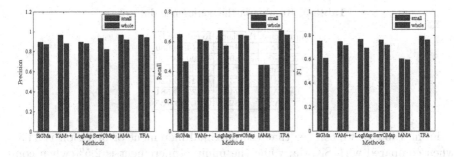

Fig. 2. Comparison on the small and whole largebio datasets.

largebio dataset. And as the results shown in Figure 2, TRA exhibits more steady performance both on the small and whole largebio datasets. It is shown that our proposed method TRA performs well on taxonomy matching task, regardless of scales of taxonomies.

Evaluation on the Different Domains of Datasets. In this section, we will evaluate the performance of our method TRA and the baseline methods on different domains. Therefore, we firstly tested these methods on the largebio dataset, then tested them on the conference dataset. Subsequently, we analyzed the adaptability of these methods both on largebio and conference datasets.

In order to see the whole performance of all the methods on the largebio dataset, we averaged the results on the small and whole largebio dataset. Table 2 presents the average results of these methods on largebio dataset. From Table 2, we can see that the maximal increment of precision is 7.5% when compared with ServOMap, while the minimal increment is 1.2% when compared with IAMA. And the maximal increment of recall is 21.8% when compared with IAMA, while the minimal is 1.7% when compared with ServOMap. And the maximal increment of F1 measure is 17.8% when compared with IAMA, while the minimal is 3.8% when compared with ServOMap. In general, the experimental results indicate that TRA outperforms all the baseline methods on the largebio dataset.

Table 2. Average results on the largebio datasets.

Approaches	Precision	Recall	F1
SiGMa	0.885	0.557	0.680
YAM++	0.924	0.606	0.732
LogMap	0.889	0.621	0.730
ServOMap	0.878	0.640	0.740
IAMA	0.941	0.439	0.600
TRA	**0.953**	**0.657**	**0.778**

In the following, we tested these methods on the conference dataset. As the conference dataset contains 7 ontologies, we took the results of these taxonomy matching pairs as a whole to evaluate the performance of TRA and the baseline methods. Table 3 presents comparison of the precision, recall and F1 measure. In Table 3, for each method, the best performance is listed in bold.

Table 3. Results on the conference dataset.

Approaches	Precision	Recall	F1
SiGMa	0.482	0.314	0.380
YAM++	0.800	0.690	0.740
LogMap	0.799	0.585	0.675
ServOMap	0.731	0.547	0.626
IAMA	0.742	0.477	0.581
TRA	**0.825**	**0.701**	**0.758**

From Table 3, we can see that the maximal increment of precision is 34.3% when compared with SiGMa, while the minimal increment is 2.5% when compared with YAM++. And the maximal increment of recall is 38.7% when compared with SiGMa, while the minimal increment is 1.1% when compared with YAM++. And the maximal increment of F1 measure is 37.8% when compared with SiGMa, while the minimal increment is 1.8% when compared with YAM++. In general, the experimental results indicate that TRA can achieve better performance than the baseline methods on the conference dataset.

Figure 3 graphically depicts the results on the largebio and conference datasets, we can see that compared with all the baseline methods, TRA obtains the highest precision, recall and F1 measure both on the largebio dataset and conference dataset. Moreover, TRA exhibits more steady performance on the test datasets. It is shown that our proposed method TRA performs well on taxonomy matching task, regardless of taxonomy domains.

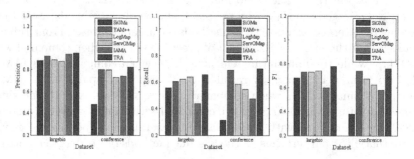

Fig. 3. Comparison on the largebio and conference datasets.

As was mentioned above, it can be seen that TRA can not only obtain better performance over taxonomies with different size, but also obtain better performance over taxonomies in different domains, which demonstrates the good adaptability of TRA. Furthermore, it is shown that the ensemble method for taxonomy matching is a very useful technique, which demonstrates the combination of an ensemble of taxonomy matchers can make full use of the advantages of each taxonomy matcher and potentially compensate for the weakness of each other. Moreover, from the accuracy of knowledge engineering perspective, a slightly improvement is of great importance to applications. It is worth noting that, since we adopt the distributed computing framework to get the taxonomy mappings generated by taxonomy matchers in parallel, the time consumption of ensemble matcher is almost the same as the single matcher, therefore, our method TRA is a useful technique for taxonomy matching.

6 Conclusions

In this paper, we studied the problem of taxonomy matching. Specifically, we proposed taxonomy metamatching, a distributed computing framework for assembling taxonomy matchers and generating an optimal taxonomy mapping. And we introduced TRA, a Threshold Rank Aggregation algorithm for this problem. The experimental results showed that our proposed ensemble method is a useful technique for taxonomy matching. Furthermore, the method performs good adaptability regardless of domains and scales of taxonomies.

Acknowledgments. This work is supported by National Grand Fundamental Research 973 Program of China (No. 2013CB329602, 2014CB340405), National Natural Science Foundation of China (No. 61174152, 61232010, 61402442, 61402022, 61402464, 61303244), Beijing Nova Program (No. Z121101002512063), and the Natural Science Foundation of Beijing (No. 4154086), National Hightech R&D 863 Program of China (No. 2015AA015803).

References

1. Ba, M., Diallo, G.: Large-scale biomedical ontology matching with servomap. IRBM 34(1), 56–59 (2013)
2. Banerjee, S., Pedersen, T.: An adapted lesk algorithm for word sense disambiguation using wordNet. In: Gelbukh, A. (ed.) CICLing 2002. LNCS, vol. 2276, pp. 136–145. Springer, Heidelberg (2002)
3. Choi, N., Song, I.Y., Han, H.: A survey on ontology mapping. SIGMOD Record 35(3), 34–41 (2006)
4. Dean, J., Ghemawat, S.: Mapreduce: Simplified data processing on large clusters. Communications of the ACM 51(1), 107–113 (2008)
5. Dwork, C., Kumary, R., Naorz, M., Sivakumarx, D.: Rank aggregation methods for the web. In: Proc. WWW (2001)
6. Fagin, R., Lotem, A., Naor, M.: Optimal aggregation algorithms for middleware. In: Proc. SIGMOD-SIGACT-SIGART Symp., pp. 102–113 (2001)
7. Grau, B.C., Dragistic, Z., Eckert, K., et al.: Results for the ontology alignment evaluation initiative 2013. In: Proc. ISWC (2013)
8. Gruber, T.R.: A translation approach to portable ontology specifications. Knowledge Acquisition 5(2), 199–220 (1993)
9. Jiménez-Ruiz, E., Grau, B.C.: LogMap: Logic-based and scalable ontology matching. In: Aroyo, L., Welty, C., Alani, H., Taylor, J., Bernstein, A., Kagal, L., Noy, N., Blomqvist, E. (eds.) ISWC 2011, Part I. LNCS, vol. 7031, pp. 273–288. Springer, Heidelberg (2011)
10. Lacoste-Julien, S., Palla, K., Davies, A., Kasneci, G., Graepel, T., Ghahramani, Z.: Sigma: Simple greedy matching for aligning large knowledge bases. In: Proc. SIGKDD, pp. 572–580 (2013)

11. Leacock, C., Chodorow, M., Miller, G.A.: Using corpus statistics and wordnet relations for sense identification. Computational Linguistics 24(1), 147–165 (1998)
12. Lee, T., Wang, Z., Wang, H., won Hwang, S.: Web scale taxonomy cleansing. VLDB Endowment 4(12) (2011)
13. Li, J., Tang, J., Li, Y., Luo, Q.: Rimom: A dynamic multistrategy ontology alignment framework. TKDE 21(8), 1218–1232 (2009)
14. Miller, G.A.: Wordnet: A lexical database for english. Communications of the ACM 38(11), 39–41 (1995)
15. Ngo, D., Bellahsene, Z.: Yam++-results for oaei 2013. In: Proc. ISWC (2013)
16. Rahm, E., Bernstein, P.A.: A survey of approaches to automatic schema matching. The VLDB Journal 10(4), 334–350 (2001)
17. Shvaiko, P., Euzenat, J.: Ontology matching: state of the art and future challenges. TKDE 25(11), 158–176 (2013)
18. Shvaiko, P., Euzenat, J.: A survey of schema-based matching approaches. Data Semantics 4, 146–171 (2005)
19. Suchanek, F.M., Abiteboul, S., Senellart, P.: Paris: Probabilistic alignment of relations, instances, and schema. VLDB Endowment 5(3), 157–168 (2011)

Distributed XML Twig Query Processing Using MapReduce

Xin Bi, Guoren Wang, Xiangguo Zhao, Zhen Zhang, and Shuang Chen

College of Information Science and Engineering,
Northeastern University, Liaoning, Shenyang, China 110819
edijasonbi@gmail.com

Abstract. Twig query processing is one of the core operations of XML queries. Centralized holistic twig algorithms suffer great efficiency losses when large-scale XML documents are partitioned and stored in the cloud. Previous work on distributed twig query processing have some limitations, e.g., utter dependence on priori knowledge of query patterns, iteration of MapReduce jobs, etc. In this paper, our arbitrary XML partitioning and storage strategy require no knowledge of query pattern; twig queries can be efficiently processed in a single-round MapReduce job with good scalability. Extensive experiments are conducted to verify the efficiency and scalability of our algorithms.

1 Introduction

Scalable data management in the context of Big Data has emerged as a new research frontier[6]. As a semi-structured data, XML has become an important de facto standard format for data storage and exchange. The rapid growth of XML data scale boosts the demands of distributed XML query processing.

To avoid a large number of intermediate results of binary structural join[1], *holistic* twig algorithm[2] utilizes node streams and chained stacks to realize the examination of whether each matched node contributes to the final results, so that all the nodes pushed into the corresponding stacks can be guaranteed to be a part of solution. TwigStack[2] achieves optimal when the twig pattern contains only ancestor-descendant relationship. Follow-up work [10,4,11,3] lead to better query processing performance and illustrate that holistic twig algorithms are the most efficient processing methods for twig queries.

Cloud computing is currently a major technology for processing large-scale data. Since centralized holistic twig algorithms cannot be applied directly to shared-nothing distributed settings, especially to the cloud environment, two *challenging tasks* have to be addressed: 1) partition and storage strategies of large-scale XML data on shared-nothing machines; 2) parallelized local computing paradigms on each separated machine.

This paper focuses on distributed twig query processing in the case that large-scale XML data are partitioned and stored under the mechanism of the cloud computing framework, such as Hadoop. That is, our distributed XML storage strategy ensures the *arbitrariness* of partition *without any knowledge of incoming queries*. Distributed TwigStack (DTS) is proposed based on MapReduce[8]

© Springer International Publishing Switzerland 2015
R. Cheng et al. (Eds.): APWeb 2015, LNCS 9313, pp. 203–214, 2015.
DOI: 10.1007/978-3-319-25255-1_17

and TwigStack[2], which processes distributed twig queries *regardless of how the XML data are partitioned*. ComMapReduce[9] framework with a coordinator is applied, which refines global keys *in a lightweight manner*. In summary, the contributions of this paper are as follows:

1. A *distributed XML storage strategy* is designed to maintain structural information, without knowledge of incoming queries.
2. A *distributed twig query processing algorithm* based on TwigStack is proposed, regardless of how the XML data are partitioned.
3. Extensive experiments are conducted to verify the effectiveness and efficiency of the proposed algorithms.

The rest of the papers is organized as follows. Section 2 presents the related work. Section 3 proposes the partitioning strategy, based on which the distributed holistic twig algorithm is proposed in section 4. Experimental results are presented in section 5. Section 6 draws the conclusion of this paper.

2   Related Work

MapReduce[8] realizes powerful parallel computing ability with high scalability, including three major computing phases: 1) *map* receives a unit of input data each time and processes it into a list of $\langle key, value \rangle$ pairs; 2) *shuffle* combines each group of the *values* corresponding to the same *key* into $\langle key, list \langle value \rangle \rangle$; 3) *reduce* aggregates all the *values* of each *key* to produce final results.

Machdi et al. in [13] proposed XML data partition strategies based on GMX [12], taking the advantages of both inter and intra query parallelism on static and dynamic data distribution. Wu in [15] proposed a polynomial-based workload distribution algorithm. Inverted list labels were distributed rather than raw XML documents. However, in both of these two papers, XML fragments have to be stored with *priori knowledge of the query pattern*.

Damigos et al. in [7] decomposed queries into single paths to realize distributed XPath processing based on MapReduce. However, it required each XML partition to contain *full prefix path* from the root node. As a consequence, the flexibility would also be limited.

Choi et al. designed a system HadoopXML in [5] to process multiple twig queries simultaneously. The input scans and intermediate results were shared, which saved many I/Os, and improved runtime load balance. However, despite of preprocessing, HadoopXML needs *two* consecutive MapReduce jobs for query processing, which we aim to avoid in this paper.

3   Distributed XML Storage

3.1   Intuition of Arbitrary Partition

Different from traditional distributed environment, we pursue the *arbitrariness* of XML partitioning and storage in the cloud *without knowledge* of query patterns. To guarantee the *completeness of results*, we represent an XML node as a triple $\langle label, rcode, type \rangle$, in which *label* is the label name, *rcode* is the region code represented as $\langle start, end, level \rangle$, *type* indicates the partition type of the edges corresponding to this node. We also define that for node v, *parent edge* is the edge between v and its parent; an edge between v and one of its children is denoted as a *child edge*. Edge-cuts can be summarized in four *partition types*:

- no-cut (type 0): neither the parent edge nor a child edge is cut;
- child-edge-cut (type 1): at least a child edges is cut, but not the parent edge;
- parent-edge-cut (type 2): the parent edge is cut, but not the child edges;
- all-cut (type 3): both the parent edge and at least one child edge are cut.

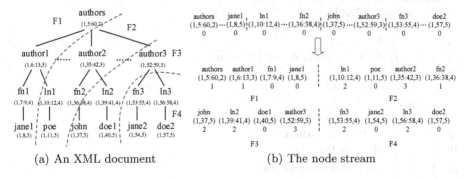

(a) An XML document   (b) The node stream

Fig. 1. Example of XML arbitrary partitioning with partition types

Example 1. Given an XML document[2] to be partitioned into four fragments as shown in Figure 1(a), while scanning the node stream, since *fragAnc* is empty, the first four nodes $\langle authors, (1,5:60,2),0 \rangle$, $\langle author, (1,6:13,3),0 \rangle$, $\langle fn1, (1,7:9,4),0 \rangle$ and $\langle jane1, (1,8,5),0 \rangle$ will not be processed. Then the stream is cut between $\langle jane1, (1,8,5),0 \rangle$ and $\langle ln1, (1,10:12,4),0 \rangle$, that is, node $\langle jane1, (1,8,5),0 \rangle$ belongs to the first fragment and node $\langle ln1, (1,10:12),4 \rangle$ is partitioned into the second fragment. According to case 1, the *type* of node $\langle ln1, (1,10:12,4),0 \rangle$ is set to 2, which will be $\langle ln1, (1,10:12,4),2 \rangle$, and of which the ancestor nodes $\langle authors, (1,5:60,2),0 \rangle$ and $\langle author, (1,6:13,3),0 \rangle$ are saved in *fragAnc*. Then the scan continues from node $\langle ln2, (1,39:41,2),0 \rangle$, of which the *type* is reset to 2; the *type* of node $\langle author3, (1,52:59,4),0 \rangle$ is reset to 3; the *type* of node $\langle fn3, (1,53:55,2),0 \rangle$ is reset to 2. The ancestor node $\langle author3, (1,52:59,4),0 \rangle$ is pushed into *fragAnc*. When the scan restarts from node $\langle jane2, (1,54,5),0 \rangle$, the *type* of node $\langle ln3, (1,56:58,4),0 \rangle$ is reset to 2. The final partitioned node stream is shown in Figure 1(b).

3.2 Partition Type Determination

Once the partition size is set by some criteria (e.g., Hadoop block size), during the traversal of the whole node stream, the *partition type* of the nodes in the previous partitions have to be re-checked, because they may have child-edge-cuts with the nodes in the following partitions. Therefore, an array named *fragAnc* is maintained by the following strategies: 1) all the ancestors of the next node after each partition cut will be pushed into $fragAnc$; 2) all the nodes at the same level as the node being traversed will be popped out from $fragAnc$.

Note that for the first node of each partition, only their ancestors may have cut edges with the remaining nodes have not been traversed; All the nodes at the same level with the node being traversed, let's say node v, will not have edges with v and nodes following v.

Algorithm 1. Arbitrary XML partitioning

Input: XML document T, block size S
Output: Partitions $parsCollection$
1 nodeStream=SAX(T);
2 **for** $i = 1$ *to nodeStream.size* **do**
3   v = nodeStream[i];
4   **if** *v is the first node of the partition* **then**
5   v.setType(2);
6   fragAnc.push(v.allAncestors());
7   **else**
8   **if** *partition.size()==S* **then**
9   **if** *!v.isLeaf()* **then**
10   v.setType(1);
11   parsCollection.add(partition);
12   v.checkFragAnc();

13 **function** checkFragAnc() **begin**
14   **if** *v.parent.isIn(fragAnc)* **then**
15   **if** *v.parent.type==0* **then**
16   v.parent.setType(1);
17   **if** *v.parent.type==2* **then**
18   v.parent.setType(3);
19   fragAnc.pop(nodes.atLevel(v.level()));

As described in Algorithm 1, given an XML document T, we first transform T into a node stream (Line 1). While scanning this node stream, let v be the current node being processed (Line 3). If v is the first node of the partition, which means the parent edge of v is cut, set type of v to 2 and push all the ancestors of v into $fragAnc$ (Lines 4-6). If v is the last node of the partition (Lines 8-11), a new partition of XML is generated. Before saving this partition, if v is not a leaf node, set the type to 1 (Line 10), meaning it will have child-edge-cut for sure; else 0. Then we check whether v has a parent in $fragAnc$ (Line 12).

Function checkFragAnc is designed to reset the type values of the nodes which have cut edges with the nodes being traversed. During the check of $fragAnc$, if v has a parent of type 0 (meaning no edge-cut), reset its type to 1 (meaning has a child-edge-cut); if type 2 (indicating having a parent-edge-cut), reset it to 3 (all-cut) (Lines 13-18). If v doesn't have a parent in $fracAnc$, we do nothing. At last, we pop out all the nodes in $fragAnc$ at the same level as v (Line 19).

4   Distributed Twig Query Processing

MapReduce is now one of the most efficient computing models in the cloud environment. In this section, based on MapReduce, we present a naïve solution to distributed twig processing, and propose a distributed holistic twig query processing algorithm on the basis of TwigStack.

4.1   Naïve Solution

Since no data exchange among mappers is allowed, and neither among reducers, arbitrarily partitioned XML data brings possibilities of losing matches across multiple partitions. A naïve solution is to run two rounds of MapReduce:

- Round 1 map: find all the matches to the query nodes on the local machine, emit key-value pairs using the match node as value and its highest ancestor which is also a match of root of the query as key;
- Round 1 reduce: collect the key-value pairs for the next round. Each final result can be derived from a sub-tree rooted by a key;
- Round 2 map: run a twig algorithm, e.g., TwigStack, to generate final results;
- Round 2 reduce: collect all the final results.

4.2   Distributed TwigStack

Intuition. The naïve method generates many unnecessary intermediate results in the first round, in addition, the start-up of a MapReduce job is time-consuming. Our framework Distributed TwigStack (DTS) aims at using one MapReduce job and reducing intermediate results as well.

In DTS, map function finds all the partial path matches on the partition stored on this machine, emits key-value pairs with globally refined keys by the coordinator of ComMapReduce. The improvement of DTS can be summarized in two aspects:

1. The matching rules are *relexed* in DTS, that is, some incomplete matching paths in TwigStack will also be considered as potential results, so that results across partitions can be retained;
2. The introduced coordinator guarantees that each group of matches need to be processed in the same reducer will have *the same keys*.

Algorithm DTS. More specifically, as described in Algorithm 2, the map function gets query Q from distributed cache (Line 2), and generates node streams *qStream* on this fragment F according to the nodes of Q (Line 3). Each map function invokes Relaxed Local TwigStack with Q, *qStreams* and partition nodes *partitionNodes*, i.e., nodes with cut edges, of fragment F as parameters (Line 6). Each machine running RL-TwigStack will send keys to the coordinator, and then the map function of Distributed TwigStack resets the keys of its own intermediate results, of which the key is the reset region encoding and the value is the partial matches to the query (Lines 7-10). The reduce function collects all the intermediate matching results corresponding to the same key and combines them into final results (Lines 12-14). Due to space constraints, the processing of ComMapReduce[9] and refining global keys are not presented.

Algorithm 2. Distributed TwigStack

```
1  function map(Q,F) begin
2      q=DistributedCache.get(Q);
3      qStream=getNodeStream(F,Q);
4      partitionNodes=getPartitionNodes(F);
5      if qStream!=null then
6          results=RL-TwigStack(Q,qStreams,partitionNodes);
7          keyArray=getKeysFromCoordinator();
8          refineKeys(results,keyArray);
9          foreach resultsᵢ ∈ results do
10             emit(resultsᵢ.key,resultsᵢ.value);

11 function reduce(key,list[value]) begin
12     resultsFinal=mergeAllResult(list[value]);
13     foreach resultsFinalᵢ ∈ resultsFinal do
14         emit(key,resultsFinalᵢ);
```

RL-TwigStack (Algorithm 3) invokes RL-GetNext function to guarantee that: 1) for any node $v_i \in child(v)$, a node of the XML document $h(v)$ has a descendant node $h(v_i)$ in stream T, or $h(v)$ has a cut child edge; 2) each $v_i \in child(v)$ satisfies the first point recursively. The *major difference* between RL-GetNext and getNext in TwigStack is that: RL-GetNext outputs not only the nodes having exact matches along the path from themselves to the leaf nodes of the query, but also the nodes with one or more cut child edges.

With a returned node q_{act} by RL-GetNext (Line 2), the clean stack operation (Lines 3, 4) ensures that the node in the parent stack of q_{act}, i.e., $Sparent(q_{act})$, is an ancestor of the node in node stream Tq_{act}. If q_{act} is the root node of the twig pattern, or the parent stack of q_{act} is not null, push q_{act} in to the stack Tq_{act} after popping the top node in Tq_{act} which is not an ancestor of q_{act} (Lines 5-7). If q_{act} is a leaf node of the twig pattern, or the descendant streams of q_{act}, i.e., $Tdesq_{act}$, have no descendant nodes of q_{act}, RL-ShowSolution (as Algorithm 5) is invoked (Lines 8-10) to produce intermediate results, which will be further explained in algorithm 5. If q_{act} is not the root node and the nearest ancestor stack is empty, the nodes in stream Tq_{act} with cut parent edge is pushed into stacks (Lines 11-17).

Compared with TwigStack, the examination criteria of whether a node can be pushed into chained stacks in RL-TwigStack is relaxed, that is, even a node in the node streams does not satisfy the getNext function, if it has one or more cut child edge, it will also be pushed into its stack. Therefore, the criteria of generating a partial solution is also relaxed in algorithm Relaxed Local ShowSolution (RL-ShowSolution), which is presented as Algorithm 5.

Algorithm 3. RL-TwigStack(q,T,partitionNodes)

---

1 **while** *!end(q)* **do**
2   q_{act}=RL-GetNext(q);
3   **if** *!isRoot(q_{act})* **then**
4   CleanStack(Sparent(q_{act}),nextBegin(Tq_{act}));
5   **if** *isRoot(q_{act})‖!empty(Sparent(q_{act}))* **then**
6   CleanStack(Sq_{act},nextBegin(Tq_{act}));
7   moveStreamToStack(Tq_{act},Sq_{act},pointerToTop(Sparent(q_{act})));
8   **if** *isLeaf(q_{act})‖!Tq$_{act}$.isAncestorOf(Tnxtodesq_{act})* **then**
9   RL-ShowSolutions(Sq_{act},1);
10   pop(Sq_{act});
11   **else if** *Tq$_{act}$.hasAncestorIn(partitionNodes)* && *(isLeaf(q_{act}) ‖ !Tq$_{act}$.isAncestorOf(Tdesq_{act}))* **then**
12   CleanStack(Sq_{act},nextBegin(Tq_{act}));
13   moveStreamToStack(Tq_{act},Sq_{act},pointerToTop(Sparent(q_{act})));
14   RL-ShowSolutions(Sq_{act},1);
15   pop(Sq_{act});
16   **else**
17   advance(Tq_{act});

In RL-ShowSolutions, each index[i] (i=1...n, n is the number of stacks) indicates the position of the node being processed in its corresponding stack (Line 2). SN is the stack ID and SP is the node position in its stack. The keys of all the path solutions are set to the highest region encode. The value of the solutions is

---

Algorithm 4. RL-GetNext(q)

---

1 **if** *isLeaf(q)* **then**
2   return q;
3 **foreach** $q_i \in children(q)$ **do**
4   n_i=RL-GetNext(q_i);
5   **if** $n_i != q_i$ **then**
6   return n_i;
7 n_{min}=minarg(nextBegin(Tn_i)); n_{max}=maxarg(nextBegin(Tn_i));
8 **while** $nextEnd(T_q) < nextBegin(Tn_{max})$ && *!Tq.hasDescendantIn(partitionNodes)* **do**
9   advance(q);
10 **if** $nextBegin(Tq) < nextBegin(Tn_{min})$ **then**
11   return q;
12 **else**
13   return n_{min};

---

a key-value pair $\langle pathType, path \rangle$. $pathType$ can be set to four values, including
a full path type "r-m-l", a path with only a root node "r-*", a path with only
a leaf node "*-l" and a path with only the middle nodes "*-m-*". There are
four cases of generating a solution: 1) if the relative leaf node of this solution
index[$originalSN$] has a cut child edge, and the relative root node index[SN]
has a cut parent edge, solutions of types "*-l", "r-*" and "r-m-l" will be generated (Lines 6 and 7); 2) if the relative leaf node has a cut child edge, and the
relative root node index[SN] has no cut parent edge, solutions of types "r-m-l"
and "r-*" will be generated (Lines 8-9); 3) if the relative leaf node has no cut
child edge, and the relative root node index[SN] has a cut parent edge, solutions
of types "r-m-l" and "*-l" will be generated (Lines 10-11); 4) if the relative leaf
node has no cut child edge and the relative root node index[SN] has no cut
parent edge, only solutions of type "r-m-l" will be generated (Lines 12-13). All
the intermediate solutions will be generated recursively (Lines 15-17).

Algorithm 5. RL-ShowSolutions(SN,SP)

```
1  originalSN=SN;
2  index[SN]=SP;
3  if SN==1 || (SN-1).size()==0 then
4      hasDes=index[originalSN].hasDescendantIn(partitionNodes);
5      hasAnc=index[SN].hasAncestorIn(partitionNodes);
6      if hasDess && hasAnc then
7       |  results=output("*-l","r-*","*-m-*");
8      else if hasDes && !hasAnc then
9       |  results=output("r-m-l","*r-*");
10     else if !hasDes && hasAnc then
11      |  results=output("r-m-l","*-l");
12     else
13      |  results.put(key,output(index[SN],...,index[originalSN]));
14     sendKeyToCoordinator(results);
15 else
16     for i=1 to index[SN].pointer do
17      |  RL-ShowSolutions(SN-1,i);
```

4.3   An Example of DTS

Example 2. Given a query "//author[fn/text()="jane" $\bigwedge$ ln/text()="doe"]", on
fragment F1, RL-TwigStack recursively invokes method RL-GetNext. Since RL-
GetNext(ln) returns $null$, RL-GetNext(fn) returns node fn, so that invocation
RL-GetNext(a) returns node a. Because stack Sa is empty and the current node
in stream Ta is node $a1$, node $a1$ is pushed into Sa. In the next round, RL-
GetNext(q) returns node fn. The parent stack of Sfn is not empty, of which
the top element $a1$ is the parent of the current node in Tfn, therefore, node $fn1$
is pushed into stack Sfn, and a pointer is set from $fn1$ to $a1$ in the chained

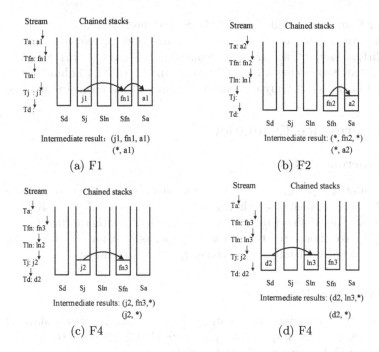

Fig. 2. Intermediate solutions on F1, F2 and F4

stack. In the third round, RL-GetNext(q) returns $j1$, which is pushed into Sj and linked to $fn1$ in stack Sfn. Since $j1$ is a leaf node and the $type$ value of $a1$ is 1, thus two key-value pairs are generated: $\langle 6 : 13, (j1, fn1, a1)\rangle$ and $\langle 6 : 13, (*, a1)\rangle$.

On fragment F2, nodes $ln1$, $a2$ and $fn2$ are stored in $fragAnc$. During RL-GetNext(ln) in the first round, the child stream of Tj is empty, and the current node in stream Tln has no cut child edge, thus RL-GetNext(ln) returns $null$. But the current node $a2$ in stream Ta has a cut child edge, thus $a2$ is pushed into stack Sa. In the second round, node $fn1$ is pushed in to stack. Thus, two intermediate results are produced, which are $\langle 35 : 42, (*, a2)\rangle$ and $\langle 35 : 42, (*, fn2, *)\rangle$.

On fragment F4, nodes $fn3$ and $ln3$ are saved in $fragAnc$. The first round of RL-GetNext(q) returns fn. The current node of stream Tfn has a cut parent edge, node $fn3$ is pushed into stack. In the second round, node $j2$ is pushed into stacks. Thus, the intermediate results of F4 are $\langle 53 : 55, (j2, fn3, *)\rangle$, $\langle 53 : 55, (j2, *)\rangle$, $\langle 56 : 58, (d2, ln3, *)\rangle$ and $\langle 56 : 58, (d2, *)\rangle$.

The computation on F3 is similar to F2, producing intermediate results $\langle 39 : 41, (d1, ln2, *)\rangle$, $\langle 39 : 41, (d1*)\rangle$ and $\langle 52 : 59, (*, a3)\rangle$.

After the local computation, coordinator receives all the keys sent from all the fragments, which are 6:13, 35:42, 53:55, 56:58, 39:41 and 52:59. Coordinator checks whether each two of these keys have ancestor-descendant relationships, and send back the refined keys to each machine running map functions. As a result, the keys on F1 and F2 remain the same; 39:41 on F3 is set to 35:42; all the keys on F4 are set to 52:59. Then all the refined intermediate results corresponding to a same key are sent to the same machine running reduce functions. That is, (j1, fn1, a1) and (*, a1) are sent to machine 1; (*, a1), (*, fn1, *), (d1, ln2, *) and (d1,*) are sent to machine 2; (*,a3), (j2,fn3,*), (j2,*), (d2,ln3,*) and (d2, *) are sent to machine 3. Only machine 3 produces a final result (a3,fn3,ln3,j2,d2).

5 Performance Evaluation

5.1 Experiments Setup

The distributed experiments are conducted on a Hadoop cluster of 1 master node and 8 slave nodes, each of which has an Intel Core 2.66GHZ CPU, 4GB memory, CentOS 5.6 and Hadoop 0.20.2. We use a 5.4GB synthetic dataset based on XMark[14]. We choose the twig pattern "//person[//country="US"]//zipcode" as Q1, "//people/person[/address/country="US"]/creditcard" as Q2.

5.2 Evaluation Results

The *scalability* comparison is presented in Figure 3. The number of slaves changes from 1 to 8. Figures 3(a) and 3(b) indicate that DTS outperforms the naïve method, especially with relatively fewer slave nodes or more complex queries. Figures 3(c) and 3(d) indicate that the efficiency of the naïve method drops more dramatically than DTS when the size of dataset increases.

Figure 4(a) and 4(b) presents the *speedup* comparison, indicating that along with the increment of the slaves number, DTS has a better scalability due to its

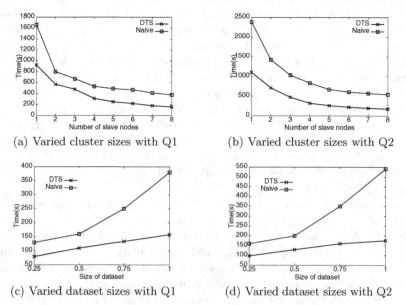

(a) Varied cluster sizes with Q1   (b) Varied cluster sizes with Q2

(c) Varied dataset sizes with Q1   (d) Varied dataset sizes with Q2

Fig. 3. Query times comparison

one-round MapReduce design, intermediate results matching and distribution mechanism. Figures 4(c) and 4(d) present the *scaleup* comparison, indicating that:1) the naïve method has a better speedup at first, because the startup of MapReduce jobs is the major part of the query time when the dataset is relatively small; 2) DTS gains a better sizeup when the scale of XML data increases.

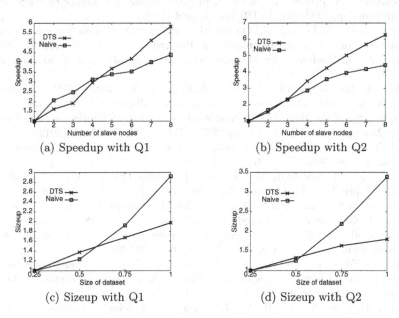

(a) Speedup with Q1 (b) Speedup with Q2

(c) Sizeup with Q1   (d) Sizeup with Q2

Fig. 4. Scalability comparison

6   Conclusion

In order to answer twig queries over large-scale XML documents in the cloud, we propose an arbitrary XML partitioning strategy without priori knowledge of the query pattern, and a distributed query processing algorithm DTS based on MapReduce, regardless of how the XML data are partitioned. Experimental results show that DTS achieves good efficiency and scalability.

Acknowledgment. This research is partially supported by the National Natural Science Foundation of China under Grant Nos. 61272181 and 61173030; the National Basic Research Program of China under Grant No. 2011CB302200-G.

References

1. Al-Khalifa, S., Jagadish, H., Koudas, N., Patel, J., Srivastava, D., Wu, Y.: Structural joins: A primitive for efficient XML query pattern matching. In: Proceedings of the 18th International Conference on Data Engineering, pp. 141–152. IEEE Computer Society, Washington, DC (2002)

2. Bruno, N., Koudas, N., Srivastava, D.: Holistic twig joins: Optimal XML pattern matching. In: Proceedings of the 2002 ACM SIGMOD International Conference on Management of Data, pp. 310–321. ACM, New York (2002)

3. Chen, S., Li, H.G., Tatemura, J., Hsiung, W.P., Agrawal, D., Candan, K.S.: Twig2stack: Bottom-up processing of generalized-tree-pattern queries over XML documents. In: Proceedings of the 32nd International Conference on Very Large Data Bases, pp. 283–294. VLDB Endowment (2006)

4. Chen, T., Lu, J., Ling, T.W.: On boosting holism in XML twig pattern matching using structural indexing techniques. In: Proceedings of the 2005 ACM SIGMOD International Conference on Management of Data, pp. 455–466. ACM, New York (2005)

5. Choi, H., Lee, K.H., Kim, S.H., Lee, Y.J., Moon, B.: HadoopXML: A suite for parallel processing of massive XML data with multiple twig pattern queries. In: Proceedings of the 21st ACM International Conference on Information and Knowledge Management, CIKM 2012, pp. 2737–2739. ACM, New York (2012)

6. Cui, B., Mei, H., Ooi, B.C.: Big data: the driver for innovation in databases. National Science Review 1(1), 27–30 (2014)

7. Damigos, M., Gergatsoulis, M., Plitsos, S.: Distributed processing of XPath queries using MapReduce. In: Catania, B., et al. (eds.) New Trends in Databases and Information Systems. AISC, vol. 241, pp. 69–77. Springer, Heidelberg (2014)

8. Dean, J., Ghemawat, S.: Mapreduce: Simplified data processing on large clusters. In: Proceedings of the 6th Conference on Symposium on Opearting Systems Design & Implementation, Berkeley, CA, USA, vol. 6, p. 10 (2004)

9. Ding, L., Wang, G., Xin, J., Wang, X., Huang, S., Zhang, R.: Commapreduce: An improvement of mapreduce with lightweight communication mechanisms. Data & Knowledge Engineering 88, 224–247 (2013)

10. Jiang, H., Wang, W., Lu, H., Yu, J.X.: Holistic twig joins on indexed XML documents. In: Proceedings of the 29th International Conference on Very Large Data Bases, vol. 29, pp. 273–284. VLDB Endowment (2003)

11. Lu, J., Ling, T.W., Chan, C.Y., Chen, T.: From region encoding to extended dewey: On efficient processing of XML twig pattern matching. In: Proceedings of the 31st International Conference on Very Large Data Bases, pp. 193–204. VLDB Endowment (2005)

12. Machdi, I., Amagasa, T., Kitagawa, H.: Gmx: An XML data partitioning scheme for holistic twig joins. In: Proceedings of the 10th International Conference on Information Integration and Web-based Applications & Services, iiWAS 2008, pp. 137–146. ACM, New York (2008)

13. Machdi, I., Amagasa, T., Kitagawa, H.: XML data partitioning strategies to improve parallelism in parallel holistic twig joins. In: Proceedings of the 3rd International Conference on Ubiquitous Information Management and Communication, ICUIMC 2009, pp. 471–480. ACM, New York (2009)

14. Schmidt, A., Waas, F., Kersten, M., Carey, M.J., Manolescu, I., Busse, R.: Xmark: A benchmark for XML data management. In: Proceedings of the 28th International Conference on Very Large Databases, San Francisco, pp. 974–985 (2002)

15. Wu, H.: Parallelizing structural joins to process queries over big XML data using MapReduce. In: Decker, H., Lhotská, L., Link, S., Spies, M., Wagner, R.R. (eds.) DEXA 2014, Part II. LNCS, vol. 8645, pp. 183–190. Springer, Heidelberg (2014)

Sentiment Word Identification
with Sentiment Contextual Factors

Jiguang Liang, Xiaofei Zhou, Yue Hu, Li Guo, and Shuo Bai

National Engineering Laboratory for Information Security Technologies,
Institute of Information Engineering, Chinese Academy of Sciences
{liangjiguang,zhouxiaofei,huyue,guoli,baishuo}@iie.ac.cn

Abstract. Sentiment word identification (SWI) refers to the task of automatically identifying whether a given word expresses positive or negative opinion. SWI is a critical component of sentiment analysis technologies. Traditional sentiment word identification techniques become unqualified because they need seed sentiment words which leads to low robustness. In this paper, we consider SWI as a matrix factorization problem and propose three models for it. Instead of seed words, we exploit sentiment matching and sentiment consistency for modeling. Extensive experimental studies on three real-world datasets demonstrate that our models outperform the state-of-the-art approaches.

1 Introduction

In recent years, sentiment analysis, also known as opinion mining, has been a hot research area and has attracted much attention from many fields (e.g., data mining, information retrieval and information security). The task of sentiment analysis is to classify a subjective text as positive or negative according to the sentiment expressing in it. Sentiment word always plays a decisive role in sentiment classification. In this paper, we concentrate on this task of sentiment word identification (SWI). Most existing automatic approaches for sentiment word identification are either dictionary-based or corpus-based. Approaches in both classes have in common that they typically exploit a set of labeled words called seed words. By calculating semantic similarities between seed words and candidate words, the orientation of each candidate word is derived. However, those seed words are always manually selected. This selection process is very subjective. Any missing key words may lead to poor performance. Therefore, those seed words based approaches have low robustness.

To tackle this problem, Yu et al. [19] and Liang et al. [9] have proposed two solutions without seed words. They respectively utilize sentiment matching and sentiment consistency for modeling. They assume that subjective corpora fit the two sentiment phenomena but have not proved it. In fact, the hypotheses is an ideal condition. Real-world corpora can not fit them well.

Inspired by the two models, we explore to combine the two sentiment contextual factors for SWI in this paper. We firstly investigate whether sentiment matching and sentiment consistency exist on real-world corpora. We find that the

© Springer International Publishing Switzerland 2015
R. Cheng et al. (Eds.): APWeb 2015, LNCS 9313, pp. 215–226, 2015.
DOI: 10.1007/978-3-319-25255-1_18

two phenomena are complementary sentiment contextual factors in real-world corpora. Then, we discuss how the sentiment contextual information could be modeled and utilized for SWI. Finally, we conduct extensive experiments to verify the proposed models. Even inspired by [19,9], this paper is different from them. The main contributions of this paper can be summarized as follows:

- This paper is the first to verify the existences of sentiment matching and sentiment consistency on real review datasets. We also demonstrate that the two phenomena are complementary sentiment contextual factors.
- We propose three novel matrix factorization based models to automatically identify sentiment words without seed words, which simultaneously exploit the sentiment labels of documents and integrate sentiment consistency information as regularization terms into the objective functions. It is the first work that utilizes both sentiment matching and sentiment consistency for modeling.
- We propose a new similarity function from both co-occurring and semantic perspectives, which combines pointwise mutual information and pre-trained word vectors.
- We conduct extensive empirical studies to compare our models with other state-of-the-art approaches.

The rest of this paper is organized as follows. We introduce the motivation in Section 2. Section 3 defines our effective sentiment word identification models. We present our experimental setup and results in Section 4, while describing the related work in Section 5. Finally conclusion appears in Section 6.

2   Motivation

In this section, we introduce our motivation exploiting sentiment contextual relations for SWI. At first, we would like to demonstrate the existence and significance of sentiment contextual factors (including sentiment matching and sentiment consistency) on three real datasets which will be introduced in Section 4.1. We classify the sentiment words into positive and negative according MPQA[1].

In order to describe sentiment matching, we calculate the percentage of sentiment words that match to the documents' polarities and define sentiment matching ratio as

$$sm_j = \frac{\sum_{i=1}^{M} I_{ij} f(d_i, c_j)}{\sum_{i=1}^{M} I_{ij}}, \tag{1}$$

where M is the document number in the corpus, d_i is the ith document and c_j is the jth word in the vocabulary list with size N. I_{ij} is the indicator function that is equal to 1 if $c_j \in d_i$ and equal to 0 otherwise. $f(d_i, c_j) \in \{0, 1\}$ indicates whether polarities of d_i and c_j are the same.

We define the sentiment consistency ratio of c_j as

---

[1] http://mpqa.cs.pitt.edu/

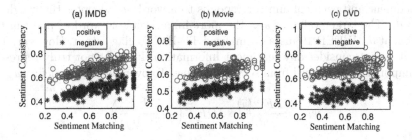

Fig. 1. Correlations between sentiment matching and sentiment consistency on three datasets.

$$sc_j = \frac{\sum_{i=1}^{M}\sum_{k=1}^{N} I_{ij}I_{ik}f(c_j, c_k)}{\sum_{i=1}^{M}\sum_{k=1}^{N} I_{ij}I_{ik}}. \tag{2}$$

We plot part of the pairs (sm_j, sc_j) as points w.r.t. sentiment matching and sentiment consistency in Fig.1. It can be observed that (1) only a very few of sentiment words and documents absolutely fit to sentiment matching, (2) more than half of sentiment words have $sm_j \geq 0.5$ (3) positive sentiment words have higher sentiment consistency ratios than negative sentiment words and they are linearly separable, (4) sentiment matching and sentiment consistency are two complementary sentiment contextual factors because $(sm_j \geq 0.5 \,||\, sc_j \geq 0.5)$ can cover the vast majority of words, and (5) *IMDB* matches the two phenomena best, then the *DVD* corpus, and the *Movie* corpus last.

The first three observations show that either sentiment matching or sentiment consistency exists in subjective corpora. Both of them can be modeled to address SWI problem. However, only one of them can not cover most of the words because $sm_j \geq 0.5$ or $sc_j \geq 0.5$ can only cover slightly more than half words. Observation (4) provides the possibility of combining them for modeling. But there are still a relatively small number of words in the area where $(sm_j < 0.5 \,\&\, sc_j < 0.5)$. One possible reason is that the pattern "negation word + negative word" is used to express positive sentiment while "negation word + positive word" is used to express negative sentiment. Negation word (e.g., not, never, don't, can't) could reverse the polarity of word. Therefore, it is necessary to do the negation preprocessing by concatenating first word after the negation word that should not be a stop word. For example, *"not a good* idea" becomes to *"not_good* idea" after negation handling. Observation (5) shows that different corpora differ in matching the two sentiment phenomena. Here, we make a prediction that the more matching the higher sentiment word identification accuracy.

3   Methodology

3.1   Problem Formulation

The M documents and N words in the corpus are represented in an $M \times N$ document-word matrix $W = [w_1, w_2, \cdots, w_N]$, in which each row corresponds to a document and each column corresponds to a word. w_{ij} denotes the weight of c_j to d_i. The larger value of w_{ij}, the more important role word c_j plays in the document d_i. WEED[19] uses normalized TF-IDF to claculate it. However, TF-IDF does not consider any sentiment information. It might not truly represent the importance of a word from sentiment view. To overcome it, we define w_{ij} as

$$w_{ij} = \frac{TF^{(i)}(j)}{|d_i|} \cdot IDF_j \cdot \left(\frac{CF^{(+)}(j)}{CF^{(-)}(j)} \right)^{l_i} \tag{3}$$

where $TF^{(i)}(j)$ is the frequency of c_j in d_i, $CF^{(+)}(j)$ and $CF^{(-)}(j)$ are the collection frequencies of c_j in positive and negative corpora, and IDF_j is the inverse document frequency of c_j. $l_i \in \{+1, -1\}$, if d_i is a positive document, then $l_i = +1$; otherwise $l_i = -1$. $|d_i|$ is the length of d_i.

Let U be an $M \times K$ document-feature matrix in which each column u_k corresponds to a latent feature and each row $\bar{u}_i = [u_{i1}, u_{i2}, \cdots, u_{iK}]$ stands for the representation of document d_i in the latent feature space. Suppose that there are M^+ positive documents and M^- negative documents in the corpus where $M^+ + M^- = M$. So U can be split into 2 sub-matrices $U = [U^+, U^-]^T$ according to documents' sentiments where $U^+ = [\bar{u}_1, \bar{u}_2, \cdots, \bar{u}_{M^+}]^T$ and $U^- = [\bar{u}_{M^++1}, \bar{u}_{M^++2}, \cdots, \bar{u}_M]^T$. $V \in \mathbb{R}^{K,N}$ is the feature-word matrix, where column v_n stands for the representation of word c_n in the latent feature space.

For each word c_n, we calculate its sentiment polarity value by

$$y_n = \frac{1}{M^+} ||U^+ v_n||_1 - \frac{1}{M^-} ||U^- v_n||_1 \tag{4}$$

where $y_n > 0$ indicates c_n is a positive word while $y_n < 0$ indicates c_n is a negative word.

3.2 Modeling Content

This paper considers sentiment word identification as a matrix factorization problem. More precisely, the document-word matrix W is approximated as UV. It amounts to solving the following optimization problem:

$$\min_{U,V} \frac{1}{2}||W - UV||_F^2 + \frac{\gamma}{2}||U||_F^2 + \frac{\lambda}{2}||V||_F^2 \tag{5}$$

where $||\cdot||_F^2$ denotes the Frobenius norm, $\gamma, \lambda > 0$ and the last two regularization terms are used to avoid overfitting. In fact, W is a sparse matrix. We only need to factorize the nonzero entries. Hence, we change Equation 5 to

$$\min_{\{\bar{u}_i\},\{v_j\}} \sum_{i=1}^{M} \sum_{j=1}^{N} I_{ij}(w_{ij} - \bar{u}_i v_j)^2 + \frac{\gamma}{2} \sum_{i=1}^{M} ||\bar{u}_i||_F^2 + \frac{\lambda}{2} \sum_{j=1}^{N} ||v_j||_F^2 \tag{6}$$

Low rank matrix factorization models, such as singular value decomposition (SVD) and non-negative matrix factorization (NMF), are always utilized to solve this problem.

3.3 Modeling Sentiment Matching

In this section, we discuss how to model sentiment matching phenomenon for SWI. Yu et al. [19] believe that document and its most component sentiment words share the same sentiments, It be true in the situation that a document only express one sentiment. Obviously, in real-world subjective documents, there might be more than one sentiments expressed in them. Based on this intuition, we hold that, for each document, the average of the sums of all its contained words' polarity values is closed to its polarity value. We formulate it as the following optimization problem

$$\min_{U,\{v_j\}} \sum_{i=1}^{M} (\frac{1}{|d_i|} \sum_{j=1}^{N} I_{ij} \frac{||U^x v_j||_1}{M^x} - l_i)^2 \tag{7}$$

where $x \in \{+, -\}$, if $l_i = 1$ then $x = +$; else $x = -$. Then the objective function exploiting sentiment matching can be formulated as

$$
\begin{aligned}
\min_{\{\overline{u}_i\},\{v_j\}} \mathcal{J}(W,U,V) = & \frac{1}{2} \sum_{i=1}^{M} \sum_{j=1}^{N} I_{ij}(w_{ij} - \overline{u}_i v_j)^2 \\
& + \frac{\alpha}{2} \sum_{k=1}^{M^+} (\frac{1}{|d_k|} \sum_{j=1}^{N} I_{kj} \frac{\sum_{i=1}^{M^+} \overline{u}_i v_j}{M^+} - 1)^2 \\
& + \frac{\alpha}{2} \sum_{k=M^++1}^{M} (\frac{1}{|d_k|} \sum_{j=1}^{N} I_{kj} \frac{\sum_{i=1+M^+}^{M} \overline{u}_i v_j}{M^-} + 1)^2 \\
& + \frac{\gamma}{2} \sum_{i=1}^{M} ||\overline{u}_i||_F^2 + \frac{\lambda}{2} \sum_{j=1}^{N} ||v_j||_F^2
\end{aligned}
\tag{8}
$$

where $\alpha > 0$. The second and the third terms are sentiment matching regularization terms with respect to U^+ and U^-. We call this model SM. In this paper, we use the gradient search method to solve the problem.

3.4 Modeling Sentiment Consistency with Average-Based Term

According to sentiment consistency, two frequently co-occurring sentiment words are more likely to have the same polarity than those of two randomly selected words. So let's further assume that, c_i' polarity should be close to the expectation polarity value of all the co-occurring words. Based on it, we propose another regularization term to impose constraint on sentiment consistency

$$\min_{v_j} ||v_j - \sum_{i=1}^{N} R_{ij} \frac{Sim(c_i, c_j)}{\sum_{k=1}^{N} R_{jk} Sim(c_j, c_k)} v_i||_F^2 \tag{9}$$

where $R_{ij} \in \{0, 1\}$, if c_i and c_j have once co-occurred then $R_{ij} = 1$; otherwise $R_{ij} = 0$. Similarity function $Sim(c_i, c_j)$ allows to treat c_i' co-occurring words differently. CONR [9] uses PMI to capture the similarity. PMI considers co-occurring frequency as the main measure. Factually, even some words seldom co-occur, they might similar from semantic perspective. For example, "great" and "greatly" hold the same meaning although they may seldom co-occur. PMI fails to capture those semantic information. To solve this problem, we present a novel similarity between c_i and c_j defined as

$$Sim(c_i, c_j) = lg \frac{p(c_i, c_j)}{p(c_i)p(c_j)} \cdot q(c_i, c_j) \cdot cos(\vec{c_i}, \vec{c_j}) \tag{10}$$

where the first term is PMI scoring, $q(c_i, c_j)$ is the co-occurring frequency in the original corpus and cos is a semantic similarity function. We use cosine similarity with word vector trained by Word2Vec[2] to calculate semantic similarity between c_i and c_j. Word2Vec is a neural network implementation that learns distributed and meaningful representation for words. Each word is represented by a fixed-length vector. Word2Vec is the state-of-the-art word vector model. It can enable words with similar semantic properties close to each other in the latent space. $\vec{c_i}$ is the vector representation of c_i by Word2Vec.

Then, the optimization formulation, which integrates sentiment matching and average-based consistency term into the learning process, could be defined as

$$
\begin{aligned}
\min_{\{\overline{u_i}\}, \{v_j\}} \mathcal{J}(W, U, V) = {} & \frac{1}{2} \sum_{i=1}^{M} \sum_{j=1}^{N} I_{ij}(w_{ij} - \overline{u}_i v_j)^2 \\
& + \frac{\alpha}{2} \sum_{k=1}^{M^+} (\frac{1}{|d_k|} \sum_{j=1}^{N} I_{kj} \frac{\sum_{i=1}^{M^+} \overline{u}_i v_j}{M^+} - 1)^2 \\
& + \frac{\alpha}{2} \sum_{k=M^++1}^{M} (\frac{1}{|d_k|} \sum_{j=1}^{N} I_{kj} \frac{\sum_{i=1+M^+}^{M} \overline{u}_i v_j}{M^-} + 1)^2 \\
& + \frac{\beta_1}{2} \sum_{j=1}^{N} ||v_j - \sum_{i=1}^{N} R_{ij} \frac{Sim(c_i, c_j)}{\sum_{k=1}^{N} R_{jk} Sim(c_j, c_k)} v_i||_F^2 \\
& + \frac{\gamma}{2} \sum_{i=1}^{M} ||\overline{u}_i||_F^2 + \frac{\lambda}{2} \sum_{j=1}^{N} ||v_j||_F^2
\end{aligned}
\tag{11}
$$

where β_1 is a positive regularization parameter. We name this model SMC_1.

3.5   Modeling Sentiment Consistency with Individual-Based Term

SMC_1 imposes a consistency regularization term to constrain sentiments among words. However, this approach is insensitive to those words that convey diverse

---

[2] http://word2vec.googlecode.com/svn/trunk/word2vec.c

sentiments but appear in the same document. It will cause information loss problem, which will result in inaccurate predicting of y. Hence, we impose constraints between one word and its co-occurring words individually

$$\min_{v_j} R_{ij} \sum_{i=1}^{N} \frac{Sim(c_i, c_j)}{\sum_{k=1}^{N} R_{jk} Sim(c_j, c_k)} ||v_j - v_i||_F^2 \qquad (12)$$

Under this scenario, the task of sentiment word identification can be mathematically formulated as solving the following optimization problem

$$\min_{\{\overline{u}_i\},\{v_j\}} \mathcal{J}(W, U, V) = \frac{1}{2} \sum_{i=1}^{M} \sum_{j=1}^{N} I_{ij}(w_{ij} - \overline{u}_i v_j)^2$$
$$+ \frac{\alpha}{2} \sum_{k=1}^{M^+} (\frac{1}{|d_k|} \sum_{j=1}^{N} I_{kj} \frac{\sum_{i=1}^{M^+} \overline{u}_i v_j}{M^+} - 1)^2$$
$$+ \frac{\alpha}{2} \sum_{k=M^++1}^{M} (\frac{1}{|d_k|} \sum_{j=1}^{N} I_{kj} \frac{\sum_{i=1+M^+}^{M} \overline{u}_i v_j}{M^-} + 1)^2 \qquad (13)$$
$$+ \frac{\beta_2}{2} \sum_{j=1}^{N} \sum_{i=1}^{N} R_{ij} \frac{Sim(c_i, c_j)}{\sum_{k=1}^{N} R_{jk} Sim(c_j, c_k)} ||v_j - v_i||_F^2$$
$$+ \frac{\gamma}{2} \sum_{i=1}^{M} ||\overline{u}_i||_F^2 + \frac{\lambda}{2} \sum_{j=1}^{N} ||v_j||_F^2$$

where $\beta_2 > 0$. For convenience sake, this model is called SMC_2.

4   Experimental Analysis

4.1   Experiment Setup

Datasets. We evaluate our method on the three real-world corpora: the Internet Movie Database[3] (IMDB), Movie review dataset collected by Pang et al.[13] and DVD reviews from NLP&CC2013[4]. The related data settings are the same as for [9].

Evaluation Metrics. Following [19,9], we consider the precision of the top K sentiment words in our experiments and we use $P@K$ and $MAP@K$ to evaluate the top-K ranked words. More precisely, $P_1@K$ is used to evaluate the ability of recognizing sentiment words from non-sentiment words while $P_2@K$ assesses the performance of classifying sentiment words as positive or negative words. $MAP_1@K$ and $MAP_2@K$ consider the order for evaluating the ranked words. Concrete definitions can be found in [9].

---

[3] http://ai.stanfor.edu/amaas/data/sentiment/
[4] http://www.datatang.com/data/44115/

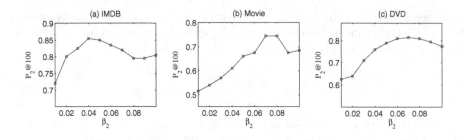

Fig. 2. P_2@100 vs. α on (a) IMDB, (b) Movie and (c) DVD.

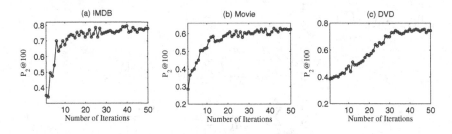

Fig. 3. P_2@100 vs. $IterationNumber$ on (a) IMDB, (b) Movie and (c) DVD.

4.2   Parameter Settings

Here we focus on parameter settings. In our experiments, we set dimension $K =$ 10. Next, we investigate the performances when the parameters change.

Tradeoff Parameters: The parameters α, β_1 and β_2 in our models play the role of learning rates and adjusting the strengths of different constraint terms in the objective functions as shown in Equation (8), (11) and (13). The impacts of α and β_1 generally share the similar trend as the impact of β_2. Hence we only illustrate the results of β_2 for SMC_2 here due to the space limitation. In the extreme case, if we use a very small value of β_2, only content term is employed. On the other side, if we employ a very large value of β_2, the sentiment consistency term will dominate the learning processes. As a result, we conduct experiments with β_2 ranging from 0.01 to 0.1. Fig.2 shows the impacts of β_2 on P_2@K. We observe that the values of β_2 impact the identification results significantly. From Fig.2, we can find that when $\beta_2 = 0.04$, SMC_2 obtains the best performance on IMDB dataset. Hence, we choose this parameter in our experiments. Similarly, we choose the parameter $\beta_2 = 0.07$ for Movie and DVD dataset. Although they may be not the best ones, the following experiments demonstrate they are adequate.

Number of Iterations: Fig.3 shows the impacts of iteration number on P_2@100 in our model. From the results, we can see that no matter using which dataset, as iteration number increases, the P_2@100 values increase gradually at first. With further increasing, P_2@100 finally turns to a relatively stable value. In order to reach converged results with an acceptable time cost, it is better to run more than 30 iterations in the following experiments.

Table 1. $P_1@K$ performance comparisons.

Dataset	$P_1@K$	SO-PMI	SVD	NMF	CONR	WEED	SM	SMC_1	SMC_2
IMDB	Top10	0.818	0.969	0.976	0.983	0.980	0.990	0.990	**0.995**
	Top20	0.780	0.975	0.962	0.975	0.965	0.975	**0.995**	**0.995**
	Top50	0.792	0.946	0.948	0.975	0.957	0.970	**0.976**	0.970
	Top100	0.793	0.937	0.931	0.962	0.948	0.950	0.955	**0.965**
Movie	Top10	0.818	0.686	0.658	0.947	0.916	0.909	0.950	**0.960**
	Top20	0.780	0.707	0.622	0.853	0.877	0.853	0.875	**0.894**
	Top50	0.792	0.737	0.673	0.826	0.826	0.834	**0.847**	0.834
	Top100	0.783	0.761	0.751	0.820	0.808	0.813	0.825	**0.840**
DVD	Top10	0.625	0.884	0.892	0.931	0.909	0.924	0.935	**0.940**
	Top20	0.595	0.904	0.880	0.916	0.913	0.921	0.924	**0.930**
	Top50	0.627	0.862	0.872	0.928	0.901	0.910	0.930	**0.936**
	Top100	0.608	0.841	0.821	0.910	0.876	0.901	0.896	**0.919**

Table 2. $P_2@K$ performance comparisons.

Dataset	$P_2@K$	SO-PMI	SVD	NMF	CONR	WEED	SM	SMC_1	SMC_2
IMDB	Top10	0.512	0.884	0.891	0.938	0.894	0.950	0.990	**0.995**
	Top20	0.553	0.836	0.870	0.917	0.861	0.875	0.950	**0.975**
	Top50	0.508	0.814	0.814	0.878	0.850	0.870	0.870	**0.894**
	Top100	0.518	0.790	0.795	0.846	0.828	0.836	0.843	**0.855**
Movie	Top10	0.512	0.634	0.681	0.833	0.744	0.807	0.850	**0.865**
	Top20	0.553	0.651	0.560	0.780	0.695	0.775	0.775	**0.820**
	Top50	0.528	0.608	0.583	0.762	0.708	0.701	**0.821**	0.800
	Top100	0.518	0.593	0.581	0.735	0.668	0.674	0.733	**0.745**
DVD	Top10	0.562	0.848	0.833	0.908	0.806	0.819	0.855	**0.920**
	Top20	0.523	0.772	0.785	0.880	0.738	0.825	0.875	**0.910**
	Top50	0.490	0.784	0.788	0.784	0.774	0.842	0.847	**0.870**
	Top100	0.445	0.717	0.712	0.762	0.732	0.800	0.814	**0.818**

4.3 Comparative Analysis

We use SO-PMI [17], WEED [19], SVD [4], NMF [8] and CONR[9] as baselines for comparisons with our models. As mentioned above, WEED and CONR are the-state-of-art methods for SWI. SO-PMI needs seed words, we randomly select 20% words for it following [9]. Other models do not need seed words.

$P_1@K$ and $P_2@K$ results of the methods on the three datasets are respectively reported in Table 1 and Table 2 with 30 iterations. As a further comparison, Fig.4 shows the $MAP@K$ results. By comparing the results of different methods, we can draw the following observations:

(1)Both SMC_1 and SMC_2 can generate better results than the state-of-the-art algorithms. It demonstrates that sentiment matching and sentiment consistency can be applied as two complementary sentiment contextual factors.

(2)SMC_1 could achieve reasonable results but do not perform as well as SMC_2 which shows that individual-based term is generally more accurate in capturing sentiment words' distributions in the corpus.

(3)Compared to WEED, SM achieves consistently better performance on three datasets. It indicates that w_{ij} could better describe the importance from word to document from the view of sentiment than TF-IDF.

(4)SM obtains better performance than SVD and NMF shows that sentiment matching could be modeled and optimized for SWI.

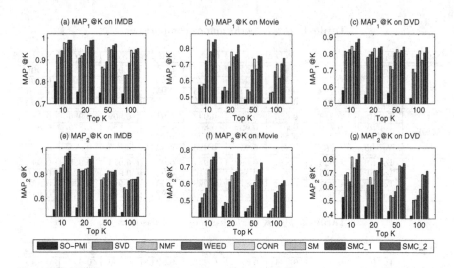

Fig. 4. MAP@K performance comparisons.

Table 3. Case Study: Top-8 Ranked Sentiment Words

SO-PMI	Pos	great, wonderful, perfect, favorite, strong, romantic, emotional, well
	Neg	generations, **suffer, stereotypical**, masters, refreshing, gems, fridge, respect
SVD	Pos	favorite, great, best, wonderful, beautifully, love, excellent, wonderfully
	Neg	bad, worst, waste, awful, pointless, not_waste, worse, like
NMF	Pos	favorite, great, best, wonderful, beautifully, love, excellent, beautiful
	Neg	worst, waste, bad, awful, not_waste, pointless, just, like
CONR	Pos	pointless, best, amazing, favorite, great, excellent, wonderful, love
	Neg	worst, waste, bad, awful, not_waste, horrible, not_even, terrible
WEED	Pos	well, beautiful, true, love, perfect, best, great, recommend
	Neg	plot, even , not_even, poor, supposed, horrible, bad, just
SM	Pos	beautifully, wonderful, perfect, powerful, perfectly, great, amazing, miss
	Neg	bad, poorly, poor, horrible, supposed, insulting, not_even, badly
SMC_1	Pos	beautifully, perfect, amazing, wonderful, true, best, great, simple
	Neg	bad, poor, horrible, poorly, not_even, terrible, supposed, waste
SMC_2	Pos	beautifully, perfect, wonderful, love, best, great, perfectly, excellent
	Neg	worst, bad, waste, awful, not_even, terrible, pathetic, lame

(5)SMC_1 performs better than CONR from which we can seen that word vectors make a contribution to similarities among words.

(6)As the size of K increases, both $P@K$ and $MAP@K$ of all methods falls accordingly. It indicates that all methods can rank the most probable sentiment words in the front of the word list.

(7)Models exploiting sentiment contextual factors generate the best results on IMDB, followed by DVD and then Movie. It proves the prediction that the identification accuracy is proportional to the matching degree of the two sentiment phenomena as discussed in Section 3.

4.4   Case Study

We conduct a case study on IMDB dataset to demonstrate the effectiveness of our proposed model in this section. Table 3 shows top-8 positive and negative words ranked by each method where the bold words are the ones with correct polarity values. SO-PMI obviously yields the worst results. Our three models perform comparable or better than WEED and CONR. Especially SMC_1 and SMC_2 — positive words and negative words are absolutely correct.

5 Related Works

Sentiment word identification is a very common theme in sentiment analysis research. A number of approaches have been proposed to address this problem [14] [16] [18]. Most of them are either dictionary-based or corpus-based.

Dictionary-Based Approaches: There is a common assumption that sentiment words hold the same sentiment polarities with their synonyms and different polarities with their antonyms. With this assumption, lots of approaches [14] [6] [10] which rely on rules and lexical databases such as WordNet [12] are proposed. In [7] and [2], negation/conjunction/disjunction rules and manually created syntactic dependency rules are used to automatically derive morphological and synonymy/antonymy relationships from WordNet. However, these dictionary-based approaches can not identify domain dependent sentiment words.

Corpus-Based Approaches: Corpus-based approaches investigate the semantic relatedness between unlabeled candidate words and seed words, and then obtain sentiment polarities of unlabeled words according to these relations. [3] and [17] calculate similarity measures, such as pointwise mutual information (PMI) and latent semantic analysis (LSA), to obtain the polarities of words. A further corpus-based approach [14] [5] is to infer the semantic relatedness of words by exploiting linguistic clues. In these approaches, negation and conjunction rules (e.g., "not", "and" or "but") are exploited. However, those approaches are sensitive to prior knowledge. Any missing rules or seed words may lead to low identification accuracy.

Recently, several methods [19,9] without seed words have been proposed for SWI. In [19], SWI has been tackled via an optimization model exploiting sentiment matching while Liang et al. [9] converts it to a matrix factorization problem exploiting sentiment consistency. Sentiment matching assumes that polarities of the document and its most component sentiment words are the same [19]. Sentiment consistency considers that two co-occurring sentiment words are more likely to share the same sentiments [9]. The two models are the state-of-the-art methods without seed words. However, they only propose and utilize the two assumptions but do not investigate whether they exist in real-world corpora. Besides, they only respectively use one factor for modeling. Could the two sentiment contextual factors be combined for the learning process? In this paper, we give an affirmative answer.

6 Conclusions

In this paper, we address the problem of sentiment word identification. We present three matrix factorization based models for SWI. Our models predict the polarities of words by exploiting the phenomena of sentiment matching and sentiment consistency. Experiments on three real datasets show that our models outperform the state-of-the-art methods without seed words.

Acknowledgments. This work was supported by Strategic Priority Research Program of Chinese Academy of Sciences (XDA06030600) and National Nature Science Foundation of China (No.61202226).

References

1. Bross, J., Ehrig, H.: Automatic construction of domain and aspect specific sentiment lexicons for customer review mining. In: CIKM, pp. 1077–1086 (2013)
2. Ding, X., Liu, B., Yu, P.: A holistic lexicon-based approach to opinion mining. In: WSDM, pp. 231–240 (2008)
3. Ganesan, K., Zhai, C.X., Viegas, E.: Micropinion generation: an unsupervised approach to generating ultra-concise summaries of opinions. In: WWW, pp. 869–878 (2012)
4. Golub, G., Reinsch, C.E.: Singular value decomposition and least squares solutions. Numerische Mathematik 14(5), 403–420 (1970)
5. Hatzivassiloglou, V., McKeown, K.R.: Predicting the semantic orientation of adjectives. In: ACL, pp. 174–181 (1997)
6. Hu, M., Liu, B.: Mining and summarizing customer reviews. In: 10th Proceeding of ACM SIGKDD, pp. 168–177 (2004)
7. Kamps, J., Marx, M., Mokken, R.J., De Rijke, M.: Using WordNet to measure semantic orientation of adjectives. In: LREC, pp. 1115–1118 (2004)
8. Lee, D.D., Seung, H.S.: Learning the parts of objects by non-negative matrix factorization. Nature 401(6755), 788–791 (1999)
9. Liang, J.G., Zhou, X.F., Hu, Y., Guo, L., Bai, S.: CONR: A novel method for sentiment word identification. In: CIKM, pp. 1943–1946 (2014)
10. Lu, Y., Castellanos, M., Dayal, U., Zhai, C.X.: Automatic construction of a context-aware sentiment lexicon: An optimization approach. In: WWW, pp. 347–356 (2011)
11. Maas, A.L., Daly, R.E., Pham, P.T., Huang, D., Ng, A.Y., Potts, C.: Learning word vectors for sentiment analysis. In: ACL, pp. 142–150 (2011)
12. Miller, G.A.: WordNet: A lexical database for English. Communication of the ACM 38(11), 39–41 (1995)
13. Pang, B., Lee, L.: A sentimental education: Sentiment analysis using subjectivity summarization based on minimum cuts. In: ACL, pp. 271–278 (2004)
14. Popescu, A., Etzioni, O.: Extracting product features and opinions from reviews. In: EMNLP, pp. 339–346 (2005)
15. Qiu, G., Liu, B., Bu, J., Chen, C.: Expanding domain sentiment lexicon through double propagation. In: IJCAI, pp. 1199–1204 (2009)
16. Rao, D., Ravichandran, D.: Semi-supervised polarity lexicon induction. In: EACL, pp. 675–682 (2009)
17. Turney, P.D., Littman, M.L.: Measuring praise and criticism: Inference of semantic orientation from association. ACM Transactions on Information Systems 21(4), 315–346 (2003)
18. Vechtomova, O., Suleman, K., Thomas, J.: An information retrieval-based approach to determining contextual opinion polarity of words. In: ECIR, pp. 507–512 (2014)
19. Yu, H., Deng, Z.H., Li, S.: Identifying sentiment words using an optimization-based model without seed words. In: ACL, pp. 855–859 (2013)

Large-Scale Graph Classification
Based on Evolutionary Computation with MapReduce

Zhanghui Wang[1], Yuhai Zhao[1,2,*], Guoren Wang[1], and Yurong Cheng[1]

[1] College of Information Science and Engineering, Northeastern University, China
[2] Key Laboratory of Computer Network and Information Integration (Southeast University),
Ministry of Education, China
zhaoyuhai@ise.neu.edu.cn

Abstract. Discriminative subgraph mining from a large collection of graph objects is a crucial problem for graph classification. Several main memory-based approaches have been proposed to mine discriminative subgraphs, but they always lack scalability and are not suitable for large-scale graph databases. Based on the MapReduce model, we propose an efficient method, MRGAGC, to process discriminative subgraph mining. MRGAGC employs the iterative MapReduce framework to mine discriminative subgraphs. Each map step applies the evolutionary computation and three evolutionary strategies to generate a set of locally optimal discriminative subgraphs, and the reduce step aggregates all the discriminative subgraphs and outputs the result. The iteration loop terminates until the stopping condition threshold is met. In the end, we employ subgraph coverage rules to build graph classifiers using the discriminative subgraphs mined by MRGAGC. Extensive experimental results on both real and synthetic datasets show that MRGAGC obviously outperforms the other approaches in terms of both classification accuracy and runtime efficiency.

1 Introduction

Applications in many scientific areas such as bioinformatics [9,17], computational chemistry [3,7], medical informatics [2], and social networks [19] produce a large amounts of data modeled as graphs. There is a great need for building an automated classification model for predicting the large number of graphs between different classes. For instance, chemical compounds can be stored as graphs, and chemists are interested in predicting which chemical compounds are active and which compounds are inactive.

The graph classification framework being widely used is first to select a set of subgraph features from graph databases, and then to build a generic classification model using the set of subgraph features selected. Discriminative subgraphs that are frequent in one class labeled graph set but infrequent in the other class labeled graph sets are more suitable for classification requirement. Therefore, the key problem of graph classification is how to mine and select the discriminative subgraph features efficiently.

Several main memory-based approaches [21,16,12,13,11] have been proposed to mine discriminative subgraphs in small-scale graph databases, but they are both time

---

* Corresponding author.

© Springer International Publishing Switzerland 2015
R. Cheng et al. (Eds.): APWeb 2015, LNCS 9313, pp. 227–243, 2015.
DOI: 10.1007/978-3-319-25255-1_19

and memory costly when applied to process large-scale graph databases. Even worse, a single machine may not be able to handle such a process due to its limited memory. Cloud computing and the widespread MapReduce framework can be used to solve the scalability and computationally-intensive problems in discriminative subgraph pattern mining.

Many data analysis techniques require iterative computations in MapReduce. HaLoop [4], a modified version of the Hadoop framework can be used for iterative data analysis applications. Evolutionary computation [6,18] is an efficient iterative approach to obtain global solutions which can be used for discriminative subgraph mining in large-scale graph databases.

In this paper, we employ the HaLoop MapReduce framework and evolutionary computation techniques to find discriminative subgraphs efficiently and propose a large-scale graph classification algorithm with MapReduce, named MRGAGC[1]. To our best knowledge, this work is the first attempt to design an algorithm for discriminative subgraph mining based on evolutionary computation with MapReduce. We summarize our contributions as follows:

- We propose a large-scale discriminative subgraph mining algorithm based on evolutionary computation with MapReduce.
- We propose three evolutionary strategies to generate a set of locally optimal discriminative subgraphs in the worker nodes efficiently.
- We propose to use subgraph coverage rules as graph classifiers using the mined discriminative subgraphs.
- Extensive experimental results on both real and synthetic datasets show that our proposed approach obviously outperforms the other approaches.

The rest of the paper is organized as follows. Section 2 describes the discriminative subgraph mining problem and introduces the MapReduce framework. In Section 3, we describe the iterative programming model and mechanisms of the evolutionary computation. Section 4 explains how we can employ the subgraph coverage rules to build graph classifiers. In Section 5, we evaluate our experiment results. Finally, we briefly review the related work in Section 6 and make a conclusion in Section 7.

2 Preliminaries and Problem Statement

In this section, we first describe the concept of discriminative subgraph, and then Section 2.2 provides a brief MapReduce primer and formalizes the problem statement.

2.1 Discriminative Subgraph

An undirected graph can be modeled as $G = (V, E, L)$ where V is a set of vertices and E is a set of edges connecting the vertices. Both nodes and edges can have labels and L is the labeling function on vertices and edges. Each graph can be attached with an unique class label.

---

[1] MapReduce Genetic Algorithm based Graph Classification.

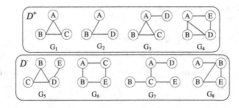

Fig. 1. A graph database $\mathcal{D}$ with 4 positive graphs and 4 negative graphs

In this paper, we use the two classes graph database as a discussible example (we emphasize that our method can be used to process multi-classes graph database with a little modification, not only two classes graph database). Fig. 1 shows a two classes graph database $\mathcal{D}$ contains 4 positive graphs and 4 negative graphs. The set of positive graphs in graph database $\mathcal{D}$ denoted as $\mathcal{D}^+$ and the set of negative graphs denoted as $\mathcal{D}^-$.

Definition 1 (Subgraph Isomorphism). *Given two graphs $G' = (V', E', L')$ and $G = (V, E, L)$, G' is subgraph isomorphic to G, if there exists an injective function $\varphi : V' \to V$, such that (1) for $\forall u, v \in V'$ and $u \neq v$, $\varphi(u) \neq \varphi(v)$, (2) for $\forall u \in V'$, $\varphi(u) \in V$ and $L'(u) = L(\varphi(u))$, and (3) for $\forall(u,v) \in E'$, $(\varphi(u), \varphi(v)) \in E$ and $L'(u,v) = L(\varphi(u), \varphi(v))$.*

If graph G' is *subgraph isomorphic* to graph G, we call that G' is a subgraph of G, denoted as $G' \subseteq G$. Here, G is a supergraph of G' or G supports G'. In this paper, a subgraph is also called a pattern.

Example 1. In Fig. 1, G_2 is a subgraph of G_4 and G_4 is a supergraph of G_2.

Definition 2 (Frequency). *Given a graph database $\mathcal{D} = \{G_1, G_2, \cdots, G_n\}$ and a graph pattern G, $\mathcal{D} = \mathcal{D}^+ \cup \mathcal{D}^-$, the supporting graph set of G is $\mathcal{D}_G = \{G_i \mid G \subseteq G_i, G_i \in \mathcal{D}\}$. The support of G in $\mathcal{D}$ is $|\mathcal{D}_G|$, denoted as $sup(G, \mathcal{D})$, the support of G in $\mathcal{D}^+$ and $\mathcal{D}^-$ denoted as $sup(G, \mathcal{D}^+)$ and $sup(G, \mathcal{D}^-)$, respectively; the frequency of G is $|\mathcal{D}_G|/|\mathcal{D}|$, denoted as $freq(G, \mathcal{D})$, meanwhile, the frequency of G in $\mathcal{D}^+$ and $\mathcal{D}^-$ denoted as $freq(G, \mathcal{D}^+)$ and $freq(G, \mathcal{D}^-)$, respectively.*

Example 2. In the two classes graph database $\mathcal{D}$ in Fig. 1, the supporting set of subgraph A-B-C is $\{G_1, G_3\}$. That is, the support of subgraph A-B-C is 2, and its frequency in $\mathcal{D}$ is 0.25, meanwhile, its frequency in $\mathcal{D}^+$ is 0.5 and its frequency in $\mathcal{D}^-$ is 0, respectively.

Definition 3 (Discriminative Subgraph). *Discriminative subgraph G is a subgraph pattern that occur with disproportionate frequency in one class versus others. The discriminate power score $d(G)$ can be calculated by a given discrimination function.*

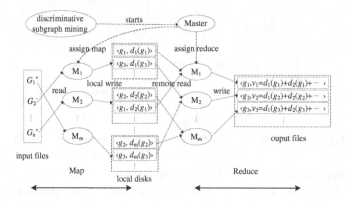

Fig. 2. The framework of MapReduce for discriminative subgraph mining

2.2 MapReduce

MapReduce which is a distributed framework for processing large-scale data contains three phases: map, shuffle, and reduce. In the map and reduce phases, users can generally specify the map function and the reduce function. The three phases as follows:

- *Map*. In this phase, each worker node reads data from a *distributed file system* (DFS), and applies a map function on the input data to produce any number of intermediate key-value pairs.
- *Shuffle*. In this phase, the key-value pairs with the same and similar keys are aggregated together, and then sent to one of the worker nodes.
- *Reduce*. In this phase, each worker node examine the keys and corresponding list of values that receive in the shuffle phase, and then applies a reduce function on each list of values. The reduce function transforms each list into new key-value pairs, which are then stored in the DFS.

With the MapReduce framework, users can implement a map function and a reduce function to process their applications. Fig.2 shows our MapReduce framework for discriminative subgraph mining. First, the master node assign the map and the reduce tasks to each worker. Then, discriminative subgraphs was searched based on evolutionary computation and the discriminative subgraph results was aggregated in each worker. At last, each worker outputs the results. We employ the HaLoop MapReduce framework because our method needs iterative computations to obtain the discriminative subgraph results. In order to accommodate requirements of iterative data analysis applications. HaLoop made several changes to the Hadoop MapReduce framework. These changes as follows:

- First, HaLoop exposes a new application programming interface to users for iterative MapReduce programs.
- Second, the master node of HaLoop contains a new loop control module that repeatedly starts new map-reduce steps that compose the loop body, until the user-specified stopping condition is met.

- Third, HaLoop uses a new task scheduler for iterative applications that leverages data locality in these applications.
- Fourth, HaLoop caches and indexes application data on worker nodes.

HaLoop not only extends MapReduce with programming support for iterative applications, it also dramatically improves their efficiency by making the task scheduler loop-aware and by adding various caching mechanisms. Fig.3 shows the discriminative subgraph mining framework of HaLoop that contains two map-reduce pairs.

Problem Statement: Let $\mathcal{D} = \{G_1, G_2, \cdots, G_n\}$ be a graph database that consists of graph G_i, for $1 \leq i \leq |\mathcal{D}|$. Large-scale discriminative subgraph mining problem is to find a set of subgraph patterns with which are more discriminative and can be built for graph classifiers.

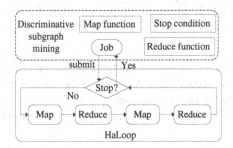

Fig. 3. The discriminative subgraph mining framework of HaLoop

3   Discriminative Subgraph Mining with Evolutionary Computation

In this section, we present the solution for large-scale discriminative subgraph mining. As shown in Fig.2, we next describe the iterative MapReduce programming model.

3.1   Evolutionary Computation Iterative Model

The iterative programs that HaLoop support can be distilled in the following core construct in Eq. (1).

$$\mathcal{R}_{i+1} = \mathcal{R}_0 \cup (\mathcal{R}_i \times \mathcal{L}) \tag{1}$$

Where $\mathcal{R}_0$ is an initial result data and $\mathcal{L}$ ia an invariant relation. A program in this form terminates when a *fixpoint* is reached when the result does not change from one iteration to the next, i.e. $\mathcal{R}_{i+1} = \mathcal{R}_i$, and the *fixpoint* is typically defined by exact equality between iterations. But HaLoop also supports an approximate *fixpoint*, where the computation terminates when either the difference between two consecutive iterations is less than a user-specified threshold, or the maximum number of iterations has been reached.

HaLoop extends MapReduce and is based on two simple intuitions. The first one is that a MapReduce worker node can cache the invariant data in the first iteration, and then reuse them in later iterations. Second, a MapReduce worker node can cache reducer outputs, which makes checking for a *fixpoint* more efficient, without an extra MapReduce job.

In order to use HaLoop program for discriminative subgraph pattern mining, we specify the loop body (as two map-reduce pairs) and rewrite the map function, reduce function and the termination condition.

Map Function: In the map step, each worker node $M_i(i = 1, \cdots, m)$ reads a subset $\mathcal{D}_i$ of $\mathcal{D}$ in the first iteration, and identifies a set of discriminative subgraphs $\mathcal{R}_i$ based on evolutionary computation to represent the set of positive graphs in $\mathcal{D}_i$ that stored in M_i. Then, for each discriminative subgraph $G \in \mathcal{R}_i$, M_i outputs a key-value pair where the key is G and the value indicates the corresponding discrimination score $d_i(G)$.

Reduce Function: With the key and value list obtained from the shuffle phase, M_i inspects the list of values with key G and computes the sum discrimination scores. Then, M_i outputs a key-value pair with key G and value equal to the sum of discrimination scores.

Stop Condition: With the key-value pairs outputs from the reduce phase, we can obtain the difference between two consecutive iterations. If the value of the difference is bigger than the δ threshold, the iteration continues. we also assign a maximum number of iterations θ to terminate the discriminative subgraph mining program.

3.2 Subgraph Pattern Evolutionary Computation Strategy

In the map step, randomly divide graph database $\mathcal{D}$ into m disjoint subsets $\mathcal{D}_i(i = 1, \cdots, m)$ and sent $\mathcal{D}_i$ to worker node M_i. The first goal of our work is to find a set of discriminative subgraphs in each worker node $M_i(i = 1, \cdots, m)$, each positive graph in $\mathcal{D}_i$ that stored in M_i must have at least one representative subgraph for classification. We achieve this goal in each worker node by exploring candidate subgraph patterns in a process resembling biological evolution (evolutionary commutation) which consist of several evolutionary mechanisms such as reproduction, selection and competition. Evolutionary computation begins with a set of sample points in the search space and gradually biases to the regions of high quality fitness [6,18].

In the problem of discriminative subgraph mining with MapReduce, we need choose a fitness function to measure the potential discriminative power of a subgraph pattern. Larger subgraph pattern with stronger discriminative power can be generated by subgraph pattern reproduction. We use a discrimination score definition as the fitness function in Eq. (2) for MapReduce.

$$d_i(G) = sup(G, \mathcal{D}_i^+) - sup(G, \mathcal{D}_i^-) \tag{2}$$

The discrimination score in Eq. (2) is used to measure the fitness of subgraph patterns in each worker node M_i, the bigger the score is, the more discriminative subgraph pattern is.

In the MapReduce framework, for the positive graph set $\mathcal{D}_i^+$ that stored in M_i, we should find a set of discriminative subgraphs as the representative subgraphs efficiently. With a given discrimination score function, we can rank all subgraph patterns by their scores. We first enumerate all subgraph patterns with 2 nodes which the discrimination scores are equal or greater than 0 as the representative subgraph candidates (a set of sample points in the search space in evolutionary computation).

(1) Subgraph Pattern Reproduction Strategy: All the representative subgraph candidates should have a probability of being selected for subgraph pattern reproduction to generate a larger subgraph. During each HaLoop iteration, we give a reproduction threshold α ($0 < \alpha \leq 1$) to randomly select a subset of the representative subgraph candidates for reproduction. When a subgraph pattern g was selected to reproduce a new larger subgraph pattern by extending one edge, it also have probabilities to generate different larger subgraph patterns. The probability of generated a larger subgraph pattern $g' \supset g$ can be calculated by Eq. (3). We only consider the larger subgraph patterns g' that have the nonnegative discriminative power.

$$p(g') = \frac{d_i(g')}{\sum_{p \text{ is in the union of all the } g'} d_i(p)} \tag{3}$$

The probability is always between 0 and 1. This reproduction strategy is commonly used in evolutionary algorithms [6]. The intuition here is that candidate subgraph patterns with higher discriminative scores are more likely be extended to larger subgraph patterns with high scores. If a subgraph pattern g reproduce a new larger subgraph pattern g' that the discrimination score is less than that of g, then the pattern reproduction is undesirable since it can not produces a more strong subgraph pattern that is more fit for survival in the representative subgraph candidates. Among the representative subgraph candidates, there are too many competitions for survival.

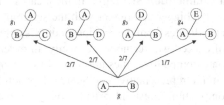

Fig. 4. An example for subgraph pattern reproduction probability

Example 3. Subgraph graph pattern g in the graph database $\mathcal{D}$ in Fig. 1 have four potential larger subgraph patterns (g_1, g_2, g_3, g_4) to reproduce, $d(g_1) = 2$, $d(g_2) = 2$, $d(g_3) = 2$ and $d(g_4) = 1$. The probability of each larger subgraph pattern reproduced from graph pattern g is shown in Fig. 4.

Extensions of different subgraph patterns can produce the same subgraph pattern because a subgraph pattern can be directly extended from all of its supergraphs. There, we should determine whether a subgraph pattern has already been generated to avoid repetitive examination of the same subgraph pattern. We use the maximal CAM code [10] for avoiding the same subgraph pattern generation.

(2) Subgraph Pattern Selection Strategy: With the new representative subgraph candidates generated from subgraph pattern reproduction step in M_i, the goal of subgraph pattern selection is to find a subset of discriminative subgraph patterns among which each positive graph in $\mathcal{D}_i$ can have at least one representative pattern for graph classification. We fist introduce the subgraph-cover definition.

Definition 4 (Subgraph-Cover). *Given a graph database* $\mathcal{D} = \{G_1, G_2, \cdots, G_n\}$ *and a subgraph pattern* g, *a graph* $G_i \in \mathcal{D}$ *is covered by subgraph pattern* g *if there is a subgraph isomorphic from* g *to* G_i. *We denote the set of graphs in* $\mathcal{D}$ *covered by subgraph pattern* G *as* $c(g, \mathcal{D})$.

For the positive graph set $\mathcal{D}_i^+$ that stored in M_i and a set of new representative subgraph candidates, we should select a subset of representative subgraphs $\mathcal{R}_i = \{g_1, g_2, \cdots, g_k\}$ to cover the set of positive graphs in $\mathcal{D}_i$. $\mathcal{R}_i$ and the value k should simultaneously satisfy Eq. (4) and Eq. (5). So, the values k can not be the same for different workers.

$$c_i(g_1, \mathcal{D}_i) \cup c_i(g_2, \mathcal{D}_i) \cup \cdots \cup c_i(g_k, \mathcal{D}_i) = \mathcal{D}_i^+ \qquad (4)$$

We use a heuristic algorithm to select $\mathcal{R}_i$ from the set of representative subgraph candidates. First, the representative subgraph candidates should be sorted in descending order of their discrimination scores, and we choose some highest scores subgraph patterns which satisfy Eq. (4) as the select result $\mathcal{R}_i$. Apparently, when we choose the top-k highest scores subgraph patterns, Eq. (5) is satisfied.

$$max\{\{d_i(g_1, \mathcal{D}_i) + d_i(g_2, \mathcal{D}_i) + \cdots + d_i(g_k, \mathcal{D}_i)\}/k\} \qquad (5)$$

(3) Subgraph Pattern Competition Strategy: In most cases, the representative subgraph candidate set become bigger and bigger with the subgraph pattern reproduction and the input subgraph candidates from the previous reduce's output, sooner or later the number of patterns will exceed worker node's capacity. In order to stabilize the subgraph candidate set, some competitive rules are needed to determine which subgraph pattern should survive in the representative candidate set or not. During each HaLoop iteration, we give a competition threshold β $(0 < \alpha \leq 1)$ to delete the vulnerable subgraph candidates.

First, a subgraph pattern that has already been reproduced should not live in the candidate set any longer because it has served its role in generate a new stronger subgraph pattern. Second, some subgraph patterns with low discrimination scores that are lower than the competition threshold ratio will be deleted from the candidate set. The competition strategy commendably reflect the biological evolutionary phenomenon: "Survival of the fittest".

$$sc(t) = \mathcal{R}(t) = \sum_{i=1}^{m} \mathcal{R}_i \qquad (6)$$

With the three evolutionary strategies, we could quickly identify a set of locally optimal discriminative subgraph results $\mathcal{R}_i$ based on evolutionary computation to cover the set of positive graphs in $\mathcal{D}_i$ that stored in M_i. Then, Map function showed in algorithm 1 output the key-value pairs that the key is a subgraph pattern and the value is the corresponding discrimination score. Next, the key and value list got from the shuffle phase, M_i inspects the list of values with the key and computes the sum discrimination scores. At last, M_i outputs a key-value pair with key equal to the subgraph pattern and the value equal to the sum of discrimination scores.

Algorithm 1. Map Function

Input: Graph dataset $\mathcal{D}_i$ and representative subgraphs, θ, α, β
Output: Representative subgraphs and corresponding discrimination scores
1 θ: maximum number of iterations;
2 α: reproduction threshold;
3 β: competition threshold;
4 Worker node $M_i(i = 1, \cdots, m)$ reads a subset $\mathcal{D}_i$ of $\mathcal{D}$;
5 t: iterative number;
6 **if** $t \leq \theta$ **then**
7 $\quad\mid\quad$ subgraph pattern reproduction strategy;
8 $\quad\mid\quad$ subgraph pattern selection strategy;
9 $\quad\mid\quad$ subgraph pattern competition strategy;
10 **else**
11 $\quad\mid\quad$ Return;
12 Output $\mathcal{R}_i$ and corresponding discrimination scores;

The stop condition calculate by the key-value pairs outputs from the Reduce phase, the sum discrimination scores of $\mathcal{R}_i = \{g_1, g_2, \cdots, g_k\}(i = 1, \cdots, m)$ in Eq. (6) was defined to determine the difference $\{sc(t + 1) - sc(t)\}$ between two consecutive iterations is less than the user-specified δ threshold or not. we also assign a maximum number of iterations θ to terminate the discriminative subgraph mining program.

We introduce an algorithm named MRGAGC for discriminative subgraph mining with MapReduce, and MRGAGC consists of two main steps. The fist step is the map step showed in algorithm 1. Through three evolutionary strategies, the map function outputs the representative subgraphs and their corresponding discrimination scores efficiently. The next reduce step showed in algorithm 2 aggregates the key-values pairs from the shuffle step and to determine whether the iteration stop or not according to the iterative stop condition, meanwhile, outputs the results.

Algorithm 2. Reduce Function

Input: Representative subgraphs $\mathcal{R}_i$ and corresponding discrimination scores, δ
Output: Union of the local optimal representative subgraphs $\mathcal{R}$
1 δ: stop condition threshold;
2 t: iterative number;
3 Aggregate the key-value pairs;
4 **if** $\{sc(t + 1) - sc(t)\} > \delta$ *and* $t \leq \theta$ **then**
5 $\quad\mid\quad$ Output $\mathcal{R}$;
6 **else**
7 $\quad\mid\quad$ Return;

Example 4. Fig. 5 shows an integrated example about discriminative subgraph mining for MapReduce with two worker nodes and three iterations in graph database shown in Fig. 1. First, the graph database was randomly divided in two graph sets, each worker reads one graph set and execute evolutionary computation to obtain a set of discriminative subgraphs. Then, the two worker nodes aggregate the discriminative subgraphs and outputs the results that will be feed in the next map-reduce pair. At last, the difference sc values (sum of subgraph discriminative scores) between two consecutive iterations meet the stop condition threshold ($\delta = 0$) and the result discriminative subgraphs (A-B-C and A-B-D) were obtained.

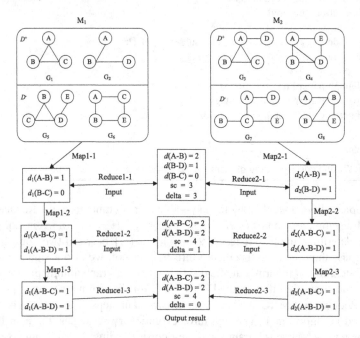

Fig. 5. Discriminative subgraph mining for MapReduce with two worker nodes in Fig. 1

Table 1. Experimental Parameter Values

parameter	values
σ	**0.1**
θ	**1000**
m	2, 4, 6, 8, **10**
δ	**0**, 5, 10, 15, 20
α	0.1, 0.2, **0.3**, 0.4, 0.5
β	0.1, 0.2, **0.3**, 0.4, 0.5

3.3 Reducing Communication Overhead

Discriminative subgraph mining with MapReduce produces a large number of key-value pairs which must be transferred via network communication. This may incur significant overhead. But, benefit for the HaLoop MapReduce framework, many invariant data could be cached in the local disk. Caching the invariant data will reduce the I/O cost and network communication cost. Meanwhile, we also use the same compression technique mentioned in [14] to reduce the communication cost.

4 Graph Classification

In this section, we will introduce the subgraph coverage rule and then build graph classifiers using the discriminative subgraphs mined by MRGAGC.

Definition 5 (Subgraph Coverage Rule). *Given a subgraph set $\mathcal{R}$, a subgraph coverage rule is defined as a classification rule in the form of $G \rightarrow L$, where $G \in \mathcal{R}$ and L is the class label. If a graph has the subgraph pattern G, then it is classified as L.*

When the graph database only have positive and negative class labels, L is always positive since we only mining subgraph patterns whose positive supports are higher than their negative supports. If a graph G' has a subgraph pattern in subgraph coverage rules indicates G' is a positive graph, whereas, it is a negative graph. We measure the classification accuracy by normalized accuracy, which is defined as follows:

$$Sensitivity = \frac{number\ of\ predictions\ that\ are\ positive\ and\ correct}{number\ of\ graphs\ that\ are\ positive}$$

$$Specificity = \frac{number\ of\ predictions\ that\ are\ negative\ and\ correct}{number\ of\ graphs\ that\ are\ negative}$$

$$Normalized\ accuracy = \frac{Sensitivity + Specificity}{2} \tag{7}$$

We generate subgraph coverage rules with the discriminative subgraphs mined by MRGAGC, and build the classifiers for large-scale graph classification. For some unlabeled graphs, we use the subgraph coverage rules and Eq. (7) to calculate the normalized accuracy about the classifiers and evaluate the quality of the classifiers.

5 Experimental Evaluation

In this section, we first evaluate the efficiency of MRGAGC on the synthetic dataset and next evaluate experimentally MRGAGC against two competitors on both synthetic and real-world graph datasets.

5.1 Experimental Setup

We set up a cluster of 11 commodity PCs in a high speed Gigabit network, with one PC as the Master node, the others as the worker nodes. Each PC has an Intel Quad Core 2.66GHZ CPU, 4GB memory and GentOS Linux 5.6. We use HaLoop (a modified Hadoop 0.20.201) to run MRGAGC and the default configuration of Hadoop, i.e., *dfs.replication* = 3 and *fs.block.size* = 64 MB.

Table 2. Summary of the NCI Data Sets

Bioassay ID	Tumor description	Actives	Inactives
NCI 1	Lung cancer	2047	38410
NCI 33	Melanoma	1642	38456
NCI 41	Prostate cancer	1568	25967
NCI 47	Nervous Sys. tumor	2018	38350
NCI 81	Colon cancer	2401	38236
NCI 83	Breast cancer	2287	25510
NCI 109	Ovarian tumor	2072	38551
NCI 123	Leukemia	3123	36741
NCI 145	Renal cancer	1948	38157
NCI 167	Yeast anticancer	8894	53622

As competitors, we consider two main memory-based discriminative subgraph mining algorithms LEAP [21] and LTS [11], and we adapt the two competitors namely MRLEAP and MRLTS to run on MapReduce, respectively. For the LEAP and LTS algorithms, each worker performs many times algorithm in the map step until all the corresponding positive graphs could be covered. So, we can obtain a set of locally discriminative subgraphs mined by these two algorithms in each worker. And, in the reduce step, the two algorithms aggregate the union of discriminative subgraphs as the results. We perform several types of measurements for MRGAGC, MRLEAP and MRLTS, and run each experiment three times for the average result. The parameter values used in the three algorithms are shown in Table 1, with default values in bold. According to the rules of the biological evolution process, $\alpha = 0.3$ and $\beta = 0.3$ are the most reasonable values.

5.2 Data Description

(1) **Synthetic:** We use a synthetic graph data generator[2] to generate 100000 undirected, labeled and connected graphs, and randomly select half of the graphs to be positive graphs. We set the number of unique nodes labels to 20, the average number of nodes and edges in each graph to 20 and 30 respectively.

(2) **NCI:** The NCI cancer screen data sets are widely used for graph classification evaluation [21,16]. We download ten NCI data sets from the PubChem database[3]. Each data set belongs to a bioassay task for anticancer activity prediction, if a chemical compound in a data set is active against the corresponding cancer, its label is positive. Table 2 lists the summary of the ten data sets. Since the data sets are unbalanced (about 5% positive graphs), for each bioassay, we randomly select a negative data set with the same size as the positive one (e.g., 2047 poistive and 2074 negative graphs in NCI 1). As a result, we obtain ten balanced data sets, which comprise 56000 graphs in total. We combine them to construct a new NCI-A dataset (balanced dataset). Meanwhile, we also use all the graphs from the ten data sets (400000 graphs) to construct a new NCI-B dataset (unbalanced dataset) in our experiment.

---

[2] http://www.cais.ntu.edu.sg/~jamescheng/graphgen1.0.zip
[3] http://pubchem.ncbi.nlm.nih.gov/

5.3   Efficiency of MRGAGC

In this subsection, we study the efficiency of MRGAGC on the synthetic dataset with respect to three parameters: α (reproduction threshold), β (competition threshold), and δ (stop condition threshold).

Fig. 6 shows the running times vs three parameters. From the result, we can discover that when the reproduction threshold α increases, the running time of MRGAGC observably increases, but with the competition threshold β increases, the time cost decreases. In addition, we observe that the bigger the stop condition threshold δ is, the less time will be cost.

The phenomena above demonstrate that the most time susceptible operation is the subgraph pattern reproduction in evolutionary computation, which is the same as the traditional main memory-based subgraph pattern mining algorithms. The more subgraphs are explored, the more time will be cost.

Table 3. Average Normalized Accuracy Between MRLEAP, MRLTS and MRGAGC

Graph datasets	MRLEAP	MRLTS	MRGAGC
Synthetic	0.704	0.712	0.74
NCI-A	0.728	0.743	0.776
NCI-B	0.652	0.672	0.705
Average	0.695	0.709	0.74

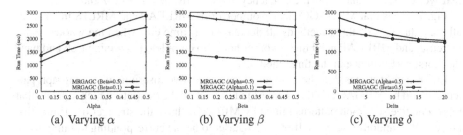

| (a) Varying α | (b) Varying β | (c) Varying δ |

Fig. 6. Running time vs α, β and δ

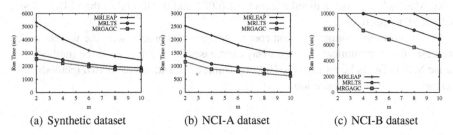

| (a) Synthetic dataset | (b) NCI-A dataset | (c) NCI-B dataset |

Fig. 7. Running time vs Number of workers m

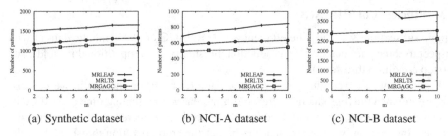

Fig. 8. Number of subgraph patterns vs Number of workers m

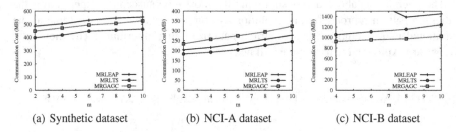

Fig. 9. Communication cost vs Number of workers m

5.4 Comparison with Other Methods

We compare MRGAGC with two other representative discriminative subgraph mining methods: MRLEAP and MRLTS. We run the three algorithms on both synthetic and real-world datasets and show the efficiency and effectiveness in Fig. 7 and Table 3.

Fig. 7 shows that MRGAGC clearly outperforms MRLEAP and MRLTS in terms of all the datasets. The three methods all decrease their time cost when the worker nodes increase, and MRLEAP need more worker nodes to complete the mining task and have the worst performance in the three methods.

Fig. 8 demonstrates that the number of representative discriminative subgraph patterns mined by MRGAGC is the smallest in terms of all the datasets, which indicate that representative subgraph patterns mined by MRGAGC have the stronger discriminative powers. We randomly select half of each dataset to be the corresponding training graph set, and the remainder half of graphs to be the testing graph set. Table 3 intuitively demonstrates the quality of representative discriminative subgraphs mined by the three methods from the aspect of average normalized classification accuracy. On average, MRGAGC achieves normalized accuracy of 0.031 and 0.045 higher than MRLTS and MRLEAP respectively.

Fig. 9 shows the communication cost vs number of workers, with the worker nodes increase, the communication cost always increases slowly for the three methods because MRGAGC is benefit for the HaLoop framework and the other two methods only perform simple map-reduce jobs.

6   Related Work

Discriminative graph pattern mining aims to mine the patterns that occur with disproportionate frequency in some classes versus others. Various efficient algorithms have been developed, such as LEAP [21], GAIA [13], LTS [11], GraphSig [16], and COM [12]. Due to the computation and I/O intensive characteristic of graph pattern ming in large-sale graph database, more and more efforts are geared towards solving it with parallel techniques. MapReduce [20,5] has emerged as a popular alternative for large-scale parallel data analysis and Hadoop is an open-source implementation of MapReduce. Several frequent subgraph mining approaches [14,8] have been proposed in MapReduce. The work in [14] proposes a two-step filter-and-refinement approach to mine the frequent subgraphs which is more suitable to massive parallelization. The work in [8] takes an iterative approach for frequent subgraph mining and it is the first one to employ MapReduce for mining a large collection of graphs. However, many data analysis techniques require iterative computations in MapReduce, such as PageRank [15], recursive relational queries [1] and so on. HaLoop [4], a modified version of the Hadoop MapReduce framework that is designed to serve the iterative applications can be used for iterative data analysis applications. Evolutionary computation [6,18] is a subfield of artificial intelligence that can be used for discriminative subgraph pattern mining. In this paper, we integrate MapReduce with evolutionary computation to mine discriminative subgraph pattern mining efficiently.

7   Conclusion

In this paper, we first propose the problem for large-scale graph classification base on evolutionary computation with MapReduce. We propose an efficient iterative MapReduce method base on HaLoop framework to mine discriminative subgraphs efficiently and explore candidate subgraph pattern space in a biological evolution way. Through evolutionary computation and three effective evolutionary strategies, the method could obtain a set of representative discriminative subgraphs efficiently, meanwhile, the iterative HaLoop framework could reduce significant communication cost. Experiments on both real and synthetic datasets show that our method obviously outperforms the other approaches in terms of classification accuracy and runtime efficiency.

Acknowledgment. This work was supported by National Natural Science Foundation of China (61272182, 61100028, 61173030 and 61173029), 973 program (2011CB302200-G), State Key Program of National Natural Science of China (61332014, U1401256), 863 program (2012AA011004), New Century Excellent Talents ($NCET$-11-0085), Fundamental Research Funds for the Central Universities N130504001).

References

1. Bancilhon, F., Ramakrishnan, R.: An amateur's introduction to recursive query processing strategies. ACM (1986)
2. Bilgin, C., Demir, C., Nagi, C., Yener, B.: Cell-graph mining for breast tissue modeling and classification. In: 29th Annual International Conference of the IEEE Engineering in Medicine and Biology Society, EMBS 2007, pp. 5311–5314. IEEE (2007)
3. Borgelt, C., Berthold, M.R.: Mining molecular fragments: Finding relevant substructures of molecules. In: Proceedings of the 2002 IEEE International Conference on Data Mining, ICDM 2003, pp. 51–58. IEEE (2002)
4. Bu, Y., Howe, B., Balazinska, M., Ernst, M.D.: Haloop: Efficient iterative data processing on large clusters. Proceedings of the VLDB Endowment 3(1-2), 285–296 (2010)
5. Cui, B., Mei, H., Ooi, B.C.: Big data: the driver for innovation in databases. National Science Review 1(1), 27–30 (2014)
6. De Jong, K.: Evolutionary computation: a unified approach. In: Proceedings of the Fourteenth International Conference on Genetic and Evolutionary Computation Conference Companion, pp. 737–750. ACM (2012)
7. Deshpande, M., Kuramochi, M., Wale, N., Karypis, G.: Frequent substructure-based approaches for classifying chemical compounds. IEEE Transactions on Knowledge and Data Engineering 17(8), 1036–1050 (2005)
8. Hill, S., Srichandan, B., Sunderraman, R.: An iterative mapreduce approach to frequent subgraph mining in biological datasets. In: Proceedings of the ACM Conference on Bioinformatics, Computational Biology and Biomedicine, pp. 661–666. ACM (2012)
9. Huan, J., Wang, W., Bandyopadhyay, D., Snoeyink, J., Prins, J., Tropsha, A.: Mining protein family specific residue packing patterns from protein structure graphs. In: Proceedings of the Eighth Annual International Conference on Resaerch in Computational Molecular Biology, pp. 308–315. ACM (2004)
10. Huan, J., Wang, W., Prins, J.: Efficient mining of frequent subgraphs in the presence of isomorphism. In: Third IEEE International Conference on Data Mining, ICDM 2003, pp. 549–552. IEEE (2003)
11. Jin, N., Wang, W.: Lts: Discriminative subgraph mining by learning from search history. In: 2011 IEEE 27th International Conference on Data Engineering (ICDE), pp. 207–218. IEEE (2011)
12. Jin, N., Young, C., Wang, W.: Graph classification based on pattern co-occurrence. In: Proceedings of the 18th ACM Conference on Information and Knowledge Management, pp. 573–582. ACM (2009)
13. Jin, N., Young, C., Wang, W.: Gaia: graph classification using evolutionary computation. In: Proceedings of the 2010 ACM SIGMOD International Conference on Management of Data, pp. 879–890. ACM (2010)
14. Lin, W., Xiao, X., Ghinita, G.: Large-scale frequent subgraph mining in mapreduce. In: 2014 IEEE 30th International Conference on Data Engineering (ICDE), pp. 844–855. IEEE (2014)
15. Page, L., Brin, S., Motwani, R., Winograd, T.: The pagerank citation ranking: Bringing order to the web. Technical Report (1999)
16. Ranu, S., Singh, A.K.: Graphsig: A scalable approach to mining significant subgraphs in large graph databases. In: IEEE 25th International Conference on Data Engineering, ICDE 2009, pp. 844–855. IEEE (2009)
17. Sharan, R., Suthram, S., Kelley, R.M., Kuhn, T., McCuine, S., Uetz, P., Sittler, T., Karp, R.M., Ideker, T.: Conserved patterns of protein interaction in multiple species. Proceedings of the National Academy of Sciences of the United States of America 102(6), 1974–1979 (2005)

18. Storn, R., Price, K.: Differential evolution-a simple and efficient adaptive scheme for global optimization over continuous spaces. ICSI Berkeley (1995)
19. Tang, L., Liu, H.: Graph mining applications to social network analysis. In: Managing and Mining Graph Data, pp. 487–513. Springer (2010)
20. Tao, Y., Lin, W., Xiao, X.: Minimal mapreduce algorithms. In: Proceedings of the 2013 International Conference on Management of Data, pp. 529–540. ACM (2013)
21. Yan, X., Cheng, H., Han, J., Yu, P.S.: Mining significant graph patterns by leap search. In: Proceedings of the 2008 ACM SIGMOD International Conference on Management of Data, pp. 433–444. ACM (2008)

Multiple Attribute Aware Personalized Ranking

Weiyu Guo[1,2], Shu Wu[1], Liang Wang[1], and Tieniu Tan[1]

[1] Center for Research on Intelligent Perception and Computing, National Laboratory of Pattern Recognition, Institute of Automation, Chinese Academy of Sciences, Beijing 100190, China
[2] College of Engineering and Information Technology, University of Chinese Academy of Sciences, Beijing 100049, China
`weiyu.guo@ia.ac.cn`, {`shu.wu,wangliang,tnt`}`@nlpr.ia.ac.cn`

Abstract. Personalized ranking is a typical task of recommender systems. It can provide a set of items for specific user and help recommender systems more correctly direct each item to its user. Recently, as the dramatically increasing social media, an entity, i.e., user and item, usually associates with multiple kinds of characterized information, e.g., explicit ratings, implicit feedbacks, and multi-type attributes (such as age, sex, occupation, or posts of user). Intuitively, comprehensively considering these information, we can obtain better personalized ranking results. However, most conventional methods only take collaborative information (explicit ratings or implicit feedbacks) or single type attributes into account. In this work, we investigate how to combine multiple attribute and collaborative information to learn the latent factors for entities and the attribute-aware mappings. As a result, we propose a novel Multiple-attribute-aware Bayesian Personalized Ranking model, Maa-BPR, for personalized ranking, which can learn reliable latent factors for entities as well as effective mappings for multiple attribute. The experimental results show that, compared with the state-of-the-art methods, Maa-BPR not only provides better ranking performance, but also is robust to new entities and the incomplete attributes.

Keywords: Personalized Ranking, Multiple Attribute, Cold Start.

1 Introduction

Due to the dramatically increasing content in the Web, users are now suffering from the information overload. To cope with such heavy burden, recommender systems, which embed personalized ranking techniques, have attracted a significant attention from both academic and industrial communities. For example, Taobao and Amazon use personalized ranking techniques to improve their recommender systems in order to attract customers for more purchasing.

Among a variety of recommendation methods, the collaborative factorization is one of the most successful approaches, which learns low dimensional latent representations for both users and items. Most of work [1,2,3] in this field is based on explicit ratings. However, explicit ratings are generated by users actively interacting with the systems, and are hard to be obtained in practice. For instance,

© Springer International Publishing Switzerland 2015
R. Cheng et al. (Eds.): APWeb 2015, LNCS 9313, pp. 244–255, 2015.
DOI: 10.1007/978-3-319-25255-1_20

users are encouraged to provide ratings for the movies on the site of MovieLens, and then the recommender system provides recommendation services based on these explicit ratings. In many scenarios, the systems can only obtain implicit interactions between users and items, e.g., whether a user viewed a web page or whether a customer purchased an item. These binary signals are called as implicit feedbacks. Recently, some factorization methods [4,5,6] are proposed to exploit the implicit feedbacks for personalized ranking. Typical examples, such as Bayesian Personalized Ranking (BPR) [7] and its extensions [8,5,1], are popular for personalized ranking by assuming that users are interested in items they had selected than the remaining items.

The above factorization methods indeed promote personalized ranking a lot. However, they are easily suffering from the cold start problem because of lacking enough collaborative information for entities. Recently, with the increasing of social media, apart from collaborative information, there is much attribute information associated with entities, such as the profiles, posts of users, and the description of items. By utilizing such information, researchers have proposed some methods [8,9] to deal with the lack of enough collaborative information. These methods usually treat the attribute information as complete supplementary to collaborative information. However, in real-word applications, the attribute information is usually noisy, incomplete and having multiple types. How to cope with these issues is still a challenge in personalized ranking. Besides, since the implicit feedbacks are widespread and easy to be collected, how to systematically combine vast multiple types of attributes and sparse implicit feedbacks to achieve better recommendation performance is still a thorny problem.

In this paper, for enhancing the performance of personalized ranking, we propose a novel model, which can fuse multiple attribute, and jointly learn latent vectors for entities as well as effective mappings for multiple attribute. Different from the traditional attribute-aware factorization models [10,8], which usually assume that the attribute information of entities is complete and only has single type, our proposed model is robust to the incomplete attributes and multiple attribute. Moreover, we present an advanced parameter learning algorithm, which is different from the traditional methods that segment the processes of learning latent vectors and learning attribute mappings into two independent parts. Through a unified parameter learning framework of systematically combining multiple attribute and implicit feedbacks, our parameter learning algorithm can learn more reliable latent vectors for entities as well as obtain the mappings from the multiple attribute spaces to the latent feature space.

In a nutshell, our contributions in this paper are listed as follows:

1. To systematically combine collaborative information and attribute information for personalized ranking, we propose a multiple-attribute-aware method, Maa-BPR, which can learn more reliable latent vectors for entities and obtain the attribute-aware mappings.

2. For fusing multiple attribute, we bring structured regularizers into our model, and provide a reasonable solution for jointly learning parameters.

3. For investigating the performance of our method, we conduct a series of experiments, and the results show that our method can combine multiple attribute and implicit feedbacks for improving personalized ranking.

2 Related Work

In this section, we discuss related work in two branches, i.e., collaborative methods and attribute-aware methods.

Collaborative methods [4,7,11] are usually based on a mass of users interactions with items, which are called as collaborative information. These methods attempt to factorize collaborative information, and map users and items into a shared latent space. Matrix factorization (MF) [12] is a classical factorization method to be used for dealing with explicit ratings. There are various extensions of MF for personalized ranking. For instance, for dealing with implicit feedbacks, implicit MF (iMF) [4] extends the basic MF by introducing adaptive confidence weights for each user-item pair. Although MF methods, e.g., iMF, can be extended to deal with implicit feedbacks, the phenomenon of data skew commonly exists in the datasets of implicit feedbacks (the number of positive samples is usually less than one percent of the total number), which causes the MF based methods to easily suffer from the over-fitting problem. Recently, Rendle et. propose a framework for personalized ranking, i.e., Bayesian Personalized Ranking (BPR) [7], which can cope with data skew in implicit feedback datasets. BPR and its extensions [5,1,6] make a pair-wise assumption that users are interested in items they had selected than the remaining items, which results in a pair-wise ranking object that tries to discriminate between a small set of selected items and an extremely large set of irrelevant items. Since massive training instances will be derived by the assumption, the learning of parameters is typically based on a stochastic gradient descent (SGD) with uniformly drawn pairs [6].

On the other hand, since the attribute information can indicate many characteristics of entities, attribute-aware methods [8,2,13] are also a kind of mainstream approaches in the field of personalized ranking. For example, in Factorization Machines (FM) [2], all kinds of attribute information are concatenated into a feature matrix, and then factors associated with attributes are learned by a process of rating regression. However, in real-world applications, the attribute information usually is multiple types, noisy and redundant. For fusing multiple attribute, [13] and [14] have been proposed, which utilize techniques of structured sparsity to handle multi-type attributes. Besides, to alleviate the cold start problem, Map-BPR [8] improve BPR framework to learn attribute-aware mappings from both collaborative information and attribute information. Note that, unlike Map-BPR that only treats the attributes as complete information with a single type, our work takes multiple, noisy and redundant attributes into account. Moreover, due to the limitation of parameter learning, the latent vectors of observed entities learned by Map-BPR only encode the collaborative information of entities, while our learning algorithm can learn the latent vectors of combining collaborative information and attribute information for entities.

3   Proposed Model

In real applications, recommender systems mainly obtain two kinds of information, i.e., collaborative information and attribute information. The information linked to a user-item pair is called as collaborative information, e.g., explicit ratings and implicit feedbacks. The information pertaining to one entity is the attribute information, e.g., profile and posts of a user, or genre, cast and description of movies, etc.

In this section, we explore these two kinds of information, and propose a new model to incorporate the collaborative information and the attribute information into a unified personalized ranking framework. Before diving into the details, we first present our model from a sketched view. Then, we will introduce how to model the implicit feedbacks and the multiple attribute respectively. Finally, we formally propose our model and give the algorithm of parameter learning.

3.1   The Sketch of Model

In traditional factorization models [4,7,11], every entity is represented by a latent vector, which can be learned if the entity occurs in the (collaborative) training set. However, when the models are suffering from an entity associated with few collaborative data, the latent vector of the entity could not be well learned. Furthermore, new entities without any collaborative information could be added into the real systems at any time. To employ factorization models for the entities of lacking collaborative information, a common method is to learn the attribute mappings from the attribute space to the latent feature space, and then estimate the latent vectors for these entities of lacking collaborative information by their existing attribute information.

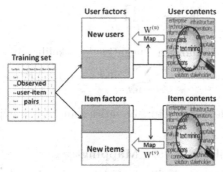

Fig. 1. The framework of combining collaborative information and attribute information into a factorization model.

Figure 1 illustrates the sketch of our factorization model. Our model is driven by three kinds of data: the collaborative training data (left), the latent factors of entities (middle) and the attributes of entities (right). The rectangles in the middle part represent the factor matrices, where the entities of lacking collaborative data have no latent factors. These unknown latent factors will be estimated using the corresponding attribute mappings. Thus, taking the attribute information into account, the training of our model consists of the following alternant learning steps: learning the latent vectors of entities from the collaborative information and the attribute information; then, learning the attribute-aware mappings from the latent vectors of observed entities and their attributes.

3.2 Learning Latent Factors From Implicit Feedbacks

BPR [7] is a popular personalized ranking framework for dealing with the implicit feedbacks. In the BPR framework, if the user u_m has selected the item v_i but not selected the item v_j, then the work of BPR assumes that, u_m prefers v_i over v_j, and defines the pairwise preference of of u_m as

$$p(i \succ_m j) := \Phi(x_{mij}), \tag{1}$$

where $\Phi(x) = 1/(1 + exp(-x))$ and $x_{mij} := r(u_m, v_i) - r(u_m, v_j)$. $r(\cdot, \cdot)$ can be any kind of scoring functions which indicate the relevance between two entities.

Because this work mainly focuses on studying the implicit feedbacks, we simply follow the pair-wise assumption of BPR [7] to create training data $D_S := \{(m,i,j)|v_i \in I_{u_m}^+ \wedge v_j \in I \setminus I_{u_m}^+\}$ from collaborative training data, where $I_{u_m}^+$ denotes the set of the observed items which are linked to the user u_m, and the triple $t = (m,i,j)$ in D_S represents the user u_m is relevant to the item v_i but irrelevant to the item v_j. For simplicity, we call v_i as a positive item of u_m, while v_j is a negative item of u_m. It should be noted that the training triples can be easily created from any datasets with explicit ratings. For example, if the user u_m has given a higher score to the item v_i than the item v_j, then we can treat v_i as the positive item of u_m and v_j as the negative one.

After obtaining the training triples, the goal of BPR is to maximize the likelihood of all pair-wise preference:

$$\arg\max_{\Theta} \prod_{(m,i,j)\in D_S} p(i \succ_m j), \tag{2}$$

which is equivalent to minimize the negative log likelihood:

$$L_{feedback} = - \sum_{(m,i,j)\in D_s} \ln \Phi(x_{mij}) + \lambda||\Theta||^2, \tag{3}$$

where Θ is the latent feature vectors of entities and λ is a hyper-parameter.

3.3 Learning Attribute-Aware Mappings

In the above sections, we have illustrated how to learn the latent factors of entities only considering implicit feedbacks. However, in the real recommender systems, entities are not always having enough collaborative information, e.g., new entities will be added into the systems at any time. Therefore, apart from the collaborative information, we also need to take the attribute information of entities into account for learning latent vectors of entities. For the sake of simply bridging the attribute space and the latent space, as shown in Figure 2, we concatenate the multiple attribute of an entity into an attribute vector, and

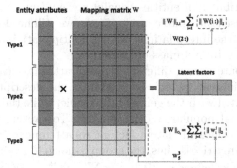

Fig. 2. The schematic diagram of mapping from the attribute space to the latent feature space, where we assume an entity contains three different types of attributes. We concatenate all attributes into an attribute vector, and map the attribute vector to be the latent factors by the attribute-aware mapping matrix. Besides, for dealing with the multiple attribute which is noisy, redundant and consists of multiple types, G_1-norm and $l_{2,1}$-norm are used for regularization on the mapping matrix.

then leverage a linear mapping to map the attribute vector to be a latent vector. Thus, our task is in turn to learn the linear mappings for attribute information.

Here, we formally present an unsupervised solution for learning the linear mappings. In this work, we consider the classical scenario of recommendation, which has two types of entities, i.e., users (e.g., customers) and items (e.g., movies, books and songs). We use u and v to denote a user and an item, respectively. For simplicity, we use e to denote an abstract entity, which can be a user or an item. $y_i^{(e)}$ denotes the latent vector of the entity e_i and the matrix $Y^{(e)} = [y_1^{(e)}, y_2^{(e)}, y_3^{(e)} ...]$ is the latent vectors of all the users/items. For example, $y_m^{(u)}$ denotes the latent vector of user u_m and $Y^{(u)}$ denotes the latent matrix of user. Besides, we use the matrix $A^{(e)} = [a_1^{(e)}, a_2^{(e)}, a_3^{(e)}, ...]$ to denote the multiple attribute of entities. $a_i^{(e)} \in \mathbb{R}^d$ is an attribute vector concatenating k types of attributes, where d is the total number of attributes of an entity, each type j has d_j attributes, and $d = \sum_{j=1}^k d_j$. Then, we can present the objective function to learn the attribute-aware mappings from the attributes of k types:

$$L_{attribute} = ||A^{(e)}W^{(e)} - Y^{(e)}||_F^2, \qquad (4)$$

where $W^{(e)} = [w_1^1, ..., w_c^1; ...; w_1^k, ..., w_c^k] \in \mathbb{R}^{d \times c}$ denotes a mapping matrix, and c is the dimension of latent vectors. w_p^q indicates the weights of all attributes in the q-th type with respect to the p-th latent factor.

However, the optimization problem expressed by Eq. (4) has infinite solutions because of lacking supervised information. Fortunately, using the BPR framework, we can obtain an approximate latent matrix $\widetilde{Y^{(e)}}$ for $Y^{(e)}$ from the collaborative information. As a result, we can treat $\widetilde{Y^{(e)}}$ as pseudo labels, and substitute it into Eq. (4). With the pseudo labels, we can solve the optimization problem, and jointly learn the mapping matrix $W^{(e)}$ from both the collaborative information and the attribute information.

However, the attributes of entities usually are multiple types, noisy and redundant. Taking these issues into considerations, we still need to design a proper scheme to deal with the interrelations among multiple attribute, and select the informative attributes from a mass of attributes.

In multiple attribute fusion, different types of attributes can be more or less discriminative for different factors. For instance, the description of a movie is usually more associated with the genre of the movie, while the cast of this movie indicates the potential popularity of the movie. To capture the interrelations between multiple attribute and latent factors, we attempt to conduct structured sparsity on mapping matrices. Thus, the group l_1-norm (G_1-norm) [15] is used for regularization, which is defined as $||W||_{G_1} = \sum_{i=1}^{c} \sum_{j=1}^{k} ||w_i^j||_2$. As illustrated in Figure 2, the G_1-norm uses l_2-norm within each type of attributes and l_1-norm between types. In this way, the sparsity between different types is enforced [15], i.e., if attributes of one type are not discriminative for the latent factors of a certain group, the elements corresponding to those latent factors in $W^{(e)}$ will be assigned with zeros (in practical case, they are usually very small values), otherwise, their weights should be large.

Besides, in certain cases, even if most attributes of one type are not discriminative for the latent factors of any groups, there are still a small number of attributes in this type to be highly discriminative. Such important attributes should be shared by all latent factors. Thus, the $l_{2,1}$-norm [16] is also used for regularization, which is defined as $||W||_{2,1} = \sum_{i=1}^{d} ||W(i :)||_2$, where $W(i :)$ is the i-th row of matrix W. Thus, the linear mappings can be learned by

$$L_{attribute} = \sum_{e \in \{u,v\}} \lambda^{(e)} (||W^{(e)} A^{(e)} - \widetilde{Y^{(e)}}||_F^2 + ||W^{(e)}||_{G_1} + ||W^{(e)}||_{2,1}), \qquad (5)$$

where $\lambda^{(e)}$ is a hyper-parameter for tuning the weight of users or items.

3.4   The Unified Model and Parameter Learning

To systematically incorporate collaborative information and attribute information into a solution, our model for learning latent vectors and attribute-aware mappings can be expressed as

$$\arg \min_{\Theta,W} \ L_{feedback} + L_{attribute} =$$
$$- \sum_{(m,i,j) \in D_s} \ln \Phi \left((y_i^{(v)} - y_j^{(v)}) y_m^{(u)} \right) + \lambda ||\theta||^2$$
$$+ \sum_{e \in \{u,v\}} \lambda^{(e)} \left(||W^{(e)}||_{G_1} + ||W^{(e)}||_{2,1} + ||A^{(e)} W^{(e)} - Y^{(e)}||_F^2 \right). \qquad (6)$$

To learn the parameters $Y^{(u)}$, $Y^{(v)}$, $W^{(u)}$ and $W^{(v)}$ in Eq. (6), we design an alternative optimization algorithm, which uses SGD with uniformly drawn training triples to learn the latent vectors and implements matrix decomposition to learn the mapping matrices.

In each iteration, when we are updating the latent factor matrix $Y^{(e)}$, we set the mapping matrix $W^{(e)}$ to be a constant and $L_{attribute}$, the entire optimization objective of attribute information, to be a regularizer. Thus, the gradient of an arbitrary latent parameter θ is

$$
\frac{\partial L}{\partial \theta} = \sum_{(m,i,j) \in D_s} \left(1 - \Phi \left((y_i^{(v)} - y_j^{(v)}) y_m^{(u)} \right) \right) \frac{\partial ((y_i^{(v)} - y_j^{(v)}) y_m^{(u)})}{\partial \theta}
$$
$$
+ \frac{\partial \sum_{e \in \{u,v\}} \lambda^{(e)} \left(|| A^{(e)} W^{(e)} - Y^{(e)} ||_F^2 \right)}{\partial \theta} + \lambda \theta. \tag{7}
$$

The updating rule for parameter θ is $\theta = \theta + \eta \frac{\partial L}{\partial \theta}$, where η is the learning rate.

On the other hand, given a latent factor matrix $Y^{(e)}$, we view $Y^{(e)}$ as pseudo labels and treat $L_{feedback}$ as a constant. Thus, the optimization objective of Eq. (6) is equal to $L_{attribute} = 0$ and the updating rule for $W^{(e)}$ can be derived as

$$
w_i = (A^{(e)} (A^{(e)})^T - \lambda D^i + \gamma \widetilde{D})^{-1} A^{(e)} Y^{(e)} (i :), \tag{8}
$$

where $Y^{(e)}(i :)$ is the i-th row of matrix $Y^{(e)}$. D^i is a block diagonal matrix with the j-th diagonal block as $\frac{1}{2||w_j^i||_2} I_j$. I_j is an identity matrix with size of d_j. $\widetilde{D}$ is a diagonal matrix with the j-th diagonal element as $\widetilde{D}(j,j) = \frac{1}{2||W^{(e)}(j:)||_2}$. Note that both $D^i (1 \leq i \leq c)$ and $\widetilde{D}$ are dependent on $W^{(e)}$. Thus, they are also unknown variables but can be approximatively calculated by the value of $W^{(e)}$ in last iteration.

In a nutshell, the iterative algorithm for learning parameters of Maa-BPR is summarized in Algorithm 1. The algorithm mainly repeats two learning steps until the parameters reach convergence, i.e., it first learns the latent factors of entities from implicit feedbacks, by SGD with uniformly drawn training triples, then it learns the attribute mapping matrices by given latent factors.

4   Experiment

In this section, we perform experiments to validate our proposed model by comparing with other approaches on real world datasets. In the following experiments, we first investigate the comprehensive prediction quality by evaluating Area Under the ROC Curve (AUC) and Mean Average Precision (MAP). Then, we study the performance of our method on Top-N recommendation. Finally, we study the cold start problem by simulating the recommendation of new items.

4.1   Datasets

For evaluating our method, we use two real world datasets, i.e., DBLP[1] and MovieLens[2], and carry out training and testing on randomly split training (80%) and testing (20%) data.

DBLP contains 2,084,055 papers with 2,244,018 citations. Each paper may be associated with abstract, authors, published year, and title. We preprocess the DBLP data to be an experimental data set in a similar way as [6]. More

---

[1] http://arnetminer.org/citation
[2] http://grouplens.org/datasets/movielens

Algorithm 1. Learn parameters for Maa-BPR

Input:

    The triple set D_s;

    The content feature of entities $A := \{A^{(u)}, A^{(v)}\}$;

Output:

    $\Theta := \{Y^{(u)}, Y^{(v)}\}$, $W := \{W^{(u)}, W^{(v)}\}$;

1: Initialize Θ and W;

2: **repeat**

3:    uniformly draw a triple (m, i, j) from D_s;

4:    **for** each latent vector $\theta \in triple\,(m,i,j)$ **do**

5:        $\theta \leftarrow \theta + \eta \frac{\partial L}{\partial \theta}$;

6:    **end for**

7:    **for** each $W^{(e)} \in W$ **do**

8:        Calculate the block diagonal matrix $\widetilde{D}$;

9:        Calculate the block diagonal matrices $D^i (1 \le i \le c)$;

10:      **for** each column $w_i (1 \le i \le c) \in W^{(e)}$ **do**

11:        $w_i \leftarrow (A^{(e)}(A^{(e)})^T - \lambda D^i + \gamma \widetilde{D})^{-1} A^{(e)} Y^{(e)}(i\,:)$;

12:      **end for**

13:    **end for**

14: **until** convergence

---

specifically, we sample 1,000 authors who have published no more than 5 papers and cited 5-100 papers from the DBLP data. Thus, we obtain 1,000 authors, 16,313 papers and 23,506 author-paper pairs. Each author-paper pair denotes a relation of the author cited the paper. In the experiments, we treat texts in published papers of an author as the content information of this author, and paper text as content information of the paper. We use the term-frequency over texts as features of content information. Our task is to predict the personalized ranking of citing papers for each author.

MovieLens includes 100,000 ratings by 943 users on 1682 movies. Each user has rated at least 20 movies. In our experiments, the age, gender and occupation of a user are used as the multiple attribute of the user, and the genre of movie and key words in title are viewed as the multiple attribute of an item. Using the same processing method in [17], we do not use the rating values but just binary rating events by assuming that users tend to rate movies they have watched. For a specific user, our task is to predict the potential ranking list of movies.

In addition, comparing these two experimental datasets, we can observe that: 1) the implicit rating matrix of DBLP data set is sparser than that of Movie-Lens; 2) each entity in MovieLens has its corresponding attributes, but entities in DBLP always lack corresponding texts. Thus, in DBLP, many authors and papers have incomplete attribute information.

4.2 Compared Methods

Table 1 shows the characteristics of compared methods. Our method, i.e., Maa-BPR is first compared with two basic methods, i.e., BPR-MF [7] and iMF

[6], which only consider implicit feedbacks. Then, we investigate two advanced attribute-aware methods, i.e., Map-BPR [8] which extends BPR with attribute mappings and FM [2] which is an attribute-aware framework. In our experiments, since the datasets we used existing data skew, i.e., the number of positive feedbacks is far less than the number of negative feedbacks, we train FM and MF with the training data sets which contain randomly drawn negative feedbacks, i.e., the proportions of negative feedbacks and positive feedbacks are 50 : 1 on DBLP and 100 : 1 on MovieLens.

Table 1. The characteristics of compared methods

Method	Attribute	Feedback	Cold start
BPR-MF	no	implicit	no
iMF	no	implicit	no
FM	yes	implicit/explicit	yes
Map-BPR	yes	implicit	yes
Maa-BPR	**yes**	**implicit**	**yes**

4.3   Results and Discussion

Figure 3 presents the ranking performance of methods evaluated by MAP and AUC. Since the attribute information is noisy and consists of multiple types, the traditional attribute-aware methods, e.g., FM, usually could not get reliable results. On the other hand, due to the lack of enough collaborative information, the methods only considering implicit feedbacks, e.g., iMF and BPR-MF, also could not get ideal results. Owing much to combining multiple attribute and implicit feedbacks, Maa-BPR consistently outperforms other methods. Furthermore, we observe that, although both Map-BPR and Maa-BPR combine attribute information and collaborative information to learn the attribute mappings, the performance of Maa-BPR on different datasets is more stable than these of Map-BPR. This result demonstrates that, our method is more robust than Map-BPR in terms of incomplete attribute information. This is because that our method fuses multiple attribute, and jointly learns latent vectors from implicit feedbacks and multiple attribute.

Figure 4 shows the precision of different methods with varying numbers of recommendations. Our method achieves the best performance in most cases, especially in recommending a small set of items. Since in real-world scenarios, users only care about several items which are listed on the top places, this experiment of Top-N recommendation in turn shows that our method can well fit recommender systems. In addition, with the number of recommendations increasing,

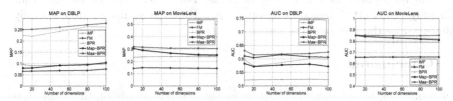

Fig. 3. The comprehensive prediction quality, i.e., Mean Average Precision (MAP) and Area under the ROC curve (AUC), on DBLP and MovieLens.

the performance of attribute-aware methods decline more quickly than the methods only taken collaborative information into account. This phenomenon may indicate that the attribute information usually has more noise than collaborative information. Thus, it is valued for designing proper schemes to alleviate the noise of attributes, when modeling the attributes of entities.

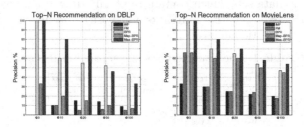

Fig. 4. The performance of Top-N recommendation on DBLP and MovieLens.

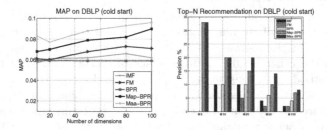

Fig. 5. The performance of new items recommendation on the DBLP dataset. We only evaluate the user-item pairs when the items are not occurring in the training set.

For investigating the cold start problem, we assess the performance of new items recommendation in Figure 5. In the experiments, We only evaluate the user-item pairs which their items are not occurring in the training set. Due to the lack of collaborative information, the traditional methods without considering attribute information can only provide random recommendation results, while most attribute-aware methods have better performance regardless of comprehensive prediction and Top-N recommendation. Moreover, Maa-BPR obtains the best results among attribute-aware methods. Besides, we can observe that FM has poor performance on the experiment of Top-N recommendation, especially in recommending a small set of items. This result may be because that FM only takes attributes into account, and Top-N recommendation is easily influenced by the noise in attributes.

5 Conclusions

For promoting personalized ranking by fusing multiple attribute, this paper has proposed a novel personalized ranking model, Maa-BPR, which combines multiple attribute and interactions between entities to learn latent vectors for entities and attribute-aware mappings for multiple attribute. Comprehensive experiments have shown that our model achieves better predictive performance and is robust to both the cold start problem and incomplete attributes. However, the strategy of randomly drawn training triples would cause a slow convergence. In the future, to speed up the parameter learning process and further promote the performance of our model, we plan to improve the sampling strategy by taking attribute information of items and social information of users into account.

Acknowledgments. This work is jointly supported by National Basic Research Program of China (2012CB316300), and National Natural Science Foundation of China (61175003, 61202328, 61420106015, U1435221, 61403390).

References

1. Qiu, S., Cheng, J., Yuan, T., Leng, H.: Item group based pairwise preference learning for personalized ranking. In: SIGIR, pp. 1219–1222 (2014)
2. Rendle, S.: Factorization machines with libFM. TIST 3(3), 57:1–57:22 (2012)
3. Shi, Y., Larson, M., Hanjalic, A.: List-wise learning to rank with matrix factorization for collaborative filtering. In: RecSys, pp. 269–272 (2010)
4. Hu, Y., Koren, Y., Volinsky, C.: Collaborative filtering for implicit feedback datasets. In: ICDM, pp. 263–272 (2008)
5. Pan, W., Chen, L.: Gbpr: Group preference based bayesian personalized ranking for one-class collaborative filtering. In: IJCAI, pp. 2691–2697 (2013)
6. Rendle, S., Freudenthaler, C.: Improving pairwise learning for item recommendation from implicit feedback. In: WSDM, pp. 273–282 (2014)
7. Rendle, S., Freudenthaler, C., Gantner, L.: Bpr: Bayesian personalized ranking from implicit feedback. In: UAI, pp. 452–461 (2009)
8. Gantner, Z., Drumond, L., Freudenthaler, C., Rendle, S., Schmidt-Thieme, L.: Learning attribute-to-feature mappings for cold-start recommendations. In: ICDM, pp. 176–185 (2010)
9. Pilászy, I., Tikk, D.: Recommending new movies: even a few ratings are more valuable than metadata. In: RecSys, pp. 93–100 (2009)
10. Chen, T., Zheng, Z., Lu, Q., Zhang, W., Yu, Y.: Feature-based matrix factorization. arXiv preprint arXiv:1109.2271 (2011)
11. Yu, X., Ren, X., Sun, Y., Sturt, B., Khandelwal, U., Gu, Q., Norick, B., Han, J.: Recommendation in heterogeneous information networks with implicit user feedback. In: RecSys, pp. 347–350 (2013)
12. Koren, Y., Bell, R., Volinsky, C.: Matrix factorization techniques for recommender systems. Computer 42(8), 30–37 (2009)
13. Tang, J., Hu, X., Gao, H., Liu, H.: Unsupervised feature selection for multi-view data in social media. In: SDM, pp. 270–278 (2013)
14. Wang, H., Nie, F., Huang, H.: Multi-view clustering and feature learning via structured sparsity. In: ICML, pp. 352–360 (2013)
15. Wang, H., Nie, F., Huang, H., Ding, C.: Heterogeneous visual features fusion via sparse multimodal machine. In: CVPR, pp. 3097–3102 (2013)
16. Obozinski, G., Taskar, M.I.: Joint covariate selection and joint subspace selection for multiple classification problems. ASC 20(2), 231–252 (2010)
17. Gunawardana, A., Meek, C.: A unified approach to building hybrid recommender systems. In: RecSys, pp. 117–124 (2009)

Knowledge Base Completion
Using Matrix Factorization

Wenqiang He, Yansong Feng*, Lei Zou, and Dongyan Zhao

Peking University, Beijing, China
{hewenqiang,fengyansong,zoulei,zhaodongyan}@pku.edu.cn

Abstract. With the development of Semantic Web, the automatic construction of large scale knowledge bases (KBs) has been receiving increasing attention in recent years. Although these KBs are very large, they are still often incomplete. Many existing approaches to KB completion focus on performing inference over a single KB and suffer from the feature sparsity problem. Moreover, traditional KB completion methods ignore complementarity which exists in various KBs implicitly. In this paper, we treat KBs completion as a large matrix completion task and integrate different KBs to infer new facts simultaneously. We present two improvements to the quality of inference over KBs. First, in order to reduce the data sparsity, we utilize the type consistency constraints between relations and entities to initialize negative data in the matrix. Secondly, we incorporate the similarity of relations between different KBs into matrix factorization model to take full advantage of the complementarity of various KBs. Experimental results show that our approach performs better than methods that consider only existing facts or only a single knowledge base, achieving significant accuracy improvements in binary relation prediction.

Keywords: Knowledge Base Completion, Matrix Completion, Matrix Factorization.

1 Introduction

Over the last few years, several large-scale knowledge bases have been developed. Notable endeavors of this kind include YAGO [1], DBpedia [2], Freebase [3], and NELL [4]. These KBs contain a great deal of facts about people, places, events and the relations between them. For example, (*Barack Obama, president Of, USA*) and (*The True Story of Ah Q, author, Lu Xun*).

However, these knowledge bases are often incomplete. For example, 93.8% of persons from Freebase have no place of birth, and 78.5% have no nationality [5]. There is a need to increase their coverage of facts to make them more useful in practical applications.

A strategy to increase coverage might be to infer new relationship simply by examining the knowledge base itself. For example, since we have known the

---

* Corresponding author.

© Springer International Publishing Switzerland 2015
R. Cheng et al. (Eds.): APWeb 2015, LNCS 9313, pp. 256–267, 2015.
DOI: 10.1007/978-3-319-25255-1_21

fact $birthPlace(X, Y)$ in DBpedia, we can infer $nationality(X, Y)$ to a great degree. The task is also called knowledge base completion. Previous methods approach this goal by examining a single knowledge base [6][7]. To overcome the insufficiency of the available inference information in a single knowledge base, surface-level textual patterns are often used in the completion process [8][9]. However, the textual patterns also bring the feature sparsity problem [10][11].

As mentioned above, there are many existing knowledge bases constructed using various methods. As a substitution of textual pattern, we can take full advantage of the complementarity which exists across various KBs implicitly to complete the knowledge base. For example, since DBpedia contains the fact $almaMater(CleveMoler, StanfordUniversity)$, we can infer $graduatedFrom$ $(CleveMoler, StanfordUniversity)$ in YAGO [1] as long as we know these two relations are similar. In this way, we will have more inference information for the completion of YAGO.

According to these observations, in this paper, we propose a novel matrix factorization model for knowledge base completion, which integrates the similarity information of different relations in various knowledge bases and reduces the data sparsity effectively. The main contributions of our work can be summarized as follows: (1) we reveal the complementarity which exists across various KBs implicitly, and integrate various KBs to complete themselves simultaneously. (2) we utilize the constraints of argument type of relations, when constructing the matrix, to decrease the data sparsity. (3) we coin the relations similarity regularization to represent the similarity constraints to matrix factorization model, and systematically illustrate how to design a matrix factorization objective function with similarity regularization.

The remainder of this paper is organized as follows. In Section 2, we provide an overview of several major approaches for knowledge base completion and related work. Section 3 describes the construction of knowledge base matrix. Section 4 details the framework of our method. The results of an empirical analysis are presented in Section 5, followed by the conclusion in Section 6.

2   Related Work

Knowledge Base Completion. In the early stage of NELL [4], its inference method was applying first order Horn clause rules to infer new facts from current facts. This method cannot scale to make complex inferences from large knowledge bases, because the computing cost of learning first-order Horn clauses is expensive–not only the search space is large, but also some Horn clauses can be costly to evaluate [12]. [6] first introduces random walk inference over knowledge base based on the Path Ranking Algorithm (PRA) [13]. In this method,

---

[1] Actually, Cleve Moler received his bachelor's degree from California Institute of Technology and a Ph.D. from Stanford University. There are two facts $almaMater(Cleve\ Moler, California\ Institute\ Of\ Technology)$, $almaMater(CleveMoler, Stan\text{-}ford\ University)$ in DBpedia, but just a fact $graduatedFrom(CleveMoler, CaliforniaInstituteOfTechnology)$ in YAGO.

the knowledge base is viewed as a graph and inferring new facts in knowledge base is formulated as a ranking task. Random walks are used to find paths which connect the source and target nodes of relation instance. These paths are used as features in a logistic regression classifier. This work improves significantly over traditional Horn-clause learning and inference method. However, it is limited by the connectivity of the knowledge base graph. To overcome this limitation, [8][10] integrate a large corpus into original knowledge base graph to improve the connectivity. But these methods dramatically increase feature sparsity caused by the textual patterns which have a large degree of synonymy. On this basis, [11] makes a remarkable improvement in reducing the feature sparsity inherent in using surface text, by incorporating vector space similarity into random walk inference over KBs.

The common characteristic of these methods is inferring new facts over a single knowledge base. Although textual sources improve the connectivity of knowledge base, they also result in feature sparsity problem because of the nature of textual patterns. In this paper, we integrate different knowledge bases rather than knowledge base with textual source, for the reason that knowledge base relations are semantically coherent.

Matrix Factorization. Factorization method is widely used in predicting new facts among knowledge base. In RESCAL[14], knowledge base which consists of n entities and m binary relations is transformed automatically into a three-way tensor K of size $n \times n \times m$. This method approximates each slice of the KB tensor K as a product of an entity matrix, a relation matrix, and the entity matrix transposed. [7] employs RESCAL to large-scale learning on the Semantic Web that honors the sparsity of LOD Data. The main advantage of RESCAL is that it can exploit a collective learning. [15][9] flattened the KB tensor into a matrix, combining the two entity models into a single model containing entity pairs. Furthermore, these methods include surface patterns into the matrix.

In this paper, we make use of the strengths of the tensor and matrix factorization method. We also use low-rank matrix factorization over the KB matrix to find latent representations for relations and entities in the knowledge base. In addition, we incorporate the similarity of relations between differen KBs into matrix factorization model in order to take full advantage of the complementarity of various KBs.

3   Matrix Construction

Our method for knowledge base completion, described in Section 4, performs matrix factorization with the complementarity of various KBs to get latent factor. Prior to detailing that technique, we first describe how to translate various knowledge bases to a matrix and formulate knowledge base completion as a matrix completion task.

Similar to [16], we will loosely follow the W3C Resource Description Framework (RDF) standard. Facts of knowledge base are represented in the form of binary relationships, in particular $(subject, predicate, object)$ (SPO) triples, where *subject* and *object* are entities and their *relation* is formulated as *predicate*. In this paper, *relation* is identical to *predicate*. After the preprocess of entity linking, we set E to the set of entity pairs $(subject, object)$ contained in various knowledge bases, R to the set of binary relations, and F to all the observed triples in various KBs. Assuming that we index an entity pair with $e \in E$, a relation with $r \in R$, and $r(e)$ is true when $(e_{subject}, r, e_{object}) \in F$, false otherwise. Producing a matrix X from various knowledge bases, which consist of m predicates and n entity pairs, is straightforward: each row in the matrix corresponds to a relation r and each column an entity pair e. Each matrix cell is denoted as X_{er} and the size of the matrix is $m \times n$. Now we need to figure out how to define the value of X_{er} .

In classical movie recommendations, an entry in user-movie matrix is the score that a user rate for a given movie. In a similar way, we can initialize each cell variable of the knowledge base matrix X_{er} to 1 when $r(e)$ is true. To fill the matrix, we get the value of multiplication of latent representations for relations and entities in the knowledge base, as the probability of a user r "like" a movie e. If the probability is greater than a threshold, it means that we have found a new relationship between the corresponding entity pairs. It is also the goal of knowledge base completion.

The knowledge bases matrix constructed by the above method is very sparse. The density of the matrix is determined by the number of existing facts in the knowledge base. However, in Table 1, we can find that the average facts of each predicate are fairly small compared with the total entity pairs and the knowledge base matrix density is less than 2%. To decrease the sparsity of matrix, we should first understand the missing data in the knowledge base matrix.

In movie recommendations, the user-movie matrix is sparse because there are many missing data caused by the user not rating the movie. The goal of recommendation is to get a prediction value for each missing data and complete the matrix. In our knowledge base matrix, the missing data is caused by the fact that there does not exist a relation r between entity pair e at present. But our goal is not to get a prediction value for each missing data.

In RDF standard, the predicate of knowledge base has a domain and a range. The domain limits the scope of subjects where such predicate can be applied and the range is the set of possible values of the predicate. For example, *Birth-Place* is a predicate using the class *owl:Person* as domain and its range is *owl:Place*. This is also called *type consistency constraint*. Following this characteristic, we can define the true-missing and false-missing data in knowledge base matrix as follows.

Definition 1. *We say an entity pair $e(subject, object)$ is consistent with relation r iff the type of subject is equivalent to or the subset of the domain of r and the type of object meets the same requirement with the range of r . For any triples $(e_{subject}, r, e_{object}) \notin F$, which are missing data in knowledge base, if $e(subject, object)$ is consistent with relation r, it is **true-missing data** in knowledge base matrix, otherwise it is **false-missing data**.*

Based on this definition, the triple *BirthPlace(Barack-Obama, Havaii)* can be treated as true-missing data, but the triple *BirthPlace(BarackObama, MichelleObama)* cannot be, even though they are all missing data in knowledge base matrix. In other words, true-missing data are just the potential target we need to predict.

Consequently, we can regard all the false-missing data as negative facts and add them to the knowledge base matrix by just setting X_{er} to 0 for these triples. This is also consistent with the intuition that X_{er} means the probability of a predicate r holds a entity pair e as mentioned above. This strategy gets rid of these false-missing data for all the missing data and increases the density obviously, as we can see in Table 1.

4   Matrix Factorization Framework

In this section, we discuss how to complete the knowledge base using matrix factorization with similarity regularization.

4.1   Low-Rank Matrix Factorization

The assumption behind a low-dimensional factor model, which is frequently used in movie recommendations, is that there are only a small number of factors influencing the preferences, and that a user's preference vector is determined by how each factor applies to that user [17]. Inspired by this, we can factorize the knowledge base matrix, and utilize the low-dimensional predicate-specific and entity pair-specific matrices to predict how likely an entity pair holds a certain relation. Our goal is sought to approximate the $m \times n$ matrix X by a multiplication of L-rank factors:

$$X \approx U^T V \tag{1}$$

where $U \in R^{l \times m}$ and $V \in R^{l \times n}$ with $l < min(m, n)$, respect to predicate-specific and entity pair-specific matrices, respectively.

Traditionally, the matrix factorization method is utilized to approximate the matrix X by minimizing:

$$\frac{1}{2} \left\| X - U^T V \right\|_F^2 \tag{2}$$

where $\|\cdot\|_F^2$ denotes the Frobenius norm. However, due to the reason that X contains a large number of missing value, we only need to factorize the existing data in matrix X. Moreover, the value of X_{er} corresponds to the probability of how an entity pair e matches the predicate r, the range of which is $[0, 1]$. Employing $U_r^T V_e$ to predict the missing value X_{er} can make the prediction outside of the range of valid values. Hence, instead of using a simple linear factor model, we adjust the inner product between predicate-specific and entity pair-specific feature vectors through a nonlinear logistic function $g(x) = 1/(1 + \exp(-x))$, which bounds the range of the predictions into $[0, 1]$. Hence, we change Equation (2) to the following optimization problem:

$$\min_{U,V} \frac{1}{2} \sum_{r=1}^{m} \sum_{e=1}^{n} I_{er}(X_{er} - g(U_r^T V_e))^2 + \frac{\lambda_1}{2} \|U\|_F^2 + \frac{\lambda_2}{2} \|V\|_F^2 \qquad (3)$$

where I_{er} is the indicator function that is equal to 1 if $X_{er} = 1$ or $X_{er} = 0$, which means $(e_{subject}, r, e_{object})$ is a existing or negative facts, and equal to 0 otherwise. In order to avoid overfitting, the norms of U and V are added as regularization terms and $\lambda_1, \lambda_2 > 0$, as the regular parameter.

The optimization problem in Equation (3) minimizes the *sum-of-squared-errors* objective function with quadratic regularization terms. Gradient based approaches can be applied to find a local minimum. In the following sections, we will introduce how to incorporate the similarity of relations, which are from different KBs, into the matrix factorization model.

4.2 Similarity-Based Regularization

In order to model the similarity information among predicates realistically, we first need to understand where the similarity comes from. Actually, in the construction of various knowledge bases, some of the predicates may be different on the lexical-level even though they represent the same relationship. Moreover, the phenomenon that they represent the same relationship means that these entity pairs of them can be shared. In this paper, we call it the *complementarity* of various knowledge bases. This is an important reason that we integrate various knowledge base into one matrix.

Here we take two popular knowledge bases, YAGO and DBpedia, as an example. The predicate $yago : graduatedFrom$ has 30,389 entity pairs and the predicate $dbpedia : almaMater$ has 64,928 entity pairs. We find that the former has 27,523 (90%) entity pairs that appear together with the latter. Hence, if a predicate r_i is more similar to predicate r_j, it means that there are more common entity pairs between r_i and r_j.

Based on the interpretation above, if predicate r_i is similar to predicate r_j, we can assume that the latent feature representations U_i and U_j of these two predicates are close in the feature space. Following this intuition, we minimize the objective function:

$$\min_{U} \sum_{i=1}^{m} \sum_{f \in S(i)} Sim(i, f) \|U_i - U_f\|_F^2 \qquad (4)$$

where $S(i)$ is the set of predicates that are similar to predicate r_j, $Sim(i, f) \in [0, 1]$ is the similarity function to indicate the similarity between predicate r_i and r_f. Hence we have the overall objective function:

$$\min_{U,V} L(X,U,V) = \frac{1}{2}\sum_{r=1}^{m}\sum_{e=1}^{n} I_{er}(X_{er} - g(U_r^T V_e))^2$$

$$+ \frac{\varphi}{2}\sum_{i=1}^{m}\sum_{f \in S(i)} Sim(i,f)\|U_i - U_f\|_F^2 \qquad (5)$$

$$+ \frac{\lambda_1}{2}\|U\|_F^2 + \frac{\lambda_2}{2}\|V\|_F^2$$

where $\varphi > 0$, which controls the similarity weight in the factorization model. A local minimum of the objective function given by Equation (5) can be found by performing gradient descent in feature vectors U_r and V_e:

$$\frac{\partial L}{\partial U_r} = \sum_{e=1}^{n} I_{er} g'(U_r^T V_e)(g(U_r^T V_e) - X_{er})V_e + \lambda_1 U_r$$

$$+ \varphi \sum_{f \in Sim(r)} Sim(r,f)(U_r - U_f) \qquad (6)$$

$$\frac{\partial L}{\partial V_e} = \sum_{r=1}^{m} I_{er} g'(U_r^T V_e)(g(U_r^T V_e) - X_{er})U_i + \lambda_2 V_e \qquad (7)$$

4.3   Similarity Function

In Section 4.2, the proposed similarity regularization term requires the weight presentation of similarities between different predicates from various knowledge bases. For the knowledge base matrix, the evaluation of similarity between two predicates can be obtained by calculating the similarity of each corresponding row vector. There are many popular methods about this. In this paper, we use a simple yet effective method: Vector Space Similarity(VSS). VSS is employed to define the similarity between vectors based on the consist of items they have in common:

$$Sim(i,f) = \frac{\displaystyle\sum_{j \in I(i) \cap I(f)} r_{ij} \cdot r_{fj}}{\sqrt{\displaystyle\sum_{j \in I(i) \cap I(f)} r_{ij}^2 \cdot \sum_{j \in I(i) \cap I(f)} r_{fj}^2}} \qquad (8)$$

where j belongs to the subset of items which predicate r_i and predicate r_f both related. r_{ij} is the probability of how entity pair j matches the predicate r_i. From the above definition, we can see that VSS similarity in $Sim(i,f)$ is within the range $[0,1]$, and a larger value means predicate r_i and r_f are more similar.

4.4   Prediction

After the low-dimensional latent feature vector U and V are learned, the next step is to predict whether the entity pair e_j hold a relationship r_i for the missing triples. Given a threshold, we could predict $r_i(e_j)$ is true if the probability value $\widehat{X}_{ij} = g(U_i{}^T V_j)$ is greater than threshold, and false otherwise.

5   Experiment

5.1   Dataset Description

We conduct experiments on the YAGO 2s and DBpedia 3.9 knowledge bases. For YAGO 2s, we use the facts about instances contained in the *yagoFacts* dataset. For DBpedia 3.9, we use the *person data* and *raw infobox properties* datasets. In order to solve the problem of aligning entity pairs among two knowledge bases, we use a simple string-matching method to find this linking.

To verify our method, we filter all existing triples as source input for the construction of knowledge base matrix according to two criteria: the object of the triple is not a literal and the number of each relation instances must be greater than 400. In the process of initializing negative data, type consistency information is obtained from *yagoSchema* and *dbpedia:owl*. After the preprocessing, the characteristics of these knowledge bases are shown in Table 1. We split the known entity pairs of these predicates (include existing and negative triples) into 80% training and 20% testing parts. We tune the parameters for our methods using a coarse, manual grid search with cross validation on the training data.

Table 1. Statistics of the data used in our experiments.

	YAGO	DBpedia
Total entity pairs	5.7M	19.3M
Total predicates	75	1358
Total Facts	7.8M	21.3M
average facts of each predicates	67K	27K
Predicates tested	37	409
Entity pairs tested	2.5M	11.4M
KBs matrix density	1.9%	
KBs matrix density(with negative data)	13.4%	

5.2   Metrics

For each testing triple $r(e)$, we will get a prediction value $\widehat{X}_{er}$ which means the probability of the entity pair e having a relationship r. Given a threshold, we can utiliaze the precision and recall to measure the prediction quality of our proposed approach.

The precision metric is defined as:

$$Precision = \frac{\left| \widehat{X}_{er} | X_{er} = 1 \wedge \widehat{X}_{er} > Threshold \right|}{\left| \widehat{X}_{er} | \widehat{X}_{er} > Threshold \right|} \tag{9}$$

The recall metric is defined as:

$$Recall = \frac{\left|\hat{X}_{er}|X_{er} = 1 \wedge \hat{X}_{er} > Threshold\right|}{|X_{er}|X_{er} = 1|} \tag{10}$$

5.3 Methods

In our experiments we seek to answer the following questions:

1. Whether the strategy of initializing negative data by using type consistency constraint has a positive effect on the results.
2. Whether incorporating similarity regularization into the matrix factorization model help to improve the performance.
3. Whether integrating various knowledge bases to predicate new facts has a better performance than using a single knowledge base.

To address these questions, we design several groups of compared methods. As shown in Table 2, **MF_basic** means using only existing facts, **MF_negative** means adding false-missing data as negative data into knowledge base matrix, **MF_similarity** means adding similarity regularization into matrix factorization model on the basis of **MF_negative**. In Table 3, **MF_yago**, **MF_dbpedia** means employing MF_similarity method just in YAGO or DBpedia, respectively. **MF_integration** means integrating two knowledge bases as the input of MF_similarity method.

5.4 Results

From Table 2, we can observe that **MF_basic** performs worse than **MF_negative**. This indicates that initializing negative data in knowledge base matrix has a positive effect on the result. Among these methods, **MF_similarity** method generally achieves better performance than the others. This demonstrates that incorporating the similarity information between predicates into the matrix factorization model improves the performance.

In addition, Table 3 shows F1 scores for the single or integration knowledge base model on some tested relations. From the data in Table 3, it is apparent that integrating two knowledge bases has a better performance than using any one of them. These results confirm that the integration of two knowledge bases will give us more available inference information in the processing of completing knowledge base.

5.5 Impact of Parameters Threshold and φ

As mentioned in Section 4.4, the parameter threshold determines the final prediction results. Figure 1 illustrate the impact of threshold on precision. It turns out that the precision increases simultaneously with the threshold. This phenomenon coincides with the intuition that a high threshold will result in more reliable prediction.

Table 2. Comparison of performance of different methods.

	Precision	Recall	F1
MF_basic	0.424	0.240	0.306
MF_negative	0.601	0.343	0.437
MF_similarity	0.693	0.416	0.520

Table 3. F1 performance of different methods for some relations tested.

	MF_yago	**MF_dbpedia**	**MF_integration**
ya : graduatefrom	0.293	-	0.610
db : almamater	-	0.264	0.492
ya : livesin	0.316	-	0.537
db : residence	-	0.301	0.544
ya : isinterestedin	0.253	-	0.425
db : maininterest	-	0.248	0.390
ya : ismarriedto	0.332	-	0.651
db : spouse	-	0.364	0.635

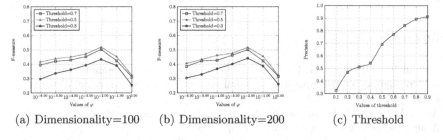

(a) Dimensionality=100   (b) Dimensionality=200   (c) Threshold

Fig. 1. Impact of parameter threshold and φ

In addition to this, the similarity weight parameter φ also plays a very important role. It controls how much our methods should incorporate the predicate similarity information between various knowledge bases. On one hand, if we use a very small value of φ, we ignore the similarity information between different predicates and simply employ the predicates' own information in making predictions. On the other hand, if we employ a very large value of φ, the similarity information will dominate the factorization processes. In many cases, we do not want to set φ to these extreme values since they will potentially hurt the prediction performance.

Figure 1 shows the impacts of φ on F-measure in our model. We observe that the value of φ impacts the prediction results significantly. From the results, we can see that no matter what the threshold and dimensionality we use, when φ increases, the F-measure value also increases at first, but after φ exceeds a certain threshold like 0.01, the F-measure value decreases with further increments of the value φ. The results in Figure 1 further support the idea that integrating the knowledge base matrix and the predicate similarity information for prediction achieves better performance than using only one of them.

6 Conclusion

In this paper, we formulate the knowledge base completion task as a matrix completion problem and develop a factor analysis approach based on matrix factorization by employing various knowledge bases. In the construction of the knowledge base matrix, we increase the density of matrix by using the type consistency constraint. In the factorization process, we integrate the similarity of different relations as regularization into the factorization model. Empirical results show that both the two strategies can help improve the performance of predicting new relations between existing entity pairs.

Acknowledgments. This work was supported by the National High Technology R&D Program of China (Grant No.2014AA015102, 2015AA015403), National Natural Science Foundation of China (Grant No.61272344, 61202233, 61370055) and the joint project with IBM Research. Any correspondence please refer to Yansong Feng.

References

1. Suchanek, F.M., Kasneci, G., Weikum, G.: Yago: a core of semantic knowledge. In: Proceedings of the 16th international conference on World Wide Web, ACM (2007) 697–706
2. Auer, S., Bizer, C., Kobilarov, G., Lehmann, J., Cyganiak, R., Ives, Z.: DBpedia: A nucleus for a web of open data. Springer (2007)
3. Bollacker, K., Evans, C., Paritosh, P., Sturge, T., Taylor, J.: Freebase: a collaboratively created graph database for structuring human knowledge. In: Proceedings of the 2008 ACM SIGMOD international conference on Management of data, ACM (2008) 1247–1250
4. Carlson, A., Betteridge, J., Wang, R.C., Hruschka Jr, E.R., Mitchell, T.M.: Coupled semi-supervised learning for information extraction. In: Proceedings of the third ACM international conference on Web search and data mining, ACM (2010) 101–110
5. Min, B., Grishman, R., Wan, L., Wang, C., Gondek, D.: Distant supervision for relation extraction with an incomplete knowledge base. In: HLT-NAACL. (2013) 777–782
6. Lao, N., Mitchell, T., Cohen, W.W.: Random walk inference and learning in a large scale knowledge base. In: Proceedings of the Conference on Empirical Methods in Natural Language Processing, Association for Computational Linguistics (2011) 529–539
7. Nickel, M., Tresp, V., Kriegel, H.P.: Factorizing yago: scalable machine learning for linked data. In: Proceedings of the 21st international conference on World Wide Web, ACM (2012) 271–280

8. Lao, N., Subramanya, A., Pereira, F., Cohen, W.W.: Reading the web with learned syntactic-semantic inference rules. In: Proceedings of the 2012 Joint Conference on Empirical Methods in Natural Language Processing and Computational Natural Language Learning, Association for Computational Linguistics (2012) 1017–1026
9. Yao, L., Riedel, S., McCallum, A.: Universal schema for entity type prediction. In: Proceedings of the 2013 workshop on Automated knowledge base construction, ACM (2013) 79–84
10. Gardner, M., Talukdar, P.P., Kisiel, B., Mitchell, T.M.: Improving learning and inference in a large knowledge-base using latent syntactic cues. In: EMNLP. (2013) 833–838
11. Gardner, M., Talukdar, P., Krishnamurthy, J., Mitchell, T.: Incorporating vector space similarity in random walk inference over knowledge bases. In: Proceedings of EMNLP. (2014)
12. Cohen, W.W., Page, C.D.: Polynomial learnability and inductive logic programming: Methods and results. New Generation Computing **13**(3-4) (1995) 369–409
13. Lao, N., Cohen, W.W.: Relational retrieval using a combination of path-constrained random walks. Machine learning **81**(1) (2010) 53–67
14. Nickel, M., Tresp, V., Kriegel, H.P.: A three-way model for collective learning on multi-relational data. In: Proceedings of the 28th international conference on machine learning (ICML-11). (2011) 809–816
15. Yao, L., Riedel, S., McCallum, A.: Probabilistic databases of universal schema. In: Proceedings of the AKBC-WEKEX Workshop at NAACL 2012. (June 2012)
16. Maximilian Nickel, Kevin Murphy, V.T.E.G.: A review of relational machine learning for knowledge graphs: From multi-relational link prediction to automated knowledge graph construction. In: Prediction to Automated Knowledge Graph Construction, IEEE (2015) arXiv:1503.00759
17. Rennie, J.D., Srebro, N.: Fast maximum margin matrix factorization for collaborative prediction. In: Proceedings of the 22nd international conference on Machine learning, ACM (2005) 713–719

MATAR: Keywords Enhanced Multi-label Learning for Tag Recommendation

Licheng Li, Yuan Yao, Feng Xu, and Jian Lu

State Key Laboratory for Novel Software Technology,
Nanjing University, Nanjing 210046, China
{jimmyli,yyao}@smail.nju.edu.cn, {xf,lj}@nju.edu.cn

Abstract. Tagging is a popular way to categorize and search online content, and tag recommendation has been widely studied to better support automatic tagging. In this work, we focus on recommending tags for content-based applications such as blogs and question-answering sites. Our key observation is that many tags actually have appeared in the content in these applications. Based on this observation, we first model the tag recommendation problem as a multi-label learning problem and then further incorporate keyword extraction to improve recommendation accuracy. Moreover, we speedup the proposed method using a locality-sensitive hashing strategy. Experimental evaluations on two real data sets demonstrate the effectiveness and efficiency of our proposed methods.

Keywords: Tag recommendation, multi-label learning, keyword extraction, locality-sensitive hashing.

1 Introduction

Tags have been widely used to categorize and search online content in many online applications such as blogs, question-answering sites, and online newspapers. For example, the question-answering site Stack Overflow[1] uses tags to categorize programming questions so that the questions can be easily found by appropriate answerers. Although tags are important for these applications, it also brings additional burdens to users. For example, when a user posts a programming question on Stack Overflow, he/she is required to add several appropriate tags for this question, which is not usually an easy task. This motivates the research to automatically recommend appropriate tags for users in these online applications.

Roughly, existing tag recommendation methods can be classified into two categories: collaborative-filtering method and content-based method (see Section 2 for a review). Methods in the first category rely on users' historical behavior, and these methods are more suitable to tag a relatively fixed set of items (e.g., music and movies). On the other hand, methods in the second category mainly use the content information, and they are more suitable for content-based applications (e.g., blogs and question-answering sites).

---

[1] http://stackoverflow.com/

© Springer International Publishing Switzerland 2015
R. Cheng et al. (Eds.): APWeb 2015, LNCS 9313, pp. 268–279, 2015.
DOI: 10.1007/978-3-319-25255-1_22

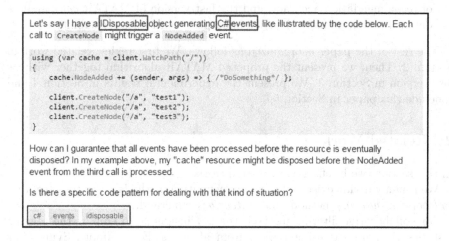

Let's say I have a |Disposable| object generating |C#| events, like illustrated by the code below. Each call to `CreateNode` might trigger a `NodeAdded` event.

```
using (var cache = client.WatchPath("/"))
{
    cache.NodeAdded += (sender, args) => { /*DoSomething*/ };

    client.CreateNode("/a", "test1");
    client.CreateNode("/a", "test2");
    client.CreateNode("/a", "test3");
}
```

How can I guarantee that all events have been processed before the resource is eventually disposed? In my example above, my "cache" resource might be disposed before the NodeAdded event from the third call is processed.

Is there a specific code pattern for dealing with that kind of situation?

    c#   events   idisposable

Fig. 1. A programming question from Stack Overflow.

In this work, we focus on recommending tags for content-based applications. Our key observation is that in content-based applications, many tags actually have appeared in the content, and these tags are potential keywords in the content. Fig. 1 shows an illustrative example from Stack Overflow. There are three tags (i.e., 'c#', 'events', and 'idisposable') for the given question in this example. They all appeared in the question content, and they all revealed key information of the question. Actually, we empirically found that around 70% tags have appeared in the corresponding questions on Stack Overflow. Based on this observation, we aim to boost the accuracy of tag recommendation by combining two content-based methods: multi-label learning and keyword extraction. Specially, we first model the tag recommendation problem as a multi-label learning problem, and then incorporate a keyword extraction method into multi-label learning. Further, since one of the challenges of tag recommendation is from the large scale of the content data, we also speedup the proposed method by employing the locality-sensitive hashing strategy.

The main contributions of this paper are summarized as follows:

- We propose a tag recommendation method (MATAR) for improving recommendation accuracy in content-based applications. The proposed MATAR method combines the multi-label learning method and the keyword extraction method.
- We propose a fast version of MATAR, which employs approximation methods to reduce the time complexity of MATAR.
- Experimental results on two real data sets show that the proposed MATAR method outperforms several existing tag recommendation methods. In particular, MATAR improves the best competitor by 3.7% - 18.9% in terms

of recommendation accuracy, and the fast version of MATAR can handle a data set including more than 1 million posts with linear scalability.

The rest of the paper is organized as follows. We first review related work in Section 2. Then, we present the proposed MATAR algorithm together with its fast version in Section 3. We present the experimental results in Section 4 and conclude this paper in Section 5.

2   Related Work

In this section, we briefly review related work.

We roughly divide existing tag recommendation methods into two categories: *collaborative-filtering* method and *content-based* method.

For collaborative-filtering method, the key insight is to employ the tagging histories (i.e., user-item-tag tuples) from all users. For example, Symeonidis et al. [18] adopt tensor factorization model on the user-item-tag tuples; Rendle et al. [11,12] further model the pairwise rankings into tensor factorization method. Other examples in this category include [4,7,15,3]. Overall, methods in this category are more suitable to tag a relatively fixed set of items (e.g., music and movies), and they are not able to recommend tags for the new content that has not been tagged yet.

In the second category of content-based method, the content itself is used as input. For example, some studies focus on the feature aspect by finding useful textual features (such as tag co-occurrence [2,21] and entropy [8]). In the algorithm aspect, classification models and topic models are widely used for tag recommendation. For example, Saha et al. [13] train a classification model for each tag and recommend tags based on multiple classifiers; Krestel et al. [5] use topic models to find latent topics from the content and then recommend tags based on these latent topics; Si and Sun [14] further propose a variant LDA model to link tags with the latent topics. Other content-based methods include [16,17,20]. In this work, we focus on content-based methods as we target at recommending tags for content-based applications.

There are also some other lines of research that are potentially useful for tag recommendation. Tag recommendation can be viewed as a multi-label learning problem by treating tags as labels. For example, TagCombine [21] is one of the few methods that apply multi-label learning for tag recommendation. However, TagCombine suffers from the class imbalance problem (i.e., each tag is only used by a very small portion of posts). Keyword extraction is another potentially useful tool for tag recommendation, as tags may appear as keywords in the content. For instance, keywords may be directly used as tags [10]; association rules between keywords and tags are also considered [19]. In this work, we propose incorporating keyword extraction into multi-label learning for better tag recommendation. To the best of our knowledge, we are the first to combine multi-label learning and keyword extraction for tag recommendation in content-based applications.

3   The Proposed Methods

In this section, we present our algorithms for tag recommendation. The proposed MATAR algorithm combines multi-label learning with keyword extraction to enhance the recommendation accuracy. We further propose MATAR-fast to speedup the computations of MATAR.

3.1   Problem Statement

Before presenting our algorithms, we first introduce some notations and define the tag recommendation problem. We use $\mathbf{X}_{m \times d}$ to denote the feature matrix where each row contains the features for an instance. Without loss of generality, we assume that there are m instances and d features for each instance. We use $\mathbf{Y}_{m \times T}$ to denote the associate tag matrix where T is the number of tags. We use subscripts to indicate the entries. That is, $\mathbf{X}_i$ and $\mathbf{Y}_i$ indicate the feature vector and the tag vector of the ith instance, respectively. Then, element $\mathbf{Y}_i(t) \in \{0, 1\}$ indicates whether the ith instance is assigned with the tth tag ($\mathbf{Y}_i(t) = 1$ indicates that the instance is assigned with the tth tag and $\mathbf{Y}_i(t) = 0$ indicates otherwise). Based on the above notations, the tag recommendation takes the existing $\mathbf{X}$ matrix and $\mathbf{Y}$ matrix as input, and aims to predict the tag vector $\mathbf{Y}_r$ for a given new instance $\mathbf{X}_r$.

3.2   The Proposed MATAR Algorithm

Next, we describe the MATAR algorithm. To make the descriptions of our algorithms easier, we further define the following notations. For a given instance $\mathbf{X}_i$, we define $\mathcal{N}(\mathbf{X}_i)$ as the set of its K-nearest neighbors. Based on the tag vectors of these neighbors, we define a *counting* vector for $\mathbf{X}_i$ as

$$S_i(t) = \sum_{\mathbf{X}_j \in \mathcal{N}(\mathbf{X}_i)} \mathbf{Y}_j(t), t = 1, 2, \ldots, T. \tag{1}$$

Basically, $S_i(t)$ counts the number of instances that have been assigned with the tth tag among the neighborhood of $\mathbf{X}_i$.

Next, for a given new instance $\mathbf{X}_r$, we define two families of probability events. The first one is U_g^t ($g \in \{0, 1\}$), where U_1^t is the event that tag t belongs to $\mathbf{X}_r$ and U_0^t is the event that tag t does not belong to $\mathbf{X}_r$. The other one is V_k^t ($k \in \{0, 1, \ldots, K\}$), which means that there are exactly k instances that have tag t among the K-nearest neighbors of $\mathbf{X}_r$. Based on these two families of probability events, we aim to estimate the probability $P_r(t)$ by which the new instance $\mathbf{X}_r$ would be tagged with tag t (i.e., the tth tag). Following the K-nearest neighborhood multi-label learning framework [22], we can define $P_r(t)$ as

$$P_r(t) = P(U_1^t | V_{S_r(t)}^t), t = 1, 2, \ldots, T. \tag{2}$$

In this work, we use the cosine similarity (i.e., the cosine distance between $\mathbf{X}_r$ and $\mathbf{X}_i$) to find the neighborhood.

Applying the Bayesian rule and decomposing the $V^t_{S_r(t)}$ event, we can rewrite Eq. (2) as

$$P_r(t) = \frac{P(U^t_1)P(V^t_{S_r(t)}|U^t_1)}{P(V^t_{S_r(t)})} = \frac{P(U^t_1)P(V^t_{S_r(t)}|U^t_1)}{\sum\limits_{g\in\{0,1\}} P(U^t_g)P(V^t_{S_r(t)}|U^t_g)}. \tag{3}$$

Computing $P(U^t_g)$ and $P(V^t_k|U^t_g)$. As shown in Eq. (3), we need to estimate the prior $P(U^t_g)$ ($t = 1, 2, \ldots, T$ and $g \in \{0, 1\}$) and the posterior $P(V^t_k|U^t_g)$ ($k \in \{0, 1, \ldots, K\}$) to compute the probability $P_r(t)$. We resort to keyword extraction and frequency counting to compute these prior and posterior probabilities.

As mentioned in introduction, the key intuition of our method is that many tags actually appeared as keywords in the content. Let w^t be the corresponding word of tag t, we have

$$P(U^t_g) = P(U^t_g|w^t)P(w^t) + P(U^t_g|\neg w^t)P(\neg w^t), \tag{4}$$

where $P(w^t)$ is the probability of which the word w^t is a keyword in the content, and $P(\neg w^t)$ is the probability of which the word w^t is not a keyword in the content. Here, $P(\neg w^t) = 1 - P(w^t)$.

Then, computing $P(U^t_g)$ becomes computing $P(U^t_g|w^t)$, $P(U^t_g|\neg w^t)$ and $P(w^t)$. For $P(w^t)$, various keyword extraction methods can be used here. In this work, we adopt the TextRank [9] method. The basic idea of TextRank is to model the text as a weighted graph. Each vertex in this graph represents a unique word in the text, and an edge between two words indicates that these two words are close to each other in the text. Then, pagerank algorithm is applied on this graph to find the keywords. The output of TextRank is a score vector in which each word is assigned with a score to indicate the importance of this word. In other words, the score for each word indicates the probability by which this word could be a keyword. Therefore, we use this score as $P(w^t)$.

To estimate $P(U^t_g|w^t)$, we can approximate the result by making a few statistics on the training data. For each tag t, we first define two variables c_w and $c_{w,t}$. We first find the instances where the corresponding word of tag t appears. Then, c_w counts the number of these instances and $c_{w,t}$ counts the number of these instances that are tagged with t. Based on these two variables, we can estimate probability $P(U^t_g|w^t)$ as

$$P(U^t_1|w^t) = \frac{c_{w,t} + s}{c_w + s \times 2}$$
$$P(U^t_0|w^t) = 1 - P(U^t_1|w^t), \tag{5}$$

where $s = 1$ is a smoothing parameter.

Similarly, to estimate $P(U^t_g|\neg w^t)$, we also define two variables $c_{\neg w}$ and $c_{\neg w,t}$ for each tag t. We first find the instances where the corresponding word of tag t does not appear, and then $c_{\neg w}$ counts the number of these instances and $c_{\neg w,t}$

Algorithm 1. The Training Stage of Our MATAR Algorithm

Input: X, Y, and K
Output: $P(U_g^t)$, $P(V_k^t|U_g^t)$, $P(U_g^t|w^t)$, and $P(U_g^t|\neg w^t)$
1: **for** $i \in \{1, 2, \ldots, m\}$ **do**
2:     Compute $P(w^t)$ for each tag in $\mathbf{X}_i$;
3:     Identify the neighborhood $\mathcal{N}(\mathbf{X}_i)$ for $\mathbf{X}_i$;
4: **end for**
5: **for** $t \in \{1, 2, \ldots, T\}$ **do**
6:     Compute $P(U_g^t|w^t)$ and $P(U_g^t|\neg w^t)$ using Eq. (5) and Eq. (6);
7:     Compute $P(U_g^t)$ using Eq. (7);
8:     Compute $P(V_k^t|U_g^t)$ using Eq. (8);
9: **end for**

---

counts the number of these instances that are tagged with t. Based on these two variables, we can estimate probability $P(U_g^t|\neg w^t)$ as

$$P(U_1^t|\neg w^t) = \frac{c_{\neg w,t} + s}{c_{\neg w} + s \times 2}$$
$$P(U_0^t|\neg w^t) = 1 - P(U_1^t|\neg w^t). \tag{6}$$

Notice that our computations of $P(U_g^t|w^t)$, $P(U_g^t|\neg w^t)$ and $P(w^t)$ are all based on the cases when the tags appeared in the content. If the tags did not appear in the content, we may directly estimate $P(U_g^t)$ as

$$P(U_1^t) = \frac{\sum_{i=1}^m \mathbf{Y}_i(t) + s}{m + s \times 2}$$
$$P(U_0^t) = 1 - P(U_1^t). \tag{7}$$

Next, we show how we compute $P(V_k^t|U_g^t)$. For each tag t, we first define two arrays c and c' of length $K + 1$. We first find the training instances whose K-nearest neighbors contain exactly k instances with tag t, and then $c[k]$ and $c'[k]$ count the number of these training instances that are tagged with t and without t, respectively. Based on these two arrays, we can estimate the posterior probabilities as

$$P(V_k^t|U_1^t) = \frac{c[k] + s}{\sum_{j=0}^K c[j] + s \times (K + 1)}$$
$$P(V_k^t|U_0^t) = \frac{c'[k] + s}{\sum_{j=0}^K c'[j] + s \times (K + 1)}. \tag{8}$$

The overall MATAR algorithm is summarized in Alg. 1 and Alg. 2 (training stage and predicting stage, respectively). Steps 1-4 in Alg. 1 compute $P(w^t)$ and $\mathcal{N}(\mathbf{X}_i)$ for each instance. Steps 5-9 estimate $P(U_g^t|w^t)$, $P(U_g^t|\neg w^t)$, $P(U_g^t)$ and $P(V_k^t|U_g^t)$. Based on the results of the training stage, the predicting stage is shown in Alg. 2. After computing $\mathcal{N}(\mathbf{X}_r)$ and $P(w^t)$ for $\mathbf{X}_r$, Steps 3-9 compute the probability $P_r(t)$ for each possible tag. During the iterations, we update

---

Algorithm 2. The Predicting Stage of Our MATAR Algorithm

Input: X, Y, K, $P(U_g^t)$, $P(V_k^t|U_g^t)$, $P(U_g^t|w^t)$, $P(U_g^t|\neg w^t)$, and $\mathbf{X}_r$
Output: $P_r(t)$
 1: Identify $\mathcal{N}(\mathbf{X}_r)$;
 2: Compute $P(w^t)$ for $\mathbf{X}_r$;
 3: **for** $t \in \{1, 2, \ldots, T\}$ **do**
 4:     Compute $S_r(t)$ using Eq. (1);
 5:     **if** $P(w^t) \neq 0$ **then**
 6:         Update $P(U_g^t)$ using Eq. (4);
 7:     **end if**
 8:     Compute $P_r(t)$ using Eq. (3);
 9: **end for**

$P(U_g^t)$ if $P(w^t) \neq 0$ and use the input $P(U_g^t)$ otherwise. Finally, we can recommend a ranking list based on the $P_r(t)$ score for each possible tag.

Algorithm Analysis. Here, we present some algorithm analysis for MATAR. Iterations of Step 2 in Alg. 1 require $O(md)$ time. The main time consumption is from Step 3. Iterations of this step require $O(m^2d)$ time. From Step 5 to Step 9, we need $O(mTK)$ time. In general, the time complexity of the training stage of our MATAR algorithm is $O(m^2d + mTK)$. Since K is usually much smaller than m and T, and the feature dimension d is a fixed constant, the time complexity of the training stage can be rewritten as $O(m^2 + mT)$.

For Alg. 2, Step 1 requires $O(md)$ time. Step 2 requires $O(d)$ time. Steps 3-9 compute the probabilities $P_r(t)$, which requires $O(TK)$ time. To sum up, the time complexity of the predicting stage of our MATAR algorithm is $O(md + TK)$, which can be rewritten as $O(m + T)$.

3.3   The Proposed MATAR-Fast Algorithm

Next, we present a fast version of MATAR. The main time consumption of our MATAR algorithm is from the nearest neighbor computations. That is, we need to compute the similarities between any two instances. To speedup MATAR, we employ an approximate method to linearly identify neighbors. Specially, we use locality-sensitive hashing [1] whose basic idea is to hash the input instances so that similar instances are mapped to the same bucket with high probability (the number of the buckets is usually much smaller than the number of the input instances). Therefore, we only need to compute the similarities within each bucket.

We consider the family of hash functions defined as follows. For each hash function in this family, we first choose a random vector $v_{d \times 1}$ where d is the dimension size, and then define the following hash function h_v:

$$h_v(a) = \begin{cases} 1 & \text{if } v \cdot a \geq 0 \\ 0 & \text{if } v \cdot a < 0 \end{cases} \tag{9}$$

Table 1. The statistics of SO, SO-Small and Math data sets.

Data	# Posts	# Tags	# Tag Assignments
SO	1,942,249	777	4,358,738
SO-Small	10,000	777	22,335
Math	19,953	501	36,813

Further, we randomly pick L hash functions from this family where L determines the number of buckets. Applying these L hash functions to one input instance would output a bit vector composed of 1s or 0s (the bit vector is regarded as the encoding of a bucket). Then, we can compute the similarities within each bucket as an approximation for similarity computations. The detailed algorithm is omitted for simplicity.

Algorithm Analysis. Finally, we analyze the complexity of MATAR-fast. As we stated above, the locality-sensitive hashing strategy is applied to find the neighborhood in Alg. 1 to reduce the complexity. The time complexity of MATAR-fast mainly depends on the L parameter, which is $O(\frac{m^2}{2^L})$. Typically, we can set $L = O(\log m)$ (e.g., let $L = \log \frac{m}{500}$, there are averagely 500 instances in each bucket). Therefore, the similarity computation for each $\mathbf{X}_i$ requires $O(1)$ time as it only involves the instances in the same bucket. Overall, the time complexity of the training stage of our MATAR-fast algorithm is $O(md + mTK)$, which can be rewritten as $O(mT)$.

4 Experiments

In this section, we present the experimental evaluations. The experiments are designed to answer the following questions:

- *Effectiveness*: How accurate are the proposed algorithms for recommending tags?
- *Efficiency*: How scalable are the proposed algorithms?

4.1 Experiment Setup

Data Sets. We use the data from two real world sites, i.e., Stack Overflow (SO) and Mathematics Stack Exchange (Math), to evaluate our algorithms. They are popular CQA sites for programming and math, respectively. For both data sets, they are officially published and publicly available[2]. For features, we adopt the commonly used 'bag of words' model. For tags, we put our focus on common tags and remove some rare tags. A tag is rare if it appeared less than ϵ times (ϵ is equal to 0.5% of the maximum value of tag appearance in the data set). To compare with the existing methods, we also randomly select a small subset of the SO data. The statistics of the three data sets are summarized in Table 1.

---

[2] http://blog.stackoverflow.com/category/cc-wiki-dump/

Table 2. The effectiveness comparisons. Higher is better. The proposed algorithms (MATAR and MATAR-fast) outperform all the other methods.

<table>
<tr><td colspan="3">(a) Result on Math data</td><td colspan="3">(b) Result on SO-Small data</td></tr>
</table>

Method	Recall@5	Recall@10	Method	Recall@5	Recall@10
SVMSim	0.5903	0.6770	SVMSim	0.3219	0.3803
MLKNN	0.5247	0.6267	MLKNN	0.3563	0.4476
LDASim	0.5139	0.6191	LDASim	0.3447	0.4402
TagCombine	0.5908	0.6429	TagCombine	0.4490	0.4936
Snaff	0.2904	0.3671	Snaff	0.2148	0.2586
MATAR	**0.6123**	**0.7122**	**MATAR**	**0.4758**	**0.5871**
MATAR-fast	**0.5941**	**0.6965**	**MATAR-fast**	**0.4504**	**0.5531**

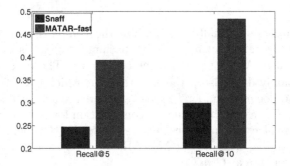

Fig. 2. The result on the entire SO data. Higher is better. Our MATAR-fast is much better than Snaff.

Evaluation Metrics. For effectiveness evaluation, recall is more important than precision in our problem setting. Therefore, we adopt the recall@k metric as defined below.

$$Recall@k = \frac{1}{m} \sum_{i=1}^{m} \frac{|Rec_i \cap Tag_i|}{|Tag_i|}, \tag{10}$$

where m is the number of test instances, Tag_i is the actual tags for instance i, Rec_i is the top-k ranked tags recommended for instance i.

For efficiency, we report the wall-clock time. All the efficiency experiments were run on a machine with eight 3.4GHz Intel Cores and 32GB memory.

4.2 Experimental Results

Effectiveness Results. We first compare the effectiveness of the proposed algorithms (MATAR and MATAR-fast) with SVMSim [13], TagCombine [21], LDASim [5], MLKNN [22], and Snaff [6] on Math data and SO-Small data. We randomly choose 90% data for training and use the rest 10% data for testing. We set $K = 60$ for Math data and $K = 70$ for SO-Small data. The results on the two data sets are shown in Table 2.

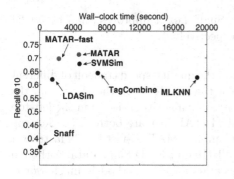
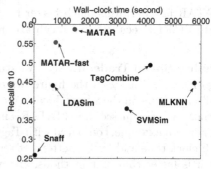

(a) Tag Recommendation on Math data   (b) Tag Recommendation on SO data

Fig. 3. The quality-speed tradeoff. The proposed MATAR-fast achieves a good balance between the recommendation accuracy and the efficiency (in the left-top corner).

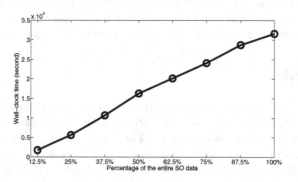

Fig. 4. Scalability of MATAR-fast. Our MATAR-fast scales linearly wrt the number of instances.

We make several observations from Table 2. First, the proposed MATAR algorithm performs the best. For example, on Math data, MATAR improves the best competitor (i.e., SVMSim) by 3.7% and 5.2% for recall@5 and recall@10, respectively; on the SO-Small data, it improves the best competitor (i.e., Tag-Combine) by 6.0% and 18.9% for recall@5 and recall@10, respectively. Second, MATAR also performs much better than MLKNN. For example, on Math data, MATAR improves the MLKNN method by 16.7% and 13.6% for recall@5 and recall@10, respectively. Since MATAR can be seen as an extension of MLKNN, this result indicates that the keyword extraction indeed helps in tag recommendation. Third, although the performance of the proposed MATAR-fast algorithm is not as good as MATAR, it still better than the compared methods.

We also give the results on the entire SO data in Fig. 2. Here, we only show the results of MATAR-fast and Snaff as only these two algorithms can return results within 24 hours on the entire SO data. As we can see from the figure, our

MATAR-fast performs much better than Snaff, i.e., it improves Snaff by 59.2% and 61.6% for recall@5 and recall@10, respectively.

Quality-Speed Tradeoff. Next, we study the quality-speed tradeoff of different algorithms in Fig. 3. In the figure, we plot the recall@10 on the y-axis and the wall-clock time on the x-axis. Ideally, we want an algorithm sitting in the left-top corner. As we can see, our MATAR and MATAR-fast are both in the left-top corner. For example, compared with TagCombine, MATAR-fast is 5.3x faster wrt wall-clock time and 18.9% better wrt recall@10 on the SO-Small data. Although Snaff is faster than our methods, the accuracy of this method is much worse. Overall, our MATAR-fast achieves a good balance between the recommendation accuracy and the efficiency.

Scalability Results. Finally, we study the scalability of our proposed MATAR-fast algorithm on the whole SO data. We vary the size of training data, and report the wall-clock time in Fig. 4. As we can see from the figure, the running time of our MATAR-fast algorithm scales linearly wrt the number of training data size, which is also consistent with our algorithm analysis in Section 3.3.

5   Conclusions

In this paper, we have proposed a novel tag recommendation algorithm MATAR. The proposed MATAR combines multi-label learning and keyword extraction to enhance the recommendation accuracy. The basic intuition behind MATAR is the fact that many tags actually have appeared in the content in content-based applications. We have further proposed a fast version for MATAR based on the locality-sensitive hashing strategy. Experimental evaluations on two real data sets demonstrate the effectiveness and efficiency of the proposed methods.

Acknowledgment. This work is supported by the National 973 Program of China(No. 2015CB352202), and the National Natural Science Foundation of China(No. 91318301, 61321491, 61100037).

References

1. Andoni, A., Indyk, P.: Near-optimal hashing algorithms for approximate nearest neighbor in high dimensions. In: 47th Annual IEEE Symposium on Foundations of Computer Science, FOCS 2006, pp. 459–468. IEEE (2006)
2. Belém, F., Martins, E., Pontes, T., Almeida, J., Gonçalves, M.: Associative tag recommendation exploiting multiple textual features. In: SIGIR, pp. 1033–1042. ACM (2011)
3. Feng, W., Wang, J.: Incorporating heterogeneous information for personalized tag recommendation in social tagging systems. In: Proceedings of the 18th ACM SIGKDD International Conference on Knowledge Discovery and Data Mining, pp. 1276–1284. ACM (2012)

4. Jäschke, R., Marinho, L., Hotho, A., Schmidt-Thieme, L., Stumme, G.: Tag recommendations in social bookmarking systems. AI Communication 21(4), 231–247 (2008)
5. Krestel, R., Fankhauser, P., Nejdl, W.: Latent dirichlet allocation for tag recommendation. In: Proceedings of the Third ACM Conference on Recommender Systems, pp. 61–68. ACM (2009)
6. Lipczak, M., Milios, E.: Learning in efficient tag recommendation. In: Proceedings of the Fourth ACM Conference on Recommender Systems, pp. 167–174. ACM (2010)
7. Liu, R., Niu, Z.: A collaborative filtering recommendation algorithm based on tag clustering. In: Park, J.J.(J.H.), Stojmenovic, I., Choi, M., Xhafa, F. (eds.) Future Information Technology. LNEE, vol. 276, pp. 177–183. Springer, Heidelberg (2014)
8. Lu, Y.T., Yu, S.I., Chang, T.C., Hsu, J.Y.J.: A content-based method to enhance tag recommendation. In: IJCAI, vol. 9, pp. 2064–2069 (2009)
9. Mihalcea, R., Tarau, P.: Textrank: Bringing order into texts. Association for Computational Linguistics (2004)
10. Murfi, H., Obermayer, K.: A two-level learning hierarchy of concept based keyword extraction for tag recommendations. In: ECML PKDD Discovery Challenge (DC 2009), p. 201 (2009)
11. Rendle, S., Balby Marinho, L., Nanopoulos, A., Schmidt-Thieme, L.: Learning optimal ranking with tensor factorization for tag recommendation. In: Proceedings of the 15th ACM SIGKDD International Conference on Knowledge Discovery and Data Mining, pp. 727–736. ACM (2009)
12. Rendle, S., Schmidt-Thieme, L.: Pairwise interaction tensor factorization for personalized tag recommendation. In: Proceedings of the Third ACM International Conference on Web Search and Data Mining, pp. 81–90. ACM (2010)
13. Saha, A.K., Saha, R.K., Schneider, K.A.: A discriminative model approach for suggesting tags automatically for stack overflow questions. In: Proceedings of the 10th Working Conference on Mining Software Repositories, pp. 73–76. IEEE Press (2013)
14. Si, X., Sun, M.: Tag-lda for scalable real-time tag recommendation. Journal of Computational Information Systems 6(1), 23–31 (2009)
15. Sigurbjörnsson, B., Van Zwol, R.: Flickr tag recommendation based on collective knowledge. In: Proceedings of the 17th International Conference on World Wide Web, pp. 327–336. ACM (2008)
16. Song, Y., Zhang, L., Giles, C.L.: Automatic tag recommendation algorithms for social recommender systems. ACM Transactions on the Web (TWEB) 5(1), 4 (2011)
17. Subramaniyaswamy, V., Pandian, S.C.: Topic ontology-based efficient tag recommendation approach for blogs. International Journal of Computational Science and Engineering 9(3), 177–187 (2014)
18. Symeonidis, P., Nanopoulos, A., Manolopoulos, Y.: Tag recommendations based on tensor dimensionality reduction. In: Proceedings of the 2008 ACM Conference on Recommender Systems, pp. 43–50. ACM (2008)
19. Wang, J., Hong, L., Davison, B.D.: Rsdc'09: Tag recommendation using keywords and association rules. In: ECML PKDD Discovery Challenge, pp. 261–274 (2009)
20. Wang, T., Wang, H., Yin, G., Ling, C.X., Li, X., Zou, P.: Tag recommendation for open source software. Frontiers of Computer Science 8(1), 69–82 (2014)
21. Xia, X., Lo, D., Wang, X., Zhou, B.: Tag recommendation in software information sites. In: Proceedings of the 10th Working Conference on Mining Software Repositories, pp. 287–296. IEEE Press (2013)
22. Zhang, M.L., Zhou, Z.H.: Ml-knn: A lazy learning approach to multi-label learning. Pattern Recognition 40(7), 2038–2048 (2007)

Reverse Direction-Based Surrounder Queries

Xi Guo[1,2], Yoshiharu Ishikawa[3], Aziguli Wulamu[1,2], and Yonghong Xie[1,2]

[1] School of Computer and Communication Engineering,
University of Science and Technology Beijing, China,
[2] Beijing Key Laboratory of Knowledge Engineering for Materials Science, China,
[3] Nagoya University, Japan
{xiguo,xieyh}@ustb.edu.cn, ishikawa@is.nagoya-u.ac.jp,
zdzchina@126.com

Abstract. This paper proposes a new spatial query called the reverse direction-based surrounder (RDBS) query, which retrieves a user who is seeing a point of interest (POI) as one of their direction-based surrounders (DBSs). According to a user, one POI can be dominated by a second POI if the POIs are directionally close and the first POI is farther from the user than the second is. Two POIs are directionally close if their included angle with respect to the user is smaller than an angular threshold, θ. If a POI cannot be dominated by another POI, it is a DBS of the user. We also propose an extended query called the *competitor RDBS query*. POIs that share the same RDBSs with another POI are defined as competitors of that POI. We design algorithms to answer the RDBS queries and competitor queries. The experimental results show that the proposed algorithms can answer the queries efficiently.

1 Introduction

In spatial databases, given the location of user o, a traditional nearest neighbor query retrieves the top-k nearest points of interests (POIs) to the user. The spatial closeness between the user and the POIs is typically used for the query. However, in real applications, *directional closeness* is also important. Recently, some studies [8][9][14][15] have focused on spatial queries considering both distance and direction. These direction-based spatial queries can provide information about the area around the user. Given a set of POIs, $\mathcal{P}$, and user o in two-dimensional Euclidean space, two POIs, p and p', are *directionally close* if their included angle, $\angle pop'$, with respect to user o is smaller than an angular threshold, θ. POI p *dominates* p' if the distance from p to o is shorter than the distance from p' to o. A DBS is a point that cannot be dominated by any other POIs. A DBS query is used to retrieve all the DBSs from the database [8][9].

Example 1. Fig. 1 shows an example of the DBS query. The circles represent users labeled a to g. The squares represent POIs, such as the restaurants, labeled A to H. Assuming that user a wants to find nearby restaurants, a issues a DBS query with $\theta = 45°$. POIs E and D are directionally close to user a. Because they are directionally close and E is closer to a than D, E dominates D. In the

© Springer International Publishing Switzerland 2015
R. Cheng et al. (Eds.): APWeb 2015, LNCS 9313, pp. 280–291, 2015.
DOI: 10.1007/978-3-319-25255-1_23

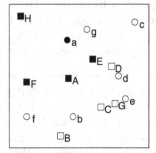

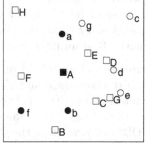

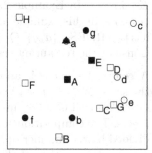

Fig. 1. DBSs of a **Fig. 2.** RDBSs of A **Fig. 3.** Shared RDBSs

same way, E dominates C and G, and A dominates B. POIs $\{A, E, F, H\}$ cannot be dominated by any other POIs and they are the DBSs of user a (solid circles).

Problem 1. The **RDBS query** finds all users that see a query POI as one of their DBSs.

Example 2. Fig 2 shows an example of an RDBS query. POI A (solid square) wants to find the users that see it as a DBS, given $\theta = 45°$. According to Example 1, user a sees A as a DBS; therefore, user a is an RDBS of A. However, another user, g, does not see A as a DBS because E can dominate A with respect to g ($\angle EgA = 33°$ and E is closer to g than A). In the same way, we also find that users c, d, and e do not see A as a DBS[1]. In summary, the RDBSs of A are $\{a, b, f\}$ (solid circles).

The query point in an RDBS query is usually a POI, whereas the query targets are the users. When the query point and the query targets are two different kinds of objects, the query is called *bichromatic*. When the query point and the query targets are the same kind of object, the RDBS query is called *monochromatic*. In this paper, we focus on bichromatic RDBS queries, although the proposed techniques can answer monochromatic queries with a few changes.

Motivation. RDBS queries are suitable for location-based services. A typical scenario is a restaurant, q, that issues an RDBS query to retrieve every customer who may visit q. Every customer retrieved should see q as their nearest restaurant in all different directions. In Fig. 2, it is assumed that the direction preference of the customers is unknown. Customers $\{a, b, f\}$ may visit A because it is the nearest restaurant in a specific direction. In other words, they are possible customers of A and A should do something to attract them, such as reach them with an advertisement.

We propose an extended query based on the RDBS query, called the *competitor RDBS query*.

Problem 2. A **competitor RDBS query** (hereafter referred to as the competitor query) retrieves the competitors of query point q and sorts the competitors by their competition scores for q.

---

[1] $\angle DcA = 22°$, $\angle EdA = 35°$, $\angle GeA = 43°$.

Example 3. In Example 2, we know that the query point A has three RDBSs, $\{a, b, f\}$. In this example, we consider another query point, E. As Fig. 3 shows, E has two RDBSs, $\{a, g\}$. Queries A and E share the same RDBS, a (solid triangle). Consider the restaurant scenario. Restaurant E is a competitor of restaurant A.

A *competitor query* is an extension of an RDBS query. Two points may share the same RDBSs. For the two restaurants, q and q', the customer a can visit q or q'. We evaluate the amount of competition between q and q' by *competition scores*. The competition score that q' has for q is the number of the same RDBSs divided by the number of RDBSs of q, whereas the competition score that q has for q' is the number of the same RDBSs divided by the number of the RDBSs of q'. We retrieve the competitors for query point q and rank them by sorting their competition scores.

Technical Overview. To answer a bichromatic RDBS query, a straightforward approach is to check every user $o \in \mathcal{O}$ and determine whether they are an RDBS of q. Observing that the users close to q are likely to be RDBSs, we propose an *RDBS candidate polygon*, in which the points are possible RDBSs. Using the candidate polygon, we can avoid performing checks for the users that cannot be RDBSs. To answer a competitor query, a straightforward approach is to check every POI in $\mathcal{P} - q$ and to determine whether it is a competitor of q. Intuitively, two POIs share the same RDBSs only if their RDBS candidate polygons intersect. We calculate the polygons of all POIs beforehand and organize them by a spatial index. Because R*-tree cannot index polygons directly, we approximate the polygons by bounding boxes.

Experimental Results. We conduct experiments to evaluate the RDBS queries and the competitor queries. In the RDBS query experiments, the results show that the users enclosed by the RDBS candidate polygon are a small portion of the whole user set. Because the polygon helps us prune many non-candidate users, our approach can answer RDBS queries very quickly. The average speed is about 100 times faster than the straightforward approach. In the competitor query experiments, the results show that when we use bounding boxes to approximate the polygons, we can filter out many POIs that cannot be competitors. Filtering out many irrelevant competitors decreases the retrieval time.

Contributions. The contributions of the paper are summarized below.

- We propose a new problem called an RDBS query and design algorithms to answer the query efficiently. (Section 3)
- We propose an extension problem, called the competitor query problem, and design algorithms to answer competitor queries efficiently. (Section 4)
- We conduct experiments to test the performance of the proposed algorithms. (Section 5)

2   Related Work

Reverse Nearest Neighbor Queries. Given a query point, q, a reverse nearest neighbor (RNN) query ([10],etc.) finds the points that see q as their nearest neighbors. RDBS queries retrieve points considering distances and directions, whereas RNN queries only consider the distances.

Direction-Based Spatial Queries. In spatial databases, the study of queries that emphasize direction is limited. The RDBS query is different from the DBS queries in previous work [8][9] because it retrieves the POIs that see the query point as a DBS rather than retrieving the POIs that are the DBSs of the query point. In Refs. [6], [14], and [15] the problem definitions and retrieval algorithms of the DBS queries are improved. The DBRNN problem has been studied previously [11][12]. The DBRNN query first retrieves the users that are RNN of q and then filters out the users that are not moving towards q and the customers that are too far away from q. It is totally different from the RDBS query. In Refs. [3][13] spatial points are retrieved considering keywords as well as distances and directions.

Reverse Skyline Queries. Skyline queries also retrieve points to optimize two or more objectives [2]. Reverse skyline queries [4] retrieve the points that see q as a relative skyline point [5]. RDBS queries are different from reverse skyline queries because whether two points are directionally close depends on the positions of two points and q. In reverse a skyline query, an attribute value of a point depends on itself and q.

3   RDBS Queries

We define RDBS queries in two-dimensional Euclidean space. In the space, there are a set of POIs, $\mathcal{P}$, and a set of users, $\mathcal{O}$. According to user $o \in \mathcal{O}$ and angular threshold θ, POI $p_1 \in \mathcal{P}$ has a dominance relationship with POI $p_2 \in \mathcal{P}$.

Definition 1. *Given a user, o, and a threshold, θ, two POIs, p_1 and p_2, are directionally close if the included angle of the vector, op_1, and the vector, op_2, are smaller than θ, that is $\angle p_1 o p_2 < \theta$.*

Definition 2. *Given a user, o, and a threshold, θ, POI p_1 dominates POI p_2 if they are directionally close and p_1 is nearer than p_2.*

Definition 3. *Given a user, o, and a threshold, θ, a POI, p, is a DBS of o if it cannot be dominated by any other POIs in $\mathcal{P}$.*

Definition 4. *Given a query POI q, and a threshold, θ, user $o \in \mathcal{O}$ is an RDBS of q if o sees q as a DBS.*

An RDBS query retrieves the RDBSs of query POI q as Problem 1 defines.

Naïve Approach. A naïve approach to retrieving the RDBSs of a query POI q, is to check every user, $o \in \mathcal{O}$, and to determine whether it is an RDBS. User o is an RDBS if o sees q as a DBS. If q is a DBS, it should not be dominated by any other points in $\mathcal{P}$. According to the location of o, point p can dominate q only if p and q are directionally close and p is closer than q^2. However, the naïve

---

^{2} Please refer to Ref. [7] for the detailed algorithm.

approach is not efficient enough. Intuitively, the users close to q are likely to be RDBSs, as Fig. 2 indicates. User o far from q may see that q is dominated by other POIs. A POI that can dominate q should be directionally close to q and be closer than q. It is easy to find such a POI near o. Based on this observation, next we analyze the area where user o must not be an RDBS of q. In other words, we try to find the area where user o could be an RDBS.

Negative Area. Consider where user o is if they do not see q as their DBS. Assuming p can dominate q with respect to o, two conditions must be satisfied according to Definition 2. One condition is that p should be nearer than q. Another condition is p and q should be directionally close. To find o that makes p and q satisfy these two conditions, we use two geometries as shown in Fig. 4.

- **Mid-Perpendicular Line.** The mid-perpendicular line, $l(p, q)$, of segment pq can separate the space into two areas. The half-plane where p is located is called $area(l, p)$. The other half-plane is called $area(l, q)$. If user o is in $area(l, p)$, p is nearer than q.
- **θ-Circle.** θ-circle $c(p, q)$ is a circle to guarantee that $\angle poq > \theta$ if user o is in the circle. According to the inscribed angle, the center of the circle should be on the mid-perpendicular line, $l(p, q)$, and the radius of the circle should be

$$r = \frac{|pq|}{2 \sin \theta}. \tag{1}$$

There would be two such θ-circles, c_1 and c_2, as shown in Fig. 4. The area enclosed by the two circles is denoted as $area(c_1, c_2)^3$. If user o is outside of $area(c_1, c_2)$, p is directionally close to q.

Using the mid-perpendicular line and the θ-circles, we separate the space into three areas: $area(l, p)$, $area(l, q)$, and $area(c_1, c_2)$. The reference points in the three areas can make p and q satisfy different conditions. Because $area(c_1, c_2)$ intersects with $area(l, p)$ and $area(l, q)$, the space is divided into four non-overlapping areas. We show the four areas and the corresponding satisfied conditions below.

1. $area(l, p) \cap \neg area(c_1, c_2)$: $|op| < |oq|$ and $\angle poq < \theta$;
2. $area(l, p) \cap area(c_1, c_2)$: $|op| < |oq|$ and $\angle poq > \theta$;
3. $area(l, q) \cap area(c_1, c_2)$: $|op| > |oq|$ and $\angle poq > \theta$;
4. $area(l, q) \cap \neg area(c_1, c_2)$: $|op| > |oq|$ and $\angle poq < \theta$.

We call area 1 the *negative area* (according to p), denoted as $area_{neg}(p)$, because the users in this area are guaranteed not to be RDBSs of q because they see p as a dominator of q. To find RDBSs efficiently, we should prune such negative areas beforehand. However, negative areas are not regular shapes. To make the negative area easily pruned, we simplify the area as a half-plane that is separated by a negative line, $l_{neg}(p)$. $l_{neg}(p)$ is in area 1 and is tangent to the two θ-circles. The gray area in Fig. 4 shows the simplified negative area separated by $l_{neg}(p)$.

---

[3] $area(c_1, c_2)$ is short for $area(c_1(p, q), c_2(p, q))$.

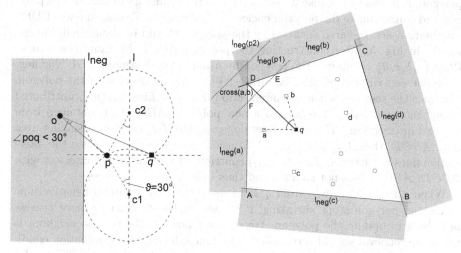

Fig. 4. Negative area of query q

Fig. 5. Candidate polygons of query q

Given a query POI q, and a POI, $p \in \mathcal{P}$, user o cannot be an RDBS of q if it is in $area_{neg}(p)$, the negative area according to p.

RDBS Candidate Polygon. We introduce the concept of an *RDBS candidate polygon*. In Fig. 5, where a group of POIs, $\mathcal{P}^* = \{a, b, c, d\}$, correspond to negative lines $l_{neg}(a)$, $l_{neg}(b)$, $l_{neg}(c)$, and $l_{neg}(d)$. The four lines form a polygon area in Fig. 5. If a user is in the polygon, like the hollow circles, they can be an RDBS. We call such an area an *RDBS candidate polygon*.

Definition 5. *Given a query POI, q, and a group of POIs, $\mathcal{P}^*$, the RDBS candidate polygon of q according to P^* is the area enclosed by the negative lines corresponding to the POIs in $\mathcal{P}^*$.*

Theorem 1. *Given a candidate RDBS polygon of query POI q, if a user is in the polygon, it could be an RDBS, otherwise, it is not an RDBS.*

We can arbitrarily select a group of POIs, $\mathcal{P}^*$, to construct an RDBS candidate polygon of the query POI q. However, a better way to construct the polygon is to use the k-nearest POIs of q because a POI closer to q corresponds to a negative line that is also closer to q. The line closer to q can prune larger negative areas and can form a smaller RDBS candidate polygon. For example, in Fig. 5, we construct the polygon by using the nearest POIs, $\{a, b, c, d\}$, where a is the first nearest neighbor of q, b is the second nearest neighbor of q, and so on. This construction order has an additional benefit. We can terminate the construction of the candidate polygon when the enclosed area does not shrink. Next, we introduce details of the termination condition.

Termination Condition. A POI p contributes a *useful negative line*, $l_{neg}(p)$, if the line intersects with the current polygon. If $l_{neg}(p)$ does not intersect with the polygon, it is useless because it cannot change the boundary of the current polygon and cannot make the polygon enclose a smaller area. To obtain fewer RDBS candidate users, the area enclosed by the polygon should be minimized. For example, in Fig. 5, after constructing the polygon $ABCD$ by using the nearest POIs $\{a, b, c, d\}$ we update the polygon by using the fifth-nearest POI. The negative line contributed by the fifth-nearest POI may intersect with the polygon. Consider the scenario when the intersection happens. Line $l_{neg}(p_1)$ contributed by p_1[4] intersects with $ABCD$ and a new polygon, $ABCEF$, is obtained from the two intersections, E and F. New polygon $ABCEF$ encloses a smaller area than $ABCD$; thus, $l_{neg}(p_1)$ is useful. Consider another scenario when the intersection does not happen. Line $l_{neg}(p_2)$ contributed by p_2 does not intersect with $ABCD$, $ABCD$ does not change, and thus Line $l_{neg}(p_2)$ is useless.

When a POI is far away from q, it cannot contribute a useful negative line. When the polygon stops shrinking, it is meaningless to retrieve more nearest neighbors to update the polygon. We consider how far the nearest neighbor is when the retrieval should terminate. Before introducing the termination condition, we define *angular adjacent* first. Assuming that there is an auxiliary polar coordinate with the origin as POI q and the axis as vector qq', where q' is $(q_x + 1, q_y)$ to make qq' parallel to the x-axis. In the polar coordinate, each POI p has a polar angle, $\omega(p)$, and a polar radius, $\rho(p)$.

Definition 6. *Given two POIs, p^- and p^+ ($\omega(p^-) \leq \omega(p^+)$), the two POIs are angular adjacent if $\nexists \omega(p^-) \leq \omega(p) \leq \omega(p^+)$, where $p \in \mathcal{P}^* - p^- - p^+$.*

In the definition, $\mathcal{P}^*$ denotes the nearest neighbors that have been checked already. For example, in Fig. 5, nearest neighbors a and b are angular adjacent because $\omega(d) < \omega(b) < \omega(a) < \omega(c)$. There are no other POIs between b and a in the list sorted by the increasing order of their polar angles. When there is no ambiguity, we use p^- and p^+ to denote two angular adjacent points. We define the *angular interval* below.

Definition 7. *An* angular interval *is the range defined by the polar angles of two adjacent POIs, p^- and p^+, that is $[\omega(p^-), \omega(p^+)]$.*

The intersection of their negative lines, $l_{neg}(p^-)$ and $l_{neg}(p^+)$, is denoted as $cp(p^-, p^+)$. For example, in Fig. 5, the intersection of $l_{neg}(a)$ and $l_{neg}(b)$ is $cp(a, b)$. The theorem below indicates when a nearest neighbor cannot contribute a useful negative line.

Theorem 2. *If the polar angle of p is between the polar angles of two angular adjacent POIs, p^- and p^+ (i.e., $\omega(p) \in [\omega(p^-), \omega(p^+)]$) and the distance from q to p is*

$$|q, p| \geq \frac{2|q, cp(p^-, p^+)|\sin\theta}{1 + \sin\theta}, \tag{2}$$

POI p cannot contribute to a useful negative line.

---

[4] For simplicity, the nearest POIs, p_1 and p_2, are not shown in Fig. 5.

For simplicity, we call the right side of Eq. 2 the *distance threshold* of the interval $[\omega(p^-), \omega(p^+)]$ and denote it as $dt(q, p^-, p^+)$. The proof of Theorem 2 is in [7]. According to Theorem 2, we can define the *closed interval*.

Definition 8. *Angular interval* $[\omega(p^-), \omega(p^+)]$ *is* closed *if there is a nearest neighbor, p, that satisfies $\omega(p) \in [\omega(p^-), \omega(p^+)]$ and $|q, p| \geq dt(q, p^-, p^+)$.*

When an angular interval is closed, it is meaningless to retrieve the next nearest POIs, which have polar angles in the interval. Assuming we are checking nearest POI p and it is in the interval $[\omega(p^-), \omega(p^+)]$. On the one hand, if $|q, p|$ is larger than or equal to $dt(q, p^-, p^+)$, p cannot contribute a useful negative line. According to Eq. 2, in the interval, all the POIs further away than p cannot contribute useful negative lines. Thus, we should terminate retrieving nearest POIs in the interval. In other words, the interval is closed. On the other hand, if $|q, p|$ is smaller than $dt(q, p^-, p^+)$, p can contribute a negative line. Thus, we compute $l_{neg}(p)$ and update the boundary of the polygon. After updating the polygon, we split interval $[\omega(p^-), \omega(p^+)]$ into two new intervals, $[\omega(p^-), \omega(p)]$ and $[\omega(p), \omega(p^+)]$. Theorem 3 summarizes the termination condition.

Theorem 3. *After retrieving nearest POIs $\{p_1, p_2, \ldots, p_m\}$ to construct the polygon, if the intervals $[\omega(p_1), \omega(p_2)]$, $\ldots$, $[\omega(p_{m-1}, p_m)]$, and $[\omega(p_m), \omega(p_1)]$ are all closed, the area enclosed by the polygon will not be smaller and retrieval can be terminated.*

To answer the RDBS queries more efficiently than the naïve approach, first we use the RDBS candidate polygon of q to obtain the users that could be RDBSs. The polygon is constructed incrementally by retrieving the nearest POIs of q. To retrieve the nearest POIs efficiently, we organize POIs $\mathcal{P}$ by a spatial index such as R*-tree [1]. After the polygon is determined, we check whether the candidate users in the polygon are real RDBSs by using the naïve approach. To retrieve the candidate users efficiently, we use another R^*-tree to index the users $\mathcal{O}^5$.

4   Competitor Queries

Based on the definition of an RDBS query, we define a competitor query. The competitors are POIs of query point q.

Definition 9. *Given a set of POIs, $\mathcal{P}$, a set of users, $\mathcal{O}$, and a threshold, θ, a* competitor *of query POI q are POIs $p \in \mathcal{P}$ that share the same RDBSs with q, that is $RDBS(q) \cap RDBS(p) \neq \emptyset$.*

$RDBS(\cdot)$ denotes the RDBS set of a POI. The same RDBSs shared by two POIs are denoted by $share(\cdot, \cdot)$. Each competitor of query POI q has a *competition score* with q.

---

[5] Please refer to Ref. [7] for the detailed algorithm.

Definition 10. *Given query POI q and its competitor p, the* competition score *that p has for q is given as*

$$score(p) = \frac{|share(q,p)|}{|RDBS(q)|}, \tag{3}$$

where $|\cdot|$ denotes the number of members in a set. The competition score can describe the effect of the competition strength that p exerts on q. According to Definitions 9 and 10, we formulate the competitor query problem in Problem 2. The competitor query problem is finding the competitors of query POI q and sorting the competitors by their competition scores.

Naïve Approach. To answer competitor queries, a naïve approach is to find $RDBS(q)$ and $RDBS(p_i)$ $(p_i \in \mathcal{P})$ for identifying the competitors from $\mathcal{P}$ that share same RDBSs with q, and then to sort the competitor by their competition scores. The RDBS set of q and $p_i \in \mathcal{P}$ can be found by using the algorithm proposed in Section 3. The competition score of a competitor can be calculated according to Eq. 3. However, the naïve algorithm is not efficient enough. One reason is that we have to calculate the RDBSs of every POI in $\mathcal{P}$, even though most POIs do not share any RDBSs with q. Another reason is that we have to calculate the RDBSs of POIs repeatedly for different queries q, although the RDBSs of a POI are independent of the queries and never change for different query POIs. To improve the naïve algorithm in terms of two aspects, we propose an index-based approach.

Bounding Boxes. Intuitively, if the RDBS candidate polygons of two POIs intersect, the two POIs are likely to share RDBSs. In other words, if the polygons of two POIs do not intersect, they do not share any RDBSs. Therefore, we calculate polygons p_{poly} of all POIs $p \in \mathcal{P}$ beforehand. When query q is issued, we find polygons that intersect with q_{poly}. The corresponding POIs of these intersecting polygons may be the competitors of q. Finally, we identify the real competitors from these candidate competitors. To retrieve the polygons intersecting with q_{poly} efficiently, we use an R*-tree to index the polygons of the POIs in $\mathcal{P}$. R*-tree is designed to index points or boxes, but it cannot index polygons directly. We approximate the polygons as boxes by a bounding approach. The bounding approach can calculate the smallest bounding box of a polygon. We use R*-tree from the Boost C++ Libraries[6]. In the library, the bounding approach has already been implemented.

5   Experiments

Datasets. We use synthetic datasets to evaluate the RDBS queries (Section 3) and the competitor queries (Section 4). The POIs and users in the datasets are two-dimensional and are generated randomly. We create six datasets containing different numbers of POIs (users): $\mathcal{P}_0$ ($\mathcal{O}_0$), $\mathcal{P}_1$ ($\mathcal{O}_1$), $\mathcal{P}_2$ ($\mathcal{O}_2$), $\mathcal{P}_3$ ($\mathcal{O}_3$), $\mathcal{P}_4$ ($\mathcal{O}_4$),

---

[6] http://www.boost.org/

and $\mathcal{P}_5$ ($\mathcal{O}_5$), which contain 100, 2000, 4000, 6000, 8000, and 10,000 points, respectively. For simplicity, the POI dataset, $\mathcal{P}_i$, and the user dataset, $\mathcal{O}_i$, are the same set of points. We randomly select 20 POIs from the experimental dataset as the query points. The results are the average results of 20 queries.

Environments. All the experiments are conducted on a machine with an Intel Core i5-2415M CPU with 4 GB of memory running Windows 7 (32 bit). All the algorithms proposed are implemented in C++. The development environment is Visual Studio 2013. The R*-tree used in the experiments is from the Boost C++ Libraries of release version 1.57.0.

RDBS Query Experiments. First, we conduct experiments to evaluate the performance of the proposed algorithms with respect to different amounts of data. We use datasets $\mathcal{P}_1$ ($\mathcal{O}_1$), $\mathcal{P}_2$ ($\mathcal{O}_2$), $\mathcal{P}_3$ ($\mathcal{O}_3$), $\mathcal{P}_4$ ($\mathcal{O}_4$), and $\mathcal{P}_5$ ($\mathcal{O}_5$). Threshold θ is 30°. Fig. 6 and 7 show the experimental results. The label "2" on the x-axis corresponds to dataset $\mathcal{P}_1$ ($\mathcal{O}_1$), which contains 2000 points. We label the other x-coordinates in the same way. In Fig. 6, the y-axis represents the numbers of RDBSs and users that have been checked. In Fig. 7, the y-axis shows the elapsed time (milliseconds) for retrieving the RDBSs on a log scale. In Fig. 6, the number of users checked is not much more than the number of RDBSs, which indicates that the filtering function of the candidate polygons works well. In Fig. 7, "naive" denotes the naïve approach, whereas "opt" denotes our approach. The experimental results show that opt runs about 100 times faster than naive because checking whether a user is an RDBS is time consuming and opt can filter out large numbers of users who cannot be RDBSs. Fig. 7 also indicates that when the dataset size increases, the time advantage of opt becomes more obvious.

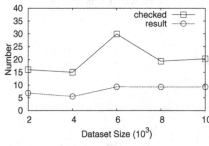
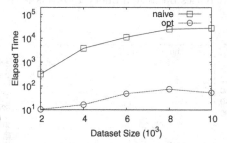

Fig. 6. Number of RDBSs as a function of dataset size

Fig. 7. Elapsed time (milliseconds) as a function of dataset size

Second, we conduct experiments to evaluate the algorithms when θ varies. We use dataset $\mathcal{P}_1$ ($\mathcal{O}_1$) and θ varies in $\{15°, 30°, 45°, 60°, 75°\}$. In Fig. 8 and 9, the x-axis represents different θ values. In Fig. 8, the y-axis represents the number of RDBSs and the number of users checked. In Fig. 9, the y-axis represents the time (milliseconds) taken to retrieve RDBSs. In Fig. 8, the number of RDBSs decreases when θ increases because q is more likely to be dominated when θ is larger. The number of users checked also decreases when θ increases because

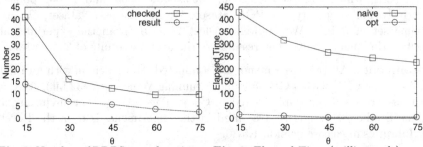

Fig. 8. Number of RDBSs as a function of θ

Fig. 9. Elapsed Time (milliseconds) as a function of θ

the candidate polygon for q decreases as θ increases. In Fig. 9, the time cost decreases when θ increases for naive and opt; as θ increases, q is more easily dominated.

Competitor Query Experiments. First, we evaluate the performance for retrieving the competitors of a query POI for different dataset sizes. We use datasets $\mathcal{P}_1$ ($\mathcal{O}_1$), $\mathcal{P}_2$ ($\mathcal{O}_2$), $\mathcal{P}_3$ ($\mathcal{O}_3$), $\mathcal{P}_4$ ($\mathcal{O}_4$), and $\mathcal{P}_5$ ($\mathcal{O}_5$). θ is set as 30°. "bb" approximates the candidate polygons by bounding boxes. "ab1" and "ab2" approximate the polygons by angular boxes, which are proposed in Ref. [7]. For brevity, we do not introduce the techniques ab1 and ab2. The average results are shown in Fig. 10. The x-axis represents the dataset sizes. Second, we evaluate the performance of retrieving competitors for different θ values. We use dataset $\mathcal{P}_1$ ($\mathcal{O}_1$) and conduct experiments on the dataset when θ is 15°, 30°, 45°, 60°, or 75°. In Fig. 11 the x-axis represents different θ values. The y-axis represents the time (seconds) taken to process the competitors. The time cost decreases as θ increases because a query POI has fewer competitors when θ is larger.

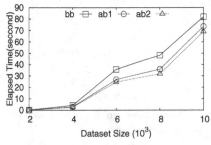

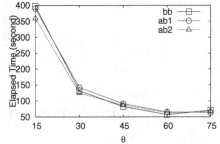

Fig. 10. Elapsed time as a function of dataset sizes

Fig. 11. Elapsed time as a function of θ

6   Conclusions and Future Work

In this paper, we proposed an RDBS query that retrieves the users seeing the query POI q as one of their DBSs. We also proposed a competitor query that retrieves the POIs that share the same RDBSs with q. We designed algorithms to answer the queries and experiments showed that the algorithms are efficient. In future studies, we intend to consider the direction preference of query q and to consider moving target points $\mathcal{O}$.

Acknowledgement. This work is supported by National Key Technology R&D Program in 12th Five-year Plan of China (No. 2013BAI13B06). This work is also supported by Fundamental Research Funds for the Central Universities (No. FRF-TP-14-025A1). This research was partly supported by KAKENHI (25280039).

References

1. Beckmann, N., Peter Kriegel, H., Schneider, R., Seeger, B.: The R*-tree: an efficient and robust access method for points and rectangles. In: SIGMOD, pp. 322–331 (1990)
2. Börzsönyi, S., Kossmann, D., Stocker, K.: The skyline operator. In: ICDE, pp. 421–430 (2001)
3. Chen, Z.-J., Zhou, T., Liu, W.Y.: Direction aware collective spatial keyword query. Journal of Chinese Computer Systems 35(5), 999–1004 (2014)
4. Dellis, E., Seeger, B.: Efficient computation of reverse skyline queries. In: VLDB, pp. 291–302 (2007)
5. Deng, K., Zhou, X., Shen, H.T.: Multi-source skyline query processing in road networks. In: ICDE, pp. 796–805 (2007)
6. El-Dawy, E., Mokhtar, H.M.O., El-Bastawissy, A.: Directional skyline queries. In: Xiang, Y., Pathan, M., Tao, X., Wang, H. (eds.) ICDKE 2012. LNCS, vol. 7696, pp. 15–28. Springer, Heidelberg (2012)
7. Guo, X., Ishikawa, Y.: Reverse direction-based surrounder queries (long version), pp. 1–30 (2015), http://www.db.is.nagoya-u.ac.jp/~ishikawa/tmp/apweb05-long.pdf
8. Guo, X., Ishikawa, Y., Gao, Y.: Direction-based spatial skylines. In: MobiDE, pp. 73–80 (2010)
9. Guo, X., Zheng, B., Ishikawa, Y., Gao, Y.: Direction-based surrounder queries for mobile recommendations. The VLDB Journal 20(5) (2011)
10. Korn, F., Muthukrishnan, S.: Influence sets based on reverse nearest neighbor queries. In: SIGMOD, pp. 201–212 (2000)
11. Lee, K.-W., Choi, D.-W., Chung, C.-W.: DART: An efficient method for direction-aware bichromatic reverse k nearest neighbor queries. In: Nascimento, M.A., Sellis, T., Cheng, R., Sander, J., Zheng, Y., Kriegel, H.-P., Renz, M., Sengstock, C. (eds.) SSTD 2013. LNCS, vol. 8098, pp. 295–311. Springer, Heidelberg (2013)
12. Lee, K.-W., Choi, D.-W., Chung, C.-W.: Dart+: Direction-aware bichromatic reverse k nearest neighbor query processing in spatial databases. Journal of Intelligent Information Systems 43(2), 349–377 (2014)
13. Li, G., Feng, J., Xu, J.: Desks: Direction-aware spatial keyword search. In: ICDE, pp. 474–485 (2012)
14. Saravanan, A., Balasundaram, S.: Enhancing context-aware services in mobile environment. In: Proc. of International Conference on Emerging Trends in Science, Engineering and Technology, pp. 39–44 (2012)
15. Saravanan, A., Balasundaram, S.: Pruning strategies to refine resource list for nn queries in location based services. In: Proc. of 2nd International Conference on Communication, Computing & Security, pp. 598–605 (2012)

Learning to Hash for Recommendation with Tensor Data

Qiyue Yin, Shu Wu, and Liang Wang*

Center for Research on Intelligent Perception and Computing,
National Laboratory of Pattern Recognition
Institute of Automation, Chinese Academy of Sciences, Beijing 100190, China
{qyyin,shu.wu,wangliang}@nlpr.ia.ac.cn

Abstract. Recommender systems usually need to compare user interests and item characteristics in the context of large user and item space, making hashing based algorithms a promising strategy to speed up recommendation. Existing hashing based recommendation methods only model the users and items and dealing with the matrix data, e.g., user-item rating matrix. In practice, recommendation scenarios can be rather complex, e.g., collaborative retrieval and personalized tag recommendation. The above scenarios generally need fast search for one type of entities (target entities) using multiple types of entities (source entities). The resulting three or higher order tensor data makes conventional hashing algorithms fail for the above scenarios. In this paper, a novel hashing method is accordingly proposed to solve the above problem, where the tensor data is approached by properly designing the similarities between the source entities and target entities in Hamming space. Besides, operator matrices are further developed to explore the relationship between different types of source entities, resulting in auxiliary codes for source entities. Extensive experiments on two tasks, i.e., personalized tag recommendation and collaborative retrieval, demonstrate that the proposed method performs well for tensor data.

Keywords: Learning to hash, Tensor data, Recommendation.

1 Introduction

To reduce the information overload and help users select potential appealing products, recommender systems play an important role by recommending relevant products to users. Till now, promising progresses have been made from both academia and industry. Generally, recommendation needs to compare a large number of items to determine the users' most preferred ones and this process will be very time consuming in the context of large user and item space. Efficiency in recommendation is accordingly becoming a challenging problem.

Several works have been proposed dealing with the efficiency problem, which can be roughly classified into data structure based methods and hashing based methods [22]. As for the data structure based methods, researchers try to shrink

---

* Corresponding author.

© Springer International Publishing Switzerland 2015
R. Cheng et al. (Eds.): APWeb 2015, LNCS 9313, pp. 292–303, 2015.
DOI: 10.1007/978-3-319-25255-1_24

the search space to reduce the comparison [20] [18] [7]. Typical examples are the kd-tree methods and some other partition methods [6] [8] [7]. Recently, hashing based methods show more superior than the data structure based methods [11]. For example, they can obtain low time complexity that is irrespective of the size of datasets and they need little memory space for the storage of the binary hash codes. Few researchers have tried using the hashing technology for fast recommendation [13] [25] [24]. Readers can refer to Section 2 for more details.

To the best of our knowledge, exiting hashing based recommendation methods only learn binary codes for users and items, and they mainly focus on matrix data. For example, they learn hash codes for users and items so that their similarity in Hamming space can approach the rating matrix. In practice, recommendation can be rather complex to be represented by matrix data, for example, recommending items to users under specific queries in collaborative retrieval [17] [4] and recommending tags to images by specific users in personalized tag recommendation [10]. All these scenarios results in three or higher order tensor data with each dimension indicating one type of entities. Taking personalized tag recommendation as an example, the resulting three order tensor has values 1 or 0 indicating whether a tag is related to an image by a specific user or not.

In summary, the above scenarios involve fast search for one type of entities (target entities) using multiple types of entities (source entities) and usually tensor data is given representing the implicit or explicit relationship between the source entities and target entities. Then conventional hashing based recommendation methods may fail for their disability dealing with tensor data. Besides, traditional hashing methods developed for multiple types of entities, i.e., cross-model hashing methods [9] [15], usually model the relationship between each two types of entities, which is different from our scenario and cannot be applied directly. Readers can refer to Section 2 for more details. Hence, in this paper, a novel hashing method for tensor data is proposed. We learn hash codes for each dimension of the tensor data. To preserve the relationships between source entities and target entities, their similarity calculation in the Hamming space is properly designed by bringing in operator matrices. Finally, extensive experiments on two complex recommendation tasks validate our proposed method.

Our Contributions: 1) To the best of our knowledge, this is the first time to learn hash codes for tensor data and meanwhile it is successfully applied to complex recommendation problems. 2) We design operator matrices to explore the relationship between different types of source entities and auxiliary codes are accordingly constructed to enhance the recommendation performance. 3) We test our model in terms of personalized tag recommendation and collaborative retrieval, and extensive experimental results validate our proposed method.

2   Related Work

In the context of large user and item space, it becomes urgent to improve the recommendation efficiency when compares user interests and item characteristics. As described before, the methods can be roughly classified into data structure based methods [20] [18] [19] [7] and hashing based methods [25] [24]. As for the former ones, some researches utilize simple partitioning based methods, which partition the items or users into groups [6] [8]. These methods can reduce the item space or user space, but they may inevitable harm the recommendation performance. Recently, some special data structures, such as kd-tree and its variants are utilized for fast search [7] [5]. However, these methods are still time consuming when dealing with high dimensional and large scale datasets [3].

Recently, hashing technology has been brought to recommendation field and shows superior than the data structure based methods. Generally, hashing is one kind of the most popular methods for large scale nearest neighbor search problems, which learns binary codes as the new representations of entities [11]. Initiated by Locality Sensitive Hashing (LSH) [1], which is a family of hash functions mapping similar points to the same hash codes with high probability, some machine learning methods are now employed to design more effective hash codes [16] [2] [23]. As in recommendation field, Zhou and Zha [25] gave the first try and learned binary codes for users and items for fast item ranking. Zhang et al. [24] learned hash codes given the implict or explicit feedback matrix to preserve the users' preferences over items. Furthermore, Wang et al. [13] [12] proposed to learn compact hashing codes for efficient tag completion and recommendation. Generally, previous hashing based recommendation methods mainly focus on the matrix data, e.g., user-item rating matrix and image-tag tag matrix. However, recommendation scenarios are more complex and usually tensor data is given, making conventional hashing methods failed. This is the motivation that promotes us to propose new hashing methods for tensor data.

Since we are learning to hash for multiple modalities, the research of cross-model hashing proposed to meet the need of similarity search across different types of entities is one of the most related works [9] [15]. However, exiting cross-model hashing methods mainly focus on modeling the pairwise relationship, i.e., the relationship between two types of entities. For example, when people uses texts, images and videos for cross-model retrieval, the pairwise relationships between texts and images, between texts and videos and between images and videos are developed. Instead we are concentrating on the scenarios that use multiple types of entities to retrieve one type of entities and typical examples are personalized tag recommendation and collaborative retrieval. Generally, different scenarios lead to distrinct hashing methods to adjust their data structures, which makes previous cross-model hashing methods cannot be applied directly to solve the problem we are focusing on. Thus, we develop new hashing methods for multiple modalities and meanwhile enable for fast recommendation.

3   Model

3.1   Notations

We are given n order tensor data R with each dimension representing a type of entities and their values indicating the relationships between the $n - 1$ types of entities and the other one type of entities. We call the $n - 1$ types of entities source entities and the other one type of entities target entities. Our goal is to learn hash codes for each type of entities to preserve their relationship and meanwhile enable for fast similarity search. For example, for personalized tag recommendation, we have three types of entities: users and images as source entities and tags as target entities. Accordingly, a three order tensor is given indicating whether a tag is annotated to an image under a specific user or not. We need to learn hash codes for each user, each image and each tag so that tags can be rapidly recommended to images given specific users.

Suppose the $n - 1$ source entities and the target entities are represented as $X^{(k)}$ ($k = 1, ..., n - 1$) and $X^{(n)}$ respectively and $m_k(k = 1, ..., n)$ denotes the number of entities of the k-th type. Then we need to learn hash codes $H^{(k)}$ ($k = 1, ..., n$) for all of them. Apart from the binary constraint on $H^{(k)}$, the key issue is to keep the similarity between the source entities and the target entities in Hamming space so that their relationship is consistent with the n order tensor data R. In the following, we will elaborate our strategy achieving this goal.

3.2   Similarity Preserving

Since the tensor data R indicates the implict or explicit relationship between all the source entities and the target entities, a natural way to calculate similarity between them is to compute the similarity between each type of source entities and the target entities and then sum them. By doing so, the relationship between the source entities and the target entities can be explored in a straight-forward manner. Then for the $s_1, ... s_n$-th value of the tensor data R, its predicted value $\hat{R}_{s_1, ... s_n}$ can be obtained by:

$$\hat{R}_{s_1, ... s_n} = \sum_{k=[1:n-1]} sim(H^{(k)}_{s_k}, H^{(n)}_{s_n}) \tag{1}$$

where s_k is an indictor of an entity of the k-th type and $H^{(k)}_{s_k}$ is the s_k-th hash code for the k-th entity in the k-th type. sim is an operator calculating the similarity between two hash codes to be introduced later.

In the above formulation, we directly consider the relationship between each type of source entities and the target entities. However, the interaction between different types of source entities are ignored, which may be essential to explore the relationship between the source entities and the target entities. For example, for collaborative retrieval, users behave quite different under different queries and clearly queries can influence users. To model the influence, we propose operator matrix for each source entity. For example, $T^{(j)}_{s_j}$, served as the operator matrix of the s_j-th entity in the j-th type, represents the influence this entity will bring to other types of source entities. Furthermore, given an entity $R_{s_1, ... s_n}$, we model the influence the other source entities to the s_k-th source entity of the k-th type as $\sum_{j \neq k} T^{(j)}_{s_j}$. By multiplying this matrix with the hash codes of the s_k-th entity in the k-th type and binary it, we can obtain a new hash codes as auxiliary codes for $H_{s_k}^{(k)}$:

$$aux(H_{s_k}^{(k)}) = sign \left(\sum_{j=[1:n-1], j \neq k} T_{s_j}^{(j)} H_{s_k}^{(k)} \right) \qquad (2)$$

where $sign$ is an element-wise symbolic operation for a vector with each element returning 1 or -1 based on whether its value is bigger than 0 or not.

Taking the interaction between different types of source entities into consideration, we modify Equation 1 as:

$$\hat{R}_{s_1,...s_n} = \sum_{k=[1:n-1]} sim(H_{s_k}^{(k)}, H_{s_n}^{(n)}) + \mu \sum_{k=[1:n-1]} sim \left(aux(H_{s_k}^{(k)}), H_{s_n}^{(n)} \right) \qquad (3)$$

where μ is a positive value balancing the effect of the auxiliary similarity term.

In Equation 3, a natural way to define the similarity between two hash codes is the fraction of common bits between them, which is widely used in many hashing based methods [25] [13]. As an example, given hash codes $H_{s_k}^{(k)}$ and $H_{s_n}^{(n)}$, their similarity is defined as:

$$sim \left(H_{s_k}^{(k)}, H_{s_n}^{(n)} \right) = \frac{1}{B} \sum_{i=1}^{B} I \left(\left(H_{s_k}^{(k)} \right)_i, \left(H_{s_n}^{(n)} \right)_i \right) = \frac{1}{2} + \frac{1}{2B} \left(H_{s_k}^{(k)} \right)^T H_{s_n}^{(n)} \qquad (4)$$

where B is the number of bits of the hash codes and $\left(H_{s_k}^{(k)} \right)_i$ is the i-th bit of $H_{s_k}^{(k)}$. I is an indicator function returning 1 if the two operator numbers are the same otherwise 0.

After the calculation of similarity between the source entities and the target entities, many kinds of loss functions can be applied to construct the objective. Here we just use the simple square loss and it is written as

$$\min_{\{H^{(k)}\}_{k=1}^{n}, \{T^{(i)}\}_{i=1}^{n-1}} \sum_{(s_1...s_n) \in O} \left(R_{s_1...s_n} - \sigma \hat{R}_{s_1...s_n} \right)^2 + \lambda \sum_{i=[1:n-1]} \sum_{j \in \{s_j\}} ||T_j^{(i)}||^2 \qquad (5)$$

where O is the observed entities of the tensor data R. σ is a scaler parameter to limit the predicted similarity to be in the interval $[0,1]$. The last term is a regular regularization on the operator matrices with a positive tradeoff parameter λ.

As for the constraints imposed on the hash codes, apart from the constraint that the elements of all the hash codes are $\{-1, +1\}$, it is necessary to restrict the hash codes to be balanced, which makes sure that each bit carries as much information as possible. In summary, the constraints are defined as $H^{(k)} \in \Omega^{(k)}, \forall k = 1, ..., n$ with $\Omega^{(k)}$ being:

$$\Omega^{(k)} = \left\{ H^{(k)} \in \{-1, 1\}^{B \times m_k}, H^{(k)} \mathbf{1} = 0 \right\} \qquad (6)$$

where the constraint $H^{(k)} \mathbf{1} = 0$ is used to preserve the balance of each bit.

3.3 Final Objective and Optimization

Bringing Equation 3, 4 and 6 into Equation 5, we can obtain the final cost function. Because of the binary constraint on the variables, the cost function in Equation 5 is not differentiable. Moreover, the balance constraint and the *sign* operator make the optimization of the objective a non-trivial problem. To solve these problems, we firstly relax the *sign* operator to real values as most hashing leaning methods have done [21] [14]. Furthermore, the two constraints are converted into soft penalty terms as in [9]. Specifically, we add the following two terms to the cost function.

$$\theta_1(\{H^{(k)}\}) = \sum_{k=1}^{n} ||H^{(k)} \odot H^{(k)} - E||^2 \qquad \theta_2(\{H^{(k)}\}) = \sum_{k=1}^{n} ||H^{(k)}1||^2 \qquad (7)$$

where $E \in R^{B \times m_k}$ is an all-one matrix and $1 \in R^{m_k \times 1}$ is an all one vector.
And the final cost is written as:

$$L = \sum_{(s_1 s_2 \ldots s_n) \in O} (R_{s_1 \ldots s_n} - \sigma \hat{R}_{s_1 \ldots s_n})^2 + \\ \lambda \sum_{i=[1:n-1]} \sum_{j \in \{s_j\}} ||T_j^{(i)}||^2 + \theta_1(\{H^{(k)}\}) + \theta_2(\{H^{(k)}\}) \qquad (8)$$

Since the final cost is in Equation 8 is not jointly convex with respect to all the variables, we use the stochastic gradient descent method to obtain a local optimal solution and the gradients are calculated as

$$\frac{\partial L}{\partial H_{s_k}^{(k)}} = \sum_{\{s_j\}_{j=1}^{n-1}, j \neq k} -2\sigma(R_{s_1 \ldots s_n} - \sigma \hat{R}_{s_1 \ldots s_n}) \frac{\partial (\hat{R}_{s_1 \ldots s_n})}{\partial H_{s_k}^{(k)}} \\ + 4\theta_1(H_{s_k}^{(k)} \odot H_{s_k}^{(k)} - E_{s_k})H_{s_k}^{(k)} + 2\theta_2 H^{(k)}1 \qquad (9)$$

$$\frac{\partial L}{\partial H_{s_n}^{(n)}} = \sum_{\{s_j\}_{j=1}^{n-1}} -2\sigma(R_{s_1 \ldots s_n} - \sigma \hat{R}_{s_1 \ldots s_n}) \frac{\partial (\hat{R}_{s_1 \ldots s_n})}{\partial H_{s_n}^{(n)}} \\ + 4\theta_1(H_{s_n}^{(n)} \odot H_{s_n}^{(n)} - E_{s_n})H_{s_n}^{(n)} + 2\theta_2 H^{(n)}1 \qquad (10)$$

$$\frac{\partial L}{\partial T_{s_j}^{(j)}} = \sum_{\{s_k\}_{k=1}^{n-1}, k \neq j} -2\sigma(R_{s_1 \ldots s_n} - \sigma \hat{R}_{s_1 \ldots s_n}) \frac{\partial (\hat{R}_{s_1 \ldots s_n})}{\partial T_{s_j}^{(j)}} + 2\lambda T_{s_j}^{(j)} \qquad (11)$$

where the gradient components in the above equation are given as

$$\frac{\partial (\hat{R}_{s_1 \ldots s_n})}{\partial H_{s_k}^{(k)}} = \frac{1}{2B} H_{s_n}^{(n)} + \frac{u}{2B} \sum_{j=[1:n-1], j \neq k} (T_{s_j}^{(j)})^T H_{s_n}^{(n)} \qquad (12)$$

$$\frac{\partial (\hat{R}_{s_1 \ldots s_n})}{\partial H_{s_n}^{(n)}} = \frac{1}{2B} \sum_{k=[1:n-1]} H_{s_k}^{(k)} + \frac{u}{2B} \sum_{k=[1:n-1]} \sum_{j=[1:n-1], j \neq k} (T_{s_j}^{(j)})H_{s_k}^{(k)} \qquad (13)$$

$$\frac{\partial (\hat{R}_{s_1 \ldots s_n})}{\partial T_{s_j}^{(j)}} = \frac{u}{2B} H_{s_n}^{(n)} \sum_{k=[1:n-1], k \neq j} (H_{s_k}^{(k)})^T \qquad (14)$$

After solving the relaxed optimization problem as in Equation 8, we can obtain real-valued representations for all the entities denoted as $\widetilde{H}^{(k)}(k = 1, ..., n)$. Finally, using the constraints of Equation 6, we can obtain the final hashing codes as:

$$H_{ij}^{(k)} = \begin{cases} 1 & \widetilde{H}_{ij}^{(k)} > median(\widetilde{H}_{tj}^{(k)} : t \in 1 : m_k) \\ -1 & Otherwise \end{cases} \qquad (15)$$

where $H_{ij}^{(k)}$ is the j-th bit of the i-th entity in type of k. The whole algorithm is summarized in Algorithm 1.

Algorithm 1. Learning to Hash for Recommendation with Tensor Data

---

Input:
    Observed tensor data R, parameters λ, θ_1, θ_2 and μ.
1: Initialize $H^{(k)}$ $(k = 1, ..., n)$ and $T^{(k)}$ $(k = 1, ..., n - 1)$;
2: Optimize Equation 8 using stochastic gradient method with their gradients calculated by Equation 9, 10 and 11;
3: Obtain the binary hashing codes using Equation 15 with the relaxed solution obtained from the above step;
4: Obtain the auxiliary codes using Equation 2;
Output:
    Hash codes for all the entities $H^{(k)}$ $(k = 1, ..., n)$, auxiliary codes for all source entities $H^{(k)}$ $(k = 1, ..., n - 1)$.

---

4   Experiments

4.1   Datasets

We report experimental results on two widely used recommendation datasets and the statistics of the databases are summarized in Table 1.

Last_fm: It contains music artist listening information and tagging information sampled from Last.fm online music system. Each user has tagged some music artists and data tuples [user, artist, tag] are given. We use this dataset to test the performance of our method in terms of personalized tag recommendation. Similar to [10], we use a p-core[1] for Last_fm to filter the dataset and p is chosen as 20 here.

MovieLens: It is published by GroupLens research group and contains personalized ratings to movies. Besides, the user tagging information is also provided and accordingly data tuples [tag, user, movie] can be obtained. We use this dataset for the experiments of collaborative retrieval. Similar to [4], the most common 50 tags are selected as the genre of the movies and [genre, user, movie] triple are utilized to mimic [query, user, item] triples for collaborative retrieval. Besides, the data preprocessing is the same as in [4].

---

[1] The p-core of a dataset D is the largest subset of D with the property that each entity has to occur in at least p posts. And a post is defined as a combination of different types of source entities [10].

Table 1. Information of the Last_fm and MovieLens datasets.

Last_fm			MovieLens		
# of posts	# of pairs	sparsity	# of posts	# of pairs	sparsity
19, 938	54, 019	99.95%	2, 111	20, 583	99.54%

4.2 Experimental Settings

To evaluate the performance of the proposed model, we compare our method with the following state-of-the-art hashing based recommendation algorithms.

CFCodeReg: Zhou and Zha [25] proposed to learn binary codes for collaborative filtering. **BCE-FIT:** Wang et al. [13] learned binary codes embedding for tag recommendation. It should be noted that the above two methods learn hash codes for matrix data, i.e., rating matrix and tag matrix respectively. However, for personalized tag recommendation and collaborative retrieval problems, three order tensor data is provided. So we may compress the tensor data into matrix form for a comparison. Like in traditional recommendation, the factors of user and query are ignored in personalized tag recommendation and collaborative retrieval respectively. **LCR-B:** Weston et al. [17] proposed the first latent collaborative retrieval algorithm (LSR). However, the learned latent representations are real values, which makes the comparison with our method unfair. Usually, the learned latent representations are in the interval [-1,+1], so we binary the real values with the constraints in Equation 6 that we have used. The learned binary codes is then compared with our method. **LHTD:** Our proposed **L**earning to **H**ash for recommendation with **T**ensor **D**ata with the similarity calculated in Equation 1. **LHTDi:** Our proposed **L**earning to **H**ash for recommendation with **T**ensor **D**ata with the similarity calculated considering the interaction between different types of source entities as in Equation 3.

In all the experiments, we use Recall as in [4] as the evaluation metric. For a given test triple $(X_{s_1}^{(1)}, X_{s_2}^{(2)}, ..., X_{s_{n-1}}^{(n-1)}, X_i^{(n)})$, we calculate the similarity for all i and sort the target entities in descending order. Then, we measure Recall@k, which is 1 if the target entity in the top k and 0 otherwise. Finally, the mean Recall@k over the whole test dataset is reported. In the parameter setting, we empirically set $\lambda = 1$ and $\theta_1 = \theta_2 = 0.001$ in all the datasets. As for μ, it controls the importance of the interaction term and is searched in the interval $[0, 1]$. For all the compared methods, we follow the suggests their authors have given to achieve their best performance. All the experiments are run 10 times with randomly choosing 80% of the observed entities as the training set and the remaining as the testing set.

4.3 Numerical Results and Analysis

From Figure 1, it can be seen that our method outperforms all the compared methods in both databases in terms of personalized tag recommendation and collaborative retrieval. Compared with LHTD, LHTDi considers the interaction

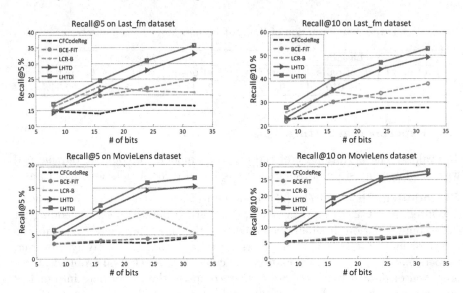

Fig. 1. Recommendation results on Last-fm and MovieLens datasets.

between different types of source entities and the resulting auxiliary codes can well explore the relationship between the source entities, thus its performance is better than that of LHTD.

Compared with our method, CFCodeReg and BCE-FIT cannot consider hashing for all types of entities and this may inevitable harm the relationship between them. So the hash codes learned by their methods can not properly approach the tensor data and their performance is relatively poor. As for LCR-B, it learns latent representations for all types of entities and considers one-side interaction between different types of source entities. Compared with it, our method takes all pairwise interactions between different types of source entities into consideration and can better reflect their relationship. Besides, the hash codes learning of LCR-B is a two-stage process, and more information will be lost compared with our method that considers learning the hash codes in one objective. And these may be the reasons that our method performs better than that of LCR-B.

4.4   Parameter Selection

In our model, λ is a regularization parameter used to avoid over-fitting and θ_1 and θ_2 are parameters controlling the soft penalty terms. All these variables are empirically set as in Section 4.2. There is also an important parameter μ that balances the effect of the auxiliary similarity term. And in this section, we test how parameter u influences the performance of our method. We choose the number of hash codes to be 8 and 32 and vary u in the interval $[0, 1]$ on both datasets. The results are shown in Figure 2. When the number of hash codes is small, the performance of LHTDi is becoming better with the increasing of u. However, once the number of hash codes is big enough, we can see that LHTDi

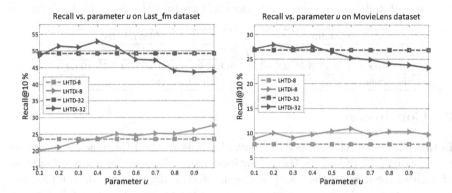

Fig. 2. Recommendation results vs. parameter u on Last_fm and MovieLens datasets. The numbers behind LHTD and LHTDi are the number of bits selected.

reaches the best performance when u is relatively small. This is reasonable because the basic similarity between source entities and target entities can not well approach the tensor data using very few bits and the interaction term is accordingly becoming important. When the number of hash codes is relatively big, the basic hash codes can well embed the information of different entities and the effect of auxiliary codes is accordingly inapparent.

4.5 Operation Ability of Different Types of Operator Matrices

In our method, we propose operator matrices to model the interaction between different types of source entities, resulting in auxiliary codes to further improve the recommendation performance. In this section, we quantize the effect of the operator matrices and observe their relationship. For an entity of one type, we firstly calculate the change between its operator matrix multiplying the hash codes of the other types of entities and the hash codes of other types of entities themselves. Then we average the changes obtained by all this type of entities as the final operating effect of this type of entities. The comparison between different types of source entities are listed in Table 2.

From Table 2, it can be seen that operator matrices of different types of source entities are not equally important. 1) In personalized tag recommendation, user operator matrices are more important than that of artist. Compared with conventional tagging system, personalized tag recommendation adds the factor of user, which influences the tagging for the same artist. This may be the reason that user operator matrices are the main factors for the interaction. 2) As in collaborative retrieval, we observe that query operator matrices are more important than that of user. Since collaborative retrieval has the entities of queries compared with traditional recommendation, it shares the similar reason as in personalized tag recommendation for the comparison.

Table 2. Influence of operator matrices on Last_fm and MovieLens databases.

Last_fm					MovieLens				
Bit	8	16	24	32	Bit	8	16	24	32
User Operator	3.17	6.59	9.54	12.87	Query Operator	3.59	7.17	10.63	14.31
Artist Operator	2.76	5.80	8.18	11.55	User Operator	3.06	6.34	8.91	12.11

5   Conclusion

In this paper, we have proposed a novel hashing method for tensor data, which retrieve one type of entities (target entities) using the query of many other types of entities (source entities) and has been successfully applied for complex recommendation problems, i.e., personalized tag recommendation and collaborative retrieval. In our model, we learn hash codes for each dimension of entities and properly design the similarity between them to approach the tensor data. Furthermore, to explore the relationship between different types of source entities, operator matrices are developed, resulting in auxiliary codes for all the source entities to further enhance the recommendation performance. Extensive experiments have demonstrated the effectiveness of our method compared with several state-of-the-art hashing algorithms. Since different source entities of the same type may share common characteristics, it may be necessary to build an operator tensor as a dictionary instead of a list of operator matrices to reduce the computation. We leave this as future work.

Acknowledgments. This work is jointly supported by National Basic Research Program of China (2012CB316300), and National Natural Science Foundation of China (61175003, 61420106015, U1435221, 61403390).

References

1. Gionis, A., Indyk, P., Motwani, R., et al.: Similarity search in high dimensions via hashing. In: International Conference on Very Large Data Bases, vol. 99, pp. 518–529 (1999)
2. Gong, Y., Lazebnik, S.: Iterative quantization: A procrustean approach to learning binary codes. In: Computer Vision and Pattern Recognition, pp. 817–824 (2011)
3. Grauman, K., Fergus, R.: Learning binary hash codes for large-scale image search. In: Cipolla, R., Battiato, S., Farinella, G.M. (eds.) Machine Learning for Computer Vision. SCI, vol. 411, pp. 55–93. Springer, Heidelberg (2013)
4. Hsiao, K.J., Kulesza, A., Hero, A.: Social collaborative retrieval. In: International Conference on Web Search and Data Mining, pp. 293–302 (2014)
5. Koenigstein, N., Ram, P., Shavitt, Y.: Efficient retrieval of recommendations in a matrix factorization framework. In: International Conference on Information and Knowledge Management, pp. 535–544 (2012)
6. Linden, G., Smith, B., York, J.: Amazon.com recommendations: Item-to-item collaborative filtering. Internet Computing 7(1), 76–80 (2003)
7. Liu, T., Moore, A.W., Yang, K., Gray, A.G.: An investigation of practical approximate nearest neighbor algorithms. In: Advances in Neural Information Processing Systems, pp. 825–832 (2004)

8. Ntoutsi, E., Stefanidis, K., Nørvåg, K., Kriegel, H.-P.: Fast group recommendations by applying user clustering. In: Atzeni, P., Cheung, D., Ram, S. (eds.) ER 2012 Main Conference 2012. LNCS, vol. 7532, pp. 126–140. Springer, Heidelberg (2012)
9. Ou, M., Cui, P., Wang, F., Wang, J., Zhu, W., Yang, S.: Comparing apples to oranges: a scalable solution with heterogeneous hashing. In: International Conference on Knowledge Discovery and Data Mining, pp. 230–238 (2013)
10. Rendle, S., Balby Marinho, L., Nanopoulos, A., Schmidt-Thieme, L.: Learning optimal ranking with tensor factorization for tag recommendation. In: International Conference on Knowledge Discovery and Data Mining, pp. 727–736 (2009)
11. Wang, J., Shen, H.T., Song, J., Ji, J.: Hashing for similarity search: A survey. arXiv preprint arXiv:1408.2927 (2014)
12. Wang, Q., Ruan, L., Zhang, Z., Si, L.: Learning compact hashing codes for efficient tag completion and prediction. In: International Conference on Information & Knowledge Management, pp. 1789–1794 (2013)
13. Wang, Q., Shen, B., Wang, S., Li, L., Si, L.: Binary codes embedding for fast image tagging with incomplete labels. In: Fleet, D., Pajdla, T., Schiele, B., Tuytelaars, T. (eds.) ECCV 2014, Part II. LNCS, vol. 8690, pp. 425–439. Springer, Heidelberg (2014)
14. Wang, Q., Si, L., Zhang, D.: Learning to hash with partial tags: Exploring correlation between tags and hashing bits for large scale image retrieval. In: Fleet, D., Pajdla, T., Schiele, B., Tuytelaars, T. (eds.) ECCV 2014, Part III. LNCS, vol. 8691, pp. 378–392. Springer, Heidelberg (2014)
15. Wei, Y., Song, Y., Zhen, Y., Liu, B., Yang, Q.: Scalable heterogeneous translated hashing. In: International Conference on Knowledge Discovery and Data Mining, pp. 791–800 (2014)
16. Weiss, Y., Torralba, A., Fergus, R.: Spectral hashing. In: Advances in Neural Information Processing Systems, pp. 1753–1760 (2009)
17. Weston, J., Wang, C., Weiss, R., Berenzweig, A.: Latent collaborative retrieval. International Conference on Machine Learning (2012)
18. Yin, H., Cui, B., Chen, L., Hu, Z., Huang, Z.: A temporal context-aware model for user behavior modeling in social media systems, pp. 1543–1554 (2014)
19. Yin, H., Cui, B., Chen, L., Hu, Z., Zhou, X.: Dynamic user modeling in social media systems. ACM Transactions on Information Systems 33(3), 10 (2015)
20. Yin, H., Sun, Y., Cui, B., Hu, Z., Chen, L.: Lcars: a location-content-aware recommender system. In: ACM SIGMOD International Conference on Knowledge Discovery and Data Mining, pp. 221–229 (2013)
21. Zhai, D., Chang, H., Zhen, Y., Liu, X., Chen, X., Gao, W.: Parametric local multimodal hashing for cross-view similarity search. In: International Joint Conference on Artificial Intelligence, pp. 2754–2760 (2013)
22. Zhang, D., Yang, G., Hu, Y., Jin, Z., Cai, D., He, X.: A unified approximate nearest neighbor search scheme by combining data structure and hashing. In: International Joint Conference on Artificial Intelligence, pp. 681–687 (2013)
23. Zhang, D., Wang, J., Cai, D., Lu, J.: Self-taught hashing for fast similarity search. In: International Conference on Research and Development in Information Retrieval, pp. 18–25 (2010)
24. Zhang, Z., Wang, Q., Ruan, L., Si, L.: Preference preserving hashing for efficient recommendation. In: International Conference on Research & Development in Information Retrieval, pp. 183–192 (2014)
25. Zhou, K., Zha, H.: Learning binary codes for collaborative filtering. In: International Conference on Knowledge Discovery and Data Mining, pp. 498–506 (2012)

Spare Part Demand Prediction
Based on Context-Aware Matrix Factorization

Jianwei Ding[1,2], Yingbo Liu[2], Yuan Cao[2], Li Zhang[2], and Jianmin Wang[2]

[1] Department of Computer Science and Technology, Tsinghua University, China
[2] Institute of Information System and Engineering,
School of Software, Tsinghua University, China
{dingjw09,caoyuan12}@mails.tsinghua.edu.cn,
{csliuyb,lizhang,jimwang}@mail.tsinghua.edu.cn

Abstract. Maintenance spare part is used to replace and update the damaged and old components in the equipment. Forecasting spare part demand is notoriously difficult, as demand is typically intermittent and lumpy. Meanwhile, with the development of the sensor and internet technology, numerous condition monitoring systems are used to monitor the working condition of equipment, generating a large variety of monitor data at runtime. In this paper, we propose a Spare Part Demand (SPD) model based on a context-aware matrix factorization approach. The SPD mode incorporates historical spare part demands, the correlation between spare part demands and working places, and the correlation between spare part demands and monitor data. We evaluate our method based on extensive experiments using historical spare demands of one important component from more than 10000 concrete pump trucks and monitor data generated by part of these pump concrete pump trucks over a period of 9 months. The results demonstrate the advantages of our method over the previous studies, validating the contribution of our method.

Keywords: Spare Part Demand Prediction, Monitor Data, Matrix Factorization.

1 Introduction

Maintenance spare part[10,18] (hereafter referred to as spare part) is ubiquitous in modern societies. Its demand arises whenever a component fails or requires replacement in the equipment. Whether a spare part demand arises is totally decided by whether a component fails in the equipment. Meanwhile, an occurrence of the component failure is directly related with the working condition of the equipment. Effective prediction for the spare part demand can give strong support to maintenance activities of the equipment in time, and also reduce the capital for spare part inventory[18].

Most of the previous studies on the spare part demand prediction methods forecast spare part demands[2,3,14] mainly according to the historical spare part demands, because the working condition of equipment is difficult to grasp in the past. However, with the rapid development of the *Internet of Things*[5,17] technology, condition monitoring systems, e.g. *KOMTRAX*[1] and *IEM*[2] have been widely applied to monitor the working

---

[1] *KOMTRAX*: http://www.komatsuamerica.com/komtrax
[2] *IEM*: http://www.sanygroup.com/group/en-us/

© Springer International Publishing Switzerland 2015
R. Cheng et al. (Eds.): APWeb 2015, LNCS 9313, pp. 304–315, 2015.
DOI: 10.1007/978-3-319-25255-1_25

condition of equipment. A large variety of monitor data are generated simultaneously, which can help us better understand the working condition of equipment. Combining monitor data generated in the equipment, it can improve the prediction of spare part demands efficiently.

However, combining monitor data for spare part demand prediction faces three main challenges. 1) Huge monitor data volume. As the sensor technology is becoming ubiquitous, increasing numbers of sensors are installed in the equipment to monitor the working condition of different components thereof. For example, the volume of monitor data generated by the condition monitoring system *IEM* exceeds 30 GB in a workday, and the count is still growing. 2) Data sparseness. The working circumstances of equipment are unforeseen. There are a variety of unforeseen issues, e.g. illegal operations and adverse working circumstances, which prevent or affect the production of monitor data. Consequently, the data sparseness problem exists in the monitor data. 3) From monitor data to spare part demand. Although the spare parts demand is decided by the working condition of equipment, the working condition of equipment depends on multiple factors, such as working time, working strength and the number of equipment, as well as weather conditions of working places. Unfortunately, it is not easy to obtain enough training data to learn the mapping between working conditions and these factors. Hence, we can not simply use monitor data as a training set to learn the mapping.

Given the aforementioned challenges, existing methods for forecasting the spare part demand do not work well. In this paper, we propose a Spare Part Demand (SPD) model based on a context-aware matrix factorization approach. The SPD model takes not only historical spare part demands but also monitor data into consideration. Next, we use a context-aware matrix factorization method to forecast the real spare part demands, and to tackle the data sparsity problem as well. In order to validate our proposed method, we conduct experiments on the real world data sets. The experimental results show that, our method outperforms the conventional methods when combining a large amount of monitor data.

Our work presents a context-aware matrix factorization method for spare part demand prediction, taking monitor data into consideration as well. Our main contributions are shown as follows:

- We infer the spare part demand for multiple regions simultaneously, using a model (titled SPD). The SPD model incorporates the historical spare part demands, and the correlation between spare part demands and monitor data, as well as the characteristics of regions.
- We apply a context-aware matrix factorization method for the SPD model inference, in a framework of collaborative filtering, to forecast the demands for the spare part and to tackle the data sparsity problem as well.
- We evaluate the effectiveness and efficiency of our method by conducting extensive experiments, using a real-world data set containing monitor data collected over a period of 9 months. Our method clearly outperforms the baseline methods.

The rest of the paper is organized as follows. After reviewing related work in Section 2, Section 3 overviews our method. We detail the methodology in Section 4 and evaluate our approach in Section 5. Finally, we conclude this paper in Section 6.

2   Related Work

The spare part mainly consist of two types: productive spare part and maintenance spare part. The former spare part is used in the manufacturing phase of the equipment, and the latter one is used in the maintenance phase of the equipment. In this paper, we mainly talk about the maintenance spare part (hereafter referred to as spare part).

2.1   Spare Part Prediction

Illustrated in Figure 1, the spare part demand prediction[1,18,20] (also called spare part forecasting in the literature) mainly contains two types: initial spare part demand prediction and subsequent spare part demand prediction. The initial spare part demand prediction methods forecast the spare part demand according to different maintenance strategies from the beginning. As the initial spare pare prediction methods lack of enough data, they mainly forecast the spare part demand using reliability model[2,6,9,19]. The subsequent spare part prediction[18] (also called mixture spare part demand prediction) mainly forecast the spare part demand according the historical demand and other information, which consists of four types of methods in the literature: time series based, regression analysis, exponential smoothing and others.

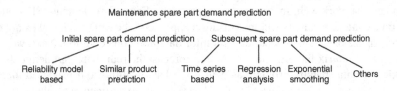

Fig. 1. A hierarchy for the spare part demand prediction in the literature.

- Time series based methods[15,16] forecast the spare part demand according to the variation rules of the historical spare part demands. These methods request that the historical spare part demands have stable variation trend. If the spare part demand is uncertain or intermittent, the prediction results will become unreliable.
- Regression analysis[7,11] finds the relation between the spare part demands and multiple factors. However, it always lacks of enough training data to train the model in the real circumstances.
- Exponential smoothing[8] is a specific technique that can be applied to the historical spare part demand to make forecasts.
- Other methods, such as neural network[4], take the historical spare part demands and other related factors as the input, to make forecasts.

2.2   Matrix Factorization

In a framework of collaborative filtering, a matrix of observed matrix is represented as $X \in \mathbb{R}^{n \times m}$. The matrix factorization technology[12,13] aims to factorize the matrix of observed values X into two matrices $Y \in \mathbb{R}^{n \times d}$ and $Z \in \mathbb{R}^{m \times d}$ such that $X' := YZ^T$

approximates X as much as possible. In most cases, the regular processing methods is to define and minimize a loss function $L(X : X')$ between observed and predicted values, to get the optimal results, as well as tackling the data sparsity problem.

3 Terminology and Notation

In this section, we first give some definitions to help the readers better understand the problem settings.

Definition 1 (Preparation cycle). *An **preparation cycle** is a time period, that is used to count the spare part demand and to manufacture the spare part.*

The preparation cycle of the spare part is usually determined by the manufacturers, e.g. the preparation cycle is set to ten days in our experiment. The setting of preparation cycle reflects manufacturing and management techniques of the spare parts, which are beyond the scope of this paper. In this paper, we make a reasonable assumption that the manufacturer is able to manufacture enough spare parts to satisfy all kinds of spare part prediction demands.

Definition 2 (Monitor data series). *A **monitor data series** is a sequence of data points, sampled and captured by a sensor typically at successive points over a preparation cycle of the equipment.*

A monitor data series is akin to a time series, which reflects the working condition of one component in the equipment. As a monitor data series is corresponding to a sensor (one component of the equipment), we can combine a group of different monitor data series together to represent the working condition of the equipment approximately, which is closely related with the spare part demand in the same preparation cycle.

Definition 3 (Region characteristics). *The **region characteristics** of the equipment reflects the geographic position characteristics in the working place of the equipment, e.g., the altitude.*

The spare part demands of different regions are totally different, because different numbers of the equipment have different working conditions in the different regions. Meanwhile, the working condition of the equipment is also affected by the real region characteristics. Take the concrete pump truck as an example, the wear degree of the concrete pump truck in the high altitude is larger in a pumping concrete, as compared with that in the low altitude. In the real cases, the region characteristics consist of the weather characteristics, the number of equipment and so on.

4 Methodology

4.1 Spare Part Demand Prediction Problem

Generally speaking, the answers to the two following questions in the spare part demand prediction are essential:

1. how often is the spare part demand checked?
2. how many spare parts are ordered in the next cycle?

The first question is to determine the order cycle of spare parts and the statistical cycle of parts that need replacement. As we analyzed before, the determination of the preparation cycle is related with the manufacture of the spare part, which exceeds the scope of this paper. This paper focuses on the second question, which is to forecast the real spare part demands in the present preparation cycle, with the help of historical spare part demands and monitor data collected in the equipment.

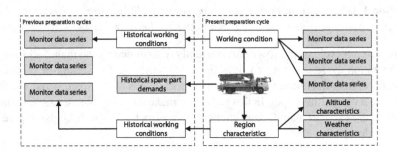

Fig. 2. Illustration for related factors in the spare part demand prediction. The factors in grey boxes are measured in the real cases, which are mainly divided into four parts: present monitor data series, region characteristics, historical spare part demands and historical monitor data series.

Figure 2 shows the all the related factors in the spare part demand prediction. For one part of the equipment in the present preparation cycle, the spare part demand is related with working conditions and region characteristics for all the pieces of the equipment. The working conditions are approximately reflected by the monitor data series generated in the equipment. The region characteristics consists of altitude characteristics, weather conditions and so on. Meanwhile, the spare part demand in the present preparation cycle is also related with the historical spare part demands and historical working conditions. The previous spare part demands reflect the variation of spare part demands in the previous preparation cycles, and the historical working conditions reflect the usage of the equipment in the previous preparation cycle.

4.2   The SPD Model

Let $M_1, M_2, ..., M_K$ represent K monitor data series, which are used to reflect the working conditions of the equipment in the $r^{th} (r = 1, ..., R)$ region. Suppose that there are T preparation cycles and the present preparation cycle is the Tth preparation cycle.

Definition 4 (Working condition matrix). *A working condition matrix, denoted as* $\mathbf{M} \in \mathbb{R}^{R \times T \times K}$, *is a three dimensional matrix. Its element* $m^{r \times t \times k}$ *represents the characteristics of the* k^{th} *monitor data series for the* r^{th} *region in the* t^{th} *preparation cycle.*

In the real cases, the element $m^{r \times t \times k}$ consists of two characteristics for the kth monitor data series for convenience–the average value and the standard deviation of all sampled values in the rth region. Because there exist some missing monitor data series for some regions, therefore the working condition matrix is actually sparse.

For one spare part, its spare part demand is also affected by its historical spare part demands. Hence, we can use a spare part demand matrix to represent the variation of historical spare part demands.

Definition 5 (Spare part demand matrix). *A spare part demand matrix, denoted as* $\mathbf{C} \in \mathbb{R}^{R \times T}$, *is a two dimensional matrix. Its element* $c^{r \times t}$ *represents the real amount of the spare part for the* r^{th} *region in the* t^{th} *preparation cycle. The element* $c^{r \times T}$ *in the last column of the spare part demand matrix is the target of our prediction, which represents the prediction of the spare part demand for the* r^{th} *region in the* T^{th} *preparation cycle.*

To sum up, these spare part demand related factors, such as working conditions, are variables that are varying with time, or with regions, or with both time and regions. As shown in Figure 3, we can use two matrices–preparation cycle dimensional matrix and region dimensional matrix, to represent spare part demand related factors respectively.

Definition 6 (Preparation cycle dimensional matrix). *A **preparation cycle dimensional matrix**, denoted as* $\mathbf{Y} := [\bar{M} \ \bar{C}] \in \mathbb{R}^{(K+1) \times T}$, *is a two dimensional matrix, where* $\bar{M} = \sum_{r=1}^{R} m^{r \times t \times k}$ *and* $\bar{C} = \sum_{r=1}^{R} c^{r \times t}$. *Its element* $y^{k \times t}(k = 1, ..., K; t = 1, ..., T)$ *represents the characteristics of the* k^{th} *monitor data series in the* t^{th} *preparation cycle. The element* $y^{(K+1) \times t}$ *in the last row represents the average spare part demands for all the regions in the* t^{th} *preparation cycle.*

The preparation cycle dimensional matrix summarizes all the time-dependent factors, which contains the monitor data series and historical spare part demands. Meanwhile, element $y^{k \times t}(k = 1, ..., K; t = 1, ..., T)$ is calculated from the aggregation of the working condition matrix. Hence, the characteristics also consist of the average and the standard deviation of all the kth monitor data series in the tth preparation cycle.

Definition 7 (Region dimensional matrix). *A **region dimensional matrix**, denoted as* $\mathbf{Z} := [\hat{M} \ \hat{D} \ \hat{C}] \in \mathbb{R}^{(K+D+1) \times R}$, *is a two dimensional matrix, where* $\hat{M} = sum_{t=1}^{T} m^{r \times t \times k}$ *and* $\hat{C} = \sum_{t=1}^{T} c^{r \times t}$. *The element* $y^{k \times r}(k = 1, ..., K; r = 1, ..., R)$ *represents the characteristics of the* k^{th} *monitor data series in the* r^{th} *region. The element* $y^{K+d \times r}(d = 1, ..., D; r = 1, ..., R)$ *represents the* d^{th} *region characteristics of the* r^{th} *region. The element* $y^{(K+D+1) \times r}$ *in the last row of the region dimensional matrix represents the average spare part demands for all the previous preparation cycles in the* r^{th} *region.*

The region dimensional matrix summarizes all the region-dependent factors, which contains the monitor data series, region characteristics and historical spare part demands in the regions. Meanwhile, element $y^{k \times r}(k = 1, ..., K; r = 1, ..., R)$ is calculated from the aggregation of the working condition matrix. Hence, the characteristics also consist of the average and the standard deviation of all the kth monitor data series for all the

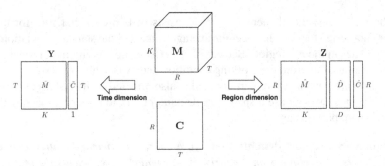

Fig. 3. Illustration for the SPD model with two matrices: preparation cycle dimensional matrix and region dimensional matrix.

previous preparation cycles in the rth region. The region dimensional matrix $\hat{D}$ also contains D region characteristics for each region, e.g. the altitude, the number of the equipment and the weather condition.

4.3   The Inference of SPD Model

As illustrated in Figure 3, we formulate the preparation cycle dimensional matrix **Y** and the region dimensional matrix **Z**, which recode the related factors from the time dimension and from the region dimension respectively. $\bar{M} \in \mathbb{R}^{T \times K}$ and $\hat{M} \in \mathbb{R}^{R \times K}$ are both matrices built based on the working condition matrix **M**, which represents the working conditions for all the regions in all the preparation cycles. $\bar{C} \in \mathbb{R}^{T \times 1}$ and $\hat{C} \in \mathbb{R}^{R \times 1}$ are also matrices built based on the spare part demand matrix **C**. $\hat{D} \in \mathbb{R}^{R \times D}$ represents the region characteristics matrix. Specially, all the matrices are sparse as we analyzed before, and the last column of the matrix C is the target of the prediction. Given the above settings, the inference can be converted into filling the missing values in the the matrices **Y**, **Z** and **C**, with the help of the existing values in the matrices.

Actually, we can achieve this goal through solely factorizing the matrix **C** into the product of two low-rank matrices based on the existing entries in the matrix **C**. However, in the real cases, the spare part demand matrix **C** does not have enough existing entries due to the data sparsity problem, especially as well as the last column is missing. In order to tackle this problem, we incorporate another two matrices **Y** and **Z**. Intuitively, $\bar{M}$ and $\hat{M}$ represent the working conditions, and $\bar{C}$ and $\hat{C}$ represent historical spare part demands. Putting **Y** together with **Z** reveals the relation between the related factors and the spare part demand in the present preparation cycle. As a result, the formulation of **Y** and **Z** can help factorize the matrix **C**.

Precisely, we can factorize the three matrices **C**, **Y** and **Z** as follows:

$$\mathbf{Y} \approx \mathbf{T} \times \mathbf{W}^T \tag{1}$$

$$\mathbf{Z} \approx \mathbf{R} \times \mathbf{F}^T \tag{2}$$

$$\mathbf{C} \approx \mathbf{R} \times \mathbf{T}^T \tag{3}$$

where **T**, **W**, **R** and **F** are all low-rank matrices representing latent factors. **Y** and **C** share the latent factor **T**. **Z** and **C** share the latent factor **R**. The collaborative

---

Algorithm 1. Illustration for the inference of the SPD model

Input: Incomplete spare part demand matrix $\mathbf{C}$, Preparation cycle dimensional matrix $\mathbf{Y}$,
        Region dimensional matrix $\mathbf{Z}$, The count of maximum iterations N, The threshold value ϵ
Output: Complete spare part demand matrix $\mathbf{C}$

1. $i = 1$;
2. Initialize the matrices $\mathbf{T}_1, \mathbf{W}_1, \mathbf{F}_1$ and $\mathbf{R}_1$
3. **while** $i < N$ and $\mathcal{L}_{i+1} - \mathcal{L}_i > \epsilon$ **do**
4.   Get the gradients $\nabla_{\mathbf{T}}\mathcal{L}, \nabla_{\mathbf{W}}\mathcal{L}, \nabla_{\mathbf{R}}\mathcal{L}$ and $\nabla_{\mathbf{F}}\mathcal{L}$ by Equation (4) $\sim$ (7);
5.   **while** $\mathcal{L}(\mathbf{T}_i - \gamma\nabla_{\mathbf{T}_i}, \mathbf{W}_i - \gamma\nabla_{\mathbf{W}_i}, \mathbf{F}_i - \gamma\nabla_{\mathbf{F}_i}, \mathbf{R}_i - \gamma\nabla_{\mathbf{R}_i}) > \mathcal{L}(\mathbf{T}_i, \mathbf{W}_i, \mathbf{F}_i, \mathbf{R}_i)$ **do**
6. $\gamma = \frac{\gamma}{2}$; // search for the maximum step size
7. $\mathbf{T}_{i+1} = \mathbf{T}_i - \gamma\nabla_{\mathbf{T}_i}, \mathbf{W}_{i+1} = \mathbf{W}_i - \gamma\nabla_{\mathbf{W}_i}$ and
8. $\mathbf{F}_{i+1} = \mathbf{F}_i - \gamma\nabla_{\mathbf{F}_i}, \mathbf{R}_{i+1} = \mathbf{R}_i - \gamma\nabla_{\mathbf{R}_i}$;
9.   **end while**
10. $i = i + 1$;
11. **end while**
12. **return** $\mathbf{C} \approx \mathbf{R}\mathbf{T}^T$

---

factorization of $\mathbf{C}$, $\mathbf{Y}$ and $\mathbf{Z}$ improves the accuracy of the approximation, as compared with the sole factorization of $\mathbf{C}$. After the factorization, we can recover the spare part demand matrix $\mathbf{C}$ through the product of $\mathbf{R}$ and $\mathbf{T}$. In order to optimize the above factorization, we define a loss function $\mathcal{L}(\mathbf{T}, \mathbf{W}, \mathbf{R}, \mathbf{F})$ as an objective function as shown in the following Formula.

$$\mathcal{L}(\mathbf{T}, \mathbf{W}, \mathbf{R}, \mathbf{F}) = \frac{1}{2}\|\mathbf{Y} - \mathbf{T} \times \mathbf{W}^T\|^2 + \frac{\lambda_1}{2}\|\mathbf{Z} - \mathbf{R} \times \mathbf{F}^T\|^2 +$$

$$\frac{\lambda_2}{2}\|\mathbf{C} - \mathbf{R} \times \mathbf{T}^T\|^2 + \frac{\lambda_3}{2}(\|\mathbf{T}\|^2 + \|\mathbf{W}\|^2 + \|\mathbf{R}\|^2 + \|\mathbf{F}\|^2)$$

where $\|\cdot\|$ denotes the Frobenius norm[3]. The first three terms in the objective function manage the loss in matrix factorization, and the last term controls the regularization over the factorized matrices so as to prevent over-fitting. The parameters λ_1, λ_2 and λ_3 are artificially set according to the real experimental cases. Hence, we have the gradients of the objective function for each matrix as follows:

$$\nabla_{\mathbf{T}}\mathcal{L} = [\mathbf{TW} - \mathbf{Y}]\mathbf{W} + \lambda_2[\mathbf{RT} - \mathbf{C}]^T\mathbf{R} + \lambda_3\mathbf{T} \qquad (4)$$

$$\nabla_{\mathbf{W}}\mathcal{L} = [\mathbf{TW} - \mathbf{Y}]^T\mathbf{T} + \lambda_3\mathbf{W} \qquad (5)$$

$$\nabla_{\mathbf{R}}\mathcal{L} = \lambda_1[\mathbf{RF} - \mathbf{Z}]\mathbf{F} + \lambda_2[\mathbf{RT} - \mathbf{C}]\mathbf{T} + \lambda_3\mathbf{R} \qquad (6)$$

$$\nabla_{\mathbf{F}}\mathcal{L} = \lambda_1[\mathbf{RF} - \mathbf{Z}]^T\mathbf{R} + \lambda_3\mathbf{F} \qquad (7)$$

However, we can find that, the objective function is not jointly convex for all the four matrices $\mathbf{T}$, $\mathbf{W}$, $\mathbf{R}$ and $\mathbf{F}$. Hence, we cannot get closed-form solutions to minimize the objective function. As a consequence, we can get the approximate optimal solutions by iteratively minimizing the objective function according to the gradient descent of the four matrices. Finally, we can get the complete spare part demand matrix through the product of $\mathbf{R}$ and $\mathbf{T}$. The last column of the spare part demand matrix is the prediction result of the spare part. The details are shown in Algorithm 1.

---

[3] http://en.wikipedia.org/wiki/Matrix_norm#Frobenius_norm

5   Evaluation

We conducted the experiments on a real-world data set from a Chinese well-known construction machinery manufacturer. The data set, collected from $R = 30$ regions, consists of two parts: the spare demands of one main component for more than $10,000$ concrete pump trucks[4] in the previous successive 9 months (a preparation cycle is set to ten days), and $K = 10$ categories of monitor data series collected from more than $1,000$ in the successive 9 months. Meanwhile, we collected $D = 3$ region characteristics for each region, which represent the average altitude, the number of concrete pump trucks and the average temperature of each region respectively.

5.1   Baselines

In the experiments, we evaluated four groups of experiments, which used the data set in 9 months ($T = 27$).

- Exponential Smoothing (**ES**) just uses the spare demands data of $T - 1$ preparation cycles to forecast the spare demand of the Tth preparation cycle for each region respectively. Its prediction formula is $c'^{r \times t} = \alpha c^{r \times t} + (1 - \alpha)c'^{r \times t-1}$, where $t = 2, ..., T; r = 1, ..., R; c'^{r \times 1} = c^{r \times 1}$ and α is smoothing factor.
- Moving Average (**MA**) uses the spare demands data of the last 3 preparation cycles to forecast the spare demand of the next preparation cycle for each region respectively. Its prediction formula is $c'^{r \times t} = \frac{c'^{r \times t-3}+c'^{r \times t-2}c'^{r \times t-1}}{3}$, where $t = 4, ..., T; r = 1, ..., R$
- Regression Analysis (**RA**) takes monitor data series together as the independent variables, and the spare demands as the dependent variables to fit a regression model for each region. Its prediction model is $\mathbf{C} = A \times \bar{M}^r$, where $\bar{M}^r$ is the aggregation of the working condition matrix for the rth region.
- Matrix Factorization (**MF**) takes both spare demands data and monitor data into consider. The details are shown in Algorithm 1.

5.2   Evaluation Metrics

The evaluation metrics of the spare part prediction consists of two parts: the evaluation of the accuracy for the prediction and the spare part satisfied rate. The first metric reflects the real prediction ability for each method. In the real cases, we use the coefficient of variation (CoV, denoted as V) between the real spare demands $c^{r \times T}$ and the prediction spare demands $c'^{r \times T}$ in all the regions to evaluate the accuracy.

$$V = \frac{\sqrt{\sum_{r=1}^{R}(c^{r \times T} - c'^{r \times T})^2/R}}{\sum_{r=1}^{R} c^{r \times T}/R} \tag{8}$$

---

[4] The concrete pump truck is a type of construction machinery, which is used to pump the concrete up to the specified destinations with the help of the mounted concrete pump in the truck.

Table 1. Statistics for the real and prediction spare part demands (unit: piece) in the Tth preparation cycle. All the prediction results are shown rounded to the nearest whole number.

Region	real	ES	MA	RA	MF	Region	real	ES	MA	RA	MF	Region	real	ES	MA	RA	MF
1	28	37	34	38	36	11	4	5	6	7	7	21	64	78	74	66	63
2	46	73	69	60	55	12	26	40	47	35	36	22	12	40	41	12	32
3	34	19	23	39	22	13	30	22	21	31	29	23	14	27	26	22	35
4	12	41	34	15	19	14	18	19	20	25	14	24	57	55	55	63	40
5	83	37	40	92	55	15	63	63	71	70	82	25	23	37	32	27	26
6	22	17	20	23	19	16	38	27	26	41	29	26	29	23	24	35	22
7	14	13	15	16	15	17	25	54	52	33	37	27	14	40	48	19	41
8	2	4	5	5	7	18	28	37	34	32	26	28	10	17	18	14	20
9	22	26	24	27	26	19	4	11	9	6	11	29	46	46	51	48	42
10	12	30	28	21	30	20	10	7	9	15	6	30	12	13	13	14	9

The CoV is the rate between the mean square error between and the average of the real spare demands. The less the CoV, the better the accuracy of the prediction. The second metric is the spare part satisfied rate (SPR, denoted as S), which is the rate between the prediction spare demands and the real spare demands.

$$S = \frac{\sum_{r=1}^{R}(c'^{r \times T}/c^{r \times T})}{R} \qquad (9)$$

If the SPR is closer to 100%, the prediction is better. Meanwhile, the prediction with more than 100% SPR is better than the prediction with less than 100% SPR.

5.3 Experimental Results

Table 1 shows the detailed statistics for the real and four prediction spare part demands in the $T^{th} = 30^{th}$ preparation cycle. As the evaluation metrics shown in Figure 4, **MF** consistently outperforms the other approaches for all the metrics, demonstrating the importance and effectiveness of combining historical spare part demands and monitor data together to forecast the future spare part demands.

As illustrated in Figure 4, the CoVs for the prediction methods **ES**, **MA**, **RA** and **MF** are 0.605, 0.582, 0.458 and 0.449 respectively, and the SPRs for the prediction methods **ES**, **MA**, **RA** and **MF** are 1.481, 1.501, 1.840 and 1.418 respectively. As the prediction results of **ES** and **MA** totally rely on the previous spare part demands, the prediction results may have significant fluctuations when the spare part demands changes greatly, such as the spare part demands in 2th region as shown in Figure 4. Conversely, **RA** and **MF** have less significant fluctuations when the spare part demands changes greatly, because they both takes the working conditions of equipment into consideration, such as the spare part demands in 5th region as shown in Figure 4. Although **RA** and **MF** have approximate CoVs, the SPR of **MF** is much less than that of **RA**. The reason is that, although **RA** and **MF** both takes monitor data series into consideration, **RA** can not deal with the data sparsity problem. Hence, **RA** does not take enough monitor data series to fit the regression model, which makes the SPR of **RA** is much larger than that of the other methods.

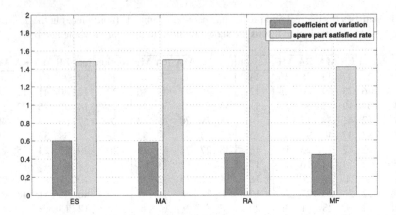

Fig. 4. Statistics of the *CoV* and *SPR* for the four experimental methods.

5.4 Discussion

Because all the experimental methods forecast the spare part demands much or less according to the historical spare part demands, the performance of the prediction results will improve when more historical data (9 months' historical data in the experiment) are used to train the model, especially **RA** and **MF**. When the period of the preparation cycle shortens, the spare part prediction will rely more on the monitor data and less on the historical data, because the spare part demands become more intermittent and more lumpy in a shorter preparation cycle. If so, **MF** will have a better performance as compared with the other methods. Especially, **MF** is even able to forecast the spare part demands per day or per hour, if enough monitor data are collected in the equipment.

6 Summary

Accurate and efficient spare part demand prediction is important to the manufacturer. In this paper, we proposed a SPD model based on a context-aware matrix factorization approach, which combines the historical spare part demands and monitor data together to forecast the spare part demand, and tackles the data sparsity problem in the framework of collaborative filtering as well. Our approach has promising performance in experimental studies, which leads us to conclude that our proposed SPD can indeed improve the spare part demand prediction in the real cases. Our future work will include optimization of the SPD model, selection of more factors and extensive applications of the SPD model.

References

1. Babai, M.Z., Syntetos, A., Teunter, R.: Intermittent demand forecasting: An empirical study on accuracy and the risk of obsolescence. International Journal of Production Economics 157, 212–219 (2014)

2. Barabadi, A., Barabady, J., Markeset, T.: Application of reliability models with covariates in spare part prediction and optimization–a case study. Reliability Engineering & System Safety 123, 1–7 (2014)
3. Cao, Y., Li, Y.: A two-stage approach of forecasting spare parts demand using particle swarm optimization and fuzzy neural network. Journal of Computational Information Systems 10(15), 6785–6793 (2014)
4. Chen, F.-L., Chen, Y.-C., Kuo, J.-Y.: Applying moving back-propagation neural network and moving fuzzy neuron network to predict the requirement of critical spare parts. Expert Systems with Applications 37(6), 4358–4367 (2010)
5. Chui, M., Löffler, M., Roberts, R.: The internet of things. McKinsey Quarterly 2, 1–9 (2010)
6. Hu, Q., Bai, Y., Zhao, J., Cao, W.: Modeling spare parts demands forecast under two-dimensional preventive maintenance policy. Mathematical Problems in Engineering (2015)
7. Hua, Z., Zhang, B.: A hybrid support vector machines and logistic regression approach for forecasting intermittent demand of spare parts. Applied Mathematics and Computation 181(2), 1035–1048 (2006)
8. Hyndman, R., Koehler, A.B., Ord, J.K., Snyder, R.D.: Forecasting with exponential smoothing: the state space approach. Springer Science & Business Media (2008)
9. Jardine, A.K.S., Tsang, A.H.C.: Maintenance, replacement, and reliability: theory and applications. CRC Press (2013)
10. Kennedy, W.J., Patterson, J.W., Fredendall, L.D.: An overview of recent literature on spare parts inventories. International Journal of Production Economics 76(2), 201–215 (2002)
11. Kim, K.-R., Yong, H.-Y., Kwon, K.-S.: Optimization for concurrent spare part with simulation and multiple regression. Journal of the Korea Society for Simulation 21(3), 79–88 (2012)
12. Koren, Y., Bell, R., Volinsky, C.: Matrix factorization techniques for recommender systems. Computer (8), 30–37 (2009)
13. Langville, A.N., Meyer, C.D., Albright, R., Cox, J., Duling, D.: Algorithms, initializations, and convergence for the nonnegative matrix factorization. arXiv preprint arXiv:1407.7299 (2014)
14. Mao, H.L., Gao, J.W., Chen, X.J., Gao, J.D.: Demand prediction of the rarely used spare parts based on the bp neural network. Applied Mechanics and Materials 519, 1513–1519 (2014)
15. Matsumoto, M., Ikeda, A.: Examination of demand forecasting by time series analysis for auto parts remanufacturing. Journal of Remanufacturing 5(1), 1–20 (2015)
16. Moon, S., Hicks, C., Simpson, A.: The development of a hierarchical forecasting method for predicting spare parts demand in the south korean navy–a case study. International Journal of Production Economics 140(2), 794–802 (2012)
17. Perera, C., Zaslavsky, A., Christen, P., Georgakopoulos, D.: Context aware computing for the internet of things: A survey. IEEE Communications Surveys & Tutorials 16(1), 414–454 (2014)
18. Qu, L., Zhang, Q.: An overview of inventory management for spare parts. Research and Exploration in Laboratory 7, 044 (2006)
19. Wang, Z., Hu, C., Wang, W., Kong, X., Zhang, W.: A prognostics-based spare part ordering and system replacement policy for a deteriorating system subjected to a random lead time. International Journal of Production Research, 1–17 (2014) (ahead-of-print)
20. Li, W.-S.: An overview of spare parts demand prediction technology. Logistics Technology 8, 30–33 (2007)

Location Sensitive Friend Recommendation in Social Network

Xueqin Sui, Zhumin Chen, and Jun Ma

School of Computer Science and Technology,
Shandong University, Jinan, 250101, China
suixueqin1010@gmail.com, {chenzhumin,majun}@sdu.edu.cn

Abstract. How to recommend friends in social network has attracted many research efforts. Most current friend recommendation methods are just based on the assumption that people will become friends if they have common interests which are usually estimated with the contents of their published posts and following relationships. However, friends recommended by these methods are only suitable for virtual social space instead of the real world. In this paper, we propose a new method to recommend friends in social network from the perspective of not just common interests, but also real-life needs. That is, we focus on finding friends that they can communicate with each other by social network and participate in some real-life activities face to face. The central idea of our approach is that we suppose people are more likely to be friends if their lives share more location overlaps besides the common interests. Currently, most people publish posts containing their real-time location information at any time, which makes it possible to detect and use the location information to recommend friends. Thus, our method combines users' published posts, their location sequences detected from the posts and how active they are in Sina Weibo to estimate whether they can become friends in not only social network but also the real world. Experiments on Sina Weibo dataset demonstrate that our method can significantly outperform the traditional friend recommendation methods in terms of *Precision*, *Recall* and $F1$ measures.

Keywords: Recommend Friends, Common Interests, Real-life Needs, Location Overlaps.

1 Introduction

Social networks such as Sina Weibo and Twitter have become very popular because they are regarded by people as the reflection and extension of their real-life. To most people, the friendship is an indispensable part of their lives in both virtual and real social space. How to recommend friends in social networks has received substantial interest from both academia and industry. Traditional friend recommendation methods in social network mainly focus on finding friends with common interests estimated with the contents of their published posts and the following relationships [1,2]. However, these methods may only suggest 'similar'

© Springer International Publishing Switzerland 2015
R. Cheng et al. (Eds.): APWeb 2015, LNCS 9313, pp. 316–327, 2015.
DOI: 10.1007/978-3-319-25255-1_26

people to be friends, which is only suitable for virtual social space instead of the real world. For example, if friends locate in some places faraway from each other, it is difficult for them to take part in a local party or meet together to do some interesting things. Therefore, it is easy to recommend friends in social network, but it is relatively hard to make friends not only in social network but also in real world. There are many scholars studying how to predict locations [3,4], which makes it possible to recommend friends considering location information besides contents. Considering locations into our recommendation ensures that recommended friends can meet the real world needs.

Our method is based on the following assumptions: (i) people are more likely to be friends if their lives share more location overlaps besides the common interests; and (ii) the more active users are in social network, the more likely they are recommended [5]. In this paper, the common interests between two users are estimated with the content similarity of their published posts. Users' location sequences with frequencies are detected from their posts and they are used to compute the location overlaps between them. If the common interests scores of two users are the same to the target user (the user we want to recommend friends to him), the one with more location overlaps with the target user will be more likely to be recommended to him. In addition, without lose of generality, how active users are in Sina Weibo is evaluated with the number of their published posts. Finally, we compute the recommendation scores of users combining above three factors and recommend the top n users as potential friends to the target user.

The rest of this paper is organized as follows. Section 2 gives some related work. Section 3 describes the friend recommendation method. Section 4 discusses the corresponding experiments and some metrics to evaluate our method and baselines. Section 5 makes conclusions and introduces our future work.

2   Related Work

There are many social networks which have employed recommendation systems to provide user experiences. Such as Facebook [2], it employed recommendation system by friends-of-friends method to recommend friends to users. The intuition is derived from the idea that it is more possible for a person to know friends of their friends rather than a random person [6]. Chen et al. [7] compared different friend recommendation algorithms in an enterprise social network and found that the algorithms based on information about the social network can find more known users while the algorithms checking for content similarity are more suited for discovering new friends.

Friend recommendation in social networks can be viewed as a link prediction problem [8] in complex networks. LibenNowell et al. firstly discussed link prediction in social networks and compared lots of methods based on node proximity [9]. These methods mainly utilize the structural information of a given social network. Kwon et al. proposed a method in [10] consisting three stages; (i) computing the friendship score using physical context; (ii) computing the friendship score using social context; and (iii) combining all of the friendship scores and recommending friends by the scoring values. The paper proposed by Hannon et al. in [1] is another typical friend recommendation method based on contents.

Just using contents to recommend friends can't meet our requirement. There are papers using locations for recommending friends. For example, Wan et al. proposed a method to recommend friends according to the informational utility, which stood for the degree to which a friend satisfied the target user's unfulfilled informational need, called informational friend recommendation in [11], and Chu et al. proposed a brand-new friend recommendation approach in [12]. The main concept is to recommend friends who have the similar interests or another thing with self to users. This paper is to recommend friends that are close in the real life. After recommending a friend to a user, there are scholars studying group recommendation such as [13,14,15] and so on. Bian et al. proposed a novel model for team recommendation to help users find interesting teams in social network [13], Deng et al. focused on studying how to improve group recommendations by making research on people interact with each other depending on their personalities or their closeness in [14]. At the same time, Nguyen et al. studied how to improve group recommendation procedure in [16].

3   Friends Recommendation Model

Without loss of generality, we use Sina Weibo as the social network to recommend friends in this paper. One post refers to an original microblogging including some words published by a user in Sina Weibo. The user here may have three kinds of posts: (i) original posts published by him directly; (ii) forwarded posts with some comments of him; and (iii) forwarded posts without any comments of him. For the (ii), we keep the comments and remove the forwarded words. For the (iii), we remove these posts. Before introducing our model, we introduce symbols used in this paper and their corresponding descriptions in Table 1.

In order to recommend friends to a user suitable for not only in social network but also the real world, we propose a novel approach which combines users' published posts, their location sequences detected from the posts and how active they are in Sina Weibo to predicate whether they can become friends. Our method supposes that people are more likely to be friends if their lives share more location overlaps besides the common interests and the more active the recommended users are, the more likely they are recommended [5]. Based on above assumptions, we give the model of our friend recommendation as follows.

Given a user u and $\{(t_1, b_1, l_1)..., (t_i, b_i, l_i)...,(t_n, b_n, l_n)\}_u$, where t_i represents a post's published time, b_i represents the post's contents and l_i is a published location (the location may be unknown). For this sequence, our method attempts to calculate the recommendation scores between the other users and user u and recommend the top n users to him. We use $P(u, r)$ to represent the probability we recommend user r to user u. r represents a recommended user and it has the same form with u. It is based on the common interests, location overlaps and how active users are in Sina Weibo. So, $P(u, r)$ is defined as

Table 1. Table of symbols we use in this paper

Notation	Description
u	the target user (we want to recommend friends to him)
r	the recommended user (we want to recommend him to a target user)
U	user set
$pop(r)$	how active user r is in Sina Weibo
D	document set of a user (a document refers all posts published in a month)
d	a document in D
$\Delta(u, l)$	the times user u went to location l in the past
E	user location frequency matrix
N_r	published post number of user r
L	Chinese location library

$$P(u,r) = \begin{cases} \frac{I(u,r)*L(u,r)*pop(r)}{\sum_{r' \in U \backslash u} I(u,r')*L(u,r')*pop(r')} & u \neq r \\ 0 & u = r \end{cases} \quad (1)$$

Where $I(u, r)$ evaluates the common interests between the user u and r, $L(u, r)$ measures the location overlaps between u and r not considering the time factor, and $pop(r)$ means how active user r is in Sina Weibo. We use $pop(r)$ not $pop(u)$ because we want to recommend user r to user u. In the following sections, we will discuss them respectively. In this equation, $I(u, u)$ and $L(u, u)$ are both 1 and we don't recommend a user to himself, so user u is excluded from the denominator.

3.1   Common Interests

As stated earlier, our method bases on a premise that friends must have some common interests. If we want to recommend user r to user u, we should first evaluate their common interests. Time factor is important for calculating the common interests, because users' interests may drift over time. In a general sense, a user does not change his interests frequently, that is, his interests usually do not drift in a short period. So, we merge all posts published in one month into a document d and get the document set D. Then $I(u, r)$ is defined as:

$$I(u,r) = \frac{1}{M(u)*M(r)} \sum_{i=1}^{M(u)} \sum_{j=1}^{M(r)} e^{-|t_i - t_j|} \frac{d_u^i \cdot d_r^j}{|d_u^i| \cdot |d_u^j|} \quad (2)$$

where d_u^i and d_r^i are the words vectors of d_u^i (document of user u that is in the ith month and $d_u^i \in D_u$) and d_r^i, and we use the tool ICTCLAS[1] to get these words. t_i and t_j are the published time of documents d_u^i and d_r^j which are represented in months. $|t_i - t_j|$ measures the gap and when the gap between t_i and t_j is large, the decay is fast. $M(u)$ and $M(r)$ represent how many months user u and user r publish posts in Sina Weibo.

---

[1] http://ictclas.org/

3.2   Location Overlaps

Location overlaps we mention here refer to the same locations between two users not considering time factor. If two users have common interests, the more overlaps between their locations, the more likely they will become friends. In this situation, the recommended users can participate in some parties or meetings together with the target user in real life.

Before getting location overlaps between users, we need to detect their past locations and corresponding frequencies. According to the location detection method proposed in the previous work [3], we first get users' locations. Paper [3] defines a Chinese location library and trains words distribution over locations based on the data getting from Wiki[2]. That means we get all locations in the sequence $\{(b_1, l_1), ..., (b_i, l_i), ...\}_u$ for user u. Then, we merge these same locations and get the frequency a user goes to a location. e_{ij} shows the probability of user u_i (a target user or a recommended user) went to location l_j according to his past locations. We use the e_{ij} to predict the probability the user u_i goes or will go to location l_j now or future. For a user u_i, the method we get e_{ij} is defined as

$$e_{ij} = \begin{cases} \frac{\Delta(u_i,l_j)}{\sum_{l_k \in L} \Delta(u_i,l_k)} & \Delta(u_i, l_j) > 0 \\ \frac{1}{\sum_{l_k \in L} \Delta(u_i,l_k)} & \Delta(u_i, l_j) = 0 \end{cases} \tag{3}$$

Where $\Delta(u_i, l_j)$ is the times user u_i goes to location l_j. For the locations where the user u_i never go there in our dataset, we consider the attendance time as once. After getting frequencies of all users go to all locations, we have the user location matrix E_U.

$$E_U = \begin{pmatrix} e_{11} & e_{12} & \cdots & e_{1|L|} \\ e_{21} & e_{22} & \cdots & e_{2|L|} \\ \vdots & \vdots & & \vdots \\ e_{|U|1} & e_{|U|2} & \cdots & e_{|U||L|} \end{pmatrix}$$

Where L is the Chinese location library we defined before. Under the common interests, if the more location overlaps, the more likely they get to be friends in real life. The matrix L_S is the location frequency probability matrix and it is defined as

$$L_S = E_U * E_U^T \tag{4}$$

$L(u, r)$ measures the location overlaps between user u and user r not considering the time factor and it is just one element in the matrix L_S.

3.3   How Active Users Are

How active users are in Sina Weibo is significant for friend recommendation especially for real-life needs. The more active users in the virtual social network space are, the more likely they will participate in some real activities. So, we need

---

[2] http://en.wikipedia.org/wiki/China#Geography

to know how active the recommended users are. $pop(r)$ is just used to measure how active the user r is in Sina Weibo and it can be defined as

$$pop(r) = \frac{N_r}{\max_{r' \in U} N_{r'}} \qquad (5)$$

where N_r is the quantity of posts published by user r, and $\max_{r' \in U} N_{r'}$ denotes the max number of posts published by users in the user set U. The largest post number in U is 185457 in our dataset.

Fig. 1 shows the relative activity of users in Sina Weibo. The X axis represents user id and the Y axis represents relative activity of users. This figure confirms that the most and the least users are small.

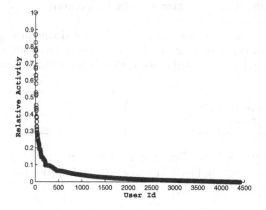

Fig. 1. The distribution of users' active in Sina Weibo.

Now, we give the algorithm of our method for recommending top n friends to the user $u \in U$. The algorithm is shown in Algorithm 1. At the beginning of the algorithm, we need to know the location sequences of all users in the user set U, then, for two users u and r where u represents the target user and r represents the recommended user, the algorithm contains three steps: (i) compute the common interests between u and r; (ii) compute location overlaps between u and r; (iii) compute how active the user r is in our dataset; and (iv) compute the recommendation score between user u and user r. At last, we recommend the top n users to the target user u according to the recommendation scores.

4   Experiments

In this section, we detail an experimental study of our method. The goals of the experiments are to understand: (i) whether friend recommendation based on common interests, location overlaps and how active users are in Sina Weibo can not only meet social network needs but also real-life needs; (ii) whether our method can significantly outperform the traditional friend recommendation

---

Algorithm 1. Location-sensitive friends recommendation

Input:
   target user set U and their published post sequences;
   where $u \in U$ represents a target user that we want to recommend friends to him;
   $r \in U \backslash u$ represents a recommended user;
Output:
   top n friends for u;
1: **for** each $u \in U$ **do**
2:    detect u's location sequence according to [3];
3:    construct u's location vector L_u;
4: **end for**
5: **for** each $r \in U$ **do**
6:    **if** $r \neq u$ **then**
7:        compute the common interests $I(u,r)$ between u and r according to Function 2;
8:        compute location overlaps $L(u,r)$ between u and r according to Function 4;
9:        compute the active of a user $pop(r)$ according to Function 5
10:       compute the recommendation score $P(u,r)$ according to Function 1;
11:   **end if**
12: **end for**
13: rank U in terms of $P(u,r)$;
14: **return** top n of U;

---

methods in terms of *Precision*, *Recall* and $F1$ measures and (iii) whether considering how active users are and location overlaps can improve the recommendation results.

4.1   DataSet

We have to collect the dataset since there is no standard corpus for evaluating our recommendation results. We collect real data from Sina Weibo. We first select 1000 active users and crawl all their posts published from Aug. 2009 to Apr. 2014. If the number of a user's published posts is smaller than 10, we remove the user. After that, there remains 769 users. When the dataset is large, we can use the common interests to exclude some users and then use our method to recommend friends. For each user, we randomly collect 10 carers' posts if the user has more than 10 carers in Sina Weibo. We use user u to show a user and user r to show another user, if user u cares user r in Sina Weibo, we say user r is user u's carer. These carers and the original users consist of user collection. Although we collect users according to this rule, the relationship is complex and it is a network. In the experiment, we assume that we don't know the relationship between users in our collection. The statistical information of the dataset is shown in Table 2.

For a post, if it contains GPS information, we use the GPS information directly. Otherwise, we use the method proposed in [3] to detect whether it contains a location and if it contains we use the detected location. If not, the location is unknown.

There is one point we want to stress, although we use Sina Weibo dataset to evaluate our experiment results, our model can be used for other datasets if they contains locations.

Table 2. Dataset from Sina Weibo

item	number
original users	769
carers	7,524
users	8,293
users' posts	39,594,373
locations	372
posts with GPS	149,021
posts with locations	6,757,750

4.2 Baseline Methods

There are five baseline methods we compared in this paper. The first method is a random recommendation method, which is defined as RR. The second method uses the location information for friend recommendation proposed by Wan et al. in [11], represented as LO. The third method is based on the common interests proposed by Hannon et al. in [1], denoted as TS. The fourth and fifth baseline methods are extended from LO and TS considering how active users are. They are represented as ALO and ATS separately.

4.3 Evaluation Measures

We use standard measures *Precision, Recall* and *F*1 to evaluate friends recommendation results. If a recommended friend generated by the methods agrees with the true friend in Sina Weibo, we view it as a correct recommendation.

Precision is the fraction of recommended friends that are correct.

Recall is the fraction of correct recommendations that are detected.

F1-score is calculated using following function: $F1 = 2*(Precision*Recall) / (Precision+Recall)$.

In this paper, we don't recommend only one friend to a target user. If we use n to measure how many friends we recommend to a user, $p@n$ means the *Precision* we recommend top n friends.

$$p@n = \frac{\sum_{i=1}^{n} correct_i}{n}; correct_i = \begin{cases} 1 & \text{ith recommendation is true} \\ 0 & others \end{cases} \quad (6)$$

Where $\sum_{i=1}^{n} correct_i$ means how many friends we recommended are right. Note that all measures above are macro-average for all users.

4.4 Results and Discussion

For each user u in the dataset, our method recommends n ($n = 1, 2, ..., 10$ in the experiments) possible friends in descending order of scores. Fig. 2 reports

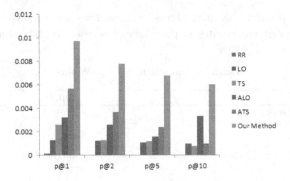

Fig. 2. Experimental result in terms of $p@n$.

the $p@1$, $p@2$, $p@5$ and $p@10$ ($p@n$ means the precision for the top@n recommendation.) for the original baseline friend recommendation approaches and our method considering common interests, location overlaps and how active users are in Sina Weibo. First, note the strong positive impact of considering how active users are and location overlaps. With considering how active users are alone, ALO and ATS methods have big improvements than the LO and TS methods separately. The gaps indicate that the active users in social network are more likely to make friends with other users. So, taking how active users are in social network into consideration can improve our recommendation results. Considering the location overlaps, our method has a big improvement than the ATS method which just consider how active users are and common interests. This result is encouraging and it also shows that our method improves the recommendation results besides meeting the social network needs and real life needs. As a negative result, we can see the poor performance of random recommendation method, which has *Precision* about 0.00012 in our dataset. This indicates that common interests and location overlaps help users become friends in social network.

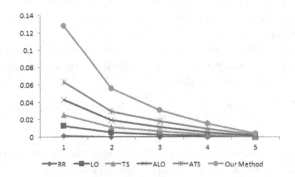

Fig. 3. Distribution of the number of users over $p@n$

Just considering the average of $p@n$ means that the·system can recommend friends better on average, but it doesn't give strong insight into the distribution of $p@n$. Hence, we have a distribution of $p@n$ in Fig. 3, which shows the user distribution of $p@n$. For example, in Fig. 3, there is one point $(2, 0.026)$, which means that when we recommend 10 friends to all users, there are about 2.6% users making more than two friends in our recommendation. From the figure, we can see that there is no user making more than five friends in our top 10 recommendation. Compared our method with baselines, our method is the best and the improvement is huge. Although the LO, TS ALO and ATS models are worse than our method, they are better than the RR model. Fig. 2 and Fig. 3 show that no matter how many friends to recommend, our method has a big improvement than baselines on average and distribution.

When we see the $Recall$ and $F1$ metrics, we don't consider RR method because we can't calculate the $Recall$ and $F1$ values for this method. Fig. 4 shows the recommendation results for all baselines and our method in $Precision$, $Recall$ and $F1$ for the top 1 recommendation. This figure demonstrates that our method can significantly outperform the baselines according to these three measures. For the $Recall$ and $F1$ metrics, our method is not high, but it is better than the four baselines which are almost 0. We further analyze these friends

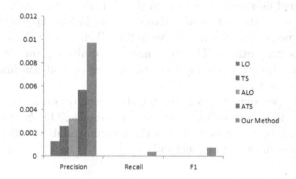

Fig. 4. The top 1 recommendation results of $Precision$, $Recall$ and $F1$

we recommend false. They have following features: (i) some users don't have enough posts; and (ii) there are not enough locations. These two reasons further confirms that common interests and location overlaps are both important in our experiment. Furthermore, since location overlaps are important, we study whether users in Sina Weibo have geographic features. Five groups are divided based on the frequency a user goes to one location and it is shown in Fig. 5. From this figure, we observe that there are 364 users in our dataset go to a lot of locations. On the other hand, there are 1321 users stay in one location all the time. There are about 32% users in our dataset have a frequency larger than 60% which confirms regional factor exists in our dataset.

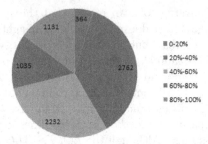

Fig. 5. 0−100% represents the highest frequency a user goes to a location. The number in the circle represents how many people in our dataset belong to this range.

5   Conclusion

In this paper, we propose a new method to recommend friends in social network from the perspective of not just common interests, but also real-life needs. We consider three characteristics to select friends for a user. First, we use posts similarity to evaluate users' common interests which is a basic factor between friends. Second, we use users' location overlaps which promise the target users and recommended users can be friends in the real life. Third, we use the amount of published posts to simulate how active users are in Sina Weibo which implies that the recommended friends are active in Sina Weibo and they are more likely to make friends with others. The experimental results on Sina Weibo dataset show that our method significantly outperforms the current baseline methods in *Precision, Recall* and *F*1.

In the future, we plan to consider the distances between users in real life for any given time to further improve the performance. On the other hand, we can use users' hometown information in the recommendation which can further improve the performance of our method.

Acknowledgement. This work is supported by the Natural Science Foundation of China under Grant No. 61272240 and 61103151, the Doctoral Fund of Ministry of Education of China under Grant No. 20110131110028, the Natural Science Foundation of Shandong Province under Grant No. ZR2012FM037, the Excellent Middle-Aged and Youth Scientists of Shandong Province under Grant No. BS2012DX017 and the Fundamental Research Funds of Shandong University.

References

1. Hannon, J., Bennett, M., Smyth, B.: Recommending twitter users to follow using content and collaborative filtering approaches. In: Proceedings of the Fourth ACM Conference on Recommender Systems, pp. 199–206. ACM (2010)

2. Xie, X.: Potential friend recommendation in online social network. In: 2010 IEEE/ACM Int'l Conference on Green Computing and Communications (Green-Com) & Int'l Conference on Cyber, Physical and Social Computing (CPSCom), pp. 831–835. IEEE (2010)
3. Sui, X., Chen, Z., Wu, K., Ren, P., Ma, J., Zhou, F.: Social media as sensor in real world: Geolocate user with microblog. In: Zong, C., Nie, J.-Y., Zhao, D., Feng, Y. (eds.) NLPCC 2014. CCIS, vol. 496, pp. 229–237. Springer, Heidelberg (2014)
4. Cheng, Z., Caverlee, J., Lee, K.: You are where you tweet: a content-based approach to geo-locating twitter users. In: Proceedings of the 19th ACM International Conference on Information and Knowledge Management, pp. 759–768. ACM (2010)
5. Zhang, C., He, Y., Ji, Y.: Temporal pattern of user behavior in micro-blog. Journal of Software 8(7), 1707–1717 (2013)
6. Hruschka, D.J., Henrich, J.: Friendship, cliquishness, and the emergence of cooperation. Journal of Theoretical Biology 239(1), 1–15 (2006)
7. Chen, J., Geyer, W., Dugan, C., Muller, M., Guy, I.: Make new friends, but keep the old: recommending people on social networking sites. In: Proceedings of the SIGCHI Conference on Human Factors in Computing Systems, pp. 201–210. ACM (2009)
8. Lü, L., Zhou, T.: Link prediction in complex networks: A survey. Physica A: Statistical Mechanics and its Applications 390(6), 1150–1170 (2011)
9. Liben-Nowell, D., Kleinberg, J.: The link-prediction problem for social networks. Journal of the American Society for Information Science and Technology 58(7), 1019–1031 (2007)
10. Kwon, J., Kim, S.: Friend recommendation method using physical and social context. International Journal of Computer Science and Network Security 10(11), 116–120 (2010)
11. Wan, S., Lan, Y., Guo, J., Fan, C., Cheng, X.: Informational friend recommendation in social media. In: Proceedings of the 36th international ACM SIGIR Conference on Research and Development in Information Retrieval, pp. 1045–1048. ACM (2013)
12. Chu, C.-H., Wu, W.-C., Wang, C.-C., Chen, T.-S., Chen, J.-J.: Friend recommendation for location-based mobile social networks. In: 2013 Seventh International Conference on Innovative Mobile and Internet Services in Ubiquitous Computing (IMIS), pp. 365–370. IEEE (2013)
13. Bian, J., Xie, M., Topaloglu, U., Hudson, T., Eswaran, H., Hogan, W.: Social network analysis of biomedical research collaboration networks in a ctsa institution. Journal of Biomedical Informatics (2014)
14. Deng, S., Huang, L., Xu, G.: Social network-based service recommendation with trust enhancement. Expert Systems with Applications 41(18), 8075–8084 (2014)
15. Tejeda-Lorente, Á., Porcel, C., Peis, E., Sanz, R., Herrera-Viedma, E.: A quality based recommender system to disseminate information in a university digital library. Information Sciences 261, 52–69 (2014)
16. Nguyen, T.T., Hui, P.-M., Harper, M., Terveen, L., Konstan, J.A.: Exploring the filter bubble: the effect of using recommender systems on content diversity. In: Proceedings of the 23rd International Conference on World Wide Web, pp. 677–686. International World Wide Web Conferences Steering Committee (2014)

A Compression-Based Filtering Mechanism in Content-Based Publish/Subscribe System

Qin Liu, Yiwen Zheng, and Kaile Wang

Tongji University, Shanghai, China
qin.liu@tongji.edu.cn, zyw9096@126.com, wangkaile@outlook.com

Abstract. Recently publish/subscribe (pub/sub) has been a popular paradigm in Internet-scale applications to decouple information publishers and subscribers. The main task of pub/sub is to find those subscriptions that match a given publication. Typically a matching algorithm utilizes a subscription indexing structure for higher efficiency. Unfortunately, given the diverse interests of publishers and subscribers, the semantic space of pub/sub becomes high dimensional and very sparse, leading to nontrivial space cost and inefficient matching algorithm. Existing work paretically tackled the both issues but at cost of scarifying another one, and few can meet the both goals. To overcome this issue, in this paper, we propose a novel coding approach. The approach can not only reduce the space cost of subscription index, and meanwhile helps optimizing the matching efficiency. Our experiments successfully demonstrate the proposed algorithm can achieve not only the low space cost of our proposed subscription indexing but also low running time of matching algorithm.

1 Introduction

Recently, the content-based publish/subscribe (pub/sub) has become a popular paradigm in Internet to decouple information producers and consumers (i.e., *publishers* and *subscribes*, respectively) with help of *brokers*. In such a system, subscribers declare their personal interests by defining subscription conditions, and publishers produce publication messages. On receiving publication messages from publishers, brokers match publications with registered subscriptions, and forward each matched publication to needed subscribers. Due to the excellent decoupling property and fine-grained expressiveness, pub/sub has been widely used for many real applications, such as event-based streaming processing, information-centric network and e-Commerce.

As the core task of pub/sub, the matching of each received publication against subscriptions typically utilizes a subscription indexing structure for higher efficiency. For example, by building a R-tree structure to index registered subscriptions, pub/sub systems can correctly find all matched subscriptions by processing a subset of subscriptions instead of all subscriptions. Unfortunately, due to the diverse interests of publishers and subscribers, the semantic space associated with publications and subscriptions becomes high dimensional and very sparse.

© Springer International Publishing Switzerland 2015
R. Cheng et al. (Eds.): APWeb 2015, LNCS 9313, pp. 328–339, 2015.
DOI: 10.1007/978-3-319-25255-1_27

For example, in e-commence applications [2], there exists a variety of product items (as publications) and each item contains significantly different attributes; meanwhile subscribers define subscription conditions based on their very personalized interests. Such publications and subscriptions lead to high dimensional and very sparse semantic space. Consequently, the full-space indexing structure such as R-Tree is inefficient to match publications against indexed subscriptions. Though the very recent work OpIndex [2] partially tacked the issue caused by the high dimensional and very sparse space, the base of OpIndex is fully built upon an assumption that pivot attributes needed by OpIndex must be known before the construction of OPIndex. Due to the decoupling and dynamic properties of pub/sub, it is practically impossible to acquire the pivot attributes before subscriptions are registered and publications are published.

To overcome the above issue, we propose a novel coding approach to save the space cost subscription index and improve publication matching efficiency. As the main contribution of our work, the proposed coding approach can unify the optimization of subscription indexing and publication matching. By using a single approach, we can achieve the both goals without scarifying any of the both goals.

The rest of paper is organized as follows. Section 2 first gives the preliminaries and Section 3 reports the overview of our approach. After that, Sections 4 and 5 present the detail of subscription index and matching algorithm, respectively. Section 6 shows the evaluation and Section 7 finally concludes the paper.

2   Preliminaries and Related Work

In this section, we introduce the data model and review related work.

2.1   Data Model

There are three roles in a content-based pub/sub system: *publishers*, *subscribers* and *brokers*. Publishers are information provider, and publish information via *publications*. Subscribers are information consumer, and declare user interests via *subscriptions*. Brokers are responsible for the matching between subscriptions and publications, and next forward matching publications to subscribers of interest.

A publication consists of a set of attribute-value pairs. Each pair contains an attribute with its data type and a single value, and the *size* of a publication is the amount of pairs. A subscription consists of a collection of predicates. The size of a subscription is the amount of predicates. There are usually three elements in a predicate: an attribute A with its data type, an operator op and an operand m. The example predicate *int price > 5* contains the attribute *price*, operator $>$ and operand *5*. Table 1 gives a list of 7 subscriptions.

In this paper, we treat every predicate as a Boolean function f. It receives a publication e as an input parameter to evaluate the publication e. Given a

Table 1. Based on the size of subscriptions, we divide the 7 subscriptions into 3 groups: $\{S_5\}$ in the group of size 1, $\{S_2, S_4, S_7\}$ in the group of size 2, and $\{S_1, S_3, S_6\}$ in the group of size 3.

SID	Subscription Conditions
S_1	$A \in [1, 7] \land B \in \{S, M, L\} \land C > 100$
S_2	$A \in [5, 9] \land D \leq 3.5$
S_3	$A \in [3, 200] \land B \in \{XS, S\} \land D \in [1.2, 4.8]$
S_4	$0 < C < 20 \land B \in \{M\}$
S_5	$D \geq 5.7$
S_6	$A \leq 4 \land B \in \{L, XL\} \land C \in [10, 50]$
S_7	$C \in [20, 30] \land D \in [5, 8]$

subscription s with the size s, the subscription s matches the publication e on the condition that $f(e) = true$ for all predicates in s. For example, a publication $price = 10$ successfully matches a predicate $price > 5$. Formally, [4] gives the matching problem definition: "Given a publication e and a set S of subscriptions, the matching problem is to find all subscriptions that are satisfied by e".

2.2   Related Work

The classic work [4] divides the set of subscriptions into distinct predicates and builds indexes respectively based on equality predicates and non-equality predicates. It maintains a bit vector to record whether a predicate is matched by an incoming publication. The key point of [4] is to pick out as few candidate subscriptions as possible. By the idea of schema-based subscription clustering, it organizes subscriptions into clusters by the size of subscriptions. In order to find clusters quickly, it defines a hash configuration to locate clusters using hash functions.

Sharing the very similar idea, [3] and [9] design a graph-based indexing structure. The terminal node of the graph is the matching result. In the binary decision diagram[3], the non-terminal node represents a Boolean function (stand for a predicate), and each has two edges marked 0 and 1 to indicate whether the predicate is matched, and every terminal node is the final matching result for one subscription. To realize a matching tree[9], the first step is to divide a predicate into a test part and a result part, and the test part represents the non-terminal node; result represents the tree edge as well as the subscription itself represents the leaf node of the tree. For each incoming event, it will go through the tree. Each path that goes to the leaf node indicates the subscription in the leaf node matches the event.

The very recent work [2] is designed to tackle the sparse and high-dimensional subscription database. The aim of the index is to pick out as few subscriptions to process the matching algorithm as possible for each incoming event. The index mechanism uses two-level partition scheme, first is to partition subscriptions based on pivot attributes into subscription lists and second is to further partition the list to predicate lists based on predicate operators$(=, \leq, \geq)$. The definition for pivot attribute is to choose an attribute $\acute{A}$ in a subscription that occurs the least times in event stream. The paper provides two optimizing approaches bit vector and hash function in second partition to quickly find the targeted predicates. There is also an counter referencing each subscription to judge whether it is fully matched to the incoming event.

3   Solution Overview

We first highlight the challenges of pub/sub systems. When given a huge amount of registered subscriptions, the pub/sub systems need to maintain a data structure to index such subscriptions. Due to the huge amount of subscriptions, we would like to save the *space* cost of such an index. Meanwhile, when publications arrive, the pub/sub systems match the incoming publications against the subscription index in order to find those matching subscriptions. Considering a high arriving rate of publications, the matching *time* is critical to optimize the throughput of pub/sub systems.

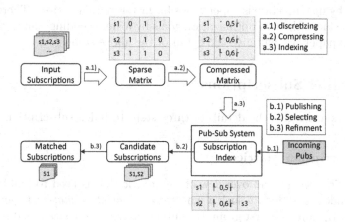

Fig. 1. Solution Overview

Until now, we have two optimization goals: the space cost of subscription index and running time of matching algorithm. Given the two goals, we propose a scheme to unify the two goals. The key of the proposed scheme is to design a coding approach to meet the two goals. First, the approach can compress the subscription index for lower space cost, and meanwhile the fast bitwise operations atop the coding approach can greatly speedup the matching algorithm for less running time.

Before going into the details in the following sections, we would like to give an overview of the proposed scheme Fig. 1 as follows. First, after subscribers register subscriptions into a pub/sub system, the system requires three steps to maintain an index for the registered subscriptions. First in step $(a.1)$, the system builds multiple matrices based on registered subscriptions. This step is to discrete the continuous ranges in subscriptions and then binary elements in the matrices. $(a.2)$ Due to the diverse interests of subscribers, the matrices in the previous step could be very sparse and lead to high space cost. Then we would like to compress the matrices for lower space cost. The key of this step is to design a coding approach for compression and we will focus on this part. $(a.3)$ Based on the compressed matrices, we finally build an indexing structure for registered subscriptions and such a structure uses less space cost than the compressed matrices.

Next, when a publisher publishes messages into a pub/sub system (step $b.1$), the system then evaluates the publications with help of the subscription index $(b.2)$ to select the candidates of matched subscriptions. The candidate selection algorithm can leverage machine bitwise operations atop the subscription index to avoid the expense scan of such an index. After that, the system finally refines such candidates for the matched subscriptions $(b.3)$.

During the above steps, the proposed coding approach not only helps compressing sparse matrices for lower space cost, but only optimizes the matching algorithm by machine bitwise operations for faster matching time. Therefore, we expect to achieve the two optimization goals by a single coding approach. In the rest of this paper, we will give the detail of the above steps.

4   Indexing Subscriptions

In this section, we give the details of three steps to index subscriptions.

4.1   Subscription Matrices

In this section, we give an overview of the matric that is used to build the subscription index, and denote such a matrix as *subscription matrix*. Given a set of input subscriptions, we first group the subscriptions by the size of subscriptions. For example, given the 7 subscriptions in Table 1, we group such subscriptions by their size.

Now for each attribute appearing in a set of grouped subscriptions, we build a corresponding subscription matrix. In the subscription matrix associated with an attribute say A_k ($1 \leq k \leq K$) where K is the dimensionality of the pub/sub system semantic space), each row represents a subscription containing A_k, and a column indicates a range of attribute values of A_k. Given the columns of the subscription matrix, we discretize the allowable range of A_i and set the element values of the matrix. In the i-th row and j-th column of a subscription matrix, the element e_{ij} is a binary value by either 1 or 0. That is, if the predicate represented by the i-th row overlaps the range represented by the j-th column, we set $e_{ij} = 1$ and otherwise $e_{ij} = 0$. For example, suppose that the allowable range of attribute A is $[1, 200]$, and we *discretize* the range into 10 columns each of which has the granularity width 20. Then given the predicate $\{1 \leq A \leq 7\}$ in

S_1 of Table 1, we set the first elements by 1 and others by 0, as shown in Table 2(1). It is because the range $[0, 20)$ associated with the first column overlaps the predicate $1 \leq A \leq 7$. Instead, for the predicate $3 \leq A \leq 200$ in S_3, we similarly set all elements from 1st column to 10-th column to be 1 and none of them zero.

4.2 Tuning Granularity

During the above discretization, we note that the granularity decides the number of columns in the subscription matrix and thus the associated space cost. In this section, we focus on how to tune a reasonable granularity that is used to divide the allowable range of a given attribute A_k.

Table 2. Subscription Matrix

Col	0	1	2	3	4	5	6	7	8	9
Span	1	20	40	60	80	100	120	140	160	180
s_1	1	0	0	0	0	0	0	0	0	0
s_2	1	0	0	0	0	0	0	0	0	0
s_3	1	1	1	1	1	1	1	1	1	1

(1) Main column partition

Col	0	1	2	3	4	5	6	7	8	9
Span	1	3	5	7	9	11	13	15	17	19
s_1	1	1	1	1	0	0	0	0	0	0
s_2	0	0	1	1	1	1	0	0	0	0
s_3	0	1	1	1	1	1	1	1	1	1

(2) Sub column partition

Col	0	1	2	3	4	5	6	7	8	9	10
Span	LB	1	20	40	60	80	100	120	140	160	180
s_1	0	1	0	0	0	0	0	0	0	0	0
s_2	0	1	0	0	0	0	0	0	0	0	0
s_6	1	1	0	0	0	0	0	0	0	0	0
s_3	0	1	1	1	1	1	1	1	1	1	1

(3) Add lower bound signal

We would like to first give a baseline solution to set the granularity before the proposed approach. Let us consider three subscriptions S_1, S_2, S_3 in Table 1. For the predicates on attribute A, we have $s_1 : \{1 \leq A \leq 7\}$, $s_2 : \{5 \leq A \leq 9\}$ and $s_3 : \{3 \leq A \leq 200\}$. Based on the predicates, we have the allowable range $1 \leq A \leq 200$. Next, we might simply utilize the machine word length denoted by B (e.g., 32 bits or 64 bits) and set the granularity $G = 200/B$. Just for simplicity and illustration, we consider the case that $B = 20$ and set $G = 200/20 = 10$. Following the setting, we have 10 columns in the subscription matrix of attribute A and the interval width is 20 as shown in Table. 2(1).

In the above figure, the three subscriptions S_1, S_2 and S_3 all set the elements in the first column by 1. It is not hard to find that two of them cover the interval of only one columns. We will show that such a case will harm the matching efficiency. Depending on the tradeoff between the space cost of subscription matrix and matching time, we set a threshold t (e.g., $t = 2$) to set the minimal number of columns, such that any subscription covers the associated intervals.

To overcome the above issue, we propose to further divide the intervals of those columns covered by too many subscriptions. Still following the above example, we further divide the interval of the first column into 10 sub-columns. Here the key is to ensure that the number of sub-columns is just equal to the one of the parent columns. As shown in Table 2(2), we have 10 sub-columns that are divided from the first column in Table 2(1). Based on the sub-columns, we can set the elements in the sub-columns when a subscription covers the interval associated with the sub-columns. Now, in Table 2(2), we find that the number of covered sub-column by any subscription is not smaller than the threshold $t = 2$.

Finally, we consider the predicate in S_6: $A \leq 4$. Differing from the above three predicates, we have an infinity point in S_6. Therefore, we define two special columns: UB and LB (i.e, short name for upper bound and lower bound. Then we can transform $A \leq 4$ into $LB \leq A \leq 4$, and have Table 2(3).

4.3   Compressed Subscription Matrices

Given the above subscription matrix, we note that the matrix could be very sparse. It is particularly true when subscribers define personalized and diverse subscription conditions. Due to high space cost caused by a sparse matrix, we would like to compress the matrix for low space cost.

The basic idea of the proposed compression is to treat the elements in each row of the above subscription matrix as binary bits and then encode the bits into integer numbers. For example, given Table 2(2), the row of S_1 is associated with the binary bits 0011100000. Given the 10 binary bits, given a base number $b = 4$, we then divide the 10 binary bits into three parts 0011, 1000 and 0000, which can be respectively encoded by three integers 3, 8 and 0. Now, instead of maintaining 10 elements for each subscription, we would like to maintain three integers, leading to less space cost. Besides the encoded numbers, we need to maintain the starting ID of the non-zero encoded numbers, such that we can easily restore the original rows. Thus, the bits 0011100000 are encoded by three pairs $\langle 0, 3 \rangle$, $\langle 4, 8 \rangle$ and $\langle 8, 0 \rangle$. For less space cost, we do not maintain the encoded numbers equal to zero, and thus maintain only two pairs $\langle 0, 3 \rangle$ and $\langle 4, 8 \rangle$.

We note that by setting a larger base number for example $b = 8$, we then treat the entire 10 bits as the binary format of two integers $56 = (00111000)$ and 0. Since we not maintain the pairs with the encoded numbers equal to zero, the 10 bits can be maintained by only one pair $\langle 0, 56 \rangle$. Thus, a larger base number could help optimize the space cost for subscription matrix. Depending on the physical machine and operating system, we typically set $b = 32$ or 64.

4.4   Subscription Index

In this section, we leverage the similarity of subscriptions to merge similar subscriptions (rows in the matrix) into single rows. In this way, we have chance to reduce the number of rows in a subscription matrix, and thus optimize the space cost of subscription matrices. Based on the compressed matrices with fewer number of rows, we design a subscription index.

Recall that each row (i.e., a subscription s_i) of a subscription matrix is associated with a set of binary bits. Given two subscriptions S_i and $S_{i'}$, we denote the associated bits to be B_i and $B_{i'}$, respectively. In case that the bits in B_i are a subset of those in $B_{i'}$, we then merge the rows of such subscriptions together into a single one by adding subscription S_i to the subscription list of $S_{i'}$.

Given an input matrix, we show how to build a subscription index by an iterative manner. First, for the matrix in the current iteration, we conduct the *subscription merging* operation on pairwise rows and form the new matrix in the next iteration. If the bits B_i of s_i are a subset of $B_{i'}$, then the two subscriptions S_i and $S_{i'}$ are merged together. We continue the merging process until no rows can be merged, and have the final subscription index.

By the merging operation over the 7 subscriptions in Table 1, we build a subscription index as shown in Table 3. The column partition for each attribute is not the same due to the different value ranges. In our design, there are 11 main columns for attribute A(details in Table 2), 5 main columns for attribute B and each stands for one enumerate value, 7 main columns for attribute C and value span is 10 for each column, as well as 8 columns for attribute D with value span of 1 and upper/lower bound signals. Therefore the start column ID for each attribute is 0,11,16 and 23 respectively. The start column ID stands for key in each key-value pair. As for each subscription, it will mark 0 or 1 in each column and get a compressed value for each attribute. Specifically, there has occurred sub columns division in column of ID 1 in attribute A as well, and it results in 2 key-value pairs for attribute A. The index format for s_5 is $< 23, 3 >$, and it can definitely put into the relating list of s_7 .

Table 3. Subscriptions Index

SIZE	SID	LIST	RELATE
3	S_1	$\langle 0, 512, 1, 1920 \rangle, \langle 11, 14 \rangle, \langle 16, 1 \rangle$	
	S_3	$\langle 0, 1023, 1, 1022 \rangle, \langle 11, 24 \rangle, \langle 23, 60 \rangle$	
	S_6	$\langle 0, 1536, 1, 1536 \rangle, \langle 11, 3 \rangle, \langle 16, 62 \rangle$	
2	S_2	$\langle 0, 512, 1, 448 \rangle, \langle 23, 120 \rangle$	
	S_4	$\langle 11, 4 \rangle, \langle 16, 96 \rangle$	
	S_7	$\langle 16, 24 \rangle, \langle 23, 3 \rangle$	S_5

5 Matching Algorithm

Based on the above coding scheme and subscription index, we highlight the proposed matching by using the very fast machine bitwise operation as follow. Consider subscription S_2 with the predicate $A \in [5, 9]$ and a publication with attribute value $A = 8$ and the base number $b = 8$. If we follow Table 2(2) to divide the matrix, the predicate $A \in [5, 9]$ is associated with the bits 0011100000. Since the point value 8 falls inside the interval $[7, 9)$, we then mark the 3-th column to 1 and all others 0, and have the binary bits 0001000000. Next, we conduct the machine bitwise operation AND: 0011100000 AND 0001000000 = 0001000000. We then determine that the publication successfully matches the predicate. Similarly, given another publication with attribute value $A = 3$ encoded by the bits 0100000000, we again conduct the bitwise operation 0011100000 AND 0100000000 $\neq$ 0100000000 and infer that the publication with $A = 3$ does not match the predicate.

Based on the above example, we find that the bitwise operation can help the matching algorithm. Specifically, for a specific attribute A, we denote $B_A(s)$ and $B_A(s)$ to be the bits encoded for the attribute value of a given publication e and predicate of a given subscription s, respectively. Then, it not hard to find that $B_A(e)$ AND $B_A(s) = B_A(e)$ is the necessary condition of the claim that s successfully matches e. As a result, if $B_A(e)$ AND $B_A(s) \neq B_A(e)$, we then claim that s must fail to match e.

Algorithm 1: Filtering Algorithm

   Input: Publication e and Subscription Index
   Output: Subscriptions matched with e
1 Initialize result set C_S for the publication e;
2 Encode e to a group of $\langle col_e, val_e \rangle$;
3 **for** *each row in subscription matrix* **do**
4   **for** *each pair $\langle col_{idx}, val_{idx} \rangle$ in the row* **do**
5   **if** $col_{idx} == col_e$ **then**
6   **if** $(val_{idx} \; AND \; val_e) == val_e$ **then**   move to the next pair ;
7   **else** Match fails, move to next row of index ;
8   **end**
9   **end**
10   **if** *all pairs in the row match successfully* **then**
11   add the subscription and its relating list of the row to C_S;
12   move to next row of index;
13   **end**
14 **end**
15 **for** *each candidate subscription in C_S* **do**
16   Do further exact matching process;
17 **end**

Alg. 1 gives the pseudocode of the proposed matching algorithm. In the initiated state, there is an empty container for each event that is prepared to store all the matched candidate subscription IDs(Line 1). Given a set of events $\mathcal{E}$ and for every event of the set, it first converts each event to the same format with subscription index, then comes the machine bitwise operation AND between each subscription index and the event(Line $3-9$). Only if all the pairs in the index matches then the index as well as its relating list will put into the candidate container.

Events in the stream will first transform to key-value pairs before bitwise operation. Given an event: $e_1 : A = 2 \land B = \{L\} \land C = 30 \land D = 10$. It is transformed to $e_1 :< 0, 512, 1, 1024 >, < 11, 2 >, < 16, 8 >, < 23, 1 >$ according to the column division in Table 3. Its start column ID is 0, 11 and 16. So bitwise operation is conducted on the three pair groups. The first pair group is $< 0, 512, 1, 1920 >$, and bitwise operation on $< 0, 512 >$ for start ID 0 is 512. We further check the pair of sub columns with start ID 1, the bitwise operation on $< 1, 1920 >$ and $< 1, 1024 >$ is 1024. The first pair group is matched in such case. The next pair group is matched and the third is not in this way. After the three checks, the index is not regarded as the candidate for e_1 in the end. At last, s_6 and s_7 are matched, and s_5 will also put into the candidate container for it is in the relating list of s_7.

6   Experiment

We evaluate the proposed compression-based matching scheme (COMPME for short) and compare it with the recent work OPIndex [2]. The implementation of OPIndex is kindly provided by the author of the paper. For fairness, we implement COMPME in C and conduct the experiment on ubuntu 14.04. Following OPIndex [2], we generate all subscriptions and publications by uniformaly distributed datasets as OPIndex does. In our experiments, we vary the number of subscriptions, attributes, maximal size of subscriptions(**MSS** for short) and events(**MES** for short) respectively, and compare COMPME with OPIndex. The experiments are conducted in high dimension scene(i.e. attribute number=20000)(**H** for short) as well as low dimension scene(i.e. attribute number is equal to **MES**)(**L** for short). In our setting, **MES** is always five times of **MSS**. Obviously, larger MSS and MES result in more memory cost, matching time and index building time. We conduct at least two cases of **MSS** and **MES** in each experiment to improve the experiment results.

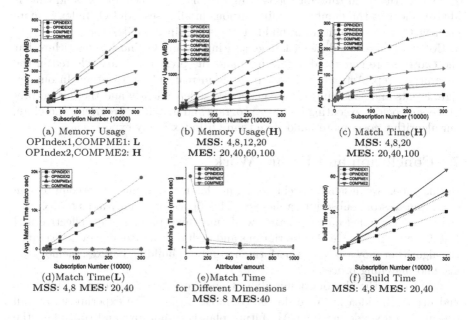

(a) Memory Usage
OPIndex1,COMPME1: **L**
OPIndex2,COMPME2: **H**

(b) Memory Usage(**H**)
MSS: 4,8,12,20
MES: 20,40,60,100

(c) Match Time(**H**)
MSS: 4,8,20
MES: 20,40,100

(d)Match Time(**L**)
MSS: 4,8 **MES**: 20,40

(e)Match Time
for Different Dimensions
MSS: 8 **MES**:40

(f) Build Time
MSS: 4,8 **MES**: 20,40

Fig. 2. Experiment Results.

① **Memory Usage**: Fig. 2(a) shows the comparison of the memory usage between COMPME and OPIndex by measuring the space cost of the source data of subscriptions and the index in high and low dimension scenes by setting the **MSS**(4) and **MES**(20). Fig. 2(b) is the experiment result of OPIndex,COMPME in high dimension scene to further approve the experiment result. Both figures show that the memory usage grows linearly with the number of subscriptions, and the memory usage of OPIndex is much larger than COMPME.

② **Matching Time**: Fig. 2(c) is the comparison of the two algorithms in high-dimensional space. From the figure we can see that average match time is almost linear to the number of subscriptions. OPIndex is more efficient with extremely small subscription size in very high dimension scene. Under the case of **MSS**(20) and **MES**(100), the time cost of OPIndex is much more than the other two cases while ours are more stable. We can conclude from the figure that as the subscription size becomes larger, the matching performance of OPIndex decreases sharply and COMPME becomes much more efficient than OPIndex.

Fig. 2(d) is the experiment result in low dimensional scene. As shown, the lines of COMPME almost overlap and its cost is much lower compared to OPIndex. The matching time in different dimensions varies a lot between the two algorithms. Fig. 2(e) directly approves that our algorithm is quite stable whenever in high and low dimensional space while OPIndex is much more efficient in high dimensional space.

③ **Index Building Time**: Building time in both algorithms includes reading subscription items from the external file and constructing relevant indexes. Fig. 2(f) shows the build time for the two algorithms. As shown, index build time is almost linear to the number of subscriptions in all cases. And OPIndex is more efficient than our algorithm in building the subscription indexes.

However, there is a drawback in generating data in OPIndex's algorithm that the two files are generated accordingly and the pivot nodes are selected in the generation step. Actually it is not possible in real pub/sub system. In our design, events and subscriptions are generated independently and there is no preprocessing before executing building program and matching program. Therefore, our algorithm costs more build time than OPIndex in my experiment.

7 Conclusion and Future Work

In this paper, we design a novel filtering mechanism that integrates compression operation to store subscription indexes. The algorithm mainly aims to decrease space cost for index data structure while matching process stays effective and fast. Given a large scale of subscriptions and high rate of incoming event streams, the algorithm shows excellent performance in both building indexes, memory usage and matching process.

We propose the idea of granularity to transform interval value to a point value and utilize the idea in [5] to do the compression job. The experiments are all executed in a single machine. My future plan is to improve and transplant the filtering mechanism to distributed system.

References

1. Eugster, P.T., Felber, P.A., Guerraoui, R., Kermarrec, A.-M.: The many faces of publish/subscribe. ACM Computing Surveys (CSUR) 78, 114–131 (2003)
2. Zhang, D., Chan, C.-Y., Tan, K.-L.: An Efficient Publish/Subscribe Index for E-Commerce Databases. Proceedings of the VLDB Endowment 7(8) (2014)

3. Campailla, A., Chaki, S., Clarke, E., Jha, S., Veith, H.: Efficient filtering in publish-subscribe systems using binary decision diagrams. In: Proceedings of the 23rd International Conference on Software Engineering, ICSE 2001, pp. 443–452. IEEE (2001)

4. Fabret, F., Jacobsen, H.A., Llirbat, F., Pereira, J., Ross, K.A., Shasha, D.: Filtering algorithms and implementation for very fast publish/subscribe systems. ACM SIGMOD Record 30, 115–126 (2001)

5. Rao, W., Chen, L., Hui, P., Tarkoma, S.: Bitlist: New full-text index for low space cost and efficient keyword search. Proceedings of the VLDB Endowment 6(13), 1522–1533 (2013)

6. Rao, W., Chen, C., Hui, P., Tarkoma, S.: Replication-based load balancing in distributed content-based publish/subscribe. In: 2013 IEEE 27th International Symposium on Parallel & Distributed Processing (IPDPS), pp. 765–774. IEEE (2013)

7. Whang, S.E., Garcia-Molina, H., Brower, C., Shanmugasundaram, J., Vassilvitskii, S., Vee, E., Yerneni, R.: Indexing boolean expressions. Proceedings of the VLDB Endowment 2(1), 37–48 (2009)

8. Sadoghi, M., Jacobsen, H.-A.: BE-Tree: An index structure to efficiently match Boolean expressions over high-dimensional discrete space. In: Proceedings of the 2011 ACM SIGMOD International Conference on Management of Data, pp. 637–648. ACM (2011)

9. Aguilera, M.K., Strom, R.E., Sturman, D.C., Astley, M., Chandra, T.D.: Matching events in a content-based subscription system. In: Proceedings of the Eighteenth Annual ACM Symposium on Principles of Distributed Computing, pp. 53–61. ACM (1999)

10. Fabret, F., Llirbat, F., Pereira, J., Shasha, D.: Efficient matching for content-based publish/subscribe systems. In: Scheuermann, P., Etzion, O. (eds.) CoopIS 2000. LNCS, vol. 1901, pp. 162–173. Springer, Heidelberg (2000)

Sentiment Classification for Chinese Product Reviews Based on Semantic Relevance of Phrase

Heng Chen, Hai Jin*, Pingpeng Yuan, Lei Zhu, and Hang Zhu

Services Computing Technology and System Laboratory
Cluster and Grid Computing Laboratory
School of computer Science and Technology
Huazhong University of Science and Technology
Wuhan, 430074, China
hjin@hust.edu.cn

Abstract. The emotional tendencies of product reviews on web have an important influence. Analysis of the sentiment of reviews on the Internet became very necessary. In this paper, a new sentiment analysis algorithm is utilized to analyze sentiment of Chinese product reviews. At training stage, a model based on skip-gram is proposed to train phrase vectors respectively on positive and negative reviews, which represent the semantic relationship of phrases. The predication of emotional tendencies of reviews based on the phrase vectors. The model does not need any modeling and feature extraction for the review data, thus it is applicable for massive data. Experimental results show that when dealing with massive data, the algorithm is better than traditional algorithms on both accuracy and learning time.

Keywords: Distributed Representation, Chinese Product Reviews, Emotional Tendency, Sentiment Classification.

1 Introduction

Shopping online has become increasingly popular, which severely impacts physical stores. User comments play an important role in the sale model of online stores. Since potential customers cannot see the actual product items, other users reviews about an item are an important third party opinion they can use for reference.

On *Taobao*[1] or *Jingdong Mall*[2], users can buy products that recommended by other consumers or from brands which have rave reviews. When they receive their products, they comment and rate the sellers based on their satisfaction with the purchases and on the seller's service.

Though there are some works that analyze blogs or forum articles to obtain the opinions [1,2], these works are based on the long text in blogs and forums which

---

* Corresponding author.
[1] http://www.taobao.com/
[2] http://www.jd.com/

© Springer International Publishing Switzerland 2015
R. Cheng et al. (Eds.): APWeb 2015, LNCS 9313, pp. 340–351, 2015.
DOI: 10.1007/978-3-319-25255-1_28

have more semantic information than user comments which are restricted to 140 words or only one sentence. The following features of comments differ them from blogs or forum articles: First, the language of comments is informal with short length. Second, the texts of comments are with the strong subjectivity. Third, there are so many emoticon such as smiling face and crying face etc. Fourth, the new expressions on web such as *xidapubeng, renjianbuchai* in Chinese and *gooooood, cooool* in English, are frequently used and quickly evolving.

There are a number of studies on sentiment analysis on short text. The approaches could fall into two categories. One is lexicon-based or knowledge-based [3,4,5], and the other is based on corpus-driven learning [6,7,8].

Lexicon-based approach predicts the emotional tendencies of sentence through a lexicon together with *TF/IDF*. Experiments show that this method has achieved a satisfying result. But there are still problems: First, different domains have different emotion express ways. Second, emotional expression in web environments is evolving, and a large number of new neologisms have sprung up every year.

For sentimental analysis of short text, corpus-driven learning is the predominant trend. The approach extracts a set of hand-designed features from the sentence, then feeds the features to a classification algorithm. Pang et al. [6] took the lead in introducing machine learning into the problem of sentiment classification. Three machine learning methods (Naive Bayes, Maximum Entropy, SVM) were employed on sentiment classification. The results showed that machine learning methods do not perform as well as on traditional topic-based categorization. A conclusion was made that sentiment classification problem is more challenging.

Based on conventional methods, [9] added the punctuation mark such as !, ?, :-), and [8] considers the external features such as star rating, emoticons and hashtags to assist the sentiment classification for Tweets in *Twitter*[3]. However, reviews of product rarely contain these features. [10] uses Recursive Neural Tensor Network to train the corpus. However, the method contains the following request: first, phrases as features need sentiment labels; second, the corpus training should with fully labeled parse trees.

In this paper, based on semantic relevance of phrases, we analyze the sentiment of Chinese reviews. The sentiment analysis for English product reviews has been widely studied in recent years, followed with many important achievements. Due to the special language traits of Chinese, the study on Chinese product reviews is much more difficult than the former.

We propose a novel approach to deal with sentiment classification. In preprocessing, we do as little as possible. In training process, a neural network architecture based on skip-gram is used to train a positive model and a negative model respectively. At this stage, the relevance of phrases is encoded in phrase vectors. The phrase vectors make up two phrase vector sets corresponding to the two models. The predication of sentences emotional tendencies is based on the phrase vectors.

---

[3] https://twitter.com

Compared with previous works, our contributions can be summarized as follows:

1. We introduce Skip-gram model to train Chinese phrases, which could obtain the semantic relevance of phrases. These phrases which frequently occur together could have much similar representation.

2. Comparing to existing sentiment classification methods, our method does not need extracting a set of complex hand-designed features from sentence.

3. Vectors of positive data and negative data are trained independently, so the model is convenient to parallel extension.

The rest of the paper is organized as follows: Section 2 gives related work. Section 3 presents how to construct the model, and the training of phrase vector. Section 4 presents our model of estimating the emotional polarity of review sentence. Section 5 reports our experiments and Section 6 concludes this paper.

2   Related Work

The model we present in this paper draws in spiration from prior work on both sentiment analysis and distributed representations.

Sentiment Analysis. [3] proposed a lexicon-based algorithm to determine the sentiment orientation of a sentence. It was based on a sentiment lexicon generated using a bootstrapping strategy with some given positive and negative sentiment word seeds and the synonyms and antonyms relations in WordNet. The sentimental orientation of a sentence was determined by summing up the orientation scores of all sentiment words in the sentence. A positive word was given the sentiment score of +1 and a negative word was given the sentiment score of -1. In [4], a similar approach was also used. Their method of compiling the sentiment lexicon was also similar. However, they determined the sentiment orientation of a sentence by multiplying the scores of the sentiment words in the sentence. Again, a positive word was given the sentiment score of +1 and a negative word was given the sentiment score of -1.

In [5], Nigam and Hurst applied a domain specific lexicon and a shallow NLP approach to assess the sentence sentiment orientation. In [7], a semi-supervised learning algorithm was used to learn from a small set of labeled sentences and a large set of unlabeled sentences. The learning algorithm was based on *Expectation Maximization* (EM) using the Naive Bayes as the base classifier. This work performed three-class classification, positive, negative, and other (no opinion or mixed opinion). In [8], sentiment classification of Twitter postings (or tweets) was studied. Each tweet is a single sentence. The authors took a supervised learning approach. Apart from the traditional features, the method also used hashtags, smileys, punctuations, and their frequent patterns. These features were shown to be quite effective.

Distributed Representations. Distributed representations for words, where words are represented as continuous vectors have a long history. It was first proposed in [11] and has become a successful paradigm, especially for statistical language modeling. A very popular model architecture for estimating *Neural Network Language Model* (NNLM) was proposed in [12], where a feed forward neural network with a linear projection layer and a non-linear hidden layer was used to learn jointly the word vector representation and a statistical language model.

Another interesting architecture of NNLM was presented in [13,14], where the word vectors are first learned using neural network with a single hidden layer. The word vectors are then used to train the NNLM. Thus, word vectors are learned even without constructing the full NNLM. In this work, we directly extend this architecture, and focus just on the first step where the phrases vectors are learned using a simple model. Representing Sentences and Document [15] are a recent trend and received much attention.

3   Construction of Training Model

Traditional text analysis considers the whole text as a vector, and treats words as dimensions. Besides the high feature dimensionality and feature sparsity, considering each word as a dimension has a drawback: the model cannot obtain the relevance of each word. The model treats terms as atomic units and there is no notion of semantic relevance between terms.

Distributed representations for terms use an elegant way to solve the problem of semantic between terms of text. For distributed representation, a term is treated as a vector. The dimensionality of a term usually set to 50 or 100, and the value of dimension is not necessary to be non-negative. Through appropriate training, the term vector can be improved, and more semantic related terms can have more similar vectors. So, using cosine function or Euclidean distance can quantitatively calculate the degree of the relevance. In this paper, we learn from distributed representation to represent phrases of Chinese product reviews.

Before the emotional tendency analysis, we give the process of training phrases vector. It is necessary to emphasize that our model trains phrase vectors for positive data and negative data respectively, in other words that there are two vector sets.

The model to construct should guarantee that a positive sentence has the maximum generating probability in positive model, and a negative sentence has the maximum generating probability in negative model. Expressed mathematically as: if S is a positive sentence, $p(S|model(P)) > p(S|model(N))$; else $p(S|model(P)) < p(S|model(N))$. Here $model(P)$ and $model(N)$ are positive and negative model respectively.

For any model, given a sentence with T phrases: $t_1, t_2, ..., t_T$, here $t_i \in V$, and V is the vocabulary. $p(S)$ can be defined as: $p(S) = p(t_1, t_2, ..., t_T)$. The equation above can be expanded as follow: $\prod_{i=1}^{T} p(t_1, ...t_{i-1}, t_{i+1}, ..., t_T|t_i) = \prod_{i=1}^{T} p(t_1, ...t_{i-1}|t_i) \cdot p(t_{i+1}, ..., t_T|t_i)$. By taking the logarithm and simplifying, we obtain the objective as follow: $\sum_{i=1}^{T} (log\ p(t_1, ...t_{i-1}|t_i) + log\ p(t_{i+1}, ..., t_T|t_i))$.

Since the long-distance words are usually less related to the current word than those closed to it, we only consider small number phrases before and after the center phrase. The number can set to $2c$, so the formula can be simplified to $\sum_{i=1}^{T}(log\ p(t_{i-c+1},...t_{i-1}|t_i) + log\ p(t_{i+1},...,t_{i+c-1}|t_i))$. Assuming the variables are independent and identically distribution (i.i.d.), $p(t_{i-c+1},...t_{i-1}|t_i) = p(t_{i-c+1}|t_i) \cdot ... \cdot p(t_{i-1}|t_i)$, and ignoring the coefficient, for farther ease of calculation, we obtain the following objective:

$$\sum_{i=1}^{T} \sum_{-c \leq j \leq c, j \neq 0} log\ p(t_{i+j}|t_i) \tag{1}$$

So, we should maximize the log probability (1) to approximately achieve our goal.

Skip-gram algorithm [13] is a preferred algorithm for us. The objective function of Skip-gram is to maximize Formula (2), which is mathematically equivalent to our objective function.

$$p(Context(t)|t) = \prod_{u \in Context(t)} p(u|t) \tag{2}$$

In our work, we refer to [13] to construct our training model and train the phrase vectors respectively. The module of training model is shown as left side of Fig. 1.

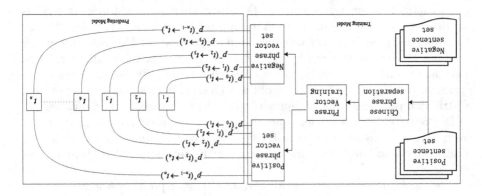

Fig. 1. The overall flow chart

4   Predicting the Emotional Polarity

4.1   Definition and Theorem

In order to better express our thoughts, we give some definitions and theorems as following.

Definition 1: Generating Probability. Define the probability that phrase B occurred after A in a model as the generating probability of B given condition A: $p(A \to B)$; define the generating probability of sentence s as p_s: $\prod_n p(t_{n-1} \to t_n)$.

Definition 2: Define the generating probability of B given condition A in positive model as $p^+(A \to B)$. Define the generating probability of B given condition A in negative model as $p^-(A \to B)$; define a sentence s generating probability in positive model p_s^+ as $\prod_n p^+(t_{n-1} \to t_n)$, define a sentence s generating probability in negative model p_s^- as $\prod_n p^-(t_{n-1} \to t_n)$.

Definition 3: Define the relevancy of phrase A and phrase B as the cosine of phrase vector C_A and phrase vector C_B in a model: $\cos\langle C_A, C_B\rangle$; define the relevancy of phrase A and phrase B in positive model as positive relevancy: $\cos^+\langle C_A, C_B\rangle$, and define the relevancy of phrase A and phrase B in negative model as negative relevancy: $\cos^-\langle C_A, C_B\rangle$.

Definition 4: Define $p^+(A \to B)$ as the positive relevancy of phrase A and B; define $p^-(A \to B)$ as the negative relevancy of phrase A and B.

Theorem 1: For a sentence in a training set, the generating probability is largest in the corresponding model.

Proving: As in previous section, training of phrase vectors in a corpus is to maximize the log probability $\sum_{i=1}^{T} \sum_{-c \leq j \leq c, j \neq 0} log\ p(t_{i+j}|t_i)$. So for the sentences in the training corpus, the phrase vector set of the corpus must be the optimal vectors for them. Two bordering phrases in a sentence can have much more similar vector representations in the corresponding model than in other models where the two phrases could not be adjacent in a sentences. Considering the definition of generating probability, the theorem is obvious.

We have also verified the conclusion in an experiment. The experimental data is a public sentimental dataset. The dataset will be introduced in Section 5. Here based on the positive data and negative data, we first train the positive phrase vector set and negative phrase vector set respectively, and then randomly select the training sentences to compute the generating probability based on the two vector sets (as Algorithm 1), finally choose the largest probability corresponding models emotional tendency as the sentence emotion. We calculate the accuracy of the classification. The result is given in Table 1. The result validates our theorem by experiment.

Table 1. Experimental verification of Theorem 1

Data set	precision
Dangdang_Book_4000	0.95
Jingdong_NB_4000	0.93
Ctrip_htl_ba_4000	0.89

Deduction 1: For a review sentence, the generating probability will get the maximum value on the phrase vector set which has the same emotional tendency of the review sentence.

Intuitively, for the same products, the positive reviews will trend to the similar expression, and the negative reviews have this feature too.

4.2   Construction of Prediction Model

Through the training of a mass of corpus based on training model we described previous section, the values of phrase vectors are changed iteratively, and these phrases frequently occur together can have the similar phrase vector. Finally, we have obtained two vector sets: the positive set and negative set respectively. We could use Definition 3 and 4 to quantitatively calculate the degree of relevance between two phrases. So the relevance of phrases is recording in the phrase vectors, and we could use it to predicate the emotional polarity of other sentences as Deduction 1. Based on the vector sets, we use a generative model to predicate the emotional polarity of a sentence. First, we generate the sentence based on the two sets, and then choose the emotion of the set which has the higher probability as the sentences sentiment orientation.

Based on Deduction 1, we give the algorithm of predicting the emotional tendency for a review sentence in Algorithm 1.

Algorithm 1. Predicting the emotion tendency of a review sentence

---

Begin:
>   **Input:** phrases vector model C^+ and C^- , review sentence $s = t_1, t_2, ..., t_n$
>   **Output:** the emotional tendency of s
>   **Initialize:** generating probability of s as $p_s^+ = 0$ and $p_s^- = 0$
>   if$(n = 1)$
>> Classify sentence s thought lexicon;
>> **return**;
>   **else**
>> Set loop variable $i = 2$;
>> **while**$(i \leq n)$
>> Computing generating probability of phrase t_i of sentence s as:
>>> $p^+(t_{i-1} \rightarrow t_i)$ and $p^-(t_{i-1} \rightarrow t_i)$;
>>> Update p_s^+ as: $p_s^+ = p_s^+ + p^+(t_{i-1} \rightarrow t_i)$, and p_s^- as $p_s^- = p_s^- + p^-(t_{i-1} \rightarrow t_i)$;

>>> $i = i + 1$;
>   if$(p_s^+ < p_s^-)$
>> Classify s as negative sentence;
>   **else**
>> Classify s as positive sentence.
End

For Algorithm 1, when a sentence only has one word, there is no generating probability for the sentence. However, this type of sentence is much easier to analyze, we only should resort to a lexicon. The module of prediction model is shown in the right side of Fig. 1.

5 Experiments

5.1 Experimental Dataset

In order to verify the reliability of our model, we perform experiments on real data which is product reviews crawled on *Jingdong mall*. These data contain electric appliances (*phone, notebook, household appliance*) and clothing (*clothes, shoes, accessories*). The dataset is crawled between March 1, 2014 and March 25, 2014. The reason to use the reviews of *Jingdong mall* is that *Jingdong* allows users to give star-rating when commenting the products. Specifically, when users comment about their feelings of shopping experience about the products, they need to choose 1-5 stars to express the degree of satisfaction. The more stars users give, the more satisfaction they have about the products, meanwhile the much more positive trend reviews they give.

So when crawling the reviews of products, we also record the corresponding stars reviews have. We put the reviews that have 4 or 5 stars in the positive data set, and the reviews that have only 1 or 2 stars in the negative data set. Through some preprocessing, we obtain our experimental data of *Jingdong mall*. The detail information is summarized in Table 2. For most customers, shopping is a pleasant experience. In *Jingdong mall*, positive reviews outnumbers negative reviews 22: 1.

Table 2. Summary of the data set

Data set	Total items	Positive/negative
Clothing	1,518,915	1,441,551/77,364
Electric_Appliances	1,836,520	1,768,171/68,349
Dangdang_Book_4000	4,000	2,000/2,000
Jingdong_NB_4000	4,000	2,000/2,000
Ctrip_htl_ba_4000	3,923	1,972/1,951

A publicly available sentiment dataset is also employed in our study. The dataset is provided by researcher Songbo Tan[4] in Institute of Computing Technology of the Chinese Academy of Sciences. It contains three small sample data sets: Book reviews of *Dangdang book mall*, notebook reviews of *Jingdong mall* and hotel review data of *Ctrip*. Each data set has approximately 2000 sentences for positive and negative respectively. The summary information is also listed in Table 2.

---

[4] http://sourcedb.cas.cn/sourcedb_ict_cas/en/eictexpert/fas/200909/
t20090917_2496724.html

5.2   Experimental Setting and Results

For each data set of *Jingdong mall*, we randomly choose 9/10 of sentences to construct the training data set. The rest sentences are used for testing.

Using the positive and negative training data sets, we have trained two phrase vector sets. Based on the phrase vectors sets, we predict the emotional tendency of sentences at the rest 13135 positive reviews and 809 negative reviews, the results are shown in Table 3 before slash (/).

Table 3. The results of test before and after given threshold

Data category	Total items	correct items	precision
Positive sentences	17,919	13,135/14,209	0.733/0.793
Negative sentences	943	809/829	0.857/0.879

Through analyzing the false judgment results, we find that these misclassified reviews mostly are long sentences, which have too much weak semantic relevancies between phrases. A single phrase's generating probability is small, but by incremental addition, they could exceed phrases which have large generating probability. The accumulation of weak generating probability could effect the judgment.

Fig. 2 is an example, which means *"The phone seems okay, though the color is wrong, and is enough for older parents"* in English. The review is a positive review, but the sentiment is predicted as negative tendency.

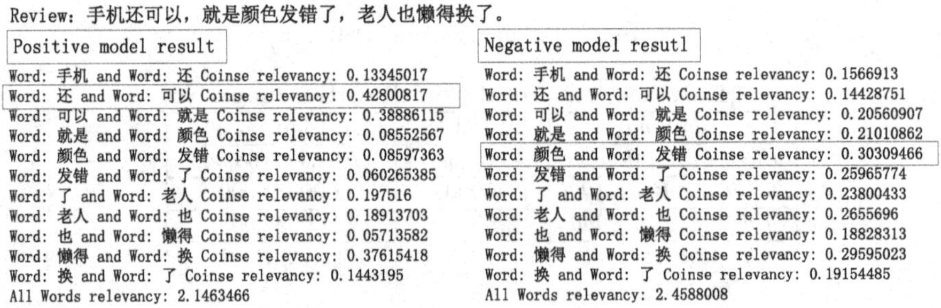

Review: 手机还可以，就是颜色发错了，老人也懒得换了。

Positive model result	Negative model resutl
Word: 手机 and Word: 还 Coinse relevancy: 0.13345017	Word: 手机 and Word: 还 Coinse relevancy: 0.1566913
Word: 还 and Word: 可以 Coinse relevancy: 0.42800817	Word: 还 and Word: 可以 Coinse relevancy: 0.14428751
Word: 可以 and Word: 就是 Coinse relevancy: 0.38886115	Word: 可以 and Word: 就是 Coinse relevancy: 0.20560907
Word: 就是 and Word: 颜色 Coinse relevancy: 0.08552567	Word: 就是 and Word: 颜色 Coinse relevancy: 0.21010862
Word: 颜色 and Word: 发错 Coinse relevancy: 0.08597363	Word: 颜色 and Word: 发错 Coinse relevancy: 0.30309466
Word: 发错 and Word: 了 Coinse relevancy: 0.060265385	Word: 发错 and Word: 了 Coinse relevancy: 0.25965774
Word: 了 and Word: 老人 Coinse relevancy: 0.197516	Word: 了 and Word: 老人 Coinse relevancy: 0.23800433
Word: 老人 and Word: 也 Coinse relevancy: 0.18913703	Word: 老人 and Word: 也 Coinse relevancy: 0.2655696
Word: 也 and Word: 懒得 Coinse relevancy: 0.05713582	Word: 也 and Word: 懒得 Coinse relevancy: 0.18828313
Word: 懒得 and Word: 换 Coinse relevancy: 0.37615418	Word: 懒得 and Word: 换 Coinse relevancy: 0.29595023
Word: 换 and Word: 了 Coinse relevancy: 0.1443195	Word: 换 and Word: 了 Coinse relevancy: 0.19154485
All Words relevancy: 2.1463466	All Words relevancy: 2.4588008

Fig. 2. The detail of a misclassified review(the left side is the detail of good review model, the right side is the detail of bad review model)

In order to further optimize the experiment and reduce the classification error, we add a threshold λ on the generating probability of phrases to filter out the weak semantic relevance, and enhance the strong semantic relevance. In our experiment, we set λ to 0.25. The optimized results are shown in Table 3, after slash. Comparing with the result before and after optimization, we can find that only add a threshold, there can be 6 percentage and 2.2 percentage improvement for predicting.

We compare our model to SVM model and Naive Bayes model. For binary classification, SVM is well known as one of the most effective methods. In spam filtering, Naive Bayes model has been successfully used. Identifying negative reviews from positive reviews can be regard as a analogy with identifying spam from normal mail, since positive reviews outnumbers negative reviews 22: 1. Considering the n-gram model is widely used in text categorization, we compare to SVM and Naive Bayes model with bi-gram features too.

The experiment is conducted on the data of electrical appliance, which includes *air-conditioning, washer, flat screen TV* reviews. These data contain 16358 positive reviews and 61661 negative reviews. The experimental results are shown in Table 4. From Table 4, we can see that, both on the precision and training time, our model is superior to the other two models. When using the all *Jingdong mall* reviews data, the results of both SVM and Naive Bayes tend to be random. What is more serious is that the training time of SVM is unusually long, hardly compared to our model.

Table 4. Comparison experiments on electrical data set

model	Training time(ms)	Test time(ms)	precision
SVM model	97,406,649	1,509,109	0.5443
Bi-gram SVM Model	123,467,453	1,794,538	0.5872
Naive Bayes Model	812,198	5,421	0.6151
Bi-gram Naive Bayes Model	1,292,322	7,482	0.6353
Our Model	2,261,238	2,512	0.7591

SVM is effective on small sample text classification. However, when the data volume is large, especially when the data do not belong to a particular domain (here the data contain *LCD TV, washer*, etc.), its emotion classification effect will reduce sharply. For Naive Bayes model, unlike its performance in spam filtering, where the more training data, the better classification effect, the accuracy here also is not high as we expected. Our model's accuracy is more than ten percentage points higher than Naive Bayes model and SVM model. The experiments illustrate that the emotion classification is different from the theme classification, more sophisticated methods are needed. The proposed model of this paper, both on the training time and accuracy has better results.

In order to more in-depth analyze the performance of our model with a variable sample space, we experiment on data from *Jingdong mall* with a number of reviews 8000, 16000, 64000, 120000, 250000, 500000, 1000000 and 2000000. Here we only display the compared results of basic SVM and Naive Bayes, since the bi-gram versions of them have similar training time and precision tendency. The results are shown in Fig. 3.

The left-hand of Fig. 3 shows the training time of the three models. The training time of our model at very low slope linear growth with the training data, though the SVM model is almost linear growth, the curve is sharply steep. Naive Bayes model has the lowest slope, but as shown in right-hand of Fig. 3, it has the least accurate. The right-hand of Fig. 3 shows how accuracy changes

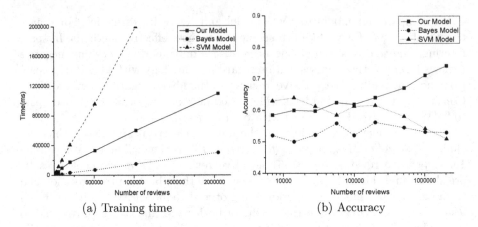

(a) Training time   (b) Accuracy

Fig. 3. The result of comparing experiments on different dataset size

with data scale changes for the three models. The predict accuracy of our model improves continually when data scale grows. However, the other two models have no obvious improvement with data increase, the SVM model even decreases.

Finally, we experiment the predict accuracy and time consuming of our model and the comparing models on the three publicly available sentiment data sets. The results are shown in Fig. 4.

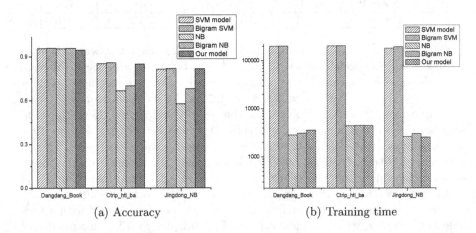

(a) Accuracy   (b) Training time

Fig. 4. The result of comparing experiments on public corpus

As shown in left hand of Fig. 4, since the data scale is small and neat, SVM model is effective. This demonstrates the ability of our model to deal with small-sample learning problems. From the figure, we can also see that our model has achieved as high accuracy as SVM model. The right hand of Fig. 4 shows that, the training and testing time of our model are much less than that of the SVM model, in an order of magnitude less time than it. That demonstrates that our model occupies preferable predication performance on small sample data too.

6   Conclusion

Based on the hypothesis that sentences with same emotional tendency often have similar language expression and collocation in Chinese, we propose a generating probability model for sentiment classification for Chinese product reviews.

In training step, two vector sets of phrases are trained respectively by the proposed model with positive and negative reviews corpus. The sentiment of a new review could be predicted by the probability of sentence generating from the two sets. We choose the emotion of set which has the highest generating probability as the new review emotional tendency.

Acknowledgments. We would like to thank all reviewers for their valuable suggestions. The research is supported by 863 Program (No.2012AA011003).

References

1. Huang, X.-J., Croft, W.-B.: A unified relevance model for opinion retrieval. In: Proceedings of CIKM, pp. 947–956 (2009)
2. Gottipati, S., Jiang, J.: Finding thoughtful comments from social media. In: Proceedings of ACL, pp. 995–1010 (2012)
3. Hu, M.-Q., Liu, B.: Mining and summarizing customer reviews. In: Proceedings of SIGKDD, pp. 168–177 (2004)
4. Kim, S.-M., Eduard, H.: Determining the sentiment of opinions. In: Proceedings of COLING, pp. 1–8 (2004)
5. Nigam, K., Matthew, H.: Towards a robust metric of opinion. In: Proceedings of AAAI, pp. 265–279 (2004)
6. Pang, B., Lee, L., Vaithyanathan, S.: Thumbs up?: sentiment classification using machine learning techniques. In: Proceedings of ACL, pp. 79–86 (2002)
7. Gamon, M., Aue, A., Corston-Oliver, S., Ringger, E.: Pulse: mining customer opinions from free text. In: Famili, A.F., Kok, J.N., Peña, J.M., Siebes, A., Feelders, A. (eds.) IDA 2005. LNCS, vol. 3646, pp. 121–132. Springer, Heidelberg (2005)
8. Davidov, D., Tsur, O., Rappoport, A.: Enhanced sentiment learning using twitter hashtags and smileys. In: Proceedings of the 23rd International Conference on Computational Linguistics: Posters, pp. 241–249 (2010)
9. Maas, A.-L., Daly, R.-E., Pham, P.-T., Huang, D., Ng, A.-Y.: Learning word vectors for sentiment analysis. In: Proceedings of ACL, pp. 142–150 (2011)
10. Socher, R., Perelygin, A., Wu, J.-Y., Chuang, J., Manning, C.-D., Ng, A.-Y.: Recursive deep models for semantic compositionality over a sentiment treebank. In: Proceedings of EMNLP, pp. 1631–1642 (2013)
11. Rumelhart, D.-E., Hinton, G.-E., Williams, R.-J.: Learning Representations by Back-Propagating Errors. Nature 323(6088), 533–536 (1986)
12. Bengio, Y., Schwenk, H., Senecal, J.-S., Morin, F., Gauvain, J.-L.: Neural Probabilistic Language Models. Innovations in Machine Learning 194, 137–186 (2006)
13. Mikolov, T., Chen, K., Corrado, G., Dean, J.: Efficient Estimation of Word Representations in Vector Space. CoRR abs/1301.3781 (2013)
14. Jurgens, D.-A., Mohammad, S.-M., Turney, P.-D., Holyoak, K.-J.: Semeval-2012 task2: measuring degrees of relational similarity. In: Proceedings of SemEval, pp. 356–364 (2012)
15. Quoc, V.-L., Tomas, M.: Distributed Representations of Sentences and Documents. CoRR abs/1405.4053 (2014)

Overlapping Schema Summarization Based on Multi-label Propagation

Man Yu[1], Chao Wang[1], Xiangrui Cai[1], Ying Zhang[1,2,*],
Yanlong Wen[1,2], and Xiaojie Yuan[1,2]

[1] College of Computer and Control Engineering, Nankai University, China
[2] College of Software, Nankai University, China
{yuman,wangchao2010,caixiangrui,zhangying,
wenyanlong,yuanxiaojie}@dbis.nankai.edu.cn

Abstract. Modern databases are usually composed of hundreds of tables. Querying an unfamiliar database is a tall order for users before they truly understand its schema. A schema summary can help to provide a succinct overview of the schema and improve the usability of databases. Existing summarization methods only focus on each element in a database belongs to one topic, ignores the fact that some elements may belong to multiple topics. This paper come up with a new method of generating overlapping summaries. It is the very first work to address the task as far as we know. We formulate overlapping schema summarization first and then introduce multi-label propagation algorithm in community detection to achieve several groups. To refine the partition, we cluster the groups additionally using hierarchical clustering algorithm. Finally, we find representative tables in each cluster to annotate the schema summary. The extensive experiments on both benchmark database and real-world database show that our approach not only achieves higher accuracy but also generates more meaningful summary.

Keywords: Multi-label schema graph, overlapping schema summarization, multi-label propagation.

1 Introduction

Real enterprise databases are usually large and complicated, which makes querying unfamiliar databases a really hard task for users. Moreover, the prevalent lack of documentation incurs significant cost in both understanding and integrating the databases. Schema summarization is a promising method to address this challenge by providing users an overview of the database relationships and schema constitution. It summarizes the schema to several clustered categories, each of which is represented to users by a topic table, making it possible to explore in depth only the relevant schema components.

Since Yu and Jagadish [13] proposed the first algorithm for automatically summarizing database schema, researches on schema summarization are conducted mainly on XML database [13] and relational database [11] [12]. This

---

* Corresponding author.

© Springer International Publishing Switzerland 2015
R. Cheng et al. (Eds.): APWeb 2015, LNCS 9313, pp. 352–364, 2015.
DOI: 10.1007/978-3-319-25255-1_29

paper focuses on the later one. Summarizing a relational database consists of two main steps. In the first step, similar tables are grouped into a category. In the second step, each category will pick up a topic table to represent it. [12] applies a weighted k-center cluster algorithm to partition the schema into specified k categories and picks up representatives by a model based on table content and random walk. [11] notices that community detection is very similar to schema summarization, so it adopts a modularity-based community detection algorithm to detect subject groups in a database and selects a representative for each category with the same method of [12].

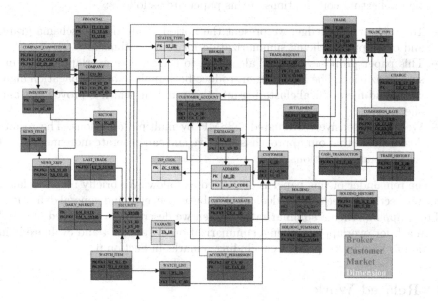

Fig. 1. TPC-E schema graph

Existing schema summarization methods just cluster a database element to one category. However, in reality, an element of a database may belong to multiple topics. Take TPC-E [7] as an example. As shown in figure 1, TPC-E models the activity of brokerage firm that executes customer trade orders and is responsible for the interactions of customers with financial markets. The database tables are pre-grouped into four categories: Customer, Broker, Market and Dimension. Table EXCHANGE records the financial exchange information of brokers, which bridges Market and Broker. However, previous work only assigns it into Market and thus results in category Broker lacking of important exchange information. Moreover, the result will confuse users by losing the connection between Market and Broker. In an enterprise-scale database, scenarios like this will be very common. A non-overlapping schema summary is not enough for users to understand a database structure. Detecting overlapping schema summarization will be necessary and helpful.

As far as we know, this paper presents the notion of overlapping schema summarization for the first time. An overlapping schema summary allows database elements belonging to multiple categories. This paper proposes an algorithm OS-SMP(Overlapping Schema Summarization based on Multi-label Propagation) to generate overlapping schema summaries. OSSMP first constructs an overlapping schema graph for a given database. And then it adopts a multi-label propagation algorithm [3] to detect overlapping groups in the graph. In addition, it utilizes hierarchical clustering to refine the groups. Similar to previous work, we choose representatives by computing importance of each table [12]. Our extensive experiments demonstrate the effectiveness of our algorithm and show that this method is really helpful for users to understand complex databases.

The intellectual contributions of this paper are as follows:

- To our best knowledge, we present the notion of multi-label schema graph and overlapping schema summarization for the first time.
- This paper proposes OSSMP algorithm to generate an overlapping schema summarization. The algorithm introduce both multi-label propagation algorithm and hierarchical clustering into this problem, which is proved fast and effective.
- We have extensively evaluated OSSMP by multiple criterions. The results demonstrate that our approach is effective and can generate more meaningful schema summarizations.

The remainder of this paper is organized as follow. We briefly review related work in section 2. Section 3 is dedicated to basic concepts and general definitions of overlapping schema summarization. Then we describe our proposed OSSMP approach for overlapping schema summarization in section 4 and evaluate it in Section 5. Finally, we present concluding remarks in Section 6.

2   Related Work

There are mainly two types of schema summarization researches according to different applicable database types. The first one focused on XML database like [13]. It proposed the notion of schema summary for the first time and developed an algorithm for generating the summary by balancing two properties, importance and coverage. Other researches all focused on the relational databases. [9] described a system iDisc to generate topical structures of databases. It partitioned the schema graph into different topic clusters via meta-clustering and took the most important tables as the representatives of topic clusters. [12] regarded the relational database schema as an undirected graph. It picked up the most important tables as the center of each cluster, and then utilized Weighted k-center algorithm to generate schema summarization. Weighted k-center is an approximation algorithm for the K-means problem which is NP-hard. When it comes to large database, the high complexity is unbearable. [8] and [10] applied community detection methods to cluster a schema into subject groups and used hierarchical clustering algorithm additionally to merge the groups. The algorithms depicted in [8] and [10] were also time-consuming as the rapid increase of database scale.

However, none of the algorithms above took overlapping schema summarization into account. To the best of our knowledge, there were no researches on overlapping schema summarization so far. Fortunately, this problem is similar to overlapping communities in graph which has much talented work these years. One kind of overlapping community detection algorithms were based on local expansion and optimization, such as [4] [5]. They expanded the community on the foundation of a local community and had a local optimization function to assess the current corporatist degree of the local community. Another kind of algorithm was to cluster links, such as [1]. It used non-overlapping community detection algorithm to detect link communities over a line graph. Since vertices can belong to multiple links, it could generate overlapping communities. The disadvantage of the above two kinds of algorithm were their high time complexity. Multi-label propagation algorithm [3] [10] was one of the fastest algorithms to detect overlapping communities. It allowed each vertex to belong to θ communities at most and can process very large networks in a short time.

3   Preliminaries

In this section, we first define of multi-labeled schema graph. Then we formulate the problem of overlapping schema summarization.

3.1   Multi-labeled Schema Graph

Given a relational database, a labeled schema graph G_L is usually defined as a triple (V, E, L), where each vertex $v \in V$ corresponds to a table and each edge $e \in E$ to a foreign key in the database, L is a labeling function that assigns a label to each table. $L(v) = l$ denotes that the label of table v is l. A multi-labeled graph G_{ML} extends the function L to multiple labels, as definition 1 interprets.

Definition 1 (Multi-labeled Schema Graph). *A schema graph in this paper is defined to be a multi-labeled weighted graph, $G_{ML} = (V, E, L_M)$, where:*

- *Each node $v \in V$ corresponds to a table in the given database.*
- *Each edge $e \in E$ is a foreign key that connects two tables. The weight of e expresses the connection strength of the two tables.*
- *L_M is a labeling function that assigns a single label or multiple labels to each table $v \in V$. In this work, a label is denoted as (c, b), where c is a category identifier and b is a belonging coefficient, indicating the membership strength of a table v and a category c.*

A table v with m labels is denoted as $L_M(v) = \{(c_1, b_1), (c_2, b_2), \ldots, (c_m, b_m)\}$. The weight function for edges is very important to generate a good schema graph, which will be discussed below.

3.2 Weight Function

A weight to each edge in the schema graph is proportional to the similarity between its two tables. Three features have been proposed to measure the similarity.

Name-Based Feature, $Sim_n(v_i, v_j)$. If two tables are similar, their table names and attribute names may be similar. We convert the similarity between two tables to the similarity between their name texts, which can be calculated by Vector Space Model.

Value-Based Feature, $Sim_v(v_i, v_j)$. If two tables are similar, their attributes may contain several similar values. Jaccard coefficient $J(A, B) = \frac{|A \cap B|}{|A \cup B|}$ is usually taken to calculate value similarity between two attributes A, B. The value similarity of two tables can be computed as the average of their attribute similarities.

Mapping-Based Feature, $Sim_m(v_i, v_j)$. If two tables are similar, there may be many mapping tuples between them. This feature reflects the connection strength of two tables' tuples and is calculated as

$$Sim_m(v_i, v_j) = \frac{q_i}{\sum fan(\tau_i)} \times \frac{q_j}{\sum fan(\tau_j)} \tag{1}$$

where τ are the tuples of table v, $fan(\tau_i)$ is the number of edges incident to tuple τ , q counts the number of tuples satisfying $fan(\tau_j) > 0$.

In this paper, we adopt a similarity kernel function introduced in [14]. It combines all of the three features to measure the similarity. The similarity of table v_i and table v_j is defined as

$$Sim(v_i, v_j) = \exp(-\frac{dist(v_i, v_j)^2}{2\sigma^2}) \tag{2}$$

σ is a parameter which can be studied by average label entropy method and $dist(v_i, v_j)$ denotes the distance between v_i and v_j, which is defined as

$$dist(v_i, v_j) = b - \beta_1 * Sim_n(v_i, v_j) + \beta_2 * Sim_v(v_i, v_j) + \beta_3 * Sim_m(v_i, v_j) \tag{3}$$

where the four parameters are estimated by machine learning method.

3.3 Problem Definition

To initialize, each table in the multi-labeled schema graph has a unique category identifier. Overlapping schema summarization is to reduce the category identifiers to user-specified k and to obtain a mapping that assigns each table to its categories. Each category is represented by a topic table t. Based on the this intuition, we formulate overlapping schema summarization as below.

Definition 2 (Overlapping Schema Summarization). *Given a multi-labeled schema graph* $G_{ML} = (V, E, L_M)$ *and a parameter* k, *an overlapping schema summarization of* G_{ML} *with size* k *is a* k-partition $C = \{C_1, C_2, ..., C_k\}$ *over the database schema, where:*

- *Each category $C_i \in C$ has a topical table, defined as $t(C_i)$.*
- *Each table v in category C_i has a label with category identifer c_i: $\forall v \in C_i$, $\exists (c_i, b_{vc_i}) \in L_M(v)$.*

4   The OSSMP Approach

Figure 2 shows the work flow of our OSSMP approach. It takes a multi-labeled schema graph as input and returns an overlapping schema summary. It is made up of three major steps: In the first step, we cluster the database tables into several groups by multi-labeled propagation algorithm. Secondly, if the groups size is larger than the specified summary size k, we use hierarchical clustering algorithm to generate a more reasonable summarization. Finally we pick up the most important table for each category as the topic tables. The details of each step will be discussed next.

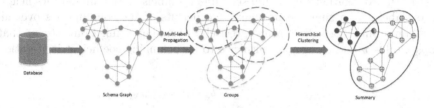

Fig. 2. Overview of OSSMP

4.1   Overlapping Groups Detection

This step aims at clustering the multi-labeled schema graph with n tables into m groups by a multi-label propagation algorithm.

Initialization. Multi-label propagation algorithm starts with a weighted multi-labeled schema graph $G_{ML}(V, E, L_M)$ defined in section 3 and a parameter θ. θ is the maximum number of categories to which any table can belong. To initialize the multi-labeled schema graph, every table in V is given a unique label pair(c, b), where the category identifier c is its table name, and the belonging coefficient b is set to 1. Therefore, there are n category identifers when initialization, where n is the number of tables in the schema graph.

Since θ influences the result of multi-label propagation directly, it also influences the final summary greatly. To set a reasonable θ, we range θ from $k-1$ to $k+3$ and choose the θ which generates the best cluster.

It assumes that a good cluster should achieve the goal of attaining high intra-category similarity. We utilize the intra-similarity of a cluster to evaluate its property.

Definition 3 (Cluster Intra-Similarity). *Let $C = \{C_1, C_2, ..., C_m\}$ be the cluster generated by multi-labeled propagation. The intra-similarity of C is denoted as $Sim(C)$:*

$$Sim(C) = \frac{1}{m} \sum_{C_i \in C} \frac{\sum\limits_{v_i, v_j \in C_i} Sim(v_i, v_j)}{C^2_{|C_i|}} \qquad (4)$$

where $Sim(v_i, v_j)$ is the similarity between v_i and v_j introduced in section 3.2, $|C_i|$ is the number of tables in C_i. Especially, if there is no edge between two tables, their similarity is 0.

The experiment result shows that, for $k = 2, 3, 4, 5$, the clusters have the highest intra-similarities when $\theta = 5$. In the following experiments, we set $\theta = 5$ in the multi-label propagation.

Propagation. The procedure of Multi-label propagation can be divided into four phases.

1) Each propagation step would set table v's labels to the union of its neighbours' labels, then sum the belonging coefficients of the communities over all neighbours, and normalize. More precisely, assuming a function $b_x(c, v)$ that maps a table v and category identifier c to its belonging coefficient in iteration x,

$$b_x(c, v_i) = \frac{\sum\limits_{v_j \in N(v_i)} b_{x-1}(c, v_j) w_{v_i v_j}}{|N(v_i)|} \qquad (5)$$

where $N(v_i)$ denotes the neighbors of v_i, and $w_{v_i v_j}$ is the weight of the edge between v_i and v_j.

2) To avoid every table owning all labels at the end, in each step, we delete the labels whose belonging coefficient is less than $1/\theta$. Category identifiers are reducing at the same time. If the belonging coefficients of all labels are less than $1/\theta$, we only retain the label with the greatest belonging coefficient.

3) After several iterations, the propagation will stop when the minimum number of tables labeled by one identifier is not changing. The concrete proof can be found in [3].

4) Let m be the number of category identifiers remaining after iterations. Tables whose labels contain identifier c_m is allocated to group C_m. Now the schema is partitioned into m groups, $C = \{C_1, C_2, ..., C_m\}$.

4.2   Group Refinement

Enterprise databases always have hundreds of tables. There may still be many groups after clustering by multi-labeled propagation. In order to generate more concise summary, we utilize an agglomerative hierarchical clustering algorithm to cluster the groups again when the group size is larger than user-specified k.

Group Similarity. Intuitively, two table groups are similar if their tables are similar. We formally define the similarity of every two groups as follow.

Definition 4 (Group Similarity). *Let C_i and C_j be two groups generated by multi-labeled propagation. The similarity of C_i and C_j is:*

$$Sim(C_i, C_j) = \frac{\sum\limits_{v_i \in C_i} \sum\limits_{v_j \in C_j} Sim(v_i, v_j)}{|C_i| |C_j|} \tag{6}$$

Hierarchical Clustering Algorithm. With similarities between any two groups, we adopt hierarchical clustering algorithm to generate abstract categories. Hierarchical clustering is one of the most common clustering algorithms. It starts with m groups generated by the multi-label propagation and puts out k categories. The algorithm first treats each group as a single category. In each iteration step, it incorporates the two categories with highest similarity. The iteration continues until the summary size k is reached. The detailed pseudo-code is shown in Algorithm 1.

Algorithm 1. Hierarchical Clustering Algorithm

Input: $C = \{C_1, C_2, ..., C_m\}$
Output: $C = \{C_1, C_2, ..., C_k\}$
1: **for** $i = m$ to k **do**
2:    find C_p, $C_q \in C$ with the maximum similarity;
3:    merge cluster C_p into C_q;
4:    delete C_p from C;
5:    **for** each C_j in C **do**
6:       compute distance between C_j and C_q;
7:    **end for**
8: **end for**
9: **return** (C);

4.3   Topic Table Selection

After getting categories of the summary, we choose the most important table in each category to be the topic table. Topic tables help users understand the subject of each category easily. We implement the importance function introduced in [12] to calculate the importance of each table. The importance of a table v is written as I_v. Each table is first given an information content to be the initial importance as

$$IC(v) = \log |v| + \sum_{v.A} H(v.A) \tag{7}$$

where $IC(v)$ is the initial information content of table v, $|v|$ is the number of tuples in v, $v.A$ ranges over all attributes in table v. $H(v.A)$ presents the entropy of the attribute A, which is defined as

$$H(v.A) = \sum_{i=1}^{m} p_i \log(1/p_i) \tag{8}$$

where m is the number of different values of attribute $v.A$. Let $v.A = \{a_1, \cdots, a_m\}$ and p_i be the fraction of tuples in v that have value a_i on attribute A.

Let n be the table number on the schema graph G. An $n \times n$ probability matrix $\prod$ reflects the information transfer between tables. Let v_i, v_j present two tables, $v_i.A - v_j.B$ be the join edge between A and B. q_A denotes the total number of join edges involving attribute $v_i.A$.

$$\prod [v_i, v_j] = \sum_{v_i.A - v_j.B} \frac{H(v_i.A)}{\log |v_i| + \sum_{v_i.A'} q_{A'} \cdot H(v_i.A')} \tag{9}$$

The importance vector $\mathbf{I}$ denotes the stationary distribution of the random walk defined by the probability matrix $\prod$. It can be computed by the iterative approach until the stationary distribution is reached.

5   Experiment

In this section, we firstly introduce the data sets and the evaluation indicators. Then we conduct experiments to evaluate the effectiveness of our proposed method. All the programs are implemented in C# using Microsoft Visual Studio.net 2010. And all experiments are carried out on a PC with Intel core i5 CPU 3.10GHz, 4GB memory and Microsoft Windows 7.

5.1   Data Sets

We evaluate our method on two databases: a benchmark single-label database (TPC-E) and a real-world multi-label database(CCEMIS [6]). TPC-E is a benchmark synthetic database which is adopted to simulate the On-Line Transaction Processing workload of a brokerage firm. And CCEMIS is a database of information management system in Nankai University. The characteristics of them are shown in table 1, where $\mathbf{T}$ is the number of non-empty tables, $\mathbf{Avg(C)}$ and $\mathbf{Max(C)}$ are the average and maximum number of columns per table, $\mathbf{Avg(R)}$ and $\mathbf{Max(R)}$ are the average and maximum number of rows per table, $\mathbf{C}$ is the number of foreign keys.

Table 1. Characteristics of data sets

	T	Avg(C)	Max(C)	Avg(R)	Max(R)	C
TPC-E	32	6	24	171127	4469625	45
CCEMIS	39	9	59	120744	4062145	73

Tables of TPC-E are pre-grouped into four categories: Customer, Broker, Market and Dimension. CCEMIS is also pre-grouped into four categories: Student, Teacher, Course and Lesson. The pre-defined categories are the ground truth and facilitate our experimental evaluation.

5.2 Evaluation Indicators

The accuracy model introduced in [12] is usually used to evaluate the non-overlapping schema summarization approaches.

Accuracy. The accuracy model measures how many tables are categorized correctly in the summary C by the following formula:

$$acc(C) = \frac{\sum_{i=1}^{k} m(C_i)}{n} \tag{10}$$

where $acc(C)$ denotes the accuracy of summary $C = \{C_1, C_2, ..., C_k\}$, k is the summary size which is specified by the users, n is the total number of database tables, $t(C_i)$ is the topic table of C_i, $m(C_i)$ denotes the number of tables in the category C_i that belongs to the same category as $t(C_i)$, in the pre-defined labeling.

Accuracy-M and F-Measure. Inspired by multi-label classification indicators in [2], we exploit Accuracy and F-measure for multi-label classification to evaluate the overlapping summaries. In order to distinguish the two accuracy, we denote the multi-label accuracy as Accuracy-M. For each table v, let T be the true categories it belongs to, S be its predicted categories in the summary. n is the total number of database tables. Then Accuracy-M and F-measure for an overlapping summary $C = \{C_1, C_2, ..., C_k\}$ are defined as follow.

$$Acc - M(C) = \frac{1}{n} \sum_{i=1}^{n} \frac{|T \cap S|}{|S|} \tag{11}$$

$$F - measure(C) = \frac{1}{n} \sum_{i=1}^{n} \frac{2|T \cap S|}{|T| + |S|} \tag{12}$$

5.3 Evaluation of Effectiveness

The baselines in this work are VHCA [11] and CWKC [12]. Both of them are non-overlapping summarization methods. To compare with them, we compute accuracy on TPC-E first. Our method achieves an overlapping summary where tables may have multiple labels. We perform additional evaluations with multi-labeled accuracy and multi-labeled F-measure on the both data sets. All of the experiments show that our method achieve better effectiveness and the generated summaries are more helpful. We will analysis the experimental results in depth.

Effectiveness over TPC-E. Fig. 3 shows the accuracy values of the three summarization approaches over TPC-E according to different k. We notice that the effectiveness of the three approaches is increasing along with the growth of k. When $k = 4$, the accuracy reaches the maximum. For $k \geq 5$, the accuracy gradually decreases. It gives a clear signal that there are only 4 categories in this database schema and it is meaningless to compute categories for $k > 5$. For each k, our OSSMP approach obtains the most accurate summary.

Fig.4 and Fig.5 show the accuracy-M and F-measure values respectively. Since TPC-E is a single-label data set, the detected overlaps will lead to the decrease of Accuracy-M and F-measure. However, OSSMP still performs better than the other two approaches.

The result shows that we find meaningful overlaps including EXCHANGE, SECURITY, ADDRESS, ZIP_CODE and STATUS_TYPE when $k = 4$. EXCHANGE bridges Marker and Broker as we mentioned above. Like EXCHANGE, SECURITY brings important information of both Marker and Customer. The other three tables belong to Dimension in original. However, it is more appropriate to divide them into other categories because they are referenced by other categories. According to our result, ADDRESS and ZIP_CODE are assigned into Costumer and Market while STATUS_TYPE is assigned into Market and Broker.

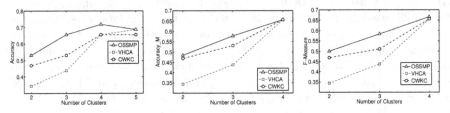

Fig. 3. Accuracy on TPC-E   **Fig. 4.** Accuracy-M on TPC-E   **Fig. 5.** F-measure on TPC-E

Effectiveness over CCEMIS. In Fig.6 and Fig.7, we plot the Accuracy-M and F-measure values for the three approaches over CCEMIS. They follow a similar trend as TPC-E with the growth of k. And the trend shows that there are also only 4 categories in CCEMIS. The graphs show that our approach performs the best. Moreover, when $k = 4$, we detect all of the four overlaps in CCEMIS. MajorClass and StudentCourse belong to both Student and Course. LessonTeacher and TeacherArrange belong to both Teacher and Lesson.

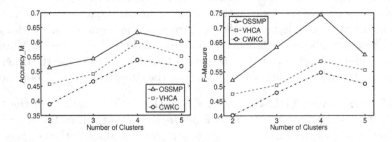

Fig. 6. Accuracy-M on CCEMIS   **Fig. 7.** F-measure on CCEMIS

6   Conclusion

In this paper, we present the notion of overlapping schema summarization for the first time. We also proposed an approach for generating overlapping schema summarization automatically. The OSSMP approach first exploits multi-label propagation algorithm to generate overlapping groups. Then it utilizes a hierarchical clustering algorithm to cluster the groups into the final categories. We have conducted an extensive study on one benchmark database and a real-world database, showing that our approach is accurate and can detect meaningful overlaps.

Acknowledgements. This work is partially supported by National Natural Science Foundation of China under Grant No. 61170184, 61402242, and 61402243, National 863 Program of China under Grant No. 2015AA015401, as well as National Key Technology R&D Program under Grant No.2013BAH01B05. This work is also partially supported by Tianjin Municipal Science and Technology Commission under Grant No. 13ZCZDGX02200, 13ZCZDGX01098, 14JC-QNJC00200, and 14JCTPJC00543.

References

1. Ahn, Y.Y., Bagrow, J.P., Lehmann, S.: Link communities reveal multiscale complexity in networks. Nature 466(7307), 761–764 (2010)
2. Godbole, S., Sarawagi, S.: Discriminative methods for multi-labeled classification (2004)
3. Gregory, S.: Finding overlapping communities in networks by label propagation. New Journal of Physics 12(10), 103018 (2010)
4. Jin, D., Yang, B., Baquero, C., Liu, D., He, D., Liu, J.: A markov random walk under constraint for discovering overlapping communities in complex networks. Journal of Statistical Mechanics: Theory and Experiment 2011(05), P05031 (2011)
5. Lancichinetti, A., Radicchi, F., Ramasco, J.J., Fortunato, S.: Finding statistically significant communities in networks. PloS One 6(4), e18961 (2011)
6. NKU: Ccemis. http://cc.nankai.edu.cn/
7. TPC: Tpc-e benchmark. http://www.tpc.org/tpce/
8. Wang, X., Zhou, X., Wang, S.: Hybrid schema summarization method of large scale database. Chinese Journal of Computers 36(8), 1616–1625 (2013)
9. Wu, W., Reinwald, B., Sismanis, Y., Manjrekar, R.: Discovering topical structures of databases. In: Proceedings of the 2008 ACM SIGMOD International Conference on Management of Data, pp. 1019–1030. ACM (2008)
10. Wu, Z.H., Lin, Y.F., Gregory, S., Wan, H.Y., Tian, S.F.: Balanced multi-label propagation for overlapping community detection in social networks. Journal of Computer Science and Technology 27(3), 468–479 (2012)
11. Wang, X., Zhou, X., Wang, S.: Summarizing large-scale database schema using community detection. Journal of Computer Science and Technology 27(3), 515–526 (2012)

12. Yang, X., Procopiuc, C.M., Srivastava, D.: Summarizing relational databases. Proceedings of the VLDB Endowment 2(1), 634–645 (2009)
13. Yu, C., Jagadish, H.: Schema summarization. In: Proceedings of the 32nd International Conference on Very Large Data Bases, pp. 319–330. VLDB Endowment (2006)
14. Yuan, X., Li, X., Yu, M., Cai, X., Zhang, Y., Wen, Y.: Summarizing relational database schema based on label propagation. Web Technologies and Applications (2014)

Analysis of Subjective City Happiness Index Based on Large Scale Microblog Data

Kai Wu, Jun Ma, Zhumin Chen, and Pengjie Ren

School of Computer Science and Technology,
Shandong University, Jinan, 250101, China
kaiwusdu@gmail.com,
{majun,chenzhumin}@sdu.edu.cn,
andyren2012@hotmail.com,

Abstract. City Happiness Index(CHI) is an important societal metric to evaluate the living status of the people in a city. In traditional method, CHI is usually calculated by the combination of other objective city indicators including economy, environment, technology and education. However, happiness is a kind of subjective feeling of people rather than the measurement on how much material wealth people have gotten. Therefore we propose a novel method to evaluate Subjective City Happiness Index(SCHI) by the analysis of public sentiment on microblogs. We carried out the analysis by mining the word distribution of the microblogs in Sina Weibo, which is the largest microblog platform in China. As an application, we used the model to calculate the SCHIs of 36 major cities of China based on 55 million microblogs posted by 0.9 million unique users. Furthermore, we investigated the variety of SCHI with time in different granularities(month, day, hour) in the year 2013.

Keywords: data mining, happiness index, microblog, city data.

1 Introduction

City happiness index is a popular indicator to reflect happiness of the residents in a city. With over half the world's population now living in urban areas, and this proportion continuing to grow, cities will only become increasingly central to human society. In this paper, we introduce a new happiness index, i.e. SCHI, to reveal the true feelings of people in cities. Expression about people's happiness can be got from many social networks. Recent progress of Web 2.0 applications has witnessed the rapid development of microblog platform in China (i.e., Sina Microblog [1]), where people can communicate with each other or express their feelings freely[1]. Sina Microblog provides similar services as other famous microblog platforms, such as Twitter[2]. Since its launch in August 2009, Sina

---

[1] http://www.weibo.com/

[2] Twitter : http://twitter.com/

© Springer International Publishing Switzerland 2015
R. Cheng et al. (Eds.): APWeb 2015, LNCS 9313, pp. 365–376, 2015.
DOI: 10.1007/978-3-319-25255-1_30

Microblog has grown into the biggest Chinese microblog with 500 million registered users by the end of 2012[2]. There are a tremendous amount of microblogs on Sina Microblog and most of them are the ture description of people's attitude towards life. Therefore, it is wise to choose Sina Microblog as our data source. In addition to the advantage of quantity, using Sina Microblog as a data source also makes us able to explore SCHI as a function of time, for the reason that every piece of microblog has a posting time. For example, we can explore which month do people feel happiest in a specific city.

Happiness is a subjective feeling of people while it is affected by objective conditions. Traditional methods to calculate city happiness index in China mainly consider the objective indicators of a city, such as GDP, healthy rate, education, green space and air quality, ignoring the subjective experience of people themselves. Recent years, the increasing amount of large scale user data on social network provides us a new way to evaluate happiness. In our micrblog-based method, we evaluated the happiness of a city through the self-reports(i.e., microblogs) of people themselves. In order to calculate SCHI, first we collected large amount of microblogs posted by people who come from the same city to form a city text. Then we gave every city a score as SCHI based on the happy and sad word frequency in the city text.

The major contribution of this paper is to propose a method to calculate SCHIs in China in terms of the analysis of public sentiment based on microblogs. After first examining happiness at the level of cities, we explored the variety of happiness with different time granularities, including month, day and hour. In this way, we showed the shift of happiness with time. At last, by analyzing the word frequency, we gave reasonable explanation of the difference between city happiness. Our study on SCHI found many useful and interesting macro-scale social phenomena. We believe that our study can provide many important social information and will do great help to both government and researchers, as well as millions of common people.

The rest of this paper is organized as follows: In section 2, we introduce some related work on social network and happiness index. In section 3, we explain the computing method of SCHI based on large amount of Sina Microblog data. Section 4 first introduces the dataset of our research and then illustrates our experimental results as well as the explanation of discoveries. Section 5 concludes our work and makes a brief description of future work.

2   Related Work

Happiness index is becoming an important research topic in various fields. Many reports on happiness have been published by sociologists and statisticians. The UN's 2012 World Happiness Report attempts to quantify happiness on a global scale with a 'Gross National Happiness' index which uses data on rural-urban residence and othder factors[3]. In the US, Gallup and Healthways produce a yearly report on the well-being of different cities, states and congressional districts[4], and they maintain a well-being index based on continual polling and survey data[5]. Other countries are also beginning to produce measures of well-being: in 2012, surveys measuring national well-being and how it relates to both health and where people live were conducted in both the United Kingdom by the Office of National Statistics[6] and in Australia by Fairfax Media and Lateral Economics[7].

While these and other approaches to quantifying the sentiment of a city as a whole rely almost exclusively on survey data, there are now a range of complementary, remote-sensing methods available to researchers. The explosion in the amount and availability of data relating to social network in the past 10 years has driven a rapid increase in the application of data-driven techniques to the social sciences and sentiment analysis of large-scale populations, such as population-level happiness measurements carried out by Facebook's internal data team[8], work focusing directly on sentiment detection based on Twitter[9,10]. Quercia et al.[11] found that monitoring tweets is an effective way to track community well-being in UK, as well as individual happiness. Detail researches by Dodds, Mitchell, Frank et al.[12,13,14,15] investigated the shift of happiness in American urban areas with the variety of geography and time, using a service named Amazon's Mechanical Turk[16].

Works related to happiness are frequent abroad, however, there are rarely such researches published in China. Inspired by the works above, we studied the Subjective City Happiness Index of China by mining large amount of Sina Microblog data.

3   Method

To evaluate the happiness degree of a city in our method, first a city text is needed. A city text is formed by a certain number of microblogs posted by sample users who come from the same city. However, before this task, we should introduce how to score the major sentiment words for a given city text. Here we use a existing affective lexicon ontology constructed by the Information Retrieval Laboratory, Dalian University of Technology[17]. Originally, there are three kinds of words in this ontology - "positive, negative, neutral". We slightly modified this ontology based on our works, including: 1)removing all the neutral words, such as "sure(确定)" and "understand(确定)". 2)adding some newly appeared common words(about 50) in microblogs which are related to happiness, such as "positive energy(正能量)" and "negative energy(负能量)". For simplification, we called this modified affective lexicon ontology as MALO in the rest of our paper. There are exactly 21984 words in MALO and every word has a discrete score which is in a set of {-9(saddest),-7,-5,-3,-1,1,3,5,7,9(happiest)}. Figure 1 shows the detail word distribution of MALO on score. Every column in it represents the number of words which lie in a certain score. From Figure 1 we can see that the numbers of happy and sad words are generally the same on every corresponding absolute score, so that there is no happy or sad bias in MALO itself.

Importantly, with this method we make no attempt to take the context of words or the meaning of a text into account. While this may lead to difficulties in accurately determining the emotional content of small texts, we find that for sufficiently large texts this approach nonetheless gives reliable results. An analogy which was used in [14] is that of temperature: while the motion of a small number of particles cannot be expected to accurately characterize the temperature of a room, an average over a sufficiently large collection of such particles nonetheless defines a durable quantity. Furthermore, by ignoring the context of words we gain both a computational advantage and a degree of impartiality; we do not need to decide a priori whether a given word has emotional content, thereby reducing the number of steps in the algorithm and hopefully reducing experimental bias.

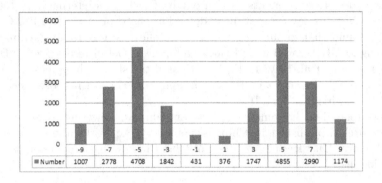

Fig. 1. Words distribution of MALO.

After having the happiness score of base words, next we selected a certain number of microblogs from every sample user, and city texts are formed by the microblogs posted by users who come from the same city. Then for a given text T containing N unique words, we calculate the subjective happiness h_{sub} by

$$h_{sub}(T) = \frac{\sum_{i=1}^{N} h_{sub}(w_i) f_{i,T}}{\sum_{i=1}^{N} f_{i,T}} = \sum_{i=1}^{N} h_{sub}(w_i) p_{i,T} \tag{1}$$

where $f_{i,T}$ is the frequency of the ith word w_i in T for which we have happiness score $h_{sub}(w_i)$, and $p_{i,T} = f_{i,T} / \sum_{i=1}^{N} f_{i,T}$ is the normalized frequency of word w_i.

Another task left is to choose an appropriate quantity of microblogs to form a city text. To reduce the personality factor of people, as well as considering the major quantity of microblogs posted by a user in a year, we selected 100 microblogs from every user. If a user's microblog number does not reach 100, we selected all. Figure 2 shows the SCHI values of different user numbers and every curve in the figure represent a city in China. The figure shows that the SCHI value becomes stable when user number reaches 20000. Recent works by Dodds et al.[18] have proved that human languages(including Chinese) reveal a universal positivity bias. Their conclusion coincides with the trend of the curves in Figure 2. We also analyzed the sentiment distribution of top N(N=200, 500, 1000, 2000, 2500) frequent words in city text of Beijing, and results are shown in Table 1. From the table we can see that the percentage of happy words is bigger than that of sad words. Therefore, we believe that with the rising of the user number, the SCHI value will keep increasing slowly. However, the result is meaningful as long as we choose the same user number for all cities, and the number is fixed at 25000 in the rest of our paper.

Table 1. Sentiment distribution of top N frequent words.

Top N	200	500	1000	2000	2500
happy words number	156	347	641	1233	1514
sad words number	44	153	359	769	986

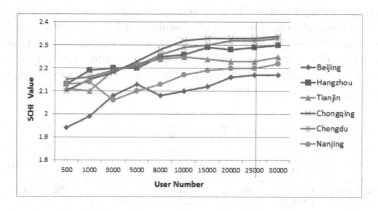

Fig. 2. SCHI values of different user numbers.

4 Dataset and Results

Our dataset contains totally 1.3 billion microblogs posted by 1.1 million users from 36 major cities, in which we selected 55 million microblogs(all posted in 2013) and 0.9 million users in our research. City information of users can be got in the user table of the dataset. In this paper, we assume that most microblogs of a user are posted in the city which he belongs. The detail of total microblog number, word frequency and SCHI value of every city are shown in Table 2. A difference of 0.1 in SCHI between cities is rather significant in our study. In order to reflect the difference between cities clearly, we map the SCHI values onto the China Map, which is shown in Figure 3. From the map we can see that cities in south China are generally happier than that of north. For the purpose of investigating the reason of such phenomenon and exploring the influencing factors of happiness, first we calculated the Spearman correlation coefficient between

SCHI and other city indicators, including GDP, GDP per capita(GDPPC), air quality(AQ)[3] and average daily leisure time(ADLT)[4]. Spearman correlation co- efficient is commonly used in calculating the relevance of two distributions. It has two values: 1) r-value(from -1 to 1) shows the degree of correlation between two distributions and larger absolute value means more relevant. The plus or minus sign of r-value represents that whether the two distributions have positive or negative correlation. 2) p-value(from 0 to 1) shows the significant degree of r-value. Usually, we are confident to say r-value is significant when p-value is under 0.05, and very significant when p-value is under 0.01.

From Table 3 we can see that SCHI has a positive correlation with other four indicators. Particularly, the p-values of (GDP,SCHI) and (ADLT,SCHI) are 0.03 and 0.0002 which are both under 0.05. First this means that people are more likely to feel happy if they have more leisure time. Meanwhile, works by Stevenson and Wolfers[19] also claim to show a direct correlation between gross domestic product and subjective well-being across American. As we all know, the economic level of south China is generally higher than that of north and recent years have witnessed a bad performance of north China's air quality. On the other hand, although west China is less developed than east, west China have a much better air quality. These factors are partly coincident with the Figure 3. After all, people's feeling about happiness can not be determined by one or two indicators, it is affected by a lot of external and latent factors.

Table 2. Data statistics of cities.

City	Microblogs Number	Total Word Frequency	SCHI	City	Microblogs Number	Total Word Frequency	SCHI
Guangzhou	1,889,999	3,089,432	2.45	Qingdao	1,656,431	2,982,986	2.22
Dongguan	1,492,970	2,282,106	2.40	Kunming	1,777,641	3,273,469	2.21
Changsha	1,764,486	3,271,602	2.38	Nanchang	1,104,904	1,960,518	2.21
Xiamen	1,723,121	2,815,117	2.34	Urumqi	1,098,313	2,001,611	2.21
Shenzhen	1,559,337	2,782,178	2.34	Changchun	1,623,619	2,901,287	2.20
Chongqing	1,482,668	2,485,645	2.33	Nanjing	1,744,841	3,055,772	2.20
Chengdu	1,898,478	3,170,007	2.32	Dalian	1,782,705	3,162,853	2.19
Wuhan	1,807,289	3,189,643	2.31	Xi'an	1,888,246	3,440,072	2.19
Hangzhou	1,893,969	3,188,413	2.29	Jinan	1,836,964	3,323,279	2.18
Shanghai	2,153,662	3,563,921	2.28	Harbin	1,787,634	3,193,679	2.18
Ningbo	1,420,197	2,450,546	2.28	ShiJZ	1,612,463	3,003,130	2.18
Guiyang	1,280,736	2,132,346	2.28	Beijing	2,146,631	3,979,412	2.17
Lhasa	710,983	1,355,670	2.28	Weifang	1,071,806	1,916,680	2.17
Xining	749,192	1,434,396	2.28	Yantai	1,100,289	1,961,539	2.17
Haikou	1,195,734	1,984,165	2.26	Yinchuan	819,094	1,554,382	2.17
Hefei	1,682,811	2,994,811	2.24	Zhengzhou	1,753,554	3,332,338	2.16
Wuxi	1,623,472	2,711,722	2.24	Huhehot	897,642	1,644,153	2.15
Tianjin	1,777,331	3,057,252	2.23	Lanzhou	1,233,654	2,364,456	2.15

---

[3] http://wenku.baidu.com/link?url=-YFd9GInkeMn9IgfLLJ8UizW2p3wNJ41pc5Yq CF804ocC5b-42ntwQGtvBnw5GDdzDjoxAzILqOIiI2GymQFZW9nEXFjC9CtaoLLjvGrW4G
[4] http://news.xinhuanet.com/finance/2012-03/20/c_122856940.htm

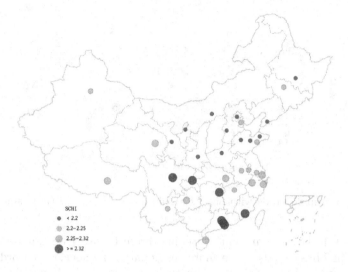

Fig. 3. Reflection of SCHI(year 2013) on China Map.

Table 3. Spearman Correlation Coefficient.

Spearman-value	(GDP,SCHI)	(GDPPC,SCHI)	(AQ,SCHI)	(ADLT,SCHI)
r-value	0.315	0.197	0.267	0.618
p-value	0.030	0.125	0.067	0.002

For further discussion, we compared two cities' SCHI values from the view of words shift contribution. We define word shift contribution $C_{T_m,T_n}(w_i)$ as follows:

$$C_{T_m,T_n}(w_i) = h_{sub}(w_i)(p_{i,T_m} - p_{i,T_n}) \qquad (2)$$

where T_m, T_n represent two city texts. Taking $Lhasa, Beijing$ as an example, Figure 4 shows the top 20 words with biggest absolute $C_{T_{Lhasa},T_{Beijing}}(w_i)$ values. All words with nonzero C values are divided into four types, including "happy word which increases", "happy word which decreases", "sad word which increases", "sad word which decreases" for which we use "+,↑", "+,↓", "-,↑", "-,↓" to represent in the figure. The increase or decrease here means the variety of $p_{i,T_{Lhasa}}$ compared with $p_{i,T_{Beijing}}$. Among the four types, "+,↑" and "-,↓" have positive contribution to the change of SCHI value. The small pie chart in Figure 4 shows the percentage of total absolute contribution of the four types, including all the words appeared. We put the original Chinese words in the bottom of Figure 4, for the reason that there may be some deviation in translation from Chinese to English.

Beijing is the capital, cultural and political centre of China, while Lhasa is a beautiful religion city in the west with a high elevation of 3650 metres. Compared with Lhasa, definitely Beijing is much more developed. However, the SCHI value

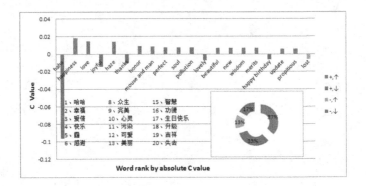

Fig. 4. Top 20 words with biggest absolute $C_{T_{Lhasa}, T_{Beijing}}(w_i)$ values.

of Lhasa is 0.11 larger than that of Beijing. From Figure 4 we can see that the happiness in Lhasa tends to be more religionary. Frequencies of words which are often appeared in religious realm such as "happiness, honor, mouse and man, soul, wisdom, merits, propitious" are larger than that of Beijing, while the frequencies of words "haha, joyful, lovely, happy birthday" are much smaller. In addition, words "haze, pollution" in Lsaha appear less than Beijing, which is coincident with the actual environment situation. Similarly, all cities in our study can be compared through words shift contribution, so that the difference between SCHIs can be explained reasonably.

Next we studied the variety of SCHI with month in 2013 by limiting the microblogs posted in a certain month, while the upper number of sampling microblogs per user is also 100 as before. Six cities including Beijing, Hangzhou, Tianjin, Chongqing, Nanjing and Chengdu are randomly selected as examples. Results are shown in Figure 5. The curves in the figure show roughly the same variation trend in the whole year and most winter months have advantages in happiness compared with summer months. The happiest month is February while the saddest month lies in April, with a difference of SCHI value from 0.3(Beijing) to 0.41(Nanjing). Forasmuch the difference between two texts is quit obvious in our study, therefore we calculated the values of $C_{T_{February}, T_{April}}(w_i)$ to explore the reasons.

Taking city Beijing as an example, Figure 6 shows top 20 words with biggest absolute $C_{T_{February}, T_{April}}(w_i)$ values. In 2013, Chinese traditional Lunar New Year lies in February, and so do the Valentine's Day and the Lantern Festival. Therefore, large amount of happy words related to the three festivals have much bigger frequencies in February, such as "joyful, lover, auspicious, reunion". In the meanwhile, there are two events happened in April: 1)an earthquake with a magnitude of 7.0 happened in Ya'an; 2)an epidemic disease named H7N9 avian influenza was spreading in China at that time. The intuitive reflection of the two events in Figure 6 is the larger frequencies of words "rescue, case(of illness), virus". However, the influence of event on SCHI has two sides. Earthquake itself is definitely a bad event for people, it will cause the decrease of happy words as well as the increasing of sad words. But in another aspect, when earthquake

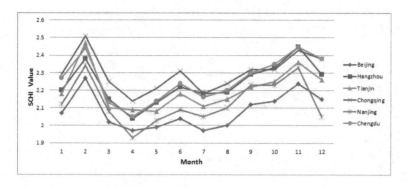

Fig. 5. Variety of SCHI with month in 2013.

happens, happy words such as "rescue, help, love" will increase too. In this light, although the calculation method of SCHI is simple, SCHI is indeed a reflection of many complicated factors. At last, the larger frequencies of sad words "haze, pollution" in February may be caused by the fireworks in Lunar New Year. In fact, more information will be revealed if more words are shown.

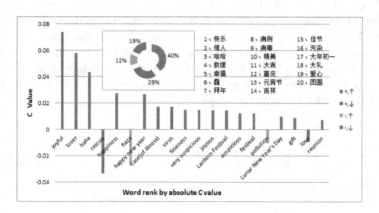

Fig. 6. Top 20 words with biggest absolute $C_{T_{February},T_{April}}(w_i)$ values.

In addition to 2013, we also made a gragh of 2012 for comparison as illustrated in Figure 7. Although the six curves in Figure 7 are not as regular as those in Figure 5, still many points are similar. For example, most cities have wave crests in June and November in both years, merely the wave crests are not so obvious in 2012. Another fact we should notice is that the traditional Lunar New Year in 2012 is in January, while the Valentine's Day and the Lantern Festival are still in February. This is why the difference between January and February in 2012 is smaller than that in 2013. In fact, many obvious or abnormal changes of the waves in two figures are caused by events, festivals, climate or other factors.

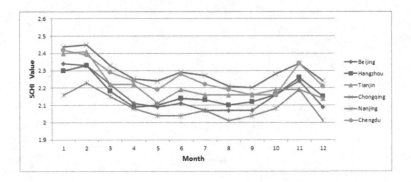

Fig. 7. Variety of SCHI with month in 2012.

Here we don't depict the reasons one by one because the limitation of space, however, they can be well explained by analyzing the word shift contribution.

Similar to month, we analyzed the variety of SCHI with day of week and hour by limiting the time when the microblogs were posted into a certain period, such as Mondays of year 2013. Results are shown in Figure 8, where the abscissa of Figure 8(a) is "day of week", while the abscissa of Figure 8(b) is "every two hours of day". Generally the SCHI doesn't show a significant variety in a week. Most cities are a little happier in Monday and Friday than other days of week. Amazingly, SCHI is relatively larger in Monday, even bigger than that of weekends in some cities. At last, most cities show the lowest SCHI in middle weekdays, including Tuesday and Wednesday. For the hours part, values in afternoon and evening are larger than that of morning, and the curves of SCHI is quite similar to the daily routine of common people.

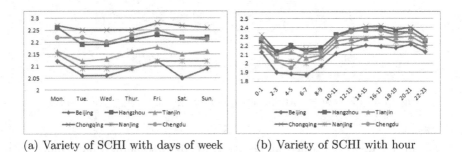

(a) Variety of SCHI with days of week   (b) Variety of SCHI with hour

Fig. 8. Variety of SCHI with days of week,hours in 2013.

5   Conclusion and Future Work

In this paper, we proposed a new index SCHI to evaluate the subjective happiness of cities in China. Large amount of microblogs posted by users from 36 cities were collected as our dataset. The major contributions of this paper are as follows: 1)reflecting the subjective happiness distribution in China by calculating the SCHI values of 36 major cities. 2)analyzing the variety of SCHI with month, day of week, hour to explore the shift of happiness with time. 3)reasoning the difference between SCHIs through word shift contribution. Compared with traditional methods, utilizing user data on social networks makes the computing of SCHI simpler and faster, but reasonable and effective at the same time.

Many useful and interesting discoveries have been found in our study. Cities in south China generally performs better on SCHI than that of north China. Meanwhile, subjective happiness was found to show positive correlation with GDP, GDPPC, AQ and ADLT, especially GDP and ADLT. For twelve months in 2012 and 2013, people feel happier in winter months. Holidays and big events do have a significant influence on SCHI. Another interesting phenomenon is that values of SCHI in Monday and Friday are larger than other days as well as weekends, although the advantage is small. Our research is meaningful because works about happiness index are quite few in China. Also SCHI information can provide government and sociologists many useful macroscopical data about the happiness status of China. For the majority of ordinary people, SCHI values can be good references when they choose working or living city.

In the future, first we plan to enlarge our dataset which includes both user and city numbers, for the purpose to make our research more robust. Besides the dataset, more analysis should be done on SHCI to give reasonable explanation about the discoveries we found. At last, we will extend our work to other group of people, such as people who have the same occupation.

Acknowledgment. This work was supported by Natural Science Foundation of China (61272240, 60970047, 61103151), the Doctoral Fund of Ministry of Education of China (20110131110028), the Natural Science foundation of Shandong province (ZR2012FM037), and Microsoft research fund (FY14-RES-THEME-25).

References

1. Gao, Q., Abel, F., Houben, G.-J., Yu, Y.: A comparative study of users' microblogging behavior on Sina Weibo and Twitter. In: Masthoff, J., Mobasher, B., Desmarais, M.C., Nkambou, R. (eds.) UMAP 2012. LNCS, vol. 7379, pp. 88–101. Springer, Heidelberg (2012)
2. Bao, P., Shen, H.W., Huang, J., Cheng, X.Q.: Popularity prediction in microblogging network: A case study on Sina Weibo. In: Proc. of WWW 2013, pp. 177–178 (2013)
3. Sachs, J.D., Layard, R., Helliwell, J.F.: World Happiness Report. Technical report, Columbia University/Canadian Institute for Advanced Research/ London School of Economics (2012)
4. Gallup-Healthways State of well-being 2011: City, state and congressional district wellbeing reports. Technical report, Gallup Inc. (2012). http://www.well-beingindex.com/files/2011CompositeReport.pdf
5. Gallup-Healthways Well-Being Index. http://www.well-beingindex.com/

6. Beaumont, J., Thomas, J.: Measuring National Well-being - Health. Technical Report July, UK Office for National Statistics (2012)
7. Lancy, A., Gruen, N.: Constructing the Herald/Age - Lateral Economics Index of Australias Wellbeing. Australian Economic Review 46, 92–102 (2013)
8. Facebook Gross National Happiness. http://apps.facebook.com/usa_gnh/
9. O'Connor, B., Balasubramanyan, R., Routledge, B., Smith, N.: From tweets to polls: Linking text sentiment to public opinion time series. In: Proc. of ICWSM 2010, pp. 122–129 (2010)
10. Bollen, J., Pepe, A., Mao, H.: Modeling public mood and emotion: Twitter sentiment and socio-economic phenomena. In: Proc. of ICWSM 2011 (2011)
11. Quercia, D., Ellis, J., Capra, L., Crowcroft, J.: Tracking "Gross community happiness" from tweets. In: Proc. of CSCW 2012, pp. 965–968 (2012)
12. Dodds, P.S., Danforth, C.M.: Measuring the happiness of large-scale written expression: Songs, blogs, and presidents. Journal of Happiness Studies, 441–456 (2009)
13. Dodds, P.S., Harris, K.D., Kloumann, I.M., Bliss, C.A., Danforth, C.M.: Temporal patterns of happiness and information in a global social network: Hedonometrics and Twitter. PLoS One 6, e26752 (2011)
14. Mitchell, L., Frank, M.R., Harris, K.D., Dodds, P.S., Danforth, C.M.: The Geography of Happiness: Connecting Twitter Sentiment and Expression,Demographics, and Objective Characteristics of Place. PLoS One 8(5), e64417 (2013)
15. Frank, M.R., Mitchell, L., Dodds, P.S., Danforth, C.M.: Happiness and the Patterns of Life: A Study of Geolocated Tweets. Scientific Reports 3 (2013)
16. Amazon Mechanical Turk Service. https://www.mturk.com/mturk/welcome
17. Linhong, X., Hongfei, L., Yu, P., Hui, R., Jianmei, C.: Constructing the Affective Lexicon Ontology. Journal of The China Society For Scientific and Technical Information 27(2), 180–185 (2008)
18. Dodds, P.S., Clark, E.M., Desu, S., Frank, M.R., Reagan, A.J., Williams, J.R., Mitchell, L., Harris, K.D., Kloumann, I.M., Bagrow, J.P., Megerdoomian, K., McMahon, M.T., Tivnan, B.F., Danforth, C.M.: Human language reveals a universal positivity bias, arXiv preprint arXiv:1406.3855 [physics.soc-ph] (2014)
19. Stevenson, B., Wolfers, J.: Economic Growth and Subjective Well-Being: Reassessing the Easterlin Paradox. Brookings Papers on Economic Activity 39, 1–102 (2008)

Tree-Based Metric Learning for Distance Computation in Data Mining*

Ming Yan, Yan Zhang, and Hongzhi Wang**

Harbin Institute of Technology, China
503365092@qq.com, {zhangy,wangzh}@hit.edu.cn

Abstract. Distance is an essential measurement of data mining. A good metric often leads to a good performance. Then how to obtain a proper metric systematically is critical. Distance metric learning is a classic method to learn distances between instances on data set with complex distributions. However, most researches on distance metric learning are based on Mahalanobis metric, which is equivalent to linear transformation on distance space that has limitation on complex data. To solve this problem, we propose a metric learning method based on non-linear transformation suitable for complex data. By using the tree model, we could address non-linearly separable data that rearrange input data and represent them to another forms, and tree model could be able to implicitly represent data to a new distance space with a non-linear activator function. Furthermore, single tree model will lead to overfit that has higher generalization errors. Therefore, we design a randomize algorithm to combining different tree models which could reduce the generalization errors in theory and practice. According to analysis, we prove the correctness and effectiveness of our algorithm in theory. Extensive experiments demonstrate that algorithm is stable and suitable for data mining.

1 Introduction

Data mining are becoming more and more important as the development of data science. In data mining, classification and clustering analysis are the core issues, in which we often need to compute the distances or similarities between two data objects. Therefore, distance metric plays an crucial role in data mining. A good metric is one of the most important key points, especially in clustering, classification and kernel methods, that reflect the relationships between data [15]. Even though the traditional distance metrics, such as Euclidean distance, are simple and easy to compute, these metrics cannot handle the data set with complex distribution, and cannot represent the latent information which cannot handle heterogeneous data well [16]. Moreover, the Euclidean distance assumes that each feature has the same importance and have no relevance between each

* This paper was partially supported by NGFR 973 grant 2012CB316200, NSFC grant 61472099,61133002 and National Sci-Tech Support Plan 2015BAH10F00.
** Corresponding author.

R. Cheng et al. (Eds.): APWeb 2015, LNCS 9313, pp. 377–388, 2015.
DOI: 10.1007/978-3-319-25255-1_31

other, but this assumption does not hold for real world data. In contrast, Distance Metric Learning (DML) [15] computes distances between instances in data set with complex distributions and reveal relevance between each feature.

In recent years, researchers found many applications of DML in many areas, such as clustering [15], classification [14] and information retrieval especially in ranking [11]. The distance metric learning was first proposed in [15] which minimize the distance between instances that are similar and subject to the constraint that dissimilar instances have a larger distances. Although this method gave a novel idea, it requires to solve a semi-definite programming problem, which is difficult in computation [4]. To avoid the problem of [15], Relevant Component Analysis (RCA) was then proposed to learn metric through linear transformation on Mahalanobis [2]. Weinberger has taken the advantages of Neighborhood Component Analysis (NCA) [7] to avoid the non-convex problem, and then proposed Large Margin Nearest Neighbor (LMNN) [14]. Davis's Information-Theoretic Metric Learning (ITML) [6] combining the information theory, expressed the DML problem as a particular Bregman optimization problem that minimize the LogDet divergence subject to linear constraints.

Even though the proposed methods could weight each feature according to its effect, the complex distribution data with non-linearly separable features, which is common in real life, is not well solved by the above methods [16]. The above methods are mostly based on a single Mahalanobis metric which is $d_M(x_i, x_j) = \sqrt{(x_i - x_j)^T A(x_i - x_j)}$, and take optimization ideas to solve the DML problems. For the constraints of optimization problem, they use the labeled data, often is pairwise similar or dissimilar. Minimize distances between similar pairs and constrain on dissimilar pairs in order to indicate pairs should or should not be grouped. The objective of Mahalanobis based metric learning is to optimize the matrix A so that it could be viewed as an appropriate linear transformation that fits the distribution of data set[9].

To overcome the shortcomings of existing methods, we propose a novel metric learning method. Instead of linear space transformation, our method is based on non-linear transformation. To represent non-linear transformation effectively, we design a tree-based model to solve this problem. Our method keep the latent structural information as the growth of a tree, and the feature sampling strategy extends internal relevance between features. In order to reduce the generalization error of single tree model, we design a combination model and develop a randomize algorithm which combine each tree generated with a randomize strategy to represent non-linear transformation. Such randomizing-combining strategy could not only keep the information of original data but also extend latent structure and probability information. We prove that the proposed method is effective and correct, and demonstrate the algorithm is stable and suitable for data mining with extensive experiments.

Our contributions in this paper are summarized as follows.

1. We propose a novel idea of tree representation-based metric learning task based.
2. We develop a leaf index feature representation algorithm to represent data by the leaves of the tree which can reveal latent information of original data.

3. We employ a Bayesian specific tree path feature representation algorithm to represent data by decision path which involves the decision process.
4. We exploit a randomized tree algorithm to reduce generalization errors and improve effectiveness.

The paper is organized as follows. In Section 2, we present the problem definition and notation. Two tree based feature representation algorithms are proposed in Section 3 which includes the leaf index feature representation algorithm and Bayesian specific tree path feature representation algorithm. We show a random trees representation strategy in Section 4 and prove the correctness of our methods. Experiment results are shown in the Section 5. At last, we draw the conclusions in Section 6.

2   Problem Definition and Notation

In this section, we define the problem as well as the notations. In the area of data mining, each record in database may have multiple attributes, such as name, topic, description, etc. Different algorithms for distance computation may take different choices of attributes. For instance, during the distance computation between tuples, one algorithm mainly takes the authorship or name, another may focus on the topic, and the description information could be used in the third algorithm. Here comes the problem, which one should be measured? Different attributes may have different semantics, thus they worth differently for the distance computation. In this paper, we aim to find out a plausible distance metric for data mining tasks.

The input of this problem is a data set with n points $\{x_i\}_{i=1}^n \subseteq \mathbb{R}^d$. When two records x_i, x_j fall into Euclidean Space, we could measure them by Euclidean distance: $d(x_i, x_j) = \|x_i - x_j\| = \sqrt{(x_i - x_j)^T(x_i - x_j)}$. General algorithms only take Euclidean distance. However, different attributes may have different semantics and weights. The goal of metric learning is to find the weight for each attribute, which could also be thought as the transformation from Euclidean Space to some other. Traditional methods [15,14,7] seek a semi-positive definite matrix A which parameterizes the Mahalanobis distance:

$$d(x_i, x_j) = d_A(x_i, x_j) = \|x_i - x_j\|_A = \sqrt{(x_i - x_j)^T A(x_i - x_j)} \qquad (1)$$

The matrix $A \subseteq \mathbb{R}^{d \times d}$ is semi-positive definite, which means A could be decomposed into $v^T v$. Then the above distance is transformed into

$$d(x_i, x_j) = \sqrt{(x_i - x_j)^T A(x_i - x_j)} = \sqrt{(x_i - x_j)^T v^T v(x_i - x_j)}$$
$$= \sqrt{(vx_i - vx_j)^T(vx_i - vx_j)} = \sqrt{\varphi(x_i, x_j)^T \varphi(x_i, x_j)} \qquad (2)$$

The equation (2) is just the general form of distance metric with an activator function $\varphi(\cdot)$. The core problem of metric learning is to find out the activator function $\varphi(\cdot)$. Then the problem is defined in Definition 1.

Definition 1. *The function $\varphi(\cdot)$ is a space transformation function for distance metric. We want to find out a proper activator function $\varphi(\cdot)$ for non-linear transformation task.*

Traditional activator function, which is often the linear, cannot represent complex data. Therefore, we aim to find out a proper activator function that has enough capabilities to explain the data which not only keeps the original information but also extends latent information and do not lose accuracy.

3   Tree Based Feature Representation

The equation (2) could be regarded as linear transformation with a vector v, and $\varphi(\cdot)$ means we times a weight within each feature. However, this kind of linear transformation may not handle the situation with data correlation and complex distribution. Therefore, we attempt to replace the $\varphi(\cdot)$ with a non-linear transformation. In this paper, we propose a robust non-linear and tuple transformations of input feature space that based on decision trees, which is one of non-linear methods usually used for classification or regression.

We build several decision trees within different selected input features, and treat individual tree as a transformation function. Here we propose two basic method to handle this situation:

1. generate new categorical features with the index of the leaf that an instance ends up.
2. generate new sequence features with the path of prediction of an instance within a tree.

In this section, we first talk about tree leaves representation method and then introduce tree path representation. The two methods represent input data to another forms from different aspects.

3.1   Tree Leaves Representation

When talking about tree-based method, people often only care about the result of classification or regression output, and ignore the structural representation information. However, the structure information is often important for representing latent factors [3]. Therefore, we can utilize the tree to do the representation task.

Our Tree Leaves Representation algorithm is described in Algorithm 1. From Line 1 to 7, we create the transformation function F_T based on tree T. We first build decision trees with different selected features, and make a prediction for an instance. In Line 8, unlike general decision tree algorithm, we sample features from the whole feature set A, and in each node, we re-sample the sub-feature set in order to increase the diversity of data representation and reveal the real meaning and distribution. The benefit is that this method could retain the original information on the maximum extent due to the randomized sample algorithm in each node. We split tree nodes based on the ID3 [8,12] which is

---

Algorithm 1. Leaf Index Feature Representation algorithm (LIFT)

Input: tree node t_node, data set $\mathbf{X} \in \mathbb{R}^d$, feature set A.
Output: Single tree Representation model F
1: **if** t_node satisfies stop criteria **then**
2:    Stop growing
3:    Create transformation function F_T based on T
4:    Keep the index idx of the leaf that an instance t falling in
5:    $F(\mathbf{t}) = F_T(\mathbf{t}) = \underbrace{0, \cdots, 0}_{idx-1}, 1, \underbrace{0, \cdots, 0}_{d-idx}$
6:    **return** F
7: **end if**
8: Random sample features $\tilde{A}$ on feature set A
9: $\tilde{a} = \arg\max_{a \in \tilde{A}} \text{Information\_Gain}(\mathbf{X}, a)$
10: Split node with $\tilde{a}$, keep the child node
11: **for** c_node in child node list **do**
12:    Add node c_node into tree T
13:    Tree grows on child c_node
14: **end for**

---

simple but effective in practice. Once the tree has been built, we keep the decision tree for transformation. Once an instance comes, the model will predict which leaf it will end up, and we should keep the index of this leaf. Then we represent the instance by the leaf all 0s but 1s which the leaf this instance falls in.

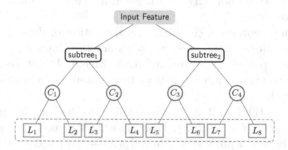

Fig. 1. Non-Linear representation with the index of tree leaves

As the growth of decision tree, the capability of explaining and representing will increase exponentially according to the tree structure. However, the single tree model will not include the whole information [3]. Therefore, we often build multiple trees to keep enough information.

For example, in Fig. 1, if an instance falls into leaf 3 in the first subtree, and falls into leaf 6 in the second subtree, then the input features of this instance will be transformed into a binary vector $[0, 0, 1, 0, 0, 1, 0, 0]$. With the growth of a decision tree, the number of leaves will increase fast, and this model will have more capability to represent input feature. On the other hand, the generated features are totally sparse representation which have been proven efficient and accurate.

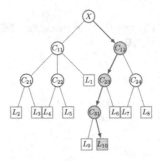

Fig. 2. Non-Linear representation with the path of prediction

3.2   Tree Path Representation

We give a strategy to represent input data based on decision path of a tree. The second method is shown in algorithm 2. As the first method, we first build several decision trees, and make prediction on each instance. Then keep the path of prediction within the tree. The pseudo code is shown in Algorithm 2, and we also take the ID3 algorithm to build decision tree. However, the mainly difference is that we do not only keep the leaf information but also the decision path information.

This algorithm represents the hierarchical relation of data, and include the latent structure information. Then we introduce the Bayesian method that remain the decision process to represent data in Line 6. A path keep not only the index but also the transition probabilities from parent node to child node, which is $p(child|parent)$.

For example, in the Fig. 2, an instance falls into the leaf 10, and we keep the path $\{C_{12} \rightarrow C_{23} \rightarrow C_{31} \rightarrow L_{10}\}$. From another tree node, we could consider this path as a bayesian process, which is $\{p(X \rightarrow C_{12}), p(C_{12} \rightarrow C_{23}), p(C_{23} \rightarrow C_{31}), p(C_{31} \rightarrow L_{10})\}$, and then add the index information each layer in the path. For this instance, when we perform the transform task, the results include a vector that includes path with index of each layer, and the probabilities of each tree node in the path. To increase the capability of this method and reduce the generalization errors, we will propose a randomize algorithm for feature representation in section 4, and prove the effectiveness.

4   Random Trees for Metric Learning

The single decision tree will lead to overfit, which perform well in training set, but weakly in testing set [12]. It usually has low bias and high variance in real

---

Algorithm 2. Bayesian Specific Tree Path Feature Representation (BSTP)

Input: tree node t_node, data set $\mathbf{X}$, feature set A.
Output: Single tree Representation model F
 1: **if** t_node satisfies stop criteria **then**
 2:    Stop growing
 3:    Create Representation function F_T based on T
 4:    For an instance t, keep the path in the tree $\{c_1, c_2, \cdots, c_k, l(t)\}$
 5:    Compute the probabilities based on Bayesian method
 6:    $F(t) = F_T(t) = \{p(root \to c_1), p(c_1 \to c_2), \cdots, p(c_{k-1} \to c_k), c_k \to l(t)\}$
 7:    **return** F
 8: **end if**
 9: Random sample features $\tilde{A}$ on feature set A
10: $\tilde{a} = \arg\max_{a \in \tilde{A}}$ Information_Gain$(\mathbf{X}, a)$
11: Split node with $\tilde{a}$, keep the child node
12: **for** c_node in child node list **do**
13:    Add node c_node into tree T
14:    Tree grows on child c_node
15: **end for**

---

application. In light of [5], an effective approach for reducing generalization errors would reduce the prediction variance, and the respective bias would be the same or not be increased too much [10]. Therefore, we will introduce a randomize algorithm for DML in this section to increase capability of the method and reduce the generalization errors, then analyze the time complexity and prove the effectiveness. In this section, we employ ensemble strategy that combines several results and then produces a new result.

4.1 Combining Random Tree Algorithm

Combining models method will solve the overfit problem in a simple way [10]. The method takes random combinations for learning process, such that they could produce different models only from a single data set $\mathcal{L}$. After that, this method combines the results to form a new results based on the generated models. In Algorithm 3, we generate feature representation based on algorithm 1

Algorithm 3. Random Trees for Feature Transformation algorithm(RTFT)

Input: data set $\mathbf{X}$, feature set A, number of combining models n.
Output: transformation function set F
 1: Initialize transformation function set F
 2: **for** $i \leftarrow 1$ to n **do**
 3:    Bootstrap instances $\tilde{X}$ on X
 4:    $F_i \leftarrow$ Generate-Feature-Transformation-Function$(\tilde{\mathbf{X}}, A)$
 5: **end for**
 6: **return** F

---

and algorithm 2. We use the transformation function set to represent input features. For each transformation function, we generate a representation vector, and combining them to become a new representation in Line 4.

Assume that there is a set of k random tree models. All of them are trained based on the same data set but built from the independent parameter t_i which is viewed as a random seed.

Theorem 1. *Our method combines the transformation results into a new representation, denoted as $F_{t_1,t_2,\cdots,t_k}$ such that the expected generalization error of the combining trees denoted by $Err(F_{t_1,t_2,\cdots,t_k})$, tends to be smaller than a single model which is $Err(F_{t_i})$.*

$$F_{t_1,t_2,\cdots,t_k} = average(F_{t_1}, F_{t_2}, \cdots, F_{t_k}) \tag{3}$$

Proof. The combination of transformation results is equivalent to the average prediction that minimizes the average errors with respect to the single result of the models. The errors of a single model are shown in formula (4)

$$\bar{E} = \frac{1}{K} \sum_i^k Err(F_{t_i}) \tag{4}$$

Similarly, the error of combining model and its variance of the single result of the combining result is:

$$\bar{V} = \mathbb{E}\{\frac{1}{K} \sum_i^k (F_{t_i} - F_{t_1,t_2,\cdots,t_k})^2\} \tag{5}$$

In statistic, the error evaluation can be viewed as:

$$Err(F_{t_1,t_2,\cdots,t_k}) = \bar{E} - \bar{V} \leq \bar{E} \tag{6}$$

While the total error of combining model is smaller than the single model because the variance $\bar{V}$ is non-negative. Therefore, the combining model leads to a more accurate model with less generalization error, and not increase the expectation bias.

4.2   Random Trees Representation

In Algorithm 1 and algorithm 2, we select sub feature space randomly, which means that each feature has the same probability to be chosen. We assume that if two instances have similarity p, then each feature has the probability p to be same [13]. First of all, we prove that our method could keep the original information, then we will show that this method will represent enough latent information include structure and probability information.

Theorem 2. *The probability of two similar instances getting the same result in one tree will approach 1. Otherwise, the probability will approach 0.*

Proof. Suppose we use k trees to represent input features, and each tree has the average height $\bar{h}$. The similarity between two instances is p. Due to the above assumption, two instances has the probability p to have the same choose for them to select split node at each layer as the tree growth. With the average height $\bar{h}$, for a single tree, the probability that two instances agree in all layers is $p^{\bar{h}}$. Then the probability that the instances do not agree in at least on layer which will lead the two end up in the different leaves will be $1 - p^{\bar{h}}$. The probability that the instances end up differently for all trees will be $(1 - p^{\bar{h}})^k$. Therefore, there will be $1 - (1 - p^{\bar{h}})^k$ probability that the instances agree in all layers of at least one tree. This form is more like the the form of an S-curve. With the $p \to 1$ which means the instances are very similar, then the probability that the two will be considered as the same in at least one tree tends to be 1. On the other hand, once the $p \to 0$, the probability tends to be 0. Once two instances has high similarity, the probability that the two will end up in same leaf for at least one tree, will approach 1, or it will approach 0.

From another point, our method will keep the original information with extremely high probability. As the growth of a decision tree, each split is based on random sampled sub-feature space. Each feature has the same probability to be chosen to extend the internal information of itself, which is an efficient non-linear representation. In the next section, we will show the experimental results to validate the effectiveness of our algorithm.

4.3   Time Complexity

Suppose we use k trees and each tree involves the whole data set $\mathbf{X}$, and we assume the size of $\mathbf{X}$ is n. To build a single tree with an average height $\bar{h}$, in each layer, the algorithm will scan the data set to decide in which the node will split. Thus, the time complexity of a single tree will be $O(n \times h)$. Generally, the height $\bar{h}$ of a single tree will be not greater than the height of a binary tree which is $\log(n)$. Then the time complexity will be $O(n \times \log(n))$. For k trees, the total complexity will be $O(k \times n \times \log(n))$. However, each tree is independent and could be built independently. Therefore, the trees could be built in parallel with non loss of accuracy.

5   Experiments

In this section, we give the experimental results on the UCI data sets. We run experiments on the Ubuntu server with the 4GB RAM and 4 cores Intel(R) Core(TM) i5-2400 CPU. We test our model with both classification and clustering tasks. We compare our approach on classical data mining methods. For classification, we compare the methods in the accuracy, while for clustering, the measure is sum square errors (SSE) [12].

5.1   Experiments on Classification

In Table 1, we test on the Movement Libras data set for k-NN classification task. The Movement Libras data set publics a data set for training, and five data sets that for testing. This data set has 90 attributes/features, and 15 classes.

Data Set	Size(K)	plain-kNN	LIFT	BSTP
movement_1	32	0.578	0.644	**0.644**
movement_5	63	0.867	0.867	**0.878**
movement_8	94	0.770	0.793	**0.800**
movement_9	32	0.467	**0.533**	0.511
movement_10	189	0.700	0.707	**0.711**

Table 1. Movement Libras data sets for k-NN classification testing

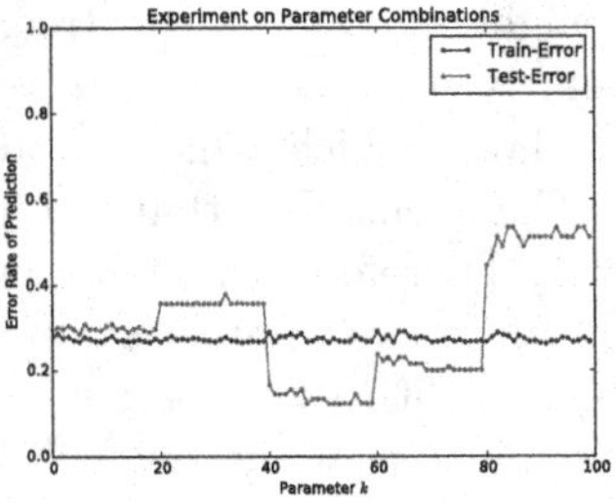

Fig. 3. Influence of Para. k

From Table 1 we observe that the accuracies of our algorithms are better than the point-based position-dependent k-Nearest Neighbors. The point-based position-dependent k-Nearest Neighbors ignore the structure representation information, but the structural information is often very important for representing latent factors. Therefore our algorithms increase the accuracies. Additionally, the BSTP algorithm has a better performance on many test sets. The BSTP algorithm keeps not only the leave information but also the decision path information of each tree.

However, not all results of the test sets are such good. It shows that even though the BSTP algorithm keeps more information, sometimes it will include noise and the influence of bad cases which often leads to overfit on training set that the training results are highly with the consistent with the distribution of training set that has been shown in Fig. 3 and Fig. 4.

Our algorithm has two essential parameters , including max tree depth h and number of trees k. The h decides the complexity of a single tree model, and the k decides the whole model. As discussed in the section 4, the capability of representation will increase while the h and k are going to be large. We conduct 100 experiments for the two parameters respectively. The results have been shown in Fig. 3 and Fig. 4. As the parameter k and h goes up, the errors first decrease then increase. When the parameter is small, the model cannot represent enough information and the errors seem to be large. While it increases gradually, the model has enough capability to represent latent information. Thus, the errors are going down. However, when parameters are large enough, the model has enough structure information of train data so that the model falls into the distribution of train data and the performance on the test set are worse and worse. As a result, our algorithm can represent data effectively by choosing a proper parameter set.

5.2   Experiments on Clustering Analysis

Clustering is one of important applications of metric learning. We experimented on drift data set of UCI data sets. The experiment takes the K-means as the base algorithm and compare the k-means result on Euclidean distance with the k-means result on our random trees distance metric, and the measure rule is the sum square errors. The implementation strategy of k-means is based on the [1], which is an effective algorithm to reduce the influence of the initialization. Another thing we should claim is that our algorithm is to represent input to another form, and the output will have more dimensions than the original. However, when the number of dimension increases, the corresponding square errors increase. Therefore, we should reduce the dimensions of representation results in order that they can have the same scale with the original input. The results are shown in Fig. 5. As observed, our algorithm has reduced the errors largely.

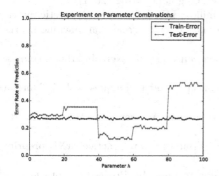

Fig. 4. Influence of Para. h

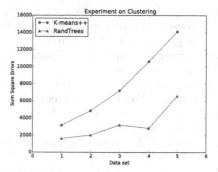

Fig. 5. Experiment on Clustering task

6   Conclusion

Distance metric plays a core role in data mining. Traditional distance metric could not handle the complex distribution situation. Metric learning has been proposed to overcome this problem. Most metric learning methods, however, are based on Mahalanobis metric that could be viewed as some kind linear transformation on distance space. Such method has many limitations. In this paper, we propose a randomize algorithm to solve the distance metric learning problem. Our algorithm, which based on combining random decision trees, has demonstrated the capability of representation of data. The mainly purpose of a representation task is to keep the original information and then to extend the latent information. The combining random trees could keep the original information as proved. And the decision trees model will reveal the latent structural and probability information. The experimental results show the effectiveness of our model is effective on both the classification task and the clustering task. For further studying, we will research on regularization in order to reduce the risk of over-fitting.

References

1. Arthur, D., Vassilvitskii, S.: k-means++: The advantages of careful seeding. In: Proceedings of the Eighteenth Annual ACM-SIAM Symposium on Discrete Algorithms, pp. 1027–1035. Society for Industrial and Applied Mathematics (2007)
2. Bar-Hillel, A., Hertz, T., Shental, N., Weinshall, D.: Learning a mahalanobis metric from equivalence constraints. Journal of Machine Learning Research 6(6), 937–965 (2005)
3. Bishop, C.M., et al.: Pattern recognition and machine learning, vol. 1. Springer New York (2006)
4. Boyd, S., Vandenberghe, L.: Convex optimization. Cambridge University Press (2009)
5. Breiman, L.: Random forests. Machine Learning 45(1), 5–32 (2001)
6. Davis, J.V., Kulis, B., Jain, P., Sra, S., Dhillon, I.S.: Information-theoretic metric learning. In: Proceedings of the 24th International Conference on Machine Learning, pp. 209–216. ACM (2007)
7. Goldberger, J., Roweis, S., Hinton, G., Salakhutdinov, R.: Neighbourhood components analysis (2004)
8. Jiawei, H., Kamber, M.: Data mining: concepts and techniques, vol. 5. Morgan Kaufmann, San Francisco (2001)
9. Kulis, B.: Metric learning: A survey. Foundations & Trends in Machine Learning 5(4), 287–364 (2012)
10. Louppe, G.: Understanding random forests: From theory to practice. arXiv preprint arXiv:1407.7502 (2014)
11. McFee, B., Lanckriet, G.R.: Metric learning to rank. In: Proceedings of the 27th International Conference on Machine Learning (ICML 2010), pp. 775–782 (2010)
12. Pang-Ning, T., Steinbach, M., Kumar, V., et al.: Introduction to data mining. In: Library of Congress (2006)
13. Rajaraman, A., Ullman, J.D.: Mining of massive datasets. Cambridge University Press (2011)
14. Weinberger, K.Q., Blitzer, J., Saul, L.K.: Distance metric learning for large margin nearest neighbor classification. In: Advances in Neural Information Processing Systems, pp. 1473–1480 (2005)
15. Xing, E.P., Jordan, M.I., Russell, S., Ng, A.Y.: Distance metric learning with application to clustering with side-information. In: Advances in Neural Information Processing Systems, pp. 505–512 (2002)
16. Xiong, C., Johnson, D., Xu, R., Corso, J.J.: Random forests for metric learning with implicit pairwise position dependence. In: Proceedings of the 18th ACM SIGKDD International Conference on Knowledge Discovery and Data Mining, pp. 958–966. ACM (2012)

GSCS – Graph Stream Classification
with Side Information

Amit Mandal[1], Mehedi Hasan[1], Anna Fariha[1],
and Chowdhury Farhan Ahmed[2]

[1] Department of Computer Science and Engineering, University of Dhaka,
Bangladesh
[2] ICube Laboratory, University of Strasbourg, France
amitducse17@gmail.com, mha_bd@yahoo.com, anna@cse.univdhaka.edu,
cfahmed@unistra.fr

Abstract. With the popularity of applications like Internet, sensor network and social network, which generate graph data in stream form, graph stream classification has become an important problem. Many applications are generating side information associated with graph stream, such as terms and keywords in authorship graph of research papers or IP addresses and time spent on browsing in web click graph of Internet users. Although side information associated with each graph object contains semantically relevant information to the graph structure and can contribute much to improve the accuracy of graph classification process, none of the existing graph stream classification techniques consider side information. In this paper, we have proposed an approach, **G**raph **S**tream **C**lassification with **S**ide information (**GSCS**), which incorporates side information along with graph structure by increasing the dimension of the feature space of the data for building a better graph stream classification model. Empirical analysis by experimentation on two real life data sets is provided to depict the advantage of incorporating side information in the graph stream classification process to outperform the state of the art approaches. It is also evident from the experimental results that *GSCS* is robust enough to be used in classifying graphs in form of stream.

Keywords: Graph Classification, Graph Streams.

1 Introduction

With the expansion of technologies that generate graph data in form of stream like Internet, social network, sensor network etc. into our everyday life, graph stream classification is gaining significant research interest in data mining and machine learning community. Graph stream classification is a two step process, where in the learning step, features are extracted from each training graph and a classification model is built, and in the classification step, class labels of each test graphs are predicted using that classification model.

© Springer International Publishing Switzerland 2015
R. Cheng et al. (Eds.): APWeb 2015, LNCS 9313, pp. 389–400, 2015.
DOI: 10.1007/978-3-319-25255-1_32

Graph is a very popular data structure, suitable for representing schema-less data like representing various activities among nodes in a network. In many graph streams, various side attributes are associated with each graph which may contain semantically relevant information to the graph linkage structure. These informations may contribute to build a more discriminative classification model. Let us look at some examples. *First*, each article residing in scientific reposi-tories(e.g. Google Scholar [1], DBLP [2]) can be represented using a bidirected authorship graph where authors are nodes and their co-author relationships con-stitute bidirectional edges among the nodes. Along with each graph, various types of side attributes like terms and keywords of the paper can be associated. *Second*, communication via message passing among users in a social network in a small time window can form a directed graph where users are nodes and messages sent and received among users make directed edges among the corre-sponding nodes of users of those messages. Different types of side attributes like user locations, user profile informations, message types (e.g. personal message, group chat) and platforms (e.g. PC, mobile) can be associated with each graph in the stream. Hence, side information should be incorporated along with graph linkage structure in the graph classification process to improve classification ac-curacy.

Features that only occur with high frequency in objects with one particular class label, used to discriminate among objects with two or more different class labels, are known as discriminative features. The aim of any classification algo-rithm is to build a classification model using discriminative features, which help to make finer distinctions among objects with different class labels, thus improv-ing classification accuracy. By considering side information along with the graph linkage structure, dimension of the feature representation of a graph object can be increased. This increased dimensionality with features related to the graph structure, can be of great use to find more discriminative features from graph objects, hence increasing classification accuracy.

One of the inherent challenges in graph stream classification is storing the received objects from the stream and extracting features from the enormous vol-ume of data. For example, there are 4.2×10^9 IPv4 addresses. So there can be 4.2×10^9 distinct nodes and 9.2×10^{18} distinct edges in the web graph stream. Storing this huge amount of data is intractable. Traditional graph classification algorithms require multiple scans over the whole data, which is not possible in stream scenario. So a summary of the graph stream is saved for future mining. As information loss is occurring while saving a summary of the graph stream, the results will be approximate rather than exact.

Various graph mining approaches have been defined in recent times to solve different tasks like classification [3], correlation mining [4], recommendation on social networks [5] etc. Though there are some approaches for graph stream clas-sification [3][6][7][8], to the best of our knowledge none of them considers side information in the mining process. Aggarwal et al. [3] proposed a probabilistic discriminative subgraph mining approach, where the received graphs are at first compressed and saved into a summary table using two random hashing schemes and then frequent and discriminative subgraphs are mined from that table to build a rule-based classifier. A clique based approach was proposed by Chi et al. [6] where the graphs are at first compressed into a fixed size node space and then discriminative clique patterns are extracted from each compressed graph to build a rule-based classifier. Discriminative hash kernel approach was proposed by Li et al. [7] to classify graph streams. Guo et al. [8] proposed an approach for graph stream classification which uses a combination of hashing and factorization of graphs. Yu et al. [9] considered side information in the graph stream clustering process and their experimental results showed the benefit of considering side information in the clustering process. One major drawback of the existing graph stream classification techniques is that, they do not consider side information in the classification process and the side information can help to improve classification accuracy.

Inspired by this drawback of the current graph stream classification techniques, in this paper we are proposing an approach *GSCS* for graph stream classification which considers side information. In *GSCS*, we have mostly followed the discriminative clique based approach proposed by Chi et al. [6] for graph structure mining. The volume of side information in the stream scenario can also be potentially infinite. So a hash based technique is used to tackle this storage problem and extract discriminative features from side information. Finally a majority voting classifier, inspired from Chi et al. [6] approach, is designed for classifying the future stream.

The rest of the paper is organized as follows: Section 2 explains our proposed approach in detail with necessary examples and algorithms. Section 3 focuses on the performance analysis of our proposed technique and shows the benefit of considering side information in the classification process. Finally we bring the paper to a close in Section 4.

2   Our Proposed Approach – GSCS

First, we introduce the problem formulation. Assume we have a stream of graphs GS denoted as $\{G_1, G_2, \ldots, G_n, \ldots\}$. E represents the set of all distinct edges, $\{(X_1, Y_1), (X_2, Y_2), \ldots, (X_n, Y_n), \ldots\}$ in the graph stream. (X_i, Y_i) represents an edge between the two nodes X_i and Y_i. Each graph G_i is drawn on the subset of massive node set V and contains a subset of edges from set E. GS also has d different types of side information associated with it, denoted by $\tau = \{T_1, T_2, \ldots, T_d\}$. Each type of side information T_l, where $l = 1, 2, \ldots, d$ has multiple values $S_l = \{S_{l1}, S_{l2}, \ldots, S_{ln}, \ldots\}$. Our aim is to incorporate side information in the graph stream classification process in order to improve the classification accuracy over the existing graph stream classification algorithms.

Now, we give an overview of our GSCS approach. GSCS is composed of four modules, showed in Figure 1. Each graph object consisting of graph structure and side information, is received from the stream and processed according to *GSCS* approach. The *first* module hashes a graph's edges into a fixed-size edge set and then maximal cliques are mined from the compressed graph. In this module, we have modified the approach in $DICH$ [6] by using a more efficient algorithm, for maximal clique detection for better performance. *Second* module is followed from the approach stated in $DICH$ [6], for managing the the exponential number of cliques into a fixed-size feature space. The *third* module is used for side information mining and its design is inspired from $DICH$ [6]. Finally, we have designed the *fourth* module to incorporate side information in the classification process.

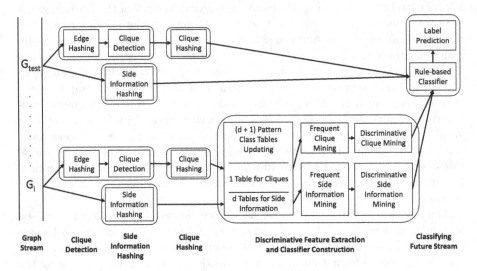

Fig. 1. $GSCS$ approach for graph stream classification with side information

As in stream scenario, due to the high incoming rate of enormous amount of data, a summary of stream needs to be saved for future mining. An in-memory data structure, enabling high speed access-update operations, is used for storing the summary. $DICH$ [6] used one in-memory table, where $GSCS$ will be using $(d + 1)$ in-memory tables. One table for the summary of the graph linkage structure, the other d tables for the summary of the d types of side information. These modules are briefly described in Section 2.1-2.4.

2.1   Clique Detection

In this module, the original edge set of each received graph G_i is hashed into a compressed edge set, and maximal cliques are mined from the compressed edge set to be used as feature representation of the linkage structure of graph. In graph stream mining, massive universe of nodes and continuous arrival of data make storing the original stream intractable. So a summary of the stream needs to be saved for mining purpose. Hence, we hash the original edge set of graph G_i to a compressed edge set. Edge compression involves hashing the two node labels of an edge into a fixed-size node space $\{1, \ldots, N\}$, and then considering the two hashed node label values as a compressed edge. Each time an edge is mapped into a compressed edge, the weight of the compressed edge is increased by 1. If multiple edges of G_i are mapped into the same index due to hash collision, the weight of that compressed edge is set to the number of edges that result into the same index after applying hash function.

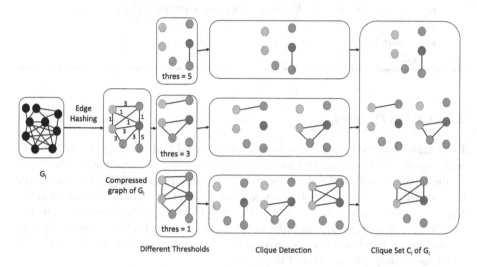

Fig. 2. Clique detection in a compressed graph

Compressed representation of the original graph is significantly smaller and it is now feasible to run clique detection algorithm. Different weights of compressed edges are taken into account and maximal cliques are mined at different edge weight threshold. Graphlet basis estimation algorithm [10] is adapted for this purpose. Though $DICH$ [6] used the naive BronKerbosch algorithm [11], for improved performance, $GSCS$ uses BronKerbosch with pivoting technique [12].

Let $\overline{G_i}$ be the compressed graph of G_i. $\overline{G_i^{(t)}}$ denotes the compressed graph which consists of only compressed edges with weight $\geq t$. Let $max(\overline{G_i})$ denotes the largest and $min(\overline{G_i})$ denotes the smallest edge weights in $\overline{G_i}$, respectively. For maximal clique detection, at first we apply threshold to the compressed $\overline{G_i}$ at different weight level t to get $\overline{G_i^{(t)}}$, where $t = \{max(\overline{G_i}), \ldots, min(\overline{G_i})\}$. BronKerbosch with pivoting algorithm [12] is used at every $\overline{G_i^{(t)}}$ to identify all the maximal cliques. All the cliques found in $\overline{G_i^{(t)}}$ at different weight levels, $\{\overline{G_i^{(t)}}\}_{t=min(\overline{G_i})}^{max(\overline{G_i})}$, are represented as the clique union set C_i of graph G_i. This procedure is shown in Figure 2 and detailed in Algorithm 1.

Algorithm 1. Clique Detection in Compressed Graph

　　　　Input : Graph G_i, size of the compressed node space N
　　　　Output: Clique set C_i of graph G_i

1 **begin**
2　　　$\overline{G_i} \leftarrow Edge - Hash(G_i, N)$
3　　　$C_i \leftarrow \phi$
4　　　**for** $t \leftarrow max(\overline{G_i})$ *to* $min(\overline{G_i})$ **do**
5　　　　　$\overline{G_i^{(t)}} \leftarrow 1(\overline{G_i} \geq t)$
6　　　　　$C_i^{(t)} \leftarrow Bron - Kerbosch - with - pivoting(\overline{G_i^{(t)}})$
7　　　　　$C_i \leftarrow C_i \cup C_i^{(t)}$
8　　　**end for**
9 **end**

2.2　Clique Hashing

Each graph received from the stream is compressed and decomposed into maximal cliques and then frequent and discriminative cliques are identified to build a rule based classifier. As there can be $3^{n/3}$ maximal cliques [13] from a graph of n vertices, so there can be exponential number of cliques generated from the compressed node space. It makes the task of quantifying clique patterns for discriminative feature mining infeasible. To tackle this problem, each clique is hashed into a fixed-size feature space and the regarding information is updated in an in-memory pattern class table.

An in-memory pattern class table Δ_0, consisting of P rows and M columns,

---

Algorithm 2. Clique Hashing

　　　　Input : Clique set C_i of graph G_i
　　　　Output: Update graph linkage structure summary table Δ_0

1 **begin**
2　　　**for** $j \leftarrow 1$ *to* $size(C_i)$ **do**
3　　　　　$H_{i,j} = hash(C_{i,j})$
4　　　　　$\Delta_0[H_{i,j}, L_i] = \Delta_0[H_{i,j}, L_i] + 1$
5　　　**end for**
6 **end**

---

where P is the size of the feature space and M is the number of distinct class labels in stream, is maintained for storing the graph structure summary. For each clique $C_{i,j}$ in clique set C_i of graph G_i, a hash value $H_{i,j} \in \{1, \ldots, P\}$ is generated. When a clique with class label L_i generates hash value $H_{i,j}$, the entry at $\Delta_0[H_{i,j}, L_i]$ is incremented by 1, so that the clique pattern $C_{i,j}$ contributes to the class label L_i. This in-memory clique pattern class table Δ_0 is continuously updated by cliques as with the progression of the stream. The steps of this procedure is given in Algorithm 2.

2.3 Side Information Hashing

As the total number of aggregated side attributes associated with each received graph from the stream can be potentially infinite and it is not feasible to store them directly, so hashing is applied to save a summary of the side information. There are d pattern class tables for storing the summary of the d different types of side information. Each table consists of P rows and M columns where P is predefined size of the feature space and M is the number of distinct class labels in the stream. All types of side information values are hashed into the same fixed-size feature space $\{1, \ldots, P\}$ using the same hash function. Each side information value S_{lj}, where $j = 1, 2, \ldots, q$ and q is a finite number, of type T_l, where $l = 1, 2, \ldots, d$, associated with each graph G_i of the graph stream is hashed to generate a hash index $H_{i,lj} \in \{1, \ldots, P\}$. The i'th graph has class label L_i. Then the entry $\Delta_l[H_{i,lj}, L_i]$, in the side information pattern-class table Δ_l is incremented by 1. These steps are shown in Algorithm 3.

Algorithm 3. Side Information Hashing

    Input   : An array of sets of all types of side information SI_i of graph G_i
    Output: Update the corresponding side information summary table Δ_l
             for each type l of side information, where $l = 1, \ldots, d$

1 **begin**
2   **for** $l \leftarrow 1$ *to* d **do**
3   **for** $j \leftarrow 1$ *to* $size(SI_i[l])$ **do**
4          $H_{i,l_j} = hash(SI_i[l]_j)$
5          $\Delta_l[H_{i,l_j}, L_i] = \Delta_l[H_{i,l_j}, L_i] + 1$
6   **end for**
7   **end for**
8 **end**

2.4 Discriminative Feature Extraction and Classifier Construction

$GSCS$ uses $(d+1)$ in-memory pattern class tables where Δ_0 is for saving graph linkage structure summary and $\Delta_1, \ldots, \Delta_d$ are for storing a sketch of each types of side information in their corresponding table. α and θ, two parameters are used for discriminative feature extraction from these tables. α is the frequent pattern threshold used for selecting frequent features and θ is the discriminative pattern threshold used for selecting frequent yet discriminative patterns [6].

At first, for identifying frequent features, the values in each row of a table are summed up and divided by the maximum row sum in that table. The resulting

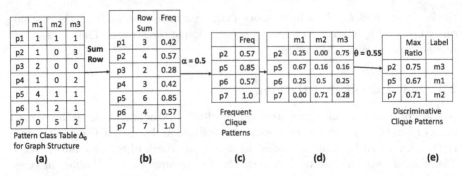

Fig. 3. A small example of frequent and discriminative feature mining. (a) Pattern class table Δ_0 which stores graph structure summary, composed of 7 clique patterns $\{p1, \ldots, p7\}$ (feature space size is 7) and 3 class labels $\{m1, m2, m3\}$. (b) The row sum of each clique pattern with corresponding occurrence frequency (i.e. each row sum in a table is divided by the maximum row sum of that table). (c) Frequent clique patterns with frequencies above the frequent pattern threshold α. (d) Each clique patterns occurrence ratio in each classes. (e) Finally, selected frequent and discriminative clique patterns where maximum ratio of each pattern is greater or equal to discriminative pattern threshold θ.

values indicate the occurrence frequency of a set of cliques in the graph stream. Frequent threshold parameter α is then used to filter out the infrequent patterns whose occurrence frequency are less than α. The remaining patterns are candidate for the discriminative feature selection.

To find the discriminative features, we need to start comparing the occurrence ratios of the features in M classes. For a candidate feature, its occurrence ratio in column i, where $i = 1, 2, \ldots, M$, indicates the probability that the feature belongs to class label i. Higher occurrence ratio of a feature on a certain class indicates a better discriminative capability. Discriminative threshold parameter θ is used to select features whose maximum ratios $\geq \theta$. The process of finding discriminating features from pattern class table using the threshold pair (α, θ) is shown in Figure 3.

Then, a majority voting rule-based classifier is constructed, which has $(d+1)$ sets of discriminative features, 1 set corresponds to the *discriminative clique patterns* and another d sets correspond to the sets of discriminative features extracted from the d side information pattern class tables. Given, a test graph G_{test}, the clique set C of G_{test} is extracted using Algorithm 1. All distinct types of side information are directly hashed and stored into the array of sets SI according to their side information type. Then, all the extracted features from graph G_{test} are given to the majority voting based classifier and for each hashed feature that has a corresponding discriminative pattern of its own type, the class label of that discriminative pattern is taken a vote for the class label of the graph G_{test}. Finally the label of the test graph G_{test} is determined from the majority of the class label votes from the classifier. This is detailed in Algorithm 4.

Algorithm 4. Graph Stream Classification with Side Information

---

Input : A test graph G_{test} from the stream
Output: Predict the class label of G_{test}

1 **begin**
2    $L \leftarrow \phi$
3    $C_{test} = Clique - Detect(G_{test})$;
4   **for** $i \leftarrow 1$ *to* $size(C_{test})$ **do**
5      $H_{0,i} \leftarrow hash(C_i)$
6      $L_{0,i} \leftarrow Find - Rule(\Delta_0, H_{0,i})$
7   **end for**
8   **for** $l \leftarrow 1$ *to* d **do**
9   **for** $i \leftarrow 1$ *to* $size(SI_l)$ **do**
10        $H_{l,i} \leftarrow hash(SI_{l,i})$
11        $L_{l,i} \leftarrow Find - Rule(\Delta_l, H_{l,i})$
12   **end for**
13   **end for**
14    $Label = Majority - Voting - Classifier(L)$
15 **end**

---

3   Experimental Results

In this section we will present the experimental results and techniques. We tested our proposed approach, GSCS for effectiveness and efficiency and compared the results with existing DICH [6] approach. We used two real word data sets, CORA [14] and IMDB [15].

CORA Data Set. The CORA data set contains scientific articles on computer science. To create a graph stream from the articles we considered the co-author relationship as edges of the graph. Research topics were used as class labels. Terms and citations were used as side information.

IMDB Data Set. The Internet Movie Database, IMDB, is a website which contains detailed information about movies and TV shows. We scraped a sample of movies from IMDB which contains 3535 movies released in United States during 2000-2015. We created graph object from each movies using actor-pair as edges. The genre of the movies were used as class labels. We extracted two types of side information: (a) plot keywords and (b) directors.

We used 90% of each data set as training data and used the other 10% as testing data. There are 4 parameters to consider while using GSCS and DICH, frequent pattern threshold α, dicriminative pattern threshold θ, the node space size N, the hash space size P. We vary the parameters and show how GSCS performs in comparison to DICH. The default values for parameters are: $\alpha = .05$, $\theta = .3$, N = 500, P = 10000.

All tests were run on an Asus K550JK running Windows 8.1 x64 with a 2.8 GHz Intel Core i5-4200H CPU and 8 GB of main memory. Both approaches were implemented with C++ and were compiled with tdm-gcc 4.9.2.

Figure 4, 5, 6 and 7 shows different effectiveness measures for GSCS and DICH in both data sets. We measured precision, recall, balanced accuracy and F1 score for multi-class classification [16]. From the graph it is apparent that GSCS performs better than DICH in terms of classification effectiveness. The extra dimensionality provided by the side information helps GSCS to be more effective.

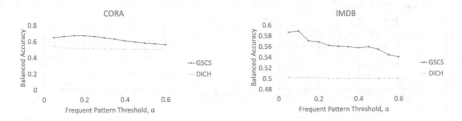

Fig. 4. Balanced accuracy comparison between GSCS and DICH by varying α

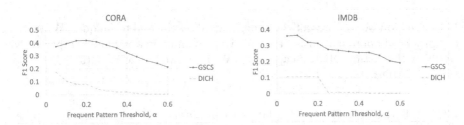

Fig. 5. F1 score comparison between GSCS and DICH by varying α

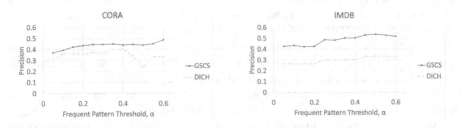

Fig. 6. Precision comparison between GSCS and DICH by varying α

Figure 8 shows that GSCS takes a little bit extra time than DICH. This is the overhead of processing the side information along with the graph structure.

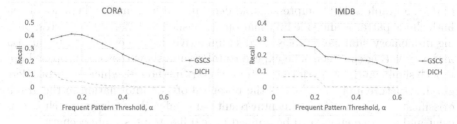

Fig. 7. Recall comparison between GSCS and DICH by varying α

GSCS spends most of its time in feature extraction from the graph and side information. GSCS can process on an average 600 graphs per second in both data sets which is quite good for the stream scenario.

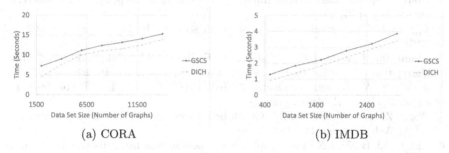

(a) CORA   (b) IMDB

Fig. 8. Time comparison between GSCS and DICH by varying data set size

Since GSCS does not store any data other than in-memory tables for graph and side information, the memory usage is almost constant. In our experiments with both data sets the maximum memory used by GSCS was 100MB. This makes GSCS ideal for the stream scenario.

From the above experimentation, we can conclude that, GSCS is effective in classification of graphs with side information and efficient in the graph stream scenario. Also GSCS outperforms the state of the art graph stream classification approach DICH by providing better performance.

4   Conclusions

In this paper we proposed the first approach which incorporates side information in the classification process of graph streams. The existing graph stream classification algorithms only consider the graph structure and do not utilize side information in the classification process. Many real life applications generate graph streams where side information is associated with each graph, which contains semantically meaningful information relevant to the graph structure, thus can help to build a more accurate classification model. Mining graph streams is a challenging problem because of the high computational cost for graph structure mining and storage difficulty in stream scenario. In our proposed approach

GSCS, a graph is first compressed and decomposed into maximal cliques. Then both clique patterns and side information are hashed and stored into corresponding in-memory summary tables for discriminative feature extraction to build a majority voting classifier. The experimental results show that our proposed approach significantly outperforms state-of-the-art method [6] which only considers graph structure and thus depicts the potential of side information in the graph classification process. Experimental results also infer that our approach is efficient and scalable enough to be applied in real life graph stream scenario.

References

1. Google scholar. Website http://scholar.google.com/ (last access: March 27, 2015)
2. Dblp: computer science bibliography. Website http://dblp.uni-trier.de/ (last access: March 27, 2015)
3. Aggarwal, C.C.: On classification of graph streams. In: SDM 2011, pp. 652–663 (2011)
4. Samiullah, M., Ahmed, C.F., Nishi, M.A., Fariha, A., Abdullah, S.M., Islam, M.R.: Correlation mining in graph databases with a new measure. In: Ishikawa, Y., Li, J., Wang, W., Zhang, R., Zhang, W. (eds.) APWeb 2013. LNCS, vol. 7808, pp. 88–95. Springer, Heidelberg (2013)
5. Wang, Z., Tan, Y., Zhang, M.: Graph-based recommendation on social networks. In: APWeb 2010, pp. 116–122 (2010)
6. Chi, L., Li, B., Zhu, X.: Fast graph stream classification using discriminative clique hashing. In: Pei, J., Tseng, V.S., Cao, L., Motoda, H., Xu, G. (eds.) PAKDD 2013, Part I. LNCS, vol. 7818, pp. 225–236. Springer, Heidelberg (2013)
7. Li, B., Zhu, X., Chi, L., Zhang, C.: Nested subtree hash kernels for large-scale graph classification over streams. In: ICDM 2012, pp. 399–408 (2012)
8. Guo, T., Chi, L., Zhu, X.: Graph hashing and factorization for fast graph stream classification. In: CIKM 2013, pp. 1607–1612 (2013)
9. Yu, P.S., Zhao, Y.: On graph stream clustering with side information. In: SIAM, pp. 139–150 (2013)
10. Soufiani, H.A., Airoldi, E.: Graphlet decomposition of a weighted network. In: AISTATS 2012, pp. 54–63 (2012)
11. Bron, C., Kerbosch, J.: Finding all cliques of an undirected graph (algorithm 457). Commun. ACM 16(9), 575–576 (1973)
12. Tomita, E., Tanaka, A., Takahashi, H.: The worst-case time complexity for generating all maximal cliques. In: Chwa, K.-Y., Munro, J.I. (eds.) COCOON 2004. LNCS, vol. 3106, pp. 161–170. Springer, Heidelberg (2004)
13. Moon, J.W., Moser, L.: On cliques in graphs. Israel Journal of Mathematics 3(1), 23–28 (1965)
14. Cora data set. Website http://people.cs.umass.edu/~mccallum/data.html (last access: March 27, 2015)
15. Imdb - internet movie database. Website http://www.imdb.com (last access: March 27, 2015)
16. Sokolova, M., Lapalme, G.: A systematic analysis of performance measures for classification tasks. Inf. Process. Manage. 45(4), 427–437 (2009)

A Supervised Parameter Estimation Method of LDA

Liu Zhenyan[1,2,3,4], Meng Dan[3], Wang Weiping[3], and Zhang Chunxia[4]

[1] Institute of Computing Technology, Chinese Academy of Sciences, BeiJing, 100190, China
[2] University of Chinese Academy of Sciences, BeiJing, 100049, China
[3] Institute of Information Engineering, Chinese Academy of Sciences, BeiJing, 100093, China
[4] School of Software, Beijing Institute of Technology, BeiJing, 100081, China
zhenyanliu@bit.edu.cn

Abstract. Latent Dirichlet Allocation (LDA) probabilistic topic model is a very effective dimension-reduction tool which can automatically extract latent topics and dedicate to text representation in a lower-dimensional semantic topic space. But the original LDA and its most variants are unsupervised without reference to category label of the documents in the training corpus. And most of them view the terms in vocabulary as equally important, but the weight of each term is different, especially for a skewed corpus in which there are many more samples of some categories than others. As a result, we propose a supervised parameter estimation method based on category and document information which can estimate the parameters of LDA according to term weight. The comparative experiments show that the proposed method is superior for the skewed text classification, which can largely improve the recall and precision of the minority category.

Keywords: LDA, parameter estimation, Gibbs sampling, skewed text classification, term weighting.

1 Introduction

Probabilistic topic model are receiving extensive attention in text mining, information retrieval, natural language processing and so on. Latent Dirichlet Allocation (LDA) proposed by Blei et al. [1] is one of the most notable and most successful probabilistic topic models for unsupervised and supervised learning. Especially for the text classification problem, LDA is a very effective dimension-reduction tool which can automatically extract latent topics and dedicate to text representation in a lower-dimensional semantic topic space.

In text classification, LDA is commonly unsupervised because the parameters of LDA are estimated without reference to category label of the documents in the training corpus. And the terms (or words) in LDA vocabulary are viewed as equally important, but the category discriminating of each term is different, especially for a skewed corpus in which there are many more samples of some categories than others. In other words, for the skewed corpus the importance of a term not only depends on the relationship between it and all categories, but also the sample size of each category. However,

R. Cheng et al. (Eds.): APWeb 2015, LNCS 9313, pp. 401–410, 2015.
DOI: 10.1007/978-3-319-25255-1_33

LDA ignores the valuable information, that is, its two default assumptions are that both the training corpus is balanced and each terms in vocabulary is equally important. Undoubtedly this will cause suboptimal categorization peformance for the skewed text classification.

To address this shortcoming, this paper will propose a novel parameter estimation method based on category and document information which can estimate the parameters of LDA according to the weight of terms. The rest of this paper is organized as follows. Section 2 will first briefly review the related work. Section 3 will analyze LDA model and its classical Gibbs Sampling parameter estimation method. Based on the analysis in section 3, a surpervised parameter estimation method will be presented in section 4, which can especially cope with the skewed text classification problem. In section 5, we will present our comparative experiments of this new method. Finally, section 6 will give the conclusions.

2   Related Work

The LDA model is still a newcomer of topic model, which is in a relatively early stage of development up to now, and most variants of LDA focus on three research directions: parameters extension, context introduction, and orienting special task [2]. However, a few researchers pay attention to term weight in LDA. In fact, the original LDA [1] didn't mention how to build vocabulary, and in subsequent sLDA (supervised Latent Dirichlet Allocation) [3] the vocabulary was chosen by TF-IDF which computed the weight of a term using the product of the term frequency (TF) and the inverse document frequency (IDF) [4]. Madsen et al. [5] proposed Dirichlet Compound Multinomial model using TF-IDF for term weighting. Similarly, Reisinger et al. [6] also made use of TF-IDF to compute term weight in their Spherical topic model.

Moreover, Zhang et al. [7] proposed a weighted LDA model in which the weight of a term is computed based on Gauss function.Wilson et al. [8] extended the LDA model by accommodate the Pointwise Mutual Information to compute term weight.The two extended LDA model aimed to reduce the negative effect of the high frequency terms for topic distribution and incorporated term weight to parameter estimation of LDA.

However, term weight computed by the above methods can only reflect the document information, not the category information. As a result, especially for a skewed corpus, most terms chosen by them may be come from a majority category, which will tend to degrade the performance of classifier directly. But at present few scholars focus on using LDA model for dimensionality reduction of the skewed corpus.

In order to tackle the skewed text classification problem, currently one of the most popular solution pursue to improve traditional term selection method, which are Document Frequency (DF), Information Gain (IG) and so on, or traditional term weighting method, such as TFIDF. For example, Wu et al. [9] proposed a novel term selection method based on category DF, Xu et al. [10] introduced Inverse Document Frequency (ICF) into IG and so on, and Zhang et al. [11] converted TF-IDF to

TF-IDF-IG. Their experiments showed these methods can largely increase the precision and recall of the minority category. Inspired by these methods, this paper will propose a new term weighting method based on TFIDF, and accommodate it to LDA for dimensionality reduction of the skewed corpus.

3   LDA and Gibbs Sampling

For text classification, LDA is a very effective dimension-reduction tool which can automatically mine topics hidden in the documents, where each topic can be viewed as a collection of correlative words, thus each document can be represented using the latent topics.

The basic idea of LDA can be thought as be origined from Latent Semantic Indexing (LSI)[12],which uses the co-occurrences of words to capture the latent semantic associations of words and constructs a lower-dimension latent semantic feature space. This derivation of LSI aims at dimensionality reducion, however there isn't a clear conception of "topic" in LSI. The mathematical basis of LSI is linear algebra, not probability theory. So LSI is not a probabilistic topic model, but lays the foundation for probabilistic topic model.

After LSI, an alternative to LSI named probabilistic LSI (pLSI) was introduced by Hoffmann. The basic idea of pLSI is a document is a mixture of topics and a topic is a mixture of words. In pLSI the concept "topic" appreares clearly, thus pLSI is regarded as the actual origin of probabilistic topic model. However, the two parameters of pLSI —the topic distributions for each document and the word distributions for each topic

— don't be treated as random variables. For this reason, pLSI is not a complete probabilistic topic model.

From Bayesian School's opinion, every parameter should be random variable and every random varialble should follow a prior distribution. So pLSI model is extended by treating the two parameters of pLSI as random variables and introducing Dirichlet prior on them. This new extended model is LDA in which the parameters of model are estimated by Bayesian method. The graphical model of LDA is depicted in Fig. 1.

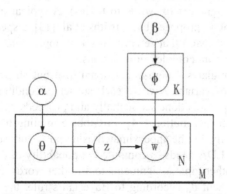

Fig. 1. Graphical model representation of LDA

Where w refers to the observed word in a document which contains N words, z refers to a latent topic, θ refers to the topic distribution for each document, ϕ refers to the word distribution for each topic, α and β are hyperparameters for Dirichlet prior distribution over both θ and ϕ respectively, M is the size of a corpus, N is the length of a document, and K is the number of latent topics in the corpus.

The generative process for a corpus under the LDA model is as follows.

1. Choose $\phi_k \sim$ Dirichlet (β), $k \in [1, K]$
2. For each document $m \in [1, M]$
    (a) Choose $\theta_m \sim$ Dirichlet (α)
    (b) For the nth word in document m, $n \in [1, N_m]$
        Choose a topic $z_{m, n} \sim$ Multinomial (θ_m)
        Choose a word $w_{m, n} \sim$ Multinomial $(\phi_{z_{m,n}})$

That is, to make a new document, at first LDA chooses ϕ_k ($k \in [1, K]$) where $\phi_{i, j} = p(w_i|z_i=j)$ refers to the probability that the jth topic is sampled for the ith word, then for each document m, chooses θ_m where $\theta_i = p(z_i=j)$ refers to the probability of word w_i under topic j, after that, for each word in the current document, chooses a topic $z_{m, n}$, and draws a word $w_{m, n}$ from that topic $z_{m, n}$.

In such LDA model, the probability of a word w_i within a document is:

$$P(w_i) = \sum_{j=1}^{K} P(w_i|z_i = j) P(z_i = j) \tag{1}$$

Furthermore, for a corpus consists of M documents and K latent topics, let $\phi = \{\phi_k\}_{k=1}^{K}$ refer to the multinomial distribution over words for each topic, and let $\theta = \{\theta_m\}_{m=1}^{M}$ refer to the multinomial distribution over topics for each document. Based on this, both ϕ and θ are the main objectives of LDA inference where ϕ represents a K×W (W is the size of the vocabulary) matrix and θ represents a M×K matrix. The parameters ϕ and θ indicate which words are important for which topic and which topics are important for a particular document, respectively.

Unfortunately, it is intractable to learn the parameters ϕ and θ directly.Instead of directly estimating them, another approach is to directly estimate the posterior distribution over z (the assignment of words to topics). A typical implement of this approach is Gibbs Sampling proposed by Griffiths et al. [13], a specific form of Markov Chain Monte Carlo (MCMC) that refers to a set of approximate iterative techniques for obtaining samples from complex distributions.

Gibbs Sampling simulates a high-dimensional distribution by sampling on lower-dimensional subsets of variables where each subset is conditioned on the value of distribution. The sampling is done sequentially and proceeds until the sampled values approximate the target distribution [14]. In Gibbs Sampling method, parameters do not be estimated directly, but be apporximated using posterior estimation of z.

Gibbs Sampler for LDA needs to compute the pobability of a topic being assigned to a word, given all other topic assignments to all other words. For an observed word w_i ($w_i \in w$) in document d_i, according to Bayesian's rule, the conditional posterior distribution for $z_i = k$ ($k \in [1, K]$) is given by

$$P(z_i = k|\mathbf{z}_{\neg i}, \mathbf{w}) \propto P(w_i|z_i = k, \mathbf{z}_{\neg i}, \mathbf{w}_{\neg i})P(z_i = k|\mathbf{z}_{\neg i})$$

$$= \frac{n_{\neg i,k}^{(w_i)}+\beta}{n_{\neg i,k}^{(\cdot)}+W\beta} \cdot \frac{n_{\neg i,k}^{(d_i)}+\alpha}{n_{\neg i,\cdot}^{(d_i)}+K\alpha} \tag{2}$$

Where $\mathbf{z}_{\neg i}$ means all topic assignment except z_i, $\mathbf{w}_{\neg i}$ means all words except w_i in the vocabulary. $n_{\neg i,k}^{(w_i)}$ is the number of times of word w_i assigned to topic k except the current assignment, $n_{\neg i,k}^{(\cdot)}$ is the total number of words assigned to topic k except the current assignment, $n_{\neg i,k}^{(d_i)}$ is the number of words from document d_i assigned to topic k except the current assignment, and $n_{\neg i,\cdot}^{(d_i)}$ is the number of words in document d_i, not including the current word w_i.

Gibbs Sampler starts by assigning each word to a random topic index in [1...K], and then assign a new topic index for every word during each iteration of Gibbs Sampling. After a burn-in period of a few hundred interations, Gibbs Sampler can reach its converged state and two matrices ϕ and θ are estimated from all topic assignment as follows.

$$\phi_{k,w_i} = \frac{n_{\neg i,k}^{(w_i)}+\beta}{n_{\neg i,k}^{(\cdot)}+W\beta} \qquad \theta_{d_i,k} = \frac{n_{\neg i,k}^{(d_i)}+\alpha}{n_{\neg i,\cdot}^{(d_i)}+K\alpha} \tag{3}$$

Here, we can see each term (or word) is equally important in calcuating the conditional posterior distribution for $z_i = k$. However, in the skewed text classification, the important of terms should be especially distinguished, or else which will largely degrade classification performance. That is, the traditional LDA model ignoring term weight must make many mistakes when classifying skewed documents. In order to overcome this limitation of LDA, we will propose an excellent term weighting method to compute term weight, which will be used to estimate the parameters of LDA.

4   A Supervised Parameter Estimation Method of LDA

The IDF factor of the traditional TFIDF is used to indicate the category discriminating power of a term, who believes that the fewer documents a term occurs in, the more discriminating power the term contributes to text classification. However, a term occurred in many documents from a category should be viewed as a strong feature, while a term occurred in fewer documents from some different categories should be viewed as a weak feature. The term weight computed by TFIDF can only reflect the document difference, not the category difference. As a result, TFIDF must be improved based on the category difference and the document difference.

Firstly, for the skewed corpus, the absolute category document frequency of term t, which is the number of documents from a category that have at least one occurrence of term t, cannot accurately measure its category discriminating power. For example, the document frequency of term t is 90 in a major category that contains 1000 instances, and the document frequency of term t is 90 in a minor category that contains

100 instances. Thus term t is more useful to identify this minor category. Therefore, a term that occurs in a minor category should be more valuable than in a major category in case of the same document occurrence number in each category. We will use Relative Category Document Frequency Difference (R-CDFD) to measure the difference of documents contain term t between category c_i and its complement category $\bar{c}_i$. The corresponding formula is given by

$$R\text{-}CDFD(t, c_i) = P(t|c_i) - P(t|\bar{c}_i) \tag{4}$$

Where $P(t|c_i)$ is the conditional probability of term t occurrence given category c_i , $P(t|\bar{c}_i)$ is the conditional probability of term t occurrence given category $\bar{c}_i$. $P(t|c_i) = \frac{D_t \cap D_{c_i}}{D_{c_i}}$, here D_t denotes the number of documents that have at least one occurrence of term t, D_{c_i} denotes the number of documents that belong to category c_i. $P(t|c_i) = \frac{D_t \cap D_{\bar{c}_i}}{D_{\bar{c}_i}}$, here $D_{\bar{c}_i}$ denotes the number of documents that belong to category $\bar{c}_i$.

Secondly, in order to give a higher score to a term occurred in a minor category, the category distribution should be taken into account. The lower the probability of the category contains term t, the higher weight term t will achieve. Moreover, another important factor is the relation between term and category which can be measured by the conditional probability of a category given that term t occurred. And then the higher this conditional probability is, the higher weight term t will achieve.

The above three factors, i.e., R-CDFD, the category distribution, the relation between term and category, can characterize respectively a profile of term weight, so the three factors should be integrated to compute term weight. Hence, an integrated factor named as Relative Category Difference (RCD) is constructed, which contains the above three sub-factors, and the corresponding formula is as follows.

$$RCD(t) = \sum_{i=1}^{|C|} |RCDFD(t, c_i)| \lg \frac{1+P(c_i|t)}{P(c_i)}$$

$$= \sum_{i=1}^{|C|} |P(t|c_i) - P(t|\bar{c}_i)| \lg \frac{1+P(c_i|t)}{P(c_i)} \tag{5}$$

Where $|C|$ denotes the total numbers of categories in the corpus, D denotes the total number of documents in the corpus, $P(c_i|t) = \frac{D_{c_i} \cap D_t}{D_t}$ is the conditional probability of category c_i given term t occurred, and $P(c_i) = \frac{D_{c_i}}{D}$ is the probability of category c_i, here D_{c_i} denotes the number of documents that belongs to category c_i and D denotes the total number of documents in the corpus.

Then, the RCD is incorporated to replace the IDF of TFIDF. In LDA the new TF-RCD term weighting schema will be used to choose the vocabulary, and estimate parameters which replace the term frequency in Eq.3 and Eq.4 with the sum of term weight as follows.

$$\phi_{k,w_i} = \frac{ws_{\neg i,k}^{(w_i)} + \beta}{ws_{\neg i,k}^{(\cdot)} + W\beta} \qquad \theta_{d_i,k} = \frac{ws_{\neg i,k}^{(d_i)} + \alpha}{ws_{\neg i,\cdot}^{(d_i)} + K\alpha} \tag{6}$$

Where $wS_{\neg i,k}^{(w_j)}$ is the weighted sum of word w_i assigned to topic k except the current assignment, $wS_{\neg i,k}^{(\cdot)}$ is the weighted sum of words assigned to topic k except the current assignment, $wS_{\neg i,k}^{(d_j)}$ is the weighted sum of words from document d_i assigned to topic k except the current assignment, and $wS_{\neg i,\cdot}^{(d_j)}$ is the weighted sum of words in document d_i, not including the current word w_i.

5   Experiment

In order to verify the new parameter estimation method of LDA, we construct experiments focused on a comparison of TFIDF and TF-RCD in LDA. We run experiments on a subset of WebKB dataset from Ana [17], which have been pre-processed that includes tokenization and stop word removal. The experiment dataset contains 4,199 documents with four categories: "project", "course", "faculty" and "student", which is a skewed corpus that 504 documents belong to "project" category, 930 to "course", 1,124 to "faculty", and 1,641 to "student". 84% of all distinct words are observed in "student" category, 80% in "faculty", 68% in "course", and 64% in "project". Fig.2 gives the category distribution and term distribution of the WebKB dataset used in our experiment.

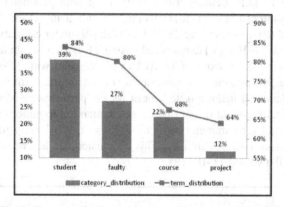

Fig. 2. Category distribution and term distribution in WebKB

On this skewed experiment dataset LDA model is trained. A 5000-term vocabulary of LDA is chosen by TFIDF or TF-RCD. And term weight is incorporated into Gibbs Sampling to assign a proper topic for the term. Then the documents are represented in latent topic space drawn by LDA. We build SVM (Support Vector Machine) classifier with LIBSVM development kit [18], in which linear kernel function is used. The reason for using SVM is that SVM has a better performance than other classification methods in text classification since it is based on the structural risk minimization principle.

Commonly the evaluation metrics for the skewed text classification are macro-averaged precision, macro-averaged recall, macro-averaged F1 [19]. Since macro-averaged scores are averaged values over the number of categories, and then the performance of classifier is not dominated by major categories. Let P be the precision, R be recall, and m denotes the total number of categories, then macro-averaged precision is $\frac{1}{m}\sum_{i=1}^{m}P_i$, macro-averaged recall is $\frac{1}{m}\sum_{i=1}^{m}R_i$, macro-averaged F1 is $\frac{1}{m}\sum_{i=1}^{m}F1_i$, where F1 is $\frac{2PR}{P+R}$.

Five-fold cross-validation is performed on the experiment dataset. For this purpose, the corpus is initially partitioned into five folds. In each experiment, four fold's data are used to train while one fold's data are used to test. The average of five experiments results is reported in Table1.

Table 1. The F1 scores comparison of four   schemes

	TF-RCD (P)	TF-RCD (V)	TF-IDF (P)	TF-IDF (V)
student	0. 9699	0. 9649	0. 9650	0. 9550
faculty	0. 9443	0. 9085	0. 8695	0. 8436
course	0. 9396	0. 9050	0. 8595	0. 8335
project	0. 9091	0. 8049	0. 7647	0. 7595
macro_ave_F1	0. 9407	0. 8958	0. 8647	0. 8479

In Table1, TF-RCD(P) denotes using TF-RCD in both parameter estimation and vocabulary choosing of LDA, TF-RCD(V) denotes only using TF-RCD in vocabulary choosing, TF-IDF(P) denotes using TF-IDF in both parameter estimation and vocabulary choosing , and TF-IDF (V) denotes only using TF-IDF in vocabulary choosing.

The macro-averaged F1 score of TF-RCD(V), compared with TF-IDF (V), is just improved   about 3%, and then we can draw a conclusion that term weight only used to build vocabulary will make a little benefit for the performance of the skewed text classifier. But if term weight doesn't pay enough attention to minor category, though term weight is used for parameter estimation, it can't also largely improve the performance of the skewed text classifier. This conclusion can be drawn from the comparison of TF-IDF (V) and TF-IDF(P).

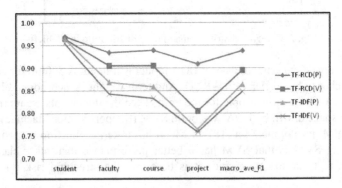

Fig. 3. The classifier performance comparison of four schemes

From Table1 we can see that the macro-averaged F1 score of TF-RCD(P), compared with TF-RCD(V), TF-IDF (P) and TF-IDF (V), is the highest and the minority categories benefit most significantly. Fig.3 gets further insights about the comparison of TF-RCD(P), TF-RCD(V), TF-IDF(P), and TF-IDF (V) with chart form. As can be seen from Fig.3, the use of TF-RCD in vocabulary choosing and parameter estimation can greatly improve the whole performance of the skewed text classifier.

6   Conclusion

TF-RCD is a superior term weighting method especially for skewed text categorization. The term weight computed by TF-RCD can not only reflect the document difference but also the category difference, while TFIDF can only reflect the document difference. The RCD of TF-RCD integrate three important factors, i.e., the relative category document frequency difference, the category distribution, the relation between term and category, can devote to measure the category discriminating power of a term.

As a result, TF-RCD can fairly choose more discriminative terms from every category to build vocabulary for LDA, and the term weights computed by TF-RCD are incorporated into parameter estimation to mine latent topics in skewed corpus. The comparative experiments show that the supervised parameter estimation method is superior for the skewed text classification, which can largely improve the recall and precision of rare category.

Acknowledgements. This work was financially supported by National Natural Science Foundation of China (61272361), also supported by Key Project of National Defense Basic Research Program of China (B11201320), National HeGaoJi Key Project (2013ZX01039 -002-001-001), National High-Tech Research and Development Program of China (2012AA011002).

References

1. Blei, D., Ng, A., Jordan, M.: Latent Dirichlet Allocation. Machine Learning Research 3(3), 993–1022 (2003)
2. Xu, G., Wang, H.: The Development of Topic Models in Natural Language Processing. Chinese Journal of Computers 34(8), 1423–1436 (2011) (in Chinese)
3. Blei, D., McAuliffe, J.: Supervised topic models. Advances in Neural Information Processing Systems 20, 121–128 (2008)
4. Salton, G., Buckley, C.: Term-Weighting Approaches in Automatic Text Retrieval. Journal of Information Processing & Management 24(5), 513–523 (1988)
5. Madsen, R., Kauchak, D., Elkan, C.: Modeling word burstiness using the dirichlet distribution. In: Proceedings of the 22nd International Conference on Machine Learning, pp. 545–552 (2005)
6. Reisinger, J., Waters, A., Silverthorn, B., Mooney, R.: Spherical topic models. In: Proceedings of the 27th International Conference on Machine Learning, pp. 903–910 (2010)

7. Zhang, X., Zhou, X., Huang, H., et al.: An improved LDA Topic Model. Journal of Beijing Jiaotong University 34(2), 111–114 (2010) (in Chinese)
8. Wilson, A., Chew, P.: Term weighting schemes for latent dirichlet allocation. In: The 2010 Annual Conference of the North American Chapter of the Association for Computational Linguistics, pp. 465–473 (2010)
9. Wu, D., Zhang, Y., Yin, F., Li, M.: Feature Selection Based on Class Distritution Difference and VPRS for Text Classification. Journal of Electronics & Information Technology 29(12), 2880–2884 (2007) (in Chinese)
10. Xu, Y., Li, J., Wang, B., Sun, C., Zhang, S.: A Study of Feature Selection for Text Categorization on Imbalanced Data. Journal of Computer Research and Development 44(suppl.), 58–62 (2007) (in Chinese)
11. Zhang, A., Jing, H., Wang, B., Xu, Y.: Research on Effects of Term Weighting Factors for Text Categorization. Journal of Chinese Information Processing 24(3), 97–104 (2010) (in Chinese)
12. Deerwester, S., Dumais, S., Landauer, T., Furnas, G., Harshman, R.: Indexing by Latent Semantic Analysis. Journal of the American Society of Information Science 41(6), 391–407 (1990)
13. Heinrich, G.: Parameter estimation for text analysis. Technical Note Version 2.9. http://www.arbylon.net/publications/text-est2.pdf (2009)
14. Steyvers, M., Griffiths, T.: Probabilistic topic models. In: Landauer, T., McNamara, D.S., Dennis, S., Kintsch, W. (eds.) Handbook of Latent Semantic Analysis. Erlbaum, Hillsdale (2007)
15. Koller, D., Sahami, M.: Hierarchically classifying documents using very few words. In: Proceedings of the 14th International Conference on Machine Learning, pp. 170–178 (1997)
16. Mladenic, D., Grobelnk, M.: Feature selection for unbalanced class distribution and Naïve Bayes. In: Proceeding of the 16th International Conference Machine Learning, pp. 258–267 (1999)
17. http://web.ist.utl.pt/~acardoso/datasets/
18. http://www.csie.ntu.edu.tw/~cjlin/
19. Manning, C.D., Raghavan, P., Schutze, H.: Introduction to Information Retrieval. Cambridge University Press (2010)

Learning Similarity Functions for Urban Events Detection by Mining Hotline Phone Records

Pengjie Ren, Peng Liu, Zhumin Chen, Jun Ma, and Xiaomeng Song

Department of Computer Science and Technology,
Shandong University, China, 250101
`chenzhumin@sdu.edu.cn`

Abstract. Many cities around the world have established a platform, entitled *public service hotline*, to allow citizens to tell about city issues, e.g. noise nuisance, or personal encountered problems, e.g. traffic accident, by making a phone call. As a result of "crowd sensing", these records contain rich human intelligence that can help to detect urban events. In this paper, we present an event detection approach to detect urban events based on phone records. Specifically, given a set of phone records in a period of time, we first learn a similarity matrix. Each element of the matrix is estimated as the probability that the corresponding pair of records describe the same event. Then, we propose an *Improved Affinity Propagation* (*IAP*) clustering approach which takes the similarity matrix as input and generates clusters as output. Each cluster is an urban event composed of several records. Extensive experiments demonstrate the great improvement of *IAP* on three standard datasets for clustering and the effectiveness of our event detection approach on real data from a hotline.

Keywords: Event Detection, Data Mining, Urban Computing, Machine Learning.

1 Introduction

The rapid progress of information and communication technologies not only modernizes people's lives, but also creates great opportunities and challenges for urban policy and management. An important part of urban management is monitoring and solving millions of urban events everyday, e.g. traffic accident, noise nuisance, etc. This problem is increasingly concerned by city administrators, especially major cities. They are calling for technologies that can automatically detect and monitor the citywide events and the distribution of events in different places.

While monitoring urban events is very difficult, some ubiquitous data sources indicating urban events are already available. For example, many cities have operated a hotline platform that allows people to make a phone call to tell about what they feel annoyed by. The phone records are actually a result of "human as a sensor" and "crowd sensing", containing rich human intelligence that can help

© Springer International Publishing Switzerland 2015
R. Cheng et al. (Eds.): APWeb 2015, LNCS 9313, pp. 411–423, 2015.
DOI: 10.1007/978-3-319-25255-1_34

detect urban events [1,2], especially emergency events [3,4]. A hotline receives a lot of phone calls everyday and many citizens tell about the same event. It is heavy work to manually find out how many events and what events these records describe. From the perspective of urban management, urban alerts must be detected early, preferably before they explode, and therefore the number of records involved may be small at the time of detection. That makes the task harder than standard topic detection, mainly due to sparsity issues.

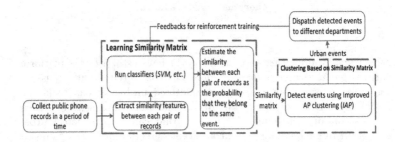

Fig. 1. Event Detection Process of *Urban-EMS*.

In this study, we present the event detection approach of our system, *Urban-EMS* (*Urban Events Monitoring System*). Generally, given a set of records, we need to identify their events and for each record, further decide which event it describes. The process of the approach is summarized in Fig. 1. We have modeled the problem as a combination of two tasks:

1. The first is learning phone record similarity: we use all types of features to learn a supervised classifier that takes two records as input and decides if the records describe the same event.

2. The second is applying an improved version of *Affinity Propagation* clustering algorithm that uses the positive confidence probability of the classifier above as a similarity measure between records and generates clusters as urban events.

Finally, the events are dispatched to different departments and their feedbacks are used to re-train the classifiers.

We detail our approach in Section 2, describe and discuss the result of our experimentation in Section 3, review related work in Section 4, and summarize our study and discuss future work in Section 5.

2   Our Approach

2.1   Modeling Similarity as a Classification Task

For our problem, many events only consist of a small number of phone records. Pure unsupervised approaches often mis-detect them. So following the methodology proposed in [5,6] for a different clustering problem, we first model our problem as a binary classification task: given a pair of records r_1 and r_2, the system must decide whether they belong to the same event (true) or not (false). Each pair of records is represented as a set of features $F(r_i, r_j)$, which is used to feed a classification model $CM(r_i, r_j)$.

$$CM(r_i, r_j) : \mathcal{PR} \times \mathcal{PR} \rightarrow \{true, false\} \tag{1}$$

$CM(r_i, r_j)$ is SVM with confidence probability estimation and default parameter settings in this paper. Once we have learned to classify records pairs, we take the positive classification confidence probability as a similarity measure.

$$S_{i,j} = P(CM(r_i, r_j) = true | F(r_i, r_j)) \tag{2}$$

where $F(r_i, r_j)$ is the set of features for records pair (r_i, r_j) that will be defined later. All similarities $S_{i,j}$ between each pair of records form the similarity matrix S, which is then used by our proposed *Improved Affinity Propagation* (*IAP*) clustering algorithm to identify urban events.

Each raw phone record r is a four-tuple (*content,time,phone,category*), where *content* is the content of the phone call recorded by telephonists, *time* is the time when the phone call is received, *phone* is the phone number, *category* is the category selected by telephonists. Note that the categories are predefined, e.g. traffic, pollution, etc. An instance is shown as follows:

content: Mr. Zhang calls: A traffic accident happens in Jiefang Road.

time: 13:35:47 2012-11-30

phone: ***

category: city issues→ city traffic→ traffic order→ traffic accident

In our study we consider a total of 13 features that capture many characteristics of phone records. Features can be divided in five families: Property Features ($F_{PF}(r_i, r_j)$), Term Position-based Features ($F_{TPF}(r_i, r_j)$), Bag-of-Word-based Features ($F_{BWF}(r_i, r_j)$), Knowledge-based Similarity Features ($F_{KSF}(r_i, r_j)$) and Neighbor-based Similarity Features ($F_{NSF}(r_i, r_j)$).

Property Features ($F_{PF}(r_i, r_j)$). The property features describe the property similarity of two records, e.g. category, address, time, etc.

Given a record, we first extract its address descriptions if exists. Then we submit extracted address descriptions to an online coordinate converter[1] to get their GPSs. Finally, the centroid of these GPSs is considered as the GPS of the record. Specifically, we adopt three approaches for address extraction, namely *Dictionary-based Extraction*, *Regex-based Extraction*, and *Named Entity Recognizer-based Extraction*. For *Dictionary-based Extraction*, we compare each segmented phase with Chinese dictionaries[2] for place names. For *Regex-based Extraction*, we define several Regex (Regular Expression) patterns to extract address descriptions, e.g. "[0-9\u4E00-\u9FA5]*?(square

---

[1] http://developer.baidu.com/map/index.php?
   title=webapi/guide/webservice-geocoding
[2] http://pinyin.sogou.com/dict/cate/index/167

|viaduct|rode|street|alley)". For *Named Entity Recognizer-based Extraction*, we use the Chinese NER[3] tool provided by Stanford NLP.

Geographical Distance Feature (f_{GDF}) measures the geographical distance of two records based on the obtained GPSs[4].

$$f_{GDF}(r_i, r_j) = \ln\left[2 \cdot 6378.137 \cdot \arcsin\left(\sqrt{\sin^2(\frac{a}{2}) + \cos(lat(r_i)) \cdot \cos(lat(r_j)) \cdot \sin^2(\frac{b}{2})}\right)\right] \tag{3}$$

where $a = lat(r_i) - lat(r_j)$, $lat(r)$ is the latitude of record r; $b = lon(r_i) - lon(r_j)$, $lon(r)$ is the longitude of r; 6378.137 is the radius of the earth in kilometers. Note that f_{GDF} is set an extreme value when GPS is not available for some records.

The other three property features are: Time Gap Feature (f_{TGF}) measures the time distance of two records in seconds. Category Feature (f_{CF}) measures the category consistency of two records. The existing category system of *Urban-EMS* is a tree structure. So f_{CF} is defined as the maximum category level that both records belong to. Person Feature (f_{PF}) indicates whether the records are from the same person. $PF = 1$ if they are, 0 if unknown, or -1 otherwise.

Term Position-based Features ($F_{TPF}(r_i, r_j)$). The first Term Position-Based Feature is called Term Match Feature (f_{TMF}).

$$f_{TMF}(r_i, r_j) = \frac{s(r_i, r_j) + s(r_j, r_i)}{2}$$

$$s(r_i, r_j) = \frac{1}{|r_i|} \sum_{p=1}^{|r_i|} \frac{|r_j| - \min\{|r_j|, |p - q||q \in Pos(r_j, r_i^p)\}}{|r_j|} \tag{4}$$

where r_i^p represents the term at the p-th position of record r_i, $Pos(r_j, r_i^p)$ is the set of all positions of record r_j where the term is r_i^p. The more words two records share, the higher f_{TMF} is. The more similar word order also indicates higher f_{TMF}.

The other three Term Position-based Features are: Longest Common Subsequence Feature (f_{LCSeqF}) measures the similarity of two records by finding the longest common subsequence. Longest Common Substring Feature (f_{LCStrF}) measures the similarity of two records by finding the longest common substring. Levenshtein Distance Feature (f_{LDF}) measures the edit distance between two sequences. Informally, the Levenshtein distance between two records is the minimum number of single-term edits (i.e. insertions, deletions or substitutions) required to change one record into the other.

Bag-of-Word-Based Features ($F_{BWF}(r_i, r_j)$). The first Bag-of-Word-based Feature, Jaccard Similarity Feature (f_{JSF}), measures the similarity of two records

---

[3] http://nlp.stanford.edu:8080/ner/
[4] http://www.movable-type.co.uk/scripts/latlong.html

by *Jaccard* metric.

$$f_{JSF}(r_i, r_j) = \frac{|T(r_i) \cap T(r_j)|}{|T(r_i) \cup T(r_j)|} \tag{5}$$

where $T(r)$ is the term set of record r.

Another Bag-of-Word-based Feature, Cosine Similarity Feature (f_{CSF}), measures the similarity of two records by *Cosine* metric.

$$f_{CSF}(r_i, r_j) = \frac{V(r_i) \cdot V(r_j)}{|V(r_i)||V(r_j)|} \tag{6}$$

where $V(r)$ is the term vector of record r.

Knowledge-Based Similarity Feature ($F_{KSF}(r_i, r_j)$). HowNet-based Similarity Feature (f_{HSF}) measures the semantic similarity of two records based on an extra Chinese knowledge base HowNet[5]. For two records r_i and r_j, suppose that $|r_i| \le |r_j|$, we have

$$f_{HSF}(r_i, r_j) = \frac{\max\limits_{p \to q} \sum\limits_{p=1}^{|r_i|} wordSim(r_i^p, r_j^q)}{|r_i|} \tag{7}$$

where r_i^p is the p-th word of r_i, similarly, r_j^q is the q-th word of r_j; the function *wordSim* is the semantic similarity of two words computed based on HowNet [7]. Records similar to each other generally tend to have same special words, or extensively high words semantic similarity. To calculate the f_{HSF} similarity between r_i and r_j, we first calculate *wordSim* on each pair of words. Then we greedily select $|r_i|$ pairs of words without repetition, which maximize the sum of their words' similarity across these two record strings. Note that $p \to q$ is a one-to-one mapping, i.e. a q-th word of r_j can only be mapped to one p-th word of r_i. Finally, to normalize this feature, we divide it by $|r_i|$.

Neighbor-Based Similarity Feature ($F_{NSF}(r_i, r_j)$). Neighbor-based Cosine Similarity Feature (f_{NCSF}) measures the semantic similarity by computing *Cosine* of the expanded term vectors.

$$f_{NCSF}(r_i, r_j) = \frac{V^n(r_i) \cdot V^n(r_j)}{|V^n(r_i)||V^n(r_j)|} \tag{8}$$

where $V^n(r)$ is the expanded term vector of record r by adding its n neighbors. In this paper, the neighbors are top $n = 5$ search results with Lucene Vector Space Model.

---

[5] http://www.keenage.com/html/e_index.html

Symmetric KL-divergence Feature (f_{SKLF}) measures the semantic similarity by computing symmetric KL-divergence between language models of the expanded records.

$$f_{SKLF}(r_i, r_j) = \frac{KL(r_i, r_j) + KL(r_j, r_i)}{2}$$

$$KL(r_i, r_j) = \sum_{w \in r_i^n} P(w|r_i^n) \ln \frac{P(w|r_i^n)}{P(w|r_j^n)}$$

(9)

where r^n is the expanded record of r by adding its n neighbors.

2.2 Improved Affinity Propagation (IAP)

Next, we detail our improved version of a famous clustering approach, *Affinity Propagation (AP)* [8]. Specifically, let $\mathcal{PR} = \{r_1, r_2, ..., r_N\}$ be a set of phone records. Let $\vartheta(i)$ associate to each r_i the index of its nearest exemplar (Each exemplar corresponds to a cluster). Then the goal of *Improved Affinity Propagation (IAP)* is to find the mapping $\theta = \{\vartheta(i)|r_i \in \mathcal{PR}\}$ maximizing the expectation:

$$E(\theta)$$
$$= \sum_{i=1}^{N} \left(\alpha S_{i,\vartheta(i)} + (1-\alpha) \frac{\sum\limits_{r_j \in \mathcal{PR}^k(C_{\vartheta(i)})} S_{i,j}}{k} \right) - \sum_{i=1}^{N} X[\vartheta(\vartheta(i)) = \vartheta(i)]$$

(10)

where $S_{i,j}$ is the similarity between r_i and r_j; $C_{\vartheta(i)} = \{r_j|\vartheta(j) = \vartheta(i); r_j \in \mathcal{PR}\}$ is the cluster whose exemplar is $\vartheta(i)$. $\mathcal{PR}^k(C_{\vartheta(i)})$ is top k representatives (excluding $r_{\vartheta(i)}$) selected from the current cluster $C_{\vartheta(i)}$. The first part in Formula 10 indicates how well $r_\vartheta(i)$ is suited to be the exemplar for r_i. The second part in Formula 10 expresses that if r_i is selected as an exemplar by some records, it has to be its own exemplar, with $X[\vartheta(\vartheta(i)) = \vartheta(i)] = \infty$ if $\vartheta(\vartheta(i)) \neq \vartheta(i)$ and 0 otherwise.

In order to detail the idea of *IAP*, we explain *IAP* with an election voting process. Specifically, when choosing an exemplar (corresponding to a cluster) for record r_i, we consider two aspects of votes balanced by a parameter α. The first aspect is $S_{i,\vartheta(i)}$, which is from the current exemplar $\vartheta(i)$. The second aspect is $\sum_{r_j \in \mathcal{PR}^k(C_{\vartheta(i)})} S_{i,j}$ which is considering other k representative records in the current cluster $C_{\vartheta(i)}$. In this paper, the k representatives are k records with highest scores in $\{S_{j,\vartheta(i)}|r_j \in C_{\vartheta(i)}, r_j \neq r_\vartheta(i)\}$. We use the second aspect to increase the robustness so as to alleviate the situation where $S_{i,\vartheta(i)}$ is incorrect or inaccurate. Also note that Formula 10 does not directly specify the number of exemplars (corresponding to the number of clusters) to be found. Similar to raw *AP*, it takes as input a real number $S_{i,i}$ for each record r_i so that records with larger values are more likely to be chosen as exemplars of their clusters. Usually, $S_{i,i}$ is set to the median value of S.

The resolution of the optimization problem defined by Formula 10 can be achieved by a *Message Passing Algorithm* [8], considering two types of messages: The "responsibility similarity" $\Phi_{i,j}$, sent from record r_i to candidate exemplar r_j, reflects the accumulated similarity for how well-suited r_j is to serve as the exemplar for r_i, taking into account other potential exemplars for r_i. The "availability similarity" $\Lambda_{i,j}$, sent from candidate exemplar record r_j to record r_i, reflects the accumulated similarity for how appropriate it would be for r_i to choose r_j as its exemplar, taking into account the support from other records that r_j should be an exemplar.

Algorithm 1. *Message Passing Algorithm* For *IAP*.

Input:
  Similarity matrix, S; Parameters, α, k; Maximum iterations, M;
Output:
  Clusters with exemplars, $\{C_{\vartheta(i)}\}$;
1: Initialize the responsibility similarity matrix Φ, availability similarity matrix Λ;
2: Initialize the integer $t = 0$;
3: **for** i=1 to S.length **do**
4:    Set $C_{\vartheta(i)} = \{i\}$;
5: **end for**
6: Run raw AP clustering to initialize $\{C_{\vartheta(i)}\}$;
7: **while** Not Convergence and $t < M$ **do**
8:    Update $\Phi_{i,j}$ and $\Lambda_{i,j}$ with Formula 11;
9:    Update $\vartheta(i)$ for each item i with Formula 12;
10:    Update $C_{\vartheta(i)}$ accordingly;
11:    t++;
12: **end while**
13: **return** $\{C_{\vartheta(i)}\}$;

All availability and responsibility similarity messages $\Lambda_{i,j}$ and $\Phi_{i,j}$ are set to 0 initially. Their values are iteratively updated according to:

$$\Phi_{i,j} = \Psi_{i,j} - \max_{j' \neq j}\{\Lambda_{i,j'} + \Psi_{i,j'}\}; i \neq j$$

$$\Phi_{i,i} = S_{i,i} - \max_{j' \neq i}\{\Psi_{i,j'}\}$$

$$\Lambda_{i,j} = \min\{0, \Phi_{j,j} + \sum_{i' \neq i,j}\max\{0, \Phi_{i',j}\}\}; i \neq j$$

$$\Lambda_{i,i} = \sum_{i' \neq i}\max\{0, \Phi_{i',i}\}$$

$$\Psi_{i,j} = \alpha S_{i,j} + (1 - \alpha)\frac{\sum\limits_{r_{j'} \in \mathcal{PR}^k(C_{\vartheta(i)})} S_{i,j'}}{k}$$

(11)

The index of exemplar $\vartheta(i)$ associated to r_i is defined as:

$$\vartheta(i) = \arg\max_j \{\Phi_{i,j} + \Lambda_{i,j} | j = 1, 2, ..., N\} \tag{12}$$

The pseudo code is shown in Algorithm 1. The algorithm stops after a maximal number of iterations or when the exemplars do not change for a given number of iterations.

3   Experiment

3.1   Evaluation of Event Detection Approach

Experimental Setup. We evaluate our event detection approach with one month public phone records from a mayor hotline[6]. We adopt two best approaches from [14] as baselines, which use *Affinity Propagation* clustering with *Cosine* similarity and *Jaccard* similarity respectively. Five groups of measures are used which are summarized in Table 1. For all measures, larger values mean better results.

Table 1. Evaluation Measures.

Groups	Symbol	Explanations
	P-J	Jaccard
Pair counting measures [9]	P-P	Precision
	P-R	Recall
	P-F	Fowlkes-mallows (flat only, non-hierarchical) [10]
Entropy based measures [11]	E-NMI-J	Joint Normalized Mutual Information
	E-NMI-S	Sqrt Normalized Mutual Information
BCubed based measures [10]	B-P	Precision
	B-R	Recall
Set Matching based measures [12]	S-P	Purity
	S-IP	Inverse Purity
Gini measures [13]	Gini	Mean of Gini coefficient

Model Performance Comparison. The comparison of our event detection approach with baselines on real public phone records dataset is summarized in Table 2. Generally, our approaches especially IAP^{CLS} achieve great improvement. As we can see, the improvement comes from two aspects. First, the proposed IAP greatly improves the clustering process, which reflects in comparison of IAP^{Cos} with AP^{Cos}, IAP^{Jac} with AP^{Jac} and IAP^{CLS} with AP^{CLS}. Second, the proposed features and supervised learning approach can approximate the real similarity between two records more accurately, which reflects in comparison of IAP^{Linear} with IAP^{CLS}. This indicates that previous annotations can be used to learn better similarity functions compared with heuristical or unsupervised approaches. We also analyze the phenomenon of the reduced measures, i.e. P-P, B-R and S-IP. The reason is that baselines tend to split large clusters into many

---

[6] The dataset is not published for privacy issues.

Table 2. Model Comparison on Real Public Phone Records Datasets.

Models	P-J	P-P	P-R	P-F	E-NMI-J	E-NMI-S	B-P	B-R	S-P	S-IP	Gini
AP^{Cos}	0.201	**0.738**	0.223	0.406	0.652	0.794	0.328	0.761	0.440	0.829	0.544
AP^{Jac}	0.163	0.720	0.174	0.354	0.635	0.784	0.274	**0.773**	0.378	**0.839**	0.523
IAP^{Cos}	0.236	0.623	0.275	0.414	0.660	0.797	0.405	0.722	0.508	0.796	0.563
IAP^{Jac}	0.216	0.681	0.240	0.404	0.655	0.796	0.352	0.747	0.460	0.812	0.549
AP^{CLS}	0.374	0.385	0.929	0.598	0.812	0.899	0.939	0.648	0.960	0.730	0.793
IAP^{Linear}	0.312	0.354	**0.996**	0.601	0.811	0.901	0.941	0.647	0.987	0.728	0.795
IAP^{CLS}	**0.492**	0.493	0.958	**0.701**	**0.839**	**0.916**	**0.989**	0.632	**0.993**	0.690	**0.811**

AP^{Cos}: Baseline. Unsupervised: raw AP clustering + Cosine similarity.
AP^{Jac}: Baseline. Unsupervised: raw AP clustering + Jaccard similarity.
IAP^{Cos}: Unsupervised: IAP clustering + Cosine similarity.
IAP^{Jac}: Unsupervised: IAP clustering + Jaccard similarity.
AP^{CLS}: Supervised: raw AP clustering + SVM + proposed features.
IAP^{Linear}: Unsupervised: IAP clustering + Linear and uniform feature combinations.
IAP^{CLS}: Supervised: IAP clustering + SVM + proposed features.

Table 3. Feature Effectiveness Analysis: *SVM Accuracy* (10-fold cross validation).

Used Features	Accuracy(%)	Used Features	Accuracy(%)
F_{PF}	79.8	$All - F_{PF}$	72.9
F_{TPF}	65.57	$All - F_{TPF}$	82.2
F_{BWF}	71.63	$All - F_{BWF}$	82.3
F_{KSF}	66.23	$All - F_{KSF}$	82.09
F_{NSF}	67.15	$All - F_{NSF}$	82.07
All	84.35		

small clusters, which results in abnormal improvement of the three measures. For example, AP^{Cos} achieves the best in terms of *Precision* (P-P). However, it's *Recall* (P-R) is very low, which leads to a 31.7% decrease in terms of *F1-measure* compared with our model IAP^{CLS}.

Feature Effectiveness. Table 3 gives the performance of different feature combinations. Unsurprisingly, combining all features gives the best choice with an accuracy of 84.35%. The accuracy is not very high which indicates that the problem is challenging. In terms of single feature, F_{PF} is the most effective. F_{PF} are some unique features for our data compared with many other short text datasets. Besides, though effectiveness of different features is varied, however drop of any feature causes an accuracy decline, which is significant according to *paired t-test* (*p-value*<0.05). This means all our proposed features are useful.

3.2 Evaluation of *IAP*: *Improved Affinity Propagation*

Experimental Setup. The improvement of *IAP* on our dataset is significant. We wonder whether *IAP* is still effective on standard clustering datasets. So we further evaluate the performance of *IAP* algorithm with three standard datasets

Table 4. Comparison of AP and IAP ($\alpha = 0.5, k = 1$) with Cosine Similarity.

	Models	P-J	P-P	P-R	P-F	E-NMI-J	E-NMI-S	B-P	B-R	S-P	S-IP	Gini
Iris	AP	0.412	**0.948**	0.422	0.632	0.488	0.682	0.422	**0.934**	0.507	**0.960**	0.678
	IAP	**0.746**	0.944	**0.780**	**0.858**	**0.649**	**0.791**	**0.780**	0.927	**0.847**	**0.960**	**0.853**
Wine	AP	0.221	0.574	0.264	0.389	0.205	0.355	0.272	**0.645**	0.365	**0.702**	0.458
	IAP	**0.323**	**0.591**	**0.416**	**0.496**	**0.227**	**0.374**	**0.430**	0.631	**0.534**	0.680	**0.530**
Zoo	AP	0.160	0.956	0.161	0.392	0.017	0.076	0.177	**0.964**	0.208	**0.980**	0.571
	IAP	**0.248**	**0.971**	**0.250**	**0.493**	**0.020**	**0.085**	**0.264**	**0.964**	**0.396**	**0.980**	**0.614**

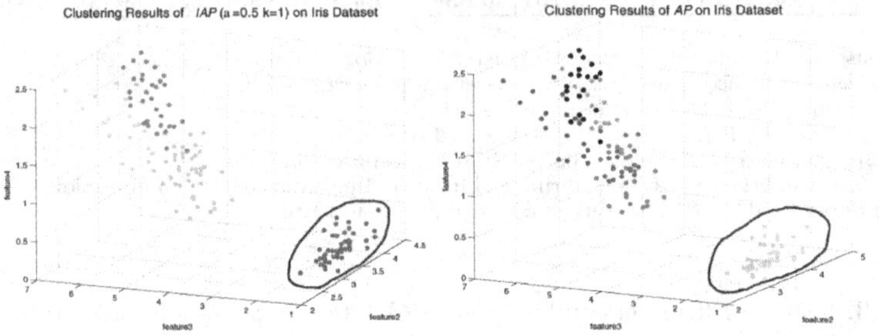

Fig. 2. Visual Comparison of AP and IAP on Iris Dataset (True clusters number : 3).

for clustering test, Iris[7], Wine[8] and Zoo[9]. The raw AP is adopted as the baseline. The same measures summarized in Table 1 are adopted.

Model Performance Comparison. The comparison of raw AP and IAP is summarized in Table 4. We report experiments with *Cosine* similarity only, other metrics have similar results e.g. *Euclidean* distance, etc. We can see that great improvement is achieved by IAP in terms of almost all measures, especially on Iris dataset. The most possible reason is, the additional consideration of representatives (corresponding to the second part of Formula 10) in the current cluster make IAP more robust. In another word, when deciding the exemplar object $\vartheta(i)$ for an object i, the wrong or inaccurate estimation of similarity $S_{i,\vartheta(i)}$ is alleviated by consideration of additional representatives during the clustering process. On the contrary, relying on a single real-value $S_{i,\vartheta(i)}$ to decide $\vartheta(i)$ tends to split large clusters. This can be confirmed by visual comparison of AP and IAP clustering results (especially the circled ones) shown in Fig. 2. We further analyze the reduced measures in Table 4. Take P-P and P-R on Iris dataset for example, the slight reduction of P-P (precision) results in great improvement of P-R (recall). The same applies to the other three pairs.

We also analyze the parameter sensitiveness of IAP. The results are shown in Fig. 3. On all three standard datasets, α is not sensitive for all tried values basically and k is also not sensitive for a wide range of values. Note that some measures decrease for the last two points because we tried $k = 25$ and $k = 30$.

---

[7] https://archive.ics.uci.edu/ml/datasets/Iris

[8] https://archive.ics.uci.edu/ml/datasets/Wine

[9] https://archive.ics.uci.edu/ml/datasets/Zoo

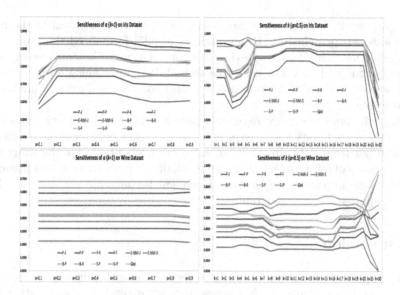

Fig. 3. *IAP* Parameters Sensitiveness Analysis. Similar results are achieved on Zoo dataset, so we omit the figures to save space.

The results mean that when applying *IAP* we do not need to worry about spend much time on tuning the parameters. Though not sensitive, the choice of parameter values indeed has a little influence on clustering results. Generally, $\alpha \in (0.2, 0.5)$ and $k \in (5, 20)$ are better choices.

4   Related Work

Recent event detection techniques include wave analysis [15], topic model [16], hierarchical dirichlet process (HDP) [17], text clustering [18] and so on [19].

AlSumait et al. [20] presented online topic model (OLDA), a topic model that automatically captures the thematic patterns and identifies emerging topics of text streams and their changes over time. Wang et al. [21] proposed a mixture Gaussian model for bursty word extraction in Twitter and then employed a novel time-dependent HDP model for new topic detection. Sakaki et al. [22] first devised a classifier of tweets based on features such as the keywords in a tweet, the number of words, and their context. Then, they produced a probabilistic spatiotemporal model for the target event that can find the center and the trajectory of the event location. Becker et al. [23] focused on online identification of real-world event and its associated Twitter messages using an online clustering technique, which continuously clusters similar tweets and then classifies the clusters content into real-world events or nonevents. Chen et al. [24] presented Non-Parametric Heterogeneous Graph Scan (NPHGS), a new approach that considers the entire heterogeneous network for event detection: they first modeled the network as a sensor network. Then, they efficiently maximized a nonparametric scan statistic over connected subgraphs to identify the most anomalous network clusters. Rozenshtein et al. [25] formalized the problem of event detection using two graph-theoretic formulations. The first one captures the compactness of an event using the sum of distances among all pairs of the event nodes. The second formulation captures compactness using a minimum-distance tree.

Although there are many researches investigating event detection, however most of them are bursty based event detection, i.e. they only focus on identifying events major from data. In addition, some are not suitable for the short text characteristics of our problem (A record is usually no more than three sentences). Besides, some important application dependent features are not considered by existing approaches, e.g. geographical features, time features, etc.

5 Discussion

In this paper, we introduce the event detection algorithm in our *Urban-EMS* system and demonstrate its effectiveness through extensive experiments. Great improvement is achieved compared with baseline approaches. In future work, we plan to further improve it by defining more features, e.g. language model features, syntax features, etc. Besides, further work is necessary on issues like streaming clustering, automatic events description, etc.

References

1. Rana, R., Chou, C., Kanhere, S., Bulusu, N., Hu, W.: Ear-phone: an end-to-end participatory urban noise mapping system. In: IPSN 2010 (2010)
2. Zheng, Y., Liu, T., Wang, Y., Zhu, Y., Chang, E.: Diagnosing new york citys noises with ubiquitous data. In: Ubicomp 2014 (2014)
3. Xie, Z., Yan, J.: Kernel density estimation of traffic accidents in a network space. Computers, Environment and Urban Systems (2008)
4. Li, J., Zhou, Y., Shang, W., Cao, C., Shen, Z., Yang, F., Xiao, X., Guo, D.: A cloud computation architecture for unconventional emergency management. In: Gao, Y., Shim, K., Ding, Z., Jin, P., Ren, Z., Xiao, Y., Liu, A., Qiao, S. (eds.) WAIM 2013 Workshops 2013. LNCS, vol. 7901, pp. 187–198. Springer, Heidelberg (2013)
5. Spina, D., Gonzalo, J., Amigó, E.: Learning similarity functions for topic detection in online reputation monitoring. In: SIGIR 2014 (2014)
6. Artiles, J., Amigó, E., Gonzalo, J.: The role of named entities in web people search. In: EMNLP 2009 (2009)
7. Xia, T.: Study on chinese words semantic similarity computation. Computer Engineering (2007)
8. Frey, B., Dueck, D.: Clustering by passing messages between data points. Science (2007)
9. Achtert, E., Goldhofer, S., Kriegel, H.-P., Schubert, E., Zimek, A.: Evaluation of clusterings–metrics and visual support. In: ICDE 2012 (2012)
10. Amigó, E., Gonzalo, J., Artiles, J., Verdejo, F.: A comparison of extrinsic clustering evaluation metrics based on formal constraints. Information Retrieval (2009)

11. Meilă, M.: Comparing clusterings by the variation of information. In: Schölkopf, B., Warmuth, M.K. (eds.) COLT/Kernel 2003. LNCS (LNAI), vol. 2777, pp. 173–187. Springer, Heidelberg (2003)
12. Zhao, Y., Karypis, G.: Empirical and theoretical comparisons of selected criterion functions for document clustering. Machine Learning (2004)
13. Rezankova, H., Loster, T., Husek, D.: Evaluation of categorical data clustering. In: Mugellini, E., Szczepaniak, P.S., Pettenati, M.C., Sokhn, M. (eds.) AWIC 2011. AISC, vol. 86, pp. 173–182. Springer, Heidelberg (2011)
14. Rangrej, A., Kulkarni, S., Tendulkar, A.: Comparative study of clustering techniques for short text documents. In: WWW 2011 (2011)
15. Weng, J., Lee, B.: Event detection in twitter
16. Diao, Q., Jiang, J., Zhu, F., Lim, E.: Finding bursty topics from microblogs. In: ACL 2012 (2012)
17. Gao, Z., Song, Y., Liu, S., Wang, H., Wei, H., Chen, Y., Cui, W.: Tracking and connecting topics via incremental hierarchical dirichlet processes. In: ICDM 2011
18. Aggarwal, C., Zhai, C.: A survey of text clustering algorithms. In: Mining Text Data 2012 (2012)
19. Gupta, M., Li, R., Chang, K.: Towards a social media analytics platform: event detection and user profiling for twitter. In: WWW 2014 (2014)
20. AlSumait, L., Barbará, D., Domeniconi, C.: On-line lda: Adaptive topic models for mining text streams with applications to topic detection and tracking. In: ICDM 2008 (2008)
21. Wang, X., Zhu, F., Jiang, J., Li, S.: Real time event detection in twitter. In: Wang, J., Xiong, H., Ishikawa, Y., Xu, J., Zhou, J. (eds.) WAIM 2013. LNCS, vol. 7923, pp. 502–513. Springer, Heidelberg (2013)
22. Sakaki, T., Okazaki, M., Matsuo, Y.: Earthquake shakes twitter users: real-time event detection by social sensors. In: WWW 2010 (2010)
23. Becker, H., Naaman, M., Gravano, L.: Beyond trending topics: Real-world event identification on twitter
24. Chen, F., Neill, D.: Non-parametric scan statistics for event detection and forecasting in heterogeneous social media graphs. In: SIGKDD 2014 (2014)
25. Rozenshtein, P., Anagnostopoulos, A., Gionis, A., Tatti, N.: Event detection in activity networks. In: SIGKDD 2014 (2014)

Answering Spatial Approximate Keyword Queries in Disks

Jinbao Wang[1], Donghua Yang[1], Yuhong Wei[2], Hong Gao[1],
Jianzhong Li[1], and Ye Yuan[1]

[1] Harbin Institute of Technology, Harbin, Heilongjiang, China
[2] ZTE Co. Ltd, Shenzhen, China
{wangjinbao,yang.dh,honggao,lijzh,yuanye}@hit.edu.cn,weiyh_zte@163.com

Abstract. Spatial approximate keyword queries consist of a spatial condition and a set of keywords as the fuzzy textual conditions, and they return objects labeled with a set of keywords similar to queried keywords while satisfying the spatial condition. Such queries enable users to find objects of interest in a spatial database, and make mismatches between user query keywords and object keywords tolerant. With the rapid growth of data, spatial databases storing objects from diverse geographical regions can be no longer held in main memories. Thus, it is essential to answer spatial approximate keyword queries over disk resident datasets. Existing works present methods either returns incomplete answers or indexes in main memory, and effective solutions in disks are in demand. This paper presents a novel disk resident index RMB-tree to support spatial approximate keyword queries. We study the principle of augmenting R-tree with capacity of approximate keyword searching based on existing solutions, and store multiple bitmaps in R-tree nodes to build an RMB-tree. RMB-tree supports spatial conditions such as range constraint, combined with keyword similarity metrics such as edit distance, dice etc. Experimental results against R-tree on two real world datasets demonstrate the efficiency of our solution.

Keywords: spatial database, approximate keyword search, index structure, query processing.

1 Introduction

More and more web sites are supporting location-based keyword search which consists of a spatial condition and a set of keywords as textual condition. Such query is formulated as the spatial keyword query, and aims to find objects with keywords description within a location region or close to a location. It requires users to query with exact keywords associated to objects, and mismatches between queried keywords and object keywords will lead to result loss. In real-life applications, such mismatch is prevalent since there are errors in object keywords and users also produce typos while inputting their queries. Research community introduces spatial approximate keyword search[8][7] to deal with the mismatch problem. Instead of looking for objects with exact keywords provided by the

© Springer International Publishing Switzerland 2015
R. Cheng et al. (Eds.): APWeb 2015, LNCS 9313, pp. 424–436, 2015.
DOI: 10.1007/978-3-319-25255-1_35

query, spatial approximate keyword query returns objects with keywords similar enough to the queried keywords. Spatial approximate keyword query enables great usability when processing massive data containing errors, and it also makes users' typos tolerant.

Recently, research community has proposed *MHR-tree* [8] and *LBAK-tree* [7] to support spatial approximate keyword search. *MHR-tree* incorporates keywords' min-hash signatures into R-tree nodes to enable search functionality over approximate keyword conditions, and it may returns incomplete query results without quality guarantee. *LBAK-tree* is a memory based index structure which stores inverted lists of q-grams into R-tree nodes. Such solution consumes too large memory to work on massive data, since it costs memory larger than the original dataset. Meanwhile, *LBAK-tree* can not be extended to disk because *LBAK-tree* node size is quite unbounded. With the rapid growth of data, spatial databases are no longer able (or cost effective) to hold all the data and indexes in main memories. Thus, it is essential to answer spatial approximate keyword queries over disk resident datasets. Up to now, there is no disk resident index to support spatial approximate keyword queries in an exact manner.

This paper proposes **R**-tree incorporated with **M**ultiple **B**itmaps (RMB-tree) summarizing keywords to support spatial approximate keyword queries. RMB-tree augments multiple bitmaps which summarize the $q - grams$ for the set of keywords inside a sub-tree, and enables searching on fuzzy keyword conditions with the summary. Given a spatial approximate keyword query, RMB-tree calculates the number of $q - grams$ absent from sub-trees for each queried keyword, and prunes sub-trees with enough absent $q - grams$. To achieve this goal, we first provide a SGB-filter with a single bitmap, and then give the design of MGB-filter based on multiple bitmaps providing better pruning effects. Compared to existing solutions, RMB-tree is able to return exact query results on disks.

This paper makes the following contributions:

1. An disk-resident index named RMB-tree is proposed. RMB-tree employs multiple bitmaps to support pruning on fuzzy textual condition.
2. We give the methods for RMB-tree utilization, including the insertion method together with the query processing method.
3. Experiments over real-life datasets are conducted, and the results validate the effectiveness and efficiency of RMB-tree.

The rest of this paper is organized as follows. Section 2 introduces the preliminary and the definition of the spatial approximate keyword query. Section 3 presents the structure and utilization methods of RMB-tree. Section 4 presents the performance evaluation, and we conclude this paper in Section 5.

2   Preliminary

This section introduces preliminaries used in the rest of this paper, including gram-based approximate string search and the formulation of the spatial approximate keyword query. Here we take edit distance as the string similarity metric, and other similarity metrics can be adopted in similar ways.

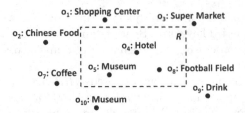

Fig. 1. An example of a spatial approximate query

2.1   Gram-Based Approximate String Search

Given a finite alphabet $\sum$, a string s is a sequence of chars in $\sum$. The ith char of s is labeled $s[i]$, and the subsequence from the ith char to the jth char of s is denoted $s[i, j]$. The length of string s is denoted by $|s|$.

Given two chars α, β which are not from $\sum$, a new string s' is built by adding $q - 1$ copies of α before s and $q - 1$ copies of β after s. The ith q-gram of string s (also referred as $g(i, s)$) is the subsequence of s' starting from the ith character with length q, and it equals to $s'[i, i + q - 1]$. A string s has $|s| + q - 1$ q-grams. The q-gram set of a string s contains $|s| + q - 1$ unique q-grams of s, and it is denoted by $GS(s)$.

The edit distance between two strings s_1 and s_2, denoted $ED(s_1, s_2)$ is defined as the minimum number of edits needed to transform one string to the other, with the allowable edit operations being insertion, deletion, or substitution of a single character.

Example 1. Given two strings $s_1 = theatre$ and $s_2 = theater$, the 2-gram set of s_1 and s_2 are $GS(s_1) = \{\#t, th, he, ea, at, tr, re, e\#\}$ and $GS(s_2) = \{\#t, th, he, ea, at, te, er, r\#\}$ respectively. The edit distance $ED(s_1, s_2) = 2$.

Existing works on string similarity search explores the relation between edit distance and q-gram sets of strings, and an intuition is that strings with small edit distance have a small number of different q-grams. The number of different q-grams of s_1 and s_2 denoted $|GS(s_1)/GS(s_2)| \leqslant q \times \theta$, if $ED(s_1, s_2) \leqslant \theta$. RMB-tree follows gram-based solutions to deal with approximate keyword conditions.

2.2   Spatial Approximate Keyword Search

Here we formulate range approximate keyword query. Suppose there is a spatial object set $O = \{o_1, ..., o_N\}$ where each object $o_i = (loc, k_1, ..., k_j)$ has a location coordinate loc and j keywords $(k_1, ..., k_j)$ describing itself.

Definition 1. *(Range Approximate Keyword Query) A range approximate keyword query Q_{RAK} is formatted as $(R, (k_1, \theta_1), ..., (k_m, \theta_m))$, where R is a spatial region and k_i is a queried keyword and θ_i is its similarity threshold for $1 \leqslant i \leqslant m$. Given a spatial object set $O = \{o_1, ..., o_N\}$, Q_{RAK} finds all the objects o satisfying that (1)$o.loc \in R$, (2)$\exists k_i \in o.keywords$ and $ED(k_i, Q_{RAK}.k_u) \leqslant Q_{RAK}.\theta_u$ for all $1 \leqslant u \leqslant m$.*

Figure 1 depicts an example of querying in a set of business objects $\{o_1, ..., o_{10}\}$. Q_{RAK} looks for objects labeled with keywords whose edit distance is no more than 2 to "$Musum$" within the square region. Object o_5 is returned.

3 RMB-Tree

This section introduces a novel disk resident index named RMB-tree. RMB-tree is able to answer queries with a spatial constraint and approximate keyword conditions. Section 3.1 describes the structure of RMB-tree, and Section 3.2 shows the utilization of RMB-tree involving specified update and query processing algorithms.

3.1 Structure of RMB-Tree

The main idea of RMB-tree is to augment R-tree with capacity of searching fuzzy keywords, and such idea is in common with recent proposed work [7][8]. RMB-tree stores a set of bitmaps called *gram bitmap* in internal nodes, and *gram bitmap* enables a tree node to judge whether there possibly exists keywords similar enough to given keywords in its descendent. Our proposed index is named RMB-tree since is augments **R**-tree nodes with **M**ultiple **B**itmaps. In next parts, we first introduce RSB-tree which is the single bitmap version of RMB-tree, then comes multiple bitmap solution of RMB-tree. The formal definition of *gram bitmap* is given as follows.

Definition 2. *(Gram Bitmap)* Given a string $s \in (\sum)^*$, an integer q and a uniform hash function $h_b : (\sum)^q \to \{1, 2, ..., L\}$, the gram bitmap of s is a sequence belong to $\{0,1\}^L$, whose the ith bit equals to 1 if and only if $\exists g \in GS(s), h_b(g) = i$. The gram bitmap of s is denoted by $GB(s)$. Given a string set S, the gram bitmap of S denoted $GB(S)$ belongs to $\{0,1\}^L$, and the ith bit of $GB(S)$ is 1 if and only if there exists a string $t \in S$ satisfying that $GB(t)[i] = 1$.

We denote the subtree rooted at an RSB-tree node N as $SubTree(N)$, and the keyword set stored in $SubTree(N)$ is denoted by $KS(N)$. An internal node N_i stores $GB(KS(N_i))$. Given a spatial approximate keyword query $Q(S, K)$, internal node N_i employs $GB(KS(N_i))$ to make a secure estimation whether $KS(N_i)$ possibly provides all necessary keywords to satisfy $Q(S, K)$'s approximate keyword constraints. This estimation may bring false positives but never loses any results. The following theorems guarantees the correctness of search space pruning at internal node N_i.

Theorem 1. *Given a string s and an edit distance threshold θ, if there exists a string $t \in KS(N_j)$ satisfying that $ED(s, t) \leqslant \theta$, then the following inequality holds.*

$$|\{k|1 \leqslant k \leqslant L, GB(s)[k] = 1, GB(N_i)[k] = 0\}| \leqslant q \times \theta$$

Proof. Since there exists a string $t \in KS(N_j)$ satisfying that $ED(s,t) \leqslant \theta$, $|GS(s)/GS(t)| \leqslant q \times \theta$. Then, inequation $|\{k|1 \leqslant k \leqslant L, GB(s)[k] = 1, GB(t)[k] = 0\}| \leqslant q \times \theta$ holds true. For all $1 \leqslant k \leqslant L$, $GB(N_j)[k] = 1$ if $GB(t)[k] = 1$. Thus $|\{k|1 \leqslant k \leqslant L, GB(s)[k] = 1, GB(N_j)[k] = 0\}| \leqslant |\{k|1 \leqslant k \leqslant L, GB(s)[k] = 1, GB(t)[k] = 0\}| \leqslant q \times \theta$.

According to Theorem 1, an RSB-tree internal node is able to prune search space among its children for spatial approximate keyword queries. Here we denote the filter with single gram bitmap by *SGB-Filter*, and it is employed to estimate whether a subtree possibly contains query results. Here is the definition of *SGB-Filter*.

Definition 3. *(SGB-Filter) Given a string t, an edit distance threshold θ and a string set S. SGB-Filter computes $|\{g|g \in GS(t), GB(S)[h_b(g)] = 0\}|$, and returns true if and only if the computed value is no larger than $q \times \theta$.*

Given a string s and a string set S, RSB-tree employs the same principle with *MHR-tree*, and they conclude there does not exist $t \in S$ that $ED(s,t) \leqslant \theta$, if $|\widehat{GS(s)/GS(S)}| > q \times \theta$. $|\widehat{GS(s)/GS(S)}|$ is the estimated value by *SGB-Filter* or min-hash signatures. *SGB-Filter* estimates whether a gram g exists in $GS(S)$ according to $GB(S)$, and it produces false positive if $g \notin GS(S)$ but $GB(S)[h_b(g)] = 1$. The hash function h_b, which is adopted in *SGB-Filter*, hashes a gram g to L positions of the bitmap with equal probability, so

$$Pr(h_b(g) = i) = \frac{1}{L}$$

Given a set of strings denoted S, and $|\bigcup_{t \in S} GS(t)| = N$, the probability $p_L = Pr(GB(S)[h_b(g)] = 1|g \notin GS(S))$ is related to the length of bitmap L. It is computed as

$$p_L = 1 - (1 - \frac{1}{L})^N \approx 1 - e^{-\frac{N}{L}}$$

when N and L are relatively large, say hundreds or thousands. In a block-based tree index, a 4KB disk page may contain hundreds of identical grams, and *SGB-Filter* fails to improve filtering effects by increasing the length of gram bitmap. Figure 2 illustrates the probability of false positive while estimating gram membership. Suppose the total gram number in a string set is around 300, we increase the single bitmap length, and find that false positive probability decreases very slowly When single bitmap length is more than 1000. Such result not only shows the limit of RSB-tree's pruning effects but also that of *MHR-tree*. In the meanwhile, increasing the length of bitmap also produces more internal tree nodes, thus costs more I/O operations during query processing. The single bitmap solution fails to further improve pruning since it takes a set of strings as a whole, and ignores the individual content of each string. Next we provide an example.

Example 2. Given string set $S = \{s_1, s_2, s_3, s_4\}$, $GS(s_1) = \{g_1, g_2, g_3\}$, $GS(s_2) = \{g_4, g_5, g_6\}$, $GS(s_3) = \{g_7, g_8, g_9\}$ and $GS(s_4) = \{g_{10}, g_{11}, g_{12}\}$, and the queried

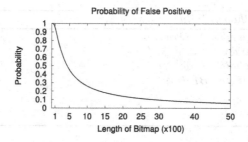

Fig. 2. Probability of Gram Membership False Positive Produced by *SGB-Filter*

string t that $GS(t) = \{g_1, g_4, g_7, g_{10}\}$. The single bitmap solution estimates $|GS(t)/GS(S)|$ as 0, and it provides no pruning effects on t since $|GS(t)/GS(s_i)|$ = 3 for $i = 1, 2, 3, 4$.

Based on the above observation, RMB-tree employs multiple bitmaps to explore the content of individual strings for improving the filtering power at tree nodes. Such strategy divides a set of strings into k groups, and builds a gram bitmap for each group. Given a string t, the multiple bitmap solution invokes *SGB-Filter* with each gram bitmap, and returns true if all the groups return true. The gram membership estimation gives a correct answer with probability

$$p_L^k = 1 - (1 - \frac{1}{\frac{L}{k}})^{\frac{N}{k}} \approx 1 - e^{-\frac{N}{k}/\frac{L}{k}} = 1 - e^{-\frac{N}{L}}$$

if the grams is nearly equally-divided into k groups. This means the multiple bitmap based solution does not decrease the accuracy of membership estimation, at the same time it provides opportunities to disseminate grams from different strings to different groups. Such dissemination increases approximate keywords pruning power since it identifies some grams do not occur in identical keywords. Next we introduce the definition of *MGB-Filter* designed for RMB-tree.

Definition 4. *(MGB-Filter) Given a string t, an edit distance threshold θ and a string set S which is divided into k partitions denoted $S_1, \ldots, S_k$, MGB-Filter invokes SGB-Filter on each partition S_i, and return true if t and θ pass a SGB-Filter on any partition.*

An RMB-tree internal node N is formatted as $< P, R, B_1, ..., B_k, CS >$, where P is the disk page id of N and R is the spatial region of $SubTree(N)$. $B_1 ... B_k$ are gram bitmaps of k keyword set partitions in $SubTree(N)$. CS is N's children set, in which each item consists of the children's page id, spatial region, and gram bitmaps. A leaf node N_L of RMB-tree is formatted as $< P, R, B_1, ..., B_k, DS >$ including N_L's disk page id, spatial region, gram bitmaps, and data set. Objects are stored in N_L's data set DS.

3.2 Utilization of RMB-Tree

This section introduces the utilization of RMB-tree, including object insertion and query processing algorithms.

Algorithm 1. *ChooseChild*

    Input: RMB-tree node N, *Object o*
    Output: $N's$ *child i*
1 **for** $j = 1$ *to* $|N.ChildSet|$ **do**
2     $c(j, o) = (ROE_j(o), BE_j(o))$;
3 $i = arg_j min\{c(j, o)\}$;
4 **return** $N.ChildSet[i]$;

Algorithm 2. *Insert*

    Input: RMB-tree node N , Object o
1 **if** N *is full* **then**
2     $Split(N, o)$
3 refine $N.SpatialRegion$ with $o.loc$;
4 **for** $kw \in o.KeywordSet$ **do**
5   **for** *each gram bitmap bm of N* **do**
6   compute $BE(bm, kw)$;
7   insert kw into bm with minimum $BE(bm, kw)$;
8 **if** N *is an internal node* **then**
9     $Child = ChooseChild(N, o)$;
10    $Insert(N, o)$;
11 **else**
12   insert o into $N.DataSet$;

Insertion of RMB-Tree Similar to R-tree's insertion, an insertion in *RMB-tree* starts from the root and selects a subtree to accomplish the insertion at each level of the tree. When an insertion reaches a leaf node finally, the object is inserted into the leaf node, then the leaf node updates its shape including spatial region and gram bitmaps. If the leaf node is full, RMB-tree processes the split from bottom to top if necessary. RMB-tree's insertion differs from R-tree in the process of subtree selection and insertion at a tree node. Algorithm 1 describes how an RMB-tree internal node N selects subtree to finish the insertion. N computes insertion cost for all of its children nodes. The insertion cost of a child node j is a two-tuple $c(j, o) = (ROE_j(o), BE_j(o))$. ROE is the increased area of spatial region overlap between the chosen child and other children after insertion. This is similar to the insertion of R*-tree. $BE_j(o)$ is the number of bits changing from 0 to 1 in j's gram bitmaps during insertion. RMB-tree chooses a child with minimum insertion cost, since smaller insertion cost leads to smaller spatial region overlap and smaller bitmap enlargement. Such selection of insertion target produces RMB-tree with better pruning power. Algorithm 2 shows how an RMB-tree inserts object o into node N. Similar to R*-tree's insertion algorithm, Algorithm 2 firstly checks whether N is full and invokes R*-tree's split algorithm to deal with node split when N is full (line 1-2). Otherwise it first

---

Algorithm 3. *RangeQueryProcessing*

    Input: RMB-tree node N, $Q(R, (k_1, \theta_1), ..., (k_m, \theta_m))$
    Output: Object set R

1 $R = \phi$;
2 **if** N *is a leaf* **then**
3   **for** $i \in N.dataSet\ inside\ R$ **do**
4   **if** i *satisfies Q's constraints* **then**
5           $R = R \cup \{i\}$;

6 **else**
7   **for** $i \in N.ChildSet$ **do**
8   boolean forward = false;
9   **if** $i.SpatialRegion$ *intersects with* $Q.R$ **then**
10   **for** *each* $k_i, 1 \leq i \leq m$ **do**
11   boolean $flag = MGB - Filter(k_i, \theta_i)$;
12   **if** $flag = true$ **then**
13                   $forward = true$;
14   break;

15   **if** $forward = true$ **then**
16           $R = R \cup RangeQueryProcessing(i, Q)$

17 return R;

---

refines N's spatial region (line 3) and gram bitmaps (line 4-7). Algorithm 2 adopts greedy strategy to minimize the loss of pruning power at N, and it chooses a gram bitmap for each of o's keyword to join. The chosen gram bitmap has fewest enlargement, thus the overall pruning power decrease is minimized. Then, Algorithm 2 finishes the insertion according to the type of N in the same manner as R*-tree (line 8-13). The method *Split* invoked in line 2 also refers to gram bitmap refinement, and it also greedily selects a bitmap for each keyword of o. The bitmap selection also aims to achieve minimum number of bits changing from 0 to 1. The detail is omitted for space.

Query Processing of RMB-Tree RMB-tree T invokes Algorithm 3 at its root node to process a range approximate keyword query. Algorithm 3 presents how an RMB-tree node N processes a query. If N is a leaf node, it checks data objects inside it, and objects satisfying Q's constraints are added into the query result. If N is an internal node, it examines all of its children with *MGB-Filter* to find which children possibly contains query results of Q (line 7-14). Then Algorithm 3 is invoked recursively at chosen children nodes, and returned results are combined (line 15-16). The query processing is similar to that of R-tree, and the difference lies in pruning with *MGB-Filter* among subtrees instead of pruning with spatial condition only. RMB-tree is able to reduce search space with spatial condition and keyword condition, thus it achieves better performance than pure R-tree solution.

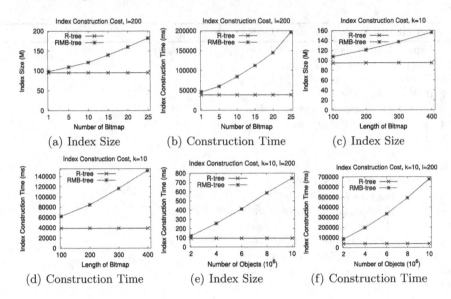

(a) Index Size   (b) Construction Time   (c) Index Size

(d) Construction Time   (e) Index Size   (f) Construction Time

Fig. 3. Index Construction

4 Performance Evaluation

RMB-tree is tested on a PC with Intel i5-2400 CPU at 3.10GHz and 4GB of memory. We use two real-world datasets named TX and CA obtained from open street map project [1]. TX and CA both contain 14 million points, where each point consist of a two-dimensional coordinate and a number of string attributes. TX and CA contain the road networks and streets for Texas and California in USA. We sample 2×10^6 to 10×10^6 objects from each of TX and CA to test the performance of RMB-tree. The sampled datasets are named TX_2 to TX_{10} and CA_2 to CA_{10} respectively. We conduct a set of 100 range approximate keyword queries as follows. For each query, we randomly select an object and generate spatial region centered at the selected object. The area of generated spatial region is determined by parameter P which is the area ratio of the generated region to the entire region. We use normalized edit distance as the similarity metric of keywords, and RMB-tree is tested with normalized edit distance including 10%, 20% and 30%. Such query list guarantees that no empty results are returned. We compare RMB-tree with R-tree, since R-tree is disk resident and it returns precise results. Only results on TX are reported due to space limit, since results on CA are quite similar. We denote the number of bitmaps by k, and the length of bitmaps by l.

We first compare RMB-tree construction costs with different settings. To illustrate the effects of bitmap numbers, we fix bitmap length at 200 and vary the number of bitmaps from 1 to 25. Figure 3(a) shows that RMB-tree size grows with the number of bitmap, and the growth is near-linear. As depicted in Figure 3(b), the time of constructing an RMB-tree with different number of

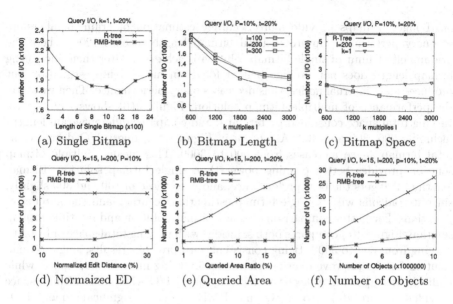

Fig. 4. Query I/O Cost

bitmaps also increases near-linearly. The reason is that increasing bitmap number decreases the fanout of internal nodes, thus produces more internal nodes. More bitmaps also take more time during object insertions, since RMB-tree seeks a proper bitmap to include each keyword of the inserted object. RMB-tree construction costs more space and time than that of R-tree, since it stores additional bitmaps in tree nodes and computes which bitmap to insert each keyword. Similar results are retrieved when we fix the number of bitmap at 10 and increase the length of each bitmap from 100 to 400. As shown in Figure 3(c) and Figure 3(d), both space and time for building an RMB-tree grow as we increase the length of each bitmap. The reason is that longer bitmaps cost more space, thus decrease internal node fanout and create more internal nodes. In the meanwhile, longer bitmaps also take more time to be operated on. We vary the number of objects inserted and test the space and time cost for building an RMB-tree as reported in Figure 3(e) and Figure 3(f). With 10 bitmaps at the length of 200, we insert 2×10^6 to 10×10^6 objects into an RMB-tree. The index size grows proportional to the number of the inserted objects, and the time of constructing index follows the same trend. The results show that RMB-tree construction cost more space and time than that of building an R-tree, but the improvement of query performance introduced later is worth the one-time investment.

Figure 4 reports the average query I/O numbers of RMB-tree and R-tree respectively, and the default number of objects is 2×10^6. To illustrate pruning effects of single bitmap solution, we build RMB-tree with a single bitmap and increase the length of bitmap from 200 to 2400. The smallest query I/O shown in Figure 4(a) depicts the limitation of single bitmap solution. As Figure 4(a) shows, increasing the length of bitmap achieves less query I/O before the length reaches 1200, then the query I/O cost increases when the length of bitmap grows on. Longer bitmaps provide more accuracy estimation of gram membership, so query performance improves until bitmap length reaches 1200, where the pruning effect limit of single bitmap solution is achieved. After then, increasing bitmap length does not provide benefit for gram membership estimation but produces more internal nodes, thus decreases query performance. Then we study the performance of multiple bitmap solution. Figure 4(b) shows results with varying total space consumed by bitmaps, and compares different gram lengths including 200, 300 and 400. As shown in Figure 4(b), query I/O decreases as total bitmap space increases from 600 to 3000. This means multiple bitmap solution provides better pruning power than single bitmap solution. Bitmap length of 200 gets the best performance, and wins those of 100 and 300 slightly, since it tradeoffs with a single bitmap estimation accuracy and the number of partitions. For better comparison of single bitmap solution and multiple bitmap solution, Figure 4(c) compares both solutions with varying total space of bitmaps ranging from 600 to 3000. Being consistent with previous results, single bitmap solution fails to improve query performance by using more bitmap space, while multiple bitmap solution keeps on reducing query I/O when more bitmap space is added. Figure 4(c) also shows that RMB-tree saves significant query I/Os compared to R-tree. Figure 4(d) shows the query performance with different keyword normalized edit distances (t) at 10%, 20% and 30%, and RMB-tree costs 69% to 84% less query I/Os than R-tree. With varying normalized edit distances, RMB-tree costs more I/Os as t increases while R-tree pays a constant I/O number. Figure 4(e) shows the impact of queried area on query performance. We vary the the area ratio from 1% to 20% and study the query I/O with different queried areas with fixed t at 20%. As shown in Figure 4(e), RMB-tree costs nearly constant number of I/Os since irrelevant subtrees are pruned by *MGB-Filter*. The query I/O of R-tree grows with the queried area, as it provides only spatial pruning effect. For all queried areas, RMB-tree costs 66% to 88% less query I/Os compared to R-tree. To test the scalability of RMB-tree, we build RMB-tree with different number of objects ranging from 2×10^6 to 10×10^6, and record the average query I/O. Figure 4(f) shows that RMB-tree query I/O grows linearly as the number of objects, and saves 80% to 84% number of I/O compared to R-tree. Thus RMB-tree scales well with the number of objects.

5   Related Work

Approximate string search is a fundamental sub-problem of spatial approximate keyword search. Existing works consist of two major types including trie-based methods and gram-based methods. Gram-based methods mainly employ a filter-and-verify framework, and more details can be referred to [6][2][11].

Spatial keyword search which returns objects with exact keywords matching query condition is well studied in [3][10][9], and so on. Most of these works employ a tree-based index structure and augment it with keyword information in each tree node. A recent excellent experiment study can be found in [5]. Our work focuses on spatial approximate keyword search and these works deal with a special case of our problem.

MHR-tree [8] and *LBAK-tree* [7] are most related to our work, since they propose index answering spatial query with approximate keyword conditions. They share the same idea as previous works to incorporate tree-based spatial index with keyword information in tree nodes. In each R-tree [4] internal nodes, *MHR-tree* employs a number of indeependent hash functions to compute a min-hash signature of the keyword set in the subtree. Given a query, *MHR-tree* estimates whether a subtree contains enough common q-grams with given keywords based on the min-hash signature. Although the estimation is unbiased, the approximate nature of min-hash signature makes *MHR-tree* miss some of the query results. *LBAK-tree* is a memory based index, which stores inverted lists of q-grams for strings in a subtree in some tree nodes. The inverted lists of q-grams help a query to judge whether a subtree possibly contains all the required keywords similar enough to the queried ones. Such solution makes a tree node size unbounded thus it can not work efficiently on disks.

6   Conclusion

In this paper, we propose a disk-resident index named RMB-tree to support spatial approximate keyword queries. We present the structure of RMB-tree together with the insertion method and the query processing method. Compared to existing works, RMB-tree is able to return exact query results on disks efficiently. We compare RMB-tree with R-tree which is the state-of-art solution for returning exact results on disk, and the comparison validates the effectiveness and efficiency of our proposed RMB-tree.

Acknowledgments. This work is funded by Project (No. 61272046) supported by the National Natural Science Foundation of China; Project supported by the Natural Science Foundation of Heilongjiang Province,China(Grant No. F201317); The Fundamental Research Funds for the Central University (Grant No. HIT.NSRIF.2015065); China Postdoctoral Science Foundation Funded Project(Grant No. 2013T60372, 2014M561351); Heilongjiang Postdoctoral Science Foundation Funded Project(Grant No. LBH-Z14118).

References

1. Open street map. http://www.openstreetmap.org/
2. Chen, L., Jiaheng, L., Yiming, L.: Efficient merging and filtering algorithms for approximate string searches. In: IEEE ICDE 2008 (2008)
3. Chen, Y.-Y., Suel, T., Markowetz, A.: Efficient query processing in geographic web search engines. In: ACM SIGMOD, pp. 277–288 (2006)
4. Guttman, A.: R-trees: A dynamic index structure for spatial searching. In: ACM SIGMOD, pp. 993–1002 (1984)
5. Jensen, C.S., Wu, D., Chen, L., Cong, G.: Spatial keyword query processing: An experimental evaluation. In: VLDB (2013)

6. Marios, H., Amit, C., Nick, K., Divesh, S.: Fast indexes and algorithms for set similarity selection queries. In: IEEE ICDE, pp. 267–276 (2008)
7. Sattam, A., Alexander, B., Li, C.: Supporting location-based approximate-keyword queries. In: ACM SIGSPATIAL, pp. 61–70 (2010)
8. Yao, B., Li, F., Hadjieleftheriou, M., Hou, K.: Approximate string search in spatial databases. In: IEEE ICDE, pp. 545–556 (2010)
9. Yinghua, Z., Xing, X., Chuang, W., Yuchang, G., Wei-Ying, M.: Hybrid index structures for location-based web search. In: ACM CIKM, pp. 155–162 (2005)
10. Zhang, D., Ooi, B.C., Tung, A.: Locating mapped resources in web 2.0. In: IEEE ICDE, pp. 521–532 (2010)
11. Zhang, Z., Hadjieleftheriou, M., Ooi, B.C., Srivastava, D.: Bed-tree: an all-purpose index structure for string similarity search based on edit distance. In: ACM SIG-MOD, pp. 915–926 (2010)

Hashing Multi-Instance Data
from Bag and Instance Level

Yao Yang, Xin-Shun Xu*, Xiaolin Wang, Shanqing Guo, and Lizhen Cui

School of Computer Science and Technology, Shandong University, Jinan, China
yangyao371@gmail.com, {xuxinshun,xlwang,guoshanqing,clz}@sdu.edu.cn

Abstract. In many scenarios, we need to do similarity search of multi-instance data. Although the traditional kernel methods can measure the similarity of bags in original feature space, the time and storage cost of these methods are so high which makes such methods cannot deal with large scale problems. Recently, hashing methods have been widely used for similarity search due to its fast search speed and low storage cost. However, few works consider how to hash multi-instance data. In this paper, we present two multi-instance hashing methods: (1) Bag-level Multi-Instance Hashing (BMIH); (2) Instance-level Multi-Instance Hashing (IMIH). BMIH first maps each bag to a new feature representation by a feature fusion method; then, supervised hashing method is used to convert new features to hash code. To utilize more instance information in each bag, IMIH regards instances in all bags as training data and apply two types of hash learning methods (unsupervised and supervised, respectively) to convert all instances to binary code; then, for a test bag, a similarity measure is proposed to search similar bags. Our experiments on four real-world datasets show that instance-level hashing with supervised information outperforms all proposed techniques.

Keywords: Learning to hash, Multi-Instance data, Image retrieval.

1 Introduction

Multi-instance learning (MIL) was first presented by Dietterich et al. [3] for the application of drug activity prediction. In MIL, each training example is treated as a bag and each bag contains multiple instances. A bag is positive if it contains at least one positive instance, and negative otherwise. The advantage of multi-instance data is that it is more natural and informable than single instance representation. In many scenarios, given a sample, we need to perform similarity search from multi-instance dataset. Traditional kernel methods that compute similarity between bags in original feature space are difficult to be used for large scale data set due to their high cost of computation and storage.

Recently, hashing method [14,13,19] is widely used to similarity search, which can tackle these two challenges by designing compact binary code in a low-dimensional space for each example. The resulting binary codes enable fast

---

* Corresponding author.

© Springer International Publishing Switzerland 2015
R. Cheng et al. (Eds.): APWeb 2015, LNCS 9313, pp. 437–448, 2015.
DOI: 10.1007/978-3-319-25255-1_36

similarity search on the basis of the hamming distance between codes. Moreover, compact binary codes are extremely efficient for large-scale data storage. Broadly, existing hashing methods can be divided into two main categories: unsupervised and supervised hashing methods. Unsupervised hashing methods design hash functions using unlabeled data to generate binary codes. A number of methods have been proposed, e.g., Locality-Sensitive Hashing (LSH) [4], Spectral Hashing (SpH) [6], Iterative Quantization (ITQ) [8], K-means Hashing (KMH) [12]. Supervised hashing methods incorporate supervised information, e.g., similar/disimilar pairs, to boost the hashing performance. Many supervised hashing methods have also been extensively studied, e.g., Semi-Supervised Hashing (SSH) [9], Minimal Loss Hashing (MLH) [11], Binary Reconstructive Embeddings (BRE) [1], Kernel-based Supervised Hashing (KSH) [10].

Both hashing methods and multi-instance learning methods have been well studied; however, few works consider how to hash multi-instance data. Thus, how to make hashing methods work on multi-instance data becomes a meaningful problem. Motivated by this, in this paper, we investigate how to apply hashing method to solve the retrieval of multi-instance data. Specifically, to hash multi-instance data, we consider the following two problems: (1) Each bag includes many instance, if we transform each bag into one hash code; then, how to merge instances in each bag together to form a new bag representation. (2) In consideration of exploiting instance information, if we convert all instance into hash code, how to measure the similarity among bags by instance hash code. To tackle these problems, we propose two multi-instance hashing methods: (1) Bag-level Multi-Instance Hashing (BMIH). BMIH first generates a set of cluster centers by K-means clustering method in the instance feature space; then, a nonlinear mapping is defined by these centers to extract a new feature representation of all bags. Based on this, supervised hashing method is proposed to hash data into binary codes. (2) Instance-level Multi-Instance Hashing (IMIH). To use more instance information in each bag, IMIH considers all instances in all bags as training data and utilizes two types of hash learning methods (unsupervised and supervised) to convert all instances to binary codes. For a new query bag, a simple metric technique is proposed to measure the similarity among bags.

The rest of this paper is organized as follows. We briefly review related work in Section 2. Section 3 presents the details of our proposed methods. Section 4 provides experimental results on several benchmark datasets. The conclusions are given in Section 5.

2   Related Work

In the following three sections, we introduce three hashing methods from unsupervised as well as supervised domains that are applied to the proposed research including Self-Taught Hashing [7], Semi-Supervised Hashing [9] and Kernel-based Supervised Hashing [10].

2.1   Self-Taught Hashing(STH)

Self-Taught Hashing (STH) [7] is the first two stage method, which first learns k-bit hash codes for all examples via an unsupervised learning step; then studies hash function via a supervised learning step.

For the hash codes learning step, STH constructs a local similarity matrix $\mathbf{W}$ and learns binary codes through the following object function:

$$\min \sum_{i,j} \mathbf{W}_{ij} \|y_i - y_j\|^2$$

$$s.t. \quad y_i \in \{-1,1\}^k, \ \sum_i y_i = 0, \ \frac{1}{n} \sum_i y_i y_j^T = \mathbf{I} \tag{1}$$

where y_i is the binary code for x_i. The equations $\sum_i y_i = 0$ and $\frac{1}{n}\sum_i y_i y_j^T = \mathbf{I}$ are the balanced and uncorrelated constraints.

For the hash function learning step, based on the binary codes learned from the previous step, STH learns it by training a set of k linear SVM classifiers. STH obtains higher hash performance than SpH [6], but STH does not utilize label information which is often available in many real world applications.

2.2   Semi-Supervised Hashing(SSH)

The semi-supervised hashing (SSH) method in [9] is proposed to solve the hash problem of the inadequate label data. It utilizes parts of pairwise knowledge between data examples for learning more effective binary codes. SSH also requires the codes to be balanced and uncorrelated. After relaxing the sign function and replacing the balancing constraint by maximizing the variance of bits, the final objection function contains two terms: a supervised term and an unsupervised term. It is easily solved as a standard eigenvalue problem.

The SSH method has shown promising results for improving hash effectiveness by leveraging the pairwise information, but it only utilizes small parts of supervised information.

2.3   Kernel-Based Supervised Hashing (KSH)

The work in [10] proposes a kernel-based supervised hashing (KSH) method for taking full advantage of the supervised information and avoiding complicated model training.

To fit linearly inseparable data, a kernel formulation $\kappa : \mathbb{R}^d \times \mathbb{R}^d \mapsto \mathbb{R}$ is employed to construct hash functions. The similarity matrix $\mathbf{S}$ is constructed by pairwise similarity among label data x_l. After that, it replaces Hamming distance by code inner product to avoid complex optimization process, and proposes the objective function $\mathcal{F}$:

$$\min_{H_l \in \{1,-1\}^{l \times r}} \mathcal{F} = \| \frac{1}{r} H_l H_l^T - \mathbf{S} \|_F^2 \tag{2}$$

where H_l denotes the code matrix of x_l. Finally, an efficient greedy optimization is designed to solve the target hash function bit by bit .

Comparing prior supervised methods [1][11] directly optimizing Hamming distances, KSH obtains both better performance and less training time cost by utilizing code inner products and a greedy optimization. At the same time, KSH also outperforms SpH [6] and SSH [9] in searching similar neighbors by incorporating more supervised information.

3   The Proposed Methods

In this section, we propose the Bag-level Multi-Instance Hashing(BMIH) and Instance-level Multi-Instance Hashing(IMIH) methods. The BMIH method maps every bag to a new feature representation and converts bag-level multi-instance hashing to a standard supervised hashing. The IMIH method treats each instance in bags as a feature vector and directly transforms them into binary codes, which takes advantage of more instance information in each bag.

3.1   Formulation

Before presenting the details, we give the formal definition of multi-instance hash learning as following. Let $\mathcal{X}$ denote the instance space. Given a data set $T = \{(X_1, L_1), ..., (X_i, L_i), ..., (X_N, L_N)\}$, where $X_i = \{\boldsymbol{x}_{i1}, ..., \boldsymbol{x}_{ij}, ..., \boldsymbol{x}_{i,n_i}\} \subseteq \mathcal{X}$ is called a bag, $L_i \in \mathcal{L} = \{-1, +1\}$ is the label of X_i and N is the number of training bags. Here $\boldsymbol{x}_{ij} \in \mathcal{X}$ is an instance $[x_{ij1}, ..., x_{ijk}, ..., x_{ijd}]^T$, where x_{ijk} is the value of $\boldsymbol{x}_{ij}$ at the kth attribute, n_i is the number of instances in X_i and d is the dimension of original space $\mathcal{X}$. If there exists $p \in \{1, ..., n_i\}$ such that $\boldsymbol{x}_{ip}$ is a positive instance, then X_i is a positive bag and thus $L_i = +1$, but the concrete value of the index p is usually unknown; otherwise $L_i = -1$. The goal is to obtain optimal binary codes $\mathbf{Y} = \{\boldsymbol{y}_1, \boldsymbol{y}_2, ..., \boldsymbol{y}_N\} \in \{-1, 1\}^{k \times N}$(for instance-level, $\mathbf{Y} \in \{-1, 1\}^{k \times n}$, n is the number of all instances) and a hash function $f : \mathbf{R}^d \rightarrow \{-1, 1\}^k$ which maps each example to its binary code with k bits.

3.2   Bag-Level Multi-Instance Hashing

One problem of hashing multi-instance data is that the bag includes many instances. In this case, we can not directly view the bag as a feature vector. Intuitively, by the feature fusion method, all instances in each bag can be transformed into a new feature representation, which contains most bag-level information.

We first line up all instances in all bags together and denote these instances as $\mathcal{I} = \{\boldsymbol{x}_1, ...\boldsymbol{x}_i, ...\boldsymbol{x}_n\}$, where n is the number of instances. Then, clustering algorithm is utilized to group all instances into a collection of clusters that similar instances are assigned into the same cluster. We represent the center of cluster as $\boldsymbol{\mu}_t$ and the set of cluster center is denoted as $\mathcal{R} = \{\boldsymbol{\mu}_1, \boldsymbol{\mu}_2, ..., \boldsymbol{\mu}_m\}$, where m is the total number of the center. Here, we apply k-means clustering method to group $\mathcal{I}$ into m clusters.

For a bag $X_i = \{\boldsymbol{x}_{ij} \mid j = 1, 2, ..., n_i\}$, the projection feature of bag, $\phi(X_i)$, can be defined as

$$\begin{cases} \phi(X_i) = [s(\boldsymbol{\mu}_1, X_i), ..., s(\boldsymbol{\mu}_m, X_i)] \\ s(\boldsymbol{\mu}_t, X_i) = \max\limits_{j=1,2,...,n_i} \exp\left(-\|\boldsymbol{x}_{ij} - \boldsymbol{\mu}_t\|^2\right), \ t = 1, 2, ..., m \end{cases} \tag{3}$$

where $s(\boldsymbol{\mu}_t, X_i)$ is interpreted as a measure of similarity between the center $\boldsymbol{\mu}_t$ and a bag X_i. $s(\boldsymbol{\mu}_t, X_i)$ is determined by the center and the closest instance in the bag. In this way, each bag will has an m-dimensional projection feature in the projection space. Since the label of bags are known, the BMIL problem is transformed into a standard supervised hash learning problem.

We represent the bags' features as $\mathcal{X}_{new} = \{\phi(X_1), ..., \phi(X_i), ..., \phi(X_m)\}$. Like most hash methods, a linear hash function is presented as follows:

$$y_i = f(\boldsymbol{x}_i) = sgn(\mathbf{W}\boldsymbol{x}_i) \tag{4}$$

where $\mathbf{W}$ is a $k \times d$ coefficient matrix and $sgn(.)$ is the sign function. For the new data set $\mathcal{X}_{new}$, the Kernel-based Supervised Hashing (KSH) [10] method can be used to learn the binary codes $\mathbf{Y}$ and coefficient matrix $\mathbf{W}$. For a query bag X_q, we first compute the bag's projection feature $\phi(X_q)$ by Eqn.3. Then, the binary code of the query bag is obtained by Eqn.4.

3.3 Instance-Level Multi-Instance Hashing

Although BMIH method solves the problem of multi-instance hashing, the instance information in each bag has not been fully taken advantage of. Thus, we want to exploit more instance information to enhance hash performance. If we view all instances as training data, it is very easy to convert all instances to binary code by presented hashing method. However, for a new query bag, we want to return similar bags rather than similar instances. In this section, a simple technique is proposed to deal with IMIH problem.

Given all instances data $\mathcal{I} = \{\boldsymbol{x}_1, ...\boldsymbol{x}_i, ...\boldsymbol{x}_n\} \subset \mathbb{R}^d$, to avoid the influence of instance order and linearly inseparable data, we employ a kernel function $\kappa : \mathbb{R}^d \times \mathbb{R}^d \mapsto \mathbb{R}$ to construct hash function. Following the Kernelized Locality-Sensitive Hashing (KLSH) [5] algorithm, we define the function $f : \mathbb{R}^d \mapsto \mathbb{R}$ with the kernel κ plugged in as follows:

$$f(x) = \sum_{j=1}^{q} \kappa(\boldsymbol{x}_{(j)}, \boldsymbol{x})w_j - b, \tag{5}$$

where $\boldsymbol{x}_{(1)}, ..., \boldsymbol{x}_{(q)}$ are q samples uniformly selected at random from $\mathcal{I}$, $w_j \in \mathbb{R}$ is the coefficient, and $b = \sum_{i=1}^{n} \sum_{j=1}^{q} \kappa(\boldsymbol{x}_{(j)}, \boldsymbol{x}_i)w_j/n$ is the bias. Note that q is fixed to a constant much smaller than the data set size n in order to maintain fast hashing. After substituting b in Eqn.5, we obtain

$$\begin{aligned} f(x) &= \sum_{j=1}^{q} (\kappa(\boldsymbol{x}_{(j)}, \boldsymbol{x}) - \tfrac{1}{n}\sum_{i=1}^{n} \kappa(\boldsymbol{x}_{(j)}, \boldsymbol{x}_i))w_j \\ &= \boldsymbol{w}^{\mathrm{T}}\bar{\boldsymbol{k}}(\boldsymbol{x}) \end{aligned} \tag{6}$$

where $\boldsymbol{w} = [w_1, ..., w_q]^T$ and $\bar{\boldsymbol{k}} : \mathbb{R}^d \mapsto \mathbb{R}^m$ is a map defined by $\bar{\boldsymbol{k}} = [\kappa(\boldsymbol{x}_{(1)}, \boldsymbol{x}) - \mu_1, ..., \kappa(\boldsymbol{x}_{(q)}, \boldsymbol{x}) - \mu_q]^T$, in which $\mu_j = \sum_{i=1}^{n} \kappa(\boldsymbol{x}_{(j)}, \boldsymbol{x}_i)/n$ can be precomputed. We want to solve k coefficient vectors $\boldsymbol{w}_1, ..., \boldsymbol{w}_k$ to construct k hash functions $\{h_i(\boldsymbol{x}) = sgn(\boldsymbol{w}_k^T\bar{\boldsymbol{k}}(\boldsymbol{x}))\}_{i=1}^{k}$.

Since the data is kernelized , most unsupervised hashing methods (such as STH [7], PCA-ITQ [8]) are easily applied to IMIH problem. We only need to substitute x by $\bar{k}(x)$ in the objective function, and the solution procedure does not change. For a query bag $\mathbf{X}_q$, we first construct kernel vector representation by Eqn.6, then compute instances binary codes. In this case, the similar instances of each instance in query bag are easily got by computing hamming distance. But we actually want to return similar bags for a query bag which are not directly received due to lacking a similarity measure. To settle this problem, a simple metric technique to measure the similarity between two bags is proposed as follows:

$$Dis(\mathbf{X}_q, \mathbf{X}_i) = \frac{1}{n_q} \sum_{q=1}^{n_q} (\min_{i=1,\dots,n_i} Ham(\boldsymbol{x}_q, \boldsymbol{x}_i)) \qquad (7)$$

where $Ham(\boldsymbol{x}_q, \boldsymbol{x}_i)$ is the hamming distance between two instances $\boldsymbol{x}_q$ and $\boldsymbol{x}_i$, and $Dis(\mathbf{X}_q, \mathbf{X}_i)$ is the distance metric between two bags $\mathbf{X}_q$ and $\mathbf{X}_i$, which first computes the minimum hamming distance between each instance in certain bag and a query bag, then averages it. If $\mathbf{X}_q$ and $\mathbf{X}_i$ have different labels, $Dis(\mathbf{X}_q, \mathbf{X}_i)$ is larger. Because the positive instance in positive bag has no matching instance and receives a larger minimum hamming distance. In the same way, if $\mathbf{X}_q$ and $\mathbf{X}_i$ are same labels, $Dis(\mathbf{X}_q, \mathbf{X}_i)$ is smller. Thus, $Dis(.)$ can reflect the similarity between two bags well. By ranking $Dis(.)$, it is easy to return similar bags of the query bag.

3.4   Embedding Supervised Information

Until now, we only learn binary code in an unsupervised manner. However, it has been shown that the hashing quality could be boosted by leveraging supervised information into hash function learning. In the multi-instance problem, we only know the bags' labels and the instances' labels are not given. But it is known that a negative bag does not contain any positive instance. Thus, we can regard all the instances in negative bags as labeled negative instances. On the other hand, since a positive bag may contain positive as well as negative instances, we can regard its instances as unlabeled ones.

As shown in [20], it treats MIL as a special case of semi-supervised learning. Similarly, IMIH problem can be viewed as a semi-supervised hash learning. We randomly select r (r is small) negative bags in which all instances are labeled negative, the left instances as unlabeled ones. In this case, similar matrix S is constructed by labeled negative instances and unlabeled instances, which contains two value $+1$ and 0. The value 0 implies that the similar/dissimilar relationship about some data pair is unknown or uncertain. Then, the Semi-Supervised Hashing (SSH) [9] method is used to learn hash function and binary codes. In the test step, we also utilize Eqn.7 to measure the similarity between two bags.

For above ISSH method, we only use the negative label, which does not fully reflect the labeled information. To exploit more label information in positive instances, in this paper, we assume that the positive instance in source bag $\mathbf{X}_i$, can be estimated and each positive only own one positive instance. This can be conducted by instance selection method such as [15,16] in an off-line manner.

After that, the selected positive instance in source bag is labeled as +1, and other instances in the same bag are unlabeled. The instances in all negative bags are labeled as −1. Thus, we define a similarity matrix $S \in \mathbb{R}^{n \times n}$ as

$$S_{ij} = \begin{cases} +1, & if\ \boldsymbol{x}_i\ and\ \boldsymbol{x}_j\ are\ similar \\ -1, & if\ \boldsymbol{x}_i\ and\ \boldsymbol{x}_j\ are\ dissimilar \\ 0, & otherwise \end{cases} \tag{8}$$

Since the similarity matrix S is obtained, we can directly employ Kernel-based Supervised Hashing (KSH) [10] method to learn hash codes and hash functions. Details can be seen in KSH. The Eqn.7 is also used to measure the similarity among bags in the test step.

4   Experiments

This section presents an extensive set of experiments to exploit which method is better for hashing multi-instance data.

4.1   Datasets

A set of datasets are utilized in evaluation as follows:

1. *Elephant, Tiger*: These datasets were acquired from [2]. Each dataset consists of 100 positive images and 100 negative images. The number of instances in each dataset is 1391, 1220. In our experiment, for each dataset, 160 bags are randomly selected as the training data, while the remaining 40 bags are used as test queries.

2. *Cars*: This dataset is original from *Graz-02*[1] and *Corel*[2]. It consists of 500 positive images and 500 negative images. All images have been preprocessed and segmented with the Blobworld system [18]. It total contains 4491 instances. In our experiment, 800 bags are randomly selected as the training data, while the remaining 200 bags are used as test queries.

3. *Airplanes*: This dataset is original from *Caltech101*[3] and *Corel*. It consists of 1330 positive images and 1300 negative images. All images have been preprocessed and segmented with the Blobworld system [18]. It total contains 11290 instances. In our experiment, 2130 bags are randomly selected as the training data, while the remaining 500 bags are used as test queries.

4.2   Experiment Settings

From bag and instance level, combining with STH [7], PCA-ITQ [8], SSH [9] and KSH [10], we propose five different methods on these datasets including BMIH-ksh, IMIH-sth, IMIH-itq, IMIH-ssh, IMIH-ksh. We also compare the five hashing methods with a typical multi-instance kernel called Normalized Set Kernels

---

[1] http://lear.inrialpes.fr/people/marszalek/data/ig02/

[2] http://archive.ics.uci.edu/ml/datasets/Corel+Image+Features

[3] http://www.vision.caltech.edu/Image_Datasets/Caltech101/

Table 1. Precision of the top 40 retrieved images on Elephant, Tiger datasets and the top 100 retrieved images on Cars, Airplanes datasets with different hash bits.

	Elephant					Tiger				
Methods	8 bits	16 bits	32 bits	48 bits	64 bits	8 bits	16 bits	32 bits	48 bits	64 bits
BMIH-ksh	0.517	0.558	0.600	0.607	0.602	0.520	0.523	0.554	0.571	0.551
IMIH-sth	0.433	0.477	0.467	0.472	0.492	0.532	0.543	0.517	0.501	0.494
IMIH-itq	0.527	0.613	0.616	0.621	0.633	0.545	0.573	0.594	0.600	0.610
IMIH-ssh	**0.654**	**0.667**	0.694	0.706	0.702	**0.602**	**0.617**	0.628	0.619	0.642
IMIH-ksh	0.451	0.641	**0.770**	**0.823**	**0.824**	0.560	0.604	**0.632**	**0.722**	**0.749**
NSK	0.840					0.832				
	Cars					Airplanes				
Methods	8 bits	16 bits	32 bits	48 bits	64 bits	8 bits	16 bits	32 bits	48 bits	64 bits
BMIH-ksh	0.505	0.504	0.593	0.649	0.665	0.457	0.509	0.528	0.545	0.564
IMIH-sth	0.701	0.651	0.633	0.655	0.660	0.473	0.565	0.599	0.586	0.618
IMIH-itq	**0.731**	0.726	0.728	0.731	0.731	0.568	0.608	0.641	0.645	0.649
IMIH-ssh	0.728	**0.769**	0.776	0.784	0.785	**0.615**	**0.685**	0.694	0.705	0.706
IMIH-ksh	0.505	0.569	**0.819**	**0.884**	**0.883**	0.547	0.569	**0.760**	**0.862**	**0.869**
NSK	0.902					0.895				

(NSK) [17]. To evaluate the quality of hashing, we use three evaluation metrics: Precision-Recall curves, Precision curves within Hamming distance 3, and Precision curves with different number of top returned samples.

The parameters m in BMIH-ksh is set to 100 for all four datasets. We will discuss more details how it affects the performance of BMIH-ksh later. Since IMIH-sth, IMIH-itq and IMIH-ksh refer to kernels, we provide them the same Gaussian RBF kernel $\kappa(\boldsymbol{x}, \boldsymbol{y}) = exp(-\|\boldsymbol{x}-\boldsymbol{y}\|/2\sigma^2)$ and the $q = 100$ for *Elephant* and *Tiger* datasets, $q = 200$ for *Cars* and *Airplanes* datasets. For IMIH-ssh, we sample 100 (400 for *Cars* and *Airplanes* datasets) random instances in negative bags from the training set to construct the pairwise constraint matrix. For IMIH-ksh, we employ the method in [15] to accomplish the instance selection. We evaluate the performance of different methods by varying the number of hash bits in the range of $\{8, 16, 32, 48, 64\}$ and calculate the average result by repeating each experiment 10 times. All our experiments were run on a workstation with a 2.67 GHz Intel Xeon CPU and 16GB RAM.

4.3 Results and Discussions

Three sets of experiments are conducted on each dataset to measure the performance of the proposed five methods to answer the following questions: (1) Which method has best performance in high-precision results such as the precision for the top retrieved samples and the precision within Hamming distance 3; (2) Which approach can outperform other methods with the precision-recall curve? (3) How is the efficiency of the proposed five methods in training and testing stage?

In the first set of experiments, we report the precision for the top 40 or 100 retrieved images with different numbers of hash bits in Table 1 for all four

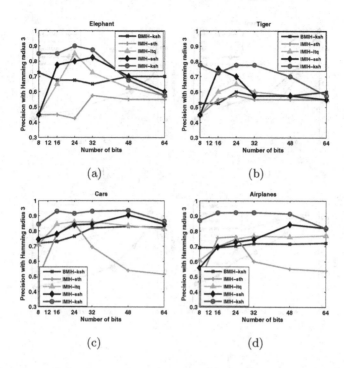

Fig. 1. Precision of the retrieved examples within Hamming radius 3 on Elephant, Tiger, Cars and Airplanes datasets with different hash bits.

datasets. From these comparison results, when hash bit is small such as 8 bits or 16 bits, IMIH-ssh obtains better performance than IMIH-ksh. But, in most cases, IMIH-ksh gives the best performance among all five hashing methods on all four datasets. We also realize that the kernel method NSK obtains higher precision than all hashing method.

The precision within hamming radius 3 are shown in Fig. 1(a)-(d). IMIH-ksh is still consistently better than other hashing methods.

In the second set of experiments, the precision-recall curves with different hash bits on all four datasets are plotted in Fig. 2(a)-(d). It can be seen that among all of these hashing methods, IMIH-ksh shows the best performance.

From these figures, we can see that IMIH-sth does not perform well in most cases. This is because IMIH-sth method does not utilize the supervised information contained in bags and leads to inefficient codes in practice. Due to the rotation operation, IMIH-itq is better than IMIH-sth in precision-recall curse. For the method BMIH-ksh, although the method uses the supervised information from labeled bags to learn hash codes, it loses instance information by converting a bag to a feature vector. Therefore, the BMIH-ksh method is superior to IMIH-sth and inferior to IMIH-ksh. IMIH-ssh achieves better results than IMIH-sth due to the incorporation of pairwise similarity constraints. However, IMIH-ssh only uses parts of negative instance label. By applying more label information, IMIH-ksh obtains better performance than IMIH-ssh in precision-recall curve.

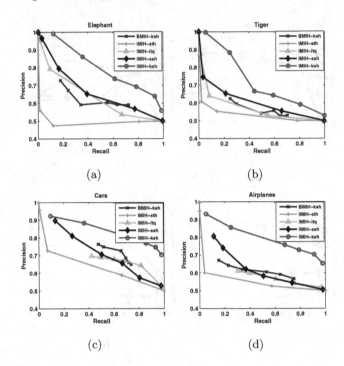

Fig. 2. Precision-Recall curve on Elephant, Tiger, Cars and Airplanes datasets with different hash bits.

In the third set of experiments, we report the training time for compressing all training bags into compact codes as well as the testing time for returning the top similar bags in Table 2. We fix the hash bit on 48 and do this experiment on for *Elephant, Cars* and *Airplanes* datasets. In Table 2, for training stage, IMIH-ssh is faster than other hashing methods and IMIH-ksh is more time consuming because it trains the hash functions one bit at a time in an iteration method. Due to converting every bag to a feature vector, BMIH-ksh is faster than IMIH-ksh. For testing stage, owing to the feature fusion, BMIH-ksh is much faster than all instance-level hashing methods. The testing time of IMIH-ksh is acceptably fast, comparable to the methods BMIH-sth, BMIH-itq and BMIH-ssh. But, for the kernel method NSK, the testing time is almost 500 times more than IMIH-ksh on *Airplanes* dataset and it will enlarge by increasing the size of dataset.

At last, we report the influence of parameter m for BMIH-ksh on *Elephant* and *Cars* datasets in Fig. 3. We chose m from 80 to 200 with step size 20. From these results, we can observe that the performance of BMIH-ksh is relatively stable with respect to m. We also have similar results on the other two datasets. But due to the limitation of space, they are not presented here.

Table 2. Training and testing time are recorded on Elephant, Cars, Airplanes datasets with 48-bit hashing code in second.

Methods	Elephant		Cars		Airplanes	
	Training	Testing	Training	Testing	Training	Testing
BMIH-ksh	8.9	0.007	76.6	0.008	497.6	0.009
IMIH-sth	49.1	1.139	516.8	3.025	12171.0	7.741
IMIH-itq	0.5	0.053	0.7	0.167	2.0	0.379
IMIH-ssh	0.4	0.051	0.5	0.136	0.5	0.527
IMIH-ksh	130.3	0.057	912.6	0.158	5841.5	0.420
NSK	–	2.700	–	34.190	–	220.170

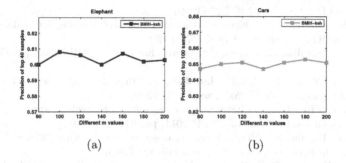

(a)   (b)

Fig. 3. Results of precision with 48 hash bits when m takes different values.

5   Conclusions

This paper explores two ways to solve how to hash multi-instance data: Bag-level Multi-Instance Hashing (BMIH) and Instance-level Multi-Instance Hashing (IMIH). For BMIH, a feature fusion method is proposed to merge a bag to a new feature vector, then the supervised hashing method (KSH) is applied to learn the hash function. For IMIH, to utilize more instance information, different type hashing methods (STH, PCA-ITQ, SSH, KSH) are used to convert all instances to binary code. Then, a similarity measure is designed to find the similar bags. Extensive experiments on four different datasets demonstrate that instance-level hashing method with supervised information (IMIH-ksh) outperforms bag-level hashing methods. Although the proposed multi-instance hashing methods are inferior to multi-instance kernel method on the precision of top retrieved samples, their search speed is much faster than kernel method.

Acknowledgments. This work is partially supported by National Natural Science Foundation of China (61173068, 61103151, 61573212), Program for New Century Excellent Talents in University of the Ministry of Education, the

Key Science Technology Project of Shandong Province (2014GGD01063), the Independent Innovation Foundation of Shandong Province (2014CGZH1106) and the Shandong Provincial Natural Science Foundation (ZR2014FM020, ZR2014FM031).

References

1. Kulis, B., Darrell, T.: Learning to hash with binary reconstructive embeddings. In: NIPS 2009, pp. 1042–1050 (2009)
2. Andrews, S., Tsochantaridis, I., Hofmann, T.: Support vector machines for multi-pleinstance learning. In: NIPS 2003, pp. 561–568 (2003)
3. Dietterich, T.G., Lathrop, R.H., Lozano-Prez, T.: Solving the multiple instance problem with axis-parallel rectangles. Artificial Intelligence 89(11), 31–71 (1997)
4. Gionis, A., Indyk, P., Motwani, R.: Similarity search in high dimensions via hashing. In: VLDB 1999, pp. 518–529 (1999)
5. Kulis, B., Grauman, K.: Kernelized locality-sensitive hashing for scalable image search. In: ICCV 2009, pp. 2130–2137 (2009)
6. Weiss, Y., Torralba, A., Fergus, R.: Spectral hashing. In: NIPS 2008, pp. 1753–1760 (2008)
7. Zhang, D., Wang, J., Cai, D., Lu, J.-S.: Self-taught hashing for fast similarity search. In: Proc. of SIGIR 2009, pp. 18–25 (2009)
8. Gong, Y.-C., Lazebnik, S.: Iterative quantization: A procrustean approach to learning binary codes. In: Proc. of CVPR 2011, pp. 817–824 (2011)
9. Wang, J., Kumar, S., Chang, S.-F.: Semi-supervised hashing for scalable image retrieval. In: Proc. of CVPR 2010, pp. 817–824 (2010)
10. Liu, W., Wang, J., Ji, R.-R., Jiang, Y.-G., Chang, S.-F.: Supervised hashing with kernels. In: Proc. of CVPR 2012, pp. 2074–2081 (2012)
11. Norouzi, M., Fleet, D.J.: Minimal loss hashing for compact binary codes. In: Proc. of ICML 2009, pp. 353–360 (2011)
12. He, K.-M., Wen, F., Sun, J.: K-means hashing: An affinity-preserving quantization method for learning binary compact codes. In: Proc. of CVPR 2013, pp. 2938–2945 (2013)
13. Wang, Q.-F., Zhang, D., Si, L.: Semantic hashing using tags and topic modeling. In: Proc. of SIGIR 2013, pp. 213–222 (2013)
14. Wang, Q.-F., Si, L., Zhang, Z.-W., Zhang, N.: Active hashing with joint data example and tag selection. In: Proc. of SIGIR 2014, pp. 405–414 (2014)
15. Xu, X., Frank, E.: Logistic regression and boosting for labeled bags of instances. In: Dai, H., Srikant, R., Zhang, C. (eds.) PAKDD 2004. LNCS (LNAI), vol. 3056, pp. 272–281. Springer, Heidelberg (2004)
16. Fu, Z.-Y., Robles-Kelly, A.: An instance selection approach to multiple instance learning. In: Proc. of CVPR 2009, pp. 911–918 (2009)
17. Gartner, T., Flach, P.A., Kowalczyk, A., Smola, A.J.: Multi-instance kernels. In: Proc. of ICML 2002, pp. 179–186 (2002)
18. Strecha, C., Belongie, S., Greenspan, H., Malik, J.: Blobworld: Image segmentation using expectation-maximization and its application to image querying. IEEE Trans. on PAMI 24(8), 1026–1038 (2002)
19. Wang, S., Huang, Z., Xu, X.-S.: A Multi-label least-squares hashing for scalable image search. In: Proc. of SDM 2015, pp. 954–962 (2015)
20. Zhou, Z.H., Xu, J.M.: On the relation between multi-instance learning and semi-supervised learning. In: Proc. of CIKM 2007, pp. 1167–1174 (2007)

A Multi-news Timeline Summarization Algorithm Based on Aging Theory

Jie Chen[1], Zhendong Niu[1,2], and Hongping Fu[1]

[1] School of Computer Science, Beijing Institute of Technology, Beijing 100081, China
[2] Information School, University of Pittsburgh, 15260 Pennsylvania, USA
zniu@bit.edu.cn

Abstract. This paper focuses on the problem of news event timeline summary in Multi-Document Summarization, which aims to summarize multi-news regarding the same event in timeline. The majority of the traditional solutions to this problem consider the text surface features and topic-related features, such as the length of each sentence, the position of the sentence in the document, the number of topic words, etc. Traditional methods ignored that every event has its life circle including birth, growth, maturity and death. In this paper, a novel approach is presented for summarizing multi-news regarding the same topic in consideration of both the traditional features and the life circle feature of each event. The proposed approach consists of four steps. First, sentences and their publishing date are extracted from each news article. Second, the extracted sentences are pretreated to reduce the influence of noises like synonyms. Third, life circle features and other four categories of features which are common used in this field are collected. Finally, SVM model is used to train these features to recognize the summary sentence of the news document. This approach have been tested on the public datasets, DUC-2002 and TAC-2010, and the results show that our approach is more effective in summarizing multi-news in timeline than existing methods.

Keywords: News event summarization, Aging theory, Timeline.

1 Introduction

Important facts about the news event summary are those which describe the developing progress of it. When people create summaries, they choose the sentences which contain key words such as major figures, places, people, dates and so on to form a short story about the event. Understandably in automatic summarization as well, it's useful to choose those key words to represent general facts and the important factors.

Everyday numerous news reporting diverse events are published on the Internet. There have been many news services(e.g. Google News) to group news into events, and then produce a short summary for each event. However, most of the news articles are not well-prepared for the progress of the event. There are lots of redundant or duplicated messages among reports about the same event.

© Springer International Publishing Switzerland 2015
R. Cheng et al. (Eds.): APWeb 2015, LNCS 9313, pp. 449–460, 2015.
DOI: 10.1007/978-3-319-25255-1_37

Multi-news summarization aims to extract the essential information about the event progress from these news articles. The timeline summarization helps to reorganize the sentences in order to get a better reading experience.

A particular challenge for multi-document summarization is how to weighing the importance of each sentence, which either depend on the words or some other latent information stored in the sentence. This requires an effective methods to analyze the hidden information. Literally, there are many synonym and polysemous words which bring lots of difficulties when computing relationships among sentences. To avoid the influence of those words, Latent semantic analysis (LSA) [10] is used to find the core words of a document. LSA is a robust unsupervised technique for deriving an implicit representation of text semantics based on observing co-occurrence of words to find semantic units of news.

Another challenge for multi-document summarization is how to deal with duplicated information among these documents about the same topic or event. Reporters usually like to review some previous issues and then tell the readers what happens now. It's helpful for readers to know the developing progress about the event and also a solution to this challenge. As stated, an event goes through a life cycle of birth, growth, maturity and death. This can be reflected from special terms utilized for describing different events that experience a similar life cycles. Aging theory [11] is a model exploited in event detection task which tracks life cycles of events using energy function. The energy of an event increases when the event becomes popular, and it diminishes with time. Similarly, it can also be used for summarization to find out the daily hot terms of events.

In this paper, the timeline summary is generated by considering both temporal and semantic characteristics from the news regarding the same event. Sentences and publishing time are extracted from the news articles. The features are extracted from five aspects to represent each sentence. Then, classification model is built with SVM. Finally, from these candidates, sentences are chosen to form the summary and displayed them with timeline, so that people can track the progress of event easily and quickly.

The remainder of this paper is organized as follows: Section 2 reviews related works on summarization and aging theory. Section 3 proposes an approach of leveraging aging theory to gain sentence feature and train the logistic regression classification model. Section 4 describes the experiments and discusses. Section 5 presents the conclusions and future plans.

2   Related Work

2.1   Multi-document Summarization

Numbers of methods about multi-document summarization have been developed recently. Generally speaking, these methods can be divided into extractive summarization and abstractive summarization. The main idea of extractive summarization is to assign scores to different words in each sentence, and then the high-score sentences will be chosen as the summary. While abstractive summarization usually do some information fusion[12], compression[13] and reformulation[14] to get the summary sentences. This paper will focus on the former method.

One of the most popular extracting method of multi-document summarization methods is MEAD[15], which represents each sentence with some sentence-level features such as term frequency, sentence position, first-sentence overlap, etc. Wan[16] proposed an extracting approach based on manifold-ranking about the information richness and novelty. News Blaster[18] clusters news into events and apply the MultiGen system to find similar senteces and reform the pieces of sentences to create the summary. Allan[19] used 'usefulness', 'novelty' to represent the sentence and provide summaries of news topics. They aimed to process the news stream and create the useful summary sentences. Wong[20] investigated a co-training method to train the model by combining labeled and unlabeled data, which used four kinds of features can be categorized as surface, content, relevance and event features.

Most recently, the multi-document timeline summarization gains enough attraction from researchers and engineers. The timeline can help readers to know the developing progress of the event. The redundancy of summary is required to be very low and the key properties of event to be retained. Lots of timeline summarization methods and applications have been developed lately. ETS[21] formulated the task as an optimization problem via iterative substitution from a set of sentences with four requirements. Tran[22] investigated five different sentence features and leveraged SVMRank to optimize the summarization task and proposed an un-biased criteria which can model timeline's characteristics. Zhao[23] involved social attention to compute the importance and capture the social attention by learning users' collective interests in the form of word distributions from Twitter. Yan[1] proposed to model trans-temporal correlations among component summaries for timelines, using inter-date and intra-date sentence dependencies. Nedunchelian[2] reused the *MEAD* and add the timestamp feature to implement their timeline summarization. Binh Tran[3] extracted the temporal information and surface features to train the regression model for predicting the summary sentences. M. Georgescu[24] offered an online system that supports the entity-centered, temporal analysis of event-related information in Wikipedia based on the detection and summarization of updates in the webpage. Binh Tran[4] put forward a framework for automatically constructing timeline summaries from web news articles. In this framework, they extracted some features for each date about articles and sentences' publishing time and reference time.

2.2 Aging Theory

In the different stages of an event, there exist different hot words to reflect the facts about the event. An effective way to extract hot topic words is that capturing variations in the distribution of different word terms occurred in the documents regarding the event. Therefore, tracking the terms to know the current stage of the life cycle is very important. It's proved that the aging theory is effective to track the current stage of news life cycle[11]. It models the news event's life cycle into four stages, birth, growth, decay, and death. These four stages can be reflected by the key words' distribution. Namely, aging theory is used to model the appearance and disappearance of terms over time. Chen[5] applied this theory to model the news event's life cycle and utilized the concept of energy to track it. In order to gain the summary of multi-documents of news domain, aging theory is considered to be effective and efficient for extracting the features of sentences.

3   Our Approach

The approach for extracting the timeline information from news regarding the same event and its corresponding reports is proposed as follows. Sentences are extracted from the training set of news articles and in a second step the features are computed for each sentence. Finally, SVM model is used to train these features with over-sampling technology to recognize the summary sentence. The pipeline for this process is depicted in Fig. 1. Hereinafter presents the key concepts and methods for the timeline summarization for news regarding the same topic.

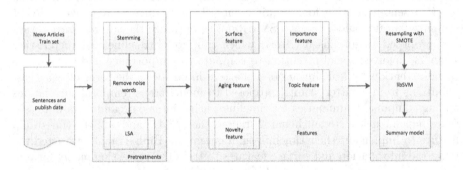

Fig. 1. The pipeline of our approach

3.1   Key Concepts

Event: News event often refers to a series of news articles about someone or something during a period of time. These articles form a news event which tell us the cause, the progress and the results about it.

   Timeline Summaries: Generally speaking, timeline is a kind of display forms for the summaries. Timeline summaries should show us the progress of this topic instand of just displaying the message according to the time sequence. Under this condition of the requirement, timeline summary of each day should describe the most important thing happened in that day regard some specific event.

The formal definition of event timeline summarization are defined as follows:

Input: Given a set of news articles $D = \{d_1, d_2, \ldots, d_n\}$ regarding the same event and these articles cover the whole progress of the event in the time span $T = \{t_1, t_2, \ldots, t_m\}$. Each news is segmented into sentences and then sentences are grouped by the publishing date to form $S = \{s_1, s_2, \ldots, s_m\}$, where s_i is a sentence group of sentences published in the i day.

Output: The multi-document timeline summarization should output the summaries along the date and each summary is the main idea of what occured in that day, i.e. $O = \{o_1, o_2, \ldots, o_m\}$, where o_i means the summary of sentences from all the sentences of that day s_i.

3.2 Sentence Feature Selection

In order to represent the most important thing happened in that day, the summary should consider the importance, the novelty, and the topic hot terms. All the features used in the experiments are shown in Table 1.

Table 1. List of features and their category

Category	Feature
Surface	length
	the count of noun words
	the count of stop words
	position in current paragraph
	position in the document
	whether it contains person name
Importance	sum of LSA scores
	sum/avg TFIDF
Topic	the count of topic words
	sum of topic words' weight
	sum of the TF top 10 words' weight
Aging	the aging score of sentence
Novelty	distance between current sentence and summary

Surface feature: This contains features computed by basic statistics, such as the length of sentence, the counts of noun words and stop words, the position in this document and paragraphs, and whether it contains person name or not.

Importance feature: This feature aims to represent the importance of the sentence. The significance of sentences is computed through linear combination of term weights with latent semantic analysis. The function is

$$Weight_{importance} = \sum_{w \in sentence} TF_w * LSA_w \tag{1}$$

where TF_w is the term frequency of word w and LSA_w is the weight of word in the LSA results.

Aging feature: This feature is used to measure the life cycle of each word. The frequency of a word changes as event going on, so we use the association between word and time interval to indicate its energy which is defined as follows:

$$E_{w,t} = F((E_{w,t-1} - \beta) + \alpha \cdot C_{w,t}) \tag{2}$$

where
$E_{w,t}$ is the energy of word w in time interval t;
α is the transfer factor;
β is the decay factor;
$C_{w,t}$ is the contribution degree of word at the time interval t;
F denote the the energy function which can convert the sum of similarities into a bounded extent. The sigmoid function is used as energy function and it is defined as:

$$F(y) = \frac{y}{1+y} \tag{3}$$

The $C_{w,t}$ function is defined as :

$$C_{w,t} = nf * \frac{TF_{w,t} \cdot IDF_w}{\sum_i^W TF_{i,t} \cdot IDF_i} \tag{4}$$

where nf is an empirical constant, $TF_{w,t}$ is the term frequency of word w in the time span t, IDF_w is the IDF value of word w, W is the whole word set during time span t.

In reality, no words describing a special event point will retain popular forever, they will decay over time. In other words, each word has a different energy value during the time span. In order to represent the word's life circle realistically, we cut down the energy of word by a decay factor β at the end of every time interval. Once the energy value became negative, it will be replaced by zero and stop decaying.

Topic feature: Each topic contains lots of sentences, i.e. each sentence contributes some information to the whole set to expand the topic. This paper uses the link analysis to compute the latent semantic between sentences. First, topic terms and topic elements can be found through the word frequency analysis. Then, the similarity among sentences construct a indirect graph of link relationship. The similarity value is computed with the cosine function and the function is defined as:

$$Similarity(s_a, s_b) = cosine(v_a, v_b) \tag{5}$$

where s_i denotes the i sentence and v_i denotes the word vector of sentence i. Since similarity value can be computed between every two sentences, the relationship forms a similarity graph. Last, pageRank algorithm is used to assign the weight to each node in this graph. The pageRank value is treated as the topic feature of the sentence. The function is :

$$Weight_{topic} = PageRank(sentence_i) \tag{6}$$

Novelty feature: In order to avoid some sentences with the same meanings be selected as the summary, the novelty of sentences should be taken into consideration. The novelty value is computed by the distance with the summary of last time span. The larger the distance, the more novel the sentence. In this research, the Jaccard similarity is used to gain this value. The function is :

$$Weight_{novelty} = 1 - Jaccard(sentence_i, summary_{ex}) \tag{7}$$

where $summary_{ex}$ is the summary sentence in the last time span. Since the Jaccard similarity is used to model the difference over words appearance between the current sentence and the summary sentences, the novelty between the current sentence and summary sentences can be measured.

3.3 Model Training

In the case of annotation data, this summarization task is considered as a classification problem. The positive data are sentences labeled as summary sentences, otherwise are negative. However, in each document only a few sentences will be labeled as the summarization sentence, i.e. the dataset is imbalanced. When the classifier is created over such an imbalanced dataset, the classifier will prefer the majority side [7]. In order to improve the precision of minority class, SMOTEBoost[7] method is used to sampling data for training the classification model. This method combines SMOTE [8] and boost technology. It has been proved effective for imbalanced data set.

4 Experiments

4.1 Datasets and Design

Since 2001, the US National Institute of Standards and Technology (NIST) has organized large-scale shared tasks for automatic text summarization within the Document Understanding Conference[1] (DUC) and the Summarization track at the Text Analysis Conference[2] (TAC). This paper choose one generic summarization and one guided summarization dataset to test the proposed approach from these datasets.

DUC-2002 dataset is used for generic summarization task in 2002. This dataset consist of 59 document sets, each in two formats - one with just the original TREC tagging, and one with an additional rough sentence tagging. Each document set contains around 12 documents from different day about the same event. Each event is also provided about 6 labeled summary sentences with the limit of 200 words. This limit is also set in our experiments.

TAC-2010 dataset is used for guided summarization task. The task aims to encourage summarization systems to make a deeper linguistic (semantic) analysis of the source documents. Thus, the previous method can be replaced that

---

[1] http://duc.nist.gov/
[2] http://www.nist.gov/tac/

relying only on document word frequencies to select important concepts and create the summary covered the predefined category. This dataset contains 906 documents around 46 topics from the New York Times, the Associated Press, the Xinhua News Agency newswires, and the Washington Post News Service. Each topic has two sets of documents, A and B, each containing 10 documents. The difference is that documents in B were published later than that in A. It's also an update summarization task, since it's required to create an 100 words updated summarization about documents in B. This dataset is good for the timeline summarization, because the summary should contains the important facts about the predefined category and show the update information. The summary sentences are relabeled based on the results from NIST in order to adapt the timeline summarization.

4.2   Annotation Method

The original annotation results of DUC-2002 and TAC-2010 are not prepared for the timeline summarization. In other words, the timeline summary sentences should not only contain those sentences with the true facts but also should along the time. We annotated these two datasets with three steps and the details are as follows: First, all the candidate sentences with publishing date are ranked in a descending order of importance. Second, the top 1 sentence is selected as the summary sentence and sentences published in the same day will be removed from the queue. Third, the next sentence in the queue will be selected and do this loop until the summary sentences can cover the whole time span of the event. The improved datasets are named as DUC-2002T and TAC-2010T.

4.3   Baselines

The experiment is started with the preprocessing like indexing, filtering out the stop words and segmenting news documents into sentences. Then the proposed method is used to the data set and generates a timeline for each chosen event. Some widely used multi-document summarization methods are implemented as the baselines.

Random The selecting method for random is that one sentence will be chosen as the summary sentence randomly from the whole document.

MEAD[3] is a famous multi-document summarization tool which extracts summary sentence based on the centroid value, positional value and first-sentence overlap.

Cluster considers that there are different themes in an event, so it divides similar sentences together into different clusters and then selects one representative sentence from each main cluster.

Allan is a similar timeline system from different aspects, which dividing sentences into on-event and off-event while ranking them with useful and novelty.

Wong combined the supervised and semi-supervised learning and used co-training method to train the labeled data and unlabeled data.

---

[3] http://www.summarization.com/mead

4.4   Evaluation Metric

In this part, ROUGE toolkit[9] is adopted, which is officially applied by DUC for document summarization performance evaluation, to evaluate the experimental results and compare these algorithms with each other. It's a recall based measure which compute unigram or bigram words matching between the auto-summarization summary and human-labeled summary sentences. All the summaries were truncated to the words limit size without removing the stop words.

4.5   Results

LibSVM[4] is used to train the classification model. In order to reduce the affect of imbalanced training data, over-sampling technology is used. First, some instances are sampled from the minority. Second, new instances are created by linear combining these selected instances. Third, these new instances are put into the training set. The model in the proposed approach 1 is trained without over-sampling, while the approach 2 is trained with over-sampling technology. Rouge-1 and Rouge-2 are used to measure the performance about recall on both datasets. Results are shown in Table 2 and Table 3. Precision value measures the percentage of labeled summary sentences among all the summary sentences created by the classifier.

Table 2. Performance on DUC-2002T dataset

Method	Precision	Rouge-1	Rouge-2
Random	0.023	0.024	0.009
MEAD	0.121	0.105	0.038
Cluster	0.075	0.061	0.018
Allan	0.077	0.061	0.021
Wong	0.144	0.245	0.089
Our approach 1	0.204	0.273	0.093
Our approach 2	**0.221**	**0.279**	**0.110**

Experiments on DUC-2002T. From Table 2, it can be seen that our approach 2 using SVM and over-sampling obtains the best results, followed by the approach 1 then *Wong* and *MEAD*. There is no surprise that Random provides the worst results. The approach 2 performs better than approach 1 is well understanded, since the resampling can gain more training samples which enhance the classifier. The *Allan*'s approach is close to the timeline summarization, but the results are not good due to the fact that only useful and novelty features are not enough to maintain the information of sentence. The *Cluster* method's bad performance is due to the random selecting strategy from the cluster center. *Wong*'s method combined the supervised and semi-supervised learning which

---

[4] http://www.csie.ntu.edu.tw/$\sim$cjlin/libsvm/

make it hard to turning the parameters. Both approach 1 and Wong shared most of features, except the aging features. The results show that the aging feature can improve the performance of summarization.

Table 3. Performance on TAC-2010T dataset

Method	Precision	Rouge-1	Rouge-2
Random	0.037	0.029	0.017
MEAD	0.129	0.076	0.028
Cluster	0.057	0.045	0.025
Allan	0.143	0.083	0.032
Wong	0.214	0.185	0.073
Our approach 1	0.211	0.279	0.102
Our approach 2	**0.233**	**0.291**	**0.121**

Experiments on TAC-2010T. Table 3 shows the results on the TAC-2010T dataset. The *Allan* method performs better than *MEAD*. The reason is that events in this dataset have more than 20 documents, which provide enough information about useful and novelty compared to the previous dataset.

From this two experiments, we can get the information that summarization methods based on machine learning are performance better than others. The result of *MEAD* are better than *Cluster*, mainly because this method use some surface feature. The *Cluster* method is the worst in our experiments, since this method cluster the same meaning sentences and choose one from cluster randomly, which introduce lots of uncertainty. The *Allan* method does not perform stable, that it will depend on the amount of documents and the content of each sentence. Our approach 2 performed better than *Wong* and approach 1 on precision and Rouge-1 metric which mainly due to over-sampling method and the new feature about aging. The over-sampling technology make the minority class can be classified better and this is proved in our experiments.

5   Conclusion and Future Work

In this paper, a novel approach involved aging theory is proposed to create timeline summarization, which focus on the news reports regarding the same event. After basic pre-precessing work, features of each sentence are extracted, including surface features, importance features, aging features, novelty features and topic features. Then the multi-document summarization task is considered as pair-wise classification task and the training data is well prepared. The final classification model is trained with SVM and over-sampling technology is used to improve the performance. Experiment results show that the proposed approach performs better compared with other widely used methods. In the future, the learning to rank framework will be tried with this group of features and we will do more experiments on different feature combinations.

Acknowledgments. This work is supported by the National Natural Science Foundation of China (No. 61370137, 61272361) and the 111 Project of Beijing Institute of Technology.

References

1. Yan, R., Kong, L., Hunag, C., Wan, X., Li, X., Zhang, Y.: Timeline generation through evolutionary trans-temporal summarization. In: EMNLP (2011)
2. Ramanujam, N.: Centroid based summarization of multiple documents implemented using timestamps. In: ICETET (2008)
3. Tran, B.: Structured summarization for news events. In: WWW (2013)
4. Tran, B., Alrifai, M., Nguyen, D.Q.: Predicting relevant news for timeline summaries. In: WWW (2013)
5. Chen, K.-Y., Luesukprasert, L., Chou, S.-C.T.: Hot topic extraction based on timeline analysis and multidimensional sentence modeling. Knowledge and Data Engineering, 1016–1025 (2007)
6. Russell Swan, James Allan: Automatic generation of overview timelines. In: ACM, pp. 49-56 (2000)
7. Chawla, N.V., Lazarevic, A., Hall, L.O., Bowyer, K.W.: SMOTEBoost: Improving prediction of the minority class in boosting. In: Lavrač, N., Gamberger, D., Todorovski, L., Blockeel, H. (eds.) PKDD 2003. LNCS (LNAI), vol. 2838, pp. 107–119. Springer, Heidelberg (2003)
8. Chawla, N.V., Bowyer, K.W., Hall, L.O., Kegelmeyer, W.P.: SMOTE: synthetic minority over-sampling technique. Journal of Artificial Intelligence Research, 321–357 (2002)
9. Lin, C.-Y.: Rouge: A package for automatic evaluation of summaries. In: Text Summarization Branches Out: Proceedings of the ACL 2004 Workshop, pp. 74–81 (2004)
10. Deerwester, S., Dumais, S.T., Harshman, R.: Indexing by latent semantic analysis. JASIS (1990)
11. Chen, C.C., Chen, Y.-T., Sun, Y., Chen, M.C.: Life cycle modeling of news events using aging theory. In: Lawry, J., G. Shanahan, J., L. Ralescu, A. (eds.) ECML 2003 (LNAI), vol. 2873, pp. 47–59. Springer, Heidelberg (2003)
12. Barzilay, R., McKeown, K.R., Elhadad, M.: Information fusion in the context of multi-document summarization. In: Proceedings of the 37th Annual Meeting of the Association for Computational Linguistics on Computational Linguistics, pp. 550–557 (1999)
13. Kevin, K., Daniel, M.: Summarization beyond sentence extraction: A probabilistic approach to sentence compression. Artificial Intelligence, 91–107 (2002)
14. McKeown, K.R., Klavans, J.L., Hatzivassiloglou, V., Barzilay, R., Eskin, E.: Towards multidocument summarization by reformulation: Progress and prospects. In: AAAI/IAAI, pp. 453–460 (1999)
15. Radev, D.R., Jing, H., Stys, M., Tam, D.: Centroid-based summarization of multiple documents. Information Processing & Management, 919–938 (2004)
16. Wan, X.: Manifold-ranking based topic-focused multi-document summarization. In: IJCAI, pp. 2903–2908 (2007)
17. Wan, X., Yang, J., Xiao, J.: Multi-document summarization using cluster-based link analysis. In: 31st Annual International ACM SIGIR Conference on Research and Development in Information Retrieval, pp. 299–306 (2008)

18. McKeown, K., Barzilay, R., Chen, J., Elson, D., Evans, D., Klavans, J., Nenkova, A., Schiffman, B., Sigelman, S.: Columbia's newsblaster: new features and future directions. In: The 2003 Conference of the North American Chapter of the Association for Computational Linguistics on Human Language Technology: Demonstrations, pp. 15–16 (2003)
19. Allan, J., Gupta, R., Khandelwal, V.: Temporal summaries of new topics. In: 24th Annual International ACM SIGIR Conference on Research and Development in Information Retrieval, pp. 10–18 (2001)
20. Wong, K.-F., Wu, M., Li, W.: Extractive summarization using supervised and semi-supervised learning. In: 22nd International Conference on Computational Linguistics, pp. 985–992 (2008)
21. Yan, R., Wan, X., Otterbacher, J., Kong, L., Li, X., Zhang, Y.: Evolutionary timeline summarization: a balanced optimization framework via iterative substitution. In: Proceedings of the 34th International ACM SIGIR Conference on Research and Development in Information Retrieval, pp. 745–754 (2011)
22. Tran, G.B., Tran, T.A., Tran, N.-K.: Leveraging learning To rank in an optimization framework for timeline Summarization. In: SIGIR (2013)
23. Zhao, W.X., Guo, Y., Yan, R., He, Y., Li, X.: Timeline generation with social attention. In: 36th International ACM SIGIR Conference on Research and Development in Information Retrieval, pp. 1061–1064 (2013)
24. Georgescu, M., Pham, D.D., Kanhabua, N., Zerr, S., Siersdorfer, S., Nejdl, W.: Temporal summarization of event-related updates in wikipedia. In: Proceedings of the 22nd International Conference on World Wide Web Companion, pp. 281–284 (2013)

Extended Strategies for Document Clustering with Word Co-occurrences

Yang Wei[1,2], Jinmao Wei[1,2], and Zhenglu Yang[1,2]

[1] College of Computer and Control Engineering, Nankai University,
Weijin Rd. 94, 300071 Tianjin, China
[2] College of Software, Nankai University
Weijin Rd. 94, 300071 Tianjin, China
weiyang_tj@outlook.com, {weijm,yangzl}@nankai.edu.cn

Abstract. To tackle the sparse data problem of the bag-of-words model for document clustering, recent strategies have been proposed to enrich a document with the relatedness of all the words in a corpus to the document, where the relatedness is estimated by the weighted sum of word co-occurrences. However, the relatedness is overestimated without eliminating the overlaps between word co-occurrences. This paper demonstrates that the weighted sum strategy gives the upper bound of the theoretic degree of relatedness. Two strategies are further proposed to approach the theoretic degree of relatedness. The first strategy is established under the extreme assumption that all the words in a document co-occur with each other. By considering the specificities of words, the second strategy gives several extended versions of the weighted sum strategy. Substantial experiments verify that the document clustering incorporated with the extended strategies achieve a significant performance improvement compared to the state-of-the-art techniques.

Keywords: document enriching, word co-occurrences, specificity.

1 Introduction

Representing documents as fixed-length vectors with the bag-of-words (BOW) model [8] is a common approach in document clustering for its simplicity, efficiency and acceptable accuracy. The dimensions of the BOW feature space are simply generated by the distinct words in a document collection. However, the sparse data weakens the effect of BOW on identifying the similarities between documents composed of different words.

Our intuitive idea is to address the sparse data problem by enriching a document with the relatedness of all the words in a corpus to the document, which is inspired by the semantic representations for words [4], [9], [13]. Since the relatedness between words cannot be obtained directly from the dimensionality reduction methods of word representation [2], [5] or the distributed representation for words [13], the distributional representation in the feature space of BOW [4], [9] is used as the statistical foundation to measure the relatedness of all the words to the original words in a document, where the relatedness between words

© Springer International Publishing Switzerland 2015
R. Cheng et al. (Eds.): APWeb 2015, LNCS 9313, pp. 461–472, 2015.
DOI: 10.1007/978-3-319-25255-1_38

is measured by their co-occurrence times in the whole corpus. The relatedness of words to a document can be obtained through the combination of the generated word vectors.

Regarding each word vector as an abstract concept, the idea of combining word vectors to represent documents is similar to those mapping documents into lower feature spaces [2], [6], [7], [11] based on the assumption that documents with similar meanings are supposed to share similar concepts. Significant improvements have been achieved with those methods. Nevertheless, the parameters, especially dimension of the space, are often difficult to be decided.

In fact, representing a document with word relatedness has been long studied [1], [10], [15]. In the previous studies, researchers mainly focus on the constructing of word vectors. The combination scheme is confined to the weighted sum of word vectors. The relatedness of a word to a document is the average of all the relatedness of the word to the original words in the document. However, such combination scheme has already been proved to hinder its usage on representing the semantic meanings of phrases [14]. By introducing three criterions to restrict the document enriching process, we deduce the theoretic degree of relatedness of a word to a document through the combination of word vectors. We further demonstrate that, the weighted sum of word vectors gives the upper bound of the theoretic degree of relatedness, and the lower bound has also been established.

Unfortunately, the theoretic degree of relatedness cannot be obtained directly. Instead, we define the specificity of a word to re-weight the weights of all the word vectors. As a result, several strategies have been proposed to extend the current weighted sum strategy. The main contribution of the newly proposed schemes is to approach the theoretic degree of relatedness by eliminating the overlaps between word co-occurrences. The experimental results agree with our analysis that significant improvements have been achieved with the proposed strategies. Furthermore, the guideline of how to select the proper strategy according to different datasets has been discussed in this paper.

2   Related Work

2.1   Word Vector

The idea that word usage can reveal semantics was claimed by Harris [8], that words occur in similar contexts tend to have similar meanings. Following this instinct, the co-occurrence times of the context words near a target word have been used to reveal the semantic meanings of the target word [3], [16]. Formally, if there is a corpus D with n documents and m distinct words in it, the set of the m words constitute the vocabulary V, and each word in V is denoted as v. Then the context of a target word v_i is defined as:

$$\mathbf{v}_i = \left(\frac{c_{i,1}}{c_i}, \frac{c_{i,2}}{c_i}, \cdots, \frac{c_{i,m}}{c_i} \right), \tag{1}$$

where $c_{i,j}$ is the co-occurrence times between v_i and v_j in D, and c_i is the total times v_i occurs in D. Generally, the meanings of words should be independent of the corpus size, so c_i is introduced here to give the basic context of each word [4]. The defined context is called *word vector*. The values in the word vector measure the relatedness of the dimensions to the target word.

The co-occurrence times can be estimated with several methods [4], [9], [12]. We choose the one used in the weighted sum of the word vectors for document representation [1]:

$$\frac{c_{i,j|D}}{c_{i|D}} = \frac{\sum_{k=1}^{n} \frac{c_{i|d_k}}{\sum_{a=1}^{m} c_{a|d_k}} \cdot \frac{c_{j|d_k}}{\sum_{a=1}^{m} c_{a|d_k}}}{\sum_{k=1}^{n} \left(\frac{c_{i|d_k}}{\sum_{a=1}^{m} c_{a|d_k}} \sum_{a=1, a \neq i}^{m} \frac{c_{a|d_k}}{\sum_{b=1}^{m} c_{b|d_k}} \right)}, \tag{2}$$

where $c_{i,j|D}$ and $c_{i|D}$ emphasize that the values are generated from the whole corpus D, and $c_{i|d_k}$ refers to the occurrence times of the word v_i in the document d_k. Together with all the word vectors produced by (2), an $m \times m$ matrix is obtained, which is called *context matrix*:

$$\mathbf{V} = (\mathbf{v}_1, \mathbf{v}_2, \cdots, \mathbf{v}_m) \ . \tag{3}$$

2.2 Context Vector Model

Given a target document d, the document vector produced by BOW is:

$$\Phi_{\text{bow}} : \mathbf{d} = (\nu_{1|d}, \nu_{2|d}, \cdots, \nu_{m|d}) \in \mathbb{R}^m, \tag{4}$$

where $\nu_{i|d}$ is the weight of the word v_i in $\mathbf{d}$. The popular weighting scheme is the Term Frequency and Inverse Document Frequency (tf $\cdot$ idf):

$$\nu_{i|d} = \text{tf}_{i|d} \cdot \text{idf}_i = \frac{c_{i|d}}{\sum_{j=1}^{m} c_{j|d}} \cdot \left(1 + \log_2 \left(\frac{n}{n_i} \right) \right), \tag{5}$$

where n_i is the number of documents in which v_i occurs.

Since BOW cannot figure out similar documents composed of different words, the Context Vector Model (CVM) tries to reveal the meanings of documents with a set of weighted word vectors [1], [10], [15]. Given the generated context matrix, the new document vector could be obtained by:

$$\Phi_{\text{cvm}} : \mathbf{d}' = \mathbf{dV} = \sum_{i=1}^{m} \nu_{i|d} \mathbf{v}_i \in \mathbb{R}^m \ . \tag{6}$$

As a result, the value of any word in $\mathbf{d}'$ is defined as the weighted sum of the relatedness of the word to the original words in d. Billhardt [1] has compared the effects of using the tf and tf $\cdot$ idf schemes to estimate the weights of the word vectors, and further exploited the deviations of the co-occurrence times in the word vectors to estimate their weights:

$$\forall \mathbf{v}_i \in \mathbf{V}, \text{dev}_i = 1 + \frac{\sum_{j=1}^{m} \left| \frac{c_{i,j|D}}{c_{i|D}} - 1 \right|}{m}, \tag{7}$$

where mean($\mathbf{v}_i$) is the average of all the values in $\mathbf{v}_i$. Then $\nu_{i|d} = \text{tf}_{i|d} \cdot \text{dev}_i$.

Experimental results demonstrate that $\text{tf} \cdot \text{idf}$ and $\text{tf} \cdot \text{dev}$ improve the performance of CVM. By revealing the improvement is achieved through eliminating the overlaps between word co-occurrences, we extend the two weighting schemes to re-weight the values for each dimension in the document vector generated by CVM directly. Nine subversions of the traditional CVM are hence generated in total.

The other symbols used for statistics throughout this paper include:

- The symbol $v_i \cdot v_j|d$ refers to the event that the words v_i and v_j co-occur in the document d. The number of the occurrence times is denoted as $c(v_i \cdot v_j|d)$, and for simplicity, it is also denoted as $c_{i,j|d}$. Generally, the co-occurrence of any k words in d is represented as $v_1 \cdot v_2 \cdots v_k|d$, and the number of the occurrence times is denoted as $c(v_1 \cdot v_2 \cdots v_k|d)$.
- The union of the two events $v_i \cdot v_j|d$ and $v_j \cdot v_k|d$ is represented as $v_i \cdot v_j \cup v_j \cdot v_k|d$, which refers to the event that v_j co-occurs with v_i or v_k in d. The number of the occurrence times of v_j in the union of the two events is denoted as $c(v_i \cdot v_j \cup v_j \cdot v_k|d)$. The union of the k events that any k words co-occur with v_j in d is represented as $\bigcup_{i=1}^{k} v_i \cdot v_j|d$, and the number of the occurrence times of v_j is denoted as $c(\bigcup_{i=1}^{k} v_i \cdot v_j|d)$.
- V_d stands for the vocabulary composed of the distinct words in the document d, and of course, $V_d \subseteq V$.

3   Document Enrichment

3.1   Document Enriching

Inspired by the success of word vectors that word usage can reveal semantics [4], [9], [12], our intuitive idea is to find the relatedness of all the words in a corpus to a document in terms of their relatedness to the original words in the document. Since the values of the word vectors generated from a corpus have been proved effective on evaluating the relatedness between words, they are adopted as standards to specify whether the relatedness of a word to a document is well defined. As a result, the enriched representation of d, denoted as d', must satisfy the criterion that the word vectors of the words in d generated from d' are equivalent to the corpus-level word vectors generated from D.

Criterion 1 $\forall v_i \in V_d, \mathbf{v}_{i|d'} = \mathbf{v}_{i|D}$.

Considering the definition of a word vector in (1), we can get the following result with Criterion 1:

$$\forall v_i \in V_d, v_j \in V, \mathbf{v}_{i|d'} = \mathbf{v}_{i|D} \Rightarrow \frac{c_{i,j|d'}}{c_{i|d'}} = \frac{c_{i,j|D}}{c_{i|D}} \Rightarrow c_{i,j|d'} = \frac{c_{i,j|D}}{c_{i|D}} \cdot c_{i|d'} \quad . \quad (8)$$

Clearly, $c_{i,j|d'}$ will be identified if $c_{i|d'}$ is given. Since only the relatedness of all the words in V to the original words in d should be considered, it'll make sense to force $c_{i|d'}$ equal to $c_{i|d}$, and to omit the other co-occurrences between the words in $V - V_d$, which correspond to the following two criterions:

Criterion 2 $\forall v_i \in V_d, c_{i|d'} = c_{i|d}$.

Criterion 3 $\forall v_i, v_j \in V - V_d, c_{i,j|d'} = 0$.

With Criterion 2, $c_{i,j|d'}$ will depend merely on $\frac{c_{i,j|D}}{c_{i|D}}$ and $c_{i|d}$, where $\frac{c_{i,j|D}}{c_{i|D}}$ is the inherent relatedness between v_i and v_j, and $c_{i|d}$ makes the degree of the relatedness between v_i and v_j decided by the occurrence times of v_i in d; According to Criterion 3, the final relatedness of v_j to d will not be affected by the words not in d. As a result, the two constraints make d' connected to d only, which distinguishes d' from the representations for other documents.

Criterion 3 makes the words in $V - V_d$ merely show up with the words in V_d:

$$\forall v_j \in V - V_d, c_{j|d'} = c(\bigcup_{v_i \in V_d} v_i \cdot v_j | d') . \tag{9}$$

Similarly, the occurrence times of the words in V_d could be represented as:

$$\forall v_j \in V_d, c_{j|d'} = c(\bigcup_{v_i \in V} v_i \cdot v_j | d') = c(\bigcup_{v_i \in V_d} v_i \cdot v_j | d') + c(\bigcup_{v_i \in V - V_d} v_i \cdot v_j | d') . \tag{10}$$

In (10), for each of the words in V_d, its occurrences are separated into two parts. In the first part, it co-occurs with the words in V_d; in the second part, it co-occurs with the words not in V_d. According to the definition of word vectors, the co-occurrences in the first part act as descriptors to describe the meanings of the words in V_d; while those in the second part are the same as the co-occurrences in (9). Only the first part is useful to estimate the relatedness between the words both in V_d. Together with (9) and the first part in (10), we get the occurrence times of all the words used to measure their relatedness to the words in V_d. These occurrence times are used to construct the new document vector:

$$\mathbf{d'} = (c(\bigcup_{v_i \in V_d} v_i \cdot v_1 | d'), c(\bigcup_{v_i \in V_d} v_i \cdot v_2 | d'), \cdots, c(\bigcup_{v_i \in V_d} v_i \cdot v_m | d')) . \tag{11}$$

We stress again that $\mathbf{d'}$ implies the total occurrence times of the words in V_d is the same as those in d, for the values of the words in $V - V_d$ can only be valid when the second part in (10) is established. To distinguish the values of each word in $\mathbf{d'}$ and the real occurrence times of the words in d', the former are called *semantic occurrence times* in the following.

Compared (11) with (4), the significant difference between BOW and the proposed method is that the values of the document vector generated by BOW are mainly depend on the local occurrences of the words in d, while the relatedness of all the words to the original words in d with the solid corpus-level statistics are used to fill the new document vector.

3.2 Calculation of the Enriched Document

Substitute Criterion 2 into (8), the required co-occurrence times between the words in d' could be obtained:

$$c_{i,j|d'} = \frac{c_{i,j|D}}{c_{i|D}} \cdot c_{i|d'} = \frac{c_{i,j|D}}{c_{i|D}} \cdot c_{i|d} . \tag{12}$$

Then the issue becomes how to get the semantic occurrence times through the co-occurrence times. As shown in Example 1, the occurrence times of v_2 could be obtained by eliminating the repeat accumulations for v_2 during the sum of the co-occurrence times. Hence $c_2 = c_{1,2} + c_{2,3} - c_{1,2,3}$. For simplicity, the repeat accumulations for a word during the sum of its co-occurrence times with other words are named as *overlaps between co-occurrences* for the word.

Example 1. For a word segment $v_1 - v_2 - v_3$, the sum of the co-occurrence times $c_{1,2}$ and $c_{2,3}$ is two, while the real occurrence times of v_2 is one. Repeat accumulations exist during the sum of the co-occurrence times.

Without loss of generality, assume the first k words in V are specified as those in V_d. The general formula to get the semantic occurrence times of each word in $\mathbf{d}'$ by eliminating the overlaps between co-occurrences could be obtained inductively as:

$$\forall v_j \in V, c\left(\bigcup_{v_i \in V_d} v_i \cdot v_j | d' \right) = \sum_{i=1}^{k} c(v_i \cdot v_j | d') - \sum_{1 \le i < a \le k} c(v_i \cdot v_a \cdot v_j | d')$$

$$+ \sum_{1 \le i < a < b \le k} c(v_i \cdot v_a \cdot v_b \cdot v_j | d') + \cdots + (-1)^{k-1} c(v_1 \cdot v_2 \cdots v_k \cdot v_j | d') . \quad (13)$$

According to (13), the semantic occurrence times will not be calculable unless the co-occurrence times among more than two words are given. Unfortunately, this information cannot be obtained under current conditions. Alternatively, several approximate approaches are proposed in the following.

Boundaries of the Semantic Occurrence Times. Ideally, if there is no co-occurrence among more than two words, the sum of the co-occurrence times between two words will contain no repeat accumulations. Namely, assume,

$$\forall v_a, v_b, v_i \in V_d, v_j \in V, c(v_a \cdot v_i \cdot v_j | d') = 0 .$$

Then all the items in (13) will equal zero except the first one:

$$\forall v_j \in V, c\left(\bigcup_{v_i \in V_d} v_i \cdot v_j | d' \right) = \sum_{v_i \in V_d} c(v_i \cdot v_j | d') = \sum_{v_i \in V_d} \frac{c_{i,j|D}}{c_{i|D}} \cdot c_{i|d} . \quad (14)$$

This is the maximum value the semantic occurrence times can approach. Conversely, by assuming every word in d' co-occurs with all the other words, the minimum value of the semantic occurrence times can be achieved. Namely, $\forall V' = (v_1, v_2, ... v_a) \subseteq V_d$, where $v_i \in V'$ is randomly selected from V_d, assume,

$$c(v_1 \cdot v_2 \cdots v_a \cdot v_j | d') = \min\{c_{1|d'}, c_{2|d'}, \cdots, c_{a|d'}, c_{j|d'}\} .$$

Specifically, if $|V'| = 1$, $c(v_a \cdot v_j | d') = c_{a,j|d'} = \min\{c_{a|d'}, c_{j|d'}\}$. Hence we have:

$$c(v_1 \cdot v_2 \cdots v_a \cdot v_j | d') = \min\{c_{1|d'}, c_{2|d'}, \cdots, c_{a|d'}, c_{j|d'}\}$$

$$= \min\{\min\{c_{1|d'}, c_{j|d'}\}, \min\{c_{2|d'}, c_{j|d'}\}, \cdots, \min\{c_{a|d'}, c_{j|d'}\}\}$$

$$= \min\{c_{1,j|d'}, c_{2,j|d'}, \cdots, c_{a,j|d'}\} . \quad (15)$$

Taken (15) into (13), the items in (13) will offset each other and leave the maximum one:

$$\forall v_j \in V, c\left(\bigcup_{v_i \in V_d} v_i \cdot v_j | d' \right) = \max_{v_i \in V_d} \{ c_{i,j|d'} \} = \max_{v_i \in V_d} \left\{ \frac{c_{i,j|D}}{c_{i|D}} \cdot c_{i|d} \right\} . \tag{16}$$

Example 1 demonstrates that for any word segment whose length is longer than two, repeat accumulations will exist with the simple sum of co-occurrence times. Therefore, the semantic occurrence times of all the words will be overestimated by (14) in practice. On the other hand, the assumption required by (16) seems too farfetched to force all the words co-occur in such a long enriched text. Pragmatically, the upper and lower bounds of the semantic occurrence times are just the approximate results of the real semantic occurrence times.

Approximation with Specificities. Since the deviations between the upper bounds and the real semantic occurrence times for words are caused by the overlaps between co-occurrences, the parameter γ is introduced to eliminate the overlaps:

$$\forall v_j \in V, c\left(\bigcup_{v_i \in V_d} v_i \cdot v_j | d' \right) = \sum_{v_i \in V_d} \gamma_{i,j} c(v_i \cdot v_j | d') = \sum_{v_i \in V_d} \gamma_{i,j} \frac{c_{i,j|D}}{c_{i|D}} c_{i|d}, 0 \le \gamma_{i,j} \le 1 .$$
$$\tag{17}$$

Theoretically, with proper γ parameters, the real semantic occurrence times can be obtained. We define the *specificities* of words to estimate the γ parameters. If a word mainly co-occurs with several specific words along the whole corpus, it is regarded as high specificity; otherwise, it is regarded as low specificity. The popular Inverse Document Frequency (idf) defined in (5) and the deviation (dev) defined in (7) can be used to estimate the specificities of each word.

Usually, words with high idf or dev values must mainly co-occur with several specific words, and therefore is regarded as high specificity. Such words are more possible to appear in different places. If not, these words tend to co-occur with all their specific words repeatedly in the same place, which has little chance to happen. Therefore, the overlaps between co-occurrences for the words with high specificities tend to be low, and they will be given a large value of γ. Since the overlaps between the co-occurrence $v_i \cdot v_j$ and other co-occurrences are affected by the specificities of both v_i and v_j, the γ parameter should be decided by both of the two factors. Hence we have:

$$\forall v_i, v_j \in V, \gamma_{i,j} = \alpha_i \cdot \beta_j, 0 \le \alpha_i \le 1, 0 \le \beta_j \le 1, \tag{18}$$

where α_i and β_j evaluate the specificities of v_i and v_j respectively.

We can define the parameters α and β to either idf or dev optionally. With either of them, α and β will stay unchanged all over the corpus. For all the words in V, their α and β parameters are organized as the vector $\boldsymbol{\alpha}$ and $\boldsymbol{\beta}$ respectively. Then (17) could be rewritten in the vector form:

$$\Phi_{\text{cvm}} : \mathbf{d}' = \mathbf{d} \times (\boldsymbol{\alpha} \times \boldsymbol{\beta} \cdot \mathbf{V}) . \tag{19}$$

Table 1. Versions of CVM

Version Name	α	β	Version Name	α	β	Version Name	α	β
CVM-UpperBound	1	1	CVM-1-IDF	1	idf	CVM-1-DEV	1	dev
CVM-IDF-1	idf	1	CVM-IDF-IDF	idf	idf	CVM-IDF-DEV	idf	dev
CVM-DEV-1	dev	1	CVM-DEV-IDF	dev	idf	CVM-DEV-DEV	dev	dev

Table 2. Statistics of the TDT2 and Reuters Datasets

	No. docs	No. clusters	Avg. cluster size	Avg. doc. length	Avg. word. freq
TDT2	10021	56	179	182	51
Reuters	8213	41	200	68	30

Generally, α and β should be normalized to make the parameters within the range defined by (18). However, the normalization will be non-necessary when the cosine similarity is used to evaluate the similarities between document vectors. In this case, normalization or not will produce the same result, for an overall normalization is performed during the calculation of the cosine similarity.

All the proposed strategies are regarded as the extended versions of CVM, for they all based on the context matrix. These versions could be divided in two branches. The first branch is the lower bound version of CVM defined by (16), CVM-LowerBound, which eliminates the overlaps between co-occurrences under a unified assumption that all the words in d' co-occur with each other. The other branch is the versions incorporated with the specificities of words, which are listed in Table 1. Specifically, by setting both α and β to one, CVM-UpperBound is the upper bound version of CVM defined by (14). By setting α to idf and β to one, CVM-IDF-1 corresponds to the classical CVM where the values in the document vectors generated by BOW are weighted with the tf $\cdot$ idf scheme. By setting α to dev and β to one, CVM-DEV-1 corresponds to the one proposed in [1]. By setting β to idf or dev, the weighting schemes used in CVM are extended to consider the specificities of all the dimensions.

4 Experiments

4.1 Experimental Setup

In this section, we evaluate our algorithms on document clustering problem. The algorithms to be compared include: 1) the traditional BOW method with the tf $\cdot$ idf weighting scheme; 2) the Non-negative Matrix Factorization based clustering (NMF in short) [6]; 3) the existing versions of CVM, including, CVM-UpperBound, CVM-IDF-1, and CVM-DEV-1; 4) the other versions of CVM proposed in this paper.

Document vectors generated by the algorithms listed above are used as the input of the k-means clustering algorithm. Two metrics are adopted in our experiments to evaluate the performance of document clustering, i.e., the accuracy (AC) and the normalized mutual information (NMI) [17]. The average similarity (avg. sim) and the average standard deviation (avg. std) of all the document vectors generated by the versions of CVM are also calculated to further demonstrate the differences between each version, where the cosine similarity is used to measure the similarity between two document vectors; and the standard deviation of a document vector is calculated as follows:

$$\text{std}(\mathbf{d}') = \left(\frac{1}{m} \sum_{i=1}^{m} (\nu_i - \bar{\nu})^2 \right)^{\frac{1}{2}}, \tag{20}$$

where ν_i is the value in $\mathbf{d}'$, and $\bar{\nu}$ is the average of all the values in $\mathbf{d}'$.

We conduct the performance evaluation using the TDT2 and Reuters datasets[1]. TDT2 consists of data extracted from six sources, including two newswires (APW, NYT), two radio programs (VOA, PRI) and two television programs (CNN, ABC). Those documents appearing in two or more categories are removed, and the categories with more than 10 documents are kept, thus leaving us with 10,021 documents in total. Reuters contains 21,578 documents which are grouped into 135 clusters. Those documents with multiple category labels are discarded, and the categories with more than 10 documents are selected. This leaves us with 8,213 documents in total. Reuters is more difficult for clustering than TDT2. In TDT2, the content of each cluster is narrowly defined, whereas in Reuters, documents in each cluster have a broader variety of content [17]. Table 2 provides the statistics of the datasets.

4.2   Performance Evaluations

The clustering results are shown in Table 3. Besides the existing versions of CVM, the results of the proposed versions with the best performances are also shown in Table 3, specifically, CVM-IDF-DEV for TDT2 and CVM-LowerBound for Reuters. The evaluations were conducted with the cluster numbers ranging from two to ten. For each given cluster number, 50 test runs were performed on different randomly chosen clusters and the average performance is reported in the table. These experiments reveal a number of interesting points:

1) The versions of CVM achieve the best performances for most cases on both datasets. This demonstrates that by introducing word co-occurrences, they can generate a better representation by enriching the target document with the relatedness between words. One possible reason for more improvement of CVM on the TDT2 dataset than on the Reuters dataset is that the average word frequency on TDT2 is much higher than that on Reuters. The high frequencies of words on the TDT2 dataset implies that the co-occurrence data is sufficient to produce more effective word vectors on the TDT2 dataset; 2) The proposed versions of CVM have competitive performance on both datasets, indicating the effectiveness on eliminating the overlaps between co-occurrences; and 3) None of the versions of CVM can perform the best on all the sub-tasks with different cluster numbers, wherefore it's more sensible to select appropriate methods in different situations rather than to find the best method for all cases.

---

[1] http://www.cad.zju.edu.cn/home/dengcai/Data/TextData.html

Table 3. Clustering Results on the TDT2 and Reuters datasets

No. cluster	2	3	4	5	6	7	8	9	10	Avg.
TDT2-Accuracy										
BOW	0.938	0.882	0.834	0.778	0.762	0.708	0.653	0.651	0.613	0.758
NMF	0.952	0.896	0.877	0.821	0.807	0.794	0.729	0.738	0.689	0.811
CVM-UpperBound	0.967	0.916	0.889	0.840	0.838	0.823	0.780	0.783	0.754	0.843
CVM-IDF-1	0.964	0.923	0.885	0.849	0.838	**0.826**	0.783	**0.785**	**0.775**	0.848
CVM-DEV-1	**0.972**	0.921	**0.895**	**0.857**	**0.847**	0.824	0.791	0.778	0.756	0.849
CVM-IDF-DEV	0.972	**0.925**	0.889	0.850	0.839	0.824	**0.792**	0.784	0.774	**0.850**
TDT2-NMI										
BOW	0.807	0.771	0.739	0.691	0.716	0.668	0.629	0.648	0.622	0.699
NMF	0.850	0.791	0.783	0.726	0.743	0.733	0.684	0.696	0.669	0.742
CVM-UpperBound	**0.883**	0.809	0.792	0.737	0.766	0.757	0.720	**0.731**	0.709	0.767
CVM-IDF-1	0.875	0.813	0.786	0.737	0.762	0.749	0.710	0.718	0.710	0.762
CVM-DEV-1	0.881	0.811	**0.799**	**0.747**	**0.769**	**0.757**	**0.725**	0.727	0.708	**0.770**
CVM-IDF-DEV	0.883	**0.814**	0.786	0.738	0.762	0.749	0.713	0.721	**0.711**	0.764
Reuters-Accuracy										
BOW	0.825	0.692	0.652	0.588	0.589	0.539	0.469	0.447	0.480	0.587
NMF	0.868	0.756	0.752	**0.698**	**0.677**	0.606	**0.598**	0.568	**0.590**	**0.679**
CVM-UpperBound	0.840	0.758	0.705	0.633	0.631	0.591	0.528	0.543	0.540	0.641
CVM-IDF-1	0.847	0.761	0.725	0.638	0.628	0.578	0.537	0.567	0.541	0.647
CVM-DEV-1	0.844	0.760	0.717	0.629	0.634	0.592	0.524	0.538	0.556	0.644
CVM-LowerBound	**0.870**	**0.764**	**0.761**	0.666	0.668	**0.619**	0.562	**0.587**	0.578	0.675
Reuters-NMI										
BOW	0.428	0.409	0.483	0.428	0.482	0.460	0.393	0.393	0.478	0.439
NMF	0.503	0.452	0.553	0.485	0.519	0.485	**0.447**	0.445	**0.513**	0.489
CVM-UpperBound	0.531	0.517	0.548	0.473	0.519	0.488	0.415	**0.447**	0.500	0.493
CVM-IDF-1	0.547	0.536	0.581	0.476	0.513	0.476	0.405	0.437	0.492	0.496
CVM-DEV-1	0.533	0.525	0.568	0.473	0.517	0.485	0.414	0.445	0.504	0.496
CVM-LowerBound	**0.565**	**0.544**	**0.614**	**0.486**	**0.541**	**0.491**	0.419	0.434	0.507	**0.511**

4.3   Method Selection

According to the above discussion, it's essential to find the rule of selecting the proper version of CVM on different situations. Figure 1 shows how the performances of different versions of CVM vary respectively. The results of different evaluation metrics are normalized into the same scale with the following function:

$$\forall \mathbf{e} = (e_1 - \bar{e}, e_2 - \bar{e}, \cdots, e_{10} - \bar{e}), \mathbf{e} = \frac{\mathbf{e}}{\|\mathbf{e}\|_1}, \tag{21}$$

where $e_1, e_2, \cdots, e_{10}$ are the evaluation results of all the ten versions of CVM for Avg. sim, Avg. std, Avg. AC, or Avg. NMI, $\bar{e}$ is the average result of the evaluation, and $\|\mathbf{e}\|_1$ is the one-norm of $\mathbf{e}$.

As we can see, the rankings of the ten versions of CVM according to Avg. sim are the same on both datasets, and along with the rising of Avg. sim, Avg. std decreases. Since the standard deviation measures the specificities of each document vector, the opposite rankings agree with the common sense that high

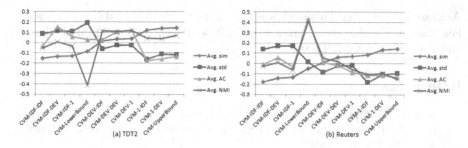

Fig. 1. The evaluation results of all the versions of CVM are arranged in ascending order of the Avg. sim values.

specificity corresponds to low similarity. By eliminating the additional overlaps between co-occurrences, the proposed version of CVM make the document vectors more specific. The rankings of the ten versions of CVM demonstrate the degrees of the overlaps between co-occurrences have been eliminated.

Figure 1 also shows that the branch incorporated with the specificities of words perform relatively stable on the two datasets, while CVM-LowerBound achieves contrary performances compared with the other versions on the two datasets. Since the semantic occurrence times of each word produced by CVM-LowerBound is decided by a single value, the maximum co-occurrence times with the original words in a document, CVM-LowerBound is more sensitive to the variance of statistics of datasets. Therefore, its performance compared with the other versions will be unstable. CVM-LowerBound has a relatively bad performance on TDT2 may attribute to the long lengths of documents as shown in Table 2. Keeping the maximum value only will lose too much information contained by the other values; on the contrary, the short lengths of documents on Reuters makes the assumption that all words co-occur with each other in the same document more sensible. So with an abundant corpus, the versions incorporated with the specificities of words are expected to perform better; with a small corpus, especially in the case as Reuters, where the lengths of documents are short, CVM-LowerBound is worth trying.

5 Conclusions

In this paper, we present a novel document representation framework to enrich a document with the relatedness of words to the document. The framework demonstrates that the overlaps between word co-occurrences hinder the performance of the traditional Context Vector Model (CVM). By defining the specificities of words, several extended strategies based on CVM are proposed to eliminate the overlaps between co-occurrences. In addition, a novel method which keeps the maximum values of word vectors on the same dimensions is proposed. The substantial experiments demonstrate that, with an abundant corpus, the branch incorporated with the specificities of words are expected to perform better; while with a small corpus, it's worth trying the maximum-value-keeping strategy.

Acknowledgments. This work was supported by the National Natural Science Foundation of China under grant 61070089, the Science Foundation of TianJin under grant 14JCYBJC15700.

References

1. Billhardt, H., Borrajo, D., Maojo, V.: A context vector model for information retrieval. Journal of the American Society for Information Science and Technology 53(3), 236–249 (2002)
2. Blei, D.M., Ng, A.Y., Jordan, M.I.: Latent dirichlet allocation. The Journal of machine Learning research 3, 993–1022 (2003)
3. Blunsom, P., Grefenstette, E., Hermann, K.M., et al.: New directions in vector space models of meaning. In: Proceedings of the 52nd Annual Meeting of the Association for Computational Linguistics (2014)
4. Bullinaria, J.A., Levy, J.P.: Extracting semantic representations from word co-occurrence statistics: A computational study. Behavior Research Methods 39(3), 510–526 (2007)
5. Bullinaria, J.A., Levy, J.P.: Extracting semantic representations from word co-occurrence statistics: stop-lists, stemming, and svd. Behavior Research Methods 44(3), 890–907 (2012)
6. Cai, D., He, X., Han, J.: Locally consistent concept factorization for document clustering. IEEE Transactions on Knowledge and Data Engineering 23(6), 902–913 (2011)
7. Deerwester, S.C., Dumais, S.T., Landauer, T.K., Furnas, G.W., Harshman, R.A.: Indexing by latent semantic analysis. JASIS 41(6), 391–407 (1990)
8. Harris, Z.S.: Distributional structure. Word (1954)
9. Iosif, E., Potamianos, A.: Unsupervised semantic similarity computation between terms using web documents. IEEE Transactions on Knowledge and Data Engineering 22(11), 1637–1647 (2010)
10. Kalogeratos, A., Likas, A.: Text document clustering using global term context vectors. Knowledge and Information Systems 31(3), 455–474 (2012)
11. Le, Q.V., Mikolov, T.: Distributed representations of sentences and documents. In: Proceedings of the 31st International Conference on Machine Learning, W&CP, vol. 32. JMLR (2014)
12. Lund, K., Burgess, C.: Producing high-dimensional semantic spaces from lexical co-occurrence. Behavior Research Methods, Instruments, & Computers 28(2), 203–208 (1996)
13. Mikolov, T., Chen, K., Corrado, G., Dean, J.: Efficient estimation of word representations in vector space. arXiv preprint arXiv:1301.3781 (2013)
14. Mitchell, J., Lapata, M.: Composition in distributional models of semantics. Cognitive Science 34(8), 1388–1429 (2010)
15. Rungsawang, A.: Dsir: The first trec-7 attempt. In: TREC, pp. 366–372. Citeseer (1998)
16. Turney, P.D., Pantel, P., et al.: From frequency to meaning: Vector space models of semantics. Journal of Artificial Intelligence Research 37(1), 141–188 (2010)
17. Xu, W., Liu, X., Gong, Y.: Document clustering based on non-negative matrix factorization. In: Proceedings of the 26th Annual International ACM SIGIR Conference on Research and Development in Informaion Retrieval, pp. 267–273. ACM (2003)

A Lightweight Evaluation Framework for Table Layouts in MapReduce Based Query Systems

Feng Zhu[1,2], Jie Liu[2,3], Lijie Xu[1,2], Dan Ye[2], Jun Wei[2,3], and Tao Huang[2]

[1] University of Chinese Academy of Sciences, Beijing, China
[2] State Key Laboratory of Computer Science, Beijing, China
[3] Institute of Software, Chinese Academy of Sciences, Beijing, China
{zhufeng10,ljie,xulijie09,yedan,wj,tao}@otcaix.iscas.ac.cn

Abstract. Table layout determines the way how the relational row-column data values are organized and stored. In recent years, considerable candidates have been developed in MapReduce based query systems; they differ on storage space utilization, data loading time, query performance and so on. In most time, users are confronted with the problem of choosing the comprehensive optimum table layout given the workloads and the schema of tables. The straightforward way to run queries on generated data and compare the results is time consuming, and incurs the inaccuracy due to the MapReduce's nondeterministic execution runtime. In this paper, we propose a lightweight framework to evaluate table layouts without running the query. The framework adopts the *black box method* to test critical metrics, and the *query aware strategy* that extracts table-layout-related operations from query. Based on the metrics and operations, the framework makes suggestions to users. We conduct extensive experiments to empirically study the popular table layouts. Through the results illustration, we discover that column projection and compression are the most two prominent factors for general cases. Moreover, we discuss optimization chances for the intermediate tables produced in high level language systems.

Keywords: Table Layout, MapReduce, Query Aware, Black Box, Performance.

1 Introduction

The MapReduce [6] programming model and its open-source implementation Hadoop [1] has now become the *de facto* framework to store and process massive data. However, it is usually too complicated to express complex analytical tasks (e.g., the business intelligence) as primitive MapReduce jobs. To enhance productivity, high level systems, that brings the relational concepts like SQL, table, row and column, have been built, such as Pig [8] and Hive [7].

In such environment, data is managed as the structured relational table that consists of rows and columns. Table layout determines the way how the two dimensional data values will be organized in the underlying distributed file system. For example, the one based on Hadoop's built-in *textfile* encapsulates table rows

R. Cheng et al. (Eds.): APWeb 2015, LNCS 9313, pp. 473–484, 2015.
DOI: 10.1007/978-3-319-25255-1_39

into records; properties, like the field delimiter, need to be explicitly specified to store and resolve tables. Besides, table layouts with sophisticated mechanisms have also been proposed, such as Zebra [5], CFile in Llama [10], CIF [11], RCFile [9], ORCFile [2], Parquet [3], Trevni [4], Trojan [13] and SLC-Store [12].

However, the different table layouts are proposed and implemented independently; a comprehensive study to compare them has not been done [14]. In practice, many situations involve evaluation on table layouts for appropriate choice, for example the following two scenarios. (1) *Table Configuration in Design Phase.* It is common for users to configure tables' layouts under specific workloads in the design phase. Migrating tables and queries from relational database to Hadoop is a representative case. In such case, the schema of tables and the queries are known, appropriate table layouts can achieve comprehensive advantages on data loading, query performance and storage space. (2) *Adaptively Set Intermediate Table Layout.* High level language systems translate queries into MapReduce job workflows. The temporary intermediate tables produced by the previous job will be consumed as input by subsequent job in the workflow. For the performance improvement, can we adaptively set the intermediate tables' layouts rather than the default setting (for example, the *sequencefile* in Hive)? Above all, the problem can be abstracted as: *Given workloads and the schema of tables, make suggestion on appropriate table layout.*

The straightforward solution is to run queries on generated or sample data, and then compare the results. However, it has many defects. To begin with, the query's MapReduce runtime in distributed environment is nondeterministic. Even the time costs on performing the same query with the same configuration twice are always different. This nondeterministic behavior is caused by many table-layout-independent factors (like scheduling), and incurs the inaccuracy of evaluating the real performance of table layouts. Moreover, it is too time-consuming to run all queries; actually, table layouts are aware of only few operations even in a complex query. Another way for the problem is to establish fine-grained performance models that cover all factors for each table layout. The challenges, which will be illustrated later, result that common performance models are unrealistic to establish.

To address the problem, we propose and implement an evaluation framework. (1) The framework, from the perspective of a task (stack: "$query \rightarrow workflow \rightarrow job \rightarrow task$") that directly interacts with table layouts, is lightweight. (2) The framework evaluates high-level user-oriented metrics, rather than table layouts's internal implementation factors, and provides users with comprehensive evaluation report. (3) The framework adopts the testing approach which is based on the *black box method* and the *query aware strategy*. Different table layouts are treated as black boxes with configuration knobs. The query aware strategy extracts the table-layout-related operations from upper computing layer's query.

Contributions. We identify three high-level user-oriented metrics of table layouts. Based on the black box method and query aware strategy, we design and implement a practical lightweight framework to evaluate different table layouts. From the perspective of a task, the framework avoids the nondeterministic behavior and the time cost of running the query. It automatically extracts table-layout-related operations from queries and rapidly evaluates the metrics. Moreover, we conduct extensive experimental studies to make empirical analysis on representative table layouts and discuss the optimizations chances for intermediate tables.

The reminder of this paper is organized as follows. In section 2, we introduce the related works. Section 3 presents the black box method and the query aware strategy. In section 4, we describe the evaluation framework. Section 5 conducts experimental study and makes discussion. We conclude our work in section 6.

2   Related Work

As early as the development era of relational database systems, table layouts had been widely studied. For example the row-oriented store [15], the column-oriented store [16,17,18] and the hybrid PAX store [19]. The weaknesses and advantages for each table layout were intensively investigated. In [18], D.J. Abadi et al makes an in-depth discussion among them.

In Hadoop's distributed environment, table layout has also attracted a wide range of interests in both academia and industry. However, these table layouts are proposed and implemented independently or even for specific workloads. Most recently, Yin Huai et al. [14] makes a comprehensive and systematic experimental study; they define three core operations to abstract table layouts' behaviors and evaluate key factors: (1) table's horizontal logical subset size, (2) the function of mapping columns to column groups, and (3) the function of packing columns or column groups in a row group into physical blocks. Based on the evaluation, practical actions to optimize read performance are suggested. Our work differs from it as follows. To begin with, we argue that the abstraction of three core operations to describe the table layouts has its limitations. It is difficult to unify all table layouts and the three factors are not common. For example, CIF has no concept of row group and column group. Our work covers both the read and write performance. Moreover, we consider computing layer's queries and implement a lightweight evaluation framework for practical use.

Related works for query optimization are usually conducted from two layers: the translator and the MapReduce computing primitive. In the MapReduce primitive layer, the study focuses on MapReduce's performance models and some common optimization techniques. For example, H. Herodotou [20] proposes cost models to analyze and optimize MapReduce programs. MRShare [21] investigates opportunities to reduce the number of MapReduce jobs, i.e., the so-called "vertical packing"; while Stubby [22] further covers "horizontal packing" opportunities between two MapReduce job workflows. Particularly, optimizing the SQL-to-MapReduce translator also gains broad attentions and several rule-based translators has emerged in recent years, such as YSmart [23]. It applies a set of rules to use the minimal number of MapReduce jobs to execute multiple correlated operations in a complex query.

3   Black Box Method and Query Aware Strategy

3.1   Table Layout Insight

We take four representative table layouts as examples: TextFile, SequenceFile, RCFile and ORCFile, which include both unstructured and structured layouts and cover various performance-related factors.

- **TextFile.** The ASCII encoded text based file format in HDFS. The MapReduce job reads one line at a time and returns the byte offset as the *key* and the line of text as the *value*, with encapsulating a table row.
- **SequenceFile.** SequenceFile provides a persistent append-only data structure for binary key-value pairs. For table layout, the key is null and the value encapsulates a table row.
- **RCFile.** RCFile stores columns of a table in a record columnar way. It first partitions rows horizontally into row splits, and then vertically partitions each row split in a columnar way.
- **ORCFile.** The optimized version of RCFile. Compared with RCFile, it provides mechanisms such as fine-grained column encoding schemes, predicate push down an so on.

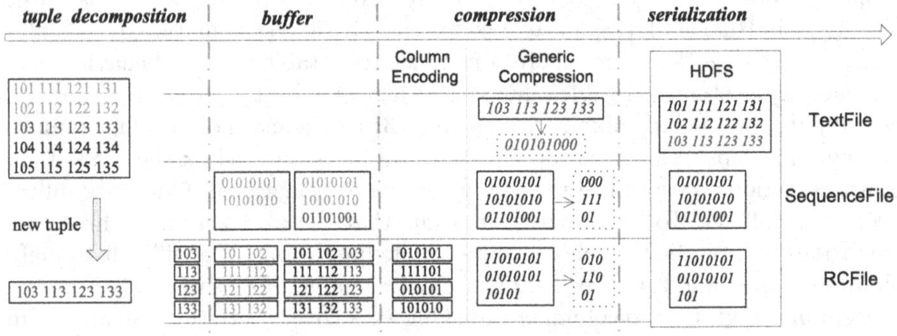

Fig. 1. Common Procedures for Writing

Fig. 1 demonstrates four optional consecutive phases (i.e., decomposition, buffer phase, compression and serialization) for writing a tuple into the file system (the case of ORCFile is similar to that of RCFile). Correspondingly, the common phases for reading records are reversed, i.e., deserialization, decompression, buffer phase and reconstruction. For simplified presentation, we only illustrate the writing behaviors here.

Tuple decomposition is the first phase for those none row-oriented table layout. In this phase, the tuple will be decomposed as different concepts like columns, column groups and so on. Before flushed to disk, data will be first managed in the buffer. The buffer not only stores data, but also maintains the metadata for data. The metadata stores the information such as column data size, column type and even statistics (like max, min and etc.) in sophisticated table layouts. When the data reaches the configured capacity, it will be written out to the disk. The compression phase consists of two levels, the column encoding schemes and the generic compression algorithms. The column encoding scheme adopts fine-grained algorithms for different column types, for example, the run length encoded algorithm for integers and the dictionary for strings. The generic compression algorithms take table as common binary files using algorithms like gzip, zlib and so on. Compression results in lower IO cost at the expense of higher CPU computing. Serialization converts in-memory data structures into bytes that can be transmitted over the network or flushed to the disk. The serialization phase is common with all table layouts.

The above insights reveal difficulties to establish fine-grained performance models for table layouts. (1) It is too complex for the models that covers all factors and table layout's configuration parameters, resulting that models are hard to be accurate. (2) Fine-grained models are difficult to validate and more importantly, the internal factors are meaningless to users and the computing layer. (3) The performance of table layout is closely related with implementations. In such case, comparisons among them are apple to an orange.

3.2 Black Box Method

Throughout this paper, we adopt the testing approach based on the black box method to evaluate the table layouts. As depicted in the left part of Fig. 2, the computing layer of MapReduce based systems write and read data row by row, respectively through the unified *write()* and *read()* interfaces.

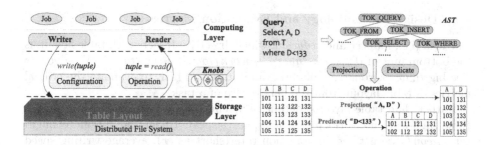

Fig. 2. Black Box Method and Query Aware Strategy

The black box method treats different table layouts as different boxes with knobs that can be adjusted by users to define configuration parameters. The knobs for writing are key/value pairs (i.e., buffer size, row group size and etc) stored in the *conf*. Based on these specified settings, the layout determines how the record data will be organized. The knobs for records reading are projection and predicate, with which the IO cost can be reduced by pushing down the operations into the storage layer. However, the two knobs are optionally implemented and not exposed by all table layouts; for example, the RCFile only supports projection.

3.3 Query Aware Strategy

It is unnecessary to evaluate a table layout through executing the query. Actually, table layouts are aware of only few operations even in a complex query; the most query execution time is spent on in-memory computations. Rather than executing the query, the query aware strategy only extracts the table-layout-related operations.

Table-Layout-Related Operations Extraction. Given a specific query, the strategy translates it into abstract syntax tree (AST) and then analyzes AST to extract the operations that can be pushed down to table layouts. In MapReduce based query systems, only the operations of *column projection* and *predicate push down* can be exploited by the computing layer. The process of extracting these two operations from AST is straightforward, the algorithm searches the tree in a depth-first behaviour and gets the operations in the nodes with the corresponding keywords (e.g, the TOK_SELECT, TOK_WHERE and so on).

As demonstrated in the right part of Fig. 2, the column projection ("A,D") and the predicate ("$D < 133$") are extracted from the query. Assume query Q has m times scans on table T, we denote the operations as the set of projection and predicate tuples, $OP(Q, T) = \{< projection_i, predicate_i > | 1 \le i \le m\}$; for the example query, the result is $OP(Query, T) = \{< (A, D), (D < 133) >\}$.

4   TLEF: Table Layout Evaluation Framework

4.1 Design Methodology and Evaluation Metrics

The task (i.e., *mapper* and *reducer*), which directly interacts with table layout, is the bottom unit of the stack "*query → workflow → job → task*". We advocate "see big things through small ones", and propose TLEF to evaluate the critical metrics of table layout from the perspective of a task.

Rather than table layouts' detailed implementation factors, TLEF evaluates three high-level user-oriented metrics, $\langle cr, ws, rs \rangle$. Under the specific configuration $conf$, we get the compression rate (short as cr), average writing speed (short as ws), average reading speed (short as rs). As the data may be stored in different node as the node of computing task, rs consists of the local reading speed rs_{local} and remote reading speed rs_{remote}.

4.2 Framework Architecture

The architecture of TLEF is demonstrated in Fig. 3. Users can input the workloads (expressed as HiveQL language), table layout configuration to be override and table specifications (including the schema of tables and other configurations

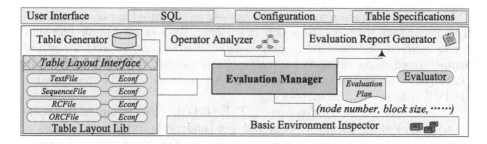

Fig. 3. The Architecture of Table Layouts Evaluation Framework.

like the range of a string column's length) through TLEF's user interface. The core of TLEF is the evaluation manager, which generates the evaluation plan and calls the evaluator to generate the read/write workloads.

The *Basic Environment Inspector* inspects the environment-specific configurations, such as the cluster's node number, the block size in HDFS and so on. The *Operator Analyzer* parses the query and extracts the table-layout-related operations (including projection and predicates) that can be pushed down to the storage layer. For the general cases, the *Table Generator* randomly generates the tables according to the table specifications. The *Table Layout Lib* not only contains built-in table layouts, but also exposes the unified interfaces. Developers can implement the interfaces for the upcoming new table layouts. The *Result Reporter* produces the evaluation metrics, based on which TLEF will make suggestions on table layout.

4.3 Configuration Space Exploration

There are a large number of variables that can be specified by users in table layout configuration and the default configuration may not be expected to be the most efficient. A straightforward exploration approach is to apply enumeration and search techniques to the full space of parameter settings. Some parameters have small and finite domains, e.g., compression type, while the domain of some parameters is unbounded. In TLEF, each implemented table layout enumerates the domains in *Econf* for individual configuration parameters. The exploration of our evaluation framework is as follows: (1) Evaluate individual parameters. For each parameter, set other parameters as default value, test the table layout with the parameter setting in *Econf* to get the metrics. (2) Heuristic search with user defined algorithm. TLEF exposes the algorithm interface for user.

4.4 Evaluation Report: Extending to Distributed Environment

Instead of solely making the choice, TLEF generates the evaluation report to users, including the contents of three respects: the high level user-oriented metrics, table layout suggestion for each query and the comprehensive analysis.

Assume the table has n rows, the row size can be estimated as rsz according to the schema; then the table size is $ts = cr \times n \times rsz$. We note the MapReduce split size as $splitsize$ and the total mapper slots number as msn. Then the length of mapper round can be calculated as $\lceil ts/(splitszie \times msn) \rceil$. For example, there are $\lceil 1204/(64 \times 16) \rceil = 2$ rounds of mappers for a 1GB table with the configuration of 64MB split size and 12 mapper slots.

Table layout suggestion for each query. TLEF lists the table-layout-related operations and calculates the new average read speed under column projection, $rs_{new} = rs \times projectedSize/rowSize$. During the query execution, the results produced by each reducer are certain and irrelevant to the table layouts; hence the write performance is determined by the metric ws. For example, when the query is data loading, table layout with better ws is preferred. As for the read performance, TLEF suggests the candidate with efficient $rs_{new} \times \lceil ts/(splitszie \times msn) \rceil$. In the distributed environment, the IO read performance may be disturbed by many factors. For the local and remote cases, we respectively define the perturbation functions ω_l and ω_r. Then the upper bound and lower bound can be estimated as the following expressions. $\lceil ts/(splitszie \times msn) \rceil \times \omega_l(rs_{local})$ and $\lceil ts/(splitszie \times msn) \rceil \times \omega_r(rs_{remote})$.

Comprehensive analysis. It is straightforward for users to make choice according to the table layout suggestions for each query. Generally, there is a tradeoff of storage space utilization and query performance, which depend on user's preference. If the previous case is the main consideration, table layout with efficient cr metric is preferred. For the latter one, it depends on different queries' weights (e.g., the one with the largest proportion or the one for the most frequent query, and etc.); we discuss some cases here. When the workload only includes data loading, table layout with better ws is preferred. However, *write-once-read-more* is frequent and query performance is usually the focus in MapReduce based query systems. For the extreme case of no extracted operation, a row-oriented table layout is appropriate for all column scans. For the general case with many table-layout-related operations, a table layout supporting them is important.

5   Experimental Study

In this section, we conduct experimental study from two dimensions. First, we empirically evaluate the popular table layouts from the perspective of a single task; we then extend the results to the distributed environment. Next we make discussion on the intermediate table layout strategy for query optimization.

Out experimental study is conducted on a local cluster of 10 DELL OptiPlex-990 nodes connected by a 1GB ethernet switch. Each node is equipped with four Intel i7-2600 3.4GHz cores, 16GB RAM and 2TB hard disk drives. Operating system is Ubuntu-11.04 x86_64. Hadoop 1.2.0 and Hive 0.12.0 are used. One node is reserved for Hadoop JobTracker and the NameNode. The other 9 nodes are used for HDFS DataNode and MapReduce TaskTracker. The HDFS block size is 256MB and each file has 2 replications. In fact, these settings can be automatically detected by basic environment inspector.

5.1   Empirical Performance Analysis

We use a synthetic relational table generated by TLEF table generator as follows: each record consists of an incremental integer ID, 6 string columns and 6 integer columns. The integers are randomly assigned values between 0 and 100000; random strings of length between 10 and 40 are generated over readable ASCII characters, similar to the schema defined in [11].

Metrics with Default Configuration. In practice, table layouts are often adopted with the default configuration, especially for the none-expert users. As the default configuration indicates the most common usage case, we first conduct experiments to study the general performance for table layouts. Table sizes are respectively 242.5MB, 263.9MB, 184.2MB and 169.7MB for TextFile, SequenceFile, RCFile and ORCFile. We write and read whole column set without the upper layer's query semantics and the results are shown in Fig. 4. It can be seen that, fine-grained column encoding schemes gain significant compression rate and row-oriented table layouts, without extra metadata maintenance and row reconstruction, outperform others for data loading and row scanning.

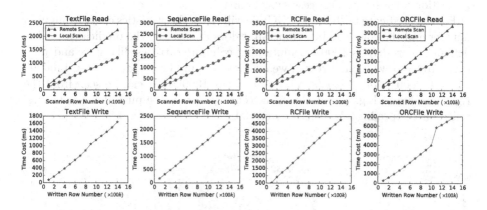

Fig. 4. Table Layout IO Performance with Default Configuration.

Configuration Space Exploration. We take two configuration parameters for ORCFile, *compression-kind* with the finite domain {None, Zlib, Snappy} and the *strip-size* with the infinite domain, to show the exploration process. The strip size varies with the values in geometrical sequences {256MB, 64MB, 16MB, 4MB, 1MB}, which is automatically generated by TLEF according to the properties defined in table layout lib. As depicted in Fig. 5, the evaluated metrics varies according to different configurations. But these differences are not very significant, diversed in the same order of magnitude.

Query Aware. As only ORCFile supports both column projection and predicate pushdown, we take it as the example to demonstrate how the query semantics can be leveraged. The client respectively reads 1400000 rows in the local mode

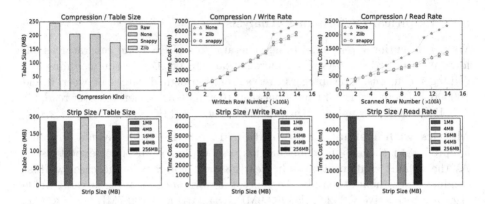

Fig. 5. Configuration Space Exploration Example: Compression and Strip Size.

with 2 integer column, 1 integer column and 1 string column, and all columns. The left and middle graphics in Fig. 6 respectively show the read time costs and the data sizes in different cases. The results indicate the effectiveness of column projection for filtering unnecessary columns.

For the predicate pushdown, ORCFile will check the statistics every configured number and the rows will be skipped if it doesn't satisfied the predicates. As the default *index-strip* is 10000 (maintaining column statistics, like max and min, every 10000 rows to skip rows), we evaluate the predicates $\sigma_1 =' id < 140000'$, $\sigma_2 =' 140000 < id < 830000'$ and $\sigma_3 =' id > 139999'$. It can be seen from the right graphic of Fig. 6 that the effectiveness of predicate push down is heavily relied on the predicate's selectivity on data.

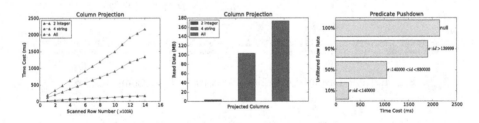

Fig. 6. Column Projection and Predicates Pushdown.

Distributed Environment. We generate the table with 56 millions rows (about 9.2GB) stored as textfile. The benchmark query is *"select count(*) from T"*, which will be translated into one MapReduce job with 39 mappers and 1 reducer in our environment. Fig. 7 demonstrates the runtime of query's execution. It can be seen that the performance of task with local data is stable, while those with the remote data differ greatly.

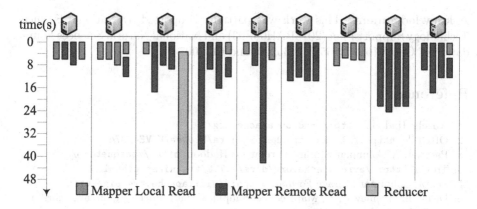

Fig. 7. Table Layout IO Performance in Distributed Environment.

5.2 Intermediate Table Layout for High Level Query Systems

Assume the producer query Q_1 has n columns $WC = \{WC_i | 1 \leq i \leq n\}$ and the consumer query Q_2 will read m columns $RC = \{RC_i | 1 \leq i \leq m\}$. To investigate whether the intermediate table layout can be adaptively set based on the Q_1 and Q_2, we discuss the characters of column projection and predicate push down.

Column projection. Generally, the column pruning (called "cp") is adopted in high level language systems' query optimizers. With this optimization, Q_2 reads all columns produced by Q_1, i.e., $m = n$ and $WC = RC$. Hence, Table layouts that support column projection can not exploit the advantage.

Predicate push down. Similarly, predicate push down (called "ppd") is also supported in query optimizers. Assume the predicate is for one column WC_k in Q_2, if and only if the WC_k is a dynamically new column generated by Q_1 (i.e., the aggregated column, calculated by user-defined functions and etc.), the predicate push down can be exploited. For instance, Q_1 is *"select col1, sum(c2) as col2 group by col1"* and Q_2 is *"select co1, col2 where col2>2"*.

In addition, the intermediate table size can only be determined after the execution of Q_1. Optimizing the intermediate table layout consequently brings not too much improvement on performance. Considering Q_2 scans all columns, setting the intermediate table layout as a row-oriented one (for example, the default *sequencefile* setting in Hive) is reasonable for most cases.

6 Conclusion

In this paper, we propose the testing approach based on black box method and the query aware strategy to evaluate the table layouts in MapReduce based environment. To assist users make the appropriate choice under specific workloads, we develop a lightweight evaluation framework. It automatically extracts table-layout-related operations from queries and rapidly evaluates the metrics.

Acknowledgement. This work was partially supported by National Science Technology Support Plan (2013BAH05F03) and National Natural Sciences Foundation of China (61202065, U1435220, and 61170074).

References

1. Apache Hadoop. http://hadoop.apache.org/
2. ORCFile. https://issues.apache.org/jira/browse/HIVE-3874
3. Parquet. A Columnar Storage Format for Hadoop. http://parquet.io/
4. Trevni. http://avro.apache.org/docs/1.7.6/trevni/spec.html
5. Zebra. Columnar Storage Format. https://wiki.apache.org/pig/zebra
6. Dean, J., Ghemawat, S.: MapReduce: Simplified data processing on large clusters. In: OSDI Conference (2004)
7. Thusoo, A., Sarma, J.S., Jain, N., et al.: Hive-a petabyte scale data warehouse using hadoop. In: ICDE Conference (2010)
8. Olston, C., Reed, B., Srivastava, U., et al.: Pig Latin: a not so foreign language for data processing. In: SIGMOD (2008)
9. He, Y., et al.: RCFile: A fast and space-efficient data placement structure in MapReduce-based warehouse systems. In: ICDE (2011)
10. Lin, Y., et al.: Llama: Leveraging columnar storage for scalable join processing in the MapReduce framework. In: SIGMOD (2011)
11. Avrilia, F., Patel, J.M., Shekita, E.J., et al.: Column-oriented storage techniques for MapReduce. In: VLDB (2011)
12. Guo, S., Xiong, J., Wang, W., et al.: Mastiff: A MapReduce-based system for time-based big data analytics. In: CLUSTER (2012)
13. Alekh, J., Quiane-Ruiz, J.-A., et al.: Trojan data layouts: right shoes for a running elephant. In: SOCC (2011)
14. Huai, Y., et al.: Understanding insights into the basic structure and essential issues of table placement methods in clusters. In: VLDB (2014)
15. Ramakrishnan, R., et al.: Database Management Systems. McGraw-Hill (2003)
16. Copeland, G.P., Khoshafian, S.: A decomposition storage model. In: SIGMOD Conference, pp. 268–279 (1985)
17. Batkin, A., Chen, X., Cherniack, M., Ferreira, M., Lau, E., Lin, A., Madden, S., et al.: C-store: A column-oriented DBMS. In: VLDB (2005)
18. Abadi, D.J., Madden, S., Hachem, N.: Column-stores vs. Row-stores: How different are they really? In: SIGMOD (2008)
19. Tsirogiannis, D., Harizopoulos, S., Shah, M.A., et al.: Query Processing techniques for solid state drives. In: SIGMOD (2009)
20. Herodotou, H., Babu, S.: Profiling, what-if analysis, and Cost-based optimization of MapReduce programs. In: VLDB (2011)
21. Nykiel, T., Potamias, M., Mishra, C., Kollios, G., et al.: MRShare: sharing across multiple queries in MapReduce. In: VLDB (2010)
22. Lim, H., Herodotou, H., et al.: Stubby: A transformation-based optimizer for MapReduce workflows. In: VLDB (2012)
23. Lee, R., Luo, T., Huai, Y., Wang, F., et al.: Ysmart: Yet another sql to mapreduce translator. In: ICDCS (2011)

An Extended Graph-Based Label Propagation Method for Readability Assessment

Zhiwei Jiang, Gang Sun, Qing Gu*, Lixia Yu, and Daoxu Chen

State Key Laboratory for Novel Software Technology
Nanjing University, Nanjing 210023, China
jiangzhiwei@outlook.com, {sungangnju,srylx}@163.com,
{guq,cdx}@nju.edu.cn

Abstract. Readability assessment is to evaluate the reading difficulty of a document, which can be quantified as reading levels. In this paper, we propose an extended graph-based label propagation method for readability assessment. We employ three vector space models (VSMs) to compute edges and weights for the graphs, along with three graph sparsification techniques. By incorporating the pre-classification results, we develop four strategies to reinforce the graphs before label propagation to capture the ordinal relation among the reading levels. The reinforcement includes recomputing weights for the edges, and filtering out edges linking nodes with big level difference. Experiments are conducted systematically on datasets of both English and Chinese. The results demonstrate both effectiveness and potential of the proposed method.

1 Introduction

Readability assessment is to evaluate the reading difficulty of documents, which can be quantified as reading levels [1]. Measuring automatically the reading difficulty of documents is helpful in areas such as foreign language education [16]. Recently, with the huge volume of documents mounting up on the web, its importance has been boosted in relevant applications, such as information retrieval [3,8].

Researches on readability assessment have a relatively long history from the early 20th century [17]. Many techniques have been used to handle it, such as readability formulae [12], logistic regression [7], SVMs [6], and many others. Text features have been developed to improve the performance of these techniques [6,10]. However, the researches on readability assessment are far from enough, and the traditional readability formulae are still the most used techniques in real world applications.

In this paper, we propose an extended graph-based label propagation method for readability assessment. In our method, documents are used as nodes, and the semantic similarities between document pairs as edges to build a graph. Label propagation is used on the graph to propagate class labels from the labeled documents (whose reading levels are known) to the unlabeled ones (i.e. the target documents). The main contributions of the paper are listed as follows:

---

* Corresponding author.

© Springer International Publishing Switzerland 2015
R. Cheng et al. (Eds.): APWeb 2015, LNCS 9313, pp. 485–496, 2015.
DOI: 10.1007/978-3-319-25255-1_40

- An extended graph-based label propagation method is proposed for the task of readability assessment. Three VSMs are used to measure the semantic similarities, and three sparsification techniques adopted to prune the edges. By incorporating the pre-classification results, four strategies are provided to reinforce the graph.
- Experiments are conducted systematically on datasets of both English and Chinese. The results demonstrate both effectiveness and potential of the proposed method.

The rest of the paper is organized as follows. Section 2 briefly introduces the background of readability assessment, plus concepts of the graph-based label propagation. Section 3 describes the details of our proposed method. Section 4 presents the experimental studies. Finally, the conclusion is made and future work is described.

2   Background

In this section, we briefly introduce the background of our work, including the concept of readability assessment and the graph-based label propagation methods.

2.1   Readability Assessment

Readability assessment is to evaluate the reading difficulty of documents, which is usually quantified as reading levels [1]. The task can be handled by regression, classification, or ranking [7,13]. At the early stage, researchers typically treated it as a regression problem, and developed many well-known readability formulae [12]. Text features were defined for these formulae to measure both lexical and grammatical complexities of the documents. More useful text features were developed recently by adopting the NLP (Natural Language Processing) techniques [6].

Recent researchers handled readability assessment by classification, and used the machine learning techniques. Collins-Thompson and Callan [4] developed statistical language models to compute the reading level of a document. Feng et al. [6] built SVMs (Support Vector Machines) to classify the reading difficulty of documents. François and Fairon [7] compared the performance of the logistic regression model and the ordinal classification model for readability assessment. Additionally, ranking techniques are also used, due to the ordinal nature of reading levels. Ma et al. [13] assigned reading levels by ranking the reading difficulty of documents. Collins-Thompson et al. [3] ranked the reading difficulty of documents for information retrieval. Jameel et al. [8] employed the Latent Semantic Indexing (LSI) for ranking the reading difficulty.

2.2   Graph-Based Label Propagation

Graph-based label propagation is a method that can be used for classification. A graph is built by entities as nodes, the inter-relations of which are represented by edges. Label propagation is used to propagate class labels from labeled entities to unlabeled ones [11]. Graph-based label propagation has been successfully applied in various applications, such as dictionary construction [11], word segmentation and tagging [19] and image parsing [18]. To our knowledge, it has not yet been applied for readability assessment.

Typically, a graph-based label propagation method consists of two main steps: graph construction and label propagation [19]. During graph construction, the similarity function is required to build edges and compute weights for the edges between pairs of the nodes [5]. Some form of edge pruning (called graph sparsification) is required to get a suitable graph for label propagation [9]. Many effective algorithms have been developed for label propagation [15,11].

3 Our Method

To determine the reading level of a document, we design an extended version of the graph-based label propagation method [15]. Firstly, we describe how to construct a graph for readability assessment. Secondly, we present the strategies to reinforce the graph by incorporating the pre-classification results. Finally, we use the label propagation algorithm to predict the reading levels. Figure 1 shows the framework of our method.

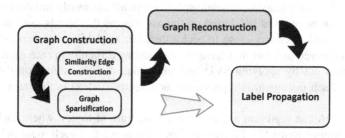

Fig. 1. Framework of the extended graph-based label propagation method

3.1 Graph Construction for Readability Assessment

As stated before, readability assessment can be viewed as a classification problem. Suppose there are l labeled documents $(d_1, c_1), \cdots, (d_l, c_l)$ and u unlabeled documents $d_{l+1}, \cdots, d_{l+u}$. Let $n = l + u$ be the total number of documents, and C be a set of labels $C = \{1, 2, \cdots, m\}$, corresponding to the m reading levels ($m = |C|$). Then, a graph $G = (V, E, W)$ for readability assessment can be built. In G, V is the set of nodes corresponding to the documents ($|V| = n$), E is the set of edges representing relations between pairs of documents, and W is a $n \times n$ weight matrix weighing the relations. From the set of documents V, the subset $V_l = \{d_1, d_2, \cdots, d_l\}$ contains the labeled documents, and subset $V_u = \{d_{l+1}, d_{l+2}, \cdots, d_{l+u}\}$ contains the unlabeled documents. Based on the graph G, the objective is to propagate (assign) labels to the nodes (documents) in V_u.

To construct an appropriate graph G for readability assessment, we have two steps. The first is building edges by similarity, the second is graph sparsification. Following gives the details.

Building Edges by Similarity. To construct edges and weights in the graph, we measure similarities between all pairs of documents, and establish an edge between each pair of node, where the computed similarity is assigned as the edge weight. By doing so, we get a complete graph. For similarity computation, we use three types of VSMs (Vector Space Models) to represent a document as a vector, and calculate the similarity (cosine similarity or Kullback-Leibler divergence). The VSMs are TF-IDF (Term Frequency-Inverse Document Frequency), LSI (Latent Semantic Indexing), and LDA (Latent Dirichlet Allocation), which are described below.

- **TF-IDF:** TF-IDF is a numerical statistic that is intended to measure how important a word is to a document in a collection or corpus. Previous studies [4] have stated that some words were more important than others in a document, and the difficulty of these words might have direct relation to the readability of the document. For example, if words measured to be important to document A are easier than words important to document B, we could conclude that A is easier to read than B. Based on the above, we compute a word-document matrix, the elements of which are computed by the TF-IDF measure described in [14]. Each document is represented as a row in the matrix.
- **LSI:** LSI defines a latent semantic space where all the words and documents in a corpus can be indexed [8], based on the assumption that words used in the same contexts tend to cluster close to each other. For readability, we can assume that words clustered together may have similar reading difficulty to each other. LSI can be carried out by applying SVD (singular value decomposition) on the TF-IDF matrix. Each document is represented as a vector (index) in the latent semantic space.
- **LDA:** LDA can represent a document as a mixture of topics, where each topic is a spectrum of terms [2]. For readability, we assume that each topic belongs to a reading level. The document having more difficult topics may be more difficult to read. Hence, we can use the probability distribution on topics to represent a document.

Graph Sparsification. For graph sparsification, we filter out edges from the complete graph based on the edge weights. We use three sparsification techniques: directed kNN (d-$k$nn), undirected kNN (u-$k$nn), and b-matching.

- **Directed kNN:** Each node keeps only k directed edge to its top-k nearest neighbors. Two nodes are nearer to each other if the corresponding edge weight is greater. This technique would lead to a directed simple graph.
- **Undirected kNN:** The procedure is the same as above, save that the edges left are treated as undirected. As a result, some nodes may have more than k neighbors.
- **b-matching:** Nearly the same as the undirected kNN, where each edge is undirected. The difference is that each node has exactly b edges to its neighbors [9].

3.2   Graph Reinforcement by Pre-Classification

For readability assessment, we add an extra step between the graph construction and the label propagation. This step focuses on reinforcing the graph to capture the ordinal relationship among the reading levels. Pre-classification is required for the graph reinforcement.

Table 1. Simple text features for Chinese and English

Features (Chinese)	Features (English)
Average number of strokes per character	Average number of characters per word
Average number of characters per word	Average number of syllables per word
Average number of strokes per sentence	Average number of syllables per sentence
Average number of characters per sentence	Average number of words per sentence
Average number of words per sentence	Average number of poly-syllable words per sentence
Complex character ratio	Average number of difficult words per sentence
Distinct character ratio	Distinct word ratio
Unique character ratio	Unique word ratio
Long word ratio	Flesch-Kincaid score
Distinct word ratio	
Unique word ratio	
Flesch-Kincaid score (adapted to Chinese)	

We build the feature-based classification models for pre-classification, which have been proved useful [6] in predicting the reading level for a document. The results of pre-classification may provide extra information to prune the edges and amplify the weights. To build the classifier, we extract simple text features [6] from documents. The features used here are listed in Table 1.

By training the classifier using the labeled documents, pre-labels (i.e. reading levels) can be predicted for the unlabeled documents, denoted as $\{\hat{c}_{l+1}, \cdots, \hat{c}_{l+u}\}$. Thus, we have the priori label of all the documents $\{z_1, \cdots, z_l, z_{l+1}, \cdots, z_{l+u}\}$, where $z_i = c_i$ ($i \in \{1, \cdots, l\}$) for labeled documents and $z_j = \hat{c}_j$ ($j \in \{l + 1, \cdots, l + u\}$) for unlabeled documents.

Based on the priori label, we propose four strategies to reinforce the graph built in Section 3.1.

- **S1:** For each edge $e_{u,v} \in E$, if $z_u = z_v$, amplify its weight $w_{u,v} \in W$ to $w'_{u,v}$ by a ($a > 1$), i.e.

$$w'_{u,v} = a \cdot w_{u,v}$$

- **S2:** For each edge $e_{u,v} \in E$, modify its weight $w_{u,v} \in W$ to $w'_{u,v}$ by the following formula:

$$w'_{u,v} = a^{(m-|z_v - z_u|)} \cdot w_{u,v}$$

where a is the same as in S1, and m is the number of reading levels.

- **S3:** For each node $v \in V$, for each node u in its neighbor set, keep the edge $e_{v,u}$ if $|z_v - z_u| \leq 1$, otherwise remove the edge.
- **S4:** The process is the same as S3, except that each weight $w_{u,v}$ is amplified as in S1 if $z_u = z_v$.

Among these four strategies, S1 amplifies the weights of edges linking nodes with identical priori labels. S2 amplifies the weight of every edge by an exponent scaled by the label difference of its two ends, which assumes an ordinal relation among the reading levels. S3 filters out edges linking nodes with dissimilar levels (the label difference bigger than 1). S4 integrates the characteristic of S1 to S3.

3.3   Label Propagation

Given a graph $G = (V, E, W)$ defined in Section 3.1, the goal of label propagation is to propagate class labels from the labeled nodes to the entire graph. To handle it, we use the label propagation method presented in [15], which is simple but effective [11]. The method iteratively updates the label distribution using Eq.1.

$$q_u^{(t)}(c) = \frac{1}{\kappa_u} \left(r_u(c)\delta(u \in V_l) + \mu \sum_{v \in \mathcal{N}(u)} w_{u,v} q_v^{(t-1)}(c) + \nu U(c) \right) \quad (1)$$

On the left side of Eq.1, $q_u^{(t)}(c)$ is the afterward probability of c (i.e. label value) on a node u at the t-th iteration. On the right side, κ_u is the normalizing constant, μ and ν are hyper-parameters, and $r_u(c)$ is the initial probability of c on the node u if u is initially labeled (i.e. belonging to the labeled set V_l). The function $\delta(x)$ returns 1 if x is true and 0 otherwise. $\mathcal{N}(u)$ denotes the set of neighbors of u. $U(c)$ is the probability of c based on the uniform distribution. The iteration stops when the change in $q_u^{(t)}(c)$ is small enough (e.g. less than e^{-3}), or t exceeds a predefined number (e.g. greater than 30).

4 Empirical Analysis

In this section, we conduct experiments based on two datasets of different languages, to investigate the following three research questions:

RQ1: During graph construction, among the three types of VSMs (Vector Space Models), which one is the best for measuring similarities between pairs of documents for readability assessment?

RQ2: During graph reinforcement, among the four strategies, which one is the most suitable for readability assessment, by using different graph sparsification techniques?

RQ3: What are the effects of varying the proportion of labeled documents on the performance of the extended graph-based label propagation method?

4.1 Corpus

The experiments are built on datasets of two languages: Chinese and English. We select Chinese because recent researches on readability assessment for Chinese are relatively few. We also select English to demonstrate that our method can be effectively applied to different languages.

For Chinese, we collect the Chinese primary school language textbooks as the dataset, which contains 6 reading levels corresponding to the 6 grades. To assure the representativeness of the Chinese dataset, we combine two widely used editions of textbooks, which cover 90% of Chinese primary schools. For English, we collect the *New Concept English* textbooks as the dataset, which contains 4 reading levels corresponding to the 4 volumes. The details of the two datasets are shown in Table 2. For Chinese, we use characters as elemental units, while for English, we use words as elemental units.

4.2 Experiment Settings

To answer the research questions, we randomly split each dataset into two sets: one is the labeled set, the other assumed unlabeled set. Firstly, a graph is built following the

Table 2. Statistics of the collected corpus

Level	Chinese			English		
	doc	char/doc	sent/doc	doc	word/doc	sent/doc
1	96	153.2	10.7	72	105.1	17.26
2	110	259.6	16.0	96	151.6	12.36
3	106	424.1	20.5	60	378.8	23.1
4	108	622.9	28.1	48	376.5	17.8
5	113	802.0	36.1	–	–	–
6	104	902.0	39.0	–	–	–

steps of graph construction described in Section 3.1, using one of the three VSMs and a graph sparsification technique. Secondly, pre-classification is made for the unlabeled set of documents, using logistic regression trained on the labeled set. Thirdly, the graph is reinforced using one of the four strategies described in Section 3.2. After that, the label propagation algorithm is applied to propagate the labels from labeled documents to unlabeled documents. For each document in the unlabeled set, its predicted label is the one with the maximum probability.

We use prediction accuracy, i.e. the ratio of documents correctly labeled in the unlabeled set, to evaluate the performances of the method with different configurations. To account for randomness, we compute the mean accuracy across the results of 100 validations for each configuration.

4.3 Comparison among the Three VSMs

For RQ1, we investigate which one of the three VSMs is the best for building edges and weights of the graph for readability assessment. We assign weights on edges using one VSM (i.e. TF-IDF, LSI or LDA) each time, along with one of the three graph sparsification techniques (i.e. d-$k$nn, u-$k$nn and b-matching). Hence, we have 9 different graphs constructed for label propagation, each denoted as a combination of the two (e.g. $TF\text{-}IDF\_d$, LSI_u and LDA_b). For each graph, we directly apply the label propagation algorithm to propagate the labels without reinforcing the graph. To make a comprehensive comparison, we conduct experiments with variant proportions of the labeled set (i.e. from 0.1 to 0.9 step by 0.1). The thresholds for sparsification (i.e. k for d-$k$nn and u-$k$nn, b for b-matching) of the graphs are set to result in well enough prediction accuracies, which will be further discussed in later sections. For the Chinese dataset, the threshold is 30, while for the English dataset, the threshold is 10.

Figure 2 shows the measured prediction accuracies for the 9 graphs in line charts. For Chinese, the performances of the graphs using LSI (i.e. LSI_d, LSI_u and LSI_b) consistently outperform the other graphs. The performances of the $TF\text{-}IDF$s and LDAs are close to each other. An exception is LDA_d, which performs better than the other two LDAs, and all the three $TF\text{-}IDF$s. As for the proportion of the labeled set, normally the performance ought to increase as the proportion increases, since more knowledge are obtained. However, for Chinese, this only holds in case of the LSIs and LDA_d, which suggests that the others are no better than random predictions.

For English, the performances of the three LSIs are again better than the other graphs. This time the three LDAs have visible better performances than the three $TF\text{-}IDF$s.

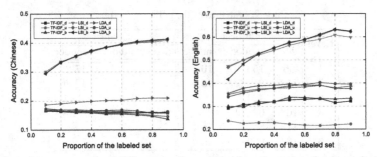

Fig. 2. Comparison among the 9 different configurations of graph construction by variant proportions of the labeled set

Again the graph sparsification techniques have little effects on the prediction accuracy. The *TF-IDF_u* performs the worst, and no better than random predictions. A phenomenon should be noted that the performances of the *LSI*s decrease when the proportion of the labeled set increases from 0.8 to 0.9. This suggests that the graphs built require further improvement.

In summary, among the three VSMs, LSI is the best in building graphs for readability assessment. The three graph sparsification techniques make little difference to the prediction performance. By the way, the graph requires careful construction, otherwise the results are no better than random predictions.

4.4 Comparison among the Four Strategies

For RQ2, we compare the performances of the four strategies of graph reinforcement on the graphs constructed by different graph sparsification techniques. Based on the findings in RQ1, we use LSI to compute edge weights. Firstly, we construct graphs with LSI, and using one of the three graph sparsification techniques (i.e. d-$k$nn, u-$k$nn and b-matching). Secondly, we reinforce each graph using one of the four strategies (i.e. $S1 \sim S4$). The graphs without reinforcement (denoted as *NS*) are also left for comparison. Finally, we apply the label propagation algorithm on the graphs to make the predictions. To make a comprehensive comparison, we conduct experiments on variant settings of the sparse threshold k (or b in b-matching). For the Chinese dataset, k is varied from 4 to 70 step by 3, and for the English dataset, k is varied from 4 to 40 step by 2. The maximum k is set according to the size of the dataset. In all the graphs, the proportion of the labeled set is 70%, the parameter a of the three strategies are uniformly set 10. Figure 3 shows the measured prediction accuracies for the graphs in line charts, where the accuracies of the pre-classification (by logistic regression, denoted as *CLF*) are also depicted as baselines.

On the Chinese dataset (the upper three charts in Figure 3, each corresponding to one sparsification technique), the accuracy of *CLF* is 0.4340. The accuracy of *NS* shows a rising trend with k, starting from 0.3427 ($k = 4$) and becoming stable at about 0.4 ($k > 25$). With b-matching, the accuracy of *NS* is slightly better than that with d-$k$nn and u-$k$nn. This means that without graph reinforcement, *CLF* is better than the graph-based label propagation for readability assessment. After applying the graph reinforcement strategies, our method can outperform *CLF* with suitable k values. As shown in the upper three figures, *S1* outperforms *CLF* when $k > 7$, while it performs worse when

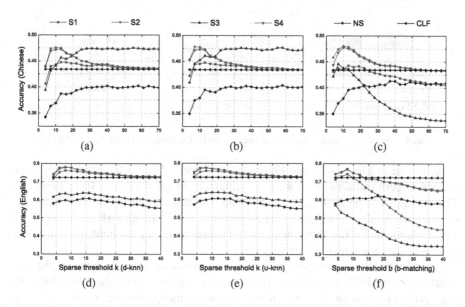

Fig. 3. Comparison among the strategies of graph reinforcement on the graphs using different sparsification techniques

$k > 25$, and with b-matching, it falls below *CLF* again. *S2* and *S4* always outperform *CLF*, where the optimal k is around 10. *S3* performs well with d-$k$nn and u-$k$nn. At starting its accuracy falls below *CLF*; but the accuracy rises quickly by k, and becomes the best (about 0.475) and remains stable from $k = 25$. Among the three graph sparsification techniques, little difference can be found between Figure 3(a) and Figure 3(b), which demonstrates that the two kNNs perform similarly. In Figure 3(c) however, both *S1* and *S3* exhibit great differences. *S1* does not always outperform *CLF* now, and *S3* drops below *CLF* at $k = 16$, and keeps going down when k increases. The reason may be that compared with kNNs, b-matching makes the nodes less different from each other, which can be alleviated by appropriate setting of a.

On the English dataset (the lower three charts in Figure 3), *CLF* still outperforms *NS*, that has a relatively stable accuracy with variant k values. This time *S3* performs poor, only slightly better than *NS* in Figure 3(d) and Figure 3(e). This suggests that *S3* is not suitable for small number of classes (The English dataset has 4 levels, while the Chinese has 6). The other three strategies have similar trends in 3(d) and Figure 3(e), which rise first, then drop, at last keep stable at larger k values, and remain better than *CLF*. *S2* and *S4* perform slightly better than *S1*. Again, b-matching makes great differences from the kNNs, where all the strategies exhibit poor performances.

As discussed earlier, the parameter a may have evident effects on the performances of the strategies (i.e. *S1*, *S2* and *S4*). We make further investigation on a. For *S1* and *S2*, we repeat the experiments in Figure 3(c) with varying a values, while for *S4*, we repeat the experiments in Figure 3(a). b-matching is repeated twice, since it performs awkward in earlier experiments, while d-$k$nn is selected for comparison, since *S2* and *S4* perform similarly. The value of a is set from 2 to 2^{11} in logarithmic scale.

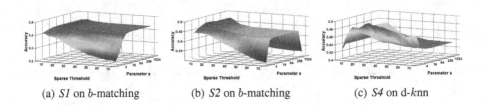

(a) *S1* on *b*-matching   (b) *S2* on *b*-matching   (c) *S4* on d-*k*nn

Fig. 4. The performances of the strategies by variant a values on the Chinese dataset

As shown in Figure 4(a) and Figure 4(b), *S1* and *S2* can achieve relatively good and stable performances when $a > 16$ (although not good at $a = 10$ in Figure 3). The strategies can perform well on *b*-matching, when the amplification parameter a is properly set. As shown in Figure 4(c), *S4* performs good on d-*k*nn when a is small, but worse when $a > 16$. The above suggests that the strategies are not limited to the *k*NNs, as long as the parameter a is carefully set.

In summary, all the strategies can improve the performances of both the graphs constructed for readability assessment, and the pre-classification, when the sparse threshold is properly set for the sparsification technique used. Among the strategies, S1 can be replaced by S2 or S4, while the latter two perform similarly, and require the amplification parameter a properly set. S3 may not be a good choice when the number of reading levels is small. We cannot find the best strategy that can significantly outperform the others for readability assessment.

4.5   Effects of the Proportion of Labeled Set

For RQ3, we study the effects of varying the proportion of the labeled set on the performance of our proposed method. We conduct experiments on both English and Chinese datasets, and vary the proportion of the labeled set across 0.1 to 0.9 step by 0.1. For the graph based method, we implement all the strategies with suitable settings. All the edges (weights) of the graphs are computed by LSI. The sparsification techniques are selected to suit the strategy, and the sparsification threshold k and amplification parameter a are set suitable for specific dataset (referring to Figure 3).

For comparison, we implement the support vector machine model (denoted as *SVM*), logistic regression model (denoted as *LR*) and the Flesch-Kincaid formula (denoted as *FK*) respectively, which are commonly used for readability assessment [7,12]. The text features used for training the former two classification models are listed in Table 1. Besides, the indices calculated by LSI are also included as features for fair comparison with the graph based methods. To develop the Flesch-Kincaid formula for Chinese, we use the number of Chinese characters and strokes instead of English words and syllables. Since the original FK formula is not built specifically on these datasets, and the resulted *FK* score does not match the reading levels here, we rebuild the formula on both datasets using linear regression. Figure 5 depicts the prediction accuracies of the implemented methods in line charts.

For Chinese, except *FK*, the prediction accuracies of all the other methods are improved when the proportion of the labeled set increases. *FK* always performs the worst. The reason is that the text features used in the formula may not be suitable to calculate directly the reading levels. The graph based method without graph reinforcement

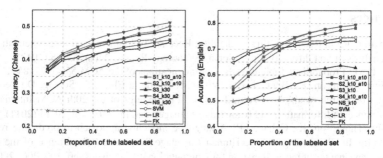

Fig. 5. Comparison among the four strategies by variant proportions of the labeled set

performs worse than the two classification models *LR* and *SVM*. After applying the graph reinforcement strategies, the graph based method (e.g. *S2_k10_a10*) can result in better performances than *SVM* and *LR*, when the proportion of the labeled set is great enough. Here *S4_k30_a2* always performs the best. However, as discussed in RQ2, *S4* can be outperformed by other strategies. This suggests that the extended graph-based label propagation method has great potential for readability assessment.

For English, again *FK* performs the worst, and may require enhancement. All the reinforcement strategies except *S3_k10* can outperform *SVM* and *LR* when the proportion of the labeled set is great. However, when the proportion is less than 0.3, *SVM* performs the best among all. This demonstrates that our method still requires improvement, especially when the proportion of the labeled set is small.

In Summary, for readability assessment, the extended graph-based label propagation method performs well when the proportion of the labeled set is great. However, when the proportion is small, the method is still in need of improvement.

5   Conclusion

In this paper, we propose an extended graph-based label propagation method for readability assessment. Three vector space models (VSMs) are employed to compute edges and weights for the graphs, along with three graph sparsification techniques. By incorporating the pre-classification results, we develop four strategies to reinforce the graphs before label propagation. Based on the datasets of both Chinese and English, we conduct experiments to investigate the effectiveness of the method. According to the experimental results, we have several findings. Firstly, among the three VSMs, the latent semantic indexing (LSI) is most suitable in building graphs for readability assessment. Secondly, all the four strategies can improve the performances of both the graphs constructed and the pre-classification, by proper setting of the graph sparsification technique and the amplification parameter. Thirdly, the method is still in need of enhancements, when the proportion of labeled documents is small.

During our future work, we plan to work over other options of the method, such as different pre-classification methods, new reinforcement strategies, and suitable graph sparsification techniques. Additionally, other high quality datasets are required to approve the effectiveness of the method.

Acknowledgments. This work is supported by the National NSFC projects under Grant Nos. 61373012, 61321491, and 91218302.

References

1. Benjamin, R.G.: Reconstructing readability: Recent developments and recommendations in the analysis of text difficulty. Educational Psychology Review 24(1), 63–88 (2012)
2. Blei, D.M.: Probabilistic topic models. Communications of the ACM 55(4), 77–84 (2012)
3. Collins-Thompson, K., Bennett, P.N., White, R.W., de la Chica, S., Sontag, D.: Personalizing web search results by reading level. In: Proceedings of the 20th ACM International Conference on Information and Knowledge Management, pp. 403–412. ACM (2011)
4. Collins-Thompson, K., Callan, J.P.: A language modeling approach to predicting reading difficulty. In: Proceedings of the Human Language Technology Conference of the North American Chapter of the Association for Computational Linguistics, pp. 193–200 (2004)
5. Daitch, S.I., Kelner, J.A., Spielman, D.A.: Fitting a graph to vector data. In: Proceedings of the 26th Annual International Conference on Machine Learning, pp. 201–208. ACM (2009)
6. Feng, L., Jansche, M., Huenerfauth, M., Elhadad, N.: A comparison of features for automatic readability assessment. In: Proceedings of the 23rd International Conference on Computational Linguistics: Posters, pp. 276–284 (2010)
7. François, T., Fairon, C.: An ai readability formula for french as a foreign language. In: Proceedings of the 2012 Joint Conference on Empirical Methods in Natural Language Processing and Computational Natural Language Learning, pp. 466–477 (2012)
8. Jameel, S., Qian, X., Lam, W.: N-gram fragment sequence based unsupervised domain-specific document readability. In: COLING, pp. 1309–1326 (2012)
9. Jebara, T., Wang, J., Chang, S.F.: Graph construction and b-matching for semi-supervised learning. In: Proceedings of the 26th Annual International Conference on Machine Learning, pp. 441–448. ACM (2009)
10. Jiang, Z., Sun, G., Gu, Q., Chen, D.: An ordinal multi-class classification method for readability assessment of chinese documents. In: Buchmann, R., Kifor, C.V., Yu, J. (eds.) KSEM 2014. LNCS, vol. 8793, pp. 61–72. Springer, Heidelberg (2014)
11. Kim, D.S., Verma, K., Yeh, P.Z.: Joint extraction and labeling via graph propagation for dictionary construction. In: Twenty-Seventh AAAI Conference on Artificial Intelligence (2013)
12. Kincaid, J.P., Fishburne Jr., R.P., Rogers, R.L., Chissom, B.S.: Derivation of new readability formulas (automated readability index, fog count and flesch reading ease formula) for navy enlisted personnel, Tech. rep., DTIC Document (1975)
13. Ma, Y., Fosler-Lussier, E., Lofthus, R.: Ranking-based readability assessment for early primary children's literature. In: Proceedings of the 2012 Conference of the North American Chapter of the Association for Computational Linguistics: Human Language Technologies, pp. 548–552 (2012)
14. Salton, G., Buckley, C.: Term-weighting approaches in automatic text retrieval. Information Processing & Management 24(5), 513–523 (1988)
15. Subramanya, A., Petrov, S., Pereira, F.: Efficient graph-based semi-supervised learning of structured tagging models. In: Proceedings of the 2010 Conference on Empirical Methods in Natural Language Processing, pp. 167–176 (2010)
16. Tanaka-Ishii, K., Tezuka, S., Terada, H.: Sorting texts by readability. Computational Linguistics 36(2), 203–227 (2010)
17. Vogel, M., Washburne, C.: An objective method of determining grade placement of children's reading material. The Elementary School Journal 28(5), 373–381 (1928)
18. Xie, W., Peng, Y., Xiao, J.: Semantic graph construction for weakly-supervised image parsing. In: Proceedings of the Twenty-Eighth AAAI Conference on Artificial Intelligence, Québec City, Québec, Canada, July 27-31, pp. 2853–2859 (2014)
19. Zeng, X., Wong, D.F., Chao, L.S., Trancoso, I.: Graph-based semi-supervised model for joint chinese word segmentation and part-of-speech tagging. In: Proceeding of the 51st Annual Meeting of the Association for Computational Linguistics, pp. 770–779 (2013)

Simple is Beautiful:
An Online Collaborative Filtering
Recommendation Solution with Higher Accuracy

Feng Zhang[1,2], Ti Gong[1], Victor E. Lee[3], Gansen Zhao[4], and Guangzhi Qu[5]

[1] School of Computer Science, China University of Geosciences, Wuhan, China
[2] Hubei Key Laboratory of Intelligent Geo-Information Processing, China University
of Geosciences, Wuhan, China
[3] Department of Mathematics and Computer Science, John Carroll University,
University Heights OH, USA
[4] School of Computer Science, South China Normal University, Guangzhou, China
[5] Department of Engineering and Computer Science,
Oakland University, Rochester MI, USA
{jeff.f.zhang,tigongcug,vlee8888,zhaogansen}@gmail.com, gqu@oakland.edu

Abstract. Matrix factorization has high computation complexity. It is
unrealistic to directly adopt such techniques to online recommendation
where users, items, and ratings grow constantly. Therefore, implementing
an online version of recommendation based on incremental matrix factorization is a significant task. Though some results have been achieved
in this realm, there is plenty of room to improve. This paper focuses
on designing and implementing algorithms to perform online collaborative filtering recommendation based on incremental matrix factorization techniques. Specifically, for the new-user and new-item problems,
Moore-Penrose pseudoinverse is used to perform incremental matrix factorization; and for the new-rating problem, iterative stochastic gradient
descentlearning procedure is directly applied to get the updates. The solutions seem simple but efficient: extensive experiments show that they
outperform the benchmark and the state-of-the-art in terms of incremental properties.

Keywords: Recommender Systems, Collaborative Filtering, Matrix Factorization, Incremental Computing.

1 Introduction

In the real world, there are many scientific and engineering fields handling high
dimension data. Typical fields include graphical modeling, video surveillance,
face recognition, latent semantic indexing (LSI) and collaborative filtering (CF)
recommendation.

These fields share the fact that their data can be represented as a matrix
in which each row represents an object, and each column denotes a dimension
(feature) of the object. However, the number of dimensions may reach to millions

© Springer International Publishing Switzerland 2015
R. Cheng et al. (Eds.): APWeb 2015, LNCS 9313, pp. 497–508, 2015.
DOI: 10.1007/978-3-319-25255-1_41

while the matrix is very sparse in general. In the fields with sparse data matrix representation, finding a lower rank approximation to the higher rank data is very important. In other words, it is significant to compute summarized dimensions (features) to explore the latent relationship between objects and features. The benefits brought by the approximation include predicting unobserved data values in data matrices, exploring latent structures underlying data matrices, smoothing errors of observed data, and finally obtaining recommendations with a higher computation performance.

Matrix factorization (MF) tries to fulfill the work of approximation. It decomposes a matrix into a product of matrices. There are many ways to perform MF. For example, QR-factorization, LU-factorization, and Singular Value Decomposition (SVD) are the most famous ones in linear algebra [5]; Non-negative factorization (NNF) is another popular one which has been used in many fields in recent years [10,17].

In the field of CF recommendation, MF has been proven to be a very successful method since the Netflix Prize competition [1,8,7,9,18,4]. The key challenge in recommendation based on MF is that the factorization process is very time-consuming. It is unrealistic to adopt such techniques to scenarios where users, items, and ratings grow constantly. In order to address this issue, the incremental version of MF seems to be a promising solution.

Inspired by the incremental nature of stochastic gradient descent (SGD) and the underlying techniques to compute incremental SVD, we introduce comprehensive methodologies to perform online CF recommendation based on incremental MF. The methodologies leverage only newcomers to incrementally update the training model, without having to rebuild the whole model. Extensive experiments have been carried out to show that the proposed methodologies perform well based on two results: the comparable recommendation accuracy and the excellent incremental properties.

1.1 Contributions

1. To our knowledge, this is the first work to apply the Moore-Penrose pseudoinverse technique [15] to implement incremental SGD-based MF and show its superior performance.
2. We propose comprehensive solutions to perform online CF from the aspects of new-user, new-item and new-rating.

2 Related Work

MF is a popular method to implement model-based CF [12,19]. In the Netflix Prize competition, Koren et al.[1,8] adopted MF methods to win one million dollars provided by the organizers. In addition to the winner, many participants there used similar MF methods to improve the predictive accuracy of recommender systems. Since then a new wave of recommender system research based on MF has taken off [7,9,18,4].

There are already some studies fulfilling the incremental training. Sarwar et al. do an early work for this [14]. They used the folding-in [3,2] method to incrementally achieve SVD-based CF. However, SVD is not considered as a practical technique to implement CF due to the sparsity problem and its poor scalability property. MF techniques can be implemented even under a very sparse data scenario and have good scalability performance, which are the major reasons why they are so popular and are utilized to be a replacement of SVD to perform CF.

Two early works in the information retrieval community [3,2] demonstrate how to use folding-in techniques to update a document index while new documents or new terms are added into the document-term space. However, the folding-in techniques are used in the incremental SVD updates in these studies and they do not consider the new-rating case. Our paper is the first one to apply the folding-in techniques to the SGD-driven MF and also consider the new-rating case.

Rendle and Schmidt-Thieme [13] make a contribution to online recommendations based on MF models. They simply and directly applied SGD to learn regularized kernel MF models. Recently, Luo et al. [11] report superior results outperforming Rendle and Schmidt-Thiemes'.

Luo et al. [11] transformed the gradient descent rules into new ones suitable to perform incremental updates. They designed two incremental regularized matrix factorization (RMF) models: the Incremental RMF (IRMF) and the Incremental RMF with linear biases (IRMF-B). Their experiments on two large, real datasets suggest positive results, which proves the efficiency of their strategy. However, their studies are based on an assumption that training examples are learned simultaneously, which may affect the recommendation accuracy, as shown in the implementation of ourselves.

3   Methodologies

3.1   Regularized Matrix Factorization

Regularized Matrix Factorization is first introduced by Webb[1] and is an implementation of latent factor models. The core idea in regularization is using some parameters such as *learning rate* and *regularization factor* to reduce the complexity of learning procedure and avoid overfitting.

Definition 1 (Regularized Matrix Factorization). Given a rating matrix $R \in \mathbb{R}^{m \times n}$, where each row vector denotes a user, each column vector denotes a column, and each entry $r_{ui} \in R$ denotes user u's preference on item i. All known entries r_{ui} are stored in the set $R_{\mathcal{K}} = \{(u,i)|r_{ui} \text{ is known}\}$. The goal of RMF is to construct a low rank approximation of R. Let f denote the dimension of the feature space, $P \in \mathbb{R}^{m \times f}$ denote the user feature matrix where each row is a user rating vector p_u, and $Q \in \mathbb{R}^{f \times n}$ denote the item feature matrix where each

---

[1] http://sifter.org/$\sim$simon/journal/20061211.html

column represents an item rating vector q_i. The approximation of the rating by user u on item i can be confined to the inner product of the corresponding user-item feature vector pair, shown as Equation (1):

$$\hat{r}_{ui} = p_u q_i = \sum_k (p_{uk} q_{ki}) \tag{1}$$

The values of p_u and q_i can be learnt by solving a regularized least squares problem, shown as Formula (2):

$$\arg\min_{p,q} \sum_{(u,i)\in\mathcal{K}} (r_{ui} - p_u q_i)^2 + \lambda(||p_u||^2 + ||q_i||^2) \tag{2}$$

where $||.||$ denotes the standard Euclidean norm, $\sum_{(u,i)\in\mathcal{K}}(r_{ui} - p_u q_i)^2$ is the loss function to evaluate the parameters p_u and q_i in order to "best" fit the original matrix R, and $\lambda(||p_u||^2 + ||q_i||^2)$ is the Tikhonov regularization [16,6] to avoid overfitting by penalizing the magnitudes of the parameters.

There are studies suggesting that incorporating three ratings, μ, the overall average rating, b_u, the observed bias of user u from the average, and b_i, the observed bias of item i from the average, could contribute to improving the prediction accuracy of models [8]. Essentially, the three parameters capture the statistical characteristics of the ratings, overall represent of rating, rating preferences of users, and popularity of items. Then the prediction rule shown as Equation (1) turns to the following:

$$\hat{r}_{ui} = \mu + b_u + b_i + p_u q_i \tag{3}$$

Correspondingly, the regularized least squares problem associated with Equation (3) becomes to:

$$\arg\min_{p,q} \sum_{(u,i)\in\mathcal{K}} (r_{ui} - \mu - b_u - b_i - p_u q_i)^2 + \lambda(b_u^2 + b_i^2 + ||p_u||^2 + ||q_i||^2) \tag{4}$$

Let $e_{ui} = r_{ui} - \hat{r}_{ui}$, then the optimization problem denoted by (4) can be easily solved by SGD techniques.

$$\begin{aligned} p_u &\leftarrow p_u + \gamma(e_{ui}q_i - \lambda p_u) \\ q_i &\leftarrow q_i + \gamma(e_{ui}p_i - \lambda q_u) \\ b_u &\leftarrow b_u + \gamma(e_{ui} - \lambda b_u) \\ b_i &\leftarrow b_i + \gamma(e_{ui} - \lambda b_i) \end{aligned} \tag{5}$$

3.2   Incremental MF

The essence of the above method is to employ MF to approximate the original matrix.

$$R \approx \hat{R} = PQ \tag{6}$$

Deduced from the above analysis, the approximation is equivalent to an optimization problem in the Euclidean space, which can be formulated as:

$$\underset{P,Q}{\arg\min} \, ||R - PQ||_F^2 \tag{7}$$

where $||.||_F$ denotes the Frobenius norm.

The process of MF shown in (6) can be explained as a space projection process. There is a dual view of the projection.

1. If R is viewed as a set of row vectors (user vectors), then Q, which is also a set of row vectors, should be viewed as base vectors in a new space. The goal of MF is to project R onto the new base of Q. In the new space, P is the coefficient matrix, in which each row consists of coefficients corresponding to the base vectors of Q.

2. If R is viewed as a set of column vectors (item vectors), then it is projected onto a new subspace of P, in which a set of column vectors form the base. In this case, Q is the coefficient matrix, in which each column consists of coefficients corresponding to the subbase vectors of P.

If the first view is taken, and a new user u signs on, then the row vector for u needs to be projected onto the new subspace with a base of Q. The challenge is to figure out the coefficients under the new subspace.

$$u^T = Q^T x \Rightarrow x = (Q^T)^+ u^T \tag{8}$$

where x is the column vector of coefficients. X^+ is the Moore-Penrose pseudoinverse of matrix X.

If the second view is taken, and a new item i is added, then the column vector for i needs to be projected onto the new space with a base of P. The challenge is to figure out the coefficients under the new space as well.

$$i = Px \Rightarrow x = (P)^+ i \tag{9}$$

Let $e = ||R - PQ||_F^2$, a measure of the quality of MF. It is obvious that the lower value of e, the higher quality of the factorization, and vice versa.

3.3 Incremental MF with Biases

Section 3.2 considers the cases where only MF is involved. This section discusses the online CF cases where biases, together with MF, are involved.

If MF incorporates the bias parameters of Equation (3), then Equation (6) can be rewritten as Equation (10):

$$R \approx \hat{R} = \mathcal{M} + \mathcal{U} + \mathcal{I} + PQ \tag{10}$$

where $\mathcal{M} = [a_{ij}]$ is a constant matrix denoting the global average, i.e., $a_{ij} = \mu$; $\mathcal{U} = [b_{u_1}, b_{u_2}, ..., b_{u_m}]^T$ is the matrix in which the user biases are stored and each b_{u_i} is a constant row vector storing user biases; $\mathcal{I} = [b_{i_1}, b_{i_2}, ..., b_{i_n}]$ is the matrix storing the item biases and each b_{i_1} stores an item bias.

Note that in $\mathcal{U}$ each row is a constant row vector, and in $\mathcal{I}$ each column is a constant column vector.

Now we consider how to perform incremental predictions based on Equation (10). To achieve this, methods to perform incremental computation of all the variables in the equation have to be found out. The problem could be separated into two parts: the first is dealing with the parameters of $\mathcal{M}, \mathcal{U},$ and $\mathcal{I}$; and the second is dealing with the MF part PQ.

In order to calculate $\mathcal{M}, \mathcal{U},$ *and* $\mathcal{I}$, formulas similar to those in [11] are adopted to compute μ, b_u and b_i.

$$\mu = \frac{\sum_{(u,i) \in T} r_{u,i}}{|T|} \tag{11}$$

$$b_u = \frac{s_u - \mu \cdot n_u - \sum_{i \in R(u)} b_i}{\beta_2 + n_i} \tag{12}$$

where T is the set of all ratings, β_1 and β_2 are two bias factors, $s_i = \sum_{(u,i) \in R(i)} r_{u,i}$, $n_i = |R(i)|$, $s_u = \sum_{(u,i) \in R(u)} r_{u,i}$, and $n_u = |R(u)|$.
and

$$b_i = \frac{s_i - \mu \cdot n_i}{\beta_1 + n_i} \tag{13}$$

Once a new rating $r_{u,i}$ is issued, μ can be updated by:

$$\mu' = \frac{\mu \cdot |T| + r_{u,i}}{|T| + 1} \tag{14}$$

b_u can be updated by:

$$b'_u = \frac{s_u + r_{u,i} - \mu' \cdot (n_i + 1) - \sum_{i \in R(u)} b'_i}{\beta_2 + n_i + 1} \tag{15}$$

and b_i can be updated by:

$$b'_i = \frac{s_i + r_{u,i} - \mu' \cdot (n_i + 1)}{\beta_1 + n_i + 1} \tag{16}$$

When a new user u signs in, then μ can be updated by:

$$\mu' = \frac{\mu \cdot |T| + \sum_{i \in R(u)} r_{u,i}}{|T| + |R(u)|} \tag{17}$$

In this case, for each $i \in R(u)$, b_i can be updated by Equation (16); and b'_u can be computed by:

$$b'_u = \frac{\sum_{(u,i) \in T} (r_{u,i} - \mu' - b'_i)}{(\beta_2 + |R(u)|)} \tag{18}$$

In order to calculate the MF, three scenarios are considered.

1. New-User signing in

When new users sign in, then based on Equation (10), Equation (8) turns to:

$$\begin{pmatrix} R \\ R_r \end{pmatrix} = \begin{pmatrix} \mathcal{M} \\ \mathcal{M}_r \end{pmatrix} + \begin{pmatrix} \mathcal{U} \\ \mathcal{U}_r \end{pmatrix} + \begin{pmatrix} \mathcal{I} \\ \mathcal{I}_r \end{pmatrix} + \begin{pmatrix} P \\ P_r \end{pmatrix} Q \Rightarrow$$
$$P_r = (R_r - \mathcal{M}_r - \mathcal{U}_r - \mathcal{T}_r)Q^+ \tag{19}$$

2. New-Item Added

When new items are added (incremental column growth), then Equation (9) turns into:

$$(R|R_c) = (\mathcal{M}|\mathcal{M}_r) + (\mathcal{U}|\mathcal{U}_r) + (\mathcal{I}|\mathcal{I}_r) + P(Q|Q_c) \Rightarrow$$
$$Q_c = P^+(R_c - \mathcal{M}_c - \mathcal{U}_c - \mathcal{T}_c) \tag{20}$$

3. New-Rating issued

In this case, the user and item involved are both the existing ones but the user now issues a new rating for the item. This case is different from the new-user and the new-item ones. Since it is observed that SGD inherently is an incremental method, this property is simply utilized to perform the incremental updates. When a new rating arrives, the SGD iterative procedure is used to achieve the update. Through this procedure, the updated μ', b'_u, b'_i, p'_u and q'_i are obtained. Then they are compared with the previous set of parameters, μ, b_u, b_i, p_u and q_i. The set of parameters leading to higher accuracy is returned as the answer.

4   Experiments

Four major experiments were carried out: 1. Experiments to judge the influence of latent number of features on the performance; 2. Experiments to determine the basis size of starting matrix; 3. Experiments to evaluate the performance of online CF: user incremental (new-user) performance, item incremental (new-rating) performance and rating incremental (new-item) performance; 4. Experiments to compare with the state-of-the-art results for the incremental recommendation precision.

Experiments have been carried out using two popular benchmark datasets: MovieLens 100K and MovieLens 1M[2], whose characteristics are described in Table 1.

All the experiments were conducted on a laptop with an Intel Core i3 CPU and 4.00GB RAM, running WIN7 32 OS. And the programming language is Python 2.6.6 Python.

The experiments involve some parameters, whose values were set as shown in Table 2.

---

[2] http://www.grouplens.org/node/73

Table 1. Dataset Description

dataset	# ratings	# users	# items	sparsity
MovieLens 100K	100,000	943	1,682	99.9403%
MovieLens 1M	1,000,209	6,040	3,706	95.5316%

Table 2. Symbol Denotation

Parameter	Meaning	Value
λ	Tikhonov Regularizing Term	0.01
γ	Learning Rate	0.002
f	Dimension of Features Space	100

4.1 Dimension of the Feature Space

The goal of the first experiment is to determine the dimension of the feature space, which is a key parameter to decide the quality of the MF. To decide such parameter, batch training and incremental training experiments were performed based on different dimension values. The results are consistent with those in [8]: increasing f benefits the accuracy of prediction in most cases. The reason may be with f increased, a better matrix approximation is performed. In our settings, it seems that for both the MovieLens 100K dataset and the MovieLens 1M dataset, 100 is a suitable f value to conduct the MF.

4.2 Basis Size

In order to perform incremental MF and online CF, the basis size of starting matrix has to be determined. The basis size may also affect the CF performance as well. Experiments were carried out to decide the basis size. For MovieLens 100K, the results show when user basis size reaches to 500, RMSE measure remains stable and changes little, so 500 was chosen as a suitable user basis size for this dataset. For the similar reason, 800 was chosen as the item basis size and 50,000 the rating basis size. For MovieLens 1M, with quite the similar reason, 2,500 was chosen as the user basis size, 1,000 the item basis size and 500,000 the rating basis size. Due to the space limitation, the above two detailed results are not shown here.

4.3 Performance of Incremental Properties

In this part of experiments, the incremental performance of MF based CF was evaluated. The performance includes the prediction precision RMSE and the elapsed time. The experiments consider the three incremental cases of new-user, new-item, and new-rating.

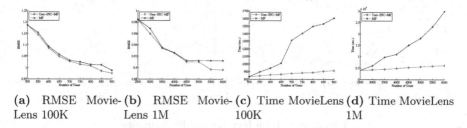

(a) RMSE Movie- (b) RMSE Movie- (c) Time MovieLens (d) Time MovieLens
Lens 100K   Lens 1M   100K   1M

Fig. 1. New-User Incremental Performance

To compare them with the batch model of matrix factorization CF, while evaluating the incremental properties, the matrix factorization CF based on batch model training using the current size of data is conducted.

1. New-User Case

For the MovieLens 100K dataset, the experiments were carried out starting with a model size of 500 users and rising up to 900 users with an increment of 50 users. And for the dataset of MovieLens 1M, the experiments were started with 2,500 users and risen up to 6,000 users with an increment of 500 users.

2. New-Item Case

For the MovieLens 100K dataset, the experiments were carried out starting with a model size of 800 items and rising up to 1,600 items with an increment of 100 items. And for the dataset of MovieLens 1M, the experiments were started with 1,000 items and risen up to 3,900 items with an increment of 500 items.

3. New-Rating Case

For the MovieLens 100K dataset, the experiments were carried out starting with a model size of 50,000 ratings and rising up to 100,000 ratings with an increment of 5000 ratings. And for the dataset of MovieLens 1M, the experiments were started with 500,000 ratings and risen up to 1,000,000 ratings with an increment of 50,000 ratings.

For the cases of new-user, new-item and new-rating, the results are respectively plotted in Fig. 1, Fig. 2 and Fig. 3. With the data size increased, better incremental performance is obtained: a slightly better RMSE and a significantly less elapsed time. The reason for the improvement in runtime is: incremental computation definitely costs much less time than batch computation. However, it seems surprising to see that the incremental version of MF works better than MF with re-computation. Overfitting could explain such results. Since the datasets contain some noises thus letting all data involve in the batch computation may not always ensure a better performance over the computation in an incremental manner. In fact, similar results can be seen in [11]. In that paper, the incremental model also outperforms the batch model in recommendation accuracy. However, our method is simpler and has a better performance, as the experimental results of the following section show.

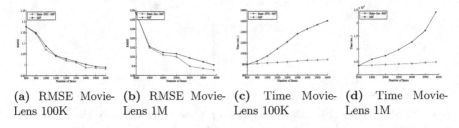

(a) RMSE Movie-Lens 100K
(b) RMSE Movie-Lens 1M
(c) Time Movie-Lens 100K
(d) Time Movie-Lens 1M

Fig. 2. New-Item Incremental Performance

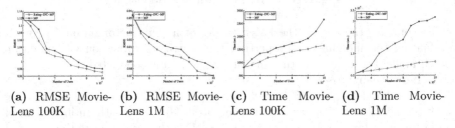

(a) RMSE Movie-Lens 100K
(b) RMSE Movie-Lens 1M
(c) Time Movie-Lens 100K
(d) Time Movie-Lens 1M

Fig. 3. New-Rating Incremental Performance

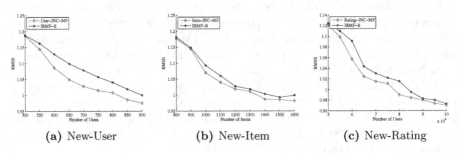

(a) New-User
(b) New-Item
(c) New-Rating

Fig. 4. Comparison With the Results in [11]: MovieLens 100K

(a) New-User
(b) New-Item
(c) New-Rating

Fig. 5. Comparison With the Results in [11]: MovieLens 1M

4.4   Comparison with State-of-the-Art

Results of this part of experiments are compared with those generated by the incremental methods proposed in [11]. It seems that [11] does not differentiate the cases into new-user, new-item and new-rating since their methods take all the three cases as rating increments. To compare their methods with ours, we reimplemented their methods under the three cases with our same controlled parameters.

The results are shown in Fig. 4 and Fig. 5, which demonstrate that the methods proposed in this paper slightly outperform the optimized one in [11], IRMF-B. We have stated the reason in the section of related work. IRMF-B assumes that each training examples are learned simultaneously, which may destroy the learning regularities and lead to less optimized results.

Also, as stated in the section of related work, the results reported in [11] are better than those in [13], so it doesn't need to replicate the experiments of [13] and compare our results with theirs. Comparing with [11] gives enough evidence to show the superiority of our methods.

5   Conclusions

Some previous studies solve the incremental CF problem only from the aspects of new-user and new-item using the classic fold-in method. For solving the new-rating problem, it seems that a different mechanism has to be invented. This paper proposes new methods to solve the problem in an integrated manner. 1. They do not need to rebuild the existing model and do true incremental recommendation. 2. The experimental results demonstrate their superior performance compared with the benchmark and state-of-the-art results. The methods proposed seem simpler than the previous ones, but they are effective and efficient, which proves an idiom: simple is beautiful.

Acknowledgments. The study is partially supported by the Natural Science Foundation of Hubei Province under Grant No. 2015CFB450, the enterprise-funded latitudinal research project under Grant No. 2014196221, and the Guangzhou Research Infrastructure Development Fund under Grant No. 2012224-12.

References

1. Bell, R.M., Koren, Y.: Improved neighborhood-based collaborative filtering. In: Proceedings of KDD-Cup and Workshop at the 13th ACM SIGKDD International Conference on Knowledge Discovery in Data Mining, pp. 7–14 (2007)
2. Berry, M.W., Dumais, S.T., OBrien, G.W.: Using linear algebra for intelligent information retrieval. Using Linear Algebra for Intelligent Information Retrieval 37(4), 573–595 (1995)
3. Dumais, S.T., Furnas, G.W., Landauer, T.K.: Indexing by latent semantic analysis. Journal of the American Society for Information Science 41(6), 391–407 (1990)

4. Ge, Y., Xiong, H., Tuzhilin, A., Liu, Q.: Cost-aware collaborative filtering for travel tour recommendations. ACM Transactions on Information Systems 32(1) (2014)
5. Golub, G.H., Loan, C.F.V.: Matrix Computations, 4th edn. Johns Hopkins University Press (2012)
6. Ito, K., Jin, B.: Inverse Problems: Tikhonov Theory and Algorithms. World Scientific Publishing Company (2014)
7. Jamali, M., Ester, M.: A matrix factorization technique with trust propagation for recommendation in social networks. In: Proceedings of the Fourth ACM Conference on Recommender Systems, pp. 135–142 (2010)
8. Koren, Y.: Factorization meets the neighborhood: A multifaceted collaborative filtering model. In: Proceedings of the 14th ACM SIGKDD International Conference on Knowledge Discovery in Data Mining, pp. 426–434 (2008)
9. Koren, Y.: Collaborative filtering with temporal dynamics. Communications of the ACM 53(4), 89–97 (2010)
10. Lee, D.D., Seung, H.S.: Learning the parts of objects by non-negative matrix factorization. Nature 401(6755), 788–791 (1999)
11. Luo, X., Xia, Y., Zhu, Q.: Incremental collaborative filtering recommender based on regularized matrix factorization. Knowledge-Based Systems 27(3), 271–280 (2012)
12. Pronk, V., Verhaegh, W., Proidl, A., Tiemann, M.: Incorporating user control into recommender systems based on naive bayesian classification. In: Proceedings of the 2007 ACM Conference on Recommender Systems, pp. 73–80 (2007)
13. Rendle, S., Schmidt-Thieme, L.: Online-updating regularized kernel matrix factorization models for large-scale recommender systems. In: Proceedings of the 2008 ACM Conference on Recommender Systems, pp. 251–258 (2008)
14. Sarwar, B., Karypis, G., Konstan, J., Riedl, J.: Incremental singular value decomposition algorithms for highly scalable recommender systems. In: Proceedings of the Fifth International Conference on Computer and Information Science, pp. 27–32 (2002)
15. Stoer, J., Bulirsch, R.: Introduction to Numerical Analysis, 3rd edn. Springer (2002)
16. Tarantola, A.: Inverse Problem Theory and Methods for Model Parameter Estimation. SIAM: Society for Industrial and Applied Mathematics (2004)
17. Wang, Y.X., Zhang, Y.J.: Nonnegative matrix factorization: A comprehensive review. IEEE Transaction on Knowledge and Data Engineering 25(6), 1336–1353 (2013)
18. Yin, H., Sun, Y., Cui, B., Hu, Z., Chen, L.: Lcars: a location-content-aware recommender system. In: Proceedings of the 19th ACM SIGKDD International Conference on Knowledge Discovery in Data Mining, pp. 221–229 (2013)
19. Zhang, Y., Koren, J.: Efficient bayesian hierarchical user modeling for recommendation systems. In: Proceedings of SIGIR 2007, pp. 47–54 (2007)

Random-Based Algorithm
for Efficient Entity Matching

Pingfu Chao[1,2], Zhu Gao[2], Yuming Li[1,2], Junhua Fang[1,2],
Rong Zhang[1,2,*], and Aoying Zhou[1,2]

[1] Institute for Data Science and Engineering
[2] Shanghai Key Laboratory of Trustworthy Computing
{51121500001,10132510331,51141500019,52131500020}@ecnu.cn,
{rzhang,ayzhou}@sei.ecnu.edu.cn

Abstract. Most of the state-of-the-art MapReduce-based entity matching methods inherit traditional Entity Resolution techniques on centralized system and focus on data blocking strategies for structured entities in order to solve the load balancing problem occurred in distributed environment. In this paper, we propose a MapReduce-based entity matching framework for Entity Matching on semi-structured and unstructured data. Each entity is represented by a high dimensional vector generated from description data. In order to reduce network transmission, we produce lower dimensional bit-vectors called signatures for those entity vectors based on Locality Sensitive Hash (LSH) function. Our LSH is required for promising cosine similarity. A series of random algorithms are designed to ensure the performance for entity matching. Moreover, our design contains a solution for reducing redundant computation by one round of additional MapReduce job. Experiments show that our approach has a huge advantages on both processing speed and accuracy compared to the other methods.

1 Introduction

Nowadays, the rapid growth of web data and User Generated Content (UGC) changes the way we used to collect and manage information. By the hands of the huge amount of web users, data become much easier to be generated. For instance, in a C2C (Customer to Customer) online business site, it becomes easy to start an online store and generate personalized structured/unstructured descriptions for its listed goods. And it is pervasive that different sellers own the same commodity with variety of descriptions together with diverse schemas. This results in the difficulty in product managing which may affect product or price comparison. Additionally, this kind of UGC is becoming large. Then it is urgent to design an efficient distributed entity matching framework by using of this UGC to identify entities that represent the same items.

Though there are already a bunch of entity matching algorithms, we face new challenges which make the traditional methods infeasible. First of all, most of the

---

* Corresponding author.

© Springer International Publishing Switzerland 2015
R. Cheng et al. (Eds.): APWeb 2015, LNCS 9313, pp. 509–521, 2015.
DOI: 10.1007/978-3-319-25255-1_42

user generated data are semi-structured or unstructured data, which come from variety data sources and in different formats/schemas. Second, a great quantity of typos occurred in UGC data dramatically reduce the data quality. Third, high computation cost occurs due to the large web data size. Traditional well designed entity matching algorithms are usually for structured data. When they confront with these three challenges including hybrid data structure without uniform schema, low data quality and huge data size, their performance is also challenged.

Lots of works have tried resolving those challenges separately: 1) Document similarity metrics[20] are introduced to measure the similarity between unstructured data, such as online documents. They provide standards of similarity measurements for unstructured data and semi-structured data. 2) Tokenization technique[18] is used to reduce the negative influence to data quality caused by typos and human mistakes. It has become an important step in data cleaning and for improving the accuracy of entity matching. 3) Data blocking strategies are designed to split data into multiple parts for parallelization and lowering computation cost. Inspired by those work, we propose a flexible parallel entity matching framework on MapReduce, which aims to resolve entity matching on unstructured data with higher efficiency and lower cost. That is for these unstructured data, our algorithm shall boost the processing speed while promise load balance and reduce network transmission cost. Those are the

The main contributions can be summarized as follows:

- We sketch out a random-based framework for entity matching based on MapReduce, and introduce our random algorithm based on this framework. We propose to generate low dimensional bit-vector signatures for entities, that are calculated from the high dimensional feature vectors by a specific Locality Sensitive Hash (LSH) function[16]. It helps to boost computation efficiency and reduce storage and network transmission costs dramatically.
- We propose a random-based permutation method inspired by PLEB[6] algorithm for increasing match accuracy, which can be well adapted on MapReduce framework. Besides, the permutation method ensures load balance during the matching process.
- We analyze the cause of redundancy problem in our algorithm, which is pervasive in many of the blocking-based algorithms. We solve this problem by adding an extra MapReduce job to reduce the redundancy rate.
- We evaluate our approaches and demonstrate their efficiency in comparison to existing blocking-based matching methods on both semi-structured and unstructured data.

2   Related Work

The idea of Entity matching (along with Record linkage and deduplication) is first introduced by geneticist Howard Newcombe in [15] who presents odds ratios of frequencies and the decision rules for delineating matches and mismatches.

Fellegi and Sunter[4] provide the formal mathematical foundations of record linkage. At the end of 20th century, since the data size grew rapidly, the main problem for entity matching changed from improving calculation accuracy to handling huge amount of data. Blocking strategy was introduced to solve this problem for it can filter majority of the entity pairs with low similarity before similarity comparison. Meanwhile, the proposal of MapReduce[3] also gave us a better platform to solve this problem.

A number of blocking-based entity matching algorithms under MapReduce framework have been presented to help dealing with big data sets[7,8,19,14]. These works are based on a assumption that there is only one key for an entity and use a map/reduce phase to handle the problem. They design different blocking strategies for map phase and then do the matching step in reduce phase. Some of the most influential works includes sorted neighborhood[12] and load-balanced entity matching[11,10]. The sorted neighborhood is a blocking technique by sorting all entities according to blocking keys, assigning a window size w and comparing entities in the window while sliding. This gives us some inspiration of the pair generation method. However, this part of work and its joint works[9] didn't mention the load balancing problem on MapReduce.

Both *BlockSplit* and *PairRange* blocking strategies mentioned in [11] focus on solving the imbalance problem. But they rely on a data analysis phase before matching job. This phase scan the input entities to collect a list of all possible pairs, then make a blocking plan to evenly divide those pairs into multiple blocks by the above two strategies. It can perfectly solve the load balancing problem on MapReduce by adding an expensive cost before matching process. Therefore, both of these two strategies are suitable for processing skewed data, but far more slow to deal with regular data or data with enormous size. Another algorithm has been presented for document-similarity computation[1] which have the same background to our work, we will compare our performance with this algorithm.

3 MapReduce-Based Entity Matching Framework

In order to get good matching accuracy for high dimensional data, the distributed entity matching framework is expected to have the following functionalities: a) Fast Entity Similarity Calculation method, b) Efficient Candidate Entity Pair Generation algorithm and c) Redundant pairs removing plan. Figure 1 shows the framework of our algorithm.

3.1 Random-Based Similarity Calculation

Cosine similarity is appropriate for measuring similarity of semi-structured and unstructured entities. However, high dimensional vectors may cause the dimensional-curse on entity matching calculation. Locality Sensitive Hashing (LSH) function which keep the property of cosine similarity, proposed by Charikar in [2], provides an option for high dimensional vectors similarity calculation distributively.

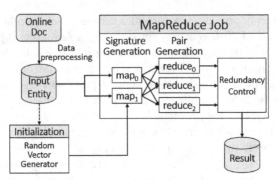

Fig. 1. Framework of random-based entity matching on MapReduce

Theorem: Suppose we have a collection of vectors in a k dimensional vector space (that is N^k). We generate a random vector r of unit length from this k dimensional space, and define a hash function h_r as Eqn.1. Then for vectors u and v, we have the corresponding relationship as calculated by Eqn.2. Goemans and Williamson[5] prove that this hash function can promise the cosine similarity in a high probability.

Since high dimensional vectors computation is time-consuming, dimension reducing is generally done for improving calculation performance. Note that the above equation is probabilistic in nature. Hence, we start to generate d numbers of random vectors from N^k and get R={r}, with $|R| = d$ and $d \ll k$. For each vector u in N^k, we can get a d-bit vector $\{h_{r1}(u), h_{r2}(u), ..., h_{rd}(u)\}$ by using hash function $h_r(u)$ for u and each vector $r_i \in R$. In such a way, we represent each vector u by a d-bit signature S_u. Then dimension k is successfully reduced to d whilst it still preserves the cosine similarity, where $d \ll k$. This d-dimension signature keeps features of the original vector. Then the huge deviation between two signatures means big difference between two entity vectors.

Cosine similarity between any two vectors is achieved by Eqn. 3. On the other hand, if we use the similarity between signatures to represent the probability in Eqn. 3, we can observe:

$Pr[h_r(u) = h_r(v)] = 1 - (hamming\ distance)/d$.

Thus, it converts the problem of finding cosine similarity between vectors to the problem of calculating hamming distance between signatures. It is more efficient in both processing speed and memory utilization. So in the following descriptions, hamming distance has the same meaning as cosine similarity.

$$h_r(u) = \begin{cases} 1 & r.u \geqslant 0 \\ 0 & r.u < 0 \end{cases} \quad (1)$$

$$Pr[h_r(u) = h_r(v)] = 1 - \frac{\theta(u, v)}{\pi} \quad (2)$$

$$\cos(\theta(u, v)) = \cos(1 - Pr[h_r(u) = h_r(v)])\pi \quad (3)$$

3.2 Entity Pairs Generation

As shown above, highly similar entities usually have similar signatures. Then sorting by lexicography will make them close to each other except the deviations occur in the first few bits of their signatures. In order to reduce this kind of deviation and increase the opportunity of getting close for similar signatures, we propose to do random permutation for these signatures as proposed by PLEB (Point Location in Equal Balls), which was first mentioned in [6] and improved in [2]. This algorithm takes random permutations to signatures and sort the permuted signatures. It aims to find vectors with short hamming distances.

A random permutation is considered as a random jumble of the bits of each signature, so the prefix deviation of two similar signatures can be prevented in some of their permutations. We apply several rounds of permutations on signature set. At each round, it generates a set of permuted signatures. Signature pairs with lower hamming distance are expected to get close in some of the sorting lists. Accordingly, we can find the top m closest neighbors for each signature and generate our entity pair candidates. Implementation details can be found in the following content.

3.3   Entity Matching on MapReduce

We aim at solving the entity matching problem on semi-structured and unstructured data. We first tokenizing the input data and generating high dimensional feature vectors based on the words frequency. We use these vectors as our data input. The goal is to find all matching pairs among these entity vectors. We define that two entities are matched when the hamming distance between their feature vectors is lower than a predefined threshold. We can also output the top N similar vectors by an additional sorting process on the result set.

Figure 1 shows matching framework on MapReduce. Before doing the matching process, we carry out three steps of preprocessing on the source data to get our expected input. Initially, we split the input entities into tokens using the Part-Of-Speech Tagger[17]. Then we generate a dictionary containing all k different tokens occurred in the data set. Finally, for each entity u, a k-dimension vector V_u is generated, in which the nth dimension represent the word frequency of the nth token in entity u. The input of our method is a set of $(key, value)$ pairs made up of entity ID E_u and its k-dimension vector V_u. In addition, another standard vector set R is introduced into the MapReduce job, which contains d ($d << k$) numbers of k bits random vectors of unit length, that is $\{r_1, r_2, ..., r_d\}$.

Figure 2 illustrates the workflow of our MapReduce job. Map phase contains steps as followings:

1. Initially, we apply $h_r(u)$ to each input entity (E_u, V_u) using R as the standard vector set. We pair V_u with every vector r_i in R ($1 \leqslant i \leqslant d$) and calculate $h_{ri}(u)$, then we get a d bit binary signature S_u for V_u represented as: $\overline{S_u} = \{h_{r1}(u), h_{r2}(u), ..., h_{rd}(u)\}$.
2. After converting the input vectors into signatures S_u, we randomly permute them t times. The permutation function can be approximated as:

$$\pi(x) = (ax + b) \bmod p \tag{4}$$

where p is prime and $0 < a < p, 0 \leq b < p$. Both a and b are chosen randomly. We apply t different random permutation for every signature (by

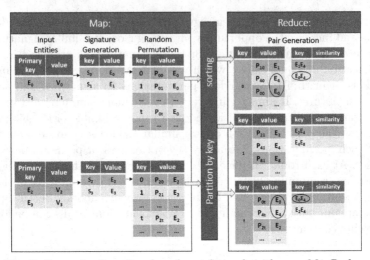

Fig. 2. Example of random-based matching algorithm on MapReduce

choosing random values for a and b, t number of times). Thus for each signature S_u, we have t different permutation results: $\{P_{u1}, P_{u2}, ..., P_{ut}\}$. We regard this result as our map output. As a consequence, we have t different map output for each entity represented as (i, P_{ui}, E_u), with i, P_{ui} and E_u referring permutation number, the corresponding permutation result and the entity ID.

In reduce phase, we expect to achieve entity pair similarity. After an automatic sorting procedure during the shuffle between map and reduce, the reduce phase faces t number of groups represented as (i, L_i) , in which L_i is the sorted list on all signatures in the i_{th} round of permutation. Then we generate matching pairs between every entity u and its closest m neighbors in the sorted list. Finally, we calculate the hamming distance of every paired entities and output those with distance below a predefined threshold. The output is formatted as $(E_u E_v, similarity)(u < v)$, which are the ID concatenation of paired entities' with its similarity value.

Overall, The map tasks change each k-dimension vector into t number of d-bit signatures. d and t are always far less than k. Generally d and t are between tens to hundreds while k are normally more than tens of thousands determined by the characteristic and size of input data. That gives a significant reduction on data volume, and also a huge cut on the network transmission cost between map and reduce. Unlike most of the blocking-based entity matching methods comprised by multiple MapReduce tasks, our matching algorithm is finished in one MapReduce job. Since each MapReduce task spends extra cost on task scheduling and network communication, the cutting on the number of MapReduce jobs can lead to performance promotion. Furthermore, since all permutations of a signature are sent to reducers evenly, which are partitioned by their permutation number i.

There are the same number of pairs for each reduce task. Therefore, our model easily solve the load balancing problem.

4   Redundancy Reduction

During the pair generation step in reduce phase, all pairs are generated in parallel from different groups. As the same entity pair may be generated from multiple groups, there can be many duplicated pairs, such as the circled pairs E_0E_4 and E_2E_4 in Figure 2. It may cause redundant computation cost, which is a pervasive problem in many MapReduce-based matching algorithms. The reason for the occurrence of these redundancy is that the features of one entity are separated into multiple parts during the map phase, and each part may possibly match any part of the other entity in the reduce phase. It may happen more frequently among these highly similar entity pairs, because of the higher possibility for each part to be matched.

In redundancy-free similarity computation model [13], it adds additional annotate on each map output to tell the reducer which reducers the rest parts of this entity will be sent to. Though it is efficient, it bases on a strong precondition that all entities sent to the same reducer will definitely be paired among each others. In our algorithm, the permuted signatures of the same entity is sent to all reducers, and for each reducer a signature is only paired with its neighborhoods. So this redundancy-free solution is inapplicable for our random-based method.

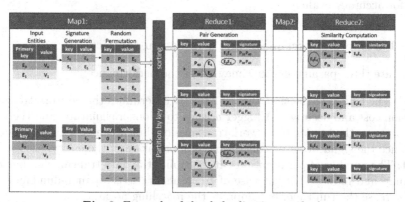

Fig. 3. Example of the deduplication method

We introduce an extra MapReduce job to reduce duplication. Figure 3 shows an example of our method. We modify the original MapReduce job in the reduce phase by cutting off the similarity computation. After generating all pairs, the reduce task terminates and outputs those pairwise information with entity IDs E_u, E_v $(u < v)$ concatenated as *key* and their ith permuted signatures P_{ui}, P_{vi} concatenated as *value*, that is $(E_uE_v, P_{ui}P_{vi})$.

The map phase of the second MapReduce job is an identity mapper which does nothing. In the following shuffle phase, all pairs with the same entity ID are grouped together. It means that all those duplicated pairs come together.

Then reducers will process the data like $(E_u E_v, list(P_{u1}P_{v1}, P_{u4}P_{v4}, ..., P_{ut}P_{vt}))$. Then we need to pick one pair of permuted signatures in the list and calculate its hamming distance. At last, we output the pair like $(E_u E_v, similarity)$ as final result.

5   Experiments

We run experiments on a 22-node HP blade cluster. Each node has two Intel Xeon processors E5335 2.00GHz with four cores and one thread per core, 16GB of RAM, and two 1TB hard disks. All nodes run CentOS 6.5, Hadoop 1.2.1, and Java 1.7.0. We evaluate the performance of our algorithm in two aspects: 1) We measure the effect on calculation performance and matching quality by using different parameters. 2) We compare the performance of our algorithm with two other state-of-the-art matching algorithms namely Document Similarity Self-Join (DSSJ)[1] and Dedoop[10].

5.1   Data Set Description

We use CiteSeerX data set. It contains nearly 1.32 Million citations of total size 2.89 GB in XML format. Each citation includes structured attributes such as *record ID, author, title, date, page, volume, publisher, etc.* It also has document *abstract*. We select a few records from CiteseerX and manually make validation sets for accuracy evaluation.

5.2   Parameter Description and Evaluation Metrics

There are three parameters that may affect the performance:

- **d**: The length of the signature. It directly determines the network transmission cost and accuracy. A bigger d leads to a longer signature, and therefore increases the burden of network transmission, but it can benefit the accuracy since the signature may contain more information of the entity.
- **t**: The number of permutations. It multiplies the data transmission between map and reduce. The increase of t can also raise the pair redundancy and increase the run-time, but improve the matching accuracy.
- **m**: The window of selecting neighborhoods. It decides the amount of pairs and also causes a change on redundancy ratio. It can influence the result accuracy and execution time as well.

We introduce four metrics to evaluate system performance:

- The **network transmission cost** is measured by summing up the size of map output since all output of map phase will be sent to reducers through network.
- The **run-time** of MapReduce jobs is recorded to compare the speed of our algorithm with different parameters.

- The **redundancy rate** is calculated as total number of generated pairs / distinct candidate pairs to show the redundancy ratio.
- The **accuracy** is also measured in this part. In order to calculate the accuracy, we prepare a validation set which contains 200 entity records for accuracy measurement. We calculate the similarity between those entities manually and generate a set of top 50 similar entity pairs as the standard result set. The accuracy is measured as the fraction of pairs that appear within the standard top 50 results.

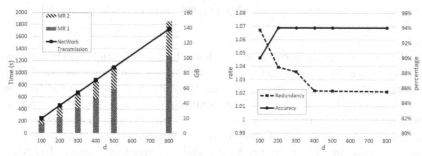

Fig. 4. Run-time and network transmission for different value of d (t=40, m=8)

Fig. 5. Redundancy rate and accuracy for different value of d (t=40, m=8)

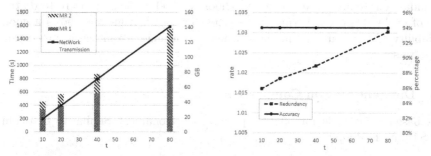

Fig. 6. Run-time and network transmission for different value of t (d=400, m=8)

Fig. 7. Redundancy rate and accuracy for different value of t (d=400, m=8)

To evaluate the performance, we first use a 200MB subset of CiteSeerX as our input to the effect of changing parameters. Figure 4 to 9 show the performance variations of our algorithm when changing one of the three parameters. Figures 4, 6, and 8 illustrate the run-time for both of two MapReduce jobs with MR1 for pair generation and MR2 for deduplication, and the transmission cost during MapReduce tasks. We can see from Figure 4 that when we increase the length of signature d, the network transmission cost together with the run-time

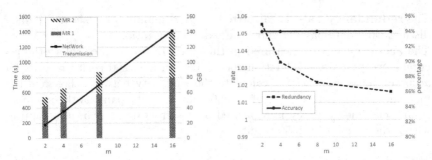

Fig. 8. Run-time and network transmis-**Fig. 9.** Redundancy rate and accuracy for sion for different value of m(d=400, t=40) different value of m (d=400, t=40)

of MapReduce jobs has a linear growth. Meanwhile, in Figure 5, as the increasing of d,the redundancy decreases steadily and the accuracy increases smoothly. The reason is that as the signature extends, the differences between entities can be found more easily. So they may have fewer chances to be paired. Therefore, the redundancy rate decreases. The performance variations with changing t and m are listed in Figure 6 to 9. Figure 6 and 8 shows similar performance with Figure 4, but it seems that in Figure 8 we spend large part of time on MR1 when $m = 2$ or $m = 4$. It is because a smaller m cannot reduce the cost on map phase when generating signatures and doing permutations. Figure 7 clearly shows that the number of permutations t can determine the redundancy rate directly. All these three parameters can strongly influence the performance. Figure4,6,8 show that when $d \geqslant 400, t \leqslant 50$ and $m \geqslant 8$, we get a better performance on both redundancy rate and accuracy. So we choose d=500, t=50 and m=10 to do the rest of evaluations. Furthermore, the linear growth of run-time shows a good scalability of our matching method, which has a huge advantage in processing big data set.

5.3 Different System Comparison

The baseline methods for our comparison are Document Similarity Self-Join (DSSJ) and Dedoop.

Table 1. Comparison of accuracy with DSSJ (d=500,t=50,m=10)

Name	Top 10	Top 20	Top 50
DSSJ	90%	95%	94%
Random-base Matching	90%	100%	94%
Dedoop	100%	100%	100%

In order to measure the accuracy, we use the validation set mentioned previously, and compare our matching result with DSSJ and Dedoop results. The standard result tests contain top 10, 20 and 50. Table 1 shows the accuracy of these top N tests. Since Dedoop compares all possible pairs of entities and calculates cosine similarity directly, the similarity result of Dedoop are always correct. However, the transmission cost is problematic that will be analyzed in the following content. When using the parameters (d=500,t=50,m=10), we can get almost the same accuracy as DSSJ method does. At the same time, we evaluate the processing speed of our method comparing with Dedoop and DSSJ using these parameters. Figure 10 shows run-time between our method and the baseline methods. We just run a small size of data since the run-time of both Dedoop and DSSJ would exceed hours if the size of data is larger than 100MB. The reason is that both of these two methods generate enormous size of pair candidates. We get nearly 100GB map output data when running Dedoop on 200MB data set. It can burden the system drastically on network transmission and is hard to be processed in memory. On the contrary, our random-based matching method shows a good scalability on data set size. The speed of our algorithm is significantly faster than Dedoop, and far more stable even when dealing with gigabytes of input data.

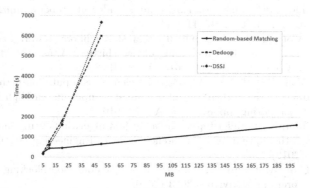

Fig. 10. Comparison of run-time with Dedoop and DSSJ (d=500,t=50,m=10)

6 Conclusion

In this paper, we study the problem of entity matching on unstructured data, which will be formatted as high-dimensional feature vectors. We take the MapReduce framework as our programming model and point out the two major challenges met on this model, which are load balancing problem and network transmission cost. We propose a random-based matching method to solve the matching problem which will help to greatly reduce the transmission cost. We use a special LSH function to generate signatures for entities, which helps to reduce entity dimensions. We take PLEB fast search algorithm to generate the candidate pairs efficiently. In addition, we propose our approach to reduce the redundancy during reduce tasks. Given the proposed algorithm, we implement it in Hadoop and analyze its performance on real data sets.

Acknowledgements. This work is partially supported by National Basic Research Program of Chi- na (Grant No. 2012CB316200), National Science Foundation of China (Grant No.61232002, 61402177 and 61332006).

References

1. Baraglia, R., De Francisci Morales, G., Lucchese, C.: Document similarity self-join with mapreduce. In: 2010 IEEE 10th International Conference on Data Mining (ICDM), pp. 731–736. IEEE (2010)
2. Charikar, M.S.: Similarity estimation techniques from rounding algorithms. In: Proceedings of the Thiry-Fourth Annual ACM symposium on Theory of Computing, pp. 380–388. ACM (2002)
3. Dean, J., Ghemawat, S.: Mapreduce: simplified data processing on large clusters. Communications of the ACM 51(1), 107–113 (2008)
4. Fellegi, I.P., Sunter, A.B.: A theory for record linkage. Journal of the American Statistical Association 64(328), 1183–1210 (1969)
5. Goemans, M.X., Williamson, D.P.: Improved approximation algorithms for maximum cut and satisfiability problems using semidefinite programming. Journal of the ACM (JACM) 42(6), 1115–1145 (1995)
6. Indyk, P., Motwani, R.: Approximate nearest neighbors: towards removing the curse of dimensionality. In: Proceedings of the Thirtieth Annual ACM Symposium on Theory of Computing, pp. 604–613. ACM (1998)
7. Kiefer, T., Volk, P.B., Lehner, W.: Pairwise element computation with mapreduce. In: Proceedings of the 19th ACM International Symposium on High Performance Distributed Computing, pp. 826–833. ACM (2010)
8. Kim, Y., Shim, K.: Parallel top-k similarity join algorithms using mapreduce. In: 2012 IEEE 28th International Conference on Data Engineering (ICDE), pp. 510–521. IEEE (2012)
9. Kolb, L., Thor, A., Rahm, E.: Parallel sorted neighborhood blocking with mapreduce. arXiv preprint arXiv:1010.3053 (2010)
10. Kolb, L., Thor, A., Rahm, E.: Dedoop: efficient deduplication with hadoop. Proceedings of the VLDB Endowment 5(12), 1878–1881 (2012)
11. Kolb, L., Thor, A., Rahm, E.: Load balancing for mapreduce-based entity resolution. In: 2012 IEEE 28th International Conference on Data Engineering (ICDE), pp. 618–629. IEEE (2012)
12. Kolb, L., Thor, A., Rahm, E.: Multi-pass sorted neighborhood blocking with mapreduce. Computer Science-Research and Development 27(1), 45–63 (2012)
13. Kolb, L., Thor, A., Rahm, E.: Don't match twice: redundancy-free similarity computation with mapreduce. In: Proceedings of the Second Workshop on Data Analytics in the Cloud, pp. 1–5. ACM (2013)
14. Lu, W., Shen, Y., Chen, S., Ooi, B.C.: Efficient processing of k nearest neighbor joins using mapreduce. Proceedings of the VLDB Endowment 5(10), 1016–1027 (2012)
15. Newcombe, H., Kennedy, J., Axford, S., James, A.: Automatic linkage of vital records (1959)
16. Ravichandran, D., Pantel, P., Hovy, E.: Randomized algorithms and nlp: using locality sensitive hash function for high speed noun clustering. In: Proceedings of the 43rd Annual Meeting on Association for Computational Linguistics, pp. 622–629. Association for Computational Linguistics (2005)

17. Toutanova, K., Klein, D., Manning, C.D., Singer, Y.: Feature-rich part-of-speech tagging with a cyclic dependency network. In: Proceedings of the 2003 Conference of the North American Chapter of the Association for Computational Linguistics on Human Language Technology, vol. 1, pp. 173–180. Association for Computational Linguistics (2003)

18. Toutanova, K., Manning, C.D.: Enriching the knowledge sources used in a maximum entropy part-of-speech tagger. In: Proceedings of the 2000 Joint SIGDAT Conference on Empirical Methods in Natural Language Processing and Very Large Corpora: Held in Conjunction with the 38th Annual Meeting of the Association for Computational Linguistics, vol. 13, pp. 63–70. Association for Computational Linguistics (2000)

19. Vernica, R., Carey, M.J., Li, C.: Efficient parallel set-similarity joins using mapreduce. In: Proceedings of the 2010 ACM SIGMOD International Conference on Management of data, pp. 495–506. ACM (2010)

20. Zobel, J., Moffat, A.: Exploring the similarity space. SIGIR Forum 32(1), 18–34 (1998)

A New Similarity Measure Between Semantic Trajectories Based on Road Networks

Xia Wu[1,2], Yuanyuan Zhu[1], Shengchao Xiong[2],
Yuwei Peng[2], and Zhiyong Peng[2,*]

State Key Laboratory of Software Engineering, Wuhan University, China
Computer School, Wuhan University, China
peng@whu.edu.cn

Abstract. With the development of the positioning technology, studies on trajectories have been growing rapidly in the past decades. As a fundamental part involved in trajectory recommendation and prediction, trajectory similarity has attracted considerable attention from researchers. However, most existing works focus on raw trajectory similarity by comparing their shapes, while very few works study semantic trajectory similarity and none of them take all the geographical, semantic, and timestamp information into consideration. In this paper, we model semantic trajectories based on road networks considering all these information, and propose a Constrained Time-based Common Parts (CTCP) approach to measure the similarity. Since the strict time constraint in CTCP may lead to many zero values, we further propose an improved Weighted Constrained Time-based Common Parts (WCTCP) method by relaxing the time constraint to measure the similarity more accurately. We conducted extensive performance studies on real datasets to confirm the effectiveness of our approaches.

Keywords: Semantic trajectory, Similarity measure, Road network.

1 Introduction

Nowadays, the smart mobile devices are very popular and the positioning technology is well developed. This technical progress leads to the accumulation of trajectories which has attracted a lot of attentions from researchers. A trajectory shows the continuous movements of an object. Usually, a raw trajectory is denoted as $\{(x_1, y_1, t_1), (x_2, y_2, t_2), \ldots, (x_n, y_n, t_n)\}$, where x_i, y_i are the geographic coordinates of the moving object at timestamp t_i and n is the number of elements in the sery. In recent years, many researchers has become interested in semantic trajectories, since additional semantic information (stops, landmarks, activities, etc.) can reflect user behavior more accurately. Generally, a semantic trajectory is a sequence consisted of timestamped locations labeled with semantic information to represent the landmarks being passed.

---

* Corresponding author.

© Springer International Publishing Switzerland 2015
R. Cheng et al. (Eds.): APWeb 2015, LNCS 9313, pp. 522–535, 2015.
DOI: 10.1007/978-3-319-25255-1_43

By finding similar trajectories, we can recommend similar users to a user (trajectory recommendation [1]), or predict the next place where a user will go (trajectory prediction [2]), etc. There are a lot of works proposed to measure the similarity of raw trajectories, but very few works focus on the similarity of semantic trajectories. Generally, a semantic trajectory contains three types of information: geographical information, semantic information, and temporal information. Fig. 1 shows three semantic trajectories of users: A, B, and C. Trajectories A and C have exactly the same stops, but they have different geographical shapes. Besides, the timestamps when they reach or leave these stops are also quite different. The geographical shapes of trajectories A and B are similar, but their stops are different. With the consideration of all the geographical, semantic and timestamp information, we can find that these three users may represent different kinds of people, i.e., they may have different occupations and living habits. Compared with raw trajectory similarity, semantic trajectory similarity can reflect how similar two users are more accurately, and thus can better support trajectory recommendation and prediction. On the other side, it is also more complicated than measuring the similarity of raw trajectories, as there are so many factors affecting the similarity need to be considered: geographical information, semantic information, and timestamp information.

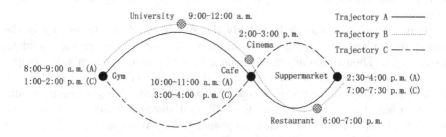

Fig. 1. An example of semantic trajectories

To the best of our knowledge, there is no existing work on semantic trajectory similarity measure which considers all these three kinds of information mentioned above. Thus, in this paper we propose a new semantic trajectory similarity measure by taking all these information into consideration. First, we model a semantic trajectory based on road networks by transforming it into a series of nodes in a graph. By doing this, we can not only simplify the computation of the similarity measure, but can also avoid the inaccurate results caused by different sampling strategies or noisy points in positioning devices. Based on this model, we propose a Constrained Time-based Common Parts (CTCP) approach of finding the common parts of two semantic trajectories to measure their similarity. However, we find that the time constraint in this approach is too strict and we may get zero values for trajectories with similar shapes and different temporal information. Therefore, we further propose an improved method called Weighted Constrained Time-based Common Parts (WCTCP) by relaxing the time constraint to measure the similarity more accurately.

The contributions of this paper are summarized below. (1)We measure the similarity of semantic trajectories by considering all the geographical, semantic and temporal information for the first time in the literature.(2)We model the semantic trajectories based on road networks and represent them as timelines. (3)We propose the CTCP approach to compute the semantic trajectory similarity and further improve the result by proposing another new approach WCTCP.(4)We conducted extensive experiments on real trajectory dataset and confirmed the effectiveness of our approaches.

The rest of this paper is organized as follows. Section 2 reviews the related work. In Section 3, we model the semantic traceries based on road networks. Section 4 gives the CTCP approach and the improved WCTCP approach. Section 5 shows the effectiveness of our approaches by extensive experiments. In Section 6 we conclude this paper.

2   Related Work

In this section, we will first briefly review the works on raw trajectories similarity, and then discuss the studies of semantic trajectory similarity closely related to this paper.

Many works were proposed to study the similarity of raw trajectory in the literature. Jonkery et al. proposed a similarity measure ED based on Euclidean distance [3]. Then dynamic time warping based measures which take temporal features into consideration are proposed, such as DTW [4] and PDTW [5]. To avoid the inaccurate results caused by noisy points, longest common subsequence based measure such as LCSS [6], edit distance based measures such as ERP [7] and EDR [8] are then proposed. In [9], Wang et al. evaluated the effectiveness of these approaches in different circumstances using a real world taxicab trajectory dataset. However, the above approaches only focus on the spatial information and ignore the temporal information of trajectories. Such problem was pointed out in [10]. In [11], Kang et al. cut the spatial space into several cells to form the raw trajectory by cell ID. Then they find the Longest Common Subsequence(LCSS) of cell sequences and the common parts of time intervals respectively and combine them together to measure the similarity. This approach can only be used in coarse-grained partition and result in low discrimination in fine-grained partition as a large number of similarity values will be zero.

In recent years, some researchers start to introduce the concept of semantic trajectory [12] [13], and several approaches are proposed to convert raw trajectories into semantic trajectories [1] [12] [14].

Up to now, a few works has studied the similarity measures of semantic trajectories. Ying et al. [15] first model the semantic trajectory on the STP-tree, then compute the similarity scores based on geographic behavior and semantic behavior respectively by STP-Tree, and combine them in a weighted way.

But this approach do not consider the time information. In [14], Li et al. proposed an algorithm to find the stops, and compared two stop sequences by matching each pair of chronologically stops extracted from two sequences respectively. In this way, two sequences are similar if each pair of stops are the same and the difference between their time intervals is less than a certain threshold. Note that this approach only consider the time interval but not the exact timestamp as we do in our approaches.

In [16], Hwang et al. measure the similarity of semantic trajectories based on road networks and they compute the distance between every pair of POIs (Points of Interest) in the trajectories. Road network model can reflect the real travel distance between two places, and it is also widely used to solve other trajectory problems such as trajectory k-nearest neighbors query [17]. However, the set of POIs in [16] contains every intersection and landmark the moving object has passed no matter it stays for a while or not, which cannot reflect real semantic stops in the trajectories. Tiakas et al. [19] also propose a new method to measure trajectory similarity based on road networks, but they only focus on the computation of raw trajectory similarity.

There are also other related works [2] [20] which measure the similarity of two trajectory patterns. The trajectory pattern is a representative trajectory of a trajectory set consisted of landmark being frequently visited. [2] mines the semantic trajectory pattern and then measures their similarity based on longest common sequence, which does not consider the temporal information. [20] computes the similarity of two trajectory patterns by assuming that a region of interest is connected by one timestamp. This is different from our approach, as we take both reach timstamp and leave timestamp into consideration which can determine the stay duration and reflect the temporal information more accurately.

3   Semantic Trajectory Model Based on Road Networks

Many approaches have been proposed to add semantic information to trajectories as stated in Section 2. In this paper we choose the algorithm in [14] to detect semantic stops from raw trajectories. Other approaches of detecting stops can also be used exchangeably. Based on the stop detection algorithm, we give the definitions of stop and semantic trajectory in the following.

Definition 1. A stop stands for a geographic region where a moving object stays over a certain time duration, and it can be formed as $< (x, y), (tin, tout) >$, where (x, y) is the geographical coordinates of the stop, tin is the reach timestamp, $tout$ is the leave the timestamp, and $tout - tin$ is larger than a threshold.

Definition 2. A semantic trajectory ST is a sequence of n_s tuples in the form of $(< (x_i, y_i), (tin_i, tout_i) >, L_i)$ for $1 \leq i \leq n_s$, where $tin_i < tout_i, tout_i \leq tin_{i+1}$, L_i is a landmark label associated with the i-th stop in landmark label set P.

Definition 3. A road network is a graph $G(V, E)$, where V is the set of nodes, and E is the set of edges. A node represents an road intersection, and an edge represents a road segment. There is an edge between two nodes iff there is a road segment connecting two intersections. Fig. 2 (a) shows a road network.

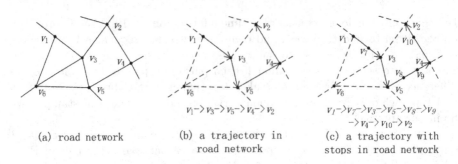

(a) road network

(b) a trajectory in
road network

v_1-> v_3-> v_5-> v_4-> v_2

(c) a trajectory with
stops in road network

v_1->v_7->v_3->v_5->v_8->v_9
-> v_4-> v_{10}-> v_2

Fig. 2. Trajectories in the road network

Definition 4. The shape of a raw trajectory T in the road network can be represented as a directed path $P_T = (v_{i_1}, v_{i_2}, ..., v_{i_n})$, where $v_{i_1} \in V$ and n is the number of road intersections passed by the moving object.

After adding timestamps to the nodes, P_T can be extended as $((v_{i_1}, t_1), (v_{i_2}, t_2), ..., (v_{i_n}, t_n))$, where t_j is the timestamp that the moving object passes node v_{i_j}, and $t_j < t_{j+1}$. Fig. 2 (b) shows the shape of a raw trajectory in the road network.

In order to integrate the semantic information into the raw trajectory, we add all stops detected as new nodes into the road network. Thus, road network $G(V, E)$ can be extended as $G_{ST}(V \cup V_{ST}, E \cup E_{ST})$, where V_{ST} is the node set representing the stops and $E_{ST} \subseteq V_{ST} \times (V \cup V_{ST})$ is the set of edges connecting two consecutive stops or a stop and its nearest intersection.

Definition 5. A semantic trajectory in road network STN can be defined as an extended directed path $P_{ST} = (v'_1, v'_2, ..., v'_{n'})$ where $v'_i \in V \cup V_{ST}$ representing stops or intersections in the trajectory, and $n' = n_s + n$ is the number of stops and intersections in the trajectory.

Fig. 2 (c) shows a trajectory with stops in the road networks. A semantic trajectory in road networks now is a new path with more nodes, where $v_1, v_3, v_5, v_4, v_2 \in V$ represent road intersections, and $v_7, v_8, v_9, v_{10} \in V_{ST}$ represent stops.

Unlike the road intersection which only need to record one timestamp, for a node in V_{ST} representing a stop, we need to record reach timestamp and leave timestamp since a moving object stays at the stop for a while. Therefore, we denote a node in V_{ST} as $(v', tin, tout)$, where tin and $tout$ are its reach time and leave time respectively. To unify the form, we also denote the node in V as $(v', tin, tout)$ with $tin = tout$. Thus, the extended semantic trajectory P_{ST} with timestamps is denoted as $((v'_1, tin_1, tout_1), (v'_2, tin_2, tout_2), ..., (v'_{n'}, tin_{n'}, tout_{n'}))$.

After we obtain the extended directed path of a semantic trajectory P_{ST}, we can represent it by a Time-Line with a sequence of stops and road intersections. Fig. 3 shows the Time-Line of the semantic trajectory in Fig. 2(c). We can see that the Time-Line is a sequence of consecutive blocks, where each block represents the travel time of a road segment. In other words, the length of each block represents the time consuming of moving from one road intersection to the next road intersection. As shown in the figure, there are five blocks $(v_1, v_3), (v_3, v_5), (v_5, v_4), (v_4, v_2)$ in the Time-Line, and $t_1, t_4, t_5, t_{10}, t_{13}$ are the

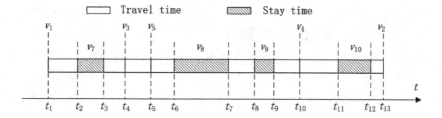

Fig. 3. A Time-Line of a semantic trajectory

timestamps when the moving object pass v_1, v_3, v_5, v_4, v_2 respectively. In the Time-Line, we also record the stops of the semantic trajectory and represent the time interval that the moving object stays at the stops by gray part, where the start of a gray block represents the reach timestamp *tin*, the end of a gray part represents the leave timestamp *tout*, and the length of a gray part shows how long the object stays at the stop. As shown in Fig. 3, the four gray parts represent four stops v_7, v_8, v_9, v_{10} of the semantic trajectory in Fig. 2(c), with time intervals $(t_2, t_3), (t_6, t_7), (t_8, t_9)$ and (t_{11}, t_{12}). In the next section, we will measure the similarity of two semantic trajectories by comparing their Time-Lines.

4 Semantic Trajectory Similarity Measures

In this section, we will first propose a basic approach Constrained Time-based Common Parts to compute the similarity of two trajectories. Then we analyze its weakness, and further propose an improved algorithm Weighted Constrained Time-based Common Parts.

4.1 Constrained Time-Based Common Parts

To measure the similarity of two semantic trajectories, we need to find their common parts in their Time-Lines. The common parts are the common or similar stops two moving objects staying in the same time interval.

To find the common parts in an efficient way, we will gradually filter out their different parts by the following steps. (1)First, we check if two trajectories have common travel time, i.e., if there is overlap between their travel intervals. (2)If two trajectories have common travel time, we then check if they are affiliated to the same edge. (3)If two trajectories have the same edge in the common travel time, we then check if there are stops on the edges have common stay time. (4)Finally, we will check wether the two stops are the same or similar. The distance of similar stops should be less than a given threshold θ.

Now we illustrate the above process by an example in Fig. 4. Here, we have two trajectories Traj A and Traj B, which are denoted as STN_a and STN_b in the road network respectively. We can see that STN_a and STN_b has a large common part of travel time. Then we check if they have common edges.

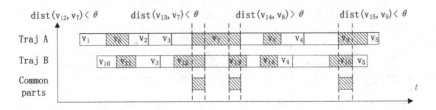

Fig. 4. Compare of two Time-Lines

Clearly, the first edges (v_1, v_2) in STN_a and (v_{10}, v_3) in STN_b are not the same, so they have no common part. The second edges in STN_a and STN_b are both (v_3, v_4), and they have 3 stops and 2 stops respectively. Since the distance between v_7 and v_{12} is less than θ, they have common part, so do v_7 and v_{13}. Although v_8 and v_{14} have the common stay time, they have no common part because their distance is larger than θ.

Next, we give the details of computing the common parts of two edges and two stops. First, we check if there is any common part between two edges by comparing their vertices. Suppose that $e_{a,i}$ is the i-th edge in STN_a containing n edges and $e_{b,j}$ is the j-th edge in STN_b containing m edges. We have the following equation.

$$e_{a,i} \cap e_{b,j} \neq \emptyset, \; if \; e_{a,i} = e_{b,j} \wedge (e_{a,i}.tout > e_{b,j}.tin \vee e_{a,i}.tin < e_{b,j}.tout)$$

where $1 \leq i \leq n, 1 \leq j \leq m$, $e_{a,i} \neq e_{b,j}$ means that $e_{a,i}$ and $e_{b,j}$ are not the same edge, $e_{a,i}.tin$ is the timestamp of passing the start vertex of $e_{a,i}$, and $e_{a,i}.tout$ is the timestamp of passing the end vertex of $e_{a,i}$.

Then, we discuss how to check if there are common parts of two stops. Suppose that $s_{a,k} \in V_{ST}$ is the k-th stop in STN_a containing g stops and $s_{b,l} \in V_{ST}$ is the l-th edge in STN_b containing h stops. The following equation holds:

$$s_{a,k} \cap s_{b,l} \neq \emptyset, \; if \; dist(s_{a,k}, s_{b,l}) < \theta \wedge (s_{a,k}.tout \geq s_{b,l}.tin \vee s_{a,k}.tin \leq s_{b,l}.tout)$$

where $1 \leq k \leq g, 1 \leq l \leq h$, $s_{a,k}.tin$ is the reach timestamp of $s_{a,k}$, $s_{a,k}.tout$ is the leave timestamp of $s_{a,k}$, and $dist(s_{a,k}, s_{b,l})$ is the Euclidean Distance between $s_{a,k}$ and $s_{b,l}$.

We say a stop s is in edge e if the start and end vertices of e are the immediate predecessor and successor of the vertex representing s. For example, in Fig. 4, v_6 is in edge (v_1, v_2) as v_1 is the immediate predecessor of v_6 and v_2 is the immediate successor of v_6. We denote $S_{a,i}$ and $S_{b,j}$ as the sets of stops in $e_{a,i}$ and $e_{b,j}$ respectively, and their stop numbers are denoted as $count(S_{a,i}) = |S_{a,i}|$ and $count(S_{b,j}) = |S_{b,j}|$ respectively. The sufficient condition that there are common parts of two edges is as follow:

$$e_{a,i} \cap e_{b,j} \neq \emptyset \wedge \; count(S_{a,i}) \times count(S_{b,j}) \neq 0 \wedge \exists s_{a,k} \cap s_{b,l} \neq \emptyset \qquad (1)$$

where $1 \leq k \leq count(S_{a,i}), 1 \leq l \leq count(S_{b,j})$.

Now we can compute the length of the common parts of two edges by the following equation:

$$|sim(e_{a,i}, e_{b,j})| = \begin{cases} \sum_{k=1}^{count(S_{a,i})} \sum_{l=1}^{count(S_{b,j})} |s_{a,k} \cap s_{b,j}| & \text{if equation (1) holds;} \\ 0, & \text{otherwise.} \end{cases} \quad (2)$$

where $s_{a,k} \in S_{a,i}$, $s_{b,l} \in S_{b,j}$, and $|s_{a,k} \cap s_{b,j}|$ is the length of their common parts. Note that the length here is the time interval but not the spatial distance.

To compute the common parts of STN_a and STN_b, we can first compute the common parts for every edge pair and get their sum as follows.

$$\boldsymbol{Sim}(STN_a, STN_b) = \frac{\sum_{i=1}^{n} \sum_{j=1}^{m} |sim(e_{a,i}, e_{b,j})|}{max\{|S_a|, |S_b|\}} \quad (3)$$

where $|S_a|$ and $|S_b|$ are the sums of lengths of all stops in STN_a and STN_b respectively. The numerator is the total length of all common parts in the stops; the denominator is maximum total length of the stops in these two trajectories. If the length of common parts equals to the length of the maximum length, the similarity is 1, if there is no common part, the similarity is 0. We name the above process of computing equation (3) the Constrained Time-based Common Parts (CTCP) approach.

However, by using this approach, we may get many similarity values of zero. It is because the time constraint is too strict for some trajectories in CTCP. For example, although two trajectories have exactly the same shape and stops, their similarity score will be 0 if they pass the same road in non-overlapped time intervals. Under this circumstance, a zero score cannot give an accurate evaluation of these two trajectories. Besides, CTCP also does not take the landmark labels into consideration. Therefore, in the next subsection, we propose an improved method to solve the above problems.

4.2   Weighted Constrained Time-Based Common Parts

To overcome the zero value problem, we need to relax time constraint to compute the similarity, i.e., we can horizontally translate the Time-Line of a trajectory. As shown in Fig. 5, before the horizontal translation, Traj A and Traj B has no common part, but they may have common parts after the translation. Note that permitting horizontal translation may undermine the affect of the timestamps in the similarity computation. So we will weight the largest value of common parts according to the distance that the Time-Line moved.

In this approach, to compute the similarity of the stops in two edges $e_{a,i}$ and $e_{b,j}$, we not only consider the common part of their stops, but also consider the landmark labels of the stops. Based on equation (2), we define the extended similarity of two edges based on stops as follows.

$$sim_s(e_{a,i}, e_{b,j}) = \alpha \frac{|sim(e_{a,i}, e_{b,j})|}{max(\sum_{s_{a,k} \in S_{a,i}} |s_{a,k}|, \sum_{s_{b,l} \in S_{b,j}} |s_{b,l}|)} + (1-\alpha) \left| \frac{L(S_{a,i}) \cap L(S_{b,j})}{L(S_{a,i}) \cup L(S_{b,j})} \right| \quad (4)$$

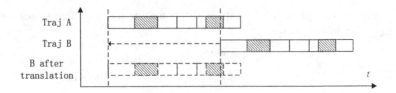

Fig. 5. Horizontal translation of a Time-Line

where $L(S_{a,i})$ and $L(S_{b,j})$ are the sets of landmark labels of all stops in $S_{a,i}$ and $S_{b,j}$ respectivley, and $0 \leq \alpha \leq 1$ is a weighting parameter. In this equation, the first part computes the ratio of the common parts to the maximum total length of stops, and the second part computes the similarity of two landmark label sets.

To compute the similarity of two edges, we will not only consider their similarity based on their stops, but also consider their distance in the road network. This can avoid getting zero value for two edges which are very close in the road network but has no common parts in their stops. Let $||STN_a||$ and $||STN_b||$ be the lengthes of STN_a and STN_b in road networks, i.e., the number of edges in STN_a and STN_b respectively. We define the distance between $e_{a,i}$ and $e_{b,j}$ as $Dist(e_{a,i}, e_{b,j})$, which representing the number of hops between the start vertex of $e_{a,i}$ and the start vertex of $e_{b,j}$ in the road network. The normalized similarity of $e_{a,i}$ and $e_{b,j}$ based on their distance in the road network is defined as follows.

$$sim_d(e_{a,i}, e_{b,j}) = \begin{cases} \frac{max(||STN_a||, ||STN_b||) - Dist(e_{a,i}, e_{b,j})}{max\{||STN_a||, ||STN_b||\}} & \text{if equation(6) holds;} \\ 0, & \text{otherwise.} \end{cases} \quad (5)$$

$$Dist(e_{a,i}, e_{b,j}) \leq max\{||STN_a||, ||STN_b||\} \quad (6)$$

$sim_d(e_{a,i}, e_{b,j})$ measures how similar two edges are by their distance in the road network. If they are the same edges, the similarity is 1; if they are very far from each other, the similarity is 0.

Now we define the overall similarity of two edges as follows.

$$sim_w(e_{a,i}, e_{b,j}) = \beta sim_d(e_{a,i}, e_{b,j}) + (1 - \beta)sim_s(e_{a,i}, e_{b,j}) \quad (7)$$

where $0 \leq \beta \leq 1$ is the weighting parameter.

Then the similarity of STN_a and STN_b can be defined as follows.

$$Sim_W(STN_a, STN_b) = \frac{\sum_{i=1}^{n} \sum_{j=1}^{m} sim_w(e_{a,i}, e_{b,j})}{||STN_a|| \times ||STN_b||} \quad (8)$$

Recall that we can gradually horizontally translate one Time-Line when we comparing two Time-Lines of trajectories. Suppose that in each step we move the Time-Line STN_b by a certain distance Δt. We denote the moved time line as STN_b^r with moved distance $r \times \Delta t$ in the r-th step. Considering the moving distance, the similarity of STN_a and STN_b can be defined as follows.

$$Sim_{W,r}(STN_a, STN_b) = \frac{\Delta T - r \times \Delta t}{\Delta T} Sim_W(STN_a, STN_b^r) \quad (9)$$

where $\Delta T = max(|e_{a,1}.tin - e_{b,m}.tout|, |e_{b,1}.tin - e_{a,n}.tout|)$, which indicates the maximum distance that one trajectory can be moved along the Time-Line. Then we can get the maximum t weight score among all steps.

$$Sim_{max}(STN_a, STN_b) = max_{r=1}^{\lfloor \Delta T/\Delta t \rfloor}\{Sim_{W,r}(STN_a, STN_b)\} \qquad (10)$$

We name the above process of computing equation (10) the Weighted Constrained Time-based Common Parts (WCTCP) approach.

5   Experiments

In this section, we conduct extensive experiments to demonstrate the effectiveness of our approaches. All the algorithms are implemented with $C\#$ and run on a computer with Intel Core i5-M430 CPU (2.27 GHz) and 6 GB memory.

Trajectory Dataset: The experiments are based on real trajectory data collected by Microsoft Research Asia. This dataset was collected in Geolife project by 178 users in Beijing in a period of over four years, which contains 17,621 trajectories with a total distance of 1,251,654 kilometers and a total duration of 48,203 hours. A GPS trajectory in this dataset is a raw trajectory represented by a sequence of time-stamped points. We will convert it into a semantic trajectory using existing semantic stop detection methods. These trajectories were recorded by different GPS loggers and GPS-phones, and thus have a variety of sampling rates. 91 percent of the trajectories are logged in a dense representation, e.g., every 1–5 seconds or every 5–10 meters per point. Fig. 6(a) shows the overview of the dataset within the fifth Ring Road of Beijing.

Road Network: We use the API interface provided by Baidu Map to find out all road segments and intersections from a vector map of Beijing. By doing this we get a road network with 171,504 vertices and 433,391 edges. We also collected the landmark labels (the label of POI) using this API. There are numerous landmark labels, among which we choose some important types of POIs given in Baidu Map, such as Restaurant, Hospital, Supermarket, Scenic spot, Bank, etc. Fig. 6 (b) shows the overview of the Bank POI set.

In our experiments, we randomly choose 5 users and 5 trajectories for each user to evaluate the algorithms. Before comparing with other methods, we first find all the semantic stops from the raw trajectory by using the algorithm in [1]. We use the default parameter setting in that paper ($\theta_d = 200$ meters, $\theta_t = 30$ minutes). Among these 25 trajectories, we detected 109 stops, and the number of stops for each trajectory is shown in Fig.7(a).

To validate that our approach WCTCP can avoid generating a large amount of zero values, we first compare our approaches with CVTI in [11]. We choose CVTI because it is also time-based measure, but designed for raw trajectories. The numbers of cell partitions in CVTI are set as 10×10 and 100×100 respectively to show the affect of different partition sizes. In WCTCP, we set $\alpha = 0.5$ and $\beta = 0.5$, and set $\Delta t = 5s$ since the sampling frequency of the dataset is every 5 seconds per point. Note that α affects the weights of stop information and land mark information, the larger α is, the more important stop information is; β

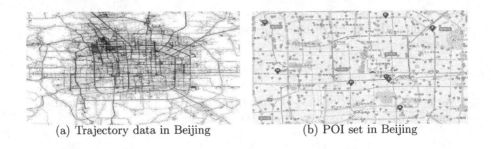

(a) Trajectory data in Beijing   (b) POI set in Beijing

Fig. 6. Data set overview

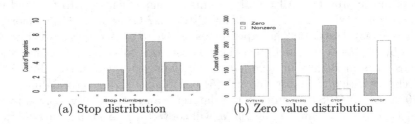

(a) Stop distribution   (b) Zero value distribution

Fig. 7. Stop and zero value distributions

affects the weights of spatial information and semantic information, the larger β is, the more important sptial information is. Here we set $\alpha = 0.5$ and $\beta = 0.5$ by considering that all stop information, land mark information and spatial information are same important. These two parameters can also be tuned by the actual demands of different users. Fig. 7(b) shows the number of zeros and number of nonzeros obtained by CVTI(10), CVTI(100), CTCP and WCTCP. We can see that CVTI(100) with 100×100 partitions results in more zeros than CVTI(10) with 10×10 partitions. Our basic approach CTCP also have many zeros since the time-constraint is too strict, but our improved approach WCTCP significantly reduces the number of zero values and it performs the best among these algorithms. Here we do not show the similarity value for each pair of trajectories for each user, as there are too many zeros in CVTI. Instead, we will choose another method HGMS [14] for further comparison.

For further evaluation of the effectiveness of the algorithms for each user, we choose HGMS in [14] as the baseline, because it measures the semantic trajectory similarity by considering geographical shape, semantic information, and travel time like we do. However, note that this travel time is only one small part of temporal information considered in our approaches. In this experiment, for each user, we compute the similarity value of each pair among its 5 trajectories and we obtain 10 similarity values. We assign each similarity value an ID, and show its values obtained by HGMS an WCTCP respectively in Fig. 8. Fig. 8(a)

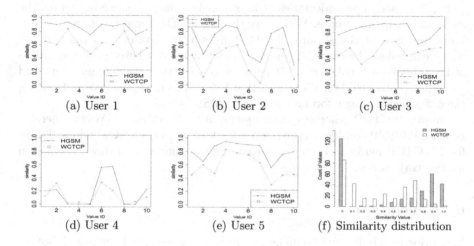

Fig. 8. Similarity comparison of HGMS and WCTCP

shows similarity values computed by HGMS an WCTCP for user 1, in which the values generated by HGMS are all very high (larger than 0.8). WCTCP generates smaller values than HGMS showing the difference between the 5 trajectories. Similar trends can also be find in Fig. 8(b), (c), and (e). Note that the similarity values obtained by HGMS and WCTCP for user 4 in Fig. 8(d) are very low. It might because user 4 have a variety of behaviors in different days. However, WCTCP still gets smaller similarity values than HGMS. Fig. 8(f) shows the distribution of similarity values among all the 25 trajectories of these 5 users. We divided [0,1] into the following intervals: 0, (0,0.1],(0.1,0.2], ...,(0.9,1] and show the number of similarity values falling in each interval. We can see that the values obtained by HGMS polarizes severely, i.e., most of the values are 0 or fall into the range of [0.7-1.0]. The values obtained by WCTCP is more balance and nearly normally distributed, where it has smaller number of zero values than HGMS and most values fall into the range of [0.4-0.7].

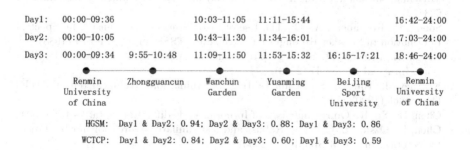

Fig. 9. Examples of trajectories of user one

Fig. 9 shows three trajectories of user one. For simplicity, we only draw the stops and their timestamps. We can see the stops of Day1 and Day2 are the same but with different timestamps. WCTCP gives a lower value 0.84 than HGMS which can reflect the time difference. WCTCP also gives lower similarity values of Day2 between Day3, and Day1 between Day3 with quite different stops and time stamps than HGMS. This shows that WCTCP is more sensitive in detecting the difference of time information than HGMS.

Overall, WCTCP generates smaller similarity values than HGMS in generally, because WCTCP considers more temporal information than HGMS does. Therefore, WCTCP performs better on measuring the difference of time information for two trajectories.

6   Conclusion

Measuring the similarity of semantic trajectories are essential and useful, but related works are very few. It is also very difficult to measure semantic trajectories as there are so many factors (geographical information, semantic information, and timestamp information) to be considered. In this paper, we first model semantic trajectories based on road networks to simplify the process of similarity computation, and then we propose a CTCP approach and an improved WCTCP approach to measure the similarity of semantic trajectories. Our extensive experiments on real datasets confirmed the effectiveness of our approaches. In our future works, we will evaluate the performance of our approaches on more datasets by comparing it with more approaches, and evaluate the affect of tuning parameters in our approaches.

References

1. Zheng, Y., Zhang, L., Ma, Z., Xie, X., Ma, W.: Recommending Friends and Locations Based on Individual Location History. ACM Tweb 5(1), February 2011
2. Ying, J., Lee, W., Weng, T., Tseng, V.: Smantic trajectory mining for location prediction. In: ACM GIS 2011 (2011)
3. Jonkery, R., De Leve, G., Van Der Velde, J.A., Volgenant, A.: Rounding Symmetric Traveling Salesman Problems With An Asymmetric Assignment Problem. OR, 623–627 (1980)
4. Soong, F., Rosenberg, A.: On The Use of Instantaneous and Iransitional Spectral Information in Speaker Recongnition, Acoustics. TSSP 36(6), 871–879 (1988)
5. Keogh, E., Pazzani, M.: Scaling up dynamic time warping for datamining applications. In: ACM SIGKDD (2000)
6. Kearney, J., Hansen, S.: Stram Editing for Animation. Technical report, DTIC Document (1990)
7. Chen, L., Ng, R.: On the marriage of Ip-norms and edit distance. In: VLDB (2004)
8. Chen, L., Ozsu, M., Oria, V.: Robust and fast similarity search for moving object trajectories. In: ACM SIGMOD (2005)
9. Wang, H., Su, H., Zheng, K., Sadiq, S., Zhou, X.: An effectiveness study on trajectory similarity measures. In: ADC (2013)

10. Frentzos, E., Grattsias, K., Theodoridis, Y.: Index-based most similar trajectory search. In: ICDE (2007)
11. Kang, H., Kim, J., Li, K.: Similarity measures for trajectory of moving objects in celular space. In: ACM SAC (2009)
12. Alvares, L., Bogorny, V., Kuijpers, B., Fernandes, J., Moelans, B., Vaisman, A.: A model for enriching trajectories with semantic geographical information. In: ACM GIS (2007)
13. Parent, C., Spaccapietra, S., Renso, C., Andrienko, G., Andrienko, N., Bogorny, V., Damiani, M., Gkoulalas-Divanis, A., Macedo, J., Pelekis, N., Theodoridis, Y., Yan, Z.: Semantic Trajectories Modeling and Analysis. ACM Computing Serveys 45(4) (2013)
14. Li, Q., Zheng, Y., Xie, X., Chen, Y., Liu, W., Ma, W.: Mining user similarity based on location history. In: ACM GIS (2008)
15. Ying, J., Lu, E., Lee, W.: Mining user similarity from semantic trajectories. In: ACM LBSN (2010)
16. Hwang, J.-R., Kang, H.-Y., Li, K.-J.: Spatio-temporal similarity analysis between trajectories on road networks. In: Akoka, J., et al. (eds.) ER Workshops 2005. LNCS, vol. 3770, pp. 280–289. Springer, Heidelberg (2005)
17. Papadias, D., Zhang, J., Mamoulis, N., Tao, Y.: Query processing in spatial network databases. In: VLDB (2003)
18. Xue, A.Y., Qi, J., Xie, X., Zhang, R., Huang, J., Li, Y.: Solving the Data Sparsity Problem in Destination Prediction. VLDB Journal (2015)
19. Tiakas, E., Papadopoulos, A.N., Nanopoulos, A., Manolopoulos, Y.: Trajectory similarity search in spatial networks. In: IDEAS (2006)
20. Chen, X., Lu, R., Ma, X., Pang, J.: Measuring user similarity with trajectory patterns: principless and new metrics. In: Chen, L., Jia, Y., Sellis, T., Liu, G. (eds.) APWeb 2014. LNCS, vol. 8709, pp. 437–448. Springer, Heidelberg (2014)

Effective Citation Recommendation by Unbiased Reference Priority Recognition

Wen-Yang Lu, Yu-Bin Yang, Xiao-Jiao Mao, and Qi-Hai Zhu

State Key Laboratory for Novel Software Technology, Nanjing University
Nanjing 210093, China

Abstract. Citation recommendation is a meaningful and challenging research problem nowadays. Most of prior researches make a simplified assumption that the citations are more preferential for the papers to cite than the uncited ones. Consequently, the unreasonable priority assertion between the cited and uncited papers derived from the above assumption makes citation recommendation prone to be biased. To address this issue, we firstly propose an instinctive assumption that the more preferential a reference is, the easier it can be recognized as a citation. Based on this assumption, we propose two methods CR and $CR+C$ aiming to find more unbiased priority between the cited and uncited papers with c-SVC. Then, a improved RankSVM model is trained for citation ranking purpose. Experimental results demonstrate that, comparing with the RankSVM model, our methods achieve 5.27% improvement on Recall@50 and 8.28% improvement on MRR. Moreover, $CR+C$ achieves advantage on efficiency by taking only 18.9% time it needs.

Keywords: citation recommendation, RankSVM, citation priority.

1 Introduction

Research papers need to cite relevant and important previous work to help readers understand their backgrounds, contexts and innovations. With the development of scientific research, more and more literatures are published every year, which makes it difficult for the researchers to go through and digest all of them. Therefore, literature search is born to solve this problem. However, traditional literature search engines, such as Google Scholar, can only retrieve a list of relevant papers using keyword-based queries. The ignorance of rich information, such as authors' personal research interests and research background, usually makes it ineffective in practice. To better address this problem, integrating diverse features for citation recommendation draws more and more attentions recently.

In recent years, some methods have been proposed for citation recommendation. He et al. [1,2] extract content-based features, which are thought to contains the specific information needs, for citation recommendation.. However, context analysis is normally time-consuming when the content is too much. Sometimes it causes ambiguity due to the lack of content. Topical similarity-based methods [3,4] find conceptually related

© Springer International Publishing Switzerland 2015
R. Cheng et al. (Eds.): APWeb 2015, LNCS 9313, pp. 536–547, 2015.
DOI: 10.1007/978-3-319-25255-1_44

papers by taking advantage of latent topic model. But topical similarity is too vague to tell which one is more suitable when a lot of papers share similar topics. It is inappropriate that these methods only focus on content information and ignore the other critical structure information. More recent work [5,6,7,8,9] utilize citation, publication and co-authorship to derive structure features which complements content-based relevance well.

However, a critical problem remaining is that existing methods make a simplified assumption that the articles actually cited by a paper are more preferential than the uncited ones. Consequently, the biased priority assertion between the cited and uncited papers derived from the assumption makes citation recommendation prone to be biased. Fig. 1 illustrates a hypothetical ranking list for a paper. For the author in Fig. 1, the uncited paper a_1 may be a better choice than the cited one c_3. During ranking learning, more unbiased priority recognition between the cited and uncited could do a great favour.

In this paper, we propose two RankSVM-based methods CR and CR+C which are easy but effective. To address priority recognition issue, we firstly propose an instinctive assumption: the more preferential one reference is, the more easily it can be recognized. Based on the assumption, two methods both train a c-SVC model to help recognizing the citation priority which can help to do better sampling. Then, a RankSVM model for citation ranking will be trained on the samples. Compared to RankSVM, our methods not only gets improvement at recall and precision but also costs less time.

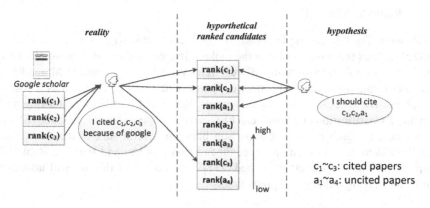

Fig. 1. A toy example showing the differences between reality and hypothesis

The rest of the paper is organized as follows. Sec. 2 gives background and problem definition. Sec. 3 introduces our proposed methods. Experimental results and discussions are provided in Sec. 4, followed by the conclusions and future work in the last section

2   Preliminary

2.1   Problem Definition

Citation recommendation is usually defined on the bibliographic network, which is described as heterogeneous networks [10]. A heterogeneous bibliographic network is represented as a directed graph G, which consists of nodes of 4 types and different types of links between them. An example is shown in Fig. 2. Formally, we introduce a definition on citation recommendation problem [11]: *Given a special query q∈Q, which comprises title, abstract, authors and target venue, a list of small subset of candidate papers d∈D need to be recommended as high quality references.*

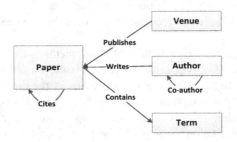

Fig. 2. Heterogeneous bibliographic network schema

2.2   RankSVM Model

As shown in Fig.1, a ranking list $rank(c_1) < rank(c_2) < rank(a_1),..., < rank(a_4)$, in which $rank(d)$ denotes the position of d in the ranking list. eg. $rank(c_1)$ is 1, is usually translated into several pairs of two different retrieval items   by RankSVM model. In RankSVM training, if $rank(d_i) < rank(d_j)$, pair $<d_i, d_j>$ will be regarded as a positive item, otherwise a negative one. By this means, RankSVM translates the ranking problem into a classification one which can be solved by Support Vector Machine (SVM).

A recent research makes a comparison between several state-of-the-art methods and RankSVM model on citation recommendation problem and find that RankSVM outperforms most of the other methods [11]. On account of this we will utilize the model for ranking training and prediction.

2.3   Feature Extraction

In order to verify the effectiveness of our methods, firstly we need to obtain some features which are extracted from either the context or structure and commonly used in the former literatures [7,8], [15,16]. The following sub-sections will describe the definition to characterize these features.

2.4   Content-Based Features

● *Topic Distribution Similarity.* Topic is a significant feature extracted from the content [13] using the unsupervised Latent Dirichlet Allocation [14] implemented by Blei et.al. We train a 100-topic models on our corpus empirically to get the topic distribution $\Gamma(d)$. The content analyzed here comprises the abstracts and titles of all the papers. We calculate the topic distribution cosine similarity for it outperforms the Kullback-Leibler divergence and the Jensen-Shannon divergence [8]:

$$\Gamma(d) = \{p(topic_1|d), p(topic_2|d),...,p(topic_{100}|d)\} \qquad (1)$$

$$TopicSimilar(q,d) = \cos(\Gamma(q), \Gamma(d)) \qquad (2)$$

● *Topic Diversity.* Some papers proposed a universal method which could draw a vast range of researchers' attention. Consequently, it is more likely to be cited by others. Entropy of the article's topic distribution is usually used to measure the topical breadth.

$$TopicDiversity(d) = \sum_{i=1}^{100} -p(topic_i|d)\cdot\log p(topic_i|d) \qquad (3)$$

2.5   Structure-Based Features

● *Recency.* Researchers normally prefer the most recent article which is relevant to their research areas [15], thus a feature describing how many years the articles have passed through is necessary:

$$Recency(q, d) = year(q) - year(d) \qquad (4)$$

● *H-Index.* H-Index is a useful measurement which can reflect the popularity of the a paper of a scientist [16].
● *Paper Reference Count.* Paper reference count, i.e. the number of times that a paper has been cited, has long been seen as an important index to measure how popular the paper is:

$$citing(d) = \{d' \in D: d' \text{ cites } d\}$$

$$PRC(d) = |citing(d)| \qquad (5)$$

● *Venue Reference Count.* Venue reference count is another important index to measure the popularity or convincement of a paper published in the venue[8].

$$VRC(v) = \sum_{d_i \in v} |citing(d_i)| \qquad (6)$$

● *Meta Path-based Features.* In bibliographic heterogeneous networks, Ref. [7] defines several features based on the schema shown in Fig.1. Three of them are defined as P-V-P, P-A-P, P-T-P. P-V-P represents whether two different papers are published in the same venue. P-A-P and P-T-P represent the amount of common authors and terms of two papers respectively.

3   Citation Recommendation Model

3.1   CR Model Overview

Fig.1 illustrates one possible citation ranking which is a best choice for one paper hypothetically. We can notice that not all cited papers rank higher than uncited ones (eg. $rank(c_3) > rank(a_1)$) and reference priority within cited or uncited papers are different (eg. $rank(c_1) < rank(c_3)$, $rank(a_1) < rank(a_4)$).

To recognize the reference priority of papers, we propose an instinctive assumption: the more preferential a reference is, the easier it can be recognized as a citation. Consequently, we think the more easily a paper is classified as cited by a classifier, the more possible it could be cited. Therefore, we train a classifier to distinguish the priority between papers.

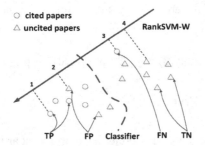

Fig. 3. 4 types of data

As Fig.3 shows, cited papers are marked as True Positive (TP) and False Negative (FN) separately by classifier. Meanwhile, uncited papers are marked as False Positive (FP) and True Negative (TN). Based on our proposed assumption, we think four different types of data could be cited with different possibility. Intutively, from the point of view of RankSVM model in Fig.3, we should set the priority between those four types of data as $rank(TP) < rank(FP) < rank(FN) < rank(TN)$. But to better fit the ground truth, our model exchange the positions of $rank(FP)$ and $rank(FN)$. That is to say, we think the ranking: $rank(TP) < rank(FN) < rank(FP) < rank(TN)$ could be a more unbiased reference priority which will help to learn a better ranking_model for recommendation. After that, we can do recommendation prediction on all candidate papers with the model.

3.2   Model learning

Firstly, we train a c-Support Vector Classification (c-SVC) classifier [17] to help recognizing reference priority between papers. In Algorithm 1, we sample part of uncited papers into A and all cited papers into C to train a c-SVC classifier. TrainClassifier() is implemented by Lin et.al [17]. All parameters for TrainClassifier are defaults. To improve classifier performance, we update classifier by iterated training. Firstly, we do classification on the training dataset with the classifier got previously. Then, to train a new classifier, we sample some TN papers randomly as negative ones into A. The iteration training makes the classifier get a higher recall but a lower precision. By doing so, we can lower the requirement that the cited papers need meet which are also the papers should be cited.

Algorithm 1. Classifier Learning

Input: Cited papers C, Uncited papers U, Loops I
Output: *Classifier*
1. $A = \{\}$
2. repeat:
3. $A = A \cup \{d : d \in U\text{-}A\}$
4. until: $|A|=|C|$
5. *Classifier* = TrainClassifier(C, A)
6. **for** i=1,2,...,I **do**
7. $A = \{\}$
8.   **for** d, d $\in C \cup U$ **do**
9.   d_label = predict(classifier, d)
10.   if (d_label==-1 & d $\in U$ & $|A|<|C|$)
11.   Update A: $A = A \cup \{d : d \in U\text{-}A\}$
12.   **end for**
13. *Classifier* = TrainClassifier(C, A)
14. **end for**
15. **Return:** Learned *Classifier*

In **Algorithm 2**, R_0, R_1, R_2 and R_3 represent 4 different types of data TP, FN, FP, TN separately. To denote the reference priority of 4 types of data, we allocate different number to different type of papers. The higher the ranking is, the bigger the number will be. TrainRanking() is implemented by Joachims et.al [12]. P_1 and P_2 represent the ratio between R_2 and $R_1 \cup R_0$, R_3 and $R_1 \cup R_0$ separately.

Algorithm 2. Ranking Model Learning

Input: *classifer*, cited papers C, uncited papers U
Output: ranking_model
1. $P=\{\}, N=\{\}, R_2=\{\}, R_3=\{\}, ranks=\{\}$
2. **for** d, d $\in$ T **do**
3.   d_label = predict(*classifier*,d)
4.   if(d_label==1):
      $P = P \cup \{d\}$
5.   else:
      $N = N \cup \{d\}$
6. **end for**
7. $R_0 = P \cap C$, $R_1 = N \cap C$
8. **repeat:**
9. $R_2 = R_2 \cup \{d : d \in P \cap U\text{-}R_2\}$
10. **until:** $|R_2|=P_1*(|R_0|+|R_1|)$

11. **repeat:**
12. $R_3 = R_3 \cup \{d{:}d \in N \cap U{-}R_3\}$
13. **until:** $|R_3|{=}P_2{*}(|R_0|{+}|R_1|)$
14. **for** $i{=}1,2,3,4$ **do:**
15.   **for** d, $d \in R_i$ **do**
16.   rank(d)= 5-i
17.   ranks = ranks $\cup$ {rank(d)}
18.   **end for**
19. **end for**
20. *ranking_model* = TrainRanking(R_0, R_1, R_2, R_3, ranks)
21. **Return:** *ranking_model*

Finishing ranking learning, we can do ranking prediction on all candidate papers directly. However, we find that classifier which is trained to get more unbiased reference priority samples can recall most of cited papers. As shown in Fig. 3, FN papers are ranked behind FP. We then propose another model called CR+C which do classification on all data firstly and then do ranking prediction on those positive labeled papers. We describe CR+C prediction procedure in Algorithm 3.

Algorithm 3. CR+C Recommendation Prediction

Input: *Classifier, ranking_model*, test data T
Output: *Ranks*
1. $D{=}\{\}$, *Ranks*={}
2.   **for** d, $d \in T$ **do**
3.   d_label = predict(*classifier*, d)
4.   if (d_label==1):
5. $D = D \cup \{d\}$
6.   **end for**
7.   **for** d, $d \in D$ **do**
8.   rank(d) = predict(ranking_model, d)
9.   *Ranks* = *Ranks* $\cup$ {rank(d)}
10. **end for**
11. **Return:** *Ranks*

4   Experiments

In this section, we apply our citation recommendation methods CR and CR+C on a DBLP citation dataset[1] extracted by Tang *et al.* [18]. Our methods will be compared with the other two methods TopicSim and RankSVM which are used as baseline in prior researches.

---

[1] http://arnetminer.org/DBLP_Citation

4.1   Data Preparation

The DBLP citation dataset extracted by Tang *et al.* covers more than 1.5 million papers from major Computer Science publication venues and gathers more than 0.9 million researchers for more than 50 years. The meta data contains normalized authors names, venues, titles, published time and citation information. Some of them are lack of abstract and complete references (eg. some papers don't have any references, some papers published in 2007 has cited the papers published in 2009). So, instead of using the entire dataset, we generated a subset based on several self-defined rules as prior research does [11], [19]. We select papers published in 2007 which has abstract, more than 15 references and their references for training, then select papers published in 2008 for validation, during 2009 and 2011 for test. Papers in both validation and test dataset should also have more than 15 references. We need also filter out some references in validation and test datasets which are absent in training dataset. Table.1 shows some details about three data subsets.

Table 1. Subset of the DBLP dataset

Subsets	Train	Validation	Test
Years	T_0=[1978, 2007]	T_1=[2008]	T_2=[2009, 2011]
Papers	25, 011	2, 011	2, 869
Paper avg citation	20.25	11.59	10.98

Keywords extraction from titles and abstracts using TF-IDF measure and the TextBlob noun phrase extractor will do help for feature PTP extraction. Both considering efficiency and effectiveness, we select implementation[2] as an alternative.

Feature normalization are always done to get a more robust model in training, and experiments results on the validation dataset shows the preprocessing increased performance. Therefore, we put all features into log-space, then scaling these log-space feature to the interval [0,1] based on the maximum and minimum values of f_i observed in the training datasets:

$$f_i\_new=scale_i(log(1+f_i\_old)) \tag{7}$$

$$scale_i(f) = \frac{f-min_i}{max_i-min_i} \tag{8}$$

No matter the value in validation or test datasets is below 0 or above 1, we set it to the endpoint. This simple zero-one scaling often performs as well as more elaborate schemes [20].

---

[2] http://textblob.readthedocs.org/en/latest/

4.2 Experimental Settings

4.2.1 Compared Methods

- *TopicSim.* It measures similarity between papers with topic modeling technique (LDA) and returns candidates which are most similar to query. We utilize feature TopicSimilar(q, d) as the ranking score which was born to measure the similarity between papers' topic distribution.
- *RankSVM[12].* It samples all cited papers into C and randomly samples some uncited papers into A. RankSVM achieves best performance when $|A|=2*|C|$.
- *CR.* After ranking_model leanring, CR do recommendation ranking prediction directly on test dataset. After tuning on validation dataset, we find CR performs best when I=5, P_1=2, P_2=0.3.
- *CR+C.* Different with CR, CR+C do classification before ranking prediction. Algorithm 3 describe the way how CR+C do recommendation prediction. CR+C has the same parameters setting as CR.

4.2.2 Evaluation Metrics

We employed common metrics Precision@M (P@M) and Recall@M (R@M) which are used to evaluate in information retrieval. They are defined as the percentage of original cited papers that appear in top-M recommended list. Besides, it is desirable that cited papers should appear earlier in the recommended list. Therefore, Mean Reciprocal Rank (MRR) was also employed over the target papers, which is defined as $\text{MRR} = \frac{1}{|Q_T|}\sum_{q\in Q_T}\frac{1}{\text{rank}(q)}$, where Q_T is the test set. Meanwhile, for citation recommendation, time efficiency is another important aspect we need to pay attention to. We utilize Average Retrieval Time (ART) which is defined as $\text{ART} = \frac{\text{TimeConsumption}}{|Q_T|}$ to measure, where TimeConsumption is the total time prediction costs.

4.2.3 Performance Comparison

First, we compare the proposed method CR with RankSVM and TopicSimilar using Precision@10, 20, Recall@20, 50, MRR and ART. Table.2 summarizes the comparison results on DBLP dataset. Due to the lack of much structural information, we can find that TopicSim is obviously worse than the other models in terms of all citation recommendation metrics. TopicSim does prediction based on feature TopicSimilar which is already known before prediction, therefore, we ignore its performance on ART here.

Table 2. Citation recommendation results comparison on DBLP

Metric	P@10	P@20	R@20	R@50	MRR	ART/s
TopicSim	0.04726	0.03952	0.06875	0.12833	0.2420	
RankSVM	0.09079	0.06819	0.12778	0.20691	0.4602	4.786
CR	**0.09965**	**0.07302**	**0.13677**	**0.21783**	**0.4982**	**4.783**
CR+C	0.09965	0.07302	0.13677	0.21783	0.4979	0.875

Even CR and RankSVM utilize same features, CR can select items with more unbiased reference priority for ranking training. Consequently, CR outperforms RankSVM on all metrics. In particular, CR obtains 5.27% improvement in Recall@50, and 8.28% growth in MRR compared to RankSVM on the DBLP dataset. For more comprehensive comparisons, we compute the precision and recall at different positions (5 to 100) to study the trends in performance change. Fig. 4 shows the comparison results of 4 methods. We can find the two methods outperforms on all positions.

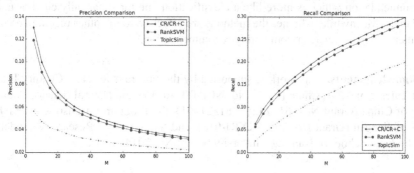

Fig. 4. Precision and recall performance comparison

CR+C and CR get almost the same performance on Precision and Recall. In prediction, CR+C filters out some cited papers during classification. Then, those filtered references cannot be recalled in following ranking prediction. Because of this, CR+C gets 0.4979 which is slightly worse than 0.4982 got by CR.

Table 3. Classification performance of CR+C

Metrics	Recall	Precision
Result	0.85014	0.00246

Even CR+C suffer somewhat loss in classification, but it outperforms CR in term of ART. CR+C costs only 18.29% as much as CR does on average. Table.3 shows that CR+C recalls 85% cited papers during classification with 0.25% precision, that is to say, CR+C needs to do ranking prediction on only 16.29% candidates rather than all like CR does. Classification costs much less time than ranking prediction needs, thus, doing classification before ranking prediction do save a lot of time.

5 Conclusion and Future Work

In this paper, we propose an easy but efficient method CR based-on RankSVM which are always used to solve citation recommendation problem. CR achieves 5.27% improvement in Recall@50 and 8.28% growth in MRR over the RankSVM. CR+C, the variation of CR, performs as well as CR does in terms of precision and recall. Meanwhile, CR+C costs only 18.29% as much as CR does. The experiments results proves that the assumption, the more preferential a reference is, the easier it can be recognized as a citation, is rational. The methods CR and CR+C we propose are feasible.

Besides, we find some interesting directions for further studying. Reference priority recognition has been proved do affect the citation recommendation. We set priority within four different types of data in a pre-defined way. If there are any other universal way that can set more precise priority within candidates without causing extra bias. RankSVM is born to solve ranking problem, how to modify it to fit the citation recommendation which is more like a classification one for it has only cited or uncited labels. Meanwhile, whether the methods are effective to the other recommendation problem, such as music or commodity recommendation.

Acknowledgments. This work is supported by the Program for New Century Excellent Talents of MOE China (Grant No. NCET-11-0213), the Natural Science Foundation of China (Grant Nos. 61273257, 61321491), the Program for Distinguished Talents of Jiangsu (Grant No. 2013-XXRJ-018), and the Scientific Research Foundation of Graduate School of Nanjing University (Grant No. 2014CL03).

References

1. He, Q., Pei, J., Kifer, D., Mitra, P., Giles, L.: Context-aware citation recommendation. In: Proceedings of the 19th International Conference on World Wide Web, pp. 421–430. ACM, April 2010
2. Huang, W., Kataria, S., Caragea, C., Mitra, P., Giles, C.L., Rokach, L.: Recommending citations: translating papers into references. In: Proceedings of the 21st ACM International Conference on Information and Knowledge Management, pp. 1910–1914. ACM, October 2012
3. Nallapati, R.M., Ahmed, A., Xing, E.P., Cohen, W.W.: Joint latent topic models for text and citations. In: Proceedings of the 14th ACM SIGKDD International Conference on Knowledge Discovery and Data Mining, pp. 542–550. ACM, August 2008
4. Tang, J., Zhang, J.: A discriminative approach to topic-based citation recommendation. In: Theeramunkong, T., Kijsirikul, B., Cercone, N., Ho, T.-B. (eds.) PAKDD 2009. LNCS, vol. 5476, pp. 572–579. Springer, Heidelberg (2009)
5. Liben-Nowell, D., Kleinberg, J.: The link-prediction problem for social networks. Journal of the American Society for Information Science and Technology 58(7), 1019–1031 (2007)
6. Strohman, T., Croft, W.B., Jensen, D.: Recommending citations for academic papers. In: Proceedings of the 30th Annual International ACM SIGIR Conference on Research and Development in Information Retrieval, pp. 705–706. ACM, July 2007
7. Sun, Y., Han, J., Yan, X., Yu, P.S., Wu, T.: Pathsim: Meta path-based top-k similarity search in heterogeneous information networks. In: VLDB 2011 (2011)
8. Bethard, S., Jurafsky, D.: Who should I cite: learning literature search models from citation behavior. In: Proceedings of the 19th ACM International Conference on Information and Knowledge Management, pp. 609–618. ACM, October 2010
9. Yu, X., Gu, Q., Zhou, M., Han, J.: Citation prediction in heterogeneous bibliographic networks. In: SDM, vol. 12, pp. 1119–1130 (2012)

10. Sun, Y., Yu, Y., Han, J.: Ranking-based clustering of heterogeneous information networks with star network schema. In: Proceedings of the 15th ACM SIGKDD International Conference on Knowledge Discovery and Data Mining, pp. 797–806. ACM, June 2009

11. Ren, X., Liu, J., Yu, X., Khandelwal, U., Gu, Q., Wang, L., Han, J.: Cluscite: Effective citation recommendation by information network-based clustering. In: Proceedings of the 20th ACM SIGKDD International Conference on Knowledge Discovery and Data Mining, pp. 821–830. ACM, August 2014

12. Joachims, T.: Optimizing search engines using clickthrough data. In: Proceedings of the Eighth ACM SIGKDD International Conference on Knowledge Discovery and Data Mining, pp. 133–142. ACM, July 2002

13. Liu, L., Tang, J., Han, J., Jiang, M., Yang, S.: Mining topic-level influence in heterogeneous networks. In: Proceedings of the 19th ACM International Conference on Information and Knowledge Management, pp. 199–208. ACM, October 2010

14. Blei, D.M., Ng, A.Y., Jordan, M.I.: Latent dirichlet allocation. The Journal of Machine Learning Research 3, 993–1022 (2003)

15. Zhou, D., Ji, X., Zha, H., Giles, C.L.: Topic evolution and social interactions: how authors effect research. In: Proceedings of the 15th ACM International Conference on Information and Knowledge Management, pp. 248–257. ACM, November 2006

16. Hirsch, J.E.: An index to quantify an individual's scientific research output. Proceedings of the National academy of Sciences of the United States of America 102(46), 16569–16572 (2005)

17. Fan, R.E., Chang, K.W., Hsieh, C.J., Wang, X.R., Lin, C.J.: LIBLINEAR: A library for large linear classification. The Journal of Machine Learning Research 9, 1871–1874 (2008)

18. Tang, J., Zhang, J., Yao, L., Li, J., Zhang, L., Su, Z.: Arnetminer: extraction and mining of academic social networks. In: Proceedings of the 14th ACM SIGKDD International Conference on Knowledge Discovery and Data Mining, pp. 990–998. ACM, August 2008

19. Yan, R., Tang, J., Liu, X., Shan, D., Li, X.: Citation count prediction: learning to estimate future citations for literature. In: Proceedings of the 20th ACM International Conference on Information and Knowledge Management, pp. 1247–1252. ACM, October 2011

20. Wu, S., Crestani, F., Bi, Y.: Evaluating score normalization methods in data fusion. In: Ng, H.T., Leong, M.-K., Kan, M.-Y., Ji, D. (eds.) AIRS 2006. LNCS, vol. 4182, pp. 642–648. Springer, Heidelberg (2006)

UserGreedy: Exploiting the Activation Set to Solve Influence Maximization Problem[*]

Wenxin Liang[**], Chengguang Shen, and Xianchao Zhang

School of Software, Dalian University of Technology, Dalian 116620, China
{wxliang,xczhang}@dlut.edu.cn, shenchengguang@gmail.com

Abstract. Influence Maximization is the problem of selecting a small set of seed users in a social network to maximize the spread of influence. Traditional solutions are mainly divided into two directions. The one is greedy-based methods and the other is heuristics-based methods. The greedy-based methods can effectively estimate influence spread using thousands of Monte-Carlo simulations. However, the computational cost of simulation is extremely expensive so that they are not scalable to large networks. The heuristics-based methods, estimating influence spread according to heuristic strategies, have low computational cost but without theoretical guarantees. In order to improve both performance and effectiveness, in this paper we propose a greedy-based algorithm, named User-Greedy. In UserGreedy, we first propose a novel concept called *Activation Set*, which is defined as a set of users that can be activated by a seed user with a certain probability under the most standard and popular independent cascade (IC) model. Based on the computation of such probabilities, we can directly estimate the influence spread without the expensive simulation process. We then design an influence spread function based on the the *Activation Set* and mathematically prove that it has the property of monotonicity and submodularity, which provides theoretical guarantee for the UserGreedy algorithm. Besides, we also propose an efficient method to obtain the *Activation Set* and hence implement the User-Greedy algorithm. Experiments on real-world social networks demonstrate that our algorithm is much faster than existing greedy-based algorithms and outperforms the state-of-art heuristics-based algorithms in terms of effectiveness.

1 Introduction

Influence maximization problem is first proposed by Dominigos and Richardson in a probabilistic perspective [6,14]. Kempe et al. [10] formulate the problem as a discrete optimization problem and propose two fundamental diffusion models, Independent Cascade (IC) model and Linear Threshold (LT) model. Given a social network, a diffusion model and a parameter k, the influence maximization problem is formally represented as choosing a set of k seed users, under the chosen diffusion model, to maximize the influence spread. Kempe et al. [10] have proven that the influence maximization problem is NP-hard under both IC and LT models. Although the problem is NP-hard, many

---

[*] This work was partially supported by NSFC (No.61272374, 61300190), SRFDP (No.20120041110046) and Key Project of Chinese Ministry of Education (No.313011).
[**] Corresponding author.

© Springer International Publishing Switzerland 2015
R. Cheng et al. (Eds.): APWeb 2015, LNCS 9313, pp. 548–559, 2015.
DOI: 10.1007/978-3-319-25255-1_45

research efforts have been made to tackle it. Traditional solutions can be mainly divided into two directions. The one is greedy-based methods and the other is heuristics-based methods.

Guaranteed by the submodular property of influence spread function, the greedy-based methods can provide a $(1 - 1/e - \varepsilon)$ approximation to the optimal solution. The basic idea of greedy-based methods is repeatedly selecting the users with largest marginal influence spread and adding them into seed users until the budget k is reached. However, computing exact influence spread under both IC and LT models is #P-hard [3]. Hence, greedy-based algorithms must run thousands of Monte-Carlo (MC) simulations for every user in each iteration to effectively estimate the influence spread. This time-consuming simulation process causes a severe performance drawback in real applications. Heuristics-based methods, in contrast, can achieve much lower computational cost by using heuristic strategies. However, the performance of heuristics-based methods is unstable because of no theoretical guarantees.

To address these issues, in this paper we propose a greedy-based algorithm, named UserGreedy. In UserGreedy, we propose a novel concept, *Activation Set*, which is defined as a set of users that can be activated by a seed user with a certain probability based on the most standard and popular independent cascade (IC) model. Using the concept of *Activation Set*, we can directly estimate the influence spread through probability calculation instead of the expensive MC simulation process. Thus, the computational cost of UserGreedy is much lower than that of traditional greedy-based algorithms. Besides, we mathematically prove that the influence spread function based on *Activation Set* has the property of monotonicity and submodularity, which provides theoretical guarantee for the UserGreedy algorithm, and hence overcomes the drawback of heuristics-based algorithms. The main contributions of this paper are summarized as follows.

- We first propose an improved greedy-based algorithm named UserGreedy for influence maximization problem in online social networks. In UserGreedy, we first present a novel concept, *Activation Set*, which is defined as a set of users that can be activated by a seed user with a certain probability under IC model. Based on the calculation of such probabilities, we can directly estimate the influence spread without the expensive MC simulation process, and therefore significantly speed up the estimation of influence spread.
- We then design an influence spread function based on the the *Activation Set* and mathematically prove that this function is still monotone and submodular. The property of monotonicity and submodularity provides theoretical guarantee for the UserGreedy algorithm, and thus effectively improve the accuracy of the estimation of influence spread.
- Besides, we propose an influence-path based method to efficiently obtain the *Activation Set* and finally implement the UserGreedy algorithm.
- Extensive experiments on real-world social networks demonstrate that UserGreedy is much faster than existing greedy algorithms and outperforms the state-of-art heuristics-based algorithms in terms of effectiveness.

The remainder of this paper is organized as follows. Section 2 presents the related work. In Section 3, we review the influence maximization problem. Section 4 depicts our UserGreedy algorithm. In Section 5 we describe our experiments and analyze the experimental results. Finally, Section 6 concludes the paper.

2 Related Work

Domingos and Richardson [6,14] firstly formulate the influence maximization problem as a ranking problem using a Markov random field model. Kempe et al. [10] formally propose two diffusion models and develop the greedy algorithm to solve the problem with approximation guarantees. However, the main drawback of their work is the scalability of their greedy algorithms. Several studies have been put to solve this issue.

Leskovec et al. [11] propose a "Lazy-Forward" optimization, called CELF. By exploiting the submodular property of influence function, CELF greatly reduces the number of evaluations on the influence spread. Chen et al. [4] generate a new smaller graph for every Monte-Carlo simulation and use a linear scan of the new graph by BFS or DFS to speed up the greedy algorithm. Wang et al. [16] exploit the community structure property of social networks and propose a community-based greedy algorithm (CGA). In [5], Cheng et al. propose a static greedy algorithm, which reuses the generated subgraphs to guarantee the submodular property of influence spread function.

Greedy-based algorithms are not scalable to large networks as it involves too many Monte-Carlo simulations. Hence, several heuristics-based algorithms are proposed under the IC model to avoid using Monte-Carlo simulations. Chen et al. [4] present a degree discount heuristic algorithm called DegreeDiscountIC. DegreeDiscountIC adds the node with the largest degree into the seed set by discounting the degrees of its neighbors in seed set. Chen et al. [3] restrict computations in the local influence regions of nodes to approximate the influence propagation and propose the PMIA algorithm. IRIE algorithm proposed in [9] integrates influence ranking with influence estimation to overcome the disadvantages of pure influence ranking methods. Heuristics-based algorithms can improve the scalability of greedy-based algorithms but without theoretical guarantees, and thus the solution quality is unstable.

Recently, several extensions of influence maximization problem have been proposed. [1,8] extend the problem into competitive settings, [2,15] study the time constrained influence maximization problem, [12] combines conformity and [7] combines novelty decay into their models.

3 Preliminaries

Generally, a social network can be seen as a social graph. A social graph $G = (V, E)$ consists of a set of vertices V representing users and a set of edges E representing relationships between users. In the Independent Cascade (IC) model, each edge $e_{u,v}$ of a social graph G is associated with a probability of $p_{u,v}$, following the probability function $p : E \leftarrow [0, 1]$. The state of each node is either active or inactive. When an initial set S of active nodes is given, the diffusion process can be triggered as follows. At each step t, each node u activated at step $t - 1$ has a single chance to activate its inactive neighbor v with probability $p_{u,v}$. If u succeeds, then v will become active at step $t + 1$. Whether or not u succeeds, it cannot make any further attempts to activate

v in subsequent rounds. The process terminates when no more activations are possible. Given S as the initial seed set, the influence spread of S, denoted as $\sigma(S)$, is the expected number of active nodes at the end of diffusion process.

Problem 1 (Influence Maximization). Given an input parameter k and a social graph $G(V, E)$, the influence maximization problem is to select a seed user set $S \subseteq V$, $(|S| = k)$ by activating them to maximize the influence spread of S, $\sigma(S)$.

It is proved that the influence maximization problem is NP-hard and the influence spread function $\sigma(\cdot)$ is monotone and submodular [10]. A set function f from sets to reals $f : 2^V \rightarrow R$ is monotone, if $f(S) \leq f(T)$ for all $S \subseteq T$. The function f is submodular, if $f(S \cup \{w\}) - f(S) \geq f(T \cup \{w\}) - f(T)$ for all $S \subseteq T$, $w \in V$ and $w \notin T$. Based on Theorem 1, a greedy algorithm can be used to solve the influence maximization problem.

Theorem 1. *In [13], if the set function f is non-negative, monotone and submodular with $f(\emptyset) = 0$, then the greedy algorithm achieves at least $(1 - 1/e)$ approximation ratio. That is, the solution found by the greedy strategy satisfies $f(S) \geq (1 - 1/e) \cdot f(S^*)$, where S^* is the optimum solution.*

4 UserGreedy

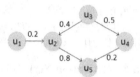

Fig. 1. A social network with the same influence probability 1.

Fig. 2. A social network with different influence probabilities.

4.1 Activation Set

To describe the *Activation Set*, let us first consider an extreme situation as shown in Fig. 1, where the influence probability is 1 for all edges. When u_3 is the only user in active status, then u_2, u_4 and u_5 will definitely activated by u_3 as the influence probability is 1. The user set activated by each user in Fig. 1 is shown in Table 1(a). According to Table 1(a), u_3 has the largest influence spread so we choose it as the first seed user. In order to calculate the marginal influence spread, the only necessary operation is to get rid of the users activated by the previously selected seed users. Thus, u_1 will be chosen as the second seed user. Figure 2 shows an example social network in IC model, in which the influence probability of each edge is in the range of $[0, 1]$. In this situation, one user can spread influence to another user via the *Influence Path* which is defined as follows.

Table 1. Activation Set of Users

	(a)			(b)
User	ActivationSet		User	ActivationSet
u_1	$\{u_1, u_2, u_5\}$		u_1	$\{(u_1, 1), (u_2, 0.2), (u_5, 0.16)\}$
u_2	$\{u_2, u_5\}$		u_2	$\{(u_2, 1), (u_5, 0.8)\}$
u_3	$\{u_3, u_2, u_4, u_5\}$		u_3	$\{(u_3, 1), (u_2, 0.4), (u_4, 0.5), (u_5, 0.388)\}$
u_4	$\{u_4, u_5\}$		u_4	$\{(u_4, 1), (u_5, 0.2)\}$
u_5	$\{u_5\}$		u_5	$\{(u_5, 1)\}$

Definition 1 (Influence Path). *Given a node $u \in V$, a simple path $\mathcal{P} =< u_1, u_2,$ ..., $u_m >$ is influence path, where $u_1 = u$ and $u_i \neq u$ for $i \neq 1$. The influence probability of path $\mathcal{P}$, $Pr(\mathcal{P}) = \prod_{i=1}^{m-1} p_{u_i, u_{i+1}}$.*

For example, in Fig. 2 the influence paths starting from u_3 are $\mathcal{P}_1 =< u_3, u_2 >$, $\mathcal{P}_2 =< u_3, u_4 >$, $\mathcal{P}_3 =< u_3, u_2, u_5 >$ and $\mathcal{P}_4 =< u_3, u_4, u_5 >$. We denote the $\mathcal{IPS}_{u,v}$ as the set of all influence paths from u to v. An influence path $\mathcal{P}$ from u to v gives a possible way for u to activate v with $Pr(\mathcal{P})$, then the probability that u activates v, $\mathcal{AP}_{u,v}$ can be calculated as Equation (1).

$$\mathcal{AP}_{u,v} = 1 - \prod_{\forall \mathcal{P} \in \mathcal{IPS}_{u,v}} (1 - Pr(\mathcal{P})) \tag{1}$$

For example, in Fig. 2, $\mathcal{IPS}_{u_3,u_2} = \{\mathcal{P}_1\}$, then $\mathcal{AP}_{u_3,u_2} = 1 - (1 - Pr(\mathcal{P}_1)) = 0.4$; $\mathcal{IPS}_{u_3,u_5} = \{\mathcal{P}_3, \mathcal{P}_4\}$, then $\mathcal{AP}_{u_3,u_5} = 1 - (1 - Pr(\mathcal{P}_3))(1 - Pr(\mathcal{P}_4)) = 1 - (1 - 0.4 \times 0.8) \times (1 - 0.5 \times 0.2) = 0.388$.

Definition 2 (Activation Set). *Given a seed user $s \in V$ that can be activated by itself with probability 1, the Activation Set of s, $AS(s)$ is defined as a set of the users that can be activated by s through the influence paths starting from s, i.e., $AS(s) = \{(v, \mathcal{AP}_{s,v}) | v \in V, \mathcal{AP}_{s,v} \in [0,1]\}$. Note that $AS(s)$ has at least one element, i.e., $(s, 1)$.*

The *Activation Set* of users in Fig. 2 are shown in Table 1(b). Based on the *Activation Set*, we can calculate the influence spread of seed users S. Firstly, we assume that each seed user independently influences v, then the user v will be influenced by the seed users S with probability $\mathbf{p}_{S,v}$ which can be calculated by Equation (2).

$$\mathbf{p}_{S,v} = 1 - \prod_{\forall u \in S} (1 - \mathcal{AP}_{u,v}) \tag{2}$$

Therefore, the influence spread of seed users S is the expected value of $\mathbf{p}_{S,v}$,

$$\sigma(S) = \sum_{\forall v \in V} 1 \cdot \mathbf{p}_{S,v}. \tag{3}$$

According to Table 1(a), the influence spread of u_3 is $\sigma(\{u_3\}) = 1.0 + 0.4 + 0.5 + 0.388 = 2.288$ and u_3 is chosen as the first seed user with the largest influence spread, $S = \{u_3\}$.

In the process of choosing the second seed user, we compute the marginal influence spread of users with Equation (3). For example, the marginal influence spread of adding u_1 into S is $\sigma(S \cup \{u_1\}) - \sigma(S) = \mathbf{p}_{S \cup \{u_1\}, u_1} + \mathbf{p}_{S \cup \{u_1\}, u_2} + \mathbf{p}_{S \cup \{u_1\}, u_3} + \mathbf{p}_{S \cup \{u_1\}, u_4} + \mathbf{p}_{S \cup \{u_1\}, u_5} - \sigma(S) = 1 + (1 - (1 - 0.2) \times (1 - 0.4)) + 1 + 0.5 + (1 - (1 - 0.16) \times (1 - 0.388)) - 2.288 = 1.218$. In the computation of $\sigma(S \cup \{u_2\})$, we notice that the influence path $\mathcal{P}_3 = < u_3, u_2, u_5 >$ contains u_2, who is already in active status and cannot be activated again. We remove the $\mathcal{P}_3$ in the calculation of $\mathbf{p}_{S \cup \{u_2\}, u_5} = 1 - (1 - 0.8) \times (1 - 0.5 \times 0.2) = 0.82$, thus $\sigma(S \cup \{u_2\}) - \sigma(S) = 1.032$, $\sigma(S \cup \{u_4\}) - \sigma(S) = 0.568$ and $\sigma(S \cup \{u_5\}) - \sigma(S) = 0.612$. Therefor, u_1 is chosen as the second seed user because of the largest marginal influence spread.

4.2 Monotonicity and Submodularity

Theorem 2. *The influence spread function $\sigma(\cdot)$ based on Activation Set is monotone and submodular.*

Proof. Let us firstly consider the set function $g(\cdot)$, which is

$$g(S) = 1 - \prod_{\forall u \in S} (1 - p_u). \tag{4}$$

Then, we can prove that $g(\cdot)$ is monotone for $S \subseteq T$ as follow.

$$\begin{aligned} g(S) - g(T) &= (1 - \prod_{\forall u \in S} (1 - p_u)) - (1 - \prod_{\forall u' \in T} (1 - p_{u'})) \\ &= \prod_{\forall u' \in T} (1 - p_{u'}) - \prod_{\forall u \in S} (1 - p_u). \end{aligned} \tag{5}$$

As $S \subseteq T$ and $p \in [0, 1]$, then $g(S) \leq g(T)$ and $g(\cdot)$ is monotone.

If $g(\cdot)$ is submodular, it should satisfy the inequation $g(S \cup \{w\}) - g(S) \geq g(T \cup \{w\}) - g(T)$, where $S \subseteq T$ and $w \notin T$. Let us firstly consider the left part of the inequality,

$$\begin{aligned} g(S \cup \{w\}) - g(S) &= (1 - \prod_{\forall u \in S} (1 - p_u) \cdot (1 - p_w)) - (1 - \prod_{\forall u \in S} (1 - p_u)) \\ &= \prod_{\forall u \in S} (1 - p_u) \cdot p_w. \end{aligned} \tag{6}$$

As $p \in [0, 1]$ and $S \subseteq T$, then

$$g(S \cup \{w\}) - p(S) \geq g(T \cup \{w\}) - p(T), \tag{7}$$

that is, $g(\cdot)$ is submodular.

As pointed out in [10], a non-negative linear combination of monotone and submodular functions is also monotone and submodular. From Equation (3), we can see that $\sigma(\cdot)$ is just the linear combination of $g(\cdot)$. Therefore, influence spread function $\sigma(\cdot)$ is monotone and submodular, which concludes the proof.   □

Algorithm 1. UserGreedy

Input: G, k
Output: seed set S
1: $S \leftarrow \emptyset, H \leftarrow \emptyset, E \leftarrow \emptyset, spread \leftarrow 0$
2: **for each** $v \in V$ **do**
3:    $AS(v) \leftarrow$ Activation Set $v.mg = \sigma(v) = \sum_i p_i$ H.insert(v) E.add(u)
4: **end for**
5: **while** $|S| \leq k$ **do**
6:    $u \leftarrow Max(H)$
7:    **if** $u \in E$ **then**
8:       $S \leftarrow S \cup \{u\}$, $E \leftarrow \emptyset, spread \leftarrow spread + u.mg$
9:    **else**
10:       $u.mg \leftarrow MG(S, AS(u), spread)$
11:       H.insert(u), E.add(u)
12:    **end if**
13: **end while**
14: return S

4.3   UserGreedy Algorithm

Based on Theorem 1 and Theorem 2, a greedy-based algorithm named UserGreedy can be proposed to solve the influence maximization problem.

Algorithm 1 shows the details of our UserGreedy algorithm. The first phase of User-Greedy is obtaining the *Activation Set* with Algorithm 2 and using Equation (3) to directly calculate the influence spread of each user. In the second phase of UserGreedy, we adopt the CELF optimization [11] to reduce the number of evaluations on the influence spread of users. Max-heap H is used to keep an ordered list of marginal gain of users from previous iteration. In lines 5-13, we iteratively choose the seed user providing the maximal marginal gain. In each iteration, a set E is used to record the examined users. We get the user u who has maximal marginal gain from the max-heap H. If u has been examined in this iteration, then it should be chosen as the next seed user. If not, subroutine MG (Algorithm 3) is called to compute the marginal gain of u with respect to S.

Algorithm 2 shows the details of obtaining *Activation Set* by enumerating influence paths. We use a tuple (v, P) to record the information of visited paths: v means the node in the path, P is the influence path starting from u. As the influence probability of each edge is $p \in [0, 1]$, the influence probability of an influence path will diminish rapidly when the length of the path becomes longer. These long paths with extremely small probabilities have few contributions to the influence spread but very time-consuming. Thus, we can choose a threshold θ to prune the paths whose $\Pr(\mathcal{P})$ are smaller than θ (Lines 5-7). $IPS_{u,v}$ records the paths from u to v (Line 4) and then we calculate the probability that u activates v by Equation (1) (Lines 9-12). We will give a heuristic value for θ and show the effect of choosing different θ in Section 5.2.

Algorithm 3 calculates the marginal gain of adding a node u into seed set S. For each node v in the *Activation Set* of seed set, as we described in Section 4.1, we recompute the activation probability from $S \cup \{u\}$ to v with Equation (2) by removing the influence

---

Algorithm 2. Obtaining Activation Set

Input: u, θ
Output: Activation Set of u
1: Initialize a queue Q and enqueue $(v = u, P = \{u\})$ into Q
2: **while** $Q \neq \emptyset$ **do**
3:     Dequeue (v, P) from Q
4:     Insert P into $IPS_{u,v}$
5:     **for each** $t \in N^{out}(v), t \notin P$ and $Pr(P) \cdot p_{v,t} \geq \theta$ **do**
6:         Enqueue $(t, P \cup \{t\})$ into Q
7:     **end for**
8: **end while**
9: **for** v in $IPS_{u,v}$ **do**
10:     $AP_{u,v} \leftarrow$ calculates by Equation (1)
11:     Insert $(u, AP_{u,v})$ into $AS(u)$
12: **end for**
13: **return** $AS(u)$

Algorithm 3. Computation of Marginal Gain

Input: $S, AS(u), spread$
Output: The marginal gain of adding u into seed set S
1: $cur \leftarrow 0$
2: **for each** node v in $AS(S \cup \{u\})$ **do**
3:     $temp \leftarrow 1$
4:     **for each** node t in $S \cup \{u\}$ **do**
5:         recompute $AP_{t,v}$ by removing the paths in $IPS_{t,v}$ with the other nodes in $S \cup \{u\}$
6:         $temp \leftarrow temp \cdot (1 - AP_{t,v})$
7:     **end for**
8:     $cur \leftarrow cur + (1 - temp)$
9: **end for**
10: **return** $mg \leftarrow cur - spread$

---

paths containing nodes in $S \cup \{u\}$. And these validate the assumption that each seed user influences v is independent in Section 4.1.

Let n (resp. m) be the number of nodes (resp. edges) in G. The time complexity of UserGreedy includes two parts. The first part is the time of obtaining *Activation Set* and the second is the time of using greedy optimization. The time complexity of the first part is O($n\bar{n}$), where $\bar{n}$ is the average number of nodes in the local region within threshold θ. The second part is $O(k^2 n |A|)$, where $|A|$ is the average size of *Activation Set*. Thus, the total time complexity of the UserGreedy algorithm is O($n\bar{n} + k^2 n |A|$).

5 Experiment

In this section, we conduct experiments on several real-world social networks to compare our algorithms with other existing algorithms. All algorithms are implemented in C++ and conducted on a server with Inter Xeon E5-2620 (2.10 GHz) and 32 GB memory.

5.1 Experiment Setup

Datasets. We use four real-world social networks, **NetPhy**[4,5][1], **Douban**[2], **Amazon**[3][3], **DBLP** [3,5,9][1], to demonstrate the performance of our algorithms comparing with other existing algorithms. The number of nodes, edges and the type of direction of the networks are summarized in Table 2.

Algorithms. We compare our algorithms with the other greedy-based algorithms and heuristics-based algorithms. The following is a list of baseline algorithms evaluated in our experiments. Since PMIA and IRIE are the two state-of-the-art heuristic algorithms, we do not implement others, such as DegreeDiscountIC and Degree Centrality.

- **CELFGreedy[CELF]**. The original greedy algorithm with the CELF optimization. Following the literature, we set $R = 10,000$, which means that in each seed selection process, 10,000 Monte-Carlo simulations are conducted to obtain the accurate estimation.
- **MixGreedy[MG]**. The mixed greedy algorithm proposed in [4] for the IC model. In MixGreedy, the first round of the seed selection uses NewGreedy and the rest rounds uses CELF optimization.
- **StaticGreedy[SG]**. A greedy algorithm reuses the generated subgraphs to reduce the number of simulations. The R, smaller than usual in other Greedy-based algorithms, is set to 500.
- **UserGreedy[UG]**. The algorithm we proposed in this paper. The parameter θ is set to 0.0005.
- **PMIA**. A heuristic algorithm uses the maximum influence paths for influence spread estimation. We use the recommended parameter setting $\theta = 1/320$.
- **IRIE**. A heuristic algorithm integrates the advantages of influence ranking and influence estimation methods. The parameters α and θ are set to 0.7 and 1/320 as suggested.
- **Degree**. A heuristic method selects k nodes according to the descending order of weighted degree.

Influence Model. In this paper, we focus on the influence propagation in the IC model and all algorithms are implemented under the IC model. We adopt the weighted cascade model for assigning propagation probabilities among edges [3,5,10]. Each edge(u,v) in weighted cascade model is assigned a propagation probability $p(u, v) = 1/d_v$, where d_v is the indegree of node v.

5.2 Experimental Results

The evaluation metrics include influence spread and running time. In order to get the ground truth of influence spread of seed users obtained by each algorithm, we conduct

---

[1] http://research.microsoft.com/en-us/
  people/weic/publications.aspx
[2] http://socialcomputing.asu.edu/pages/datasets
[3] http://snap.stanford.edu/data/

Table 2. Statistics of datasets

Datasets	#Nodes	#Edges	Direction
NetPhy	37K	231K	undirected
Douban	154K	654K	directed
Amazon	334K	925K	directed
DBLP	654K	2.0M	undirected

influence estimation using 10000 times of MC simulation with respect to the size k, from 1 to 50. For the comparison of running time, we focus on the seed size $k = 50$.

Influence Spread. Figure 3 shows the influence spread obtained by algorithms on the four datasets. As CELF algorithm is very time-consuming, we only do CELF on the small scale networks, **NetPhy** and **Douban**. MixGreedy and StaticGreedy almost perform the same results as CELF, so we only show the result of CELF.

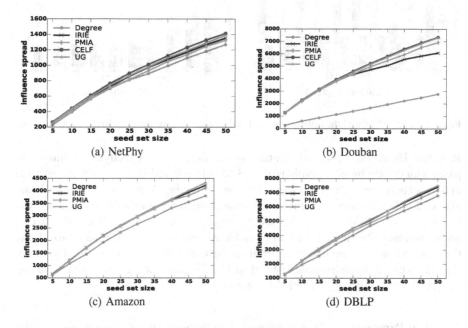

(a) NetPhy

(b) Douban

(c) Amazon

(d) DBLP

Fig. 3. Influence spread achieved by various algorithms on four datasets.

Among all these tests, our UserGreedy algorithm has stable performance and performs consistently better than IRIE, PMIA and Degree. IRIE is not stable as it has poor performance on **Douban** but performs better than PMIA on other datasets. The results show that the performance of heuristic algorithms cannot be guaranteed. PMIA is more stable than IRIE but has the worse performance than UserGreedy algorithms in all datasets. The results of Degree show that simply choosing high degree nodes is not effective to solve the influence maximization problem.

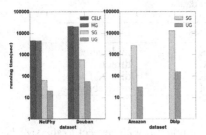
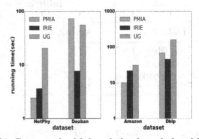

(a) Compared with greedy-based algorithms   (b) Compared with heuristics-based algorithms

Fig. 4. Running time of different algorithms on the test datasets.

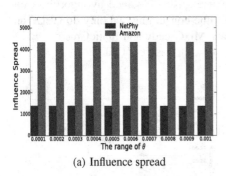
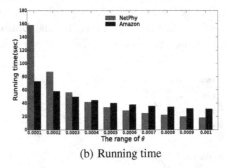

(a) Influence spread   (b) Running time

Fig. 5. Parameter θ affects the influence spread and running time on datasets NetPhy and Amazon.

Running Time. Figure 4(a) shows the running time of UserGreedy algorithms compared with greedy-based algorithms, CELF, MixGreedy and StaticGreedy. As reported in other experiments, CELF is extremely time-consuming. MixGreedy performs better than CELF, but the improvement is very small, only 0.8%. StaticGreedy is faster than CELF Greedy and MixGreedy but much slower than UserGreedy. When the size of graph increases, the running time of StaticGreedy grows very fast. Figure 4(b) shows the running time of UserGreedy algorithm compared with heuristics-based algorithms. Heuristics-based algorithms are faster than UserGreedy, but we can see that the gap between UserGreedy and IRIE,PMIA is not large.

Tuning of Parameter θ. In Algorithm 2, we propose a pruning parameter θ to control the length of influence path. Here, we study how different θ can affect the influence spread and running time. We run UserGreedy with different value of θ (0.0001 $\rightarrow$ 0.001) on datasets **NetPhy** and **Amazon**. Figure 5(a) and 5(b) show the results of influence spread and running time, respectively. In fact, different θ on both datasets **NetPhy** and **Amazon** has little effect on influence spread, but smaller θ can speed up the running time. It means that we can choose an appropriate θ to balance the influence spread and running time. We consider the θ is related with $1/R$ (R is the number of Monte-Corlo simulations) as the relationship between frequency and probability. So we suggest $\theta = [1 \sim 10] \times 1/R$.

6   Conclusion

In this paper, we proposed an improved greedy-based algorithm, UserGreedy, for solving the influence maximization problem under the independent cascade model. UserGreedy exploits the *Activation Set* to estimate influence spread instead of time-consuming Monte-Carlo simulations. We have mathematically proven that influence spread function based on *Activation Set* has the property of monotonicity and submodularity, which provides theoretical guarantee for the UserGreedy algorithm. Besides, we proposed an influence-path based method to efficiently obtain the *Activation Set*. The extensive experiments on real-world social networks demonstrate that the User-Greedy algorithm achieves better performance than the state-of-the-art heuristics-based algorithms and runs much faster than other greedy-based algorithms.

References

1. Bharathi, S., Kempe, D., Salek, M.: Competitive influence maximization in social networks. In: Deng, X., Graham, F.C. (eds.) WINE 2007. LNCS, vol. 4858, pp. 306–311. Springer, Heidelberg (2007)
2. Chen, W., Lu, W., Zhang, N.: Time-critical influence maximization in social networks with time-delayed diffusion process. In: AAAI, pp. 592–598 (2012)
3. Chen, W., Wang, C., Wang, Y.: Scalable influence maximization for prevalent viral marketing in large-scale social networks. In: KDD, pp. 1029–1038 (2010)
4. Chen, W., Wang, Y., Yang, S.: Efficient influence maximization in social networks. In: KDD, pp. 199–208 (2009)
5. Cheng, S., Shen, H., Huang, J., Zhang, G., Cheng, X.: Staticgreedy: solving the scalability-accuracy dilemma in influence maximization. In: CIKM, pp. 509–518 (2013)
6. Domingos, P., Richardson, M.: Mining the network value of customers. In: KDD, pp. 57–66 (2001)
7. Feng, S., Chen, X., Cong, G., Zeng, Y., Chee, Y.M., Xiang, Y.: Influence maximization with novelty decay in social networks. In: AAAI, pp. 37–43 (2014)
8. He, X., Song, G., Chen, W., Jiang, Q.: Influence blocking maximization in social networks under the competitive linear threshold model. In: SDM, pp. 463–474 (2012)
9. Jung, K., Heo, W., Chen, W.: Irie: Scalable and robust influence maximization in social networks. In: ICDM, pp. 918–923 (2012)
10. Kempe, D., Kleinberg, J., Tardos, É.: Maximizing the spread of influence through a social network. In: KDD, pp. 137–146 (2003)
11. Leskovec, J., Krause, A., Guestrin, C., Faloutsos, C., VanBriesen, J., Glance, N.: Cost-effective outbreak detection in networks. In: KDD, pp. 420–429 (2007)
12. Li, H., Bhowmick, S.S., Sun, A.: Cinema: conformity-aware greedy algorithm for influence maximization in online social networks. In: ICDT, pp. 323–334 (2013)
13. Nemhauser, G.L., Wolsey, L.A., Fisher, M.L.: An analysis of approximations for maximizing submodular set functions. Mathematical Programming 14(1), 265–294 (1978)
14. Richardson, M., Domingos, P.: Mining knowledge-sharing sites for viral marketing. In: KDD, pp. 61–70 (2002)
15. Rodriguez, M.G., Schölkopf, B.: Influence maximization in continuous time diffusion networks. In: ICML, pp. 313–320 (2012)
16. Wang, Y., Cong, G., Song, G., Xie, K.: Community-based greedy algorithm for mining top-k influential nodes in mobile social networks. In: KDD, pp. 1039–1048 (2010)

AILabel: A Fast Interval Labeling Approach for Reachability Query on Very Large Graphs

Feng Shuo, Xie Ning, Shen de-Rong[*], Li Nuo, Kou Yue, and Yu Ge

College of Information Science & Engineering, Northeastern University, Shenyang, China
shendr@mail.neu.edu.cn

Abstract. Recently, reachability queries on large graphs have attracted much attention. Many state-of-the-art approaches leverage spanning tree to construct indexes. However, almost all of these work require indexes and original graph in memory simultaneously, which will limit the scalability. Thus, a new interval labeling approach called AILabel (Augmented Interval Label) is proposed in this paper. AILabel labels each node with quadruples. Index construction time of AILabel is O($m+n$), which requires only one traversal through the graph. Besides, AILabel only needs index to answer the queries. We further proposed an approach D-AILabel based on AILabel to handle reachability queries on dynamic graph. Finally, experiments on real and synthetic datasets are conducted and prove that AILabel can efficiently scale to large graph and reachability queries on dynamic graph can be effectively handled by D-AILabel.

Keywords: graph, dynamic graph, reachability query, interval labeling.

1   Introduction

In recent years, graph model is getting increasingly popular with many new applications such as computational biology, computer networks and ontology management. How to efficiently answer reachability queries has attracted lots of interests[1].

There are two classical reachability computation approaches. One is to employ a depth-first search (DFS) or breath-first search (BFS) on the original graph rather than constructing indexes. Its computation complexity is O($m+n$) in the worst case, where n and m denote the number of the nodes and edges respectively. The other is to precompute and construct a full transitive closure. The reachability queries can be answered in O(1) time. However, the transitive closure requires a quadratic space complexity. Because of the large index size and the high computation complexity, the two approaches face the bottleneck of scalability for handling large graphs.

In order to efficiently solve this problem, quite a few graph approaches have been proposed[2-13]. All these approaches lie between the two classical computation methods. They focus on the tradeoff among index construction time, index size and query time. Compared with the classical approaches, existing approaches can be

---

[*] Corresponding author.

© Springer International Publishing Switzerland 2015
R. Cheng et al. (Eds.): APWeb 2015, LNCS 9313, pp. 560–572, 2015.
DOI: 10.1007/978-3-319-25255-1_46

extended to larger graphs. However, these approaches still have their own limitations. Besides, most approaches are not designed for dynamic graph.

We propose a novel interval labeling approach called AILabel and an improved approach D-AILabel handling dynamic graph. AILabel not only can scale the existing state-of-the-art reachability indexes, but also can be easily extended to large graph. D-AILabel can handle dynamic graph effectively. We make the following contributions:

— We design a novel interval labeling approach AILabel. The index construction time is $O(m+n)$ and the query time is $O(m-n)$.
— We propose four filtering optimizations to improve AILabel.
— We propose D-AILabel index to answer the reachability in dynamic graphs.
— Our experiments show the high efficiency and scalability of AILabel and effectiveness of D-AILabel.

The paper is organized as follows. Section 2 introduces related work. Section 3 gives some basic concepts. Section 4 proposes AILabel index, including index construction, querying, and optimization. Section 5 discusses dynamic graph index D-AILabel. Section 6 evaluates our proposal with empirical studies and Section 7 concludes this paper.

2   Related Work

Existing approaches processing reachability queries can be classified into two categories: hop labeling approaches and interval labeling approaches.

Hop labeling approaches encode the reachability information on the nodes. Each node u records two node sets u_{in} and u_{out}, where u_{in} maintains the nodes that can reach u and u_{out} maintains the nodes that u can reach. To answer whether u can reach v, a join operation between u_{out} and v_{in} is performed to determine whether a non-empty subset exists. 2-hop label proposed by Cohen et al.[2] is the first hop labeling approach. The approach selects some intermediary nodes as hop nodes to encode the reachability information. However, 2-hop labeling approach takes high computational complexity. Naturally, a number of improved approaches are further proposed[3-5].

The other category is interval labeling approaches. Inspired by [14], Optimal Tree-cover [6] is the first interval labeling approach by assigning each node an interval label generated by a spanning tree of the graph. Without loss of reachability information, additional interval label is required to store the information of non-tree edges. Optimal Tree-cover is hard to be applied to large graphs, although an optimization to reduce the labeling size has been proposed[7]. Dual Labeling[8] is another interval labeling approach designed for processing sparse graph. With the graph getting denser, the construction and storage of index become larger and the performance becomes worse.

GRAIL[9], Ferrari[10], GRIPP[11] and RIAIL[12] are all interval labeling approaches, and they are closely related to our method. GRAIL is an approximate interval labeling approach that has good performance dealing with negative queries. While in the case of positive queries, a large portion of the graph has to be expanded.

Inspired by GRAIL, Seufert S et al.[10] proposed Ferrari. A tradeoff between the number of labels and practical memory space is given. While both GRAIL and Ferrari need to store index and original graph in memory to keep efficiency. Further Yes-Label[13] is proposed to reduce operations on original graph. A novel indexing structure is proposed in [13] to optimize I/O efficiency for the situation that index and original graph cannot be stored in memory simultaneously. Different from GRAIL and Ferrari, in GRIPP, some nodes have replicated storage with extra intervals. Thus, when the query cannot be answered directly, the query is decomposed into a series of queries. Because of the iterative query, the query time of GRIPP is higher than O(m-n). To solve this problem, RIAIL is proposed, whose querying time is O(m-n). Besides, the optimization of GRIPP can be also leveraged in RIAIL to speed up negative queries.

Optimal Tree-cover[6] and DAGGER[15] are both designed for handling dynamic graphs. Optimal Tree-cover adds new constraint to index directly, leading to a poor efficiency after a certain update operation. But DAGGER induce the update of all the ancestor nodes of the new update nodes. Thus, each update will cost longer time. Both of the two approaches have their own limitations in dealing with dynamic graphs.

3   Basic Concepts

— Tree edge and Non-tree edge: Given a spanning tree of the graph, all the edges are divided into two components: the edges in the spanning tree (tree edges) and the edges not in the spanning tree (non-tree edges).
— Interval label: Each node is assigned an interval generated by the spanning tree. There are two main interval labeling types: [*pre, post*], where *pre* and *post* are preorder and postorder respectively; [*min_pre, post*], where *post* is the postorder and *min_pre* is the minimum *pre* of all the descendant nodes in original graph.
— Hop node and Special node: For a non-tree edge, we denote the starting node as special node and the ending node as hop node.
— The hop nodes of *u*: If *w* is a special node which is *u*'s descendant node in the spanning tree, then we denote the ending nodes of non-tree edges starting from *w* as the hop nodes of *u*.
— The special nodes of *u*: we denote all the special nodes which are the descendants of *u* in the spanning tree as the special nodes of *u*.

We use an example to demonstrate the definition. In Fig. 1, we employ a depth-first search (DFS) to the directed acyclic graph G_{dag}, where solid lines represent the tree edges and the dotted lines represent the non-tree edges. In the spanning tree, *F, I, J, H* and *C* are special nodes and correspondingly *G, H* and *K* are hop nodes. The hop nodes of *B* are *G* and *K*, and the special nodes of *B* are *F, I, J* and *H*.

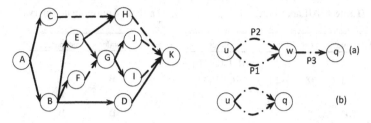

Fig. 1. Directed acyclic graph G_{dag} **Fig. 2.** Example of Theorem 1

4   AILabel Index

In this paper, we just discuss the situation of Directed Acyclic Graphs (DAGs) in the absence of special statement, because a directed graph can be transformed into DAG after folding each Strongly Connected Component (SCC) into a single node. Besides, we assume that there is at most one edge between a certain pair of nodes.

4.1   B-AILabel Index

As stated above, GRIPP only uses indexes of the graph. But because of the iterator query, GRIPP performs not satisfying. To reduce the iterative times, a directly way is to assign a set of hop nodes to each node. Thus, we propose B-AILabel. Each node is assigned a triple <*pre, post, hops*>, where *pre* and *post* constitute the interval denoted in Section 3 and *hops* is the set of hop nodes of the certain node.

To get B-AILabel index, we start at each root node. During the depth-first traversal of the graph, when v is visited through a non-tree edge from u, we put v into u's *hops*. After all u's children have been visited, u's *hops* is added into *hops* of u's parent through the tree edge. Thus, B-AILabel index can be generated through a single traversal, and the construction time is $O(m+n)$.

Unfortunately, to get *hops* set, we need to put the children's *hops* into their parents. This will increase index construction time greatly with the increase of non-tree edges.

4.2   AILabel Index

To solve the shortage of B-AILabel, we propose AILabel. Each node is assigned a quadruple <*pre, post, hops, directs*>, where *hops* has a bit difference with B-AILabel in definition. If v is a special node, we regard all the hop nodes of v that connected to v directly as *hops* (Each node in v's *hops* connects to v through a non-tree edge). For the nodes that are not special nodes, its *hops* is set null. *directs* is the same as those in B-AILabel.

Similar with B-AILabel, during the depth-first traversal of the graph, *pre* and *post* can be easily gotten. For *hops*, when a node v is visited through a non-tree edge starting from node u, we put v into u's *hops* directly. When a special node w is visited, w is added to all its ancestor nodes' *directs*. Table 1 shows AILabel index of G_{dag} in Fig. 1.

Table 1. AILabel Index of G_{dag} **Table 2.** Approximate Interval Labels

node	pre	post	hops	directs
A	0	21	[ ]	[I, J, H, F, C]
B	1	18	[ ]	[I, J, H, F]
C	19	20	[H]	[ ]
D	2	5	[ ]	[ ]
E	6	15	[ ]	[I, J, H]
F	16	17	[G]	[ ]
G	7	12	[ ]	[I, J]
H	13	14	[K]	[ ]
I	8	9	[K]	[ ]
J	10	11	[K]	[ ]
K	3	4	[ ]	[ ]

node	tree_parent
A	-1
B	A
C	A
D	B
E	B
F	B
G	E
H	E
I	G
J	G
K	D

Theorem 1. There is no duplication in *directs* of any node.

Proof: Assume that *directs* of node u contains a duplicate node q. According to the description of *directs*, there must be two different paths from node u to node q in the spanning tree. As shown in Fig. 2(a), there always exist a node w that have two incoming edges in the spanning tree, which is contrary to the property of spanning tree.

According to Theorem 1, there is no need to perform extra set merging operation. And there is no duplicate edge in original graph, so no duplicate nodes are in *hops*. The index construction time of AILabel is also $O(m+n)$. As we release the set merging operation, AILabel has better performance than B-AILabel in fact. But it is hard to accurately calculate the number of nodes in *directs*. So we cannot give an exact upper bound of index size. Lots of experiments demonstrate that index size of AILabel is less than B-AILabel in most datasets.

Following we will show how to answer reachability query whether u is reachable to v using AILabel. The process has two steps as follows:

— Answer the query by the interval containment relationship. If $u_{pre} < v_{post} \leq u_{post}$, u is reachable to v. Otherwise, carry on next step.
— Generate two types of new queries. If the answer of any query is positive, u is reachable to v. Otherwise, u is unreachable to v.
  • Type 1: The queries are whether the nodes in u's *hops* can reach v. These queries should further be answered through both steps.
  • Type 2: The queries are whether the nodes in u's *directs* can reach v. These queries just need to be answered through the second step.

For example, consider the graph G_{dag} in Figure 1. And the query is whether B can reach C denoted as $Q(B,C)$. Because $(B_{pre}=1) \leq (C_{post}=20) \leq (B_{post}=18)$ is not correct, we carry on the second step. And new queries $Q(I,C)$, $Q(J,C)$, $Q(H,C)$, $Q(F,C)$ will be generated. Because I, J, H, F are all in B's *directs*, all these queries just need be answered through the second step. After iteratively querying, the answer is got negative.

4.3   Optimization

To improve the performance, we propose the following four filtering optimizations:

— Optimization 1: Identify Duplicate Queries.

Lots of duplicate queries will be generated in the second step, thus duplicate removal is necessary. In the example above, $Q(K,C)$ generated by $Q(I,C)$, $Q(J,C)$ and $Q(H,C)$ should be computed thrice. Hence, we employ a set to store the visited nodes. Before generating new queries, search the set and check whether it has been answered before.

— Optimization 2: Reduce Interval Containment Judgment.

If a new query is generated by *directs*, the query just need to be answered through the second step. In the example of Section 5.2, when dealing with $Q(I,C)$ generated by $Q(B,C)$, we realize that I is in B's *directs*. Thus, the interval of I must be contained in B's. And because we have got that the interval of C is not contained in B's, there is no need to repeatedly query $Q(I,C)$ through the first step.

— Optimization 3: Approximate Interval Labels.

This optimization is good at dealing with the queries with negative answers. We adopt [*min_pre, post*] interval labeling method defined in Section 3. For a certain node pair u and v, if the interval of u is not contained in v's, we can conclude that u cannot reach v. Otherwise, if contained, we need to do the normal querying.

— Optimization 4: Level Filter[14].

This optimization is good at dealing with the queries with negative answers either. We assign level=0 to the root nodes. And levels of the other nodes are the maximum level of their parent nodes plus 1. For a certain node pair u and v, if level of u is smaller than v's, v cannot reach u.

5   Dynamic Graph Index D-AILabel

B-AILabel and AILabel introduced above are both working on static graphs. Nevertheless, graphs are not always static in practical applications such as twitter and Facebook. Hence, how to deal with graph update operations is a great challenge. In this section, we further propose D-AILabel, an approach that can handle dynamic graphs efficiently based on AILabel. For the reason that the updates on original graph can be transformed into the updates on DAG, we simply discuss the updates on DAG in this paper.

5.1   D-AILabel Index

In D-AILabel, each node is assigned a quintuple <*tree_parent, pre, post, hops, directs*>, where *tree_parent* denotes the parent node in the spanning tree and *pre, post, hops, directs* have the same definitions as those in AILabel. For the root nodes that have no parent nodes, its *tree_parent* sets -1. Thus, D-AILabel can be also constructed in $O(m+n)$ either. Table 2 shows part of D-AILabel index of G_{dag} in Figure 1, the value of the rest attributes is the same as those in Table 1.

5.2   Querying D-AILabel

There are four types of update: node insertion, node deletion, edge insertion and edge deletion. Because node deletion can be regard as a series of edge deletion and node insertion just need to set index on the new node, we just consider two situations: edge insertion and edge deletion.

Edge Insertion.
There are four cases when an edge (u,v) is inserted into DAG:
— Case 1: Both u and v are not isolated nodes in DAG.
  • If v has no parent node, (u,v) must be a tree edge. And:
    ○ If v is a not a special node: Add v's *directs* into *directs* of u and u's ancestor nodes in the spanning tree. That is because there have no intersection between v's *directs* and u's or u's ancestor nodes'. And the information of (u,v) is stored in the intervals.
    ○ If v is a special node: Add v into *directs* of u and u's ancestor nodes in the spanning tree. That is because there may be intersection between v's *directs* and u's.
    Next, we assign u to v's *tree_parent*. And a new traversal is recalled. The new edge does not influence non-tree edges, thus *hops* does not need modification.
  • If v has a parent node, (u,v) must be a non-tree edge. And:
    ○ If u is a special node: Add v into u's *hops*. And all the nodes in v's *directs* can be generated by v.
    ○ If u is not a special node: The new edge transforms u into a special node. Thus, add u into *directs* of u's ancestor nodes and add v into u's *hops*.
— Case 2: u is not an isolated node and v is an isolated node.
  The new edge (u,v) must be a tree edge. Thus, we add (u,v) into the spanning tree and recall a new traversal to refresh the intervals.
— Case 3: u is an isolated node and v is not an isolated node.
  • If v has a parent node, v is added into u's *hops* directly and the u's interval sets $[maxid+1, maxid+2]$, where $maxid$ is the maximum value among all intervals.
  • If v has no parent node. The operation can refer to Case 1.
— Case 4: u and v are both isolated nodes.
  We add (u,v) into the spanning tree and assign the interval of u and v as $[maxid+1, maxid+4]$ and $[maxid+2, maxid+3]$ separately. u's *tree_parent* sets v.

Edge Deletion.
We divide edge deletion (u,v) operations into following two cases:
— Case 1: The edge (u,v) is a tree edge.
  We delete v from *directs* of u and u's ancestor nodes in the spanning tree. And set u's *tree_parent* as -1, which means that v has no parent node in the spanning tree. Then we will relabel the spanning tree whose root is v.
— Case 2: The edge (u,v) is a non-tree edge. There are following two situations according to whether (u,v) is the only non-tree edge outgoing from u.
  • (u,v) is the only non-tree edge outgoing from u:
    After (u,v) is deleted, u is no more a special node. Thus, we delete u from *directs* of u's ancestor nodes and add u's *directs* into *directs* of u's ancestor nodes. At last, delete v from the hops of u.

- *(u,v)* is not the only non-tree edge outgoing from *u*:

After non-tree edge *(u,v)* deletion, there are still non-tree edges left outgoing from *u*. And we just need to delete *v* from *u*'s *hops*.

6   Experiment

In our experiments, all approaches are implemented using C++. And all the experiments are performed on the computer equipped with Intel Quad-Core i7-2600 CPU at 3.40DHz and 4GB of main memory. The operating system in use is a 64-bit installation of Ubuntu 12.04 LTS with kernel 3.2.0.

6.1   Experiment on Statics Graph

— *Datasets.*
We conduct experiments on real and synthetic graph shown in Table 3.

Table 3. Datasets

	Small Sparse Graphs							
	agrocyc	amaze	anthra	ecoo	human	kegg	Mtbrv	Nasa
nodes	12684	3710	12499	12620	38811	3617	9602	5605
edges	13657	3947	13327	13575	39816	4395	10438	6538

	Small Sparse Graphs		Small Dense Graphs				
	vchocyc	xmark	arxiv	citeseer	go	pubmed	yago
nodes	9491	6080	6000	10720	6793	9000	6642
edges	10345	7051	66707	44258	13361	40028	42392

	Large Graphs					Synthetic Graphs	
	citeseer	uniprot22m	citeseerx	cit-Patents	go-uniprot	10m2x	10m5x
nodes	693947	1595444	6540401	3774768	6967956	10M	10M
edges	312282	1595442	15011259	16518947	34770235	20M	50M

— *Approaches.*
B-AILabel: In order to reduce the index construction time of B-AILabel, we uses hash set to store *hops* of each node.

GRAIL and Ferrari: We set the relevant parameters of GRAIL and Ferrari based on the recommendation in original paper. And we got the codes of the two approaches[1].

GRIPP: Different from the other approaches, GRIPP is a database-implemented approach. It is compared with GRAIL in [7]. In order to be fair to all, we conduct experiments in the same environment as [7] to compare AILabel with GIRPP.
— *Methodology.*
The metrics we compare on are: index size, query performance and index construction time. Index size denotes the size of all the labels. The type of label is Integer in our experiments; Query performance is the querying time of 100000 queries. In our ex-

---

[1] https://code.google.com/p/grail/
    https://code.google.com/p/ferrari-index/

periment, the 100000 queries are the same to those in paper [14]; Index construction time is the time cost to establish the index.

Small Sparse Graph

As shown in Fig. 3(a), the index construction time of B-AILabel is almost twice as long as AILabel. The construction time of AILabel is the shortest, because only one depth-first traversal is required. Moreover, GRAIL requires twice traversals, and Ferrari requires an optimal tree cover algorithm to encode interval to each node firstly and assign multiple labels on each node. And the query performance is shown in Fig. 3(c). Ferrari has the best performance, because Ferrari has many labels to reduce the queries on original graphs. The performance of AILabel is much closely to Ferrari, and then comes GRIAL. That is because when the query cannot be answered directly, GRIAL needs to do a traversal on original graph, however AILabel just traverses the non-tree edges.

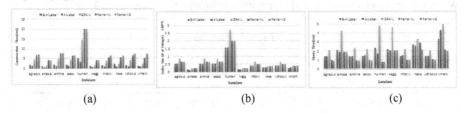

(a) (b) (c)

Fig. 3. Performance on small sparse graphs

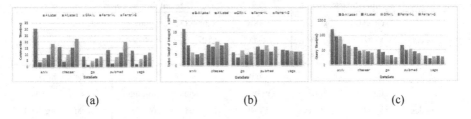

(a) (b) (c)

Fig. 4. Performance on small dense graphs

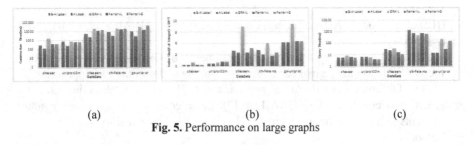

(a) (b) (c)

Fig. 5. Performance on large graphs

Small Dense Graph

From Fig. 4, AILabel has the shortest index construction time. That is because for the other approaches, more non-tree edges are required to be stored in the index and lots of duplicate removal operations is required. AILabel has the smallest size in most datasets except *arxiv* and *yago*. The average degree of the two dataset is relatively larger than the other datasets. More non-tree edges are generated, so AILabel takes more space to maintain *directs*. However, GRAIL and Ferrari need to put the index and the original graphs in memory simultaneously. Fig. 4(c) shows the query time of the four methods. Query time of AILabel is much closer to GRAIL than in sparse graphs. That is because GRAIL has multiple interval labels on each node which can answer the queries quickly.

Large Graph

Fig. 5 shows that index construction time of B-AILabel is 2-3 times longer than AILabel in large graphs. According to [7], index size of AILabel is smaller than GRIPP. *citeseer* and *uniprot22m* are large sparse graphs. In the experiment, AILabel use the multiple interval labeling optimization to further improve the performance. However GRAIL and Ferrari cannot simply utilize the optimization, because these two approaches need to put the original graph in main memory and there is no more space for additional intervals. For *go-uniprot*, the performance of AILabel is the best. That is because we cannot store the index of Ferrari and the original graph in memory simultaneously. Thus disc I/O is avoidless, resulting the longer query time.

Table 4. Index size of synthetic graphs(# of Integer)

data	B-AILabel	AILabel	GRAIL	Ferrari-L	Ferrari-G
10m2x	66400380	57238009	-m	81433672	94999993
10m5x	196823223	89050907	-m	-m	-m

Table 5. Query time of synthetic graphs(ms)

data	B-AILabel	AILabel	GRAIL	Ferrari-L	Ferrari-G
10m2x	47.01	44.21	-t	-t	-t
10m5x	-t	2237.77	-t	-t	-t

Table 6. Index construction time of synthetic graphs(ms)

data	B-AILabel	AILabel	GRAIL	Ferrari-L	Ferrari-G
10m2x	14906	8197.68	-t	117935	118310
10m5x	49281.7	13446.33	-t	-t	-t

Synthetic Graphs

The result is shown in Table 4, Table 5 and Table 6, where -t donates that the query cannot be finished in 5 minutes and -m denotes that the index cannot be constructed in 5 minutes. The index construction time of B-AILabel is 2-4 times longer than AILabel and the index size of B-AILabel is obviously larger than AILabel. Though Ferrari-L and Ferrari-G can construct indexes on 10m2x, but the query time of Ferrari-L and Ferrari-G can't be accepted. Because the original graph and the index cannot be stored in the memory together, disc I/O cannot be avoided. Especially, GRAIL failed in dealing with these two synthetic dataset.

6.2   Experiment on Dynamic Graph

We utilize some of the datasets in Section 7.1. Each dataset will generate 5000 update operations randomly. Table 7 shows the number of operations on different datasets. (*a, d, e, n* are add, delete, edge, node for short respectively.)

Table 7. Update operations

dataset	a_e	d_e	a_n	d_n
agrocyc	3004	749	1010	237
kegg	2955	762	1048	235
xmark	2984	727	1057	232
citeseer	2983	760	1019	238
go	2982	749	1036	233

Table 8. Average update time (ms)

dataset	a_e	d_e	a_n	d_n
agrocyc	0.13	0.0042	0.0050	0.0046
kegg	0.07	0.0095	0.0048	0.0037
xmark	0.10	0.0213	0.0058	0.0053
citeseer	0.01	0.0018	0.0060	0.0057
go	0.27	0.0057	0.0062	0.0067

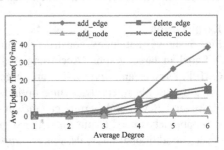

Fig. 6. D-AILabel with node amount 50k

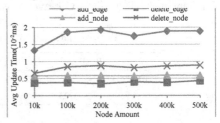

Fig. 7. D-AILabel with average degree 2

Table 8 shows the average update time. From the result, we can conclude that the cost of node insertion operation is least. The reason is node insertion operation only needs to set an interval to the new node. In comparison, the other three update operations will take longer update time owing to the extra cost of updating the labels or relabeling nodes. What's more, the average edge insertion operating time of *citeseer* is the least, although *citeseer* has a high average degree. So, average update time changes greatly for different datasets.

We further test the scalability of D-AILabel in Fig. 6 and Fig. 7. In the experiment of Fig. 6, we set the amount of nodes to a fixed number 50000 and change the average degree of the graph. For each graph, we generated 5000 updates for each type randomly. From Fig. 6, we get the conclusion that when the average degree is less than 4, the change of edge insertion is not obvious. This is because that the increase of non-tree edges brings much difficulty. Fig. 7 shows the update performance of D-AILabel with a fixed average degree 2. We range the node amount from 10k to 500k. In particular, the update time of edge insertion changes obviously. While, the increasing trend of edge deletion and node deletion is not as obvious as that of edge insertion. The update time of node insertion stays stable. Combining the result from Fig. 6, we conclude that the average update time of D-AILabel mainly depends on the average degree of the DAG.

7   Conclusion

Reachability query is a basic issue in the management of graph data. With the development of various applications, graph has changed greatly in the aspect of its scale. How to adapt to these changes and answer reachability queries accurately, rapidly and extensively is a very important issue. Therefore, we propose an approach AILabel that can answer reachability queries on static graphs and also propose D-AILabel dealing with dynamic graphs. The high efficiency and scalability of AILabel and the availability of D-AILabel are also demonstrated. But AILabel cannot be expanded to the reachability queries with label constraint. Therefore, we will focus on reachability queries with label constraint in the future work.

Acknowledgement. This work is supported by the National Basic Research 973 Program of China under Grant (2012CB316201) and the National Natural Science Foundation of China under Grant (61033007, 61472070).

References

1. Aggarwal, C.C., Wang, H.: Managing and mining graph data. Springer, Heidelberg (2010)
2. Cohen, E., Halperin, E., Kaplan, H., et al.: Reachability and distance queries via 2-hop labels. In: Proceedings of the Thirteenth Annual ACM-SIAM Symposium on Discrete Algorithms, pp. 937–946. ACM Press, Philadelphia (2002)
3. Cheng, J., Yu, J.X., Lin, X., Wang, H., Yu, P.S.: Fast computation of reachability labeling for large graphs. In: Ioannidis, Y., et al. (eds.) EDBT 2006. LNCS, vol. 3896, pp. 961–979. Springer, Heidelberg (2006)
4. Cheng, J., Yu, J.X., Lin, X., et al.: Fast computing reachability labelings for large graphs with high compression rate. In: Proceedings of the 11th International Conference on Extending Database Technology: Advances in Database Technology, pp. 193–204. ACM Press, New York (2008)
5. Jin, R., Xiang, Y., Ruan, N., et al.: 3-hop: a high-compression indexing scheme for reachability query. In: Proceedings of the 2009 ACM SIGMOD International Conference on Management of Data, pp. 813–826. ACM Press, New York (2009)
6. Dietz, P.F.: Maintaining order in a linked list. In: Proceedings of the Fourteenth Annual ACM Symposium on Theory of Computing, pp. 122–127. ACM Press, New York (1982)
7. Yıldırım, H., Chaoji, V., Zaki, M.J.: GRAIL: a scalable index for reachability queries in very large graphs. The VLDB Journal—The International Journal on Very Large Data Bases 21(4), 509–534 (2012)
8. Wang, H., He, H., Yang, J., et al.: Dual labeling: Answering graph reachability queries in constant time. In: Proceedings of the 22nd International Conference on Data Engineering, pp. 75–75. IEEE, Washington, DC (2006)
9. Yildirim, H., Chaoji, V., Zaki, M.J.: Grail: Scalable reachability index for large graphs. Proceedings of the VLDB Endowment 3(1-2), 276–284 (2010)
10. Seufert, S., Anand, A., Bedathur, S., et al.: Ferrari: Flexible and efficient reachability range assignment for graph indexing. In: Proceedings of the 2013 IEEE International Conference on Data Engineering, pp. 1009–1020. IEEE, Washington, DC (2013)

11. Trißl, S., Leser, U.: Fast and practical indexing and querying of very large graphs. In: Proceedings of the 2007 ACM SIGMOD International Conference on Management of Data, pp. 845–856. ACM Press, New York (2007)
12. Ning, X., De-Rong, S., Shuo, F., et al.: RIAIL: An Index Method for Reachability Query in Large Scale Graphs. Journal of Software 25(Suppl. (2)), 213–224 (2014)
13. Zhang, Z., Yu, J.X., Qin, L., et al.: I/O cost minimization: reachability queries processing over massive graphs. In: Proceedings of the 15th International Conference on Extending Database Technology, pp. 468–479. ACM Press, New York (2012)
14. Agrawal, R., Borgida, A., Jagadish, H.V.: Efficient management of transitive relationships in large data and knowledge bases. In: Proceedings of the 1989 ACM SIGMOD International Conference on Management of Data, pp. 253–262. ACM Press, New York (1989)
15. Yildirim, H., Chaoji, V., Zaki, M.J., Dagger, Z.M.J.: A scalable index for reachability queries in large dynamic graphs[EB/OL] (2014). http://arxiv.org/abs/1301.0977

Graph-Based Hybrid Recommendation Using Random Walk and Topic Modeling

Hai-Tao Zheng, Yang-Hui Yan, and Ying-Min Zhou

Tsinghua-Southampton Web Science Laboratory,
Graduate School at Shenzhen, Tsinghua University, Beijing, China
zheng.haitao@sz.tsinghua.edu.cn,
{yanyh13,zhou-ym14}@mails.tsinghua.edu.cn

Abstract. In this paper, we propose a graph-based method for hybrid recommendation. Unlike a simple linear combination of several factors, our method integrates user-based, item-based and content-based techniques more fully. The interaction among different factors are not done once, but by iterative updates. The graph model is composed of target user's similar-minded neighbors, candidate items, target user's historical items and the topics extracted from items' contents using topic modeling. By constructing the concise graph, we filter out irrelevant noise and only retain useful information which is highly related to the target user. Top-N recommendation list is finally generated by using personalized random walk. We conduct a series of experiments on two datasets: movielen and lastfm. Evaluation results show that our proposed approach achieves good quality and outperforms existing recommendation methods in terms of accuracy, coverage and novelty.

Keywords: hybrid recommendation, random walk, topic modeling, sparsity, novelty.

1 Introduction

With an exponential growth of information available on the Internet, it has become increasingly important to help people get personalized services. Recommendation systems, which are designed to solve the problem by analyzing users' preference, are studied extensively in prior research.

The most widely used recommendation techniques are collaborative filtering-based (CF) and content-based (CB) methods. In the user-based CF [2], target user is recommended new items that are rated by his similar-minded neighbors. While in the item-based CF [3], the items similar to target user's historical records will be recommended. User-based CF tends to recommend popular items, which, however, harms the overall diversity for all users. Item-based CF can boost long-tail items, but decrease individual diversity. CB usually calculates the similarity between constructed user profile and item contents, which is often used to help improve the performance of CF methods. By considering both the collaboration and the content, hybrid recommendation technology can achieve

© Springer International Publishing Switzerland 2015
R. Cheng et al. (Eds.): APWeb 2015, LNCS 9313, pp. 573–585, 2015.
DOI: 10.1007/978-3-319-25255-1_47

the benefits of different methods. But a simple linear combination of two factors in the user similarity calculation or the results adjustment can not perform effectively, which even decreases the precision for specific users.

Graph-based recommendation model [12, 14] is flexible, which can make a good integration of varieties of contextual information. A recommendation list is generated by using personalized random walk(RWR) or hitting time in the graph. However, in the existing graph-based methods, random walk is usually conducted in the whole graph. Even a subgraph constructed from depth-first or breadth-first search is still very large. This practice not only causes high computational complexity, but also introduces too much noise, affecting the performance.

In this paper, we propose a hierarchical graph model for hybrid recommendation, in order to combine different techniques together appropriately and maintain small computational complexity. When recommending for the target user, we only select relevant information from each factor to construct a concise graph. Thus the running time is greatly reduced and irrelevant noises are avoided. On the other hand, as the collaboration and content factors are combined more naturally by using topic modeling and iterative interaction in the well-defined structure, our accuracy is highly improved. Finally, we can alleviate the data sparsity as random walk-based ranking allows us to utilize indirect relationships between graph nodes. And we can boost long-tail items, as long as they contain the same topics target user prefers.

In conclusion, our main contributions are listed as follows:

- We propose a novel hierarchical graph model for hybrid recommendation, which integrates different factors iteratively. And we systematically study our well-defined data fusion graph structure and justify its rationality.
- Our method can alleviate the data sparsity and boost long tail recommendations that many existing methods suffer. Also, we can cover the majority of distinct items in data corpus.
- We conduct enough experiments on two datasets. Results show that our method performs better in three measures especially in sparse dataset.

The rest of this paper is organized as follows. Section 2 presents some related work on recommendation. We give a detailed description of our proposed method RWR-UIC in section 3. Experiments and analysis are included in section 4. Finally we conclude the paper with remarks of our work in section 5.

2   Related Work

Recommendation systems are basically divided into two categories: CF-based and content-based(CB). CF [1] explores user-item rating matrix and can be further classified into user-based CF and item-based CF. User-based CF [2] assumes that similar users express similar interests in future items. Items rated by similar-minded neighbors are recommended to the target user. User similarities are calculated based on ratings. [3] proposed an item-based method. They recommend new items which are similar to the items target user has rated. Items similarities are also calculated using rating matrix. In CF, usually a small subset of users or items are used as neighbors. CF is simple in training phase and can be used in many domains such as news and multimedia. The most important feature of CF is content-independent [11]. But if ratings are sparse, Standard CF methods result in poor results. Content-based methods [6] try to construct user's profile using items' contents. To generate recommendations, Similarity between candidate item's contents and the constructed user profile is calculated. CB does not suffer from rating sparsity, but a big drawback is that it ignores the implicit associations between users, which leads to poor results purely using CB. Hybrid recommendation methods [10] can obtain the advantages of both methods by considering collaboration and content at the same time, but, usually a simple weighted combination of two factors cannot perform effectively.

Latent factor models have been successfully applied to recommendation. For example, [8] proposed a three-layer aspect model in which ratings and contents are combined in a probabilistic way, to address the cold-start problem. Matrix factorization was developed in [9,20]. It decomposes user-item rating matrix into low latent space. To predict the missing score, we just need to multiply the latent vectors of candidate item and target user. However the latent space doesn't have evident interpretation for human beings. Additionally, if training set is rather sparse, factorization may suffer from overfitting.

Graph-based methods are getting more and more attention recently. By setting a biased probability of jumping to the starting nodes, RWR is very useful for personalized recommendation [13]. [14] studied a click-through bipartite graph for a series of applications including recommendation task. [16] proposed a novel recommendation method which performs random walk on an items' graph, where the edges denote similarities between items. [7] adopted a multi-layer graph model for personalized query-oriented reference paper recommendation, but incorporating too many terms may make the graph extremely large. [12] proposed a random walk-based model which combines users, items, tags, social relationship into the whole graph. There is one thing in common in the above graph-based recommendation models: To generate recommendations, random walk is usually conducted in the whole graph. This practice not only leads to high computational complexity but also brings unnecessary noise. In this work, we just construct a sub-graph, which contains information mostly related to target user. collaboration and content more integrated more naturally and fully.

3   Proposed Method

The framework of our proposed method **RWR-UIC**(**R**andom **W**alk with **R**estart which combines **U**ser-based, **I**tem-based and **C**ontent-based factors) is shown in Fig.1. In the bottom part, concept mining on items' contents and similarity calculation between users and items are done offline to save online response time. In the online procedure, when a user request recommendations, we firstly construct a concise graph as illustrated in Fig.3 for him, then top-N list is generated by using personalized random walk.

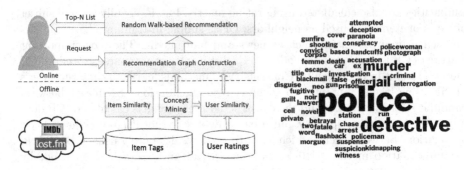

Fig. 1. System framework   **Fig. 2.** Sample concept tag cloud

3.1   Concept Mining

A document may involve multiple topics. Using the words alone, we may fail to find conceptually related documents that use different wordings. Topic models like PLSI [4] and LDA [5] treat document as a probabilistic mixture of topics and estimate the doc-topic distributions from a collection of documents unsupervisedly. When the corpus is not particularly large, LDA can effectively prevent overfitting because it adds priors to the parameters. In this work, we adopt LDA topic model and use gibbs sampling [18] to infer the topic distributions.

In our rating corpus, items contents are not available. However, annotation tags in the social media websites, such as lastfm and imdb, are good descriptions of artists' musical style or films' storyline. By crawling these tags, we can construct items' contents. However, the uncontrolled tagging behavior in websites results in tag redundancy and ambiguity. So LDA is used to help us capture the co-occurrence between related tags and extract the items' topic distributions $p(t_k|i)$. The results of this step are used for both user profiling and recommendation graph construction. A sample tag cloud is generated by using top 30 terms of an topic. We can see that words are related to "crime", as illustrated in Fig.2.

3.2   User Profiling

Formally, suppose that there exist a set of users U=$\{u_1, u_2, ..., u_M\}$ and a set of items I=$\{i_1, i_2, ..., i_N\}$. For each user u that belongs to U, a list of items are available with the corresponding rating $R_{u,i}$. To generate recommendations, we start with constructing user's preference profile as a two-attribute tuple $\overrightarrow{u} = \{T, L\}$. T represents u's interest distributions in content topics, in the format of vector $\{p(t_1|u), p(t_2|u), ..., p(t_K|u)\}$. L denotes a list of rated items $\{i_1, i_8, ...\}$ that u has shown interest in. We update T by the following procedure. T is firstly initialized to $\overrightarrow{0}$. Then $\forall$ item i which has been rated by u, we multiply $R_{u,i}$ with i's topic distribution $\{p(t_1|i), p(t_2|i), ..., p(t_K|i)\}$, and add the multiplied vector into T. Finally, $p(t_k|u)$ is normalized by $\sum_{k=1}^{K} p(t_k|u)$. Based on the constructed $\overrightarrow{u}$, we use Eq.(1) to calculate user similarity. $\delta \in [0,1]$.

$$S(u_1, u_2) = \delta \cdot sim(T_{\overrightarrow{u_1}}, T_{\overrightarrow{u_2}}) + (1 - \delta) \cdot sim(L_{\overrightarrow{u_1}}, L_{\overrightarrow{u_2}}) \qquad (1)$$

Similarity is usually calculated by cosine Eq.(2) or Pearson correlation Eq.(3). $\vec{r_a}$ and $\vec{r_b}$ are two vectors. $\bar{r}_a$ and $\bar{r}_b$ are the mean values of respective vectors. $sim_{a,b}$ is the similarity between $\vec{r_a}$ and $\vec{r_b}$.

$$sim_{a,b} = \frac{\sum_{i=1}^{N} r_{a,i} \cdot r_{b,i}}{\sqrt{\sum_{i=1}^{N} r_{a,i}^2 \cdot \sum_{i=1}^{N} r_{b,i}^2}} \tag{2}$$

$$sim_{a,b} = \frac{\sum_{i=1}^{N} (r_{a,i} - \bar{r}_a) \cdot (r_{b,i} - \bar{r}_b)}{\sqrt{\sum_{i=1}^{N} (r_{a,i} - \bar{r}_a)^2 \cdot \sum_{i=1}^{N} (r_{b,i} - \bar{r}_b)^2}} \tag{3}$$

To calculate Pearson, we need to isolate co-occured items of two vectors in advance. Additionally, if the number of co-occured items between two vectors is small, penalizing the correlation score obtained from very few evidence can improve prediction accuracy. More specifically, set a threshold ϵ. If the number of co-rated items τ is less than ϵ, we multiply the score by $\frac{\tau}{\epsilon}$ [12]. In our experiments, $sim(T_{\vec{u_1}}, T_{\vec{u_2}})$ is calculated using Eq.(2), $sim(L_{\vec{u_1}}, L_{\vec{u_2}})$ is calculated by Eq.(2) in movielen and Eq.(3) in lastfm for getting better performance. It is also common to use only a subset of users for both performance and accuracy when making predictions. In our experiments, A constant number is set.

3.3   Recommendation Graph Construction

Algorithm 1 describes the process of graph construction by using selected information which is highly related to target user. The graph has four layers which combine user-based, item-based and content-based factors together as illustrated in Fig.3. Neighbor layer contains the target user labeled with blue and his similar-minded neighbors. In candidate layer: On one hand, we add items rated by those similar-minded users(user-based); On the other hand, we also add items labeled with blue that are similar to target user's historical records from steps 8 to 14(item-based). Namely, we treat target user himself as a fake neighbor with candidate items whose number is equal to the average rated items of all real neighbors. Items similarities are calculated using Eq.(2) based on content tags. Tags' importance are weighted by classical $TF * IDF$. It is easy to see that the items from two factors may be overlapped. In candidate layer, we remove the items target user has rated and set them as history layer labeled with yellow(content-based). Every connection in graph G is bi-directional.

In the graph, the left part acts as collaboration factor; while the right part represents content factor. Collaboration factor influences candidate items by direct rating link. Since simple term matching is not precise for linking items because of existing synonyms and polysemants. We use intermediate topic layer to help content factor propagate its influence to the candidate items. Top-N recommendation list is finally generated from the candidate items in the second layer by using personalized random walk.

Algorithm 1. Recommendation Graph Construction

Input : Target user μ, Item topics $T[N][K]$, User-Item rates $R[M][N]$

Output: $G(V, E, W)$: a set of triples: $\langle node_a, node_b, weight \rangle$

1 **begin**

2   **Init** : $TotalCount \leftarrow 0$, $Candidate\_Set \leftarrow \emptyset$, $Item\_Map \leftarrow \emptyset$, $G \leftarrow \emptyset$

3   **for** $u \in \mu.Neighbors$ **do**

4   **for** $i \in R[u] \& i \notin R[\mu]$ **do**

5             $G \leftarrow G \cup \langle u, i, R[u][i] \rangle$

6             $Candidate\_Set \leftarrow i \cup Candidate\_Set$

7         $TotalCount \leftarrow TotalCount + Size(R[u])$

8   **for** $i \in R[\mu]$ **do**

9   **for** $\nu \in i.Neighbors$ & $\nu \notin R[\mu]$ **do**

            // Merge:put ν into $Item\_Map$ and sum up values of same ν

10             $Merge(< \nu, simi_{\nu,i} * R[\mu][i] >, Item\_Map)$

11     $Num \leftarrow TotalCount/Size(\mu.Neighbors)$

12   **for** $\nu \leftarrow Sort(Item\_Map.Values, descending)[1 : Num]$ **do**

13         $G \leftarrow G \cup \langle \mu, \nu, Item\_Map.get(\nu)/\sum_{i \in R[\mu]} simi_{\nu,i} \rangle$

14         $Candidate\_Set \leftarrow \nu \cup Candidate\_Set$

15   **for** $Topic$ $k = 1 : K$ **do**

16   **for** $i \in R[\mu] \cup Candidate\_Set$ **do**

17             $G \leftarrow G \cup \langle i, k, T[i][k] \rangle$

3.4 Random Walk-Based Recommendation

To generate the top-N recommended items in the proposed graph, we use personalized random walk with restart(RWR). By setting a biased probability of jumping to the starting nodes that are highly related to target user, RWR allows us to calculate the relatedness between candidate items and target user's preference. In our graph, neighbor nodes and history nodes are highly related to the target user, thus they are regarded as the starting nodes.

RWR works as follows: From the starting nodes, RWR is performed by randomly jumping to another linked node at each step, the jumping probability is proportional to the weight of outside links. Additionally, in each step there also exist probability α to force random walk restart at the starting nodes. Formally it is defined by Eq.(4).

$$r^{t+1} = (1 - \alpha)Mr^t + \alpha q \qquad (4)$$

M is the transition matrix. $M_{i,j}$ denotes the probability of j being the next state given that the current state is i. q is the initial query vector in which the elements corresponding to the starting nodes are set to 1, others are set to zero. r^t records the visiting probability of each node at step t. Update the values of vector r iteratively until convergence at step l. Finally r_i^l represents the relatedness between node i and the starting nodes.

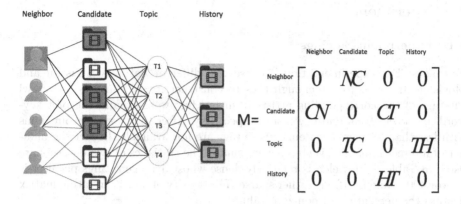

Fig. 3. Recommendation Graph

Fig. 4. Transition matrix of left graph. NC, CN, CT etc. are M's sub-matrices

In our setting, $q = (q_N, q_C, q_T, q_H)$, and the elements represent the four layers. Based on the starting nodes, we set $q = (\beta q_N, 0, 0, (1 - \beta)q_H)$. β controls the tradeoff between collaboration and content. q_N^u denotes the similarity between neighbor u and target user. Considering that target user is also put into the neighbor layer, we set his similarity equal to 1. q_H^j denotes the importance of item j in the content factors. Here, we define $q_H^j = R_{u,j} * IDF(j)$ where $IDF(j) = Log(\frac{|I|}{ItemFreq(j)})$. $|I|$ denotes the number of total items. q_N and q_H are then normalized to 1 separately. $r^{(0)}$ is the initial visiting probability which is set equal to q. To get transition probability of each node, we need to normalize each row of M to 1, as illustrated in Fig. 4. In order to stop CN(ratings) from suppressing CT(topic distributions), we consider to give equal contributions from candidate layer to its two sides. Namely we firstly normalize CN and CT to 1 for each row respectively. Then CT and CN are normalized to 1 together for each row. After getting the stable visiting probability of each node, we select the top-N items from candidate layer by sorting their values in descending order.

3.5 Complexity Analysis

The main computation of random walk-based methods is updating the values of each node in the graph. Assuming that the number of average rated items per user is P, then there are N users, (N+1)P items, and K topics nodes in our recommendation graph. For updating user nodes, the computational complexity is O(NP). The computational complexities for candidate, topic and history nodes are O(NP(N+K)), O(K(NP+P)) and O(KP) respectively. Therefore, the total complexity in one iteration is O(P(N^2+2NK+2K+N)). Usually K and N are small fixed values, which indicates that the running time of our method is proportional to the number of average rated items per users. In a long period, average ratings of the whole users will not change so much, so the complexity of our proposed method possesses good stability.

4   Experiment

4.1   Experiment Settings

Datasets. For experimentation, we use two different datasets[1]: Lastfm and Movielen. In order to get enough tags to construct items' contents, we crawl the tags of the corresponding artists from last.fm[2] website, and the plot keywords of movies from imdb[3] website respectively. After that, we filter out items with less than 20 tags and items whose tags are not available. In fact the number of filtered items in this step is very small. The statistics of two datasets are listed in Table 1. Movielen is relatively dense whose average ratings per user is almost 400, while Lastfm is rather sparse. The density of user-item rating matrix denotes the percentage of non-zero values.

Compared Methods. To better understand our proposed method, We compared with five representative algorithms. (1)UCF: User-based collaborative filtering [15]; (2)ICF: Item-based collaborative filtering [15]; (3)UICF: Hybrid collaborative filtering, which combines the predicted score from both UCF and ICF to generate the final rankings. To consider the content-based factors, users similarities are calculated by using Eq.(1) described in section 3.2. $\delta = 0.5$; (4)PureSVD: An algorithm based on Matrix Factorization. [19] conducted extensive experiments to suggest that PureSVD outperforms all other powerful models such as AsySVD , SVD++ [20]; (5)RWR-FULL: Personalized random walk in graph which contains all the users and items [12]. They use target user and his already rated items as the starting nodes.

Further Study. To further explore the rationality of the way in which we fuse different factors. We compared a set of variants of our proposed method: RWR-U only considers the user-based factor; RWR-UI considers user-based and item-based factors; RWR-UC considers user-based and content-based factors. Their respective parameters are set in the same way as RWR-UIC.

Evaluation Metrics. To measure the performance, we adopt the following metrics to cover various aspects of our consideration. Additionally, we randomly split each user' ratings into five parts equally. 20% of the items in each user's profile are put aside as T_u for testing. As default, we focus on the performance of top 200 recommendations.

(a). **Accuracy.** We use precision, recall and MAP curves to measure accuracy. $Precision = \frac{|recs \cap tests|}{|recs|}$ and $Recall = \frac{|recs \cap tests|}{|tests|}$ have been used widely in recommendation [17]. To address the bias of rank positions in the list, we also utilize MAP. If correct answers are ranked higher, MAP is higher. MAP=$\frac{1}{|U|} \sum_{u \in U} \{ \frac{1}{|T_u|} \sum_{k=1}^{n} P(k)h(k) \}$. $h(k) = 1$ if the kth recommended item belongs to T_u, 0 otherwise. P(k) denotes the precision ranked up to k.

---

[1] http://grouplens.org/datasets/hetrec-2011/
[2] www.lastfm.com
[3] www.imdb.com

(b). **Coverage.** Obviously, accuracy is not enough for evaluation. Recommendations should cover distinct items stored in database as many as possible to boost presentation and sales. Coverage=$\frac{1}{|I|}|\bigcup_{u \in U} R(u)|$. R(u) denotes the recommended items for user u. $|I|$ is the total number of unique items.

(c). **Novelty.** It is trivial to recommend popular items which are so apparent to bring few surprise to users. Thus good recommendation lists should be better in Novelty=$\frac{1}{|U|} \sum_{u \in U} \{ \frac{1}{|R(u)|} \sum_{i \in R(u)} log(item\_pop(i) + 1) \}$. The lower, the better. Here, $item\_pop(i)$ denotes the rating popularity of item i.

Parameters Setting. The threshold of similar users or similar items in our proposed method, is set equal to UCF and ICF, so that we can compare the performance between basic CF and our method fairly. In experiments, we set them as 30 in both datasets for default. β controls the balance between collaboration and content, and we tested its sensitivity on a set of limited values $\beta \in \{0, 0.01, 0.05, 0.1, 0.5, 1\}$. In the random walk-based methods, α controls the probability of jumping to the starting nodes. We set α=0.8 as proposed in RWR-FULL [12], because we also find that when α increases, MAP, precision and recall all increase. This can be explained as a more personalized model as increasing α makes the model go back to the initial query vector q more frequently. The number of topics, K is set 30 in movielen and 40 in lastfm by cross validation. Iteration of all the random walk-based methods is set 100. The reduced dimension of PureSVD is set 50 as analyzed in [19] for getting better performance.

4.2 Results and Analysis

β controls the tradeoff between collaboration and content in query vector, so we firstly conducted a series of experiments to study its influence on final performance. If $\beta = 1$, we just regard neighbors as the starting nodes; while if $\beta = 0$, starting nodes are only history items. Because of a limitation of space, we only give the findings. We find that spreading out from only one side can not perform well. The optimal β is 0.01 in movielen and 0.05 in lastfm, which are very small. Because in the right graph part, nodes are fully connected by dense edges, while the left is rather sparse. As the transmitted values to other nodes need to be divided by outdegrees, to make the influence from history layer enough significant, we must give higher weight to q_H , which means a lower β for q_N.

Then we compared our method with existing algorithms using two datasets. From Fig.5, we see that UCF Performs better than ICF especially in lastfm. Because the number of users is much smaller than that of items and we can get more information for each user than each item. However, UICF does not

Table 1. Datasets Statistics

	# users	# items	# average rates	# average tags	# total rates	density
Movielen	2113	8631	394	101	834116	4.57%
Lastfm	1890	14998	47	56	90013	0.31%

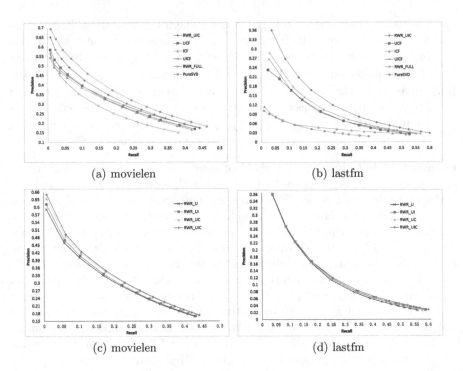

(a) movielen   (b) lastfm

(c) movielen   (d) lastfm

Fig. 5. Interpolated Precision-Recall curves for top-200 items. Fig(a),(b) compare our methods with other algorithms. Fig(c),(b) explore different variants of our approach.

gain much better performance than UCF, though it considers both content and collaboration. This phenomenon verifies our discussion that a simple weighted combination of several factors in the user similarity calculation and ranking adjustment can not perform effectively. But, our method RWR-UIC always outperforms other algorithms. The reason is that: On one hand, the random walk of our method matches people's decision-making process. Normally, people start with considering items: (1) similar to their history records (2) their similar-minded neighbors also like (3) contain the topics they prefer. The above three aspects are considered in our method at the same time in deciding an item's relatedness. And the interaction between different factors are not done once, but by iterative updates, so different factors are integrated fully. On the other hand, graph-based structure allows us to calculate the similarity between different items, even they have no direct connections. Namely, some extra latent relationships are explored.

Although both RWR-UIC and RWR-FULL use the graph-based ranking to generate top-N lists, RWR-UIC performs much better than RWR-FULL in both datasets. We argue that incorporating too much information brings some irrelevant noise which reduces the opportunity of transmitting to really related nodes, as RWR-FULL do. In contrast, We just retain the most relevant users and items in the graph and consider useful content factors apart from rating links. Another finding is that graph-based methods perform better in sparse dataset. In Lastfm

Table 2. Summary of MAP, Coverage and Novelty for all algorithms

	UCF	ICF	UICF	PureSVD	RWR-FULL	RWR-U	RWR-UI	RWR-UC	RWR-UIC
					Movielen				
MAP	0.1822	0.1707	0.1849	**0.2265**	0.1501	0.1858	0.1903	0.1983	0.2010
Coverage	0.22	0.4831	0.2649	0.3599	0.0936	0.3007	0.5796	0.4518	**0.6001**
Novelty	6.2908	6.0313	6.2611	6.0130	6.4308	6.2511	6.1828	6.1187	**5.9950**
					Lastfm				
MAP	0.1145	0.0598	0.1161	0.0517	0.1271	0.1464	0.1497	0.1487	**0.1517**
Coverage	0.4582	0.2709	0.3998	0.1937	0.4713	0.4983	0.6216	0.6504	**0.7148**
Novelty	3.3846	4.0411	3.6733	3.5219	3.8098	3.2975	3.2449	3.1428	**3.0934**

whose density is 0.39%, RWR-UIC and RWR-FULL both perform much better than all the other methods. While in the dense Movielen, RWR-FULL drops a lot in precision and recall. The advantage of our method RWR-UIC is not as significant as that in Lastfm. We explain it as follows: On one hand, graph-based methods can utilize indirect relationships to calculate similarity, which is rather useful in alleviating the sparsity of ratings; On the other hand, when density increases a lot from Lastfm to Movielen, Precision and Recall increase correspondingly. But introduced noise also weakens the growth rate.

PureSVD performs better than our method in the dense movie dataset, which suggests the good performance of utilizing latent rating patterns based on matrix factorization. However, in the sparse music dataset, PureSVD performs the worst of all. It is evident that PureSVD suffers from overfitting when the rating matrix is very sparse. In general, our method has stable performance in both datasets.

From Table 2, we see that our method performs better in novelty. Because we also consider items generated by item-based technique, as long as they are similar to user's history records. More importantly, content-based factor further boosts similar items, if they contain the topics target user prefers. If we just consider rating links, more popular items will be recommended as RWR-FULL. So we can recommend not so "apparent" items.

From Fig.5(c) 5(d) and Table.2. We find that considering more factors such as item-based or content-based techniques always improves the performance of RWR-U. But RWR-UIC performs the best. It is verified again that the intuition of our random walk reflects people's decision-making process. People can judge an item's relatedness more accurately based on all the three aspects. Meanwhile the graph structure is suitable and effective to fuse different factors.

As random walk is conducted in a concise sub-graph which only contains carefully selected relevant information. Our method has a fast speed of convergence. We tested our method(written by JAVA) in a server with 32 GB RAM. In movielen, our running time is 220 ms, while RWR-FULL needs 1.94s; In lastfm, we just need 40 ms. RWR-FULL needs 174ms. The complexity of our method is greatly reduced compared to existing graph-based methods. PureSVD can be run fast online, but SVD decomposition is rather time-consuming offline.

5 Conclusion

In this paper, we propose a graph-based method for hybrid recommendation, which can combine different factors iteratively. Our method can alleviate the data sparsity and boost long tail items. Experimental results show that our method performs well especially in sparse dataset. In the future, we plan to explore more relationship in the same layer to further improve the performance.

Acknowledgement. This research is supported by the 863 project of China (2013AA013300), National Natural Science Foundation of China (Grant No. 61375054 and 61402045) and Tsinghua University Initiative Scientific Research Program (20131089256), and Cross fund of Graduate School at Shenzhen, Tsinghua University (JC20140001).

References

1. Su, X., Khoshgoftaar, T.M.: A survey of collaborative filtering techniques. Advances in Artificial Intelligence 2009, 4 (2009)
2. Herlocker, J.L., Konstan, J.A., Borchers, A., Riedl, J.: An algorithmic framework for performing collaborative filtering. In: SIGIR 1999, pp. 230–237 (1999)
3. Sarwar, B., Karypis, G., Konstan, J., Riedl, J.: Item-based collaborative filtering recommendation algorithms. In: WWW 2001, pp. 285–295 (2001)
4. Hofmann, T.: Probabilistic latent semantic indexing. In: SIGIR 1999, vol. 24, pp. 50–57 (1999)
5. Blei, D.M., Ng, A.Y., Jordan, M.I.: Latent dirichlet allocation. The Journal of Machine Learning Research 3, 993–1022 (2003)
6. Lops, P., De Gemmis, M., Semeraro, G.: Content-based recommender systems: State of the art and trends. In: Recommender Systems Handbook, pp. 73–105 (2011)
7. Meng, F., Gao, D., Li, W., Sun, X., Hou, Y.: A unified graph model for personalized query-oriented reference paper recommendation. In: CIKM 2013, pp. 1509–1512 (2013)
8. Popescul, A., et al.: Probabilistic models for unified collaborative and content-based recommendation in sparse-data environments. In: UAI 2001, pp. 437–444 (2001)
9. Koren, Y., Bell, R., Volinsky, C.: Matrix factorization techniques for recommender systems. Computer 42(8), 30–37 (2009)
10. Balabanović, M., Shoham, Y.: Fab: content-based, collaborative recommendation. Communications of the ACM 40(3), 66–72 (1997)
11. Das, A.S., Datar, M., Garg, A., Rajaram, S.: Google news personalization: scalable online collaborative filtering. In: WWW 2007, pp. 271–280 (2007)
12. Konstas, I., Stathopoulos, V., Jose, J.M.: On social networks and collaborative recommendation. In: SIGIR 2009, pp. 195–202 (2009)
13. Haveliwala, T.H.: Topic-sensitive pagerank. In: WWW 2002, pp. 517–526 (2002)
14. Craswell, N., Szummer, M.: Random walks on the click graph. In: SIGIR 2007, pp. 239–246 (2007)
15. Karypis, G.: Evaluation of item-based top-n recommendation algorithms. In: CIKM 2001, pp. 247–254 (2001)

16. Yildirim, H., Krishnamoorthy, M.S.: A random walk method for alleviating the sparsity problem in collaborative filtering. In: RecSys 2008, pp. 131–138 (2008)
17. Shani, G., Gunawardana, A.: Evaluating recommendation systems. In: Recommender Systems Handbook, pp. 257–297 (2011)
18. Porteous, I., Newman, D., Ihler, A., Asuncion, A., Smyth, P., Welling, M.: Fast collapsed gibbs sampling for latent dirichlet allocation. In: KDD 2008, pp. 569–577 (2008)
19. Cremonesi, P., Koren, Y., Turrin, R.: Performance of recommender algorithms on top-n recommendation tasks. In: RecSys 2010, pp. 39–46 (2010)
20. Koren, Y.: Factorization meets the neighborhood: a multifaceted collaborative filtering model. In: KDD 2008, pp. 426–434 (2008)

RDQS: A Relevant and Diverse Query Suggestion Generation Framework

Hai-Tao Zheng and Yi-Chi Zhang

Tsinghua-Southampton Web Science Laboratory,
Graduate School at Shenzhen, Tsinghua University, Shenzhen, China
zheng.haitao@sz.tsinghua.edu.cn, zhangyichi12@mails.tsinghua.edu.cn

Abstract. Traditional query suggestion methods mainly leverage click-through information to find related queries as recommendations, without considering the semantic relateness between queries. In addition, few studies use click-through distribution in diversifying query suggestions. To address these issues, we propose a novel and effective framework to generate relevant and diversified query suggestions. We combine query semantics and click-through information together to generate query suggestion candidates which are highly relevant to original query , we use click-through distribution to diversify the candidates. We evaluate our method on a large-scale search log dataset of a commercial engine, experimental results indicate that our framework has significantly improved the relevance and diversity of suggested queries by comparing to four baseline methods.

Keywords: query suggestion, diversity, click-through information, query semantics.

1 Introduction

With the development of Internet, search engines have become one of the most important way to get information. User inputs queries into the search engine's search blank, then the information retrieval system returns the corresponding results according to the query user issues. However, users are seldom satisfied with the results at the first time. On one hand, users' input is always short, it can hardly reflect their search intent clearly. On the other hand, the queries themselves are ambiguous, search engines may have difficulty in giving right answers to users at the first time. Therefore, how to help users to formulate suitable queries has been recognized as a challenging problem to search engines. To solve this problem a valuable technique, query suggestion, has been employed by most commercial search engines, such as *Google*[1], *Bing*[2] and *Yahoo!*[3]

Query suggestion plays an important role in improving the usability of search engines and it has been well studied in academia as well as industry. Traditional

---

[1] http://www.google.com/
[2] http://www.bing.com/
[3] http://www.yahoo.com/

© Springer International Publishing Switzerland 2015
R. Cheng et al. (Eds.): APWeb 2015, LNCS 9313, pp. 586–597, 2015.
DOI: 10.1007/978-3-319-25255-1_48

query suggestion approaches mainly focus on recommending queries relevant to the original query. Due to the redundancy in the results of query suggestion, more and more researches have turned their attention into query suggestion diversifying. However, as an important feature to distinguish different query suggestions, click-through distribution has never been used in query suggestion diversifying work.

In addition, previous query suggestion methods measure similarity relied on either click-through information or query semantics, few of them combine the two factors together to generate query recommendations. In fact, click-through information and query semantics are both important factors in measuring the similarity between queries. Only considering one of them would reduce the relevance of recommendatory results.

In this paper, we propose a Relevant and Diverse Query Suggestion (RDQS) generation framework to produce relevant as well as diverse query suggestions. Specifically, our approach generates suggestion candidates using click-through information, re-ranks them by taking query semantics into account, then diversifies them based on click-through distribution between queries and URLs.

We evaluate our framework for query suggestion using click-through data of a commercial search engine. We measure the performance from different aspects. Empirical experimental results show that our framework can effectively generate highly relevant and diverse query suggestions.

The main contributions of our approach can be summarized as follows:

1. We combine query semantics and click-through information to generate query suggestions, thus the suggestions are not only related in click-through relations but also similar in semantics. query suggestions.
2. Since click-through distribution is an excellent feature to distinguish different kinds of queries, we are the first to apply click-through distribution in diversifying query recommendations.
3. We conduct comprehensive experiments to evaluate RDQS by comparing four baseline methods. Experiments results show that RDQS has significant performance in terms of relevance and diversity.

The rest of this paper is organized as follows: Section 2 presents the related work of query suggestion; Section 3 introduces our RDQS framework; In section 4 we conduct experiment to evaluate our proposal; We conclude this paper with future work in Section 5.

2   Related Work

Query suggestion techniques are used to help users to express their information needs. Search logs have been widely used in recent studies. Craswell[1], Wang[2], Mei[4] and Cao[8] leverage query logs first to construct a Query-URL bipartite graph, then generate suggestions using random walk model based on the click-through relations between queries and URLs. Boldi et.al.[5] utilize query logs to construct a query-flow graph, then apply a short random walk on the graph to generate query suggestions without the click-through information. Song[6] and Ozertem[7] mine a variety of characteristics of queries and use machine learning methods to find out related queries.

Since only considering relevance of query suggestion is far from satisfying users, more and more studies have focused on diversifying query suggestions. Zhu et.al. recommend diverse and relevant queries based on the intrinsic query manifold[15], and they find among the existing researches, Maximal Marginal Relevance (MMR) [11] is the most well-known method used for result set diversification. Ma et.al. [10] combine MMR with random walk to diversify query suggestions. Zhu et.al. [12] apply an absorbing random walk on the graph. In order to achieve diversity, it turns the selected object into an absorbing state and then selects the next object based on the expected number of visits to each node before absorption.

However, the approaches mentioned above cannot guarantee to produce relevant as well as diverse recommendations. These methods fail to combine query semantics and click-through information together to generate highly relevant recommendations. In addition, they all ignore click-through distribution which is an important feature in diversifying process. In this paper, we will show how to combine query semantics and click-through information to guarantee recommendations which are closely related to original query, and how to take advantage of click-through distribution in diversifying process.

3 Methodology

RDQS framework is primarily composed of three steps. The first step is to generate suggestion candidates, and we adopt random walk algorithm to generate n query suggestion candidates in this paper. After that, we re-rank the candidates based on a relevancy criterion using snippets to achieve query suggestions that are more relevant to original query. In the last step, we diversify these suggestions using click-through distribution and generate the final results. The whole process is sketched in Figure 1.

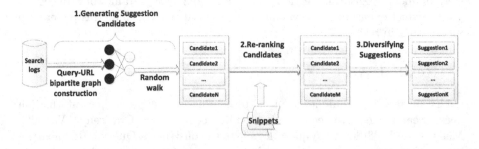

Fig. 1. The framework of RDQS

3.1   Generating Suggestion Candidates

First we use random walk to generate query suggestion candidates. We choose random walk because it's an efficient way to generate candidates as many as possible. Step 1 of RDQS focuses on generating candidates from search logs based on the forward random walk. A Query-URL bipartite graph is constructed as $G = (V, E)$, in which vertexes are composed by two parts $V = V_1 \bigcup V_2$. V_1 represents the set of all unique queries and V_2 represents the set of all unique URLs. There exits an edge from node $x \in V_1$ to node $y \in V_2$ if y is a clicked URL of query q in the search logs. The weight $w(q, u)$ is the total number of times that u is clicked when query q is issued. Note that since the edge is an undirected edge, the weight $w(q, u)$ is the same as $w(u, q)$. The transition probabilities from node i to node j in this paper is defined as follows:

$$P_{t+1|t}(j \mid i) = \frac{\omega(i, j)}{\sum_k \omega(i, k)} \tag{1}$$

where k ranges over all nodes. $P_{t+1|t}(j \mid i)$ denotes the transition probability from node i at step t to node j at time step $t + 1$.

We represent the transitions as a sparse matrix A and perform the random walk using A. We calculate the probability of transition from node i to node j in t steps as $P_{t|0}(j \mid i) = [A^t]_{ij}$. The random walk sums all the probabilities of all paths of length t between the two nodes. It gives a measure of the volume of paths between these two nodes. If there are many paths the transition probability will be high. The larger the transition probability $P_{t|0}(j \mid i)$ is, the more the node j is similar to node i. We select the top n largest $P_{t|0}(j \mid i)$ as candidates. In this paper, we set the $n = 200$ empirically, and 200 is big enough to cover all kinds of related queries.

3.2   Re-ranking Candidates

There are some noises in the generated query candidates. We find that some candidates rank high but not related to the original query, while some highly related candidates have low ranking. In order to guarantee that the results are ranked base on their semantic relevance, we choose to use query semantics to re-rank the candidates.

Query semantics is the content that are closely related to the query in semantic. As we all know, it's hard to determine the degree of semantic relevance between two queries according to their literal meaning because queries are short, generally consisted of less than ten words. However, we can extend the query semantics with related snippets.

Snippets, the few lines of text that appear under each search result, are designed to give users a sense for what's on the page and why it's relevant to their query. We utilize snippets to enrich query semantics, then we can leverage traditional documents similarity comparison methods to calculate queries similarity. In this paper we crawl top ten snippets from Google for each query.

For two queries q_i and q_j, we tokenize their related words into term vector $\overrightarrow{Q_i} = (t_{i1}, t_{i2}, ..., t_{im})$ and $\overrightarrow{Q_j} = (t_{j1}, t_{j2}, ..., t_{jn})$. t_{kl} represents the frequency of term t_l in query q_k, the cosine similarity between two vectors $\overrightarrow{Q_i}$ and $\overrightarrow{Q_j}$ $\cos(\overrightarrow{Q_i}, \overrightarrow{Q_j})$ can measure the similarity of the two queries. The similarity of two queries is calculated as follow:

$$sim(\overrightarrow{Q_i}, \overrightarrow{Q_j}) = \frac{\sum_{k=1}^{n} t_{ik} \times t_{jk}}{\sqrt{\sum_{k=1}^{n} t_{ik}^2} \times \sqrt{\sum_{k=1}^{n} t_{jk}^2}} \tag{2}$$

We compare all the candidates to original query, and re-rank the candidates according to its similarity to original query from high to low. We choose top k instead of all re-ranked results as the input of step 3, because more candidates would bring unrelated suggestions.

3.3 Diversifying Suggestions

Click-through information reflects users' click behaviour in search engines. We define query q_i and URL u_j exist a click-through relationship $R(q_i, u_j)$ when u_j being clicked after q_i being issued. Click-through distribution reflects click-through information over queries. Queries may be redundant in semantic if they have similar click-through distribution. On the contrary, a query will be diverse to an other query if the two queries' click-through distributions are different. For example, for a Query-URL bipartite shown in Figure 2, we can achieve three suggestions – 'nikeid nike', 'nike id' and 'nike shoes' – for original query 'nike'. However 'nike id' seems like a redundant suggestion with 'nikeid nike' since they have similar click-through distribution. In fact, 'nike shoes' is actually a better

Fig. 2. An example of click-through distribution

suggestion than 'nike id' if we already add 'nikeid nike' in final suggestion results.

Basically similar queries lead to similar click-through distribution, so we can diversify the query suggestion candidates according to the diverse between the click-through distribution. The main idea of the proposed diversified method is that the less click-through distribution the two queries share, the more different they are. Let C be the set of all query candidates, and S be the set of all final query suggestions. At first, S is empty and C is consisted of all the candidates generated from step 2. We first pick the candidates in C that is most similar to original query, that is the one has the highest score in step 2, as the first suggested query. After the selection of the first suggested query, we then employ the click-through distribution to calculate the diversity score of the rest queries in C, pick up the candidate which has the highest score as final query suggestion and add it into S. We pick out a candidate in C which shares least click-through information with original query and suggestions in S. The diversifying algorithm is summarized in Algorithm 1.

Algorithm 1. Diversifying Query Suggestions

Input: candidates set C and suggested queries set S
Output: A ranked list of all suggested queries L, and its size is K
Steps:

1: Initialize candidates set C and suggested queries set S: add original query and top suggestion candidate c_0 generated in re-ranking process into S, add all the results generated in re-ranking process except c_0 into C.
2: Repeat the following statements until K queries are suggested in L:
   (1) For all queries q_i in $C(q_i \in C)$, calculate diversity score:

$$d(q_i) = \sum_k div(q_i, q_k)$$

   q_k is an element in S, that is $q_k \in S$
   (2) Pick the next suggested query as $\arg\max_{q_i \in C} d(q_i)$, and add it into set S

In this paper, we propose two different ways to calculate diversity between two queries $div(q_i, q_k)$. They are described as follows:

$RDQS_1$: For a query q, we can draw a vector $\overrightarrow{q} = (\omega_1, \omega_2, \omega_3, ..., \omega_n)$ from search logs, the elements in this vector are URLs that clicked through q and w_i in $\overrightarrow{q}$ is the weight of URL_i.

$$\omega_i = \frac{count_i}{\sum\limits_{k=1}^{n} count_k} \tag{3}$$

$count_i$ represents the count of URL_i being clicked through q. Then we can calculate diversity of two queries as follows:

$$div(q_i, q_j) = 1 - \cos(\overrightarrow{q_i}, \overrightarrow{q_j})$$

$$= 1 - \frac{\sum_{k=1}^{n} \omega_{ik} \times \omega_{jk}}{\sqrt{\sum_{k=1}^{n} \omega_{ik}^2} \times \sqrt{\sum_{k=1}^{n} \omega_{jk}^2}} \tag{4}$$

$RDQS_2$: For a query q, there exists a union $U(q) = (u_1, u_2, u_3, ..., u_n)$, which u_i in $U(q)$ is the URL that clicked through query q. Then we can define the diversity of two queries q_i and q_j as follows:

$$div(q_i, q_j) = 1 - \frac{U(q_i) \cap U(q_j)}{U(q_i) \cup U(q_j)} \qquad (5)$$

The time complexity of Algorithm 1 is $O(|C| * k)$, actually the number of queries in C and k are not bigger, and the time complexity of Algorithm 1 is acceptable.

4   Experiment

4.1   Experimental Setup

In this section, we conduct empirical experiments to show the effectiveness of RDQS framework. We select AOL Search Data as our data set, which is a collection of real search log data based on real users. In summary, the data set consists of 20M web queries collected from 650K users over three months. For each query, the following details are available: a user ID, the query itself, timestamp, the clicked URL and the rank of that URL.

Since this dataset is raw data recorded by the search engine, and contains many noises. We conduct a similar method employed in [2] to clean the raw data. we clean the data by keeping those frequent well-formatted English queries (queries which only contain character 'a', 'b', ..., 'z' and space, and appear more than 5 times). After cleaning, we obtain a total of 9,752,848 records, 604,982 unique queries and 785,012 unique URLs.

We construct a query-URL bipartite graph on our data set, and randomly sample a set of 120 queries from our data set as the testing queries. For each testing query, we obtain six query suggestions. In order to evaluate the quality of the results, three experts are requested to rate the query suggestion results with '0' or '1', in which '0' represents 'irrelevant' and '1' means 'relevant'.

In order to evaluate the effectiveness of our framework, we use four methods as baselines, they are *Forward Random Walk(FRW)* [1], *Backward Random Walk(BRW)* [1], *Hierarchy Agglomerative Clustering(HAC)* [9], *and Diversifying Query Suggestions(DQS)* [10]. In order to evaluate the effect of re-ranking process, we add the results which are generated by re-ranking candidates using snippets(RCS) to compare.

4.2   Evaluation Measurements

In this paper, we adopt three metrics(*Relevance, Auto-Diversity* and *Human-Diversity*) to evaluate the precision and diversity in query suggestions.

Precision. In order to evaluate the quality of the results, three experts are requested to rate the query suggestion results with "0" and "1", where "0" means "irrelevant" and "1" means "relevant". We use the precision measurement in our experiment, i.e., precision at position n is defined as:

$$p@n = \frac{rn}{n} \qquad (6)$$

where rn is the number of relevant queries in the first n results.

Auto-diversity. Auto-diversity metric is a metric using a commercial search engine(i.e., Google), we adopt the same method used in [14]. Specifically, given two queries q_1 and q_2, we compute the proportion of different URLs among their top k ($k = 10$ in this paper) search results, the definition is as follows:

$$d(q_1, q_2) = \begin{cases} 1 - \frac{|o(q_1, q_2)|}{k} & \text{if } q_1 \neq q_2, \\ 0 & \text{otherwise.} \end{cases}$$

where $|o(q_1, q_2|$ is the number of URLs among the top k search results of the query q_1 and q_2. Then for a query q_1, the diversity of its suggestion is defined as:

$$div(q_1) = \sqrt{\frac{\sum_{q_1 \in U} \sum_{q_2 \in U} d(q_1, q_2)}{|U| (|U| - 1)}} \qquad (7)$$

Human-experts-diversity. There exists a problem in auto-diversity evaluation: if a query suggestion has nothing to do with the original query but they share some URLs, it would harm the users' experience, however it still contribute the diversity score in auto-diversity evaluation.

To solve this problem, we propose a new diversity evaluation metric: Human-experts-diversity. Manual evaluation is essential since human experts have a better understanding of the latent semantic meaning than any machines. We ask experts to label the relationship between two queries, there are three kinds of relationships: unrelated, duplicate, distinct. Experts label the query candidates based on the queries semantics and their relations to the original query.

After the accessors labeled all the suggestions, we calculate the diversity of one query suggestion method M as all the number of distinct queries dtn divided by total number of relevant queries rn which consisted of distinct ones dtn and duplicate ones dpn:

$$
\begin{aligned}
div(M) &= \frac{\sum dtn}{\sum rn} \\
&= \frac{\sum dtn}{\sum dtn + \sum dpn}
\end{aligned}
\qquad (8)
$$

4.3   Results and Discussion

Case Study. Some examples of suggestions generated by six methods for original test queries are given in Table 1. From the results shown in Table 1, we can see that some approaches, such as FRW, BRW, HAC, recommend closely related queries while introducing many redundant suggestions. For examples, for the test query 'jet blue', FRW recommends redundant suggestions 'jetblue', 'jetblue airways', 'blue jet', 'jet blue airways', 'jetblue airlines', 'jet blue airline'. BRW recommends redundant suggestions 'jetblue', 'jet blue airlines', 'jetblue

Table 1. Query suggestion comparisons among all methods

Query	jetblue	playboy	nike
FRW	jetblue jetblue airways blue jet jet blue airways jetblue airlines jet blue airline	sex play boy playboy magazine myspace playboy centerfolds my space	nike shoes nikeid nike golf nike soccer nike id niketown
BRW	jetblue jet blue airlines jetblue airlines airlines jetblue airways jet blue airways	playboy store www playboy com playboy magazine nude myspace pictures playboy plus playboy centerfolds	nikeid nike soccer nikeid nike nike id nike football nike sb
HAC	jetblue jetblue airlines jetblue airways jet blue airlines jet blue airways jet blue airline	www playboy com play boy jay hickman wackiest ship in army bing russell diana hyland	nike soccer nikeid nikeid nike nike store jim rome is burning jim rome
DQS	jetblue jet blue airlines jetblue airlines airlines jetblue airways jet blue airways	sex nicole narain my space playboy models playboy pictures kara monaco	nike shoes nike tennis apparel mens shox elevate xx jordan nike fitness apparel adidas
$RDQS_1$	airtran airlines delta air jetblue jetblue airways jetblue airlines cheap airline bargains	playboy playmates playboy plus playboy centerfolds playboy store playboy of the month playboy pictures	nike air max iconic nike yoga nike fitness apparel nikeid nike nike football nike shoes
$RDQS_2$	airtran airlines delta air airline flights cheap airline bargains jetblueairlines airtran	playboy big boobs playboy store playboy plus playboy pictures playboy lingerie playboy centerfolds	nike air max iconicst nike fitness apparel nike golf nike dunks nikeid nike nike football

airlines', 'jetblue airways' and 'jet blue airways'. HAC recommends redundant suggestions 'jetblue', 'jetblue airlines', 'jetblue airways', 'jet blue airlines', 'jet blue airways', 'jet blue airline'. For the test query 'playboy', HAC suggests redundant suggestions 'www playboy com' and 'play boy'. Since traditional methods only focus on relevancy, they do little work to diversify the results, they produce many redundant recommendations inevitably.

Meanwhile, we can easily find that DQS recommends queries with better diversity. However, there still exist some redundancy in the approach. For example, for the test query 'jet blue', 'jetblue', 'jet blue airlines', 'jetblue airlines', 'jetblue airways' and 'jet blue airways' suggested by DQS are the same meaning.

Moreover, results recommended by these baseline methods may not so closely related to the original query. For example, suggestion of 'myspace' with respect to 'playboy', and suggestion of 'jim rome' with respect to 'nike'.

Different from other methods, which only leverage search logs to generate suggestions, our method exploits semantic relationships between original query and suggestions based on snippets. We re-rank the suggestion candidates based on their relevancy using snippets. Therefore, the suggestions are guaranteed to be more semantically related to the original queries. We also use click-through information to diversify the results. Among all these approaches, we observe that $RDQS_1$ and $RDQS_2$ obtain best performance, more relevant as well as diverse queries can be found in their query suggestion results.

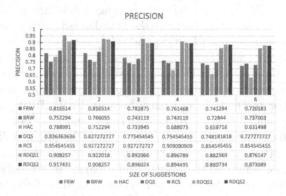

	1	2	3	4	5	6
FRW	0.816514	0.816514	0.782875	0.761468	0.741284	0.720183
BRW	0.752294	0.766055	0.743119	0.743119	0.72844	0.737003
HAC	0.788991	0.752294	0.733945	0.688073	0.658716	0.631498
DQS	0.836363636	0.827272727	0.775454545	0.754545455	0.748181818	0.727272727
RCS	0.954545455	0.927272727	0.927272727	0.909090909	0.854545455	0.854545455
RDQS1	0.908257	0.922018	0.892966	0.896789	0.882569	0.876147
RDQS2	0.917431	0.908257	0.896024	0.894495	0.880734	0.873089

Fig. 3. Precision of query suggestion over Re-ranking, $RDQS_1$ and $RDQS_2$ and other methods

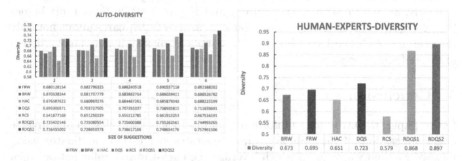

Fig. 4. Auto-diversity of query suggestion over Re-ranking, $RDQS_1$ and $RDQS_2$ and other methods

Fig. 5. Human-experts-diversity of query suggestion over Re-ranking, $RDQS_1$ and $RDQS_2$ and other methods

Experimental Results. Figure 3 shows the comparison results of precision. Figure 4 shows the comparison results of auto-diversity. Figure 5 shows the comparison results of human-experts-diversity. We have serval findings from the results:

Firstly, in Figure 3 we can see that RCS achieves the best score on precision. In fact, RCS is 13.14% higher than FRW. This is because re-ranking leverages semantics to re-rank the results generated in FRW, thus the re-ranking results are highly relevant to original queries. $RDQS_1$, $RDQS_2$ and RCS take both semantic and click-through relations into consideration, while other baselines only care about click-through relations, so we can observe that $RDQS_1$, $RDQS_2$ and RCS are about the same on precision, much better than other baselines. In addition, RCS is a litter better than $RDQS_1$ and $RDQS_2$ on precision. This is because $RDQS_1$ and $RDQS_2$ diversify the results generated by RCS, and it brings some irrelevant suggestions.

Secondly, both $RDQS_1$ and $RDQS_2$ leverage click-through distribution to diversify suggestion results, and achieve the best diversity performance. In Figure 4 and Figure 5 we can see both $RDQS_1$ and $RDQS_2$ achieve the highest diversity score, both $RDQS_1$ and $RDQS_2$ are at least 10% higher than other methods on diversity. It shows that click-through distribution is an effective feature to diversify query suggestion results. DQS is worse than $RDQS_1$ and $RDQS_2$ but better than other methods in diversity, because FRW, BRW and HAC only focus on finding related queries. $RDQS_1$, $RDQS_2$ and DQS take a step towards diversifying, thus achieve a higher diversity score. $RDQS_1$ and $RDQS_2$ obtain a much better performance than DQS shows that the overlaps of click-through distribution is more effective than hitting information in diversifying process.

Thirdly, RCS achieves the lowest performance in auto-diversity and human-experts-diversity. This is because RCS re-rank results based on the semantics of queries, the suggestions which are similar in semantics are usually representing the same meaning. The results in RCS exist redundancy.

Lastly, as can be seen from Figure 4 and Figure 5, $RDQS_2$ is a bit higher than $RDQS_1$ in diversifying score. That's because $RDQS_1$ cares about not only the overlaps of clicked URLs but also the counts of clicked, it may be overfitting in diversifying process.

5   Conclusion

In this paper, we introduce a uniform framework to generate relevant and diverse query suggestions. Our framework first generates suggestion candidates using forward random walk model which leverage the click-through information, then re-rank the candidates utilizing snippets related to queries. Finally it diversifies candidates based on click-through distribution. In this way, our framework can generate query suggestions by simultaneously considering diversity and relevance between queries in a unified way. The empirical results clearly show that our approach outperforms all the baseline methods in recommending highly diverse as well as closely related query suggestions.

Acknowledgement. This research is supported by the 863 project of China (2013AA013300), National Natural Science Foundation of China (Grant No. 61375054 and 61402045), Tsinghua University Initiative Scientific Research Program Grant No.20131089256, and Cross fund of Graduate School at Shenzhen, Tsinghua University (Grant No. JC20140001).

References

1. Craswell, N., Szummer, M.: Random walks on the click graph. In: Proceedings of the 30th Annual International ACM SIGIR Conference on Research and Development in Information Retrieval. ACM (2007)
2. Wang, X., Zhai, C.: Learn from web search logs to organize search results. In: Proceedings of the 30th Annual International ACM SIGIR Conference on Research and Development in Information Retrieval. ACM (2007)

3. Spink, A., Jansen, B.J.: A study of web search trends. Webology 1(2), 4 (2004)
4. Mei, Q., Zhou, D., Church, K.: Query suggestion using hitting time. In: Proceedings of the 17th ACM Conference on Information and Knowledge Management. ACM (2008)
5. Boldi, P., et al.: Query suggestions using query-flow graphs. In: Proceedings of the 2009 workshop on Web Search Click Data. ACM (2009)
6. Song, Y., Zhou, D., He, L.-W.: Post-ranking query suggestion by diversifying search results. In: Proceedings of the 34th International ACM SIGIR Conference on Research and Development in Information Retrieval. ACM (2011)
7. Ozertem, U., et al.: Learning to suggest: a machine learning framework for ranking query suggestions. In: Proceedings of the 35th International ACM SIGIR Conference on Research and Development in Information Retrieval. ACM (2012)
8. Cao, H., et al.: Context-aware query suggestion by mining click-through and session data. In: Proceedings of the 14th ACM SIGKDD International Conference on Knowledge Discovery and Data Mining. ACM (2008)
9. Beeferman, D., Berger, A.: Agglomerative clustering of a search engine query log. In: Proceedings of the Sixth ACM SIGKDD International Conference on Knowledge Discovery and Data Mining. ACM (2000)
10. Ma, H., Lyu, M.R., King, I.: Diversifying Query Suggestion Results. In: AAAI, vol. 10 (2010)
11. Carbonell, J., Goldstein, J.: The use of MMR, diversity-based reranking for reordering documents and producing summaries. In: Proceedings of the 21st Annual International ACM SIGIR Conference on Research and Development in Information Retrieval. ACM (1998)
12. Radlinski, F., et al.: Redundancy, diversity and interdependent document relevance. ACM SIGIR Forum 43(2) (2009)
13. Baeza-Yates, R., Tiberi, A.: Extracting semantic relations from query logs. In: Proceedings of the 13th ACM SIGKDD International Conference on Knowledge Discovery and Data Mining. ACM (2007)
14. Uhlmann, S., Lugmayr, A.: Personalization algorithms for portable personality. In: Proceedings of the 12th International Conference on Entertainment and Media in the Ubiquitous Era. ACM (2008)
15. Zhu, X., Guo, J., Cheng, X., et al.: A unified framework for recommending diverse and relevant queries. In: Proceedings of the 20th International Conference on World Wide Web, pp. 37–46. ACM (2011)

Minimizing the Cost to Win Competition in Social Network

Ziyan Liu[1], Xiaoguang Hong[1,*], Zhaohui Peng[1], Zhiyong Chen[1,2], Weibo Wang[1], and Tianhang Song[1]

[1] School of Computer Science and Technology, Shandong University, Jinan, China
[2] National Engineering Laboratory for ECommerce, Jinan, China
liuziyan@mail.sdu.edu.cn, {hxg,pzh,chenzy}@sdu.edu.cn,
{wangweibo1001,songth1202}@gmail.com

Abstract. In social network, influences are propagating among users. Influence maximizing is an important problem which has been studied widely in recent years. However, in the real world, sometimes users need to make their influence maximization through social network and defeat their competitors under minimum cost, such as online voting, expanding market share, etc. In this paper, we consider the problem of selecting a seed set with the minimum cost to influence more people than other competitors. We show this problem is NP-hard and propose a cost-effective greedy algorithm to approximately solve the problem and improve the efficiency based on the submodularity. Furthermore, a new cost-effective Degree Adjust heuristics is proposed to get high efficiency. Experimental results show that our cost-effective greedy algorithm achieves better effectiveness than other algorithms and the cost-effective Degree Adjust heuristic algorithm achieves high efficiency and gets better effectiveness than Degree and Random heuristics.

Keywords: social network, influence propagation, competition, cost, Degree Adjust.

1 Introduction

With the recent popularity of online social network such as Facebook, Twitter, Sina microblog, wechat, etc, many researchers have studied the influence diffusion problem in social network, including the diffusion of ideas, news, new products, and so on. We collectively refer to these diffusions as influence propagation. In this field, much research are related to influence maximization problem which is the problem of finding at most k nodes (seed nodes) in social network that could maximize the spread of influence.

Influence maximization problem was first formulated as a discrete optimization problem in [1]. Independence cascade model (IC) and liner threshold model (LT) are two most popular influence diffusion models. Most research do not consider the competition problem. However, competition exists extensively in the

---

* Corresponding author.

© Springer International Publishing Switzerland 2015
R. Cheng et al. (Eds.): APWeb 2015, LNCS 9313, pp. 598–609, 2015.
DOI: 10.1007/978-3-319-25255-1_49

real world. For example, two candidates want to win the election campaign, two competing companies expand market share at the same time, etc.

Some recent studies have considered competitive influence propagation. Most of them study the competitive influence maximizing problem which is the problem of finding at most k seeds in competitive social network that could maximize the spread of influence [7], while some of them study the influence blocking problem which is the problem of finding at most k seeds in competitive social network that could minimizing the spread of "bad" influence[6,10]. But sometimes, we not only need to expand our influence, but also need block competitors' influence simultaneous. Only our influence is lager than competitors, we can achieve the victory in competition. Different seed node often need different cost, so with the minimum seed set to win the competition not means the total cost is minimum. The problem of how to select a seed set with minimum cost to beat the competitor is needed to solved.

In this paper, we formulate the problem as Minimizing Cost to Win Competition problem (MCW). We prove that MCW problem under the competitive independence model(CIC) is NP-hard. Firstly, We propose cost-effective Greedy algorithm to approximately solve this problem. Because this algorithm requires Monte Carlo simulation to compute nodes' influence, which is inefficient. Then we improve the Cost-Effective Greedy algorithm's efficiency base on the submodularity. The improved cost-effective greedy algorithm runs obviously faster than original algorithm. Furthermore, to meet the need of solving the problem efficiently in large social network, we propose an efficient heuristic algorithm called Degree Adjust. It is similar to Degree heuristic but the effectiveness is much better. Our main contributions are as follows:

- We propose a new problem called minimizing cost to win competition (MCW) and prove it is NP-hard.
- We propose a considering cost performance Greedy algorithm to approximately solve MCW problem and improve the efficiency base on the submodularity.
- We propose a new Degree Adjust heuristics. Its running time is much less than greedy algorithm and effectiveness is better than other heuristics.

The rest of paper is organized as follows: in section 2, we review related work. We formulate this problems and prove it is NP-hard in section 3. In section 4, we present the Cost-Effective Greedy algorithm and Improved Cost-Effective Greedy algorithm. In section 5, we propose the Degree Adjust heuristic. In section 6, we experimentally compare the effectiveness and efficiency of those algorithms. Finally, we conclude the paper in section 7.

2   Related Work

Influence maximization is the problem of finding at most k nodes in social network that could maximize the spread of influence. This problem is first modeled as discrete optimization problem in [1] and they summarize two basic influence propagation models, independent cascade model (IC) and linear threshold model (LT). Based on the two models, [1] proposes a greedy algorithm to solve the problem. A number of studies have tried to solve this problem more efficiently [2,4,5,7,9]. But those research do not consider the competition problem.

The competitive influence propagation has been captured by a number of research in recent years [9,10,11]. Dubey, et al. studied the problem as a games in which firms compete for customers located in the social network[7], and they study the existence of Nash Equlibrium, showing that it is exist and is unique. Bharathi, S extends the IC model to competitive influence propagation and gives approximation algorithm for computing the best response to an opponent's strategy [3]. Influence blocking maximizing problem has been studied in [6,10]. They did not consider that we need to expand our influence and block competitors' influence simultaneous. Goyal, Amit consider minimizing budget problem in social network but he does not address the competition problem [8,12]. Sometimes we need maximizing our influence and blocking competitors influence simultaneous for win a competition. Further for the budget limit, we often want to win the competition with minimum cost. So we propose MCW problem.

3   Model and Problem

In this section, we define the CIC model and MCW problem. Then, we prove MCW problem is NP-Hard.

3.1   CIC Model

Independence cascade model is a basic diffusion model proposed in [1], which has been used in many works [2,4,5]. In IC model, a social network is considered as a directed graph G (V,E). Where V is the set of vertices representing users in social network and E is the set of directed edges representing relationship between users. Each edge is associated with an influence probability which means once a vertex is influenced at this step, it can influences its out-neighbors with the probability through the edge at next step. Each vertex has two states, activated or inactive, once a vertex is activated, it cannot change to inactive. At step 0 a seed set is activated while all other vertices are inactive. At any later step $t > 0$, if vertex v is activated in step t-1, it has once chance to activate each inactive out-neighbor u with probability p_{vu}. If v succeeds, then u becomes activated in step t. The process runs until no more vertices can be activated.

We now define the competitive independence model (CIC), in which every vertex has three states. Influenced by S(it means it is influenced by our selected seed Set S), influenced by C (it means it is influenced by competitor's selected seed Set C), or not be influenced. At step 0, all nodes in seed set S were influenced by S, all nodes in seed set C were influenced by C. At any step $t > 0$, if vertex v is influenced by S or C in step t-1, at step t, it has once chance to influence each current inactive out-neighbor u with probability p_{vu}. If v succeeds, then u becomes the same state with v. If there are two or more nodes trying to influence u at the same step, their attempts are sequenced in arbitrary order. Once a node is influenced by S or C, it can't change its state.

3.2 Minimum Cost to Win Competition Problem

In our problem, we use a directed graph G=(V,E) to represent the social network the same as influence maximizing problem. We use the CIC model as the propagation model in our problem. Given a target set $U \subset V$, we suppose competitor has selected seed set C and never change, We try to find seed set S to win the competition. let $InfS_U(S)$ represents the number of users influenced by S in the target set U and $InfC_U(S)$ represents the number of users influenced by C in the target set U. When the con-text is clear, we usually omit the U and let $InfS(S)$ represents the number of users influenced by S in the target set U and use $InfC(S)$ represents the number of users influenced by C in the target set U. We use $\delta(S)$ represent the number of users influenced by S minus the number of uses influenced by C in the target set U.

$$\delta(S) = InfS(S) - InfC(S) \tag{1}$$

The optimization problem we want to solve is to find the set S with minimum cost such that the influence of S is bigger than that of C. The formal problem is defined bellow.

Definition 1 (Minimizing Cost to Win Competition Problem). *The input of the problem includes the directed graph $G(V,E)$, the competitor selected seed set C and $|C| < \frac{|V|}{2}$, the influence propagation probability P, the $cost_v$ of node $v \in V$, the target set U, We use cost(S) represent the total cost of seed set S.*

$$cost(S) = \sum_{v \in S} cost_v \tag{2}$$

The target is to find a seed set S^* that can win the competition with the minimum total cost. that is

$$S^* = \arg\min_{\delta(S)>0} cost(S) \tag{3}$$

Theorem 1. *The MCW problem is NP-hard under the CIC model.*

Proof. Considering an instance of Set Cover Problem, $U = \{u_1, u_2, \ldots, u_n\}$ is the universal set. $S = \{S_1, S_2, \ldots, S_m\}$ represents sets union equals U. The covering optimization problem is to find a set of subsets whose union is equal to U and uses the fewest sets. We can encode the Set Cover instance as an special instance of MCW problem.

Given an arbitrary cover set problem instance. We define a corresponding directed bipartite graph with 2n+m-1 nodes represent a special MCW problem. We partition the nodes into three parts, the first part has n nodes and node i corresponding u_i in U, the second part has m nodes and node j corresponding to the S_j in S, the third part has n-1 nodes. The target set U is the union of part one and part three. If $u_i \in S_j$, there is a directed edge from node j to node

---

Algorithm 1. Cost-Effective Greedy Algorithm

Input:
  $G(V, E)$, seed C selected by the competitor, $cost_v$
Output:
  Seed set S
1: initialize S=$\emptyset$
2: **while** $\delta(S) <= 0$ **do**
3:    **for** each $v \in V$ **do**
4:        $p_v = \frac{\delta(S \cup v) - \delta(S)}{cost_v}$
5:    **end for**
6:    select $u = \arg\max_v \{p_v | v \in V \setminus (S \cup C)\}$
7:        $S = S \cup u$
8: **end while**
9: **return** S

---

i with a propagation probability $p_{ji} = 1$. We assume all nodes' cost is 1 and the competitor has selected the nodes in the third part as his seed set C. As he has influenced n-1 node in target set U, so if we want to win the competition with minimum cost, we must find the minimum seed set S to influence all the n nodes in the first part. Suppose we find a solution S. if $S \cap U \neq \emptyset$, we can replace each $u \in S \cap U$ with his neighborhood v and get a new seed set S^* which is also the solution to MCW problem. The solution is also a solution to set cover problem. Conversely, any solution to the Set Cover instance is also a solution to the special MCW problem. Since the set cover is NP-hard, MCW problem is NP-hard too.

4   Cost-Effective Greedy Algorithm

In this section, we try to use greedy algorithm to approximately solve the MCW problem. We have discussed the hardness of the problem in the above section. The most extensively studied algorithm for influence maximizing problem is greedy algorithm and always has a better effect than other algorithms. So we firstly try to use greedy algorithm to approximately solve the problem. For traditional influence maximizing problem is different to MCW, so traditional algorithms studied for influence maximizing problem must be altered to adapt to the MCW problem.

Algorithm 1 gives the details of Cost-Effective Greedy algorithm. $\delta(S)$ represent the number of users influenced by S minus the number of uses influenced by C. $\delta(S) > 0$ this shows that the influence of S has surpassed that of C, namely, we have won the competition. In Algorithm 1, we use Monte Carlo simulation to approximately calculate the influence of C and S under the CIC model. We use $p_v = \frac{\delta(S \cup v) - \delta(S)}{cost_v}$ to represent the cost performance of v and every round we select the node u with the highest cost performance until $\delta(S) > 0$.

Because the Monte Carlo simulation need to run many times to obtain accurate estimate, the running time is very long especially in large social network. We improve the efficiency of cost-effective greedy algorithm based on the submodularity . A fuction f(.) is submodular if it has the following property : $f(S \cup v) - f(S) \geq f(T \cup v) - f(T)$, for all $S \subseteq T$ and $v \notin S$. This means the margin gain of adding v to a set S is at least as high as that of adding the v to a superset of S.

Theorem 2. $\delta(\cdot)$ *is submodular under the CIC model.*

Proof. Since influence spreading in G(N,E) is a stochastic process under the CIC model. Considering a newly influenced node v attempts to influence his neighborhood w with probability p_{vw}. We can view that this stochastic outcome has been determined by flipping a coin with bias P_{vw}. It dose not matter whether the coin is flipped at the beginning of the whole process or at the moment of v trying influence w. So we assume that at the beginning of this process, we pre-flip all the coin to determine which edges are live (this means that the influence propagation probability in this edge is 1) and store the result. We also pre-flip all the coin to determine if C and S try to influence v at the same time, which will success.

Let $G'(N', E', NC, NS)$ represents the pre-flip result. The E' represents the set of live edges in E, and N' represents the node in N, NC(NS) represents the node will be influenced by C(S) if C and S try to influence it at the same time. We now prove that to every pre-flip $G'(N', E', NC, NS)$, the $\delta_{G'}(\cdot)$ is submodular.

Let S1 and S2 be two set of nodes and $S1 \subseteq S2$, $\delta_{G'}(S1 \cup v) - \delta_{G'}(S1)$
$= (infS_{G'}(S1 \cup v) - infC_{G'}(S1 \cup v)) - (infS_{G'}(S1) - infC_{G'}(S1))$
$= (infS_{G'}(S1 \cup v) - infS_{G'}(S1)) + (infC_{G'}(S1) - infC_{G'}(S1 \cup v))$

Let s(v,w) represents the shortest graph distance from v to w in G' and g(T,w) represents the shortest graph distance from any node in T to w in G'. If $s(v, w) < g(C, w)$ or $((s(v, w) = g(C, w))\&w \in NS)$, and v is influenced by S, v will successfully influence w, and w becomes the same state with v. Let $S(v) = \cup\{w|(w \in N')\&(s(v, w) < g(C, w)||(s(v, w) = g(C, w)\&w \in NS)\}$ then $infS(S1) = |\cup_{u \in S1} S(u)|$.

It is clear that the number in S(v) not in $\cup_{u \in S1}\{S(u)\}$ is at least as larger as the number in S(v) not in $\cup_{u \in S2}\{S(u)\}$, it means $infS_{G'}(S1 \cup v) - infS_{G'}(S1)) \geq (infS_{G'}(S2 \cup v) - infS_{G'}(S1))$

Let M represents all nodes that can be reached from any node in competitor seed set C, if $S = \emptyset$, $infC_{G'}(S) = |M|$.

Let $C(v) = \cup\{w|(w \in M)\&(s(v, w) < g(S, w)||(s(v, w) = g(C, w)\&w \in NC)\}$ Then $infC_{G'}(S1) = N - \cup_{u \in S1}C(u)$

It is clear that the number in C(v) not in $\cup_{u \in S1}\{C(u)\}$ is at least as larger as the number in C(v) not in $\cup_{u \in S2}\{C(u)\}$, it means $infC_{G'}(S1) - infC_{G'}(S1 \cup v) \geq infC_{G'}(S2) - infC_{G'}(S2 \cup v)$.

It follows that $\delta_{G'}(S1 \cup v) - \delta_{G'}(S1) \geq \delta_{G'}(S2 \cup v) - \delta_{G'}(S2)$.

It means $\delta_{G'}(\cdot)$ is submodular. $\delta(s) = \sum_{all\ G'} Prob[G'] \cdot \delta_{G'}$. since all $\delta_{G'}(S)$ are submodular, so $\delta(S)$ is submodular too.

We use p_v^i to represent the cost performance of v and use S^i to represent seed set has been selected in previous i rounds. We use p_v^j to represent the cost performance of v and S^j to represent seed set has been selected in previous j

Algorithm 2. Improved Cost-Effective Greedy Algorithm

Input:

$G(V, E)$,seed C selected by the competitor,$cost_v$

Output:

Seed set S

1: initialize S=$\emptyset$
2: **for** each $v \in V$ **do**
3:     $p_v = 0$
4: **end for**
5: **while** $\delta(S) <= 0$ **do**
6:     $maxp = 0$
7:     **for** each $v \in V$ **do**
8:         **if** $(p_v > maxp || p_v == 0)$ **then**
9:             $p_v = \frac{\delta(S \cup v) - s\delta(S)}{cost_v}$
10:             **if** $p_v > maxp$ **then**
11:                 $maxp = p_v$
12:                 $maxv = v$
13:             **end if**
14:         **end if**
15:     **end for**
16:     select $u = \arg\max_v \{p_v | v \in V \setminus (S \cup C)\}$
17:     $S = S \cup u$
18: **end while**
19: **return** S

rounds. If $j > i$ then $S^i \subseteq S^j$, as $\delta(\cdot)$ is submodular, so $\delta(S^i \cup v) - \delta(S^i) \geq \delta(S^j \cup v) - \delta(S^j)$
$p_v^i = \frac{\delta(S^i \cup v) - \delta(S^i)}{cost_v} \geq \frac{\delta(S^j \cup v) - \delta(S^j)}{cost_v} = p_v^j$,similar $p_u^i \geq p_u^j$. If $p_v^j \geq p_u^i$ then $p_v^j \geq p_u^j$
and in the j round we no need to recompute p_u^j, because node u must not be added to the seed S in this round.

Algorithm 2 gives the details of improved cost-effective greedy algorithm. Which is similarly to the origin algorithm, in the first round, it also need to compute the cost performance of all nodes. In the next round, if the cost performance of a node in the last round is not better than the best cost performance having been computed in this round, there is no need to compute it again. Our experiments show when competitor random selects a seed set with 100 nodes, the Improved Cost-Effective Greedy could achieve more than six times speedup in selecting the seed set S.

5   Degree Adjust Heuristic

Although we have improved the cost-effective greedy algorithm, the running time is still fairly long. We may need to find some heuristics to solve this problem more efficiently. Degree is a simple heuristic which in every round selecting the node with the largest degree. Degree is frequently used to select seed set in influence maximizing problem, but it does not consider the neighbors state. If a node v's out-neighbor u is influenced, node v can't influence u anymore and if u is an influenced in-neighbor of v, u may influence v in the next round.

We propose a new heuristic called Degree Adjust, which achieves better effectiveness than other heuristic and running time is much shorter than Greedy. We use $O(v)$ to represent the out-neighbors of v and $I(v)$ represent the in-neighbors of v. $E(v)$ represents the edges connected to v. Let $N(v) = \{v\} \cup O(v) \cup I(v)$, $G'(v)$ represents the subgraph of G(V,E) with N(v) as the vertices and $E(v)$ as the edges. We use the v's influence in this subgraph to forecast the overall influence and adjust the degree amount. Let O_v represents the out-neighbors number of v and I_v represent the in-neighbors number. Let I_v^C (I_v^S) represents the node number influenced by C(S) in I(v). let O_v^C (O_v^S) represents the node number influenced by C(S) in O(v). If we do not select v as the seed set of S, then expect number of C influenced in $G'(v)$ is
$c1 = I_v^C + O_v^C + ((1 - (1 - P)^{I_v^C}) \cdot (1 - P)^{I_v^S} + \frac{1}{2} \cdot (1 - P)^{I_v^C} \cdot (1 - P)^{I_v^S}) \cdot P \cdot (O_v - O_v^C - O_v^S)$.
The expected number of S influenced in $G'(v)$ is
$s1 = I_v^S + O_v^S + ((1 - (1 - P)^{I_v^S}) \cdot (1 - P)^{I_v^C} + \frac{1}{2} \cdot (1 - P)^{I_v^S} \cdot (1 - P)^{I_v^C}) \cdot P \cdot (O_v - O_v^C - O_v^S)$.
If we have selected v as the seed set of S, then expect number of C influenced in $G'(v)$ is $c2 = I_v^C + O_v^C$,
The expected number of S influenced in $G'(v)$ is
$s2 = 1 + I_v^S + O_v^S + P \cdot (O_v - O_v^C - O_v^S)$.
The expected profit of v in $G'(v)$ is

$$profit_v = (s2 - c2) - (s1 - c1)$$
$$\approx 1 + P \cdot (O_v - O_v^C - O_v^S) \cdot (1 - P(I_v^S - I_v^C))$$

The Degree heuristic thinks expected profit is $1 + P \cdot O_v$ so we use d_v to represent the new degree and use $profit_v$ to represent the expected profit of v.

$$d_v = (O_v - O_v^C - O_v^S) \cdot (1 - P(I_v^S - I_v^C)) \tag{4}$$

In the Degree Adjust algorithm in every round we find the node v with max d_v. And in the cost-effective Degree Adjust we find the node v with best cost performance $profit_v$. We use $p_v = \frac{profit_v}{cost_v}$ to represent the cost performance.

Algorithm 3 gives the details of cost-effective Degree Adjust algorithm. At each round we find the node v with the biggest $profit_v$ and recompute nodes' $profit_v$.

6   Experiments

To test the efficiency and effectiveness of our algorithms for solving MCW problem, we use three real datasets to compare the total cost and running time of different algorithms.

Algorithm 3. Cost-Effective Degree Adjust Algorithm

Input:
   $G(V, E)$,seed C selected by the competitor, $cost_v$
Output:
   Seed set S
1: initialize S=$\emptyset$
2: **for** each $v \in V$ **do**
3:    compute $I_v^C, O_v^C, I_v^S, O_v^S$
4: **end for**
5: **for** each $v \in V$ **do**
6:    $d_v = (O_v - O_v^C - O_v^S) \cdot (1 - P(I_v^S - I_v^C))$
7: **end for**
8: **while** $\delta(S) <= 0$ **do**
9:    $profit_v = \frac{1+d_v \cdot P}{cost_v}$
10:    select $u = \arg\max_v \{profit_v | v \in V \setminus S \cup C\}$
11:    $S = S \cup u$
12:    **for** each neighbor v of u **do**
13:       recompute $I_v^C, O_v^C, I_v^S, O_v^S, d_v, profit_v$
14:    **end for**
15: **end while**
16: **return** S

6.1   Experimental Setup

The real datasets used in experiments are got from Stanford Large Network Dataset Collection[1]. The details of the three datasets are listed in Table 1.

Table 1. details of datasets

Name	Nodes	Edges	Description
Epinions	7.6K	238.1K	Who-trusts-whom network of Epinions.com
Slashdot	7.7K	176.5K	Slashdot social network
Cit-HepPh	34.5K	421.5K	Arxiv High Energy Physics paper citation network

The code is written in C++ and run on a Linux server with 2.00GHz Intel Xeon E5-2620 and 32G memory. In the experiment we set $cost_v = 100 + v.outdegree$, in which v.outdegree represents the v's out-neighbor number. We set the influence propagation probability $P = 0.01$ and target set $U = V$. We run the simulation under the CIC model on the network for 10000 times and take the average value as the influence S and C. We compared the algorithms are as follows:

– Degree Adjust: in every round computing the d_v for each node in V and adding the node v with largest d_v to S until $\delta(S) > 0$.

---

[1] https://snap.stanford.edu/data/

- Cost-Effective Degree Adjust: (Algorithm 3) computing the $profit_v$ for each node in V at each round and adding the node v with largest $profit_v$ to seed set S until $\delta(S) > 0$.
- Degree : a simple heuristic that every round selecting a seed with largest out-degree until $\delta(S) > 0$.
- Random : randomly selecting node and adding to S until $\delta(S) > 0$.
- Improved Greedy :in every round adding the node with the biggest margin gain to S until $\delta(S) > 0$.
- Improved Cost-Effective Greedy: (Algorithm 2)in every round adding the node with the best cost performance to S until $\delta(S) > 0$.

6.2   Experimental Results

In the experiment, we mainly test the running time and total cost of different algorithms. We run all algorithms 10 times on every datasets and use the mean values to represent the running time and total cost.

Fig. 1 shows the total cost of different algorithms on the Epinions dataset under the CIC model. The vertical axis represents the total cost of S when S can win the competition. The horizontal axis represents the size of seed set C which has been selected by competitor randomly. From the figure we can see that the effectiveness of Random algorithm is worst. Our Degree Adjust algorithm's effectiveness is similar to the Degree algorithm. Our cost-effective Degree Adjust algorithm's effectiveness is better than Degree and Degree adjust. The effectiveness of improved Greedy is better than other heuristic but not as good as the improved cost-effective Greedy.

Fig. 2 shows the running time of different algorithms on the Epinions dataset under the CIC model. The vertical axis represents the running time and The horizontal axis represents the size of C. From the figure we can see that the running time of Degree Adjust, Cost-Effective Degree Adjust and Degree are similar, while Random often takes longer time than other heuristic algorithms because Random often needs more seeds to win the competition. Improved Greedy and improved cost-effective Greedy algorithms run very slowly comparing to other heuristic even we have used multiprocess to accelerate computation.

Fig. 3 and Fig. 4 show the experiment results in Slashdot. From Fig. 3 we can see that the effectiveness of different algorithms are similar to that in fig1. The improved cost-effective algorithm gets a much better effectiveness in this dataset while the improved cost-effective Degree Adjust does not get desired effect. Fig. 4 shows the running time of different algorithms on the Slashdot dataset . The results are similar to the results in Epinions dataset.

Fig. 5 and Fig. 6 show the experiment results in Cit-HepPh. From Fig. 5 we can see that the improved cost-effective algorithm achieves the best effectiveness and effectiveness of Random is the worst as other dataset, while the gap between different algorithms is not significant. Fig. 6 shows the running time of different algorithms on the Cit-HepPh dataset. The results are similar to the results in Epinions and Slashdot datasets.

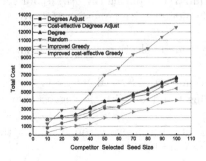

Fig. 1. Total cost of different algorithms on the epinions graph.

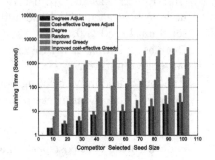

Fig. 2. Running time of different algorithms on the epinions graph.

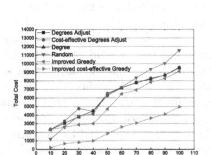

Fig. 3. Total cost of different algorithms on the slashdot graph.

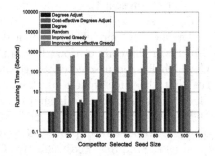

Fig. 4. Running time of different algorithms on the slashdot graph.

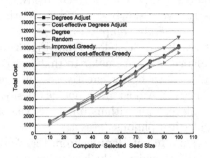

Fig. 5. Total cost of different algorithms on the Cit-HepPh graph.

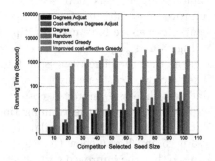

Fig. 6. Running time of different algorithms on the Cit-HepPh graph.

7   Conclusion and Future Work

Minimizing the cost to win competition is a very useful problem in the real word application. However this problem is negative by former works. In this paper, we define this problem and give some algorithms to solve this problem. Experiments in real datasets show that our cost-effective greedy algorithm achieves better effectiveness and cost-effective Degree Adjust heuristic algorithm achieves high efficiency and get better effectiveness than Degree and Random heuristics. Future work will focus on the problem of wining competition from multiple competitors in social network.

Acknowledgments. This work was supported by the National Natural Science Foundation of China (Grant No.61303005, No.61100167), the Natural Science Foundation of Shandong Province of China (Grant No.ZR2013FQ009), the Science and Technology Development Plan Project of Shandong Provience (Grant No.2014GGX101047, No.2014GGX101019), and the Fundamental Research Funds of Shandong University (2014JC025).

References

1. Kempe, D., Kleinberg, J., Tardos, E.: Maximizing the spread of influence through a social network. In: KDD, pp. 137–146 (2003)
2. Wei, C., Yajun, W., Siyu, Y.: Efficient influence maximization in social networks. In: SIGKDD, pp. 199–208 (2009)
3. Bharathi, S., Kempe, D., Salek, M.: Competitive influence maximization in social networks. In: Deng, X., Graham, F.C. (eds.) WINE 2007. LNCS, vol. 4858, pp. 306–311. Springer, Heidelberg (2007)
4. Chen, W., Wang, C., Wang, Y.: Scalable influence maximization for prevalent viral marketing in large-scale social networks. In: KDD, pp. 1029–1038 (2010)
5. Jung, K., Heo, W., Chen, W.: Irie: Scalable and robust influence maximization in social networks. In: ICDM, pp. 918–923 (2012)
6. He, X., Song, G., Chen, W., et al.: Influence Blocking Maximization in Social Networks under the Competitive Linear Threshold Model. In: SDM (2012)
7. Dubey, P., et al.: Competing for Customers in a Social Network. Social Science Electronic Publishing (2006)
8. Goyal, A., et al.: On minimizing budget and time in influence propagation over social networks. In: Social Network Analysis and Mining, pp. 179–192 (2013)
9. Borodin, A., Filmus, Y., Oren, J.: Threshold models for competitive influence in social networks. In: Saberi, A. (ed.) WINE 2010. LNCS, vol. 6484, pp. 539–550. Springer, Heidelberg (2010)
10. Budak, C., Agrawal, D., Abbadi, A.E.: Limiting the spread of misinformation in social networks. In: WWW, pp. 665–674 (2011)
11. Chen, W., Collins, A., Cummings, R., Ke, T., Liu, Z., Rincn, D., Sun, X., Wang, Y., Wei, W., Yuan, Y.: Influence maximization in social networks when negative opinions may emerge and propagate. In: SDM, pp. 379–390 (2011)
12. Zhang, P., Chen, W., Sun, X., et al.: Minimizing seed set selection with probabilistic coverage guarantee in a social network. In: KDD, pp. 1306–1315 (2014)

Ad Dissemination Game in Ephemeral Networks

Lihua Yin[1], Yunchuan Guo[1,*], Yanwei Sun[1], Junyan Qian[2],
and Athanasios Vasilakos[3]

[1] Institute of Information Engineering, Chinese Academy of Sciences, Beijing, China
[2] Guangxi Key Lab of Trusted Software(Guilin University of Electronic Technology), China
[3] Dept of Computer and Telecommunications Engineering,
University of Western Macedonia, Kozani, Greece
guoyunchuan@nelmail.iie.ac.cn

Abstract. The dissemination of ads in an ephemeral network has become an important research topic. A challenge of ad dissemination is the guarantee of robustness such that rational nodes in the ephemeral network have sufficient impetus to forward ads, despite facing the limitation of resources and the risk of privacy leakage. This paper proposes a strategy for ad dissemination in an ephemeral network. Acknowledging the assumption of incomplete information, we propose a bargaining-based game G^D that can be used by nodes to decide whether to forward an ad. There exists Bayesian Nash equilibrium in G^D with which the proposed approach provides nodes a strong impetus to disseminate the ads with higher dissemination accuracy.

Keywords: Ephemeral networks, information dissemination, games.

1   Introduction

Smartphones, equipped with low-cost and short-range communication devices (e.g., Wi-Fi and Bluetooth), high-cost and long-range communication devices (e.g., 3G and 4G devices), powerful processors, and rich operating systems have become widely available. As consumers depend on smartphones for phone calls and entertainment, merchants rely on them for capturing markets, including the ad market. FORTUNE[1] magazine reports that Facebook, one of the well-known social networks, had revenues of $1.3 billion from mobile ads, amounting to almost 60 percent of the company's overall revenue in 2013 and the global mobile ad market will reach $31 billion in 2014. Acknowledging such enormous business opportunities, multiple ad systems for mobile networks have been developed (e.g., B-MAD [1], SNMART [2] and Concierge [3]). Recently, significant effort has been expended on constructing ad dissemination systems for *ephemeral networks* (e.g., SID[4], P^3-Coupon [5] and SSD[6]).

An ephemeral network has two basic features: (1) Nodes in the network have high *mobility*. (2) Mobile nodes often use short-range communication services such as

---

[*] Corresponding author.

[1] http://tech.fortune.cnn.com/2014/04/30/the-key-to-facebooks-future-mobile-ads-everywhere-you-look/

© Springer International Publishing Switzerland 2015
R. Cheng et al. (Eds.): APWeb 2015, LNCS 9313, pp. 610–622, 2015.
DOI: 10.1007/978-3-319-25255-1_50

Wi-Fi and Bluetooth (because these services are frequently free). Consequently, interactions between nodes are often short and ephemeral. Hence, the network is called an *ephemeral network*. In recent years, information dissemination for *ephemeral networks* has attracted widespread attention from academia to industry [7]–[14].

Motivation. Although many solutions have been proposed in the literature for ad dissemination in ephemeral networks, the selfness for ad dissemination remains is a challenging problem. In an ephemeral network, nodes are frequently selfish and reluctant to participate in forwarding an ad for the following reasons: (1) *Resource Limitation*. Smartphones' resources are typically limited. Disseminating ads tends to consume energy and reduce a node's overall lifespan. (2)*Privacy Leakage*. Smartphones record significant private information, e.g., user' contacts, photos, and call records, and most users are concerned with this information being leaked when disseminating ads. Nodes may be unwilling to disseminate an ad if they do not receive sufficient incentive. Some existing incentive strategies are based on the assumption of complete information [23]] [41] (i.e., that a node knows all the information of the other parties). This assumption, however, is unrealistic in ephemeral networks because of the transient and short-lived nature of the networks.

Contribution. In this paper, we investigate the robustness of ad dissemination in ephemeral networks. In contrast to existing approaches, we consider individual rational nodes that can decide themselves to forward an ad. We begin by analyzing the factors that could affect a node's decision and encourage the node to cooperate. Based on a realistic *incomplete information* assumption, we propose a *bargaining-based dissemination* game G^D to assist nodes in making their decision whether to forward an ad. It is shown that there exists at least one Bayesian Nash equilibrium in G^D. Simulation results confirm that our approach provides appropriate incentive and efficiently trades-off the utility between disseminators and ad providers and encourages nodes to disseminate ads. The proposed scheme has also higher dissemination accuracy: the majority of the ads are disseminated to the interested nodes.

2   Related work

Information Dissemination in Ephemeral Networks. The early research landscape on ephemeral networks predominantly focused on routing. In 2005, Motani et al. [7] began to study information dissemination in delay tolerant networking (DTN) and proposed a simple, scalable and low-cost architecture for information searches. Because their approach is based on epidemic dissemination, it is inefficient under non-uniform mobility pattern. Since 2005, significantly more effort has been directed towards information dissemination in ephemeral networks. Drabkin et al. [8] reviewed the work in this field from 2005 to 2011. In our paper, we will only survey the related work of the past three years (i.e., from 2012 to today). In 2012, based on the concept of acknowledgment, Ros et al. [9] proposed a broadcast protocol for reliable and efficient data dissemination in vehicular networks. Their approach resolves propagation at road intersections and does not require intersection recognition. Its coding overhead, however, was a bottleneck. Exploiting the concurrency potential of sensor nodes to resolve the coding overhead, Gao et al. [10] extended the Deluge (a popular protocol for bulk data dissemination) and proposed a multithreaded data dissemination protocol. At the same time, based on random linear network coding, Tang et al.[11] proposed a one-sided protocol and Zeng et al. [12] proposed a segmented network coding scheme. These two approaches minimize the time that a node requires to receive a copy of an entire message and can efficiently exploit the transmission opportunity. Based on the idea of location, Fan et al. [13] and Ishihara et al. [14] proposed data broadcasting schemes for vehicular ad hoc networks and for mobile social networks, respectively. Teng et al. [5] and Ning et al. [4] designed ad dissemination systems for ephemeral networks. They focused on privacy protection and cash withdrawal, respectively.

Although significant effort has been directed towards designing efficient schemes for disseminating information in ephemeral networks, almost all of the proposals require that nodes in the network are cooperative, and it is assumed that these nodes are willing to cooperate. Such an assumption is unrealistic in ephemeral networks. A rational and selfish node may be unwilling to share its resources with other nodes if its resources are limited.

Incentive in Ephemeral Networks. To encourage selfish nodes to cooperate, multiple strategies have been proposed, primarily in two categories: credit-based [4], [5], [15]–[17] and reputation-based [18] approaches.

Zhang et al. [15] used controlled coded packets as a virtual commodity currency to induce cooperative behavior and to reduce overhead. Teng et al. [5] designed the probabilistic one-ownership forwarding algorithm to disseminate large-scale coupons in ephemeral networks. Based on a time slot, Ning et al. [4] proposed a self-interest-driven scheme to motivate cooperation among selfish nodes for ad dissemination in social networks. Refaei et al.[18] proposed a reputation-based incentive scheme to monitor the changes of node behavior quickly and accurately.

In the literature, game theory is often used to model and analyze incentive behavior. For example, Chen et al. [19] used a coalitional game to stimulate cooperation in vehicular ad hoc networks. Li et al. [20] analyzed the cooperation incentives in both the price-based mechanism and reputation-based mechanism. We [17] built a game model for cooperative authentication to assist nodes in making decisions. Raya et al. [21] and Reidt et al. [22] presented novel key-revocation game schemes for ephemeral and ad hoc networks, respectively. Manshaei et al. presented an overview of applying game-theory to solve security and privacy problems in networks. Although many schemes have been proposed in the literature for encouraging the cooperation of nodes, none of them provides solutions for encouraging nodes to forward ads to the interested nodes in ephemeral networks.

3   System Model

In this paper, we study robust ad dissemination in ephemeral networks formed by mobile smartphones (throughout this paper, we refer to smartphones as nodes or disseminators). The proposed model includes two parts: Ad providers (e.g., C2C stores) and mobile nodes.

Ad providers (i.e. merchants) are responsible for broadcasting ads to potential disseminators in their range of communication, and awarding digital cash to all disseminators. Our goal is to encourage nodes to disseminate ads to potential and interested consumers.

Mobile Nodes. Each node is a potential disseminator in the proposed model. Assumptions regarding disseminators are as follows: (1) The node's computing capability is powerful and their energy is strong but limited; thus, they can run a watch-like mechanism to receive, verify and forward ads from neighboring nodes. (2) Interactions among neighboring nodes do not rely on network infrastructures, rather, on device-to-device communications (e.g., Wi-Fi or Bluetooth) because of their low cost. However, nodes access the Internet through network infrastructures using long-range radio (e.g., 3G or 4G network) or shot-range communications (e.g., Wi-Fi connected to the Internet). (3) To prevent location or identity from being monitored, we adopt pseudonyms that are widely adopted in both industry and academia. In generally, pseudonyms are scarce resources, that is, the number of pseudonyms is often limited, and if frequently changed, they will be quickly exhausted. After exhaustion, nodes must purchase new pseudonyms from certification authorization (which is responsible for managing pseudonyms). (4) A node's digital cash is its private property and only its owner can make withdrawals using online payment services, such as PayPal and Zhifubao[2]. Once receiving ads from an ad service provider or a neighboring node, disseminators can choose one of the following actions: (1) Forward the ads to their neighboring nodes. (2) Do nothing.

4   Dissemination Game

We propose a bargaining-based game, called the dissemination game, to model the dissemination behaviors between ad providers and mobile nodes and to incent additional nodes to disseminate ads.

Before a mobile node forwards an ad to its neighbor, a bargain is first executed between the node and the ad provider. In this bargain, the mobile node who provides the dissemination service behaves as a seller. An ad provider, who requests a node to disseminate his ad, acts as a buyer. Before bargaining with a node, an ad provider determines the value of its ad and offers a bidding price. A node calculates its cost for disseminating this ad and offers an asking price. If the asking price is less than or equal to the bidding price, the bargain is established at their agreed upon price. Upon completion of the dissemination, the node will obtain the digital cash from the ad provider, with possible delay, and the bargain ends.

Note: although the dissemination of the ad may pass through multiple hops of relay nodes, each bargain is only performed between a disseminator and an ad provider. Therefore, when an ad is forwarded from one node to another, a new bargain is required. This means that for each ad, the number of successful bargains equals the number of disseminations. Next, we discuss the factors influencing the cost of disseminating an ad.

---

[2] Zhifubao is a third party payment system and widely used in China.

Costs. Two factors influence the cost of disseminating an ad: the consumed resource for dissemination, the reputation of the node and the interest. In general, resource consumption directly determines the cost; reputation and interest indirectly affect the cost. We call the three factors *resource cost* and *interest weight*.

Resource Cost. As described in Section 0, we use pseudonyms to prevent the node's identity or location from being leaked. In general, pseudonyms are not free and are a scarce resource. Once pseudonyms become exhausted, the node must purchase replacements with digital cash. The energy consumed for dissemination also affects the cost. A lengthier ad will lead to a higher key cost and a higher energy cost. Therefore, the resource cost c_{res} for the ad monotonically increases with the length of ad. For simplicity, we define $c_{res} = \gamma \times |ad|$, where $\gamma > 0$ and $|ad|$ is the length of ad.

Interest Weight. If a node has a higher interest for an ad, it is more willing to share this ad with others; thus, having a lower asking price. For example, a football fan might be willing to share a football ad with other fans. Consequently, this fan may ask a lower price. For an ad, we use $ii_{cd}(ad)$ and $ii_r(ad)$ to the interest index of the current disseminator and its receiver, respectively. $iw(ii_{cd}, ii_r, ad)$ is used to represent the *interest weight* that the current disseminator forwards the ad towards the receiver. In general, the *interest weight* should satisfy the two properties: (1) $iw(ii_{cd}, ii_r, ad)$ monotonically decreases as the receiver's interest index $ii_r(ad)$ increases. (2) If the receiver's interest indices are the same, $iw(ii_{cd}, ii_r, ad)$ decreases as the disseminator's level $ii_{cd}(ad)$ increases.   The rationale of the two properties is that it is easier to persuade interested nodes to receive an ad and the interested nodes are more willing to persuade others. Essentially, the above properties can be generalized as follows:

$$iw(il_{cd}, il_r, ad) < iw(il_{cd'}, il_{r'}, ad) \text{ if either}$$

$$il_{r'}(ad) < il_r(ad) \text{ or} \tag{1}$$

$$il_r(ad) = il_{r'}(ad) \text{ and } il_{cd'}(ad) < il_{cd}(ad) \tag{2}$$

All interest levels are denoted by a set (IL), for example, IL= {disgust, dislike, neutrality, like, fascination} with interest level disgust<dislike<neutrality<like<fascination. For simplicity, we assign a unique positive integer from the set {0...$|IL|$-1} to represent the interest index of an interest level $il \in IL$. TABLE 1 illustrates an example of interest indices used in the sequel. Both statistics and data mining [25] are efficient methods to determine the interest level.

Table 1. Interest Indices

Interest level	disgust	dislike	neutrality	like	fascination
Interest index	1	2	3	4	5

Similar to a *reputation weight*, there are several methods to calculate the interest weight. In our model, a simple but useful one is adopted:

$$iw(il_{\text{cd}}, il_{\text{r}}, ad) = \frac{|IL| \times (|IL| - ii_{\text{r}}(ad)) + (|IL| - ii_{\text{cd}}(ad))}{|IL| \times |IL| - 1} \tag{3}$$

We can prove that formula (3) satisfies (1), (2)and $0 \leq iw \leq 1$. Cost c_i for node i to disseminate an ad depends on the cost of the resources and the *interest weight* and is defined as

$$c_i = (1 + iw) \times c_{res}. \tag{4}$$

Values. It is difficult for an ad provider to evaluate the value of forwarding an ad. In practice, a rational ad provider does not bid the price that is more than the value; therefore, we can evaluate the value by analyzing its highest bidding price. The bidding price includes two parts: (1) an ad provider rewards an average bonus to each node for its contribution to forward the ad (called *dissemination bonus*, denoted by *d_bonus*). (2) For a disseminator, if a node is influenced by this disseminator and purchases goods, then the ad provider rewards this disseminator a special bonus (called *purchase bonus*, denoted by *p_bonus*). The majority of the existing schemes (e.g., [5] [26]) assume that both *dissemination bonus* for each dissemination and *purchase bonus* for each purchase are fixed. In our paper, we also obey these assumptions. Next, we discuss the value of forwarding an ad for a provider. Many purchasers exist on an ad dissemination path[3] (Figure 1 gives an example regarding a dissemination path, where the circle with blue filling indicates a disseminator and a circle with red filling represents a purchaser). Assume that the number of purchasers in the dissemination path of a disseminator is m. Ideally, the disseminator should receive the bonus with $d\_bonus + m \times p\_bonus$. However, in practice, for some reasons (e.g., the limitation of the ad budget), an ad provider prefers to limit the number of obtaining a *purchase bonus* for a disseminator [4]. For simplicity, M is used to denote the maximum number of *purchase bonuses* that the ad provider intends to reward a disseminator. Thus, the value v (i.e., the highest bidding price that an ad provider would like to offer) of forwarding an ad can be calculated, as follows.

$$v = d\_bonus + M \times p\_bonus. \tag{5}$$

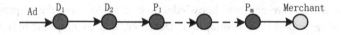

Fig. 1. Example for dissemination path.

Note: v in formula (5) is the highest price that an ad provider wishes to pay for each forwarding and is unknown to the disseminators (because a disseminator does not known the value of M).

Bargaining Procedure and Utility: Generally, c_i depends on *resource cost* and *interest weight*. Before disseminating an ad, rational node i first offers the asking price ap_i with $c_i \leq ap_i$ and the rational ad provider offers the bidding price bp with

---

[3] Every ad maintains a list of all disseminators. The dissemination path of an ad for a disseminator refers to the path from the disseminator to the last purchaser.

$bp \leq v$. Knowing the asking price and bidding price, the disseminator and ad provider negotiate the final price, as follows.

Case 1. If $ap_i \leq bp$, then the bargain is struck at the agreed-upon price P:

$$P = \epsilon \times ap_i + (1 - \epsilon) \times bp \tag{6}$$

where $0 \leq \epsilon \leq 1$. After bargain ends, node i forwards the ad to its neighbors. In formula (6), the agreed-upon price P is the weighted average of the bidding price and the asking price. When $\epsilon = 1$, formula (6) grants the ad provider the right to make only one offer that the disseminator can accept or reject. When $\epsilon = 0$, Similarly, formula (6) grants the disseminator the right to make only one offer that the disseminator can accept or reject. After forwarding this ad, node i earns digital cash (of value P) from the ad provider.

Case 2. If $ap_i > bp$, the bargain fails.
In the above two cases, it is important for the disseminator and ad provider to offer a suitable asking/bidding price. To address this problem, several methods have been proposed, one of which is the assumption of linear functions for the asking (bidding) price. This assumption has an established tradition [27]–[30] in the bargain process, because it is consistent with intuition: in the majority of cases, the asking (bidding) price should be directly proportional to the cost (value). In our work, this assumption is adopted. In detail, we define ap_i and bp as follows: ap_i $(c_i) = \alpha_a + \beta_a \times c_i$ and bp $(v) = \alpha_b + \beta_b \times v$, where α_a, β_a, α_b, $\beta_b \geq 0$.

Upon obtaining the agreed-upon price, we can evaluate the utility of the disseminators and the ad provider. The utility u_i of the disseminator i *is* $u_i = P - c_i$ and utility u_m of the ad provider *is* $u_m = v - P$.

5   Game Analysis

In this section, we study the dissemination game G^D with incomplete information. If cost c_i^D and value v are common knowledge to both node i and the ad provider, this is the *complete information* assumption. However, this assumption may be overly stringent, because:(1) Although a disseminator can know d_bonus and p_bonus in formula (5), it cannot accurately obtain the value v in formula (5), because M in formula (5)may be the confidence of an ad provider and is not provided to disseminators. (2) Although an ad provider can obtain the resource cost (i.e., the c_{res} in formula (4)) by simulating the behavior of a disseminator, it cannot calculate the c_i^D in formula (4) because it cannot know the interest level iw of the disseminator.

Therefore, the assumption of *complete information* does not apply to the proposed model. To address this problem, the assumption of *incomplete information* is adopted in our paper. That is, although a disseminator/ad provider does not know the value/cost of the other party, probability density functions of the value/cost are the common knowledge (this assumption has been widely used in the bargaining-based game for ephemeral networks [17], [22], [23], [31]). We assume that:

(1) Value v is uniformly distributed over the interval (v_{bot}, v_{top}), where v_{top} and v_{bot} are the top and bottom of the value, respectively.

(2) Cost c_i is uniformly distributed over the interval (c_{bot}, c_{top}), where c_{top} and c_{bot} are the top and the bottom of the cost respectively.

Note: The assumption of uniform distribution with respect to the value from a seller's perspective and the cost from a buyer's perspective is simple, however, was widely used in early studies [32]–[35] and more recent studies [36]–[38], because of the two advantages: (1) "it enables us to generate additional (and finer) theoretical insights"[36], (2) "it is easy to induce with dice and easy to explain to the subjects" [39]. Because uniform distribution can model the behaviors of both buyers and sellers [37], [38], it is adopted in our work. Given these two density functions, we re-define the utility u_m of an ad provider, as follows:

$$u_m = (v - ((1 - \epsilon)bp + \epsilon E[ap_i(c_i)|bp \geq ap_i(c_i)]) \times prob(bp \geq ap_i(c_i))) \tag{7}$$

where $E[ap_i(c_i)|bp \geq ap_i(c_i)]$ is the expectation of the asking price of a disseminator in the case that the bidding price is greater than or equal to the asking price. Similarly, the utility u_i of node i is:

$$u_i = (\epsilon ap_i + (1 - \epsilon)E[bp(v))|bp(v) \geq ap_i] - c_i) \times prob(bp(v) \geq ap_i) \tag{8}$$

where $E[bp(v))|bp(v) \geq ap_i]$ is the expectation of the bidding price in the case that the bidding price is greater than or equal to the asking price.

Proposition: In game G^D, there exists at least one Bayesian Nash equilibrium.

Proof. We can prove this Proposition by maximizing formula (7) and (8), as follows.

$$prob(bp \geq ap_i(c_i)) = prob(c_i \leq \frac{bp - \alpha_a}{\beta_a}) = \frac{(bp - \alpha_a) - \beta_a c_{bot}}{\beta_a(c_{top} - c_{bot})} \tag{9}$$

$$E[ap_i(c_i)|bp \geq ap_i(c_i)]$$

$$= \int_{\alpha_a + \beta_a \cdot c_{bot}}^{bp} \frac{1}{bp - (\alpha_a + \beta_a \cdot c_{bot})} x dx = \frac{bp + \alpha_a + \beta_a \cdot c_{bot}}{2} \tag{10}$$

From formula (9) and (10), we can obtain:

$$u_m = \left(v - \frac{(2 - \epsilon)bp + \epsilon(\alpha_a + \beta_a \cdot c_{bot})}{2}\right) \times \frac{(bp - \alpha_a) - \beta_a c_{bot}}{\beta_a(c_{top} - c_{bot})}$$

The first order condition of u_m for its maximum is $\dfrac{du_m}{dbp} = 0$. By solving this order condition, we can obtain:

$$bp = \frac{v}{2 - \epsilon} + (1 - \epsilon)\frac{\alpha_a + \beta_a c_{bot}}{2 - \epsilon} \tag{11}$$

Similarly, by solving the order condition of u_i for its maximum, we can obtain

$$ap_i = \epsilon * \frac{\beta_b v_{top} + \alpha_b}{1 + \varepsilon} + \frac{c_i}{1 + \varepsilon} \tag{12}$$

According to (11) and (12), we can obtain Bayesian Nash equilibriums for game G^D, as follows.

$$bp\ (v) = \frac{(\epsilon - \epsilon^2)v_{top} + (\epsilon^2 - 3\epsilon + 2)c_{bot}}{4 - 2\epsilon} + \frac{1}{2 - \epsilon}v \tag{13}$$

$$ap_i\ (c_i) = \frac{\epsilon - \epsilon^2}{2 + 2\epsilon}c_{bot} + \frac{1}{2\epsilon}v_{top} + \frac{1}{1 + \epsilon}c_i \tag{14}$$

In the sequel, we assume that the average of the bidding price and asking price is used as the agreed-upon price [33][41] (i.e., $\epsilon = 0.5$). Following this assumption, we can obtain its Bayesian Nash equilibriums, consistent with [42], as follows:

$$bp\ (v) = \frac{1}{4}c_{bot} + \frac{1}{12}v_{top} + \frac{2}{3}v \tag{15}$$

$$ap_i\ (c_i) = \frac{1}{12}c_{bot} + \frac{1}{4}v_{top} + \frac{2}{3}c_i \tag{16}$$

According to the bargaining procedure given in section 5 and the Bayesian Nash equilibriums, we can design the game algorithm. For the limitation of space, we omit it, and refer the reader to our previous work [17] for details of designing the bargain-game-based algorithm.

6   Evaluation

In our simulation, we adopt a city scenario including the GPS traces of all taxis in the Tiananmen area, in Beijing, gathered from 12:00 AM to 11:59 PM on November 30, 2012. Our scenario has an area of 12.988 km × 8.537 km in a Cartesian coordinate system. Moreover, the GPS points span from 39.8846° N to 39.9307° N in latitude and from 116.3518° E to 116.4441° E in longitude. We assume that (1) each passenger in a taxi owns a smartphone equipped with Wi-Fi over a communication range of 200 m, which is 80 percent of the range specified by the IEEE 802.11n standard4; (2) the time for each bargain (including the Wi-Fi connection time and ad forwarding time) is not more than 60 s. Assumption (2) is realistic because 802.11n can operate at a maximum net data rate from 54 Mbit/s to 600 Mbit/s. Based on these two assumptions, we define the logical encounter between two taxis as follows: two taxis logically encounter if (1) the distance between the two taxis is less than 200 m, and (2) the retention time of these two taxis, the distance of which is less than 200 m, is greater than 60 s. It is only when two taxis logically met that the passengers in them can strike a bargain. In our experiments, we gathered the GPS traces of 10,084 taxis, where 9,804 taxis encountered each other at least once. Each taxi is a disseminator. For simplicity, the resource cost C_{res} for each dissemination or reporting is set to 0.1, that is, C_{res}=0.1. We compared four strategies: the selfish strategy (i.e., nodes without incentives do not forward ads), the selfless strategy (i.e., all nodes unconditionally forward ads until their resources are exhausted), 30% selfless (i.e., 30 percent of nodes are selfless) and our dissemination game.

---

[4]   IEEE 802.11n-2009—Amendment 5: Enhancements for Higher Throughput. IEEE Standards Association. 29 October 2009.

Figure 2 presents the average number of forwards (*ANF*) for an ad. We can see that G^D is efficient at stimulating nodes. In the selfish strategy, no ad is forwarded. In both the selfless strategy and the 30%-selfless strategy, the *ANF* linearly increases with respect to the initial resource. The reason is that every disseminator consumes a fixed number of resources that cannot be effectively replenished. As a result, once the initial resource (e.g., pseudonyms) is exhausted, nodes cannot forward any ad. In game G^D, however, nodes can purchase resources by digital cash if their resource is exhausted. This means that the resource owned the node in G^D is unlimited and therefore, their *ANF*s are constant.

Figure 3 presents the dissemination accuracy, which denotes the level that an ad is forwarded to an interested node. As shown in TABLE 1, interest index 1, 2, 3, 4 and 5 denote interest level "disgust", "dislike", "neutrality", "like", and "fascination". From Figure 3 , we can see that in G^D, approximately 32.1, 32.3 and 28.9 percent of the ads are forwarded to "fascination", "like" and "neutrality" nodes, respectively and 6.7 percent is disseminated to the "disgust" or "dislike" nodes. This means that game G^D has a higher accuracy and almost no ads are regards as SPAM. The reason is that in G^D, the interest level of a receiver for ads will affect the cost of the asking price and a node with a higher interest level will have a lower cost and a higher incentive. Consequently, the bargain with a higher interest level is more likely to succeed. In both the selfless strategy and the 30%-selfless strategy, a node forwards an ad to a random receiver and does not consider the receiver's interest. Thus, nodes in the network will receive ads of all interest levels with almost the same probability. The performance under varying some parameters in the different kinds of scenarios is not included in this paper due to the limitation of space. We will present it in our future paper.

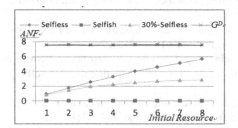 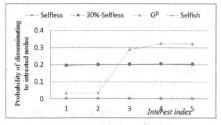

Fig. 2. ANF vs. Initial resource   **Fig. 3.** Dissemination Accuracy

7   Conclusion

In this paper, we focused on ad dissemination in ephemeral networks and proposed a bargaining-based dissemination game to assist nodes in deciding whether to forward an ad. We obtain a Bayesian Nash equilibrium for this game. Based on this equilibrium, the proposed approach encouraged additional nodes to disseminate ads.

In the future, more performance metrics should be studied, e.g. communication overhead. Additionally, the real identity of a user in ephemeral networks is often hidden from the other general users; as a result, false or even harmful ads are often disseminated and bring serious threats to our daily life. Preventing false advertisements and trace their origin has become an urgent and critical task in the future.

Acknowledgements. This work was supported by the National High Technology Research and Development Program of China (2013AA014002) and "Strategic Priority Research Program" of the Chinese Academy of Sciences (XDA06030200), Guangxi Natural Science Fundation (2014GXNSFAA118365, 2014GXNSFDA118036), the High Level of Innovation Team of Colleges and Universities in Guangxi and Outstanding Scholars Program Funding, and Program for Innovative Research Team of Guilin University of Electronic Technology.

References

1. Aalto, L., Göthlin, N., Korhonen, J., Ojala, T.: Bluetooth and WAP Push Based Location-Aware Mobile Advertising System. In: Proceedings of the International Conference on Mobile Systems, Applications, and Services - MobiSYS, pp. 49–58 (2004)
2. Kurkovsky, S., Harihar, K.: Using ubiquitous computing in interactive mobile marketing. Pers. Ubiquitous Comput. 10(4), 227–240 (2006)
3. de Castro, J.E., Shimakawa, H.: Mobile Advertisement System Utilizing User's Contextual Information. In: International Conference on Mobile Data Management, p. 91 (2006)
4. Ning, T., Yang, Z., Wu, H., Han, Z.: Self-Interest-Driven Incentives for Ad Dissemination in Autonomous Mobile Social Networks. In: Proceedings of the Annual IEEE International Conference on Computer Communications (INFOCOM), pp. 2310–2318 (2013)
5. Teng, J., Zhang, B., Bai, X., Yang, Z., Xuan, D.: Incentive-Driven and Privacy-Preserving Message Dissemination in Large Scale Mobile Networks. IEEE Trans. Parallel Distrib. Syst (2014)
6. Lee, S.-B., Park, J.-S., Gerla, M., Lu, S.: Secure Incentives for Commercial Ad Dissemination in Vehicular Networks. IEEE Trans. Veh. Technol. 61(6), 2715–2728 (2012)
7. Motani, M., Srinivasan, V.: PeopleNet: Engineering A Wireless Virtual Social Network. In: Proceedings of the International Conference on Mobile Computing and Networking (MOBICOM), pp. 243–257 (2005)
8. Drabkin, V., Friedman, R., Kliot, G., Segal, M.: On Reliable Dissemination in Wireless Ad Hoc Networks. IEEE Trans. Dependable Secur. Comput. 8(6), 866–882 (2011)
9. Ros, F.J., Ruiz, P.M., Stojmenovic, I.: Acknowledgment-Based Broadcast Protocol for Reliable and Efficient Data Dissemination in Vehicular Ad Hoc Networks. IEEE Trans. Mob. Comput. 11(1), 33–46 (2012)
10. Gao, Y., Bu, J., Dong, W., Chen, C., Rao, L., Liu, X.: Exploiting Concurrency for Efficient Dissemination in Wireless Sensor Networks. IEEE Trans. Parallel Distrib. Syst. 24(4), 691–700 (2013)
11. Tang, B., Ye, B., Guo, S., Lu, S., Wu, D.O.: Order-Optimal Information Dissemination in MANETs via Network Coding. IEEE Trans. Parallel Distrib. Syst. 25(7), 1841–1851 (2014)
12. Zeng, D., Guo, S., Hu, J.: Reliable Bulk-Data Dissemination in Delay Tolerant Networks. IEEE Trans. Parallel Distrib. Syst. 25(8), 2180–2189 (2014)
13. Fan, J., Chen, J., Member, S., Du, Y.: Geocommunity-Based Broadcasting for Data Dissemination in Mobile Social Networks. IEEE Trans. Parallel Distrib. Syst. 24(4), 734–743 (2013)

14. Ishihara, S., Nakamura, N., Niimi, Y.: Demand-Based Location Dependent Data Dissemination in VANETs. In: Proceedings of the Annual International Conference on Mobile Computing & Networking (MOBICOM), pp. 219–221 (2013)
15. Zhang, C., Zhu, X., Song, Y., Fang, Y.: C4: A New Paradigm for Providing Incentives in Multi-Hop Wireless Networks. In: Proceedings of the Annual IEEE International Conference on Computer Communications (INFOCOM), pp. 918–926 (2011)
16. Guizani, M., Rachedi, A., Gueguen, C.: Incentive Scheduler Algorithm for Cooperation and Coverage Extension in Wireless Networks. IEEE Trans. Veh. Technol. 62(2), 797–808 (2013)
17. Guo, Y., Yin, L., Liu, L., Fang, B.: Utility-based Cooperative Decision in Cooperative Authentication. In: Proceedings of the Annual IEEE International Conference on Computer Communications (INFOCOM), pp. 1006–1014 (2014)
18. Refaei, M.T., Dasilva, L.A., Eltoweissy, M., Nadeem, T.: Adaptation of Reputation Management Systems to Dynamic Network Conditions in Ad Hoc Networks. IEEE Trans. Comput. 59(5), 707–719 (2010)
19. Chen, T., Zhu, L., Wu, F., Zhong, S.: Stimulating Cooperation in Vehicular Ad Hoc Networks: A Coalitional Game Theoretic Approach. IEEE Trans. Veh. Technol. 60(2), 566–579 (2011)
20. Li, Z., Shen, H.: Game-theoretic analysis of cooperation incentive strategies in mobile ad hoc networks. IEEE Trans. Mob. Comput. 11(8), 1287–1303 (2012)
21. Raya, M., Manshaei, M.: Revocation games in ephemeral networks. In: Proceedings of the ACM Conference on Computer and Communications Security (CCS), pp. 199–210 (2008)
22. Reidt, S., Srivatsa, M., Balfe, S.: The fable of the bees: incentivizing robust revocation decision making in ad hoc networks. In: Proceedings of the ACM Conference on Computer and Communications Security (CCS), pp. 291–302 (2009)
23. Manshaei, M.H., Zhu, Q., Alpcan, T., Basar, T., Hubaux, J.-P.: Game Theory Meets Network Security and Privacy. ACM Comput. Surv. 45(3), 1–39 (2013)
24. Adrian, P., Koenig-Lewis, N.: An experiential, social network-based approach to direct marketing. Direct Mark. An Int. J. 3(3), 162–176 (2009)
25. Surana, A., Kiran, R.U., Reddy, P.K.: Selecting a Right Interestingness Measure for Rare Association. In: International Conference on Management of Data, pp. 115–124 (2010)
26. Sigala, M.: A framework for designing and implementing effective online coupons in tourism and hospitality. J. Vacat. Mark. 19(2), 165–180 (2013)
27. Spulber, D.F.: Bertrand competition when rivals' costs are unknown. J. Ind. Econ. 43(1), 1–11 (1995)
28. Bbink, K.L.A., Randts, J.O.B.: Price competition under cost uncertainty: A laboratory analysis (2002)
29. Lynch, L., Hardie, I., Parker, D.: Analyzing Agricultural Landowners' Willingness to Install Streamside Buffers
30. Ford, J.L., Kelsey, D., Pang, W.: Information and Ambiguity: Contrarian and Herd Behaviour in Financial Markets, 1–43 (November 2006)
31. Freudiger, J., Manshaei, M.H., Hubaux, J.-P., Parkes, D.C.: Non-Cooperative Location Privacy. IEEE Trans. Dependable Secur. Comput. 10(2), 84–98 (2013)
32. Wilson, R.B.: Competitive bidding with asymmetric information. Manage. Sci. 13(11), 816–820 (1967)
33. Chatterjee, K., Samuelson, W.: Bargaining under incomplete information. Oper. Res. 31(5), 835–851 (1983)
34. Holt, C.A., Sherman, R.: The loser's curse. Am. Econ. Rev. 84(3), 642–652 (1994)

35. Kagel, J.H., Levin, D.: Behavior in multi-unit demand auctions: Experiments with uniform price and dynamic Vickrey auctions. Econometrica 69, 413–454 (2001)
36. Ding, M., Eliashberg, J., Huber, J., Saini, R.: Emotional Bidders—An Analytical and Experimental Examination of Consumers' Behavior in a Priceline-Like Reverse Auction. Manage. Sci. 51(3), 352–364 (2005)
37. Lin, Z., Dc, W., Vandell, K.D.: Illiquidity and Pricing Biases in the Real Estate Market. Real Estate Econ. 35(3), 291–330 (2007)
38. Cheng, P., Lin, Z., Liu, Y.: Home Price, Time-on-market, and Seller Heterogeneity under Changing Market Conditions. J. Real Estate Financ. Econ. 41(3), 272–293 (2010)
39. Davis, D.D., Holt, C.A.: Experimental Economics. Princeton University Press, Princeton (1993)
40. Hargittai, E., Litt, E.: The tweet smell of celebrity success: Explaining variation in Twitter adoption among a diverse group of young adults. New Media Soc. 13(5), 824–842 (2011)
41. Li, Y., Yu, J., Wang, C., Liu, Q., Cao, B., Daneshmand, M.: A novel bargaining based incentive protocol for opportunistic networks. In: IEEE Globcom, pp. 5285–5289 (2012)
42. Zhang, W.: Game theory and information economics. Truth & Wisdom Press (1996)

User Generated Content Oriented Chinese Taxonomy Construction

Jinyang Li, Chengyu Wang, Xiaofeng He, Rong Zhang, and Ming Gao*

Institute for Data Science and Engineering, Software Engineering Institute,
East China Normal University, Shanghai, China
{jinyangli,chengyuwang}@ecnu.cn, {xfhe,rzhang,mgao}@sei.ecnu.edu.cn

Abstract. The taxonomy is one of the basic components in knowledge graphs as it establishes types of classes and semantic relations among the classes. Taxonomies are normally constructed either manually, or by language-dependent rules or patterns for type and relation extraction or inference. Existing work on building taxonomies for knowledge graphs is mostly in English language environment. In this paper, we propose a novel approach for large-scale Chinese taxonomy construction based on user generated content. We take Chinese Wikipedia as the data source, develop methods to extract classes and their relations mined from user tagged categories, and build up the taxonomy using a bottom-up strategy. The algorithms can be easily applied to other Wiki-style data sources. The experiments show that the constructed Chinese taxonomy achieves better results in both quality and quantity.

Keywords: knowledge graph, taxonomy, Wikipedia.

1 Introduction

In a knowledge graph, the *taxonomy* is a basic component of the entire system as it specifies the sets of classes and entities, relations between classes and entities, and the topological structure of the knowledge graph. Currently, much research work has been devoted to creating taxonomies for knowledge graphs. Taxonomies are created manually or automatically. In projects such as ReadTheWeb [1] and DBpedia [2,3], classes and their hierarchical relations are pre-defined, while automatic approaches have also been proposed by exploiting the rich semantics in unstructured texts (in Probase [4]) or semi-structured wikis (in YAGO [5,6,7] and WikiTaxonomy[8]).

Research on taxonomy construction is worthy efforts. Other than the common challenges such as sparsity, incompleteness and heterogeneity [4], taxonomy construction approaches are still highly language-dependent [9]. Taxonomy construction from Chinese Wikipedia is challenging due to the following factors:

- **Lack of Data Sources:** The construction of taxonomy heavily relies on data sources. For example, Freebase[1] provides a knowledge repository, which

---

* Corresponding author.
[1] http://www.freebase.com/

© Springer International Publishing Switzerland 2015
R. Cheng et al. (Eds.): APWeb 2015, LNCS 9313, pp. 623–634, 2015.
DOI: 10.1007/978-3-319-25255-1_51

is the backbone in Google Knowledge Graph. WordNet [10] contains rich hierarchical relations between entities and is served as a source for classes in YAGO. However, these counterparts in Chinese are not readily available.

- **Hard to Obtain Language Patterns:** Extraction patterns in English language cannot be directly extended to other languages, such as Chinese. For example, while plural forms can be used to detect concepts in English [5,8], there are no explicit singular/plural forms in Chinese nouns. In [11], Chinese language patterns are designed to extract isA relations from plain text. But they can not be applied to user generated categories, which are short text rather than complete sentences.

- **Low Coverage of Cross-lingual Approaches:** Although some cross-lingual methods have been proposed based on cross-lingual links between data sources in different languages [7,12], the coverage is relatively low. Figure 1 shows the number of articles, categories and their overlaps for Chinese and English Wikipedias. We can find only 34.66% of articles and 15.60% of categories in Chinese Wikipedia are covered in the English version. Thus, cross-lingual approaches have limited power for entities and classes that are unique in certain languages.

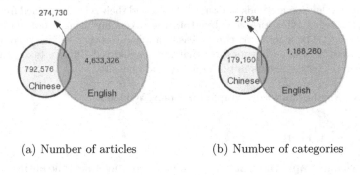

(a) Number of articles   (b) Number of categories

Fig. 1. Comparison of Chinese and English Wikipedias

In this paper, we study the problem of taxonomy construction from Chinese Wikipedia. We address the the problem based on two main ideas, namely: (a) detecting the isA relations from Wikipedia article titles and user tagged categories as accurate as possible, and (b) constructing a taxonomy in terms of all isA relations, where an isA relationship represents that one entity/class is a subclass of another. We summarize the main research contributions of this work as follows:

- We design a classification-based method to extract isA relations from Chinese Wikipedia titles and categories with an accuracy of over 95%. We also apply inference-based and mining-based approaches to generate isA relations that are not explicitly expressed in Wikipedia categories.

- We assemble these isA relations into a complete taxonomy. Specially, we construct a taxonomy from Wikipedia in a bottom-up manner via integrating these isA relations.

– We evaluate our constructed taxonomy in scale and accuracy measures. The experimental results show that our constructed taxonomy has high coverage and accuracy in entity space, class space and relation space.

The rest of the paper is organized as follows. Section 2 covers the related work. In section 3, we formulate the problem and present the stages of our approach for constructing taxonomy from Chinese Wikipedia. Section 4 and 5 elaborate the isA relation detection and taxonomy construction process, respectively. Section 6 covers our experimental studies on Chinese Wikipedia. Finally, we give the concluding remarks in Section 7.

2   Related Work

In knowledge graph area, a lot of research efforts have been focused on constructing taxonomies. There are three ways of constructing taxonomy: manual approaches [2,3], automatic approaches [8,5,6,7,4] and cross-lingual approaches [7,12,13].

Some knowledge graphs have hand-craft, fixed taxonomy with fine quality, such as NELL and DBpedia. In NELL, categories are manually arranged into a hierarchical structure so that entities are extracted from texts and mapped to certain categories by coupled training [14,1]. In DBpedia, there is a cross-lingual, universal taxonomy. Entities are mapped to the taxonomy by contributors of the project [2,3]. The major drawback of manually constructed taxonomies is relatively low coverage, especially in newly emerged areas and specific domains.

Several projects leverage the rich semantic information in Wikipedia to derivate the taxonomy automatically. WikiTaxonomy [8] utilizes methods based on the connectivity of Wikipedia network and lexicon-syntactic features to classify isA and notIsA relations. In WordNet [10], concepts (synsets) are well organized by experts with clear semantic relations. YAGO [5,6,7] combines Wikipedia categories and WordNet by mapping Wikipedia categories to WordNet concepts. Currently the largest taxonomy is Probase [4]. Instead of extracting relations from Wikipedia, it takes natural languages from Web pages as input and generates isA pairs using Hearst patterns [15]. However, these approaches focus on English sources, and cannot be easily extended to Chinese sources.

The existing taxonomy can also be leveraged to construct a taxonomy in another language. In YAGO3 [7], Wikipedias in multiple languages are used to build one coherent knowledge base with the English version. Also, Wang et al. [12,13] studied the problem of cross-lingual taxonomy derivation from English and Chinese Wikipedias and proposed a cross-lingual knowledge validation method via Dynamic Adaptive Boosting. Although cross-lingual approaches are promising when multilingual links or knowledge exist. Due to the low coverage of cross-lingual information and the significant difference between Chinese and English, these methods can not be employed to construct taxonomies with a lot of language-specific knowledge.

3 Chinese Taxonomy Construction

Constructing a Chinese taxonomy is challenging. We briefly introduce our problem and provide a sketch of our approach in this section.

Problem Description. Wikipedia is a large repository that can be modeled as a set of Wikipedia articles W. In our paper, each article is a 2-tuple $w = (e, C)$ where e is the title of the article, which is served as a candidate entity in our taxonomy and C is the set of user generated categories for e.

A taxonomy $T = (V, E)$ is a rooted, labeled tree where nodes V are entities or classes and edges E represent isA relations. Specifically, for each non-root $e \in V$, there exists a class c where (e, isA, c) holds.

However, it is a non-trivial task to identify (e, isA, c) from w because most categories express the semantic relatedness to the entity, or the topics or fields the entity belongs to. For example, in the article for *Jack Ma* in Chinese Wikipedia, categories include *1964 births, Alibaba Group, Business person in online retailing*, etc. Only *Business person in online retailing* is the suitable class for *Jack Ma*.

Besides, in the previous example, *1964 births* indicates that *Jack Ma* is a *person*. However, we do not know the isA relation between *Business person in online retailing* and *person*, while this relation is necessary to construct the high level structure of the taxonomy.

In our paper, we further divide isA relations into two types, namely *instanceOf* and *subclassOf*. The relations between entities and classes are called *instanceOf* relations. And *subclassOf* relations are used between classes. Our goal is to derive a large and accurate Chinese taxonomy T from W which contains *instanceOf* and *subclassOf* relations.

Overview of Our Approach. Our approach consists of two key stages below.

Stage 1: Generate isA relations. As shown in Figure 2, Stage 1 generates isA relations from categories by a classification model, infers isA relations from rational categories and extends existing isA relations via rule mining.

Stage 2: Construct Chinese taxonomy. In Stage 2, we derive a tree to present the Chinese taxonomy. In this stage, we take each isA relation as a subtree and propose an algorithm to construct the taxonomy in a bottom-up manner via node merging, cycle removal and subtree merging.

4 isA Relation Generation

In this section, we give a detailed description of our isA relation generation algorithm. The framework of our algorithm is shown in Figure 2. We first preprocess the Chinese Wikipedia pages by a filter to remove irrelevant pages, which will be discussed in Section 6. And then we train a classification model to detect isA relations. For the negative ones, we generate some efficient rules to infer isA relations. Finally, all the isA relations will be extended and we will obtain the whole isA relations set.

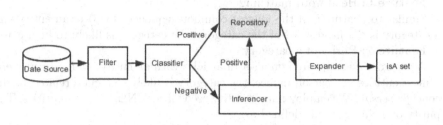

Fig. 2. The framework of generating isA relations

4.1   isA Relation Classification

Scoring Function. To distinguish isA relations from others, we carefully design a scoring function. Given an entity e, a category c, and two sets of features F_1 and F_2, the function outputs a positive number for isA relation; negative otherwise, defined as follows:

$$y(e,c) = w_1 \cdot F_1(e,c) + w_2 \cdot F_2(c) + w_0 \qquad (1)$$

where $F_1(e,c)$ considers both information of e and c (called entity-dependent features) while $F_2(c)$ (called entity-independent features) only takes the properties of c into account.

Feature Sets. We now briefly introduce our two feature sets. Features 1-4 are entity-dependent features while features 5-7 are entity-independent features.

Feature 1: Length of a category

The basic intuition is that if the length of a category is too long or short, it may be too general or too specific to describe the class of an entity.

Feature 2: POS tag

Usually a valid class is a noun or a noun phrase. We perform word segmentation and POS tagging on category names. We use the POS tag of the head word of the category as a feature.

Feature 3: Thematic category

As is described in [8], some categories, such as finance, politics, entertainment, etc. are thematics categories rather than conceptual classes. We have collected a set of themes in Chinese. We take whether a category or the head word of a category is a thematic word as a feature.

Feature 4: Language pattern

In English, a conceptual category is often in the form of *premodifier + head word + postmodifier* (see [5]). We have observed that in Chinese, the pattern is *premodifier + "de" + head word* where "de" is an auxiliary character "的" in Chinese. We then perform pattern matching on categories.

Feature 5: Common sequence of entity and category

In Chinese, entities and categories may have a common subsequence. For example, the category *political party* is a class for the entity *Labor Party*. We take the existence of the common sequence of a feature.

Feature 6: Head word matching

Similar to Feature 5, if the longest common sequence (LCS) of an entity and a category is the head word of the category, the category is likely to be a class.

Feature 7: Purity of a category

A category c is appeared on a set of articles with an entity set E_c. Intuitively, if most entities are person names, the category is likely to a class related to the concept *person*. We employ named entity recognition (NER) to tag entities. The purity of a category c is defined as:

$$purity(c) = max_{l \in L} \frac{|E_l \cap E_c|}{|E_c|} \qquad (2)$$

where L is a collection of NE tags and E_l is the collections of entities that are labeled as l. Given a pre-defined threshold τ, we define whether $purity(c) > \tau$ as a feature.

Weights Learning. Features in $\boldsymbol{F_1}(e, c)$ and $\boldsymbol{F_2}(c)$ have different discriminative powers for the isA classification task, which are represented by weight vectors $\boldsymbol{w_1}$ and $\boldsymbol{w_2}$. To learn the weights $\boldsymbol{w} = (\boldsymbol{w_1}, \boldsymbol{w_2}, w_0)$, we employ a sequential minimal optimization technique based on linear SVM. Given a training set of positive and negative samples, we optimize the objectives $\parallel \boldsymbol{w} \parallel^2 + C\Sigma_i \xi_i$ under the constraint:

$$y_i(\boldsymbol{w_1} \cdot \boldsymbol{F_1}(e, c) + \boldsymbol{w_2} \cdot \boldsymbol{F_2}(c) + w_0) \geq 1 - \xi_i \qquad (3)$$

where $\xi_i \geq 0$ and C is the turning parameter.

4.2 isA Relation Inference

The classifier can be of help to extract *conceptual* classes from Wikipedia categories, however, a large number of *relational* categories can provide useful knowledge about an entity.

Rational categories can be leveraged to extract relations or properties. For example, we may extract a relation *(A, graduateFrom, PKU)* from the category *PKU graduates* in article A. We can also extract a property *(A, diedIn, 1964)* from the category *1964 deaths*. In the following, we elaborate how to infer isA relations from relational facts and properties.

Formally, a *relation* is a triple *(subj, predicate, obj)* that the subject and object are entities of certain classes. A *property* is also a triple but the object is a literal (such as string, numerical, date, etc.) rather than an entity.

For each relation R, given a relation predicate, a subject class SC, and an object class OC, then the referred rule is shown as follows:

$$(subj, R, obj) \Rightarrow (subj, isA, SC) \wedge (obj, isA, OC) \qquad (4)$$

For each property P, given a property predicate and a subject class SC, the inferred rule is shown as follows:

$$(subj, P, obj) \Rightarrow (subj, isA, SC) \qquad (5)$$

However, not all such rules can be used to generate isA relations. The reasons are twofold: firstly, the type inference rules may not be accurate. Secondly, the classes of subjects and objects can be very diverse. As a result, it is difficult to assign a simple class (i.e., SC or OC) for the rule. To ensure high accuracy, we only consider the top most frequent rules as isA relations. See Section 6 for detailed rules and their evaluation results.

4.3 isA Relation Expansion

Classes have different levels of abstraction. Some classes are high-level and cover a broad spectrum of entities, such as *person*, while others describe a domain specific region, such as *Chinese pop music composer*. In this part, our goal is to extract isA relations between classes of different levels. For example, given two isA relations *(A, isA, Chinese pop music composer)* and *(A, isA, person)*, isA relation *(Chinese pop music composer, isA, person)* should be extended to build the taxonomy.

In this section, we introduce our relation expansion technique in the framework of association rule mining. Given a class c, the *contribution* of the class is defined as the number of distinct entities that labeled by the class c:

$$contrib(c) = |\{e.subj|e.obj = c \wedge e \in E\}| \qquad (6)$$

The *match* for two classes c_1 and c_2 is the number of matched entities:

$$match(c_1, c_2) = |\{e.subj|e.obj = c_1 \wedge e \in E\} \cap \{e.subj|e.obj = c_2 \wedge e \in E\}| \quad (7)$$

Then, the measure *confidence* between c_1 and c_2 corresponds to the ratio of the match and the contribution of c_1:

$$conf(c_1, c_2) = \frac{match(c_1, c_2)}{contrib(c_1)} \qquad (8)$$

We can see $conf(c_1, c_2)$ determines whether class c_1 is a subclass of class c_2. When $conf(c_1, c_2)$ is no less than a pre-defined threshold, isA relation (c_1, isA, c_2) is formed. The higher the confidence score $conf(c_1, c_2)$ is, the more likely it is that class c_1 belongs to class c_2. Because when $conf(c_1, c_2)$ is closer to 1, more entities in c_1 also belong to the class c_2.

Other than isA relations generated from Wikipedia categories, this approach tries to extract the potential isA relations from a higher level of inter-articles.

5 Taxonomy Construction

The naive approach is to construct a graph via combining all relations detected in Stage 1. This approach is not effective because there exist many inconsistent or noisy relations after Stage 1. To avoid these drawbacks, we effectively construct the taxonomy in a bottom-up manner via incorporating three subtree operations. In summary, the process of taxonomy construction can be divided into three phases, namely, node merging, cycle removal and sub-tree merging.

Node Merging. Initially, isA relations are considered as sub-trees, $T(x)$ with root node x. We highlight two key operations to construct the sub-trees.

- **Horizontal Merge.** Given two isA relations (a, isA, x) and (b, isA, x) (i.e., two sub-trees), we join the sub-trees together to construct a new one when two child nodes a and b share a common parent node x.
- **Vertical Merge.** Given two isA relations (a, isA, b) and (b, isA, x) (i.e., two sub-trees $T(b)$ and $T(x)$), we extend the depth of $T(x)$ via adding child node a to node b in $T(x)$.

These two operations will be repeated until no sub-tree is generated or changed.

Cycle Removal. A taxonomy T can be viewed as a directed, acyclic graph. Unfortunately, cycles may be formed when we merge nodes vertically. For example, two isA relations (a, isA, b) and (b, isA, a) will be merged into a cycle. Thus we proposed an algorithm to remove cycles.

In the cycle removal algorithm, we first create a direct graph G from the set of isA relations S. Then, we utilize the DFS algorithm to check the connectivity of G. For each connected component cc in graph G, we check whether any edge exists in cc but not in the DFS tree produced by DFS. The edges are removed and no cycle exists as a result.

Sub-tree Merging. After node merging and cycle removal, sub-trees with a root of high level classes have been produced. However, these sub-trees are not inter-connected with each other. The manual effort in the whole taxonomy construction process is that we define several classes with high level of abstraction (e.g., *animal, event, organization*, etc.) and connect these sub-trees together. Finally, we assign a common root node to the sub-trees to build the complete taxonomy. And we label isA relations between classes and entities as *instanceOf*, others as *subClassOf*.

6   Experiments and Evaluation

6.1   Data Source

In this paper, our dataset is from Chinese Wikipedia dump[2] generated from September 12, 2014. In total, we extract 677,246 candidate entities for Chinese taxonomy construction. Every title of articles in Wikipedia dump is considered as a candidate entity. We clean up the data by the following steps:

1. Convert traditional Chinese characters to simplified Chinese;
2. Filter out pages without useful information;
3. Remove list pages, redirect pages, disambiguation pages, template pages and administrative pages, which do not contain candidate entities.

---

[2] http://download.wikipedia.com/zhwiki/20140912/

6.2 Taxonomy Analysis

Size and Accuracy. In our taxonomy, there are in total 581,616 entities and 79,470 classes. Among these classes, 72,873 are extracted from Wikipedia categories, and the rest are classes of high level abstraction, generated from either inferring or mining approaches described in Section 4.

To evaluate the accuracy of extracted relations. We randomly select 2,000 relations from each set of relations (*instanceOf*, *subclassOf* and the whole *isA* relations) and manually label whether a relation is correct or not. We calculate the confidence interval of accuracy with significance level $\alpha = 0.05$. As shown in Table 1, the accuracy is over 95% for both *instanceOf* and *subclassOf* relations.

Table 1. Size and accuracy of relations

Relation type	Number	Accuracy	Samples
subClassOf	85,072	95.85% ± 2.16%	2000
instanceOf	1,233,291	97.80% ± 0.86%	2000
total	1,317,956	97.60% ± 0.71%	2000

Comparison. It is not easy to compare our taxonomy with others, especially when they are for different languages. Because each taxonomy is a part of the knowledge graph and knowledge graphs are usually based on different data sources, structures and relations.

However, to show that our taxonomy contains unique knowledge that can not be captured by knowledge graphs in English, such as YAGO. We utilize the inter-language links in Wikipedia to map entities and categories from English to Chinese. If there is a hyperlink between an English and a Chinese article describing the same entity, the mapping can be formed. We perform the mapping process on Wikipedia categories in a similar fashion. We call the Chinese version of YAGO generated by mapping approach *YAGO-C* in this paper. Table 2

Table 2. Comparison in size and coverage

	Our taxonomy	YAGO-C	Coverage
Entity	581,616	274,730	47.15%
Class from Wikipedia categories	72,873	27,934	38.33%
High level class	6,597	-	11.70% (estimated)

shows the comparison results between YAGO-C and our taxonomy. Compare to 274,730 entities and 27,934 classes in YAGO-C, there are 581,616 entities and 79,470 classes in our taxonomy, which is much larger in size.

Except for classes and entities extracted from Widipedia, we also generate high level classes by isA inference and expansion. As YAGO combines Wikipedia and WordNet to construct a knowledge base, we analyze the coverage of high level classes in our taxonomy by language translation. We sample 1,000 high level classes randomly from our taxonomy and translate them into English. And we find that only 117 high level classes are covered in WordNet. Generally, the coverage of entities, classes from Wikipedia categories and high levels classes is quite low in YAGO-C, with percentages of 47.15%, 38.33% and 11.70%, respectively.

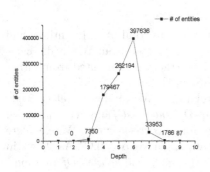

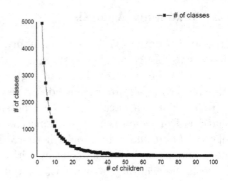

Fig. 3. Entity size distribution   **Fig. 4.** Class size distribution

Topological Structure. To understand the structure of our constructed taxonomy better, we evaluate the coverage of the taxonomy by observing the structure of the tree. We measure the depth of each leaf node and breadth of the taxonomy tree. We find that the depth ranges from 3 to 9, and the breadth ranges from 87 to 882,473.

We also evaluate the ability of the taxonomy on abstraction and expression. The depth of an entity in the tree indicates the ability of describing the entity. For example, given an entity *Lu Chen (magician)*, if the depth is 2 with the parent node *person*, we only know Lu Chen is a person. But if the depth is 5 with the path *living being, person, producer, Taiwanese television personality, Lu Chen (magician)*, we will know much more about Lu Chen.

In Figure 3, it shows that entities with the depth of 6 account for the majority of the entity set. It is normal that an entity may have different paths from it to the root, especially for people. We consider entity with multi-paths as different ones and that is why the size of entity set is larger than the entity space.

We also count the number of children for each class. As shown in Figure 4, the number of classes decreases rapidly as the number of children increases. When the number is more than about 50, the number of classes is very small. In fact, the classes with single child account for about 27.5%.

6.3 Performance Evaluation

Classification. To evaluate the performance of our classification model, We randomly select 4,600 (entity, category) pairs from the dataset and label them as positive (isA) or negative (notIsA). We randomly split 85% of the labeled data to train the classifier and test on the remainder.

We use precision, recall and F-measure to evaluate our linear SVM classifier. As as shown in Figure 5, the overall F-Measure is 97.0%, which proves the efficacy of our classification approach.

To further compare the contributions of features, we remove one feature and train the classifier with the rest at a time. As a result, we train another seven classifiers. To evaluation the contribution of each feature f, we define *Decrement in F-measure* as follows:

$$DF(f) = \frac{FM(F) - FM(F \setminus f)}{FM(F)} \tag{9}$$

where $FM(F)$ denotes the F-measure of the original classifier with feature set F and $FM(F \setminus f)$ denotes the F-measure of the classifier without feature f. From Figure 6, it is clearly observed that feature 3 is more discriminative with a higher DF score.

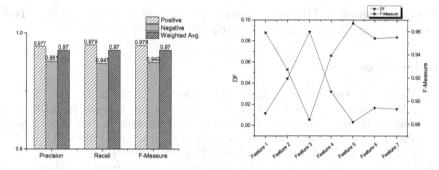

Fig. 5. Performance of linear SVM classifier **Fig. 6.** Evaluation of features and classifiers

isA Relation Inference and Expansion. As discussed in Section 4.2, we use pattern matching to leverage the semantics of relational categories. In the implementation, we use regular expressions to match Wikipedia categories to generate isA relations. In total, we design 70 regular expressions to match categories. Table 3 shows some of the regular expressions we use to perform inference. Note that when the object class does not exist, the rule can be leveraged to extract a property rather than a relation. We perform accuracy tests on each rule as well.

Table 3. Examples of inference rules

Subject Class	Object Class	Regular Expression	Num of Extraction	Accuracy
city	province	(.*省)市镇	32,091	100%
political leader	position	(.*(委员|参议员|参政员|议员))	13,881	100%
person	-	(.*? \ d{1,4}年)逝世	10,148	99%
person	-	(.*? \ d{1,4}年)出生	4,801	99%
monarch	-	(.*?)(君主|国王)	3,649	100%

As discussed in section 4.3, we expand isA relations by calculating the *confidence*. In fact, 4,707 isA relations are generated from all the existing isA relations. We set confidence to be 0.05 to filter out noisy, incorrect isA relations. We extract 3,380 isA relations with the accuracy 88% and coverage 71.8%.

7   Conclusion

In this paper, we propose a hybrid method to construct a Chinese taxonomy from user generated content. We generate a large number of accurate isA relations via directly classifying relations from categories, inferring relations from relational facts and properties, and extending existing relations by association rule mining. Furthermore, we construct the hierarchical structure to represent the taxonomy in a bottom-up manner. The experimental results illustrate that our Chinese taxonomy has a large scale and achieves a high accuracy.

However, Wikipedia is a rich knowledge repository that contains more than entities and categories. More isA relations can be mined from plain texts in Chinese. We will take effort to extract more isA relations and enlarge our taxonomy.

Acknowledgment. This work is partially supported by National Science Foundation of China (Grant No.61232002, 61402177 and 61332006).

References

1. Carlson, A., Betteridge, J., Kisiel, B., Settles, B., Hruschka Jr., E.R., Mitchell, T.M.: Toward an architecture for never-ending language learning. In: AAAI (2010)
2. Auer, S., Bizer, C., Kobilarov, G., Lehmann, J., Cyganiak, R., Ives, Z.G.: Dbpedia: A nucleus for a web of open data. In: Aberer, K., et al. (eds.) ISWC/ASWC 2007. LNCS, vol. 4825, pp. 722–735. Springer, Heidelberg (2007)
3. Lehmann, J., Isele, R., Jakob, M., Jentzsch, A., Kontokostas, D., Mendes, P.N., Hellmann, S., Morsey, M., van Kleef, P., Auer, S., Bizer, C.: DBpedia - a large-scale, multilingual knowledge base extracted from wikipedia. Semantic Web Journal 6(2), 167–195 (2015)
4. Wu, W., Li, H., Wang, H., Zhu, K.: Probase: A probabilistic taxonomy for text understanding. In: SIGMOD (2012)
5. Suchanek, F.M., Kasneci, G., Weikum, G.: Yago: A Core of Semantic Knowledge. In: WWW (2007)
6. Hoffart, J., Suchanek, F.M., Berberich, K., Weikum, G.: Yago2: A spatially and temporally enhanced knowledge base from wikipedia. Artif. Intell. 194, 28–61
7. Mahdisoltani, F., Biega, J., Suchanek, F.M.: YAGO3: A knowledge base from multilingual wikipedias. In: CIDR (2015)
8. Ponzetto, S.P., Strube, M.: Deriving a large-scale taxonomy from wikipedia. In: AAAI, pp. 1440–1445 (2007)
9. Wang, C., Gao, M., He, X., Zhang, R.: Challenges in chinese knowledge graph construction. In: ICDEW, pp. 59–61 (2015)
10. Fellbaum, C. (ed.): Wordnet, an Electronic Lexical Database. MIT Press (1998)
11. Fu, R., Guo, J., Qin, B., Che, W., Wang, H., Liu, T.: Learning semantic hierarchies via word embeddings. In: ACL, pp. 1199–1209 (2014)
12. Wang, Z., Li, J., Li, S., Li, M., Tang, J., Zhang, K., Zhang, K.: Cross-lingual knowledge validation based taxonomy derivation from heterogeneous online wikis. In: AAAI, pp. 180–186 (2014)
13. Wang, Z., Li, J., Wang, Z., Li, S., Li, M., Zhang, D., Shi, Y., Liu, Y., Zhang, P., Tang, J.: Xlore: A large-scale english-chinese bilingual knowledge graph. In: ISWC, pp. 121–124 (2013)
14. Carlson, A., Betteridge, J., Wang, R.C., Hruschka Jr., E.R., Mitchell, T.M.: Coupled semi-supervised learning for information extraction. In: WSDM (2010)
15. Hearst, M.A.: Automatic acquisition of hyponyms from large text corpora. In: COLING, pp. 539–545 (1992)

Mining Weighted Frequent Itemsets
with the Recency Constraint

Jerry Chun-Wei Lin[1], Wensheng Gan[1], Philippe Fournier-Viger[2],
and Tzung-Pei Hong[3,4]

[1] Innovative Information Industry Research Center(IIIRC)
School of Computer Science and Technology
Harbin Institute of Technology Shenzhen Graduate School, Shenzhen, P.R. China
jerrylin@ieee.org, wsgan001@gmail.com
[2] Department. of Computer Science
University of Moncton, Moncton, Canada
philippe.fournier-viger@umoncton.ca
[3] Department of Computer Science and Information Engineering
National University of Kaohsiung, Kaohsiung, Taiwan, R.O.C.
[4] Department of Computer Science and Engineering
National Sun Yat-sen University, Kaohsiung, Taiwan, R.O.C.
tphong@nuk.edu.tw

Abstract. Weighted Frequent Itemset Mining (WFIM) has been proposed as an alternative to frequent itemset mining that considers not only the frequency of items but also their relative importance. However, an important limitation of WFIM is that it does not consider how recent the patterns are. To address this issue, we extend WFIM to consider the recency of patterns, and thus present the Recent Weighted Frequent Itemset Mining (RWFIM). A projection-based algorithm named RWFIM-P is designed to mine Recent Weighted Frequent Itemsets (RWFIs) based on a novel upper-bound downward closure property. Moreover, an improved algorithm named RWFIM-PE is also proposed, which introduces a new pruning strategy named Estimated Weight of 2-itemset Pruning (EW2P) to prune unpromising candidate of RWFIs early. An experimental evaluation against a state-of-the-art WFIM algorithm on the real-world and synthetic datasets show that the proposed algorithms are highly efficient.

Keywords: weighted frequent itemset, recency constraint, weight constraint, EW2P strategy.

1 Introduction

Pattern mining is an important sub-field of data mining consisting of discovering patterns in databases such as association rules [2, 7] and sequential patterns [5, 11], among others [8, 9, 14, 16]. Frequent Itemset Mining (FIM) [2, 3, 7] has become an important data mining task having a wide range of real-world applications. However, an important limitation of FIM is that it considers the frequencies of items/itemsets in a transactional database but not other implicit factors such as their weight, interest, risk or profit. To address this issue, the

© Springer International Publishing Switzerland 2015
R. Cheng et al. (Eds.): APWeb 2015, LNCS 9313, pp. 635–646, 2015.
DOI: 10.1007/978-3-319-25255-1_52

problem of Weighted Frequent Itemset Mining (WFIM) was proposed by considering both the weight (importance) and frequency of patterns [6, 9, 13–16].

Cai et al. first defined a weighted-support model by multiplying the support of each item by its average weight [6]. Wang et al. assigned different weights to various items to mine Weighted Association Rules (WAR) [15]. Tao et al. developed the WARM (Weighted Association Rule Mining) algorithm and designed the *weighted downward closure* property to mine WFIs [13]. Vo et al. proposed a Weighted Itemset Tidset (WIF)-tree and adopted the Diffset structure to speed up WFIM [14]. Moreover, several extensions of WFIM have been proposed in many other fields, such as mining weighted association rules without pre-assigned weights [12] and weighted sequential patterns [10], etc.

Although WFIM can reveal more useful information from a database than FIM, WFIM suffers from an important limitation, which not considers how recent the patterns are. In this study, we address this issue. Our contributions are fourfold.

- A novel type of patterns called Recent Weighted Frequent Itemsets (RWFIs) is proposed to reveal more useful and meaningful WFIs by considering not only the weight and frequency of itemsets but also their recency constraint.
- A time-decay strategy is defined to automatically assign weights to transactions based on recency. A transaction is set with a higher recency value if it is more recent, which is more realistic than WFIM for real-world applications.
- The RWFIM-P algorithm is proposed to efficiently mine RWFIs. It relies on a projection-based approach and a new pruning property named *recent weighted frequent upper-bound downward closure* (*RWFUBDC*) property.
- An improved version of RWFIM-P named RWFIM-PE is also proposed. It relies on a new Estimated Weight of 2-itemset Pruning (EW2P) strategy to prune unpromising candidates early.

2 Preliminaries and Problem Statement

2.1 Preliminaries

Let $I = \{i_1, i_2, ..., i_m\}$ be a finite set of m distinct items appearing in a transactional database $D = \{T_1, T_2, ..., T_n\}$, where each transaction $T_q \in D$ is a subset of I, and has a unique identifier ccalled its *TID*. A unique existence weight $w(i_j)$ is assigned to each item $i_j \in I$, which represents its importance (e.g. profit, interest, risk). Existence weights for all items are stored in a weight table $wtable = \{w(i_1), w(i_2), ..., w(i_m)\}$. An itemset $X \in I$ with k distinct items $\{i_1, i_2, ..., i_k\}$ is of length k and is referred to as a k-itemset. An itemset X is said to be contained in a transaction T_q if $X \subseteq T_q$. Furthermore, for an itemset X, let the notation $TIDs(X)$ denotes the TIDs of all transactions in D containing X. As a running example, Table 1 shows a transactional database containing 10 transactions, which are sorted by purchased time.

Definition 1 (Item weight). The weight of an item i_j in D is denoted as $w(i_j)$, and represents the importance of this item to the user ($w(i_j) \in (0, 1]$).

Table 1. An example database

TID	Purchased time	Items
1	2015/1/08, 09:10	b, c, d, e
2	2015/1/09, 11:20	b, d
3	2015/1/11, 08:20	b, c, e
4	2015/1/12, 09:15	c, d
5	2015/1/12, 15:20	b, c, e
6	2015/1/14, 08:30	a, c, e, f
7	2015/1/14, 15:25	b, d, f
8	2015/1/15, 09:10	a, c, d, e, f
9	2015/1/16, 08:30	a, c, d, f
10	2015/1/18, 09:00	a, b, c, e, f

Table 2. Derived RWFIs

Itemset	*wsup*	*Recency*
(b)	4.2	2.8825
(c)	8.0	4.4675
(d)	3.0	3.0679
(e)	2.7	3.2404
(ac)	2.8	3.0945
(ce)	4.35	3.2404
(cf)	2.7	3.0945
(acf)	2.333	3.0945

For example, a user could define the weight table of items in Table 1 as *wtable* = $\{w(a)= 0.4, w(b)= 0.7, w(c)= 1.0, w(d)= 0.5, w(e)= 0.45, w(f)= 0.35\}$.

Definition 2 (Item weight in a transaction). The weight of an item i_j in T_q is defined as the weight of i_j in D. Thus: $w(i_j, T_q) = w(i_j), 1 \leq q \leq |D|$.

For example, the weight of (b) in T_1 is $w(b, T_1) = w(b) = 0.7$.

Definition 3 (Itemset weight). The weight of an itemset X in D $(w(X))$ is defined as the sum of the weights of all items in X divided by the number of items in X, that is: $w(X) = \frac{\sum_{i_j \in X} w(i_j)}{|X|}$, where $|X|$ is the cardinality of X.

For example, the weight of (bce) is calculated as $w(bce)=(w(b) + w(c) + w(e))/3 = (0.7 + 1.0 + 0.45)/3 = 0.7167$.

Definition 4 (Itemset weight in a transaction). The weight of an itemset X in T_q is defined as the weight of the itemset X in D, that is: $w(X, T_q) = w(X)$.

For example, the weight of (bce) in T_1 is calculated as $w(bce, T_1)=w(bce)=0.7167$.

Definition 5 (Weighted support of an itemset in D). The weighted support of an itemset X in D is denoted as $wsup(X)$, and is defined as:

$$wsup(X) = \sum_{X \subseteq T_q \wedge T_q \in D} w(X, T_q) = w(X) \times sup(X).$$

For example, the (bce) appears in transactions T_1, T_3, T_5 and T_{10}; the weighted support of (bce) is calculated as $wsup(bce)=\{w(bce, T_1) + w(bce, T_3) + w(bce, T_5) + w(bce, T_{10})\} = w(bce) \times 4 = (0.7 + 1.0 + 0.45)/3 \times 4 = 2.8668$.

Definition 6 (Weighted Frequent Itemset, WFI). Let α be a user-defined percentage value named the minimum weighted-support threshold. An itemset X in D is said to be a weighted frequent itemset (*WFI*) if its weighted support is no less than the minimum weighted-support threshold multiplied by the number of transactions in D, that is: $WFI \leftarrow \{X | wsup(X) \geq \alpha \times |D|\}$.

For example, suppose that α is set to 18%. An itemset (a) is not considered a WFI since $wsup(a) = 0.4 \times 4 = 1.6$, which is smaller than $(18\% \times 10) = 1.8$. However, the (b) is considered a WFI since $wsup(b) = 0.7 \times 6 = 4.2 > 1.8$.

Definition 7 (Recency of a transaction). Let transactions are sorted by purchased time, the *Recency* of T_q is denoted as: $R(T_q) = (1 - \delta)^{|D|-T_q}$, where δ is a user-specified time-decay factor ($\delta \in (0,1]$), $|D|$ is the current timestamp which is equal to the number of transactions in D, and T_q is the TID of the currently processed transaction.

Thus, a high *Recency* value is assigned to the most recent transactions. For example, assume that the user set the time-decay factor δ to 0.15. *The Recency* of T_1 and T_9 are respectively calculated as $R(T_1) = (1 - 0.15)^{(10-1)} = 0.2316$ and $R(T_9) = (1 - 0.15)^{(10-9)} = 0.85$.

Definition 8 (Recency of an itemset in a transaction). The *Recency* of an itemset X in T_q is defined as: $R(X, T_q) = R(T_q)$.

Definition 9 (Recency of an itemset in D). The *Recency* of an itemset X in D is denoted as $R(X)$, and defined as: $R(X) = \sum_{X \subseteq T_q \wedge T_q \in D} R(X, T_q)$.

For example, the *Recency* of (c) in T_1 is $R(c, T_1) = R(T_1) = 0.2316$. The *Recency* of (bce) in D is calculated as $R(bce) = R(bce, T_1) + R(bce, T_3) + R(bce, T_5) + R(bce, T_{10}) = 0.2316 + 0.32057 + 0.4437 + 1.0) = 1.9959$.

Definition 10 (Recent Weighted Frequent Itemset, RWFI). An itemset X in a database D is said to be a recent weighted frequent itemset $(RWFI)$ if it satisfies the following two conditions:

$$RWFI \leftarrow \{X|wsup(X) \geq \alpha \times |D| \wedge R(X) \geq \beta\},$$

where α is the user-specified minimum weighted-support threshold and β is a user-specified minimum recency threshold.

For example, consider that the minimum recency threshold β is set to 2.5. The (bc) is not a RWFI since its *Recency* is calculated as $R(bc) = 1.9959 < 2.5$. The (acf) is a RWFI since $wsup(acf) = 2.333 > 1.8$ and $R(acf) = 3.0945 > 2.5$. The full set of RWFIs for α $(= 18\%)$ and β $(= 2.5)$ is shown in Table 2.

2.2 Problem Statement

Let D be a transactional database, a weight table *wtable* indicating the weight of each item, and the user-specified minimum weighted-support threshold α and minimum recency threshold β. The problem of RWFIM is to find the complete set of RWFIs of length k, that is k-itemsets satisfying the two following conditions w.r.t weight and recency: 1) $wsup(X) \geq \alpha \times |D|$; 2) $R(X) \geq \beta$.

3 Proposed RWFIM-P Algorithm

We first propose the **R**ecent **W**eighted-**F**requent **I**temset **M**ining **P**rojected-based (RWFIM-P) algorithm to efficiently mine RWFIs. RWFIM-P utilizes a projection-based approach to explore the search space. Moreover, it introduces a novel property named *recent weighted frequent upper-bound downward closure* (*RWFUBDC*) to prune the search space. The proposed *RWFUBDC* property is presented thereafter.

3.1 Proposed Recent Weighted Frequent Upper-Bound Downward Closure (RWFUBDC) Property

A major challenge to propose an efficient RWFI mining algorithm is that the well-known *downward closure* (*DC*) property used in FIM and WFIM does not hold in RWFIM, that is subsets of a RWFI may or may not be RWFIs. To restore the *DC* property, we introduce the *RWFUBDC* property in the proposed RWFIM-P algorithm. This property can greatly reduce the size of the search space for mining RWFIs.

Definition 11 (Transactional upper-bound weight, $tubw$). The transactional upper-bound weight of a transaction T_q ($tubw(T_q)$) is defined as:

$$tubw(T_q) = max\{w(i_1, T_q), w(i_2, T_q), \ldots, w(i_j, T_q)\},$$

where j is the number of items in T_q.

For example, $tubw(T_1) = max\{w(b, T_1), w(c, T_1), w(d, T_1), w(e, T_1)\} = max\{0.7, 1.0, 0.5, 0.45\} = 1.0$.

Theorem 1. *The weight of any itemset X in a transaction T_q is always smaller or equal to the transactional upper-bound weight of T_q: $w(X, T_q) \leq tubw(T_q)$.*

Proof. Since $tubw(T_q) = max\{w(i_1, T_q), w(i_2, T_q), \ldots, w(i_j, T_q)\}$, thus:

$$w(X, T_q) = \frac{\sum_{i_j \in X \wedge X \subseteq T_q} w(i_j, T_q)}{|X|} \leq \frac{max\{w(i_j, T_q)\} \times |X|}{|X|} = tubw(T_q).$$

Definition 12 (Transactional accumulation upper-bound weight, $taubw$). The transactional accumulation upper-bound weight of an itemset X in D is denoted as $taubw(X)$ and is defined as:

$$taubw(X) = \sum_{X \subseteq T_q \wedge T_q \in D} tubw(T_q).$$

For example, the $taubw$ of (bce) is calculated as $taubw(bce) = tubw(T_1) + tubw(T_3) + tubw(T_5) + tubw(T_{10}) = 1.0 + 1.0 + 1.0 + 1.0) = 5.0$.

Definition 13 (Recent Weighted Frequent Upper-Bound Itemset, RW-FUBI). An itemset X in D is called a recent weighted frequent upper-bound itemset (*RWFUBI*) if it satisfies the following two conditions:

$$RWFUBI \leftarrow \{X | taubw(X) \geq \alpha \times |D| \wedge R(X) \geq \beta\}.$$

For example, (b) is a $RWFUBI$ since $taubw(b) = 5.4 > 1.8$, $R(b)= 2.8825 > 2.5$; but (bc) is not a $RWFUBI$ since $taubw(bc)= 3.4 > 1.8$, $R(bc)= 1.9959 <2.5$.

Theorem 2. (Downward Closure Property of RWFUBI, *RWFUBDC*)
Let X^k be a k-itemset. If X^k is a RWFUBI, then any subset X^{k-1} of X^k is also a RWFUBI, that is $taubw(X^k) \leq taubw(X^{k-1})$ and $R(X^k) \leq R(X^{k-1})$.

Proof. Since $X^{k-1} \subseteq X^k$, $TIDs(X^k \in T_q) \subseteq TIDs(X^{k-1} \in T_q)$. Thus:
$taubw(X^k) = \sum_{X^k \subseteq T_q \wedge T_q \in D} tubw(T_q) \leq \sum_{X^{k-1} \subseteq T_q \wedge T_q \in D} tubw(T_q) = taubw(X^{k-1})$,
$R(X^k) = \sum_{X^k \subseteq T_q \wedge T_q \in D} R(X^k, T_q) \leq \sum_{X^{k-1} \subseteq T_q \wedge T_q \in D} R(X^{k-1}, T_q) = R(X^{k-1})$.

From the above results, it can be concluded that if X^k is a $RWFUBI$, any subset X^{k-1} is also a $RWFUBI$. Thus the $RWFUBDC$ property holds and can be used to prune unpromising candidates w.r.t the minimum weighted-support threshold and minimum recency threshold during the search for RWFIs.

Theorem 3. (RWFIs $\subseteq$ RWFUBIs) *Let RWFUBIs be the set of recent weighted frequent upper-bound itemsets in the database D, and RWFIs be the set of recent weighted frequent itemsets in D. The recent transactional accumulation weighted frequent upper-bound downward closure (RWFUBDC) property implies that RWFIs$\subseteq$RWFUBIs. In other words, if a pattern is not a RWFUBI, it will not be a RWFI.*

Proof. $\forall X \in D$, $w(X) = w(X, T_q)$. From Theorems 1 and 2, it can be obtained that $w(X, T_q) \leq tubw(T_q)$ and $R(X^k) \leq R(X^{k-1})$. Thus:
$wsup(X) = \sum_{X \subseteq T_q \wedge T_q \in D} w(X, T_q) \leq \sum_{X \subseteq T_q \wedge T_q \in D} tubw(T_q) = taubw(X)$.
$\Rightarrow wsup(X) \leq taubw(X)$.

3.2   Procedure of RWFIM-P Algorithm

The proposed RWFIM-P algorithm works as follows. It first scans the original database to discover the set of recent weighted frequent upper-bound 1-itemsets $(RWFUBI^1)$ and recent weighted frequent 1-itemsets $(RWFI^1)$. Then, the designed **Mining-RWFI**(X, db_X, k) function is recursively called to mine the complete set of RWFIs. The RWFIM-P algorithm and the **Mining-RWFI**(X, db_X, k) procedure are respectively given in Algorithm 1 and Algorithm 2. In these pseudocode, notice that the projection of a database db by an itemset X is defined as the set of transactions from db containing X and is denoted as db_X.

4   Proposed RWFIM-PE Algorithm

Although the $RWFUBDC$ property used in RWFIM-P is effective to prune the search space, many candidates are still generated that are not RWFIs. To address this issue, an improved version of RWFIM-P named **R**ecent **W**eighted-**F**requent **I**temset **M**ining **P**rojected-based approach with **E**arly pruning (RWFIM-PE) is proposed. It improves upon RWFIM-P by introducing a novel pruning strategy name EW2P (Estimated Weight of 2-itemset Pruning) to prune more candidates and thus speed up the mining process. This pruning strategy is based on the recent weighted frequent upper-bound (RWFUB) model.

Input: D, a transactional database; *wtable*, a predefined weight table; δ, the
time-decay threshold; α, the minimum weighted-support threshold; β,
the minimum recency threshold.

Output: The set of recency weighted frequent itemsets (RWFIs).

1 **for** *each* $T_q \in D \wedge i_j \in T_q$ **do**
2 | calculate $sup(i_j)$, $tubw(T_q)$, $R(T_q)$;
3 | calculate $R(i_j)$, $taubw(i_j)$;

4 **for** *each* $i_j \in D$ **do**
5 | **if** $taubw(i_j) \geq \alpha \times |D| \wedge R(i_j) \geq \beta$ **then**
6 | | $RWFUBI^1 \leftarrow RWFUBI^1 \cup \{i_j\}$;
7 | | calculate $wsup(i_j)$;
8 | | **if** $wsup(i_j) \geq \alpha \times |D|$ **then**
9 | | | $RWFI^1 \leftarrow RWFI^1 \cup \{i_j\}$;

10 sort $RWFUBI^1$ in lexicographical order;
11 **for** *each* $i_j \in RWFUBI^1$ **do**
12 | calculate db_{i_j}, the projection of D by i_j;
13 | call **Mining-RWFI**(i_j, db_{i_j}, **1**);
14 **return** RWFIs;

Algorithm 1. RWFIM-P

Input: X, a prefix itemset; db_X; the projection of D by X; k, the length of X.
Output: The set of RWFIs of a prefix X.

1 $PC^{k+1} \leftarrow \{X'|X' = X \cup \{y\} \wedge y \in RWFUBI^1 \wedge y$ is greater than all items in
X according to the lexicographical order};
2 **for** *each* $X' \in PC^{k+1}$ **do**
3 | **for** *each* $X' \subseteq T_q \wedge T_q \in db_{X'}$ **do**
4 | | calculate $db_{X'}$, the projection of db_X by X';
5 | | calculate $taubw(X')$, $R(X')$, $sup(X')$;

6 | **if** $taubw(X') \geq \alpha \times |D| \wedge R(X') \geq \beta$ **then**
7 | | calculate $wsup(X')$;
8 | | **if** $wsup(X') \geq \alpha \times |D|$ **then**
9 | | | $RWFI^{K+1} \leftarrow RWFI^{K+1} \cup X'$;
10 | | $RWFUBI^{k+1} \leftarrow RWFUBI^{k+1} \cup X'$;
11 | | call **Mining-RWFI**(X', $d_{bX'}$, $k+1$);

12 $RWFIs \leftarrow \bigcup_k RWFI^{k+1}$;
13 **return** RWFIs of X.

Algorithm 2. Mining-RWFI(X, db_X, k)

4.1   Estimated Weight of 2-itemset Pruning

Theorem 4. *If the transactional accumulation upper-bound weight of a 2-itemset X in D is less than the minimum weighted support, X will not be a RWFUBI or RWFI. Moreover, any superset of X will not be a RWFUBI or RWFI either.*

Proof. Let be a 2-itemset X, and a k-itemset X^k ($k \geq 3$) such that $X \subset X^k$. From Theorem 2, it can be obtained that $taubw(X^k) \leq taubw(X^{k-1})$ for any subset X^{k-1} of X^k. Thus, if a 2-itemset X is not a RWFUBI, any k-itemset ($k \geq 3$) that is a superset of X will not be a RWFUBI or RWFI. Therefore, for any itemset Y, if there exists a 2-itemset $W \subset Y$ such that $taubw(W) < \alpha \times |D|$, any supersets of Y can be skipped in the mining process.

To apply this pruning technique efficiently, a constructed Estimated Weight of 2-itemset Structure (EW2S) is built for storing the $taubw$ values of all pairs of items in D. A constructed EW2S is shown in Table 3. The proposed pruning technique is efficient to prune itemsets that are not RWFIs early.

Table 3. Constructed EW2S

Item	a	b	c	d	e	f
b	1.0	-	-	-	-	-
c	4.0	4.0	-	-	-	-
d	2.0	2.4	4.0	-	-	-
e	3.0	4.0	6.0	2.0	-	-
f	4.0	1.7	4.0	2.7	3.0	-

4.2   Procedure of RWFIM-PE Algorithm

The proposed RWFIM-PE algorithm (Algorithm 3) is similar to the RWFIM-P algorithm. The difference is that the EW2S needs to be constructed initially during the first database scan. Moreover, the mining procedure for deriving RWFIs is modified to verify the new pruning property for each generated itemset. Due to the page limitation, the detailed pseudocode of the modified **Mining-RWFI** procedure is not provided.

---

1 perform steps 1 to 3 from *Mining-RWFI(X,db_X,k)* function;
2 construct *EW2S* ;
3 **for** *each $X' \in PC^{k+1}$* **do**
4   **for** $k := 1 \wedge \exists (X \cup i_j) \in EW2S \wedge taubw(X \cup i_j) < \alpha \times |D|$ **do**
5   continue;
6 perform steps 4 to 15 from ***Mining-RWFI(X, db_X,k)*** function.

Algorithm 3. *Mining-RWFI function(X, db_X, k)* of RWFIM-PE

5 Experiments

We performed extensive experiments to evaluate the proposed algorithms. Because this is the first paper to consider the recency of WFIs, no algorithm for the exact same problem was available for comparison. We thus used a state-of-the-art algorithm named PWA [9] for WFIM as a baseline algorithm. Experiments are conducted on two datasets including the real-world dataset (retail) [1] and the synthetic dataset (T10I4D100K) generated using the IBM Quest Synthetic Data Generator [4]. We assessed execution time, number of patterns and scalability. The weight of each item was randomly selected in (0,1]. The retail dataset has 88,162 transactions with 75 distinct items, the average length of transactions is 10.3. The T10I4D100K has 100,000 transactions with 870 distinct items, and the average length of transactions in it is 10.1.

5.1 Execution Time

The execution time of the three algorithms are first compared under a fixed time-decay threshold (D) with varied minimum weighted-support thresholds (MWs) and varied minimum recency thresholds (MRs), respectively. Results are shown in Fig. 1 and Fig. 2.

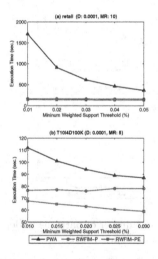

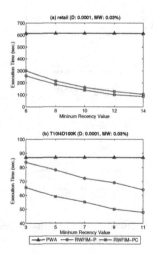

Fig. 1. Execution time under varied MWs.

Fig. 2. Execution time under varied MRs.

In Fig. 1, it can be seen that RWFIM-P and RWFIM-PE perform very well compared to PWA, and RWFIM-PE has the best performance. This result is reasonable since both the proposed RWFIM-P and RWFIM-PE algorithms consider an additional constraint that is the recency of patterns. By considering this additional constraint, a larger part of the search space can be pruned and less patterns are discovered. RWFIM-PE is faster than RWFIM-P because it adopts the novel EW2S pruning strategy to eliminate itemsets early. This allows RWFIM-PE to avoid performing the projection operation for forming the sub-database of each pruned itemset. It is interesting to observe that for the retail dataset, RWFIM-PE is still faster than PWA and RWFIM-P. The reason is that most unpromising candidates are early pruned by the EW2P properties, and thus the EW2P strategy is more effective.

From Fig. 2, it can also be seen that RWFIM-PE has better performance than RWFIM-P and PWA under various MRs. In particular, the proposed algorithms are generally from about one to two orders of magnitude faster than PWA. The reason is that when the MR value is set higher, a larger part of the search space can be pruned and there is less RWFIs. In summary, the proposed RWFIM-PE algorithm outperforms the RWFIM-P and PWA algorithms.

5.2   Patterns Analysis

The number patterns are compared under a fixed time-decay threshold (D) with varied MWs and varied MRs, respectively. Results are shown in Fig. 3 and Fig. 4, respectively.

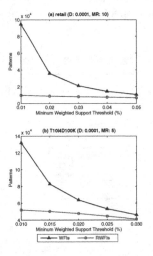

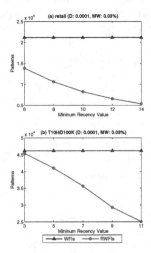

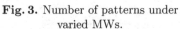

Fig. 3. Number of patterns under	Fig. 4. Number of patterns under
varied MWs.	varied MRs.

From Fig. 3, it can be seen that the number of discovered RWFIs and WFIs decrease as MW is increased, which was expected. The number of discovered RWFIs is always smaller than the number of WFIs under varied MWs in both sparse and dense datasets. It indicates that numerous WFIs are discovered but few RWFIs are produced when considering the recency of transactions. This situation happens especially when MW is set to small values since numerous irrelevant WFIs are then pruned. From Fig. 4, it can also be seen that the number of RWFIs dramatically decreases as MR increases, while the number of WFIs remains stable. The reason is the same as above. In particular, less RWFIs are found but they are more valuable for real-life applications compared to WFIs since they represent recent trends.

5.3 Scalability Analysis

Fig. 5 shows the scalability of the compared algorithms under varied dataset sizes T10I4N4KD|X|K when the time-decay threshold, MR and MW are set as 0.0001, 5, and 0.03%, respectively. We measured execution time, memory consumption, and the number of generated RWFIs and WFIs. The variable X indicates the dataset size and was varied from 100K to 500K, by increments of 100K.

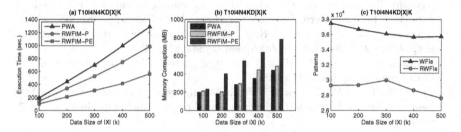

Fig. 5. Scalability of compared algorithms.

It can be seen that all compared algorithms scale well with respect to dataset size and that the proposed RWFIM-P and RWFIM-PE algorithms scale better than PWA. When the dataset size is increased, RWFIM-PE becomes more and more faster than RWFIM-P thanks to the EW2P strategy. However, RWFIM-PE consumes more memory than RWFIM-P because it uses the additional EW2S to store *taubw* values of all 2-itemsets (see Fig. 5(b)). From Fig. 5(c), it can also be seen that the number of discovered RWFIs remains quite smaller than the number of WFIs. Thus, the RWFIs can reveal a smaller but more meaningful set of recent patterns to the user compared to the whole set of WFIs.

6   Conclusion

In this paper, a novel type of patterns named recent weighted frequent itemsets (RWFIs) is proposed to solve the limitations of traditional weighted frequent itemset mining by considering both the weight and recency of patterns. A projection-based algorithm named RWFIM-P is presented to mine RWFIs based on a novel upper-bound downward closure property. Moreover, an improved algorithm named RWFIM-PE is also proposed, which introduces a new pruning strategy named Estimated Weight of 2-itemset Pruning (EW2P) to prune non RWFIs early. An experimental evaluation against a state-of-the-art WFIM algorithm on the real-world and the synthetic datasets show that the proposed algorithms are highly efficient. For future works, we will try to extend the proposed RWFIM framework and the upper-bound model to other relative issues.

Acknowledgment. This research was partially supported by the Shenzhen Strategic Emerging Industries Program under grant ZDSY20120613125016389, by the Tencent Project under grant CCF-TencentRAGR20140114, and by the Natural Scientific Research Innovation Foundation in Harbin Institute of Technology under grant HIT.NSRIF.2014100.

Bibliography

[1] Frequent itemset mining dataset repository, `http://fimi.ua.ac.be/data/`

[2] Agrawal, R., Imielinski, T., Swami, A.: Database mining: A performance perspective. IEEE Trans. on Knowledge and Data Engineering 5, 914–925 (1993)

[3] Agrawal, R., Srikant, R.: Fast algorithms for mining association rules in large databases. In: The Intern. Conf. on Very Large Data Bases, pp. 487–499 (1994)

[4] Agrawal, R., Srikant, R.: Quest synthetic data generator, `http://www.Almaden.ibm.com/cs/quest/syndata.html`

[5] Agrawal, R., Srikant, R.: Mining sequential patterns. In: The Intern. Conf. on Data Engineering, pp. 3–14 (1995)

[6] Cai, C.H., Fu, A.W.C., Kwong, W.W.: Mining association rules with weighted items. In: Intern. Database Engineering and Applications Symposium, pp. 68–77 (1998)

[7] Chen, M.S., Han, J., Yu, P.S.: Data mining: An overview from a database perspective. IEEE Trans. on Knowledge and Data Engineering 8, 866–883 (1996)

[8] Geng, L., Hamilton, H.J.: Interestingness measures for data mining: A survey. ACM Computing Surveys 38 (2006)

[9] Lan, G.C., Hong, T.P., Lee, H.Y., Lin, C.W.: Mining weighted frequent itemsets. In: The 30th Workshop on Combinatorial Mathematics and Computation Theory, pp. 85–89 (2013)

[10] Lan, G.C., Hong, T.P., Lee, H.Y.: An efficient approach for finding weighted sequential patterns from sequence databases. Applied Intelligence 41, 439–452 (2014)

[11] Srikant, R., Agrawal, R.: Mining sequential patterns: Generalizations and performance improvements. In: Apers, P.M.G., Bouzeghoub, M., Gardarin, G. (eds.) EDBT 1996. LNCS, vol. 1057, pp. 3–17. Springer, Heidelberg (1996)

[12] Sun, K., Bai, F.: Mining weighted association rules without preassigned weights. IEEE Trans. on Knowledge and Data Engineering 20, 489–495 (2008)

[13] Tao, F., Murtagh, F., Farid, M.: Weighted association rule mining using weighted support and significance framework. In: ACM SIGKDD Intern. Conf. on Knowledge Discovery and Data Mining, pp. 661–666 (2003)

[14] Vo, B., Coenen, F., Le, B.: A new method for mining frequent weighted itemsets based on wit-trees. Expert Systems with Applications 40, 1256–1264 (2013)

[15] Wang, W., Yang, J., Yu, P.S.: Efficient mining of weighted association rules (WAR). In: ACM SIGKDD Intern. Conf. on Knowledge Discovery and Data Mining, pp. 270–274 (2000)

[16] Yun, U., Leggett, J.: WFIM: Weighted frequent itemset mining with a weight range and a minimum weight. In: SIAM Intern. Conf. on Data Mining, pp. 636–640 (2005)

Boosting Explicit Semantic Analysis by Clustering Paragraph Vectors of Wikipedia Articles

Hai-Tao Zheng* and Wenzhen Wu

Graduate School at Shenzhen, Tsinghua University, China
zheng.haitao@sz.tsinghua.edu.cn,
wuwz12@mails.tsinghua.edu.cn

Abstract. Explicit Semantic Analysis (ESA) is an effective method that utilizes Wikipedia entries (articles) to represent text and compute semantic relatedness (SR) for text pairs. Analogous to ordinary web search techniques, ESA also suffers from the redundancy issues due to the on-going expansion of the amount of Wikipedia entries. Entries redundancy could lead to biased representation that lay particular emphasis on semantics from a large number of similar entries. On the other hand, original ESA for SR has a weak point that it does not consider the correlations or similarities between the Wikipedia articles of the text representations. To tackle these problems, We develop a novel method to cluster the redundant or similar entries by similarity measurement based on Paragraph Vector (PV), a neural network language model. Results of experiments on four datasets show that our framework could gain better performance in relatedness accuracy against ESA.

Keywords: Semantic Relatedness, Explicit Semantic Analysis, Paragraph Vector, Clustering.

1 Introduction

With the boom of Internet and social media, a vast volume of data has been generating and cumulating incessantly. Text content accounted for the majority among the User-generated Content [1] data in web. Automatically estimating the semantic relatedness of two text fragments is fundamental for many text mining and IR applications. In this work, we develop a SR computing framework exploiting inter-entries distance to reduce the redundancy and utilize the correlations between Wikipedia articles of text representation by ESA. Our experiments show its superiority against original ESA.

Existing approaches measuring text semantic relatedness could be classified to three categories. lexical overlap is a kind of straight-forward but weak method, for mainly text relatedness is not based on common terms of given text pair. The other kind of approaches are based on knowledge bases such as WordNet [2]

---

* Corresponding author.

© Springer International Publishing Switzerland 2015
R. Cheng et al. (Eds.): APWeb 2015, LNCS 9313, pp. 647–657, 2015.
DOI: 10.1007/978-3-319-25255-1_53

or Cyc [3]. Texts are initially mapped to taxonomy nodes in knowledge bases, then the SR of text pairs could be computed by exploiting links and paths between nodes [4][5][6][7]. Besides, LDA [8], LSA [9] and ESA [10] constitute the $3rd$ category of schemes that generate feature vector representation for text by exploring the words co-occurrence relationships or word occurrence statistics in a corpus [11]. There is also much effort [16][14] are devoted to combine a variaty of individual models together to gain better performance. Normally cosine similarities are then utilized to measure the SR.

It is known that when reading human can refer the content in text to things in real world and relations between them. Analogously, ESA served as an interpreter that refer given text to semantically related articles of Wikipedia. Hence, a semantic relatedness of a pair of texts sharing little common words could be captured by ESA. Prior work [12] shows that ESA achieve the state of the art performance within the individual methods on SR tasks. different from works devoted to combine a variaty of methods, mostly including ESA, to train a integrated model that mostly outperform the methods seperatedly, our work focuses on improvement of the ESA method, which is also able to be incorperated with other methods.

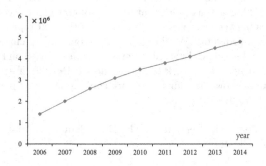

Fig. 1. The growth of Wikipedia articles [19].

For a given text fragment,ESA treat it as a collection of words disregarding the order. Those words, after stemming process, are mapping to a list of article entries and corresponding scores via a pre-built inverted indices base that link to the articles in Wikipedia. Normally the scores in the list were calculated by TFIDF scheme. By this means, ESA provide an more effective and elegant model to represent a text comparing to Bag of Words model, and then Using this model to computing semantic relatedness among text pair. However, the semantic relatedness between entries, especially for the redundancy issue, is not considered in ESA. As is shown in Figure 1, the scale of Wikipeida has experienced a huge growth during the last several years. That more and more very specific entries being established in Wikipedia may results in redundant entries in the text representation by ESA. There are two shortcomings for this situation in semantic relatedness computing. First, when text pair share a considerable amount of entries which is highly similar to each other in their representations, ESA scores may be unreasonably enhance due to the bias in computing. On the other hand, when both text representations in a pair contain some entries respectively about a specific topic, if they share no common entries, the resulting score will be underestimated. We believed that the system performance could be enhanced by overcoming those disadvantages. Hence, we devise a framework exploiting entries similarity to reduce the redundancy problems.

In the paper, we propose a SR computing framework that employ density based clustering to improve ESA accuracy. The framework incorporates Paragraph Vector, a newly devised text embedding technique that remarkably superior to Bag of Words methods, to represent Wikipedia entries. In the framework we present 1) a density based clustering method to group similar Wikipedia entries to form new text representation that bridge the gap between similar entries, and 2) a scoring mechanism by using the maximum score in each group and disregarding other entries to avoid redundancy. We conduct experiments on four dataset, and the result indicate that our system gains better performance comparing original ESA.

The structure of this paper is organized as follow. Related works is stated in Section 2. Then we outline the main idea of ESA, Section 3 is devoted to a detailed description of our framework. In which we persent two key algorithm of our framework. Preprocessing and experimental evaluation will be elaborated in Section 4. Finally, we provide concluding remarks and suggestions for future work in Section 5.

2   Preliminary

We firstly outline the ESA's principal fundamentals. ESA initially maintain a database stores the mapping between terms and their relevant Wikipedia articles. Similar to ordinary IR tasks, all the Wikipedia articles with meaningful content are indexed by terms (except the stop words) they contain. Hence each non-stop term is mapped to a list of Wikipedia articles. Namely

$$Vec(t) = \{e_1, e_2, e_3, \ldots, e_n\} . \tag{1}$$

Where $e_i = \{e_i.title, e_i.score, e_i.fulltext\}$.

TF-IDF measure are chosen as a score of the mapping in original ESA implementation. So we have

$$e_i.score = tf\_idf(t, e_i.fulltext) . \tag{2}$$

So far, $Vec(t)$, which contains a list of Wikipedia articles and their corresponding scores related to t measured by TF-IDF scheme, could be served as the semantic representation of the single term t. For longer documents, the representation is the centroid of the vectors representing the each terms.

$$Vec(d) = \sum_{t \in d} \frac{tf(t)}{|d|} Vec(t) \tag{3}$$

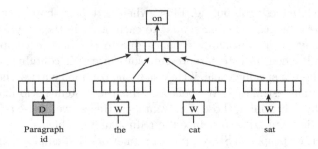

Fig. 2. The framework of Paragrah Vector [18].

Then we present the key idea of the PV [18] technique that is recently developed. PV is the extended version of the Word2Vec [17][1] that served to learn vector representation of words from a corpus. As a neural language model (Figure 2), Word2Vec is originally designed to predict word based on given context. However, what make it unique from other language model is the compact vector representation (100 dimensions by default) for every words. In the training process, both the weights of the model and the representation vectors of the words are updated simultaneously. After training is finished, every words in the corpus get a fixed-length vector representation. with the same architecture, Paragrah Vector are gained by treat the sentences or paragraphs as the a single word in the beginning of the text. Experiments [18] on sentiment analysis and informantion retrieval show that PV could ourperform other model including Bag of Words, which is also why we adopt it to mearsure the similarities of the Wikipedia articles. It is worth noting that although PV itself could be adopted to measure the text similarity for corpora, it does not apply to the scenarios that only small amount of text are given. In those cases ESA is more suitable.

3   Methodology

3.1   Overview

The proposed framework (Figure 3) is comprised of three components. 1) Ordinary ESA that generate Wikipedia articles based representation for given text pairs; 2) Wikipedia articles embedding, in which we running the PV program on all the Wikipedia articles to generate and store on database 100-dimensional vectors for each article; and 3) A clustering based algorithm for redundancy reduction (Algorithm 1). Based on the similarity between the precomputed PVs of different articles, original ESA representations of the text pairs is reconstructed to relatively more compact representations with little redundancy.

3.2   Algorithm

In this section, we present an algorithm called Semantic Relatedness via Paragraph Vector based Clustering (SRPVC) to compute SR for a given text pair.

---

[1] http://code.google.com/p/word2vec/

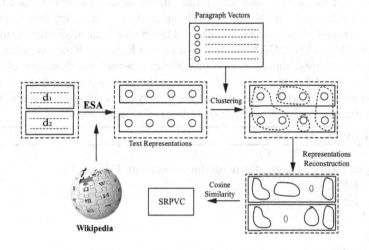

Fig. 3. The proposed framework

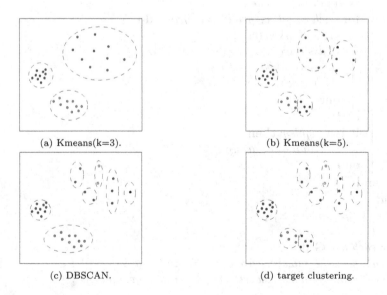

(a) Kmeans(k=3).

(b) Kmeans(k=5).

(c) DBSCAN.

(d) target clustering.

Fig. 4. The effect of Kmeans,DBSCAN and our target clustering

Comparing to existing clustering algorithms, the key point of SRPVC is that it guarantees the maximum distance between points within cluster, i.e., the Wikipedia articles of corresponding points in the same cluster are considerably related to each other. In this way, highly related or similar articles, treated as redundant entries, were merged together. The effect of our algorithm comparing to Kmeans [20] and DBSCAN [21] is illustrated in Figure 4. Kmeans assigns data points to the nearest means iteratively, without regard for distances of points within clusters. Relatively distant points may be assigned together by Kmeans. DBSCAN clusters together points that are closely packed together, without consider the diameters (i.e. maximum distances) of clusters also.

Algorithm 1. SRPVC

Input: text pair T_1, T_2
Output: score of semantic relatedness between T_1 and T_2
$E_1 \leftarrow ESA(T_1)$;
$E_2 \leftarrow ESA(T_2)$;
$S \leftarrow E_1 \cup E_2$;
for *each* $c_i \in S$ **do**
   | $c_i.group \leftarrow i$;
end
for *each* $c_i \in S$, $c_j \in S$ **do**
     $v_i \leftarrow ParaVec(c_i)$;
     $v_j \leftarrow ParaVec(c_j)$;
     if $distance_{cosine}(v_i, v_j) < \alpha$ **then**
         $isNeighbor \leftarrow true$;
         for *each* c_k *s.t.* $c_k.group = c_i.group$ **do**
             $v_k \leftarrow SenVec(c_k)$;
             if $distance_{cosine}(v_j, v_k) > \beta$ **then**
                 $isNeighbor \leftarrow false$;
                 $break$;
             end
         end
         if $isNeighbor$ **then**
             $c_i.group \leftarrow c_j.group$;
         end
     end
end
$G \leftarrow \{c.group \mid c \in S\}$;
$i \leftarrow 0$;
for *each* $g \in G$ **do**
     $g1_i.score \leftarrow max\{c.score \mid c \in E_1 \text{ and } c_i.group = g\}$;
     $g2_i.score \leftarrow max\{c.score \mid c \in E_2 \text{ and } c_i.group = g\}$;
     $i \leftarrow i + 1$;
end
return $cosine\_similarity(g1, g2)$;

We explain the Algorithm 1 as follows: Firstly, Both text's ESA representations, as a whole ($S \leftarrow E_1 \cup E_2$), are treated as the input of clustering. Every data point, which represent a single Wikipedia article, is initially assigned to a specific group. Then any two groups are iteratively merge to a larger group if their locality property satisfies certain condition. In the algorithm we set the condition that 1) the minimun distance of data points of two group respectively must not larger that α, and 2) distances of all the entries within the new cluster must be not larger than β. By this way algorithm make sure entries within the same group are related enough to eath other. In other words, for any data point c in a group, the distances of nearest point and farthest point from c in the same group should not be higher than α and β respectively. After clustering, the final groups (clusters from clustering) served as the vector space for new representation vectors of the text pair. Unrelated in previous representation but actually similar Wikipedia entries may be related to each other by being assigned to same groups. For each text and each group, the maximum ESA scores are adopted as the corresponding scores of the groups in new representations, as shown in the final part of the algorithm. By ignoring other scores in the same group, redundancy issue will be mitigated.

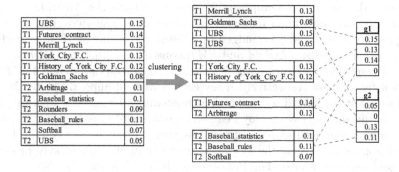

Fig. 5. An example of the algorithm SRPVC

To make the idea of the algorithm more intuitive, we present an intermediate result (Figure 5 and the Representations Reconstruction component of Figure 3) that depict a segment of ESA representation of a text pair, before and after the clustering. As depicted on the left, although only one Wikipedia entry (UBS) is shared by the text pair, there are still a lot of Wikipedia entries are considerably related. similar entries are clustered together as the dimensions of new vector space after clustering, as shown in the middle of the figure. For each groups, Picking the maximum score for both text, disregarding other entries to avoid redundancy, forming the final representation vectors on the right of the figure.

4 Evaluation

4.1 Preprocessing

We acquire the snapshot of Wikipedia dump (static documents) of December 2nd, 2013, then extract and transform data into relational database. Similar to the preprocessing in [12][15], the following phrases are involved.

- Extracting data, mainly including full text of articles, from original Wikipedia dump XML files.
- Stemming and indexing the content field of pages.
- Using index function to calculate IDF value for each term-document pair, and then insert them into the database.

As for of the PV of each Wikipedia articles, the program running on a ordinary PC took approximately 50 hours to get the output. We also store the output results in database for the convenience of future queries.

4.2 Experimental Setup and Results

Experiments on four datasets are employed to demonstrate the effectiveness of our method. *Lee.50* [13] comprise 50 documents from the Australian Broadcasting Corporation's news mail service. Those documents, covering a variety of topics, are paired and then judged by volunteer students on their relatedness to each other. *MSRpar,MSRvid* and *SMTeuroparl*[2] are stand for *Microsoft Research Paraphrase Corpus, Microsoft Research Video Description* and *WMT2008 Develoment Dataset* respectively. They each consist of approximately 750 text pairs with human-judged scores on their relatedness. Consistent with the prior works on this dataset, we employ Pearson correlation to measure the relatedness between human-judged scores and resulting score output by our system. Higher correlation indicates that the system output results are more likely in accord with gold standard scores. The Pearson correlation is defined as

$$\rho_{X,Y} = \frac{\sum_{i=1}^{n}(X_i - \bar{X})(Y_i - \bar{Y})}{\sqrt{\sum_{i=1}^{n}(X_i - \bar{X})^2}\sqrt{\sum_{i=1}^{n}(Y_i - \bar{Y})^2}} . \tag{4}$$

As is shown in Table 1, we apply our method to top 400 entries in ESA on *Lee.50* dataset to explore the effect of parameters. System gain best performance when $\alpha = 0.25$ or 0.3 and $\beta = 0.4$, hence those two configurations are adopted in the experiments on all the datasets.

Experimental results are presented in Figure 6. The parameter α indicated 1 minus the cosine similarity threshold that entries being clustered. We explore the System performance on the configurations against the number of the entries in ESA phrase. Results show that SRPVC system can significantly enhance the ESA performance on the datasets except *MSRpar*.

---

[2] http://www.cs.york.ac.uk/semeval-2012/task6/data/uploads/datasets/train-readme.txt

Table 1. The effect of parameter α and parameter β on *Lee.50* dataset

Corr \ α β	0	0.1	0.15	0.2	0.25	0.3	0.35
0.2	0.662	0.664	0.673	0.692	0.692	0.692	0.692
0.3	0.662	0.664	0.672	0.699	0.719	0.722	0.722
0.4	0.662	0.664	0.674	0.703	**0.726**	**0.724**	0.707
0.5	0.662	0.664	0.675	0.704	0.723	0.702	0.683
0.6	0.662	0.663	0.675	0.706	0.717	0.693	0.661

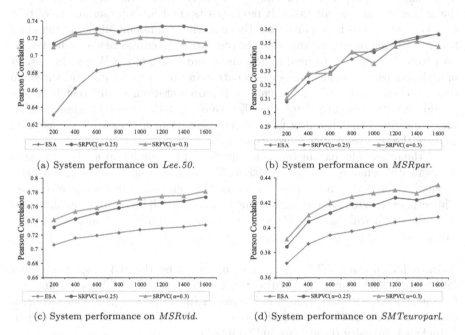

(a) System performance on *Lee.50*.

(b) System performance on *MSRpar*.

(c) System performance on *MSRvid*.

(d) System performance on *SMTeuroparl*.

Fig. 6. System performances comparing to ESA on different datasets.

It is noted that *MSRpar* is abbreviation for *Microsoft Research Paraphrase Corpus*, which indicate that *MSRpar* dataset is scored for measuring whether two texts are paraphrase to each other rather than SR. For instance, *"Amgen shares gained 93 cents,or 1.45 percent, to $65.05 in afternoon trading on Nasdaq."* and *"Shares of Allergan were up 14 cents at $78.40 in late trading on the New York Stock Exchange."*, two sentences in *MSRpar* that are about different stocks on different markets topic, although were considerably related in semantics, were scored only 1.333 (range from 0 to 5) for the extent of being paraphrases. Since ESA is prerequisite of our method, its outcome is highly related to our method's. As it is shown, ESA gains relatively low accracy in *MSRpar*, indicating that the generated entries by ESA do not capture enough semantics of the original texts. Therefore clustering on those entries also does not make sense.

In a word, system gains notable improvement comparing to ESA in the datasets except *MSRpar*. Informally, those results reveal that clustering the representation entries to reduce their redendancy could effectively enhance the semantic relatedness performance for the cases that ESA show relatively good outcomes.

5   Conclusion and Future Work

In the paper we propose a clustering scheme to reduce redundancy in ESA representations of texts for SR tasks. It reconstructs the ESA representations of text based on the groups from clustering. By clustering redundant entries in the same text are merged together. And also, the semantic relatedness beween simlilar entries from two text is captured by the maximum scores of the Wikipedia entries of each groups. By this means, surplus entries in the representations of the same text are eliminated, and similar entries in representations of different texts are considered. Experiments show the effectiveness of this method against ESA, which indicated that the new form of representations for texts after clustering capture the text semantics more accurate.

On the other side, parameters configuration of the model for a wider range of datasets still remains an open question. Moreover, frequent access to database to get PV for texts and online clustering both are time-consuming processes. Clustering the Wikipedia articles offline in advance will be a promising direction to explore. We will focus on those issues to improve our method in the future work.

Acknowledgments. This research is supported by the 863 project of China (2013AA013300), National Natural Science Foundation of China (Grant No. 61375054 and 61402045), Tsinghua University Initiative Scientific Research Program Grant No.20131089256, and Cross fund of Graduate School at Shenzhen, Tsinghua University (Grant No. JC20140001)

References

1. Krumm, J., Davies, N., Narayanaswami, C.: User-generated Content. IEEE Pervasive Computing 7(4), 10–11 (2008)
2. Miller, G.A.: WordNet: a lexical database for English. Communications of the ACM 38(11), 39–41 (1995)
3. Lenat, D.B.: CYC: A large-scale investment in knowledge infrastructure. Communications of the ACM 38, 33–38 (1995)
4. Wu, Z., Palmer, M.: Verbs semantics and lexical selection. In: Proceedings of the 32nd Annual Meeting on Association for Computational Linguistics, pp. 133–138. Association for Computational Linguistics (June 1994)
5. Leacock, C., Chodorow, M.: Combining local context and WordNet similarity for word sense identification. WordNet: An Electronic Lexical Database 49(2), 265–283 (1998)

6. Lin, D.: An information-theoretic definition of similarity. In: ICML, vol. 98, pp. 296–304 (July 1998)
7. Jiang, J.J., Conrath, D.W.: Semantic similarity based on corpus statistics and lexical taxonomy. arXiv preprint cmp-lg/9709008 (1997)
8. Blei, D.M., Ng, A.Y., Jordan, M.I.: Latent dirichlet allocation. The Journal of Machine Learning Research 3, 993–1022 (2003)
9. Dumais, S., Furnas, G., Landauer, T., Deerwester, S., Deerwester, S.: Latent semantic indexing. In: Proceedings of the Text Retrieval Conference (1995)
10. Gabrilovich, E., Markovitch, S.: Computing Semantic Relatedness Using Wikipedia-based Explicit Semantic Analysis. In: IJCAI, vol. 7 (2007)
11. Yazdani, M., Popescu-Belis, A.: Computing text semantic relatedness using the contents and links of a hypertext encyclopedia. Artificial Intelligence 194, 176–202 (2013)
12. Gabrilovich, E., Markovitch, S.: Wikipedia-based semantic interpretation for natural language processing. Journal of Artificial Intelligence Research 34(2), 443 (2009)
13. Lee, M.D., Pincombe, B., Welsh, M.: A comparison of machine measures of text document similarity with human judgments. In: 27th Annual Meeting of the Cognitive Science Society (CogSci 2005), pp. 1254–1259 (2005)
14. Banea, C., Hassan, S., Mohler, M., Mihalcea, R.: Unt: A supervised synergistic approach to semantic text similarity. In: Proceedings of SemEval 2012, pp. 635–642 (2012)
15. Zheng, H.-T., Wu, W., Jiang, Y., Xia, S.-T.: Exploiting Level-Wise Category Links for Semantic Relatedness Computing. In: Loo, C.K., Yap, K.S., Wong, K.W., Teoh, A., Huang, K. (eds.) ICONIP 2014, Part II. LNCS, vol. 8835, pp. 556–564. Springer, Heidelberg (2014)
16. Zheng, C., Wang, Z., Bie, R., Zhou, M.: Learning to Compute Semantic Relatedness Using Knowledge from Wikipedia. In: Chen, L., Jia, Y., Sellis, T., Liu, G. (eds.) APWeb 2014. LNCS, vol. 8709, pp. 236–246. Springer, Heidelberg (2014)
17. Mikolov, T., Chen, K., Corrado, G., Dean, J.: Efficient estimation of word representations in vector space. arXiv preprint arXiv:1301.3781 (2013)
18. Quoc, L., Mikolov, T.: Distributed Representations of Sentences and Documents. In: Proceedings of the 31st International Conference on Machine Learning, pp. 1188–1196 (2014)
19. http://stats.wikimedia.org/EN/TablesWikipediaEN.htm
20. MacQueen, J.: Some methods for classification and analysis of multivariate observations. In: Proceedings of the Fifth Berkeley Symposium on Mathematical Statistics and Probability, vol. 1(14), pp. 281–297 (June 1967)
21. Ester, M., Kriegel, H.P., Sander, J., Xu, X.: A density-based algorithm for discovering clusters in large spatial databases with noise. In: Proceedings of KDD, pp. 226–231 (August 1996)

Research on Semantic Disambiguation in Treebank

Lin Miao[1], Xueqiang Lv[1], Yunfang Wu[2], and Yue Wang[1]

[1] Beijing Key Laboratory of Internet Culture and Digital Dissemination Research,
Beijing Information Science and Technology University, Beijing 100101, China
[2] School of Electronic Engineering and Computer Science,
Peking University, Beijing 100871, China

Abstract. The increasingly widespread application of natural language processing technology leads parsing to play a significant role. As a result, the size and quality of treebank have become the focus of relevant research. However, there exists data sparseness when we use the treebank to parse. With the help of Cilin semantic information and words contextual information, this paper proposes a context-based lexical semantics disambiguation method. After applying this method on CTB (Chinese Treebank) 5.0 and TCT (Tsinghua Chinese Treebank), using Berkeley Parser achieved relatively good results. In Penn Chinese Treebank, the precision and recall rates reached 85.35% and 84.34% respectively, and the F value reached 84.84%. Comparing with the parsing results of using the original corpus, the correct rate increased by 1.86% and the recall rate increased by 1.02% and the comprehensive index F value increased by 1.35%. As consequence, the overall parsing error rate dropped by 8.17%.

Keywords: Treebank, Data sparseness, Semantic disambiguation, Cilin.

1 Introduction

Syntactic analysis is the process of analyzing the grammatical structure of the sentence automatically, using certain syntax rules. Acting as the fundamental technology for deep natural language analysis, syntactic analysis is a major part of natural language understanding. Technologies based on NLP are becoming increasingly widespread. For this reason, Syntactic analysis has come to play a central role in NLP applications, and is heavily used in Machine Translation, information extraction, QA system and retrieval system, etc. Treebank is a collection of natural language sentences annotated with syntactic or semantic structure. Treebanks are used to study syntactic phenomena (for example, diachronic corpora can be used to study the time course of syntactic change). Once parsed, a corpus will contain frequency evidence showing how common different grammatical structures are in use. Treebanks also provide evidence of coverage and support the discovery of new, unanticipated, grammatical phenomena. However, parsing results may be affected directly by treebanks' qualities. It will produce diverse levels of data sparseness problems in parsing process, which will affect the performance of parsing seriously, because of the limitations of treebank resources, existing analysis algorithms and optimization methods of treebank.

© Springer International Publishing Switzerland 2015
R. Cheng et al. (Eds.): APWeb 2015, LNCS 9313, pp. 658–669, 2015.
DOI: 10.1007/978-3-319-25255-1_54

In previous studies, significant improvement has been made in the English syntactic analysis because of the introduction of semantic information. In the meantime, the impact of data sparseness on syntactic analysis has been reduced as well. For example, Eugene Charniak[5], McDonald[10], S. Petrov[11] and so on have achieved great success, by using semantic information. Berkeley Parser designed by S. Petrov[11] is a non-lexicalized parser, however, the chosen features are extracted from the lexical information. The core idea of Berkeley Parser is, namely, simple word clustering, and words are represented as corresponding Category identification. This is also the use of the semantic information, such as date of classification, currency classification etc. However, the classification here is designed for English, as for Chinese, the granularity of this classification is not sufficient to address the impact of the data sparseness that treebank brought.

Eneko Agirre et al[12] shows much of the gain in parsing using semantic information to optimized treebank. Using WordNet as semantic dictionary, they firstly resolve ambiguity of word sense, and then map words into corresponding semantic coding. All of this provides significant improvement. Combining with the characteristics of Chinese parsing and the methods of Chinese semantic disambiguation, in this paper, we provide a method of semantic encoding mapping based on contextual information, which classifies, encodes and maps words that have the same or similar semantic functions according to synonymous transformation, so as to resolve data sparseness problem in parsing.

2   Data Sparseness in Parsing

In traditional parsing process, we use lexical sequence, contextual words and speech information to build syntax tree. Using these features in parsing may produce data sparseness problem. Table 1 shows three categories of data sparseness in parsing.

Table 1. Category of Data Sparseness

Example	Training sentence	Testing sentence	Reasons of data sparseness
1	这里产的土豆非常的好吃！	这里盛产马铃薯。	Different expressions of the same thing causes data sparseness.
2	"安静！开始上课了！"	屋里好安静。	The same word in different contexts, having different parts of speech and grammatical functions, causes data sparseness.
3	"你真八卦！什么都打听！"	这山洞中竟然有一个八卦图。	The same word with multiple meanings has different sense in different contexts, causing data sparseness.

From Table 1, as example 1 shows, "土豆" in training corpus and "马铃薯" in testing set represent the same one thing, but are different expression ways. However, for parser model, data sparseness may cause in testing process when there is not "马铃薯" but "土豆" in testing corpus. In this case, "马铃薯" is an unknown word. As example 2 shows, two "安静" in two sentences, the former is a verb to make something quiet; the latter is an adjective to describe the status of silent. If we don't distinguish between two "安静", the parse model will recognize "安静" as a verb, so as to cause data sparseness. As for example 3, "八卦" in sentence1 means gossip, while in sentence 2 it means the eight trigrams. The same word has different meanings when in different contexts. If parsing directly, parse model can't learn the meaning in sentence 2, which will cause serious data sparseness.

Thus, we hypothesize that the reason for data sparseness is lack of contextual semantic information. Therefore, adding semantic information, semantic disambiguation in different contexts can solve the data sparseness caused by above situations.

3   Semantic Dictionary

In this paper, we use **HIT Information Retrieval Laboratory Synonyms Cilin Extended Version** provided by Harbin Institute of Technology Information Retrieval Laboratory, which is a thesaurus of 70,000 words and arranged by semantics. This thesaurus consists of not only synonyms, but also a number of the words of the same classification, namely broadly related words.

3.1   Lexical and Semantic Representation

Table 2 shows part of the lexical coding in **HIT Information Retrieval Laboratory Synonyms Cilin Extended Version**.

Table 2. Cilin (extended version) Words Coding Table

Coding Status	1	2	3	4	5	6	7	8
Symbol Example	D	a	1	5	B	0	2	=\#\@
Symbol Character	Broad heading	Middle heading	Tiny heading		Lexical group		Tiny lexical group	
Rank	Rank 1	Rank 2	Rank 3		Rank 4		Rank 5	

A Cilin coding segment from Aa02A01= to Aa02A13@ is given in Fig. 1. Each row in Fig. 1 is a class of words with the same semantic.

```
Aa02A01= 俺 本人 个人 人家 身 斯人 我 吾 余 于 咱 咱家 依
Aa02A02= 鄙 鄙人 不才 不肖 仆 区区 小人 小子 愚 在下
Aa02A03@ 老子
Aa02A04= 老汉 老朽
Aa02A05@ 老娘
Aa02A06@ 愚兄
Aa02A07= 小弟 兄弟
Aa02A08= 民女 奴 妾 妾身
Aa02A09= 孤 寡人 朕
Aa02A10= 卑职 奴才 奴婢 下官 职
Aa02A11= 贫道 小道
Aa02A12@ 贫僧
Aa02A13@ 下臣
```

Fig. 1. Cilin coding segment

In **HIT Information Retrieval Laboratory synonyms Cilin extended version**, each word corresponds to one or more coding, and each coding corresponds to one sense of word. According to the characteristics of Cilin, we divide the words into two categories as following.

- **Univocal word**: a word corresponds to only one coding, that is, there is only one word meaning even in different contexts.
- **Polysemous word**: a word corresponds to multiple coding, that is, the word has multiple meanings in different contexts.

3.2 Cilin Representation of Set

Cilin can be represented in the form of set, according to the structure of it. We can express Cilin using "CODE TO WORD" and "WORD TO CODE" as shown following.

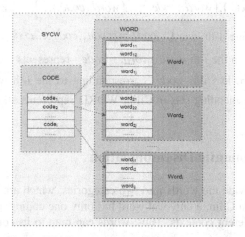

Fig. 2. Cilin Representation of CODE TO WORD

As shown in Fig. 2, $SYCW = \{(code_1, word_1), (code_2, word_2), \ldots (code_i, word_i), \ldots\}$ represents Cilin. $code_i$ represents the i-th encoding. $word_i = \{word_{i1}, word_{i2}, \ldots word_{ij}, \ldots\}$ represents the Cilin corresponding to $code_i$. $word_{ij}$ represents the j-th $word_i$ in Cilin. $CODE = \{code_1, code_2, \ldots code_i, \ldots\}$ represents set of all the encoding in Cilin. $WORD = \{word_{11}, word_{12}, \ldots word_{ij}, \ldots\}$ represents set of all the words in Cilin.

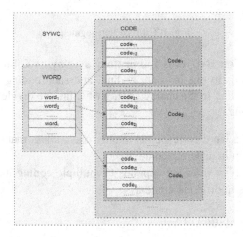

Fig. 3. Cilin Representation of WORD TO CODE

Fig. 3 shows the other way to express Cilin. $SYWC = \{(word_1, code_1), (word_2, code_2), \ldots (word_i, code_i), \ldots\}$ represents Cilin. $word_i$ represents the i-th word. $code_i = \{code_{i1}, code_{i2}, \ldots code_{ij}, \ldots\}$ represents the Cilin corresponding to the i-th word $word_i$. $code_{ij}$ represents the encoding of the j-th synonymous word in Cilin $code_i$ corresponding to the i-th word $word_i$. CODE represents all of the encoding sets in Cilin, WORD represents all of the word sets in Cilin.

4   Lexical Semantic Disambiguation

In this paper, we divide the words into two categories, which are univocal word and polysemous word. In Cilin, a univocal word has only one coding, namely, a word has one meaning in different contexts. Thus, words can map to its corresponding encoding directly.

Polysemous word has multiple meanings. Each meaning corresponds to its own syntax and structural information. Data sparseness problems stem from the ambiguity caused by multi-meaning words. For this reason, resolving ambiguity of multi-meaning words and mapping words into semantic encoding based on its surrounding context, the impact caused by word sense ambiguity can be reduced.

For convenience, SC is defined as univocal word coding set.

- **Definition 1:** Univocal word coding set (SC) is encoding set of univocal word in treebank.

4.1   Build Contextual Dictionary

Polysemous word presents a variety of semantic in different contexts. To solve semantic disambiguation, we should use Context. For ease of description, lexical context and encoding context are introduced as two concepts.

- **Definition 2:** nm_ lexical context: In the training corpus, nm_ lexical context is a set of preceding n words and succeeding m words. And n>0, m>0, if n=0, nm_ lexical context means lexical context of succeeding; if m=0, nm_ lexical context means lexical context of preceding.
- **Definition 3:** nm_ encoding context: In the training corpus, set $word_i$ in set SYCW corresponds to a code $code_i$ in Cilin. So nm_ encoding context is nm_ lexical context of all the univocal words in set $word_i$. And n>0, m>0, if n=0, nm_ encoding context means encoding context of succeeding; if m=0, nm_ encoding context means encoding context of preceding.

Extracting nm_ lexical context of all the univocal words in treebank, calculating the corresponding weights, to build contextual dictionary $CONTEXTDIC = \{(code_1, C_{P1}, C_{S1}), (code_2, C_{P2}, C_{S2}), ..., (code_i, C_P, C_{Si}), ...\}$. $code_i$ is the i-th encoding in Cilin.

Set $C_{Pi} = \{(cword_1, k_1, weight_1), (cword_2, k_2, weight_2), ...(cword_j, k_j, weight_j), ...\}$ is the set of preceding context words corresponding to $code_i$. $cword_j$ is the j-th word in context words of $code_i$, k_j represents that $cword_j$ is the k_j-th preceding context word of code $code_i$. $weight_j$ is the weight of $cword_j$. Similarly, C_{Si} is the set of succeeding context words corresponding to $code_i$, and the internal structure of the set is the same as C_{Pi}. In this paper, since context words uniquely identify the context of encoding, we increase the weights of the context words concentrated in the current encoding. Based on the idea of TF • IDF algorithm, we can calculate the weight W of context words using the following function:

$$W = TF \times IDF = TF(cword) \times \log(\frac{P}{q})$$ (1)

Among this, TF (cword) is the frequency of the current encoding preceding (succeeding) context word cword. P is the numbers of all encoding, q is the number of occurrences of word cword encoding in preceding (succeeding) context.

Because of the limitation of the Cilin, we require every encoding of the word to be disambiguated can be able to find at least one corresponding univocal word in Cilin, in order to ensure the accuracy of word sense disambiguation.

4.2   Contextual Vector Representation

nm_ lexical context and nm_ encoding context are consist of a series of words. Words are expressed by weights in the calculation process, then, word sense disambiguation can be resolved by calculation. We design a vector representation of context to represent the contextual information of a polysemous word or a encoding, by contextual vector of polysemous word or encoding in documentation set. The so-called context word here is nm_ lexical context of the words in treebank and nm_ encoding context of the code in context dictionary. Contextual vector is vector consists of every context words' weights in context dictionary, when context words arranged in a certain order.

$word_t$ represents the t-th word in treebank, $code_i$ represents a encoding of $word_t$ in Cilin. In set CONTEXTDIC, $code_i$ corresponds to set C_{Pi} and set C_{Si}. In set C_{Pi}, for each element, k_j ranges from 1 to n. k_j has n values, that is, n contextual vectors are needed to represent nm_ encoding context of $code_i$. Likewise, we also need m contextual vectors to represent nm_ encoding context of $code_i$.

$Context_{P(S)k} = \{x_{k1}, x_{k2}, ... x_{ks}, ...\}$ represents the set of context words in the k-th position of preceding (succeeding) context of $code_i$, and x_{ks} represents the s-th context word in the k-th position of preceding (succeeding) context of $code_i$. Contextual vector $\vec{V}_{P(S)k}(b_{x_{k1}}, b_{x_{k2}}, ... b_{x_{ks}}, ...)$ is built based on set $Context_{P(S)k}$, where, $b_{x_{ks}}$ represents the weight of x_{ks} which is the preceding (succeeding) context word of $code_i$. Each context word needs a contextual vector of $word_t$ to be built. Contrasting elements of $word_{t-k}$ ($word_{t+k}$) with $Context_{P(S)k}$, record weight as 1 when identical items, whereas, when different record weight as 0, to build the k-th contextual vector $\vec{V}_{P(S)k}'(0,0,...1,...)$ in preceding(succeeding) context of $word_t$ with the same dimension as $\vec{V}_{S(P)k}$.

For this, contextual vectors of polysemous word are $\vec{V}_{P1}', \vec{V}_{P2}', ..., \vec{V}_{Pn}', \vec{V}_{S1}', \vec{V}_{S2}', ..., \vec{V}_{Sm}'$; contextual vectors of encoding are $\vec{V}_{P1}, \vec{V}_{P2}, ..., \vec{V}_{Pn}, \vec{V}_{S1}, \vec{V}_{S2}, ..., \vec{V}_{Sm}$.

4.3   Semantic Similarity Calculation

If the polysemous word in treebank is not semantic disambiguated but directly mapped into semantic encoding, there may exist several semantic encodings corresponding to it. For this reason, we design an algorithm to calculate semantic similarity, that is, calculate the similarity of word in treebank and its corresponding encoding according to semantic information. If the similarity reaches a certain threshold, we consider in current context the word represents the meaning of corresponding code, namely, the word is mapped into its encoding.

We can get contextual vectors of polysemous word and contextual vectors of encoding using the method in 4.2. At the same time, we can get the similarity of context of polysemous word and context of encoding by the method of similarity calculation, as the function following:

$$SimCos_P(word_k, code_i) = \frac{1}{n}\sum_{i=1}^{n}\frac{\vec{V}_P \cdot \vec{V}_P'}{|\vec{V}_P| \times |\vec{V}_P'|}$$

$$= \frac{1}{n}\sum_{j=1}^{n}\frac{a_{w_{k-j}} \times b_{x_{ji}}}{\sqrt{a_{w_{k-j}}^2} \times \sqrt{(\sum_{i=1}^{s} b_{x_{ji}}^2)}}$$

(2)

$$SimCos_S(word_k, code_i) = \frac{1}{m}\sum_{i=1}^{m}\frac{\vec{V}_S \cdot \vec{V}_S'}{|\vec{V}_S| \times |\vec{V}_S'|}$$

$$= \frac{1}{m}\sum_{j=1}^{m}\frac{a_{w_{k+j}} \times b_{y_{ji}}}{\sqrt{a_{w_{k+j}}^2} \times \sqrt{\sum_{i=1}^{s} b_{y_{ji}}^2}}$$

(3)

Ultimately, function calculating the similarity of word w_k and code $code_i$ is following:

$$SimCos(word_k, code_i) =$$

$$\frac{n}{n+m} \times SimCos_P(word_k, code_i)$$

(4)

$$+ \frac{m}{n+m} \times SimCos_S(word_k, code_i)$$

When polysemous word is disambiguated, semantic encoding not exists in training sets may be leaded in and cause data sparseness, if handled improperly. Thus, when polysemous word is disambiguated, semantic similarity of polysemous word $word_i$ and its encoding should be computed. If the maximum value of the similarity $SimCos_{max} > Threshold$, and its encoding $code_{max} \in SC$, $code_{max}$ is the semantic encoding of $word_i$ in current context. $Threshold$ is threshold of semantic similarity, and SC is encoding set of univocal words.

5   Experiments and Results Analysis

We trained and tested on two kinds of treebanks. One is Chinese Treebank (CTB)5.0, training set is 1-270 and 400-1151, testing set is 270-300. The other is CIPS-SIGHAN2012 parsing evaluation Tsinghua Chinese Treebank (TCT), training corpus size from small to large total 5 parts, using number 5-9 to represent the training set and act as training corpus for experimental group 1,2,3,4,5. Take corpus 9 of 10 as part of the complement test set. In this paper, as we aim at optimize the treebank, we use existing parser-Berkeley Parser to process the training data and testing data. Corpus 1 is CTB5.0 unoptimized treebank, and its training, parsing results are used as Baseline method of corpus1. Corpus 2 is unoptimized TCT, and act as baseline method of corpus 2 after training.

5.1   Evaluation Indicators

In this paper, we use Precision (P), Recall (R) and F value to evaluate the results of parsing. The algorithm is as following:

$$P = \frac{The\ number\ of\ phrases\ correctly\ labeled}{The\ number\ of\ labled\ phrases} \times 100\% \tag{5}$$

$$R = \frac{The\ number\ of\ phrases\ correctly\ labeled}{The\ number\ of\ original\ phrases\ in\ treebank} \times 100\% \tag{6}$$

$$F\ value = \frac{R \times P \times (1 + \beta^2)}{R + P \times \beta^2} \times 100\% \tag{7}$$

In the function, β is weighing factor of correct rate (P) and recall rate (R). In this paper, (P) and (R) are important equally, thus, when $\beta=1$, F value is also known as F-1 value.

5.2   Selection of Similarity Threshold

In processing of polysemous word disambiguation, we decide whether the current word mapping into semantic coding depend on the similarity of semantic encoding and semantic in current context. Then we need to select similarity threshold. Threshold range is [0, 1]. In this paper, we use 0.0 as base number in steps of 0.1. To illustrate method of selecting threshold, we use corpus 1 as example and select optimal threshold.

Actually, with the increase of the threshold, the number of polysemous word to be disambiguated in the process of disambiguation gradually reduced. When the threshold is exactly equal to 0.0, there is no limitation on polysemous word sense disambiguation. Thus, words can be replaced only if they meet the polysemy replacement rules. The results of threshold selection are shown in Fig. 4.

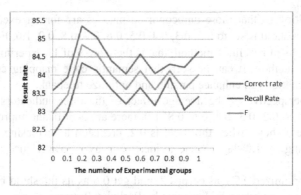

Fig. 4. 11 Groups of Experiments for Threshold Selection

After analysis of Fig. 4, integrated indicator F value is the highest, when threshold is 0.2. Therefore, we use 0.2 as threshold value in corpus 1.

5.3   Experiments and Analysis

Comparative Experiment-baseline of corpus 1 uses raw Penn Treebank corpus which is parsed by Berkeley Parser lately. The results of all experiments can be seen in Table 3.

Table 3. The Results of CTB

Groups	P (%)	R (%)	F value (%)
Baseline1	83.67	83.32	83.49
1.0	84.64	83.41	84.02
0.9	84.22	83.06	83.63
0.8	84.31	83.93	84.12
0.7	84.05	83.17	83.61
0.6	84.58	83.66	84.12
0.5	84.04	83.21	83.62
0.4	84.40	83.62	84.01
0.3	85.06	84.12	84.59
0.2	85.35	84.34	84.84
0.1	83.94	82.98	83.46
0.0	83.58	82.35	82.96

It seems that when threshold is set to 0.1, 0.2, 0.3, 0.4, 0.5, 0.6, 0.7, 0.8, 0.9, 1.0, precision is higher than baseline1 method. As it illustrates, semantic encoding is able to precisely express the word meaning in current context. Overall, it achieves an increase in accuracy of parsing, from the partial point of view, it improves the phrase recognition. When threshold is set to 0.2, 0.3, 0.4, 0.6, 0.8, 1.0, higher precision than baseline1 method shows that semantic encoding is able to precisely express grammatical elements of the word in current context. It improves recall of parsing as well as phrase recognition so that more unrecognized phrases are further excavated. Meanwhile, when threshold is set to 0.2, 0.3, 0.4, 0.5, 0.6, 0.7, 0.8, 0.9, 1.0, F value is higher than the results of baseline1 method, that is there are 9 of 11 experiments are better than baseline1 method. It can be seen that semantic code mapping combined with semantic information can enhance parsing and optimize treebank.

From the perspective of the average effect of the three indicators above, when threshold is set to 0.2, 0.3, 0.4, 0.6, 0.8, 1.0, there are significant improvements over baseline. Table 4 shows when threshold is 0.2, precision is increasing by 1.68%, recall is increasing by 1.02%, F value is increasing by 1.35%. Hence, there is some improvement.

We use unoptimized TCT as corpus 2, and get 0.6 as its threshold by experiment. Table 4 shows the results of TCT parsing by Berkeley Parser.

Table 4. The Results of TCT

Experimental Group	Baseline2			Optimized TCT		
Corpus Number	P (%)	R (%)	F (%)	P (%)	R (%)	F (%)
1	81.03	80.94	80.98	81.25	80.98	81.11
2	81.34	81.25	81.30	81.70	81.66	81.68
3	82.52	82.39	82.45	82.61	82.53	82.57
4	82.69	82.59	82.64	82.71	82.61	82.66
5	83.03	82.93	82.98	83.29	82.98	83.13

Table 4 shows a great success over baseline2 method. This proves semantic encoding will precisely express the word meaning in current context by semantic code mapping, simultaneously get greater results in parsing and phrase recognition. And we will get remarkable improvement as the increasing of corpus.

6   Conclusions and Prospects

In this paper, we optimize treebank using semantic information, and get a great success with a decrease of 8.17% over error rate in corpus 1 and an improvement of parsing result in corpus 2. Therefore, semantic information plays a significant role to treebank building and parser enhancing. At the same time, it performs well in phrase recognition. So semantic information and semantic disambiguation are crucial to phrase recognition.

We carry on our experiments using Berkeley Parser, finding out lexical information is useful to syntactic analysis of the non-lexical.

We can solve part of data sparseness problem, however, to eliminate data sparseness completely, we should improve the method of syntactic analysis and optimize treebank. The next step is carried out in two directions, one is to expand the size of the semantic treebank and improve the quality of the treebank annotation, and the other is to optimize the method of parsing to resolve data sparseness problems.

Acknowledgements. The research work was supported by National Natural Science Foundation of China under Grants No. 61271304 and Beijing Natural Science Foundation of Class B Key Project under Grants No. KZ201311232037.

References

1. Manning, C.D., Schütze, H.: Foundations of statistical natural language processing. MIT Press (1999)
2. Wang, Y., Ji, D.: Summary of Chinese Treebank. Contemporary Linguistics 11(1), 47–55 (2009)
3. Zhang, M., Zhang, Y., Che, W., et al.: Chinese Parsing Exploiting Characters. In: ACL (1), pp. 125–134 (2013)
4. Hatori, J., Matsuzaki, T., Miyao, Y., et al.: Incremental joint approach to word segmentation, pos tagging, and dependency parsing in chinese. In: Proceedings of the 50th Annual Meeting of the Association for Computational Linguistics. Long Papers, vol. 1, pp. 1045–1053. Association for Computational Linguistics (2012)
5. Charniak, E.: Statistical parsing with a context-free grammar and word statistics. In: AAAI/IAAI 2005, vol. 18, pp. 598–603 (1997)
6. Jones, B.K., Johnson, M., Goldwater, S.: Semantic parsing with bayesian tree transducers. In: Proceedings of the 50th Annual Meeting of the Association for Computational Linguistics: Long Papers, vol. 1, pp. 488–496. Association for Computational Linguistics (2012)
7. Chen, W., Zhang, M., Li, H.: Utilizing dependency language models for graph-based dependency parsing models. In: Proceedings of the 50th Annual Meeting of the Association for Computational Linguistics: Long Papers, vol. 1. Association for Computational Linguistics (2012)
8. Feng, V.W., Hirst, G.: Text-level discourse parsing with rich linguistic features. In: Proceedings of the 50th Annual Meeting of the Association for Computational Linguistics: Long Papers, vol. 1. Association for Computational Linguistics (2012)
9. Carreras, X.: Experiments with a High-order Projective Dependency Parser. In: Proceedings of the CoNLL 2007 Shared Task Session of EMNLP-CoNLL, pp. 957–961. CoNLL, Prague (2007)
10. McDonald, R., Lerman, K., Pereira, F.: Multilingual dependency analysis with a two-stage discriminative parser. In: Proceedings of the Tenth Conference on Computational Natural Language Learning, pp. 216–220. Association for Computational Linguistics (2006)
11. Petrov, S., Barrett, L., Thibaux, R., et al.: Learning accurate, compact, and interpretable tree annotation. In: Proceedings of the 21st International Conference on Computational Linguistics and the 44th Annual meeting of the Association for Computational Linguistics, pp. 433–440. Association for Computational Linguistics (2006)
12. Agirre, E., Baldwin, T., Martinez, D.: Improving Parsing and PP Attachment Performance with Sense Information. In: ACL (2008)

PDMA: A Probabilistic Framework for Diversifying Recommendation Lists

Yang-Hui Yan, Ying-Min Zhou, and Hai-Tao Zheng

Tsinghua-Southampton Web Science Laboratory,
Graduate School at Shenzhen, Tsinghua University, Beijing, China,
{yanyh13,zhou-ym14}@mails.tsinghua.edu.cn,
zheng.haitao@sz.tsinghua.edu.cn

Abstract. In this paper, we derive a probabilistic ranking framework for diversifying the recommendations of baseline methods. Unlike conventional approaches to balance relevance and diversity, we produce the diversified list by maximizing user's current marginal aspect preference, thus avoiding the hyperparameters in making the tradeoff. Before diversification, we adopt clustering to generate a much smaller set of candidate items based on three requirements: efficiency, relevance and diversity. As a result, it helps us not only reduce the search space greatly but also promote a slight increase in performance. Our framework is flexible to incorporate new preference aspects and apply new marginal aspect preference algorithms. Evaluation results show that our method can get better diversity than others and maintain comparable accuracy to baseline methods, thus a better balance between relevance and diversity.

Keywords: recommendation diversification, framework, candidate generation, marginal preference.

1 Introduction

Recommendation systems [1] are popularly used in helping online websites provide personalized services to their customers, in order to promote sales and profits. Conventional recommendation methods [2–4,6] are mainly accuracy-oriented for pursuing good predication accuracy. However better accuracy can not guarantee user's satisfaction, because recommended items often have similar contents. Such a list cannot reflect user's full spectrum of interests, which results in heavy redundancy and partial interest presentation. This problem has been addressed recently [9,11–13] by stressing on another important measure-diversity. For example, [13] diversified the lists based on item taxonomy, [12] explored the adaptive diversification level in different situations. They have studied the problem from different perspectives. Based on existing work, we suggest that good diversification method should satisfy four requirements together: efficiency, relevance, diversity and adaptability.

This paper proposes a novel probabilistic framework for diversifying the baseline recommendation lists, in order to increase the diversity of existing well-designed accuracy-oriented methods and maintain good relevance meanwhile.

© Springer International Publishing Switzerland 2015
R. Cheng et al. (Eds.): APWeb 2015, LNCS 9313, pp. 670–682, 2015.
DOI: 10.1007/978-3-319-25255-1_55

Overall, our method iteratively produces the diversified list by maximizing an objective function based on mathematical derivation. Unlike [13,14], we avoid using hyperparameters in controlling the trade-off between accuracy and diversity, since it is difficult to determine the optimal balance between them in advance. Every step, we select the item which maximally meets user's current marginal preference. To calculate the relevance between items and user's preference, we specifically model user's interests as a set of aspects, including some explicit features such as item genres and implicit factors learned from data. Once an item is selected, user's aspect preference weight is updated. If some latent aspects are heavily provided, we will diminish the corresponding weight on those specific aspects. In other words, we consider not only the target user's personal preference but also the additional influence produced by the already selected items. Consequently, a diversified recommendation list is generated.

Before diversification, we use clustering to generate a much smaller set of items as the candidates. Our candidate generation method is designed by considering three diversification requirements: efficiency, relevance and diversity. As a result, it greatly accelerate the speed of diversification process without decreasing the other performances. Finally, We can provide adaptive diversification [12] for users holding different ranges of interests. In other words, diversified items keep in accordance with user's preferred types, as we fully utilize different user's tastes both in candidates generation and diversification process.

In conclusion, our main contributions are listed as follows:

- We derive a probabilistic ranking framework for diversifying the baseline recommendation lists. By maximizing user's marginal aspects preference, we avoid using hyperparameters to balance relevance and diversity.
- Our framework can be run efficiently enough to satisfy the online requirements by using candidates generation to reduce the search space, which is important for the scalability of recommendation systems.
- We conduct a series of experiments on movie dataset using two baseline methods. Experiments show that our method performs well in terms of accuracy, diversity, and efficiency at the same time.

The rest of this paper is organized as follows. We give a detailed description of our proposed diversification framework in section 2. Then section 3 presents the experiments and analysis. Related work on recommendation diversity are included in section 4. Finally we conclude the paper with remarks of our work in section 5.

2   Proposed Method

The basic workflow of our proposed framework PDMA (Probabilistic Diversification using Marginal Aspect preference) is that: Given a list of ranked items B returned by the baseline recommendation method, we firstly assign each item into corresponding clusters. And then we initialize the candidate pool by selecting some items from each cluster, whose number is proportional to the target user's cluster preference weight. Finally we iteratively produce the diversified recommendation list D using MMAP(Maximal Marginal Aspect Preference). Once an item is selected, candidate pool and user's marginal preference are updated. The overall framework is shown in Fig.1.

2.1   Candidates Generation Using Clustering

Before applying diversification, We group similar items in the baseline recommendation list to form item clusters. In each cluster, items are ranked based on its initial relevance score.

Note that the key point of this paper is not to propose a clustering method. As the items usually have taxonomy, we directly treat genres as clusters. If an item i belongs to many genres, it can be assigned into multiple clusters. Based on user's past rating history, we calculate user's preference weight on each genre as Eq.(1). G represents a set of distinct genres, $r(u)$ denotes user u's past rated items, $r_{u,i}$ gives u's rating on item i, $h(i,g) = 1$ if i belongs to genre g, and 0 otherwise.

$$P(g|u) = \frac{\sum_{i \in r(u)} r_{u,i} \cdot h(i,g)}{\sum_{g \in G} \sum_{i \in r(u)} r_{u,i} \cdot h(i,g)} \tag{1}$$

After getting item clusters and user's cluster preference, we initialize the candidate pool based on the following three considerations. First, the number of total candidate items should be much smaller than the items in the baseline list, because a much smaller search space allows the online process to be done efficiently. For diversification, namely, we just select a few items from each cluster. Second, more representative items from each cluster should be selected at earlier time. In other words, items with higher initial relevance score are selected with higher priority. This helps us maintain the relevance of diversified list. Third, the number of selected items from each cluster is proportional to the user's preference weight. Thus if target user has rated few items from a specific cluster, items of this cluster have few opportunities to be processed. This further helps us avoid unnecessary irrelevant items in advance and maintain relevance, as we consider

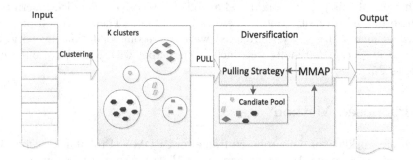

Fig. 1. Diversification Framework

user's personal preference. Finally, considering that selected items from different clusters may be replicated, we only remain unique items in the candidate pool.

2.2 Probabilistic Diversification

The derivation of our probabilistic diversification method is based on the probability ranking principle(PRP) [10], which helps us define an objective function to be optimized. Formally, let u represents target user, i denotes an item and R be a binary random variable representing relevance. For ranking, we need to estimate each item i's probability $P(R = 1|u, i)$. Further, let $A = \{a_1, ...a_{|A|}\}$ be the set of aspects that influence user's preference. Then we model user's information needs as a set of aspects $u \subseteq A$. Following this practice, the information contained in an item is also model as $i \subseteq A$.

We consider an item relevant if it contains any aspect in common with u. Namely, an item is relevant if it contains at least one aspect which also exist in user u's needs.

$$P(R = 1|u, i) = P(u \cap i \neq \emptyset) = P(\exists a_j, a_j \in u \cap i) \tag{2}$$

For simplicity, we can reasonably assume the independence of $a_j \in u$ and $a_{k \neq j} \in u$, also of $a_j \in i$ and $a_{k \neq j} \in i$. Equation 1 then can be rewritten by

$$P(R = 1|u, i) = 1 - \prod_{j=1}^{|A|}(1 - P(a_j \in u)P(a_j \in i)) \tag{3}$$

Aspect preference $P(a_j|u)$ and aspect correlation $P(a_j|i)$ can be estimated by using some methods described later. Then we can replace $P(a_j \in u)$ with $P(a_j|u)$, $P(a_j \in i)$ with $P(a_j|i)$.

$$P(R = 1|u, i) = 1 - \prod_{j=1}^{|A|}(1 - P(a_j|u)P(a_j|i)) \tag{4}$$

Applying Eq(4) to each item allows us to determine the one to be ranked first. For the subsequent items, we must view relevance in the condition of already selected items for diversification. Suppose we have selected the first $k - 1$ items in a ranked list $(i_1, ..., i_{k-1})$ and are now considering the relevance of i_k, the item at rank k. We assume that if a specific aspect appears in these first $k - 1$ items, then a repetition in i_k will bring few additional benefits, which matches the "law of diminishing returns" in economics. Let $P(a_j|u, i_1, ..., i_{k-1})$ represents u' current marginal aspect preference in the condition of the first $k - 1$ selected items. Specific $P(a_j|u, i_1, ..., i_{k-1})$ modeling will be discussed later. By abandoning high-order terms whose value is very small, we get Eq(5). An intuitive explanation for Eq.(5) is that probability $P(R_k = 1|u, i)$ is measured by how much item i satisfy u' current marginal aspect preference.

$$P(R_k = 1|u, i_k) = 1 - \prod_{j=1}^{|A|}(1 - P(a_j|u, i_1, ...i_{k-1})P(a_j|i_k))$$

$$\approx 1 - (1 - \sum_{j=1}^{|A|} P(a_j|i_k)P(a_j|u, i_1, ...i_{k-1})) \qquad (5)$$

$$= \sum_{j=1}^{|A|} P(a_j|i_k)P(a_j|u, i_1, ...i_{k-1})$$

Multiply Eq.(5) with $r_{norm}(i_k)$ which is normalized into [0,1] based on item's initial relevance score, the final diversification formulation is defined by Eq.(6).

$$P(R_k = 1|u, i_k) = \sum_{j=1}^{|A|} r_{norm}(i_k)P(a_j|i_k)P(a_j|u, i_1, ...i_{k-1}) \qquad (6)$$

Alg.1 describes the diversification procedure of our proposed framework. Each time we only select one item I^* from the candidate pool $\Re$, which gets the maximal marginal aspect relevance using Eq.(6). Based on the selected item's genres $\mathbf{G}_{I^*}$, we pull one item from each genre cluster to update the candidate pool, in which replicated items are removed.

Algorithm 1. Diversification Procedure

Input   : Candidate pool $\Re$, Item clusters $C = \{c_1, c_2, ...c_{|G|}\}$
Output: Output list D
1 **begin**
2 | **Init** : $D \leftarrow \emptyset$
3 | **while** $\Re \neq \emptyset \bigwedge |D| < N$ **do**
4 | | **for** *Item* $I \leftarrow \Re$ **do**
5 | | | calculate $S(I)$ using Eq.(6)
6 | | $I^* \leftarrow \arg\max S(I)$
7 | | $\Re \leftarrow \Re \setminus \{I^*\}$
8 | | $D \leftarrow D \cup \{I^*\}$
9 | | **for** $g \leftarrow \mathbf{G}_{I^*}$ **do**
10 | | | $I' \leftarrow c_g$
11 | | | $I' \rightarrow \Re$

2.3   Preference Aspects Modeling

In order to implement the above recommendation diversification, we need to specifically define the set of aspects A which influence user's preference. In this work, we extract a space of item features including explicit item genres and implicit latent factors. Formally, we build user's profile and item's profile as three-attribute vectors $\mathbf{u} = \{G, T, F\}$ and $\mathbf{i} = \{G, T, F\}$. Each dimension in $\mathbf{u}$ represents how much user u likes this aspect; while the corresponding dimension in $\mathbf{i}$ denotes how much item i contains this aspect.

Genre aspects G: Taxonomies exist for various domains. Take movie for example, there are about 20 genres. However people usually focus on only a small set of types such as romance, drama or fiction. This phenomenon provide us a way to distinct different user's preference. For i_G, we directly set $P(g|i) = 1$ if item contains that genre, 0 otherwise. For u_G we use Eq.(1) described before.

Content aspects T: Item tags provide us some text information which is a good characterization of the items' contents. Similar items usually have related description words. Using topic models such as LDA [7], we can model each item from the concept level to get user's semantic preference. Item's topic distribution $i_T = \{p(t_1|i), p(t_2|i), ..., p(t_{|T|}|i)\}$ is estimated by using gibbs sampling [20]. u_T is firstly initialized to $\mathbf{0}$. Then $\forall$ item i which has been rated by u, we multiply $r_{u,i}$ with i's topic distribution $\{p(t_1|i), p(t_2|i), ..., p(t_{|T|}|i)\}$, and add the multiplited vector into u_T. Finally, $p(t_k|u)$is normalized by $\sum_{k=1}^{|T|} p(t_k|u)$.

Rating aspects F: Matrix factorization has been developed in [6], which decomposes rating matrix into low latent space. $u_F = \{p(f_1|u), p(f_2|u), ..., p(f_{|F|}|u)\}$, $i_F = \{p(f_1|i), p(f_2|i), ..., p(f_{|F|}|i)\}$. The latent factors reflect user's implicit rating patterns. In order to make the formulation of probabilistic diversification meaningful, we set the feature weight equal to 0 if $p(f_k|u) < 0$ or $p(f_k|i) < 0$. Then u_F and i_F are normalized by their respective sum.

Finally, user's preference vector $\mathbf{u}$ needs to be normalized to 1. We set equal importance weight to the above mentioned three kinds of aspects, as we do not know which is better. Namely, $\mathbf{u} = \{\frac{1}{3}G, \frac{1}{3}T, \frac{1}{3}F\}$.

2.4 Marginal Preference Modeling

As discussed in section 2.2, $\forall a \in A$, we need to model user's marginal aspect preference $P(a|u, i_1, ..., i_{k-1})$ for diversification. It should be modeled by considering both user's tastes and the influence of already selected items, which need to penalize $P(a|u)$, if aspect a is heavily provided. Here we discuss two methods.

PDMA-P: If we model $P(a|u, i_1, ..., i_{k-1})$ as the probability that u still prefers a on condition that $i_1, ...i_{k-1}$ have not satisfied the aspect a. Then it can be rewritten by Eq.(7).

$$P(a|u, i_1, ..., i_{k-1}) = P(a|u) \prod_{j=1}^{k-1} (1 - P(a|i_j)) \tag{7}$$

PDMA-Q: $P(a|u, i_1, ..., i_{k-1})$ can also be modeled by Eq.(8). It denotes the quota of aspect a after some assigned items, which is motivated by an party election method Sainte-Laguë [15].

$$P(a|u, i_1, ..., i_{k-1}) = P(a|u)\{\frac{P(a|u)}{1 + 2\sum_{j=1}^{k-1} P(a|i_j)}\} \tag{8}$$

3 Experiment

3.1 Experiment Settings

Dataset. For experimentation, we use a public movie dataset[1] which contains 0.8M ratings from about 2.1K users on about 8.6K items. The number of average ratings per user is almost 400, and data density is 4.57%. In order to construct movies' contents, we crawl the plot keywords of the corresponding movies from imdb[2] website. After that, we filter out items with less than 20 tags and users with less than 20 rated items, in fact the number of filtered users in this step is very small. The number of average crawled tags per movie is about 100.

Compared Methods. To compare the performance of our proposed framework in terms of diversity and accuracy, we utilize the following methods:

1. Non-diversified: In this work, we use two popularly used recommendation methods UCF(user-based collaborative filtering) and PureSVD to produce the baseline lists. Both of them perform well in terms of accuracy. The number of neighbors in user-based CF [2] is set 30, and the reduced dimension of PureSVD is set 50 as proposed in [4] to get better performance.
2. Proposed: By using the marginal preference algorithms described in section 2.4, we instantiate our proposed framework into two diversification methods: PDMA-P and PDMA-Q. And we set the number of latent features $|T| = 30$, $|F|$=30 in our experiments, which is found to perform well.
3. STD: The method proposed by Ziegler [13] [14], also sequentially diversify the original list. Movie genres are used to calculate the items similarities. parameters $\kappa = 0.75$, $\Theta_F = 0.3$ are set according to empirical analysis in [13].
4. xQuAD: If we treat user as query and item as document, the popularly used diversification method [16] in IR can also be adjusted to our settings. The "nuggets" here are movie genres. For better tradeoff, we set $\lambda = 0.5$.

Evaluation. To compare the performance, we adopted the following metrics to cover various aspects of our consideration. Additionally, we randomly split each user' ratings into five parts equally. 20% of the items in each user's profile are put aside for testing. In user u' reserved test set, we take items whose rating is greater than u's average rate in training set as real u_{tests}. Because some users are strict raters, while others on the contrary. As default, we re-rank the top 200 items returned by the baseline method for each user, and focus on the evaluation of diversified list with N=10.

---

[1] http://grouplens.org/datasets/hetrec-2011/
[2] www.imdb.com

(a). **Accuracy**. Accuracy is still one of our considerations of performance. We intuitively use $Precision = \frac{|recs \cap tests|}{|recs|}$ and $Recall = \frac{|recs \cap tests|}{|tests|}$ which have been adopted widely. In order to consider the rank positions in the list, we also adopt Mean Average Precision(MAP). If correct answers are ranked higher, MAP is higher. $I(k)=1$, if the kth recommended movie belongs to u_{tests}, 0 otherwise. $P(k)$ denotes the precision ranked up to k.

$$MAP = \frac{1}{U} \sum_{u \in U} \frac{\sum_{k=1}^{N} P(k)I(k)}{|u_{\text{tests}}|} \tag{9}$$

(b). **Diversity**. Diversity is our another focus on performance. Here, we use $DNG@K$ to solely evaluate it without considering relevance. $DNG@K$ [12] is used to measure the number of covered item genres in the top-N list. Based on the positions of items in the list, rank bias is considered. $h_{g,k}$ is an indication function, equal to 1 if the genre g which belongs to the kth recommended movie, never appears ahead, 0 otherwise.

$$DNG@K = \sum_{k=1}^{K} \frac{\sum_{g=1}^{|G|} h_{g,k}}{log_2(k+1)} \tag{10}$$

(c). **Tradeoff**. In order to measure the tradeoff between relevance and diversity, we utilize $\alpha NDCG$. when calculating $\alpha NDCG$, movie genres are used as the "nuggets". $\alpha NDCG@K = \frac{\alpha DCG@K}{\alpha IDCG@K}$. where $q_{k-1,g}^{u}$ denotes the number of movies ranked up to position k-1 which contain genre g, in the recommended list for user u. $J_u(k,g) = r_{u,i_k}$, if the kth recommended movie contains g and belongs to u's tests, 0 otherwise. α controls the penalty of redundancy, which is set as 0.5 in our experiments. $\alpha IDCG@K$ denotes the highest value of $\alpha DCG@K$, which contains the "ideally" diversified list.

$$\alpha DCG@K = \sum_{k=1}^{K} \frac{\sum_{g=1}^{|G|} J_u(k,g)(1-\alpha)^{q_{k-1,g}^{u}}}{log_2(1+k)} \tag{11}$$

3.2   Results

Apart from modeling marginal aspect preference, clustering and aspects modeling are also two important components of our proposed framework. Here we firstly explore their influence on the final performance.

Aspect Evaluation. In our proposed method, we utilize three different yet relevant information: movie genres, content topics and rating factors to model user aspects, which are assumed to influence user's preference. Prior approaches often use one single type of features at one time. In order to test the performance of our hybrid setting, we evaluated different combinations G, GT, GF and GTF using two proposed method on two baseline recommendation list. Table 3.2 lists the comparison results. It is obvious to find that: (1) Our hybrid combination

Table 1. Comparison on using different aspect combinations in our proposed method: PDMA-P and PDMA-Q. G:genre, T:topic, F:factor

			Precision	MAP	Recall	DNG	αNDCG@5	αNDCG@10
UCF	PDMA-P	G	0.319	0.057	0.073	**27.26**	0.335	0.361
		GT	0.351	0.065	0.08	26.53	0.347	0.373
		GF	0.343	0.063	0.078	26.05	0.353	0.371
		GTF	**0.364**	**0.069**	**0.083**	25.52	**0.36**	**0.38**
	PDMA-Q	G	0.321	0.06	0.074	**22.86**	0.332	0.357
		GT	0.334	0.064	0.077	22.76	0.338	0.364
		GF	0.344	0.067	0.079	22.53	0.343	0.365
		GTF	**0.352**	**0.07**	**0.081**	22.38	**0.348**	**0.37**
PureSVD	PDMA-P	G	0.374	0.076	0.086	**27.65**	0.398	0.42
		GT	0.415	0.087	0.095	26.35	0.409	**0.43**
		GF	0.40	0.083	0.092	25.76	0.407	0.422
		GTF	**0.421**	**0.09**	**0.096**	25.12	**0.413**	**0.43**
	PDMA-Q	G	0.381	0.081	0.087	**22.15**	0.386	0.407
		GT	0.395	0.085	0.091	22.03	0.39	0.412
		GF	0.402	0.087	0.092	21.99	0.391	0.411
		GTF	**0.409**	**0.089**	**0.094**	21.91	**0.393**	**0.414**

GTF which considers all the three sets of aspects, always performs the best in terms of accuracy and $\alpha NDCG$. (2) Accuracy always improves when combining more additional features with item genres. The reason is that if two movie have the same genre assignments, taking into more effective features can help us find which is better more accurately. (3) Diversity decreases as accuracy improves. but $\alpha NDCG$ becomes better. Because irrelevant items which do not belong to user's preference, are avoided more accurately, although they satisfy the requirements that items should be diversified in the list. Thus a better balance in $\alpha NDCG$.

Clustering Evaluation. On the other hand, clustering is also important in our proposed framework. It just selects a much smaller set of items for the next diversification step. Obviously, candidates generation using clustering can accelerate the online process speed, but we still need to explore its influence on other performance measures. Table 2 presents the comparison results. We surprisingly find that clustering constantly gets a slight increase than not using it in terms of accuracy and $\alpha NDCG$. It suggests that our proposed candidate generation method is effective for diversification. This can be explained for two reasons: (1)As the selected items come from different clusters, we can guarantee the diversity of candidate items. (2) Candidate items are selected based on user's personal cluster preference. If user has not rated any item of one cluster, then items from that cluster will not be processed, this help us avoid unnecessary items in advance. Additionally, more representative items in each cluster are selected at earlier time, which helps further keep the relevance.

Comparison. It is obviously to be found that after diversification, accuracy measures such as precision and MAP both decrease. But our proposed methods

Table 2. Comparison on whether to use clustering or not, which is an important step for generating candidates. We just use the GTF aspect combination for comparison.

		cluster	Precision	MAP	Recall	DNG	αNDCG@5	αNDCG@10
UCF	PDMA-P	Yes	**0.364**	**0.069**	**0.083**	25.52	**0.36**	**0.38**
		No	0.353	0.067	0.081	**25.98**	0.355	0.375
	PDMA-Q	Yes	**0.352**	**0.07**	**0.081**	**22.38**	**0.348**	**0.37**
		No	0.339	0.067	0.078	22.35	0.339	0.364
PureSVD	PDMA-P	Yes	**0.421**	**0.09**	**0.096**	25.12	**0.413**	**0.43**
		No	0.415	0.089	0.095	**25.62**	0.412	0.428
	PDMA-Q	Yes	**0.409**	**0.089**	**0.094**	21.91	**0.393**	**0.414**
		No	0.40	0.087	0.092	**21.92**	0.39	0.409

Table 3. Performance comparison on different methods.

	Precison	MAP	DNG@5	DNG@10	αNDCG@5	αNDCG@10
UCF	**0.368**	**0.075**	13.85	22.21	0.302	0.332
PDMA-P	0.364	0.069	**19.83**	25.52	**0.36**	**0.38**
PDMA-Q	0.352	0.07	15.38	22.38	0.348	0.37
STD	0.326	0.062	19.59	**26.90**	0.303	0.331
xQuAD	0.360	0.074	15.97	22.36	0.345	0.364
PureSVD	**0.443**	**0.103**	13.54	22.16	0.375	0.399
PDMA-P	0.421	0.09	18.98	25.12	**0.413**	**0.43**
PDMA-Q	0.409	0.089	14.93	21.91	0.393	0.414
STD	0.372	0.079	**19.38**	**26.97**	0.354	0.379
xQuAD	0.439	0.10	15.32	22.38	0.403	0.421

PDMA-P and PDMA-Q remain comparable performance to baseline methods. On the other hand, diversity improves a lot especially in low ranks listed as 5. However purely pursuing good accuracy can not get the better tradeoff of αNDCG, as UCF and PureSVD perform. From the comparison results, we can see that our methods constantly perform better than baseline methods in terms of αNDCG, which is more significant in diversifying the UCF results. we explain our better performance as follows: (1)Exploiting features from both contents and ratings help us fully model user' preference aspects together with item genres. (2)When diversification, we consider both user's tastes and the influence of already selected items, and combine them appropriately based on probabilistic derivation, which guarantees the relevance, diversity and adaptability. STD performs well in diversification, but its accuracy measures are very poor. Because its diversification method does not consider different user's personal tastes. Some irrelevant items leads to the poor αNDCG, though they meet the requirements of diversity. xQuAD performs well both in accuracy and αNDCG. Because they consider both the item's initial relevance score and user's personal preference in diversification. However, it is difficult to make the optimal trade-off manually.

Table 4 lists our running time, conducted in unix system with 4G RAM and quad-core processors of 1.8GHz. To diversify the top 200 items for 2.1K users, we

Table 4. Summary of running time, as the number of baseline list increases. Two different settings are listed by using PDMA-P based on the UCF method.

	100	200	400	600	800	1000	2000
Candidates-Yes	2.6s	3.6s	5.6s	7.8s	10s	12.5s	21.6s
Candidates-No	3.1s	6.8s	21.7s	47.6s	84.2s	129.1s	446.2s

only need 3.6s. It suggests that our framework is efficient enough to satisfy the online requirements. We also find that as baseline list increases, candidate generation using clustering helps us make the running time change slowly. Because the number of candidate items maintains almost constant no matter how the baseline grows. However, directly diversifying the total baseline items increases sharply, which was similarly found in STD and xQuAD.

4   Related Work

CF(Collaborative Filtering) [1,3] has achieved great success in terms of predicative accuracy and scalability. Neighbor-based algorithms and latent factor models are the most popularly used CF methods. User-based CF [2] assumes that similar-minded users express similar interests in future items, while latent factor models [5,6] factorize user-item rating matrix into low latent space to get hidden preference factors. However better accuracy can not guarantee users' satisfaction for recommending similar redundant items.

Recently, researchers have addressed the problem that diversity is regarded as another important factor to improve user satisfaction. [13] proposed a re-ranking method to do topic diversification by using item taxonomies. They use a hyperparameter λ to control the balance between relevance and diversity. [9] proposed a set-oriented method to promote recommendation diversity, which combines relevance and diversity in a unified model in one step. [11,17] raised the issue of making diversified recommendation over time. [12] studied the problem of tackling with adaptive recommendation diversification by applying portfolio theory into latent factor models.

In information retrieval and web search, diversification has been studied extensively [15,16,19]. [8] firstly bridge the gap between search diversity and recommendation diversity. [10] studied the problem of measuring list diversity. Our work is related to [8] [10]. But we derive an more general probabilistic diversification framework by maximizing user's marginal aspects preference.

5   Conclusion

In this paper we propose a probabilistic framework for diversifying recommendation lists, which helps us avoid using hyperparameters to make the tradeoff between relevance and diversity and can be run efficiently enough to satisfy the online requirements. However, we can still do some work to improve our framework. Instead of using the same set of aspects for each user, we plan to develop a mechanism to apply different aspects into different users with individual preference. And we can design new more effective marginal preference algorithms to further improve the performance.

Acknowledgement. This research is supported by the 863 project of China (2013AA013300), National Natural Science Foundation of China (Grant No. 61375054 and 61402045) and Tsinghua University Initiative Scientific Research Program (20131089256), and Cross fund of Graduate School at Shenzhen, Tsinghua University (JC20140001).

References

1. Adomavicious, G., Tuzhilin, A.: Toward the next generation of recommender systems:a survey of the state-of-the-art and possible extensions. IEEE TKDE 17(6), 734–749 (2004)
2. Herlocker, J.L., Konstan, J.A., Borchers, A., Riedl, J.: An algorithmic framework for performing collaborative filtering. In: SIGIR 1999, pp. 230–237 (1999)
3. Sarwar, B., Karypis, G., Konstan, J., Riedl, J.: Item-based collaborative filtering recommendation algorithms. In: WWW 2001, pp. 285–295 (2001)
4. Cremonesi, P., Koren, Y., Turrin, R.: Performance of recommender algorithms on top-n recommendation tasks. In: RecSys 2010, pp. 39–46 (2010)
5. Koren, Y., Bell, R., Volinsky, C.: Matrix factorization techniques for recommender systems. Computer (8), 30–37 (2009)
6. Koren, Y.: Factorization meets the neighborhood: a multifaceted collaborative filtering model. In: KDD 2008, pp. 426–434 (2008)
7. Blei, D.M., Ng, A.Y., Jordan, M.I.: Latent dirichlet allocation. The Journal of Machine Learning Research 3, 993–1022 (2003)
8. Vargas, S., Castells, P., Vallet, D.: Intent-oriented diversity in recommender systems. In: SIGIR 2011, pp. 1211–1212 (2011)
9. Su, R., Yin, L.A., Chen, K., et al.: Set-oriented personalized ranking for diversified top-n recommendation. In: ACM Conference on Recommender Systems, pp. 415–418 (2013)
10. Clarke, C.L.A., Kolla, M., Cormack, G.V., et al.: Novelty and diversity in information retrieval evaluation. In: SIGIR 2008, pp. 659–666 (2008)
11. Lathia, N., Hailes, S., Capra, L., et al.: Temporal diversity in recommender systems. In: SIGIR 2010, pp. 210–217 (2010)
12. Shi, Y., Zhao, X., Wang, J., Larson, M., Hanjalic, A.: Adaptive diversification of recommendation results via latent factor portfolio. In: SIGIR 2012, pp. 175–184 (2012)
13. Ziegler, C.N., McNee, S.M., Konstan, J.A., Lausen, G.: Improving recommendation lists through topic diversification. In: WWW 2005, pp. 22–32 (2005)
14. Ziegler, C., Lausen, G., Schmidt-Thieme, L.: Taxonomy-driven computation of product recommendations. In: CIKM 2004, pp. 406–415 (2004)
15. Dang, V., Croft, W.B.: Diversity by proportionality an election-based approach to search result diversification. In: SIGIR 2012, pp. 65–74 (2012)
16. Santos, R., Macdonald, C., Ounis, I.: Exploiting Query Reformulations for Web Search Result Diversification. In: WWW 2010, pp. 881–890 (2010)
17. Koren, Y.: Collaborative filtering with temporal dynamics. In: KDD 2009, 447–456 (2009)

18. Hurley, N., Zhang, M.: Novelty and diversity in top-N recommendation-analysis and evaluation. TOIT 10(4), 14, 1–30 (2011)
19. Carbonell, J.G., Goldstein, J.: The Use of MMR, Diversity-Based Reranking for Reordering Documents and Producing Summaries. In: SIGIR 1998, pp. 335–336 (1998)
20. Porteous, I., Newman, D., Ihler, A., Asuncion, A., Smyth, P., Welling, M.: Fast collapsed gibbs sampling for latent dirichlet allocation. In: KDD 2008, pp. 569–577 (2008)

User Behavioral Context-Aware Service Recommendation for Personalized Mashups in Pervasive Environments

Wei He, Guozhen Ren, Lizhen Cui, and Hui Li

School of Computer Science and Technology, Shandong University, Jinan, China
hewei@sdu.edu.cn

Abstract. With the rapid development of mobile Internet and increasing amount of smart devices, Internet services have been integrated into peoples' daily lives. Due to the features of end-user-oriented mashups in pervasive environments, new challenges have been presented to conventional mashup approaches, including the complexity of user behaviors, the difficulty of predicting real-time user preference and other dynamic contexts. In this paper, we propose a new paradigm for behavioral context-based personalized mashup provision in pervasive environments by integrating mashup construction and execution into user natural behaviors. In the proposed paradigm, users with similar behavior patterns are identified and then probability distributions of potential behavior selection for user clusters are discovered from historical mashup logs, which provide supports for predicting and recommending user activities for future mashup constructions. Analysis and experiments indicate that our approach can effectively simplify personalized mashup composition, as well as improve the quality of mashup composition and recommendation based on behavioral contexts and personalization in pervasive environments.

Keywords: Pervasive computing, Behavioral context, Mashup, Personalization, Service Recommendation.

1 Introduction

With the rapid development of mobile Internet and the increasing growth of smart devices, more and more Internet-based services are being closely integrated into end-users' daily behaviors, which definitely bring better experiences for users than traditional desktop environments. Analogously, in such pervasive scenarios, the construction and execution processes of end-user-oriented mashups are also integrated into user's natural behaviors with procedure and interaction features [1, 2]. Therefore, user behavior patterns should be considered during the process of constructing and selecting mashup solutions in order to improve user experiences. Together with personalized service provisions and dynamic contexts, these new features have presented great challenges to conventional mashup approaches.

© Springer International Publishing Switzerland 2015
R. Cheng et al. (Eds.): APWeb 2015, LNCS 9313, pp. 683–694, 2015.
DOI: 10.1007/978-3-319-25255-1_56

The biggest challenge is the complexity of user behavior processes in pervasive environments. Different users always show various behavior patterns and preferences and it is difficult to perceive and predict real-time user behavior patterns in dynamic contexts. Personalized factors, including user habit and preference, have much impact on the selection of user behaviors. Even in the same context, different users have various preferences for activities and services. Furthermore, mobility of user devices and dynamic contexts increase the difficulties of user preference awareness. Let's consider an example of service mashups for dining out based on available web-based APIs, in which user activities probably include finding restaurant, reserving seats, locating and going to destination, parking and ordering, involving potential services such as LBS, map service, navigation service, reservation service and dish order service etc. Actually, there are multiple potential mashup solutions to meet the requirement and it is important to select a suitable one for the particular user. In conventional approaches, the mashup schema with involved activities is required to be defined and constructed in a mashup tool before it can be executed. However, it is both difficult and unrealistic for unprofessional users to schedule such a complete model in advance due to their limitations of professional knowledge and available information. Even though a mashup solution is pre-defined manually or automatically, probably it cannot achieve satisfactory results during the following execution period. Current context-based approaches for service composition and recommendation rarely consider the influence of user behavioral patterns and service relationships. Therefore, in order to improve user's experience, it is more crucial to help user planning a satisfying personalized mashup solution based on user behavioral contexts and preference, rather than only recommending independent services.

In this paper, we focus on the integrated cycle of both user behavior and mashup execution, instead of only recommending services. We propose a construction approach for personalized mashups based on user behavioral contexts. The main idea of this paper is as follows. First, potential preference-based user behavior patterns are discovered by applying pattern mining tasks to historical mashup logs. Then, based on the probabilistic distribution of user behaviors, according to user goal, user behavior traces and behavior patterns of similar users, target user's upcoming behaviors are predicted and the next activity is recommended. Next, the selected activity will be grounded to concrete Web-based APIs followed by executing the grounded service. By repeating the last two steps of the process, a personalized mashup schema is composed progressively until the final goal is achieved. In this proposed paradigm, mashup composition is simplified and end-user is not required to construct a schema in advance from scratch with necessary professional knowledge.

The rest of this paper is organized as follows: Section 2 gives problem definitions and the proposed system model; In section 3, we describe the detailed pattern mining approach based on historical logs. Section 4 depicts the iterative construction algorithm for end-user-oriented personalized mashups; In section 5, simulation experiments are illustrated; Section 6 gives related work; section 7 summarizes the main contributions of the paper.

2   Problem Definitions and System Model

A mashup instance represents a historical execution of mashup, which involves multiple components with specific context and execution sequence. The mashup log is the set of finite discrete mashup instances, each of which records a particular execution trace with component invocations and user contexts.

Definition 1 (*Mashup Instance*). A ***Mashup Instance*** is expressed as a tuple: $mi = <u, g, ts>$, with u denoting the end-user, g denoting user goal: $g = <In, Out, Desc>$, where $In = \{in_1, in_2, ...\}$ is the set of all input parameters, $Out = \{out_1, out_2, ...\}$ is the set of all output parameters of the mashup, and $Desc = \{kw_1, kw_2, ...\}$ is the set of keywords describing the mashup functionalities. The third attribute ts denotes task sequence: $ts = <t_1, t_2, . . ., t_n>$, where each task t_i ($1 \le i \le n$) is a tuple: $ti=<mc_i, ctx_i>$ with mc_i denoting mashup component and ctx_i denoting the context of current task. In the task sequence, the parameters required by component mc_i come from the outputs generated by one or more precursor components $mc_{i-1}, mc_{i-2}, ... mc_1$ ($1 \le i \le n$).

In the scenario of pervasive mashups, conventional context is extended with user behavior information. User behavioral context records the behavior trace (or activity sequence) he/she has performed during the period of mashup execution.

Definition 2 (*Behavioral Context*). A ***Mashup Context*** is a prefix subset of the task sequence $mi.ts$ of a mashup instance, which is expressed as: $bc = <t_1, t_2, . . ., t_m>$, $m \le n$, where each task t_i is a tuple: $ti=<mc_i, ctx_i>$ with mc_i denoting mashup component and ctx_i denoting the context of current task.

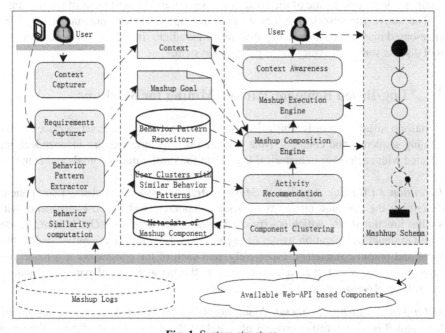

Fig. 1. System structure

The architecture of user behavioral context-based mashup construction and service recommendation is illustrated in figure 1, with multiple components including behavior pattern mining module, context awareness module, knowledgebase of behavior patterns and mashup composer & executer.

Historical logs record users' mashup activities, including behavior traces, involved web resources and the related contexts during the execution period, from which quite a few valuable information for future mashup construction can be discovered. User behavior pattern extractor generates the probability distribution for user activity selection. Based on the activity traces in which different users are involved, user behavior similarity computing module evaluates user preferences to candidate activities in particular contexts and partial activity traces, so that users with similar behaviors are identified. Component clustering module generates activities by clustering actual web-based components with similar functionalities.

Context awareness module is responsible for capturing and updating user behavioral contexts in current mashup execution, as well as other real-time contexts which may affect the selection of user activities, such as location and time.

Mashup composer and executer are the core parts of the system including activity recommendation engine, schema composing engine and service execution engine. According to user goal and user's previous behavior traces, activity recommendation engine computes user's preference values for candidate activities based on the extracted probability distribution of user behavior and recommend the next activity. Then, composition engine enhances current mashup schema by combining the recommended activity. Execution engine grounds the activity with available web-based component and perform execution. The result of execution engine will feed back to composition engine as one of the conditions to generate the next steps of the mashup.

The above framework provides an effective approach for constructing a preference-based mashup instance using an iterative paradigm. In the following sections, we will discuss some key issues in the proposed system.

3   Log-Based Behavior Pattern Mining for Mashups

A mashup instance records a historical execution trace involving user, services with a running sequence and contexts. Note that the mashup instances are independent and isolated with each other in historical logs, even for similar mashup goals and contexts.

Definition 3 (Mashup Activity). A *Mashup Activity*, also called abstract component, describing the common attributes of a set of concrete web-based components with similar functionalities, which is expressed using a tupole: $ma = <Op, In, Out, Desc>$, with Op denoting operation name, $In = \{in_1, in_2, ...\}$, $Out = \{out_1, out_2, ...\}$ denoting input and output parameters respectively, and each of the parameters is a tuple: $p = <name, Desc>$, where $p \in In \cup Out$, Desc is the keyword set of parameter annotations: $Desc=\{anno_1, anno_2, ...\}$.

Definition 4 (Mashup Schema). Assuming A is a set of activities, C is a set of contexts and B is the Cartesian product of A and C: $B= A \times C$. Let B^* denotes the set of finite sequences of B. A **Mashup Schema** σ is a finite sequence of activities in $B*$, $\sigma \in B*$, which can be expressed as: $\sigma = <(a_1,c_1), (a_2,c_2),\cdots\cdots, (a_n,c_n)>$, $\sigma[i]= (a_i,c_i)$, $|\sigma|=n$ is length of the sequence, $1 \leq i \leq n$.

A mashup schema is an abstract template describing multiple mashup instances with similar contexts and behavior patterns. As the basic component of mashup schema, activity extraction becomes one of the fundamental tasks in mashup pattern mining. At present, there have been quite a few researches on service clustering and classification. We adopt a component clustering approach based on functionality similarity and interface compatibility, which is described in detail in our previous works [11].

3.1   Similarity of User Activity Traces

Let σ be a complete mashup instance in historical logs, σ_t denotes the activity sequence of σ. We define the activity adjacency relation of σ_t as: $AAR(\sigma_t) = \{<t_i,t_{i+1}>\}$. Also, Let $ADR(\sigma_t)$ denotes the activity dependence relation: $ADR(\sigma_t) = \{<t_i,t_j>\}$, $i<j$. Therefore, the activity adjacency relation and activity dependence relation are two types of decomposition for an activity sequence with different rigorous levels.

Definition 5 (Activity Trace Similarity). Let σ_1, σ_2 represent two arbitrary activity sequences: $\sigma_1=<t_1, t_2, \ldots, t_n>$, $\sigma_2=<t'_1, t'_2, \ldots, t'_m>$, the **Activity Trace Similarity** of σ_1 and σ_2 is defined as

$$Sim_t(\sigma_1,\sigma_2) = \frac{|ANR(\sigma_1) \cap ANR(\sigma_2)|}{|ANR(\sigma_1) \cup ANR(\sigma_2)|} * w + \frac{|ADR(\sigma_1) \cap ADR(\sigma_2)|}{|ADR(\sigma_1) \cup ADR(\sigma_2)|} * (1 - w) \quad (1)$$

Where $0 \leq w \leq 1$, is a weight value for measuring the importance of activity adjacency relation and activity dependence relation.

3.2   Similarity of User Behavior Patterns

Collaborative filtering recommendation approaches based on similar users have been proven significant effectiveness in multiple domains. Similarly, introducing the factor of user preference in personalized mashup construction will definitely improve the quality of activity recommendation for end-users. In this scenario, users are considered similar if they have similar behavior patterns for the same goal.

Definition 6 (User Preference for Activity Trace). The preference of user U for an activity trace is expressed as a tuple: $EP=(P, \delta)$ with $P=\{P_1, \ldots\ldots,P_n\}$ denoting the finite set of the activity traces that user U has selected, $\delta(P_i)$ denoting the preference value for activity trace P_i which is the times that U has performed P_i in historical logs.

Based on the matrix of user preference for activity traces, the behavior pattern similarity between any users can be measured. Among current approaches for measuring user similarity, Pearson correlation coefficient [4] has been proved to be effective enough in multiple domains. In our approach, Pearson correlation coefficient is used to measure the preference similarity for activity traces between user u and v:

$$r_{uv} = \frac{\Sigma_i \in I_{uv}(P_{u,i} - \overline{P_u}) \times (P_{v,i} - \overline{P_v})}{\sqrt{\Sigma_i \in I_{uv}(P_{u,i} - \overline{P_u})^2} \sqrt{\Sigma_i \in I_{uv}(P_{v,i} - \overline{P_v})^2}}$$

Where r_{uv} denotes the preference similarity of user u and v, I_{uv} means the common activities in the traces they selected, $P_{u,i}$, $P_{v,i}$ is the preference value of user u, v for activity trace P_i, and $\overline{P_u}$, $\overline{P_v}$ denotes the average preference value of user u, v for their common activities I_{uv}.

Definition 8 (Behavior Pattern Similarity). Assuming the preferences of user U_i and U_j for activity traces are $EP_i = (P_i, \delta_i)$ and $EP_j = (P_j, \delta_j)$ respectively, the behavior pattern similarity between user U_i and U_j is defined as

$$A_{ui \Leftrightarrow uj} = \begin{cases} r_{ij}, & r_{ij} \geq 0 \\ 0, & else \end{cases} \quad (2)$$

Where r_{ij} is the Pearson correlation coefficient of the preference value for the common activity traces of user U_i and U_j.

4 Incremental Construction Algorithm for Mashup Schema

The mashup construction is a progressive procedure involving two phases in each iteration: schema construction phase and service execution phase.

4.1 Support Function for Activity Trace

Based on user partial activity traces, historical mashup instances provide different supports for recommendation of the user's future activities. That is, the more similar the trace fragment and the historical mashup instances are, the more valuable the historical patterns are to support the construction of current mashup.

Let σ be the activity trace of a mashup instance in historical logs, σ_p denotes the prefix sub-sequence of σ, ρ is a partial execution trace (i.e. mashup fragment), $G(\sigma)$ and $G(\rho)$ denote the goal of σ and ρ respectively. Then we define the **Support Function** of mashup instance σ for partial execution trace ρ, i.e. the probability that activity trace evolves into mashup instance σ:

$$S(\rho, \sigma) = Sim_t(\rho, \sigma) * w + S_g(\rho, \sigma) * (1 - w) \quad (3)$$

Where $Sim_t(\rho, \sigma)$ is the activity trace similarity defined in formula (1), and $S_g(\rho, \sigma)$ denotes mashup goal similarity which is defined as

$$S_g(\rho, \sigma) = \frac{|g(\sigma).\text{Out} \cap g(\rho).\text{Out}|}{|g(\sigma).\text{Out} \cup g(\rho).\text{Out}|}$$

The support function for activity trace is used to filter the historical mashup logs, so that the instances supporting the partial execution trace can be identified and the other unrelated mashup instances are excluded.

4.2   Mashup Construction Algorithm

In the following we describe the schema construction phase of the iterative mashup composing process, i.e. the algorithm for activity recommendation based on user past behavior trace.

Algorithm : Behavioral context-based activity recommendation
Input: mashup fragment f, mashup goal g, user u, threshold of mashup similarity h
Output: enhanced mashup fragment $\hat{f}$.

Let $f = (<s_1, c_1>, <s_2, c_2>, <s_3, c_3> \cdots\cdots, <s_k, c_k>)$;
$ssi = \{\}$; //Initialize the set of similar mashup instances of f;
For each mashup instance mi in the historical logs
    Compute similarity $S(f, mi)$ using formula (4);
    If $(S(f, mi) \geq h)$
      $ssi = ssi \cup \{mi\}$;
    End if
End for
Identify user cluster o for user u based on formula (2); //$u \in o$
$r[0..c]=0$; //The selection times of candidate activities by users in o;
For each user $\hat{u}$ in user cluster o
    For each candidate activity $ca[i]$
      $r[i] = r[i] +$ the times of $ca[i]$ that user $\hat{u}$ has selected in logs;
    End for
End for
Assign $ca[k]$ to na where $r[k]$ is the greatest value in $r[0..c]$;
$\hat{f} = f \cup <na>$;
Return $\hat{f}$;

The algorithm describes schema construction phase of the iterative mashup composition process, which aims to enhance the mashup fragment by computing and recommending the next activity. The complete mashup can be constructed progressively by performing the algorithm repeatedly. Once the recommended activity is confirmed, it should be grounded to a particular Web-API based component registered in the repository. Currently, there have been quite a few researches focusing on mashup service selection and recommendation.

5   Simulation Experiments

According to the effectiveness and efficiency of the mashup composition approach, an application scenario of service mashups are constructed, in which dining-out related services are provided for end-users. In this application scenario, a user goal may be described as "Finding a restaurant within 10 miles, getting there, parking and having dinner with my friends" with probably multiple web-based services.

5.1 Experimenting Data

1 Web-Based Components

The detailed construction process involve 2 steps. In the first step, real-world services on Internet are searched and extracted to generate the primary components in the registry from the sources of general service providers (such as Google Place APIs, Baidu mapping APIs etc.) and specific platforms(such as Yelp APIs, Dianping.com APIs etc.). The descriptive information for both functionalities and interfaces of these public services are extracted, and then annotations are generated for each component based on its native descriptions. Besides, some QoS attributes are randomly generated, including usability, performance and reliability etc. In the second step, we also created and annotated virtual services with a random number for each primary component based on its meta-data, so that the registered components come to a certain quantity. The basic statistic info of the constructed components is listed in table 2.

Table 1. The statistics of components

Total number	372		
Clusters based on functionality similarity	15		
Number of annotations	1, 3, 5		
Component classification	**Number of components**	**Primary components**	**Virtual components**
Total	372	51	321
LBS-based service	19	3	16
Navigating service	18	3	15
Restaurant finding	44	9	36
Reservation service	26	5	21
···...···...	···...	···...	···...

2. Historical Mashup Instances

Based on the meta-data of generated components, we constructed simulation data for historical user behavior traces, i.e. mashup instances. The mashup instances were created by randomly selecting components following the constraints of compatible interfaces and functionalities. Then, the generated instances were adjusted from two aspects. One adjustment is to increase clustering property for the behavior traces of similar users. On the other side, a percentage value measuring component popularity was introduced for each component and let the appearance frequency of components in the generated logs roughly follow their popularity distribution, so that the actual situation could be reflected as much as possible. The statistics of generated mashup instances is listed in table 3.

Table 2. The statistics of generated mashup instances

Set of mashup instances	
Number of mashup instances	3500
Number of involved users	60
Number of involved components	317
Estimated number of activities	29
Number of mashup goals	12
Number of contexts	5

5.2   Experimental Results

First, experiments are performed to verify the component clustering approach based on functionality similarity and interface compatibility. The experimental results of component clustering in different conditions are described in our previous works [11]. In this paper, we also compared our component clustering approach, referred to "FSandIC-based clustering" with other popular clustering algorithms. In literature [6], the authors proposed two service similarity computing approaches: Euclidean-distance and Cosine-distance measurement based on vector space for multi-dimensional properties, which is referred to "ED-based clustering" and "CD-based clustering" respectively. For similar purpose, Literature [7] proposed functionality-based and process-based similarity measurements for component clustering which is referred to "FS-based clustering". The difference degree between our clustering method and the other 3 approaches were computed based on service similarity matrices generated by the 4 methods, including difference value for service pair $<s_i, s_j>$ defined as:

$$D(s_i, s_j) = d_1(s_i, s_j) - d_2(s_i, s_j),$$

and the overall difference value define as

$$D(M_1, M_2) = \sqrt{\frac{\sum_{i=1}^{n} \sum_{j=i+1}^{n}(d_1(s_i, s_j) - d_2(s_i, s_j))^2}{n(n-1)/2}}$$

where $d_1(s_i, s_j) \in M_1$, $d_2(s_i, s_j) \in M_2$ denotes the similarity value between service s_1 and s_2 in matrix M_1 and M_2 respectively. The difference degrees of clustering between our method and the other 3 approaches, as well as the combined comparison, are shown in figure 2. The results indicate that the clustering result of our method is close to FC-based clustering with an overall difference of 1.9, and has much bigger difference with the other approaches. This is because the two closer clustering methods considered both functionality and interface compatibility, and the other 2 methods focus on multi-dimensional spaces including functional semantic, location info and QoS attributes.

The next experiment was performed to verify the effect of mashup construction. The generated mashup instances were divided into 2 parts: sample instances and benchmark instances. Experiments were carried based on sample data to generate recommended mashups, then the results were compared with the benchmark part of mashup instances with similar goal and contexts.

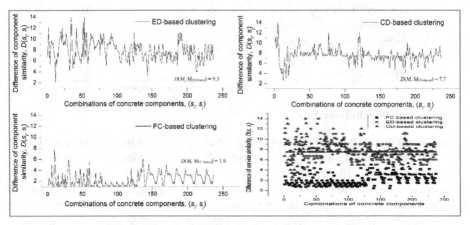

Fig. 2. Clustering differences among the approaches

In order to measure the results of mashup construction, we define matching rate according to benchmark data with mashup goal g and user u:

$$mr(g, u) = \frac{\sum_{c_i \in C} |S_b(g, r, c_i) \cap S_r(g, r, c_i)|}{\sum_{c_i \in C} |S_b(g, r, c_i)|},$$

where $S_r(g, r, c)$ denotes the set of recommended components generated in the experiments, $S_b(g, r, c)$ is the components in the benchmark instances.

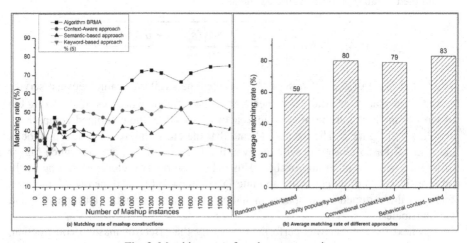

Fig. 3. Matching rate of mashup constructions

Based on the definition of matching rate, we compared our algorithm with other service composition approaches according to different numbers of mashup instances. We investigated some representative approaches of service composition, including context-awareness based [5], pervasive-environments based [3] and planning-based[8] composition. We simplified these methods and implemented the core ideas, then compared the results with our algorithm. Figure 3(a) illustrates the matching rate of the algorithms with different number of generated mashups. With the increasing number of generated mashup instances to be evaluated, the matching rate with benchmark data becomes stable. Another experiment in figure 3(b) shows the average matching rate for different approaches.

6   Related Work

In recent years, context-based service recommendation for mashup provisions have been attracted much attention. Some researchers performed role mining and service recommendation based on users' historical selection records in various physical environments [9]. Hussein et al. capture a service's requirements as two sets of scenarios, and then deal with the dynamic changes of contexts using adaptation requirement [10]. Due to the successful effects in many other domains, personalized recommendation was widely introduced into service discovery and selection. Meng et al. proposed a similar user-based CF approach to recommend services by annotating users' preferences with keywords [12]. Chen et al. adopted a visualized technology to improve recommendation comprehensibility and implemented personalized recommendation based on a CF algorithm [13]. To improve recommendation qualities in pervasive environments, quite a few researches combine contexts into collaborative filtering methods. Shin et al. proposed a CF-based recommendation approach with aggregated contexts [14]. Karatzoglou et al. constructed a multi-dimensional matrix of "user-item-context" by extending conventional "user-item" matrix for a context-aware recommendation approach [15].

7   Conclusion

In this paper, we propose a new paradigm for behavioral context-based personalized mashup provision in pervasive environments by combing user behavior and mashup instance execution into an integrated process. Analysis and experiments indicate that our approach can effectively simplify personalized mashup composition without depending on end-users' professional knowledge, as well as improve the quality of mashup composition and recommendation based on behavioral contexts and personalization in pervasive environments.

Acknowledgement. This work is supported by the National Natural Science Foundation of China under Grant No. 61303085, the Natural Science Foundation of Shandong Province of China under Grant No. ZR2013FQ014, 2014ZRFM031, the Science and Technology Development Plan Project of Shandong Province under Grant 2014GGX101019, 2014GGX101047 and Shandong Province Independent Innovation Major Special Project under Grant No. 2013CXC30201.

References

1. Daniel F, Koschmider A, et al. Toward process mashups: key ingredients and open research challenges. Proceedings of the 3rd and 4th International Workshop on Web APIs and Services Mashups, p.1-8, 2010
2. Fisichella M, Matera M. Process flexibility through customizable activities: A mashup-based approach. 2011 IEEE 27th International Conference on Data Engineering Workshops, 2011:226 - 231.
3. Zhou J, Gilman E, Palola J, et al. Context-aware pervasive service composition and its implementation. Personal and Ubiquitous Computing, 2011, 15(3):291-303
4. Good N, Schafer JB, Konstan JA, et al. Combining collaborative filtering with personal agents for better recommendations. In: Proc. of the 16th National Conf. on Artificial Intelligence. Menlo Park: AAAI Press, 1999. 439–446
5. Medjahed B, Atif Y. Context-based matching for Web service composition. Distributed and Parallel Databases, 2007, 21:5-37
6. Platzer C, Rosenberg F, Dustdar S. Web service clustering using multidimensional angles as proximity measures. ACM Transactions on Internet Technology. 2009, 9(3): 1-26
7. Sun P, Jiang C. Using service clustering to facilitate process-oriented semantic web service discovery. Chinese Journal of Computers, 2008, 31(8): 1340-1353
8. Hatzi O, Vrakas D, Nikolaidou M, et al. An Integrated Approach to Automated Semantic Web Service Composition through Planning[J]. Services Computing, IEEE Transactions on, 2012, 5(3):319 - 332.
9. Wang J, Zeng C, He C, et al. Context-aware role mining for mobile service recommendation. In: Proceedings of the 27th Annual ACM Symposium on Applied Computing. New York: ACM, 2012. 173-178
10. Hussein M, Han J, Yu J, et al. Scenario-Based Validation of Requirements for Context-Aware Adaptive Services. In: Proceedings of the IEEE International Conference on Web Services. New York: IEEE Press, 2013. 348-355
11. He W, Li Q, Cui L, et al. A Context-Based Autonomous Construction Approach for Procedural Mashups[C]. //Web Services (ICWS), 2014 IEEE International Conference on. IEEE, 2014:487 - 494.
12. Meng, S, Dou W, Zhang X, et al. KASR: A Keyword-Aware Service Recommendation Method on MapReduce for Big Data Application. IEEE Transactions on Parallel and Distributed Systems, 2014
13. Chen X, Zheng Z, Liu X, et al. Personalized QoS-aware web service recommendation and visualization. IEEE Transactions on Services Computing. 2013, 6(1): 35-47
14. Shin D, Lee J, Yeon J, et al. Context-aware recommendation by aggregating user context. In: IEEE Conference on Commerce and Enterprise Computing. New York: IEEE, 2009. 423 - 430
15. Karatzoglou A, Amatriain X, Baltrunas L, et al. Multiverse recommendation: n-dimensional tensor factorization for context-aware collaborative filtering. In: Proceedings of the fourth ACM conference on Recommender systems. New York: ACM, 2010. 79-86

On Coherent Indented Tree Visualization of RDF Graphs

Qingxia Liu, Gong Cheng, and Yuzhong Qu

State Key Laboratory for Novel Software Technology, Nanjing University, China
qxliu.nju@gmail.com, {gcheng,yzqu}@nju.edu.cn

Abstract. Indented tree has been widely used to organize information and visualize graph-structured data like RDF graphs. Given a starting resource in a cyclic RDF graph as root, there are different ways of transforming the graph into a tree representation to be visualized as an indented tree. In this paper, we aim to smooth user's reading experience by visualizing an optimally coherent indented tree in the sense of featuring the fewest reversed edges, which often cause confusion and interrupt the user's cognitive process due to lack of effective way of presentation. To achieve this, we propose a two-step approach to generate such an optimal tree representation for a given RDF graph. We empirically show the difference in coherence between tree representations of real-world RDF graphs generated by different approaches. These differences lead to a significance difference in user experience in our user study, which reports a notable degree of dependence between coherence and user experience.

Keywords: Browsing, coherence, indented tree, RDF visualization.

1 Introduction

The recent explosion of Linked Data has brought about a large volume of graph-structured RDF data from various domains. To help human users find and consume it on the Web, a broad range of Linked Data search engines [3,13] and browsers [2,12] have been developed to collect, process, and present RDF data. However, it is challenging to present such machine-friendly, graph-structured semantic data in an open-domain environment, which domain-specific visualizations [4,11] cannot inclusively support. Domain-independent visualizations like node-link diagram and indented tree are needed.

Whereas node-link diagram has been extensively studied [5] and its drawbacks have been discussed [10], what we focus on in this paper is indented tree, which has been successfully used in many applications and has shown its potential to visualize semantic data [6]. Although indented tree has been adopted by some Linked Data search engines and browsers [2,3], very little research attention has been paid to the fact that given a starting resource as root, a cyclic RDF graph (e.g. Fig. 1a) can be transformed, by duplicating vertices and reversing edges when necessary, into different tree representations (e.g. Fig. 2a and Fig. 2b) and thus be visualized as different indented trees. A question that follows is whether

© Springer International Publishing Switzerland 2015
R. Cheng et al. (Eds.): APWeb 2015, LNCS 9313, pp. 695–706, 2015.
DOI: 10.1007/978-3-319-25255-1_57

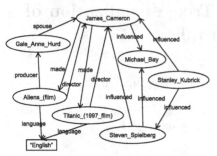

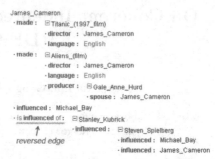

(a) Node-link diagram (with IRIs in el-
lipses and literals in rectangles)

(b) Expandable/collapsible indented
tree (with labels of reversed edges pre-
fixed by "is" and suffixed by "of")

Fig. 1. Different visualizations of the same RDF graph (with IRIs replaced by local
names)

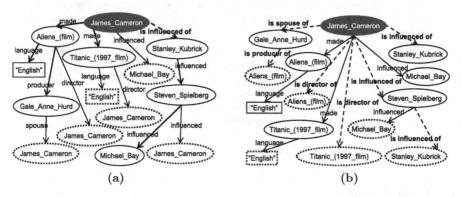

Fig. 2. Different tree representations of the same RDF graph (with starting vertex
shaded and reversed edges dashed)

and how these different representations influence the user's browsing experience,
which drives our research in this paper.

Our work is motivated by the observation that we still lack a general, effective
way of presenting reversed edges in RDF. Conventional solutions either simply
attach a special symbol (e.g. a hyphen) or words like "is ... of" to each reversed
edge, giving rise to confusing property names like "influenced-" or "is influenced
of" as illustrated in Fig. 1b, which are harmful to the user's reading experience
and frequently interrupt the user's cognitive process. In light of this, we propose to reduce the number of reversed edges as far as possible, to provide the
user with a coherent indented tree visualization of RDF graph. Specifically, our
contribution in this paper is threefold.

– To our knowledge, we are among the first to give a formalized analysis of
  indented tree visualization of RDF graphs from the coherence aspect.

- We devise a two-step approach to transform an RDF graph into a tree representation having the fewest reversed edges.
- We empirically show the difference in coherence between tree representations of real-world RDF graphs generated by our approach and two popular baseline approaches. We also carry out a preliminary user study to reveal the relationship between coherence and user experience.

The remainder of this paper is structured as follows. Section 2 formalizes the problem. Section 3 describes our approach. Section 4 reports experiments. Section 5 discusses related work. Section 6 concludes the paper with future work.

2   Problem Statement

Let I, B, and L be pairwise disjoint sets, comprising all IRIs, blank nodes, and literals, respectively, which are known collectively as resources. An RDF triple (subject-predicate-object) t is a tuple $(subj(t), pred(t), obj(t)) \in (I \cup B) \times I \times (I \cup B \cup L)$. An RDF graph $\mathbb{T}$ is a set of RDF triples, which can be represented by an edge-labeled directed graph $G = (V_G, E_G)$ where the vertex set V_G comprises the subjects and objects of all the triples in $\mathbb{T}$ (i.e. $V_G = \{subj(t) : t \in \mathbb{T}\} \cup \{obj(t) : t \in \mathbb{T}\}$), and each edge $e \in E_G = \mathbb{T}$ is directed from $subj(e)$ to $obj(e)$ labeled with $pred(e)$. Figure. 1a shows an RDF graph visualized as a node-link diagram.

The essence of an indented tree visualization is a vertex- and edge-labeled rooted tree where edges are all directed away from a particular root vertex. To transform an RDF graph G (which may contain cycles) into a tree TR, we may have to map a vertex of G to more than one vertex of TR; however, a bijection between the edge set of G and the edge set of TR is required. Since some edges of G may have to be mapped to TR in a reversed direction, we introduce a set of inverse labels I_R satisfying $I_R \cap I = \emptyset$, and define a function $inv : (I \cup I_R) \to (I \cup I_R)$ that inverses an edge label and satisfies $inv(inv(p)) = p$.

Definition 1 (Tree Representation of RDF Graph). *Given an RDF graph $G = (V_G, E_G)$ and a starting resource $s \in V_G$, a tree representation of G starting from s is a vertex- and edge-labeled rooted tree $TR = (V, E, L_V, L_E, r, f)$ where*

- V *is the vertex set,*
- E *is the edge set consisting of ordered pairs on V,*
- $L_V : V \to V_G$ *is a surjection that labels each vertex in V with a vertex in V_G, i.e., vertices of TR are labeled only with vertices of G and every vertex of G appears as the label of at least one vertex of TR,*
- $L_E : E \to (I \cup I_R)$ *labels each edge in E with an IRI or an inverse label ($e \in E$ is called a reversed edge when $L_E(e) \in I_R$),*
- $r \in V$ *with $L_V(r) = s$ is the root, and called the starting vertex of TR,*
- $f : E \to E_G$, *defined by $\forall e = (u, v) \in E$, $f(e) = (L_V(u), L_E(e), L_V(v))$ or $f(e) = (L_V(v), inv(L_E(e)), L_V(u))$ (in the latter case, $L_E(e) \in I_R$), is a bijection, i.e., there is a one-to-one correspondence between edges of TR and edges of G, and*

– $\forall u \in V_G, |\{v \in V : L_V(v) = u, d^+(v) > 0\}| \leqslant 1$ *(where $d^+(\cdot)$ returns the outdegree of a vertex), i.e., each resource can be expanded in at most one place in the indented tree visualization.*

The set of all tree representations of G starting from s is denoted by $\mathbb{TR}(G, s)$.

For instance, Fig. 2 illustrates two tree representations of the RDF graph in Fig. 1a, both with starting resource James_Cameron. Since this RDF graph contains cycles, vertices such as James_Cameron in Fig. 2a and Michael_Bay in Fig. 2b become labels of more than one vertex in tree representation. When visualizing a tree representation like Fig. 2a as an indented tree in Fig. 1b, the direction of an edge is presented from left to right. Further, both Fig. 2a and 2b contain reversed edges, and in Fig. 1b we present an inverse label $inv(p)$ by prefixing p with "is" and suffixing p with "of".

Let $\delta(TR)$ denote he number of reversed edges in a tree representation TR. We argue that the tree representation in Fig. 2a is better than that in Fig. 2b because the former is with a smaller δ and thus when being visualized as indented tree, the former appears more "coherent". We define the *coherence* of a tree representation TR as the percentage of non-reversed edges it contains:

$$coh(TR) = \frac{|E| - \delta(TR)}{|E|} . \tag{1}$$

For instance, the tree representation in Fig. 2a contains 1 reversed edge ($\delta = 1$) and its coherence is $12/13$. By contrast, the one in Fig. 2b contains 7 reversed edges ($\delta = 7$) and its coherence is $6/13$. (In both cases, $|E| = 13$.)

Given an RDF graph $G = (V_G, E_G)$ and a starting resource $s \in V_G$, we aim to find a tree representation in $\mathbb{TR}(G, s)$ that has the largest coherence:

$$\underset{TR \in \mathbb{TR}(G,s)}{\arg\max} \; coh(TR) , \tag{2}$$

which is called an *optimal* tree representation of G starting from s. For instance, Fig. 2a gives an optimal tree representation of the graph in Fig. 1a starting from James_Cameron.

3   Approach

Now we introduce an approach to generating an optimal tree representation for an RDF graph given a starting resource. RDF graphs considered in this section are assumed to be simple (i.e. without loops or parallel edges) and connected. The correctness proof of the approach and a discussion about graphs that are not simple or being disconnected are given in our technical report.[1]

---

[1] http://ws.nju.edu.cn/r2t/coherent-tree-report2015.pdf

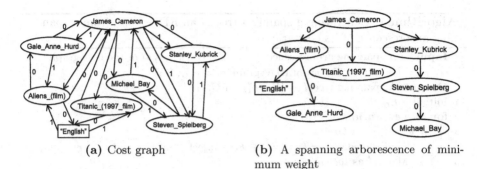

(a) Cost graph

(b) A spanning arborescence of minimum weight

Fig. 3. A cost graph and its spanning arborescence of minimum weight

3.1 Overview

We firstly present an overview of our two-step approach.

In Step 1 (c.f. Sect. 3.2), we transform not the entire but only a spanning tree (which is not necessarily a rooted tree) of the RDF graph into a tree representation, which is only a partial tree representation of the entire RDF graph. For instance, for the RDF graph in Fig. 1a and starting resource James_Cameron, the tree representation in Fig. 2a is not completely generated in this step; only its subgraph excluding dotted vertices and their associated edges is generated. This subgraph, as a tree representation of a spanning tree of the RDF graph, is actually one that has the smallest δ value among all the tree representations of spanning trees of the RDF graph starting from James_Cameron. We obtain this subgraph with the help of an edge-weighted cost graph and its spanning arborescence of minimum weight as illustrated in Fig. 3.

In Step 2 (c.f. Sect. 3.3), we complete the tree representation by transforming the remaining edges of the RDF graph. Since what we have transformed in Step 1 is a spanning tree, i.e., every vertex of the RDF graph has labeled a vertex of the partial tree representation, in Step 2 we can complete the tree representation by constantly adding new vertices and edges without introducing reversed edges, as illustrated by the dotted vertices and their associated edges in Fig. 2a.

3.2 Step 1: Transforming a Spanning Tree

In this step, we find and transform a spanning tree of the RDF graph into a tree representation that has the smallest δ value among all possibilities under a given starting resource.

A spanning tree of RDF graph $G = (V_G, E_G)$, as a spanning subgraph of G, can be denoted by $T = (V_G, E'_G)$ where $E'_G \subseteq E_G$. An arbitrary spanning tree of G can be straightforwardly transformed into a tree representation (which is a partial tree representation of G) according to Algorithm 2, where the *toBeReversed* mark on an edge of T is given in Algorithm 1 (c.f. Step 1.3) to indicate that the edge is directed to the starting resource and thus needs to be reversed

---

Algorithm 1. Generating a spanning tree of an RDF graph from a spanning arborescence of its cost graph.

Input : A spanning arborescence $T_C = (V_C, E'_C, w|_{E'_C}, s)$ of the cost graph
$\qquad\quad$ $G_C = (V_C, E_C, w)$ of RDF graph $G = (V_G, E_G)$.
Output: A spanning tree $T = (V_G, E'_G)$ of G.

1 Initialize E'_G to $\emptyset$;
2 **for** *each* $e = (u, v) \in E'_C$ **do**
3 $\quad$ **if** $w|_{E'_C}(e) = 1$ **then**
4 $\quad\quad$ Find the edge $e' = (v', p, u') \in E_G$ subject to $u' = u$ and $v' = v$;
5 $\quad\quad$ Mark e' as *toBeReversed*;
6 $\quad$ **else**
7 $\quad\quad$ Find the edge $e' = (u', p, v') \in E_G$ subject to $u' = u$ and $v' = v$;
8 $\quad$ Add e' to E'_G;

Algorithm 2. Transforming a spanning tree of an RDF graph into a tree representation.

Input : A spanning tree $T = (V_G, E'_G)$ of RDF graph $G = (V_G, E_G)$ and a
$\qquad\quad$ starting resource $s \in V_G$.
Output: A tree representation $TR = (V, E, L_V, L_E, r, f)$ of T.

1 Initialize V to $\emptyset$;
2 **for** *each* $v \in V_G$ **do**
3 $\quad$ Add a new vertex v' to V;
4 $\quad$ Define $L_V(v') = v$;
5 Let $r \in V$ be the vertex subject to $L_V(r) = s$;
6 Initialize E to $\emptyset$;
7 **for** *each* $e = (u, p, v) \in E'_G$ **do**
8 $\quad$ Find the vertex $u' \in V$ subject to $L_V(u') = u$;
9 $\quad$ Find the vertex $v' \in V$ subject to $L_V(v') = v$;
10 $\quad$ **if** *e is marked as toBeReversed* **then**
11 $\quad\quad$ Add a new edge $e' = (v', u')$ to E and define $L_E(e') = inv(p)$;
12 $\quad$ **else**
13 $\quad\quad$ Add a new edge $e' = (u', v')$ to E and define $L_E(e') = p$;
14 $\quad$ Define $f(e') = e$;

---

to ensure that TR is a rooted tree. The spanning tree to be transformed is found by the following three sub-steps.

Step 1.1: Constructing a Cost Graph. We firstly construct an edge-weighted directed graph, called cost graph, to reflect whether or not a directed edge exists between each ordered pair of vertices of the RDF graph. Every pair of adjacent vertices of the RDF graph will be connected by a pair of inverse edges in the cost graph, weighted by either 0 or 1, where 0 indicates the existence of a corresponding directed edge in the RDF graph, or 1 otherwise.

Definition 2 (Cost Graph). *Given an RDF graph* $G = (V_G, E_G)$, *its* cost
graph *is an edge-weighted directed graph* $G_C = (V_C, E_C, w)$ *where*

- $V_C = V_G$ *is the vertex set,*
- E_C *is the edge set consisting of ordered pairs on* V_C *with* $(u, v) \in E_C$ *if and
  only if* $\exists p \in I$, $(u, p, v) \in E_G$ *or* $(v, p, u) \in E_G$, *i.e.,* G_C *contains a pair of
  inverse edges for every pair of adjacent vertices of* G, *and*
- $w : E_C \rightarrow \{0, 1\}$ *is an edge labeling function defined by:* $\forall e = (u, v) \in E_C$,
  $w(e) = 0$ *if and only if* $\exists p \in I, (u, p, v) \in E_G$, *i.e., 0 indicates the existence
  of a corresponding directed edge in* G, *or* $w(e) = 1$ *otherwise.*

The strong connectivity of G_C is ensured when G is connected.

For instance, Fig. 3a shows the cost graph of the RDF graph in Fig. 1a.

**Step 1.2: Finding a Spanning Arborescence of Minimum Weight in
the Cost Graph.** In the cost graph, we aim to find a spanning arborescence
of minimum weight with the given starting resource as its root. Recall that
an arborescence is just a rooted tree, and the weight of an arborescence in an
edge-weighted graph is defined as the sum of the weights of the edges in the
arborescence. A spanning arborescence of the cost graph $G_C = (V_C, E_C, w)$, as
a spanning subgraph of G_C, can be denoted by $T_C = (V_C, E'_C, w|_{E'_C}, s)$ where
$E'_C \subseteq E_C$, $w|_{E'_C}$ is the restriction of w to E'_C, and $s \in V_C$ is the given root. Note
that for any given root, the existence of a spanning arborescence of G_C is ensured
by the strong connectivity of G_C. A spanning arborescence of minimum weight
with a given root can be found by [7] with running time $O(|V_C| \log |V_C| + |E_C|)$.
Given the cost graph in Fig. 3a and `James_Cameron` as root, Fig. 3b illustrates
a spanning arborescence of minimum weight of 1.

**Step 1.3: Obtaining a Spanning Tree from the Spanning Arborescence
of Minimum Weight.** We will show a way of generating a spanning tree of
the RDF graph from a spanning arborescence (with root s) of its cost graph,
satisfying that the δ value of the tree representation of this spanning tree starting
from s is exactly equal to the weight of this spanning arborescence. Therefore,
the desired spanning tree to be transformed in Step 1 will be the one generated
from the spanning arborescence of minimum weight.

A spanning tree T of the RDF graph can be generated from a spanning arborescence T_C (with root s) of its cost graph according to Algorithm 1. In the tree
representation of T starting from s obtained according to Algorithm 2, reversed
edges exactly come from those generated by line 4 of Algorithm 1. Therefore, the
δ value of this tree representation is exactly equal to the weight of T_C. For instance, the weight of the spanning arborescence in Fig. 3b is equal to the δ value
of the tree representation of the spanning tree it generates in Fig. 2a (i.e. the
subgraph excluding dotted vertices and their associated edges).

3.3 Step 2: Transforming Remaining Edges

Having generated a partial tree representation of the entire RDF graph, in this
step, we complete the tree representation by transforming the remaining edges of

---

Algorithm 3. Completing a tree representation by transforming the remaining edges of an RDF graph.

Input   : The tree representation $TR = (V, E, L_V, L_E, r, f)$ of a spanning tree $T = (V_G, E'_G)$ of RDF graph $G = (V_G, E_G)$.
Output: TR as a tree representation of G.

1 Initialize V_0 to V;
2 **for** *each* $e = (u, p, v) \in (E_G \setminus E'_G)$ **do**
3 $\quad$ Find the vertex $u' \in V_0$ subject to $L_V(u') = u$;
4 $\quad$ Add a new vertex v' to V;
5 $\quad$ Define $L_V(v') = v$;
6 $\quad$ Add a new edge $e' = (u', v')$ to E;
7 $\quad$ Define $L_E(e') = p$ and $f(e') = e$;

---

the RDF graph. A key observation is that the transformation of a spanning tree gives rise to every vertex of the RDF graph labeling a vertex of the partial tree representation. As described by Algorithm 3, we can transform each remaining edge (u, p, v) of the RDF graph to extend the tree representation by adding to it a new vertex v' labeled with v and a new edge (u', v') labeled with p where u' is the vertex labeled with u in the partial tree representation. It can be verified that the result is a representation tree of the entire RDF graph, and no additional reversed edges are introduced in this step. For instance, in Fig. 2a, dotted vertices and their associated edges are transformed from the remaining edges in Fig. 1a.

4   Experiments

In this section, we will show the effectiveness of the proposed approach by empirically comparing the coherence of tree representations of real-world RDF graphs generated by our approach and two popular baseline approaches. Then we will present a preliminary user study to show how the coherence of tree representation influences the user's browsing experience. Finally we will test the running time of our approach in practice.

4.1   Test Cases and Approaches

Our experiments were carried out on a sample of the Billion Triples Challenge 2011 Dataset,[2] comprising 10,000 RDF documents (i.e. RDF graphs) that were randomly selected, with 10,790 connected components as test cases. In each case, a vertex having the largest outdegree was designated as the starting resource.

Three approaches were compared in the experiments.

– **BFS** generated a tree representation of an RDF graph by following a breadth-first search (beginning with the starting resource) of its undirected underlying graph, constantly adding an edge with a new vertex to the representation for each edge traversed in the search. This approach was adopted by

---

[2] http://km.aifb.kit.edu/projects/btc-2011/

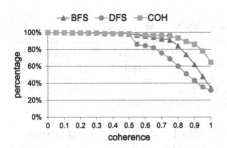

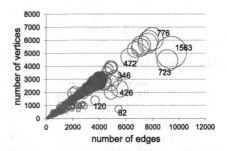

Fig. 4. Cumulative distributions of coherence of tree representations

Fig. 5. Running time (ms) of COH on each test case (bubble), in proportion to the area of bubble.

Falcons [3] for visualizing entity descriptions in search snippets. Figure 2b shows a tree representation of the graph in Fig. 1a generated by BFS.

- **DFS** was similar to BFS but followed depth-first search instead.
- **COH** generated tree representations according to Sect. 3.

4.2 An Empirical Analysis of Coherence

Figure 4 depicts the cumulative distributions of coherence of tree representations generated by different approaches for all the test cases. All the three approaches generated fairly coherent tree representations at the level of $coh \geq 0.5$ for more than 98% of the test cases. When raising the level to $coh \geq 0.8$, COH still achieved this for 93.42% of the test cases, whereas the percentages for BFS and DFS dropped significantly to 83.31% and 59.37%, respectively. Interestingly, COH generated tree representations without reversed edges (i.e. $coh = 1.0$) for 64.62% of the test cases (being almost twice as large as the percentages for BFS and DFS), indicating that reversed edges were unavoidable in the tree representations of approximately one third of the test cases.

To summarize, COH, which was designed to be the best possible approach to generating coherent tree representations, far outweighed the popular BFS and DFS in practice. BFS performed relatively well compared with DFS.

4.3 A Preliminary User Study

We carried out a user study to reveal the relationship between coherence and user experience. Eighteen university students having considerable knowledge of RDF were invited to carry out browsing tasks. In each task, the RDF graph in a random test case was visualized as an indented tree based on its tree representation generated by one of the three approaches. The subject was instructed to browse and then retell the information provided. Each subject carried out two tasks based on each of the three approaches. The subject responded to questions Q1 and Q2 in Table 1 after each task, and responded to Q3 after two

Table 1. Questions and Responses about User Experience

Question	Response: Mean (SD)			$F(2,34)$	LSD post-hoc
	BFS	DFS	COH	(p-value)	($p < 0.05$)
Q1: I can easily understand the information provided by this RDF graph.	3.42 (0.20)	3.03 (0.15)	**3.64** (0.17)	4.023 (0.027)	COH > DFS
Q2: I can easily learn the structure of this RDF graph.	3.33 (0.18)	2.97 (0.18)	**3.61** (0.20)	3.813 (0.032)	COH > DFS
Q3: Overall, I am satisfied with this way of RDF visualization.	3.61 (0.20)	2.28 (0.25)	**3.67** (0.16)	13.641 (0.000)	COH, BFS > DFS

consecutive tasks based on the same approach. Questions were responded using a five-point Likert item from 1 for strongly disagree to 5 for strongly agree.

Table 1 presents statistics about each subject's average responses to Q1 and Q2 and response to Q3 based on each approach. Repeated measures ANOVA (F and p) revealed that the differences in subjects' mean responses to the three questions were all statistically significant ($p < 0.05$). According to Q3, subjects were overall slightly more satisfied with COH than BFS, both of which significantly outweighed DFS as revealed by LSD post-hoc tests ($p < 0.05$). Similar results were observed on Q1 and Q2, capturing subjects' understanding of information and learning of graph structure, respectively. Specifically, COH was considered more helpful than BFS and DFS, although LSD post-hoc tests ($p < 0.05$) only saw significant differences between COH and DFS.

It was noticeable that the coherence ranking of the three approaches (i.e. COH>BFS>DFS) given by Fig. 4 agreed with the response ranking given by Table 1. It would be interesting to further investigate the correlation between coherence and responses about user experience. We calculated the Spearman's rank correlation coefficient (ρ) between the coherence of tree representations provided to each subject and the responses made. On Q1 and Q2, positive ρ values were observed on 15 (83.33%) and 12 (66.67%) subjects, respectively; and average ρ values were 0.26 and 0.21, respectively, indicating that coherence and user experience were notably dependent.

4.4   A Performance Test

We tested the performance of our approach on an Intel Xeon E3-1225 v2 with 2GB memory for JVM. The average running time on all the test cases was 24 ms. Figure 5 shows the running time on each test case. The maximum running time needed for an RDF graph with not more than 2,000, 4,000, 6,000, 8,000, and 10,000 RDF triples was 90, 295, 426, 776, and 1,563 ms, respectively, being reasonably fast.

5   Related Work

Popular domain-independent ways of visualizing an RDF graph (or more generally, graph-structured data) include node-link diagram, indented tree, and adjacency matrix. *Node-link diagram* straightforwardly presents a graph to the user as it is represented conceptually, but is problematic in terms of usability particularly when the graph has more than a manageable number of vertices or a complex structure [6,10]. *Indented tree* has proven to be more effective in such cases [6]. In fact, people are probably more familiar with tree-like environment because a wide variety of types of data in the world have been organized in this way. Thanks to its focus on a given root vertex to start with, indented tree particularly suits entity browsing, and has been adopted by entity browsers like Tabulator [2] and entity search engines like Falcons [3]. *Adjacency matrix* also suits large graphs better than node-link diagram [9], and it helps to identify certain patterns of data [8].

Compared with node-link diagram and adjacency matrix, indented tree has received relatively not much research attention. However, it is worth noting that a cyclic RDF graph can be transformed into *different tree representations* and thus be visualized as different indented trees. For instance, in Falcons [3], a tree representation is obtained by traversing the edges of the RDF graph in a breadth-first manner, whereas in this paper, we address the coherence of tree representations and propose to generate one that has the fewest reversed edges. We have shown in the experiments that *different tree representations lead to a significance difference in user experience, which considerably depends on coherence.* Besides, we are aware of some research that transforms an RDF graph into a tree or forest for indexing purposes [1]. However, coherence is not considered in that work because it is motivated by a different application scenario.

There are also many domain-specific visualizations [4,11], which however are out of the scope of this paper.

6   Conclusion and Future Work

We have presented a formalized analysis of indented tree visualization of RDF graphs from the coherence aspect. An indented tree visualization based on a coherent tree representation of RDF graph is expected to smooth the user's reading experience and reduce confusion and interruption to the user's cognitive process caused by unfriendly, reversed presentation of RDF triples. Our preliminary user study has shown a notable dependence between coherence and user experience. Specifically, coherent indented tree visualization is beneficial to the understanding of information as well as the learning of graph structure, leading to improved user satisfaction. However, it would be interesting to explore other aspects of tree representation (e.g. depth, branching factors) that may influence the user's browsing experience, which will be our future work.

We have proposed an approach to generating optimal tree representations in terms of coherence. In fact, the approach has been designed to handle not only RDF graphs but graphs in general, and we will apply it to other types of graph-structured data in future work. As to RDF data, we plan to extend our approach by exploiting the semantics of OWL (e.g. inverse properties) so that more reversed edges may be avoided.

Acknowledgments. The authors thank Jidong Jiang, Chenxi Qiu, and the anonymous reviewers. This work was supported in part by the NSFC under Grant 61170068 and 61223003, and in part by the JSNSF under Grant BK2012723.

References

1. Bartů, S.: Designing Indexing Structure for Discovering Relationships in RDF Graphs. In: Dateso 2004 Annual International Workshop on DAtabases, TExts, Specifications and Objects, pp. 7–17. CEUR-WS.org (2004)
2. Berners-Lee, T., Chen, Y., Chilton, L., Connolly, D., Dhanaraj, R., Hollenbach, J., Lerer, A., Sheets, D.: Tabulator: Exploring and Analyzing Linked Data on the Semantic Web. In: 3rd International Semantic Web User Interaction Workshop (2006)
3. Cheng, G., Qu, Y.: Searching Linked Objects with Falcons: Approach, Implementation and Evaluation. Int'l J. Semant. Web Inf. Syst. 5(3), 49–70 (2009)
4. Dadzie, A.-S., Rowe, M.: Approaches to Visualising Linked Data: A Survey. Semant. Web J. 2, 89–124 (2011)
5. Deligiannidis, L., Kochut, K.J., Sheth, A.P.: RDF Data Exploration and Visualization. In: 1st Workshop on CyberInfrastructure: Information Management in eScience, pp. 39–46. ACM, New York (2007)
6. Fu, B., Noy, N.F., Storey, M.-A.: Indented Tree or Graph? A Usability Study of Ontology Visualization Techniques in the Context of Class Mapping Evaluation. In: Alani, H., Kagal, L., Fokoue, A., Groth, P., Biemann, C., Parreira, J.X., Aroyo, L., Noy, N., Welty, C., Janowicz, K. (eds.) ISWC 2013, Part I. LNCS, vol. 8218, pp. 117–134. Springer, Heidelberg (2013)
7. Gabow, H.N., Galil, Z., Spencer, T., Tarjan, R.E.: Efficient Algorithms for Finding Minimum Spanning Trees in Undirected and Directed Graphs. Combinatorica 6(2), 109–122 (1986)
8. Gallego, M.A., Fernández, J.D., Martínez-Prieto, M.A., de la Fuente, P.: RDF Visualization using a Three-Dimensional Adjacency Matrix. In: 4th International Semantic Search Workshop (2011)
9. Ghoniem, M., Fekete, J.-D., Castagliola, P.: On the Readability of Graphs Using Node-link and Matrix-based Representations: A Controlled Experiment and Statistical Analysis. Inf. Vis. 4, 114–135 (2005)
10. Karger, D., schraefel, m.c.: The Pathetic Fallacy of RDF. In: 3rd International Semantic Web User Interaction Workshop (2006)
11. Katifori, A., Halatsis, C., Lepouras, G., Vassilakis, C., Giannopoulou, E.: Ontology Visualization Methods — A Survey. ACM Comput. Surv. 39(4), 10 (2007)
12. Marie, N., Gandon, F.: Survey of Linked Data Based Exploration Systems. In: 3rd International Workshop on Intelligent Exploration of Semantic Data (2014)
13. Oren, E., Delbru, R., Catasta, M., Cyganiak, R., Stenzhorn, H., Tummarello, G.: Sindice.com: A Document-oriented Lookup Index for Open Linked Data. Int'l J. Metadata Semant. Ontologies 3(1), 37–52 (2008)

Online Feature Selection Based on Passive-Aggressive Algorithm with Retaining Features

Hai-Tao Zheng and Haiyang Zhang

Graduate School at Shenzhen, Tsinghua University
Tsinghua Campus, The University Town,
Shenzhen 518055, P.R. China

Abstract. Feature selection is an important topic in data mining and machine learning, and has been extensively studied in many literature. Unlike traditional batch learning methods, online learning is more efficient for real-world applications. Most existing studies of online learning require accessing all the features of training instances, but in real world, it is often expensive to acquire the full set of attributes. In online feature selection process, when a training instance arrive, a fixed small number of features will be selected, and then the other features will be ignored. However, those ignored features may be useful and selected in later instances. If we only consider the new instances for these special features, it will lead to extreme errors. To address these issues, we improved a novel algorithm with Passive-Aggressive Algorithm and retaining features. Then we evaluate the performance of the proposed algorithms for online feature selection on several public datasets, and we can see from the experiments that our algorithm consistently surpassed the baseline algorithms for all the situations.

Keywords: Feature Selection, Online Learning, Binary Classification.

1 Introduction

Feature selection is an important topic in data mining and machine learning. It has been extensively studied in many literature [11, 12, 13]. In classification problems, it is the process of selecting a subset of the terms occurring in the training set and using only this subset as features. By removing irrelevant and redundant features, feature selection can alleviate the effect of the curse of dimensionality, improve the generalization performance, accelerate the learning process, and enhance the interpretive performance of the model, finally improve the performance of prediction models.

Most existing studies of feature selection are restricted to batch learning, which assumes the feature selection task is conducted in an off-line learning fashion and all the features of training instances are given a priori. But in real-world applications, training examples often arrive in a sequential manner, and the full information of training data is sometimes expensive to collect.

An online classifiers should involve only a small and fixed number of features for classification. It is particularly important and necessary when a real-world application

R. Cheng et al. (Eds.): APWeb 2015, LNCS 9313, pp. 707–719, 2015.
DOI: 10.1007/978-3-319-25255-1_58

has to deal with sequential training data of high dimensionality. There are two different types of online feature selection tasks: online feature selection (OFS) when the learner can access all the features of training instances, and OFS when the learner is only allowed to access a fixed small number of features for each training instance to identify the subset of relevant features. In the two tasks, the goal of OFS is to efficiently identify a fixed number of relevant features for accurate prediction. And for the second task, it is a more challenging scenario than the first task.

In OFS process, when a training instance arrive, a fixed small number of features will be selected, and then the other features will be ignored. But those ignored features may be useful and selected in later instance. If we only consider the new instance for these special features, it will lead to extreme errors. So these features should be recovered rather than ignored.

In this paper, we utilize Passive-Aggressive Algorithm to make feature selection online rather than offline, and we propose an algorithm to achieve OFS with partial inputs. We retain the features those have been ignored to improve accuracy and reduce the standard deviation. Finally we evaluate the performance of the proposed algorithms.

The main contribution of this paper is summarized as follows: (1) we propose an algorithm named Online Feature Selection based on Passive-Aggressive Algorithm with Retaining Features (RFOFS) to achieve effective online feature selection with full/partial inputs; (2) we validate the effectiveness of RFOFS by conducting an extensive set of experiments.

The rest of this paper is organized as follows. Section 2 reviews related work. Section 3 presents the problem and the proposed algorithms. Section 4 discusses our empirical studies and Section 5 concludes this work.

2   Related Work

Recently, online machine learning has become a hot research topic due to the efficiency of the algorithms and effectiveness in real-world applications. A classical online learning method is the well-known Perceptron algorithm [3, 4], which updates the model by adding a new example with some constant weight into the current set of support vectors when the incoming example is misclassified. Recently, a lot of new online learning algorithms have been proposed, in which many of them usually follow the criterion of maximum margin learning principle [5, 1, 7], for example, the Passive-Aggressive algorithm [1]. It updates a classifier that is near to the previous function while suffering less loss on the current instance. The PA algorithm is limited in that it only exploits the first order information, which has been addressed by the recently proposed confidence weighted online learning algorithms that exploit the second order information [6, 9, 8]. Most studies of online learning require the access to all the features of training instances. When the number of active features is small and fixed, the online learning problem becomes more challenging [10]. Online learning algorithms are very promising in real-world applications, especially for training large-scale datasets and data being incrementally added.

Feature selection (FS) has been the focus of interest for quite some time and has been studied extensively in the literatures of data mining and machine learning [14, 11]. The existing FS algorithms generally can be grouped into three categories: supervised, unsupervised, and semi-supervised FS. Supervised FS [15~17] selects features according to labeled training data. And when there is no label information available, unsupervised feature selection [18~20] attempts to select the important features which preserve the original data similarity or manifold structures. Semi-supervised feature selection methods [21~23], as its name says, exploit both labeled and unlabeled data information. The existing supervised FS methods can be further divided into three groups, depending on how they combine the feature selection search with the construction of the classification model: Filter methods, Wrapper methods, and Embedded methods approaches. Filter methods [10, 15, 13] choose important features by measuring the correlation between individual features and output class labels, without involving any learning algorithm; wrapper methods [16] rely on a predetermined learning algorithm to decide a subset of important features. Although wrapper methods generally tend to outperform filter methods, they are usually more computationally expensive than the filter methods. Embedded methods [17] aim to integrate the feature selection process into the model training process. They are usually faster than the wrapper methods and able to provide suitable feature subset for the learning algorithm. For unsupervised feature selection, some representative works include Laplacian Score [18], Spectral Feature Selection [19], and the recently proposed $\ell 2$-Norm Regularized Discriminative Feature Selection [20]. Feature selection has found many applications [14], including bioinformatics, text analysis and image annotation. Our OFS technique generally belongs to supervised FS.

We note that it is important to distinguish Online Feature Selection addressed in this work from the previous studies of online streaming feature selection [24]. In those works, features are assumed to arrive one at a time while all the training instances are assumed to be available before the learning process starts. This differs significantly from our online learning setting where training instances arrive sequentially, a more natural scenario in real-world applications.

3   Online Feature Selection Based on Passive-Aggresive Algorithm

3.1   Problem Setting

As defined in paper [2], we also consider the problem of online feature selection for binary classification in this paper. We denote the instance presented to the algorithm on round t by x_t, and each $x_t \in R^d$ is a vector of d dimension. We assume that x_t is associated with a unique label $y_t \in \{-1, +1\}$. The input pair(x_t, y_t) is received over the trials. In this problem, we are going to select a relatively small number of features from the d features (we assume d is a large number) for efficient linear classification. In each round t, the learner presents a classifier $w_t \in R^d$ which will be calculated as the linear function $\text{sgn}(w_t^T x_t)$ and used to classify instance x_t. We require the classifier w_t to have at most B non-zero elements, i.e.

$$\|w_t\|_0 \le B \tag{1}$$

where B>0 is a predefined constant, and consequently at most B features of x_t will be used for classification. We refer to this problem as online feature selection. Our goal is to design an effective and stable strategy for online feature selection that is able to make a small number of mistakes and a small standard deviation.

3.2   Baseline Methods

A straightforward approach to online feature selection is to modify the Perceptron algorithm by applying truncation. Specifically, in the t-th trial, truncating the classifier w_t by setting everything but the B largest (absolute value) elements in w_t to be zero. This truncated classifier is then used to classify the received instance. When the instance is misclassified, updating the classifier by adding the vector $y_t \, x_t$ where (x_t, y_t) is the misclassified training example. Although this simple algorithm selects the B largest elements for prediction, it does not guarantee that the numerical values for the unselected attributes are sufficiently small, which could potentially lead to many classification mistakes.

PA Algorithm avoid this problem by updating the classifier not only when the instance is misclassified, but also when suffering small loss (Algorithm 1). On rounds where algorithm attains a margin less than 1, the classifier will be updated and truncated.

Jialei Wang et.al proposed another efficient algorithm [2]. It also avoid this problem by exploring the sparsity property of L1 norm. And they gives the mistake bound of the Algorithm and proof.

Algorithm 1. Simple OFS via PA Algorithm

```
Input
   B: the number of selected features
   u: aggressiveness parameter
    : step size
Initialization
       w₁ = 0
for t = 1,2,,,,T do
       Receive xₜ
       Make prediction sgn(wₜᵀxₜ)
       Receive yₜ
       if yₜwₜᵀxₜ ≤ 1 then
```
$$\eta = \min\left(u, \frac{1 - y_t w_t^T x_t}{\|x_t\|_2}\right)$$
$$\tilde{w}_{t+1} = w_t + \eta y_t x_t$$
$$w_{t+1} = \text{Truncate}(\tilde{w}_{t+1}, B)$$
```
       else
```
$$w_{t+1} = w_t$$
```
       end if
   end for
end for
```

```
Algorithm 2. Truncate Algorithm
function w = Truncate(ŵ, B)
if ‖ŵ‖₀ > B then
    w = ŵᴮ where ŵᴮ is ŵ with everything but the B largest elements
set to zero.
else
    w = ŵ
end if
```

Unfortunately, several problems exist in these approaches. After the online feature selection process, some features will be ignored. However, when the new training instances arrive, the features which have been ignored may be selected, then only the new instance was considered for these features. This bias can lead to extreme errors. To see this problem, consider the case where the $w_t=(1,0,0,0,1)$, and the input pattern $x_t=(0,1,1,0,0)$. It causes a classification mistake, and then updates the w by x. But we can see the 2 to 4 elements in w_{t+1} are only depending on x_t. This is irrational, because online learning classifier w_t should associate with all the input patterns before.

3.3   RFOFS with Full Inputs

In this task, we assume the learner is provided with full inputs of every training instance (i.e. $x_1, \ldots, x_T$). We first present a state-of-art OFS algorithm and OFS with PA Algorithm as our baseline.

To solve the problem, we hold two classifiers. The one truncated is used for feature selection and making prediction. The other one named as w_all_t is used for updating when misclassified, and it saved the features that have been ignored.

Based on these ideas, we present a new approach for OFS (Online Feature Selection based on Passive-Aggressive Algorithm with Retaining Features, RFOFS) as shown in Algorithm 4. The online learner maintains a linear classifier w_t that has at most B non-zero elements for make prediction and a classifier w_all_t for updating classifiers and preserving ignored features.

When a training instance (x_t, y_t) reached, we make prediction by a linear tion sgn$(w_t^T x_t)$. If the instance suffering a loss, the classifier for updating is updated by online gradient descent. Then the classifier is projected to a L2 ball to ensure that the norm of the classifier is bounded. At last the classifier is truncated to generate the classifier which is used for feature selection.

Algorithm 3. RFOFS with Full Inputs

```
Input
   u: aggressiveness parameter
   B: the number of selected features
    : regularization parameter
    : step size
```

```
Initialization
```
$$w_1 = 0$$
$$w\_all_1 = 0$$
```
for t = 1,2,,,,T do
```
   Receive x_t

   Make prediction $\text{sgn}(w_t^T x_t)$

   Receive y_t

   if $y_t w_t^T x_t \leq 1$ then

   $\eta = \min\left(u, \frac{1 - y_t w_t^T x_t}{\|x_t\|_2}\right)$ //compute according to PA Algorithm

   $w\_all_{t+1} = w\_all_t + \eta y_t x_t$ //update

   $w\_all_{t+1} = \min\left\{1, \frac{\frac{1}{\sqrt{\lambda}}}{\|w\_all_{t+1}\|_2}\right\} w\_all_{t+1}$ //project to a L2 ball

   $w_{t+1} = \text{Truncate}(w\_all_{t+1}, B)$ //truncate

   else

   $w_{t+1} = w_t$

   end if
```
end for
```

3.4   RFOFS with Partial Inputs

In the above discussion, although the classifier w only consists of B non-zero elements, it requires the full knowledge of the instances. We can further constrain the problem of online feature selection by requiring no more than B attributes of x_t when soliciting input patterns. We note that this may be important for a number of applications when the attributes of objects are expensive to acquire.

We utilize a ε-greedy online feature selection approach with partial input information by employing a classical technique for making tradeoff between exploration and exploitation. In this approach, we spend ε of trials for exploration by randomly choosing B attributes from all d attributes, and the remaining 1−ε trials on exploitation by choosing the B attributes for which classifier has non-zero values. When a training instance (x_t, y_t) arrived, we make prediction by a linear function sgn($w_t^T x_t$). If the instance suffering a loss, the classifier for updating is updated by online gradient descent. Then the classifier is projected to a L2 ball to ensure that the norm of the classifier is bounded. At last the classifier is truncated to generate the classifier which is used for feature selection.

4   Experiment

In this section, we conduct an extensive set of experiments to evaluate the performance of the proposed online feature selection algorithms. We will evaluate the online predictive performance of the two OFS tasks on several benchmark datasets from UCI machine learning repository.

4.1   Datasets

We test the proposed algorithms on a number of publicly available benchmarking datasets. All of the datasets can be downloaded either from LIBSVM website[1] or UCI machine learning repository[2]. Besides the UCI data sets, we also adopt two high-dimensional real text classification datasets based on the bag-of-words representation: (i) the Reuters Corpus Volume 1 (RCV1)[3]; (ii) 20 Newsgroups datasets[4], we extract the "comp" versus "sci" and "rec" versus "sci" to form two binary classification tasks.

Table 1 shows the statistics of the datasets used in our following experiments. Each sample is a d+1 dimension vector $(y_t, x_t[1], x_t[2], ..., x_t[d])$.

Table 1. List of datasets in our experiments

Dataset	#Samples	# Dimensions
. magic04	19020	10
german	1000	24
splice	3175	60
spambase	4601	57
a8a	32561	123
svmguide3	1243	21
rcv1	20242	47236
20Newsgroup("rec"vs"sci")	3161	61188
20Newsgroup("comp"vs"sci")	3518	61188

4.2   Experimental Setup and Baseline Algorithms

We compare our proposed OFS algorithm against the following three baselines:

- The modified perceptron by the simple truncation step, denoted as "PE" for short;[4]
- A randomized feature selection algorithm, which randomly selects a fixed number of active features in an online learning task, denoted as "RAND" for short;

---

[1] http://www.csie.ntu.edu.tw/~cjlin/libsvmtools/
[2] http://www.ics.uci.edu/~mlearn/MLRepository.html
[3] http://www.datatang.com/data/28793
[4] http://qwone.com/~jason/20Newsgroups/

- OFS via PA Algorithm shown in Algorithm 1, denoted as "OPA" for short; [1]
- OFS via Sparse Projection proposed in paper [2] shown in Algorithm 3, denoted as "OFS" for short.[2]

To make a fair comparison, all algorithms adopt the same experimental settings. All the experiments were run over 20 times, each time with a random permutation of a dataset. OFS with full inputs in experiment 1 was evaluated against the RAND, OPA and OFS. OFS with partial inputs in experiment 2 was evaluated against the RAND, PE and OFS. Our experiment was run in matlab.

4.3   Experiment 1: RFOFS with Full Inputs

Table 2 summarizes the online predictive performance of the compared algorithms with a fixed fraction of selected features on the datasets.

Table 2. Mistakes of algorithms with full inputs

Algorithm	magic04	german	splice
RAND	8686±84.7	473± 18.8	1514±26.8
OPA	5960±1326.7	446±80.3	742±78.0
OFS	6023±1342.3	449±82.7	735±68.3
RFOFS	**5027±3.5**	**361±7.6**	**680±11.4**
Algorithm	spambase	a8a	svmguide
RAND	1831±38.0	15649±77.3	561±17.6
OPA	906±157.1	8657±1246.8	399±66.4
OFS	913±157.8	9424±2545	400±66.8
RFOFS	**712±14.4**	**8847±50.2**	**374±16.2**
Algorithm	rcv1	"rec"vs"sci"	"comp"vs"sci"
RAND	19961±18.8	2977±18.8	3300±18. 24
OPA	7656±1043.3	840±30.8	980±37.9
OFS	7325±974.4	1393±30.2	1591±30.9
RFOFS	**1746±34.3**	**754±28.5**	**893±24.9**

Several observations can be drawn from the results. First of all, we found that among all the compared algorithms, the RAND algorithm has the highest mistake rate for all the cases, and the simple OPA algorithm can outperform the RAND algorithm considerably, which shows that it is important to learn the active features in an OFS task. Second, we found the Jialei Wang's OFS algorithm's mistake rate is similar to the simple OPA algorithm, they both achieved a small mistake rate, but their standard deviation is far more than the RAND algorithms. This indicates the two algorithm is both not stable. Finally, our algorithm achieved the smallest mistake rate, and our algorithm's standard deviation is also the smallest in these four algorithms. This shows that the proposed algorithm is able to considerably improve the performance of the feature selection approach.

To further examine the online predictive performance, Figure 1(a~i) shows how the mistake rates is varied over iterations accord the entire OFS process on the chosen datasets. Similar to the previous observations, we can see that our OFS algorithm consistently surpassed the other three algorithms for all the situations. Besides, we also found that the more the training instances received, the more significant the gain achieved by our OFS algorithm over the other baselines. This is because that the more the training instances received, the more features were ignored in the OPA and OFS algorithm, then the more mistakes occurred in those algorithms. This again verifies the efficacy of our OFS algorithm.

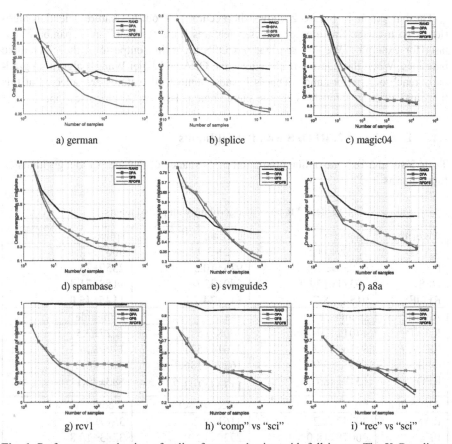

a) german   b) splice   c) magic04

d) spambase   e) svmguide3   f) a8a

g) rcv1   h) "comp" vs "sci"   i) "rec" vs "sci"

Fig. 1. Performance evaluation of online feature selection with full inputs. The X-Coordinate stands for the number of training samples while the Y-Coordinate stands for the online average rate of mistakes.

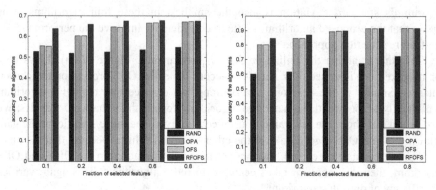

Fig. 2. Online Classification Accuracy with various fractions of selected features

The bar graph in Figure 2 shows the accuracy of online feature selection algorithms with varied fractions of selected features (take 'german' and 'spambase' dataset for example). We can see that our OFS algorithm outperform the other three baselines. When the fraction is close to 1, the OPA and Jialei Wang's algorithm gradually reach our performance. This is because almost all features are selected at this time, and all these algorithms use the same linear classifiers for classification.

4.4   Experiment 2: RFOFS with Partial Inputs

Table 3 summarizes the online prediction performance of the compared algorithms on four datasets.

Table 3. Mistakes of algorithms with partial inputs

Algorithm	magic04	german	splice
RAND	18909±11.5	992± 3.2	3065±12.1
PE	17938±26.4	923±10.5	2359±25.5
OFS	10311±179.0	580±33.6	1158±71.4
RFOFS	**9158±78.7**	**517±24.7**	**1145±45.0**
Algorithm	spambase	a8a	svmguide
RAND	4401±17.6	30083±54.3	1227±2.5
PE	3276±43.2	21614±521	1109±14.3
OFS	1831±65.9	16274±136.2	713±46.2
RFOFS	**1319±48.6**	**11734±103.2**	**583±40.1**
Algorithm	rcv1	"rec"vs"sci"	"comp"vs"sci"
RAND	20242±1.2	3161±0.2	3517±0.2
PE	20078±18.0	2930±12.4	3283±15.8
OFS	9056±164.5	1864±26.7	2096±38.5
RFOFS	**6753±160.3**	**1692±25.3**	**2020±35.0**

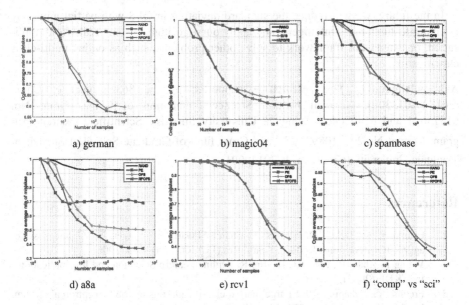

Fig. 3. Performance evaluation of online feature selection with partial inputs. The X-Coordinate stands for the number of training samples while the Y-Coordinate stands for the online average rate of mistakes.

Several observations can be drawn from the results. First, we found that the RAND algorithm suffered the highest mistake rate for all cases. This again shows that it is important to learn the active features for the inputs and the weight vector. Second, we found that the proposed RFOFS algorithms achieved the smallest mistake rates. This shows that the proposed RFOFS technique is effective for learning the most informative features under the partial input situation.

To further examine the online predictive performance, Figure 4(a~f) shows how the mistake rates is varied over iterations accord the entire OFS process on the chosen datasets. Similar to the previous observations, we can see that the proposed RFOFS algorithm consistently surpassed the other three algorithms for all the situations. Besides, we also found that the more the training instances received, the more significant gains achieved by the proposed RFOFS algorithm over the other baselines. This again verifies the efficacy of the proposed RFOFS algorithm.

5   Conclusion and Future Work

In this paper, we utilized Passive-Aggressive Algorithm and OFS Algorithm to make feature selection online rather than offline. We retained the features that have been ignored to improve accuracy and reduce the standard deviation. Then we evaluate the performance of the proposed algorithms for online feature selection on several public datasets. From the experiment results, we can see our algorithm is significantly more efficient and stable than the other algorithms.

In the future, we will reduce the standard deviation in OFS with partial inputs by retaining features more intelligent, and apply our algorithm to some real-world application, or extend our framework to solve other problems, such as online multi-class classification.

Acknowledgement. This research is supported by the 863 project of China (2013AA013300), National Natural Science Foundation of China (Grant No. 61375054 and 61402045), Tsinghua University Initiative Scientific Research Program (Grant No.20131089256) , and Cross fund of Graduate School at Shenzhen, Tsinghua University (Grant No. JC20140001).

References

1. Crammer, K., Dekel, O., Keshet, J., Shalev-Shwartz, S., Singer, Y.: Online passive-aggressive algorithms. J. Mach. Learn. Res (JMLR) 7, 551–585 (2006)
2. Wang, J., Zhao, P., Hoi, S.C.H., Jin, R.: Online Feature Selection and Its Applications. In: TKDE (2012)
3. Freund, Y., Schapire, R.E.: Large margin classification using the perceptron algorithm. Machine Learning 37(3), 277–296 (1999)
4. Rosenblatt, F.: The perceptron: A probabilistic model for information storage and organization in the brain. Psychological Review 65, 386–407 (1958)
5. Hoi, S.C.H., Wang, J., Zhao, P.: LIBOL: A Library for Online Learning Algorithms. Nanyang Technological University (2012)
6. Dredze, M., Crammer, K., Pereira, F.: Confidence-weighted linear classification. In: ICML, pp. 264–271 (2008)
7. Zhao, P., Hoi, S.C.H., Jin, R.: Double updating online learning. Journal of Machine Learning Research 12, 1587–1615 (2011)
8. Crammer, K., Dredze, M., Pereira, F.: Exact convex confidenceweighted learning. In: NIPS, pp. 345–352 (2008)
9. Crammer, K., Kulesza, A., Dredze, M.: Adaptive regularization of weight vectors. In: NIPS, pp. 414–422 (2009)
10. Bekkerman, R., El-Yaniv, R., Tishby, N., Winter, Y.: Distributional word clusters vs. words for text categorization. Journal of Machine Learning Research 3, 1183–1208 (2003)
11. Dash, M., Liu, H.: Feature selection for classification. Intell. Data Anal. 1(1-4), 131–156 (1997)
12. Saeys, Y., Inza, I., Larra ñaga, P.: A review of feature selection techniques in bioinformatics. Bioinformatics 23(19), 2507–2517 (2007)
13. Yu, L., Liu, H.: Feature selection for high-dimensional data: A fast correlation-based filter solution. In: ICML, pp. 856–863 (2003)
14. Guyon, I., Elisseeff, A.: An introduction to variable and feature selection. Journal of Machine Learning Research 3, 1157–1182 (2003)
15. Dash, M., Gopalkrishnan, V.: Distance based feature selection for clustering microarray data. In: Haritsa, J.R., Kotagiri, R., Pudi, V. (eds.) DASFAA 2008. LNCS, vol. 4947, pp. 512–519. Springer, Heidelberg (2008)
16. Kohavi, R., John, G.H.: Wrappers for feature subset selection. Artif. Intell. 97(1-2), 273–324 (1997)

17. Xu, Z., Jin, R., Ye, J., Lyu, M.R., King, I.: Non-monotonic feature selection. In: ICML, p. 144 (2009)
18. He, X., Cai, D., Niyogi, P.: Laplacian score for feature selection. In: NIPS (2005)
19. Zhao, Z., Liu, H.: Spectral feature selection for supervised and unsupervised learning. In: ICML, pp. 1151–1157 (2007)
20. Yang, Y., Shen, H.T., Ma, Z., Huang, Z., Zhou, X.: l2, 1-norm regularized discriminative feature selection for unsupervised learning. In: IJCAI, pp. 1589–1594 (2011)
21. Zhao, Z., Liu, H.: Semi-supervised feature selection via spectral analysis. In: SDM (2007)
22. Ren, J., Qiu, Z., Fan, W., Cheng, H., Yu, P.S.: Forward semi-supervised feature selection. In: Washio, T., Suzuki, E., Ting, K.M., Inokuchi, A. (eds.) PAKDD 2008. LNCS (LNAI), vol. 5012, pp. 970–976. Springer, Heidelberg (2008)
23. Xu, Z., King, I., Lyu, M.R., Jin, R.: Discriminative semi-supervised feature selection via manifold regularization. IEEE Transactions on Neural Networks 21(7), 1033–1047 (2010)
24. Wu, X., Yu, K., Wang, H., Ding, W.: Online streaming feature selection. In: ICML, pp. 1159–1166 (2010)
25. Hoi, S.C.H., Wang, J., Zhao, P., Jin, R.: Online Feature Selection for Mining Big Data. In: BigMine, pp. 93–100 (2012)

Online Personalized Recommendation Based on Streaming Implicit User Feedback*

Zhisheng Wang, Qi Li, Ye Liu, Wei Liu, and Jian Yin

School of Information Science and Technology, Sun Yat-sen University, China
issjyin@mail.sysu.edu.cn

Abstract. Since user preference is drifting over time, modeling temporary dynamic recommender system has been proven to be valuable for accurate recommendation performance. However, user feedback is continuously updating while the traditional recommender system is trained off-line in batch mode so that it cant capture user taste change in time. In this paper, we build a dynamic real-time recommendation model based on implicit user feedback stream to improve both the recommendation accuracy and training efficiency. Moreover, our model has obvious advantages over the traditional approaches in diversity, interpretability, and strong robustness to hostile attack. Finally, we conduct experiments on two real world datasets to validate the effectiveness of our proposed method and demonstrate the superior performance when compared with state-of-the-art approach.

1 Introduction

Recommender system, as a powerful tool for information filtering, has attracted a lot of interest during the past two decades. And a line of remarkable success have been made in recent years [2]. However, the era of Big Data presents some new challenges.

Recommendation Based on Implicit User Feedback. Traditional recommender systems used to be built based on explicit user feedback, e.g. ratings. However, implicit user feedback exists more widely, e.g. clicks, purchases, and various kinds of user-item interactions. Compared with ratings, these feedback can be collected more easily and inexpensively. There is a big short coming in using the standard recommendation formulation in this situation: the feedback we can observe either is positive or is missing. To tackle this problem, researchers have made great efforts to utilize this type of information and gained great achievements [7, 11]. Nevertheless, recommendation based on implicit feedback is still a new area which is valuable to be explore.

Real-Time Recommendation. Timeliness becomes more and more important in this fast changing world. Recommending items in time can greatly improve

---

* This work is supported by the National Natural Science Foundation of China (61033010, 61272065), Natural Science Foundation of Guangdong Province (S2011020001182, S2012010009311), Research Foundation of Science and Technology Plan Project in Guangdong Province (2011B040200007, 2012A010701013).

R. Cheng et al. (Eds.): APWeb 2015, LNCS 9313, pp. 720–731, 2015.
DOI: 10.1007/978-3-319-25255-1_59

satisfaction and effectiveness. Although existing dynamic models [4,8] are able to generate high quality recommendations by incorporating time factor. However, coping with fast changing trends in the presence of large scale data might be a challenge, since retraining such batch-training models is costly [6]. Moreover, the batch-training models can not utilize the latest information (e.g. the user feedback in the last few hours), which are valuable for reflecting user's current need. One alternative is to learn the recommendation model online, updating the parameters for each new observation. This calls for an online learning solution.

To address the two aforementioned challenges, this paper focuses on how to efficiently capture user dynamic preference from streaming implicit feedback over time. The main contributions of this work can be concluded as follows:

- We propose the implicit feedback recommendation model, which flexibly integrates implicit user feedback, item features, and temporal popularity in a principled manner by modeling users' personal sensitivity and conformity.
- To increase training efficiency and provide real-time recommendation results, we further propose the online implicit feedback recommendation model, namly oIFRM. By comparing user adoption probability and user confidence, oIFRM divide the collected feedback into three kinds: valuable feedback, habitual feedback, and noises. And then by adjusting the learning step for each feedback automatically, the model can not only capture user preference change but also reduce the impact of noises.
- Comprehensive experiments are conducted to validate the performance of the proposed methods. The online model outperforms the state-of-the-art RMFX [6] in both efficiency and effectiveness. Meanwhile, our proposed model has advantages on recommendation diversity, interpretability and robustness.

2 Related Work

Recommendation Based on Implicit Feedback. Regarding the observed interactions as positive samples is the common view in the implicit feedback scenarios. However, the missing ones which contain the interactions in the future can not be considered as negative samples directly. This problem was defined as One Class Collaborative Filtering ($OCCF$) in [10] and Unbalanced Class Problem in [2]. An intuitive idea is to introduce negative samples from missing data. Existing approaches can be broadly classified into two categories: those that randomly sample negative samples from the missing data [3,10], and those that treat all the missing data as negative samples with adding weights on them [10]. However, anyway it is inevitable to introduce noise (cast unobserved positive samples as negative samples) while taking advantage of the "unrated" information. 0-1MF [9] and Collaborative Ranking [3,11] are two leading models to leverage "unrated" information. 0-1MF defines the positive sample $R_{ij} = 1$ if user i has interacted with the item j, otherwise the negative sample $R_{ij} = 0$. Then the problem is converted to predict the unobserved entries based on the observed "ratings". Collaborative Ranking is the pairwise model which aims to find the correct ranking order by converting the rank task to maximize the probability that positive samples rank before randomly sampled "negative" ones [11].

Real-Time Recommendation. Extended from traditional matrix factorization model, dynamic recommendation models like timeSVD++ [8] and DMF [4] can improve recommendation accuracy by incorporating time factor. However, these dynamic models have limitations on both effectiveness and training efficiency since they are trained off-line in batch mode. Online Collaborative Filtering (OCF) [1] techniques have been explored in recent years. By applying the simple online gradient descent (OGD) algorithms, OCF avoids the highly expensive re-training cost of traditional batch matrix factorization algorithms. Based on pairwise ranking methodology, RMFX [6] follows a selective sampling strategy to perform online model updates based on active learning principles.

3 Model

In this section, we first present IFRM, which is tailor-made for implicit feedback recommendation scenarios. And then, we extend IFRM to an online version.

3.1 Implicit Feedback Recommendation Model

Following the common assumption in Collaborative Filtering, We also believe that the degree that how a user prefer an item can be measured. In the implicit feedback recommendation scenarios, we denote this "degree" as *Adoption Tendency Score* (ATS) which characterizes the tendency for a user to adopt (interact with) an items. However, such score itself is meaningless, since the users comparatively select items in practice. For example, for a given user i and item j, the tendency score that i adopts j, which is denoted $A_{i,j}$, is 5. It is hard to say i would adopt j or not. For user i, the score to item j is useful only by comparing with the scores of other items. Generally speaking, if $A_{i,j}$ is obviously larger than most ATS with respect to user i, user i may adopt item j with a high probability. Following this idea, for a common user, we assume that the behavior that he or she interacts with a target item is determined by *Relative Adoption Tendency.*

Definition 1. *The Relative Adoption Tendency* $\Delta_{ij} = \frac{A_{i,j}}{\overline{A_i}}$ *reflects the relative tendency for user i to adopt the item j, where $A_{i,j}$ is the ATS for user i to adopt item j, and $\overline{A_i}$ is the average ATS with respect to user i, which also serves as "negative prototype". The higher Δ_{ij} is, the higher probability for user i to adopt item j is.*

$A_{i,j}$ can be regarded as a function with respect to user i and item j. It can be designed differently according to different assumptions or application scenarios. Before develop it carefully, we will first present the overview of IFRM.

Inspired by [11], to transform Δ_{ij} to a probability, we normalize it to $[0,1]$ interval by utilizing sigmiod function $\phi(x) = \frac{x}{1+x}$. The probability that user i adopts item j is:

$$P_{ij} = \phi(\Delta_{ij}) = \frac{1}{1 + \Delta_{ij}^{-1}} \qquad (1)$$

Note that if $\Delta_{ij} = 1$, $P_{ij} = 0.5$ which indicates that user i shows the same adoption tendency with respect to item j compared with the most "normal" items. This property is useful to update model online (see the following subsection).

We denote $O = \{< i,j > | user\ i\ adopted\ item\ j\}$ which is the observed adoption set. By combining all the observed adoptions, we derive the likelihood of observed O as:

$$P(O|\Theta) = \prod_{<i,j>\in O} P_{ij} = \prod_{<i,j>\in O} \frac{1}{1 + \Delta_{ij}^{-1}} \qquad (2)$$

where Θ is the parameters in the model.

The Bayesian formulation of finding appropriate ATS is to maximize the following posterior probability:

$$P(\Theta|O) \propto P(O|\Theta)P(\Theta) = \prod_{<i,j>\in O} \frac{1}{1 + \Delta_{ij}^{-1}} \mathcal{N}(0, \Sigma_\Theta) \qquad (3)$$

where $P(\Theta)$ is a normal distribution with zero mean and variance-covariance matrix Σ_Θ.

The log of the total probability over the observations is given as follows:

$$lnP(\Theta|O) = ln \prod_{<i,j>\in O} \frac{1}{1 + \Delta_{ij}^{-1}} \mathcal{N}(0, \Sigma_\Theta) = -(\sum_{i,j} ln(1 + \Delta_{ij}^{-1}) + \lambda_\Theta \|\Theta\|^2)$$

To maximize the log-posterior over the observations is equivalent to minimize the proposed objective function in Eq.4 with hybrid quadratic regularization terms.

$$\underset{\Theta}{argmin} \sum_{<i,j>\in O} ln(1 + \Delta_{ij}^{-1}) + \lambda_\Theta \|\Theta\|^2 \qquad (4)$$

where λ_Θ are model specific regularization parameters. A local minimum of the objective function given by Eq.(4) can be found by performing gradient descent in model parameters Θ, which are iteratively updated.

In the rest of this subsection, we will introduce how to design *Adoption Tendency Score* $A_{i,j}$ in common and present a unified model which flexibly integrate various kinds of usual information. We intuitively design $A_{i,j}$ from three parts, including 1) the intrinsic preference which is determined by both user's latent features U_i and item's latent features V_j, 2) the influence of context which can be estimated by user's personalized sensitivity S_i and item's features F_j, and 3) the influence of social environment which can be modelled by the personal degree of user's conformity C_i and the popularity of item $Pop(j)$ in a specific social environment. Thus, $A_{i,j}$ can be formally defined as:

$$A_{i,j} = U_i \times V_j + S_i \times F_j + C_i \times Pop(j) \qquad (5)$$

For good generalization capability, we consider $Pop(j)$ in a global environment rather than a local social environment. Because the latter needs extract knowledge (e.g social relationship or social interactions) which is not always available. In this paper, we simply use *Temporal Popularity* of item j, which is denoted as $TP(j)$, to take the place of $Pop(j)$. $TP(j)$ can be estimated by counting the adoptions with respect to item j in a sliding time window.

Finally, we formally present the optimization objective of IFRM:

$$L := \underset{U,V,S}{argmin} \sum_{<i,j>\in O} ln(1 + \Delta_{ij}^{-1}) + \lambda_1(\|U\|^2 + \|V\|^2) + \lambda_2 \|S\|^2 \qquad (6)$$

From Eq.6 we can see that advioding sampling negatives, IFRM naturally address One-class problem. This property make it much more suitable for online learning, since each observation can be learnt separately.

3.2   Online Implicit Feedback Recommendation Model

Inspired by [1], we further extend IFRM to an online version by performing Online Gradient Descent(OGD) on loss function L with respect to model parameters U,V,and C. Different with standard Stochastic Gradient Descent (SGD), OGD processes samples one pass. Thus, the key issue of performing OGD is how to determine the learning step T for every sample, since OGD cant go back and learn again. State-of-the-art methods [1,6] adopt a fixed T for all samples. The drawbacks are obvious: 1)The parameter T needs to be set and tuned in practice. 2)There is not a reasonable setting of T for all feedback, since the value of user feedback is different. 3)The online model can be easily affected by noise since we equally treat all the feedback including noisy ones. Unfortunately, noise is inevitable in this setting: On one hand, implicit feedback data like clicks can be more noisy for those hostile attacks launched by a robot or crawler. On the other hand, it is hard to perform data cleaning on data stream online. To overcome the three drawbacks, we design an online learning mechanism which dynamically and automatically adjusts the learning rate for each feedback. By assuming the value of user feedback is different, oIFRM reinforces to learn the new trend while weakens to learn the habitual feedback and noise. By comparing user adoption probability and user behavior confidence, oIFRM can distinguish between valuable feedback(new trend), habitual feedback(well-learnt knowledge) and noisy feedback. To be specific, oIFRM reinforces to learn those feedback with lower occurrence probability and higher user confidence, weakens to learn those feedback with higher occurrence probability or with lower confidence. Note that the probabilistic framework of IFRM has provided the "user adoption probability". We develop a new component θ_i to describe the confidence of an observed behavior with respect to a user i.

The online learning algorithm of oIFRM is described as in Algorithm.1. At the very beginning, we finish the initialization work and perform a burn-in training process. The burn-in training process is designed to fulfill *Temporal Popularity* component and to make the online model available to give a prediction. In our experiment, the parameter W is set to 10^5. We utilize a simple method to adjust θ. For each user i, we maintain a pool to remember all the P_{ij} (Eq.1) in the past. by averaging all the P_{i*} in the pool, we can obtain the current θ_i. Note that the average ATS should be computed based on updated parameters. In practice, we can cache a few temporary variables(including average V, average F, and average popularity) for acceleration.

Algorithm 1. online Implicit Feedback Recommendation Model

Input: Implicit feedback stream, FS=$\{\cdots (User_i, Item_j)\cdots\}$; Normalized item feature vectors F; The dimension of latent factor D; The length of sliding window W; The regularization parameters for U, V and S, λ_1, λ_2; Learning Rate α; Initialization of Confidence θ;

Output: A series of real-time recommendations;

1: Initialization: Randomly construct U, V, S, load F, and construct initial confidence θ for each user

2: Normalize U, S for each user and normalize V for each item

3: For each feedback (i, j) in FS coming in sequence do:

4:    If index of incoming feedback $< W$:

5:        Burn-in training

6:    Else:

7:        Recommend Top-N items to user i (for real-time evaluation)

8:        Adjust θ_i according to P_{ij} computed by Eq.1

9:        While $P_{ij} < \theta_i$:

10:            Update $U_i \leftarrow U_i - \alpha(\frac{\partial L}{\partial U_i} + \lambda_1 U_i)$

11:            Update $V_j \leftarrow V_j - \alpha(\frac{\partial L}{\partial V_j} + \lambda_1 V_j)$

12:            Update $S_i \leftarrow S_i - \alpha(\frac{\partial L}{\partial S_i} + \lambda_2 S_i)$

13:            Update $C_i \leftarrow C_i - \alpha\frac{\partial L}{\partial C_i}$

---

4   Experiments

In this section, we conduct experiments to evaluate the performance of the proposed models by comparing with the state-of-the-art method.

4.1   Datasets

We base our experiments on two real-world datasets LastFM and Douban. The basic statistics of the datasets are summarized in Table 1.

Last.fm is a popular music website. Music listening is one of the most classic forms of implicit user feedback. Furthermore, the preference of users are drifting more casually and the new songs are emerging at all times, which calls for real-time recommendation eagerly. As a result, it is very suitable for evaluating the performance of different online recommendation methods based on implicit feedback in real-time recommendation scenarios. This dataset represents the whole

Table 1. Characteristics of the datasets

Dataset	# Users	# Items	# Features	# Total Feedback
LastFM	132	83,245	11,650	348,446
Douban	6,250	14,878	14,496	345,793

listening habits of 132 users from Feb. 14th to Aug. 1st in 2005. Over three hundreds of thousands <user, artist, song, timestamp> tuples are collected from Last.fm API(http://www.last.fm/api). We consider the songs as items and the artists as features.

Douban is one of the largest Chinese social review sites where users can review movies and provide ratings to them. Douban also allows the user to add a movie into watched list or wish list without giving an explicit rating score. We consider all the type of <user,movie> interactions as implicit user feedback. To obtain rich information, we further crawl the webpages of the relative movies from Douban website and extract 14496 features in total, including Actors(14233), Genres(37), Language(130) and Location(96). The trivial process of feature extraction is omitted for limited space.

4.2   Experimental Protocol

We design an online evaluation architecture to evaluate the performance of different models in an online fashion. For each user feedback in chronological order, following the conventional widely-used evaluation strategy in most works [5,6] for implicit feedback recommendation, we randomly sample 1000 items as irrelevant ones and construct a candidate list which contains these 1000 items together with the one really interact with at the moment. We randomly shuffle the constructed candidate list as the input of various recommendation models. The goal is to rank the item which users really interact with in front of the randomly sampled ones. Finally, we evaluate the recommendation quality based on the recommendation lists which are also the output of different models.

We use three popular evaluation metrics *Rank*, *Hit Ratio* and *Coverage* to measure and compare the performance of various recommendation models. *Rank* measures the percentile position of the relevant item in the whole recommendation list. For example, if the relevant item is ranked at the top of the recommendation list, then $Rank = 0\%$. And $Rank = 100\%$ if the relevant item is the last one in the recommendation list. *Hit Ratio* is a precision metric. If the relevant item is in the *Top-N*, then we have a *hit*, otherwise we have a *miss*. To estimate the *Coverage*, we calculate the proportion of recommended items among the whole item set. We average the *Rank*, *Hit Ratio*, and *Coverage* scores of every 5000 sequential recommendation lists to obtain a statistical evaluation result.

We implement following models as comparative methods. **batch Implicit Feedback Recommendation Model (bIFRM)** is the off-line version of our proposed model. In our experiment, for every 50000 passed samples, we add these

50000 new samples into training set and retrain IFRM. And the model will be retain to make the next 50000 times recommendations. **Temporal Popularity (TP)** is the simplest but effective method for online recommendation. By sliding a window, TP always recommends the most popular items in recent time period. Note that it is also a component in our model. **Stream Ranking Matrix Factorization (RMFX)** [6] is the state-of-the-art method to make real-time recommendation based on implicit user feedback.

4.3   Results

Accuracy and Diversity. We enforce all the comparative methods to make real-time Top-N recommendation before each user-item interaction occurred. Note that the *Hit Ratio* and *Coverage* metrics evaluate the performance of Top-N recommendation. We respectively set $N = 5, 10, 20, 30$ and conduct a series of experiments. Since the overall performances of various models are consistent for all the different N, for limited space, we only report the results when $N = 30$ in the following section.

To show the dynamic performance of different methods over time, we plot Fig.1. The observations can be summarized into the following points: (1) Online models generally achieve higher recommendation accuracy with lower *Rank* and higher *Hit Ratio*. In the scenarios like music-listening where user's interest is drifting fast and items update continually, the online recommendation models show significant superiority for capturing the temporal dynamics immediately. In the scenarios like movie-watching where user's taste is relatively stable and items update occasionally, the performance gain of implementing recommendation models online is not much obvious. On both datasets, the online version oIFRM always performs better than the batch version bIFRM. Even at the very first few evaluations, their performances are significantly different. Note that after the burn-in process, bIFRM and oIFRM are similar. However, oIFRM learns fast from the latest user feedback and utilizes more feedback information while bIFRM is static. Consequently, it is much valuable to design an online learning method for higher recommendation accuracy. (2) Our models make more diverse recommendations meanwhile retain relatively high accuracy. Personalization plays an important role. oIFRM adopts the common idea of modelling users' personal preference. Furthermore, it models users' personal sensitivity and conformity, which introduce more personality. Even though the performance of TP over accuracy is comparable to our models, the advantage of our models over diversity is obvious, since TP always generates non-personalized recommendations. RMFX is basically a pairwise ranking approach, which makes it tend to recommend popular items (the latent features of those items are generally bigger). Thus, the *Coverage* RMFX achieves is relatively low. A reasonable recommendation model always tends to recommend popular items, because they have high chances to be chosen by users. However, always predicting popular items will narrow users' horizon and hide the value of the Long Tail.

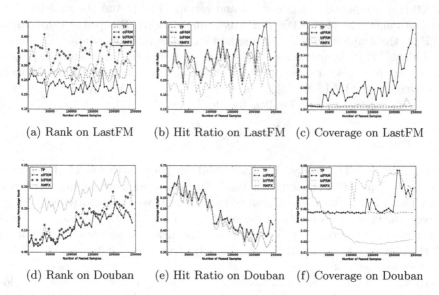

(a) Rank on LastFM   (b) Hit Ratio on LastFM   (c) Coverage on LastFM

(d) Rank on Douban   (e) Hit Ratio on Douban   (f) Coverage on Douban

Fig. 1. Performance Comparison Over Time for Real-time Top-30 Recommendation on LastFM and Douban Dataset

Interpretability and Robustness. Interpretability is another requirement of recommender system. We will show some properties of oIFRM from this perspective and verify the utility of introduced components, namely users' personalized conformity C and Confidence θ.

Fig.2 depicts the changing trends of user conformity over time. We plot the Maximum, the Average, and the Minimum value of C at each time-stamp. From these three basic statistics, we can draw some interesting conclusions. Generally speaking, the conformity of the users on Last.fm is much less than that on Douban, since the average conformity value of Last.fm is around 0.7 (and still decreasing) while that on Douban is around stable 1.0. The reason is that most people tend to watch popular movies but few people spend time on enjoining the minority (From Fig.2 (b) we observe that the average line is very close to the maximum line). On the contrary, there are still considerable part of people who tend to follow their own tastes to choose distinctive songs to listen. The difference between the Maximum line and the Minimum line increases over time show that our online recommendation model can gradually capture the degree of different users' personalized conformity.

Another advantage of our model is the robustness to hostile attack. As we mentioned before, since it is hard to process data cleaning online, the capacity to tolerate noises (especially attacks) is very important to an online recommendation model, especially the one constructed from implicit user feedback (a crawler can easily "click" a mass of items within a small time period). One common type of attack is random visit by a robot (or crawler). Thus, we conduct experiments

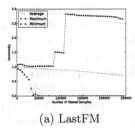
(a) LastFM

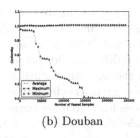
(b) Douban

Fig. 2. The Statics of Users' Personalized Conformity Change Over Time

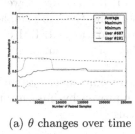
(a) θ changes over time

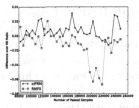
(b) Difference over Hit Ratio after adding noise

Fig. 3. Confidence Parameter θ Changes Over Time after Adding Noise on LastFM dataset

by simulating the attacks in practice. Firstly, we randomly choose two time point (around the 80000th and the 160000th sample). At each time point, we simulate the random visit behavior of one chosen user. Specifically, we insert the "random visit noises" (10000 randomly sampled interactions) into sample stream. Finally, our model and RMFX are trained(updated) by the noisy sample stream. We plot the statistics of confidence θ in Fig.3 (a) and the difference over *Hit Ratio* in Fig.3 (b). From (a), we can observe that the confidences of the two chosen users indeed decrease at the inserted time point respectively. This verifies the effectiveness of the introduced confidence parameters θ. In practice, we can detect the anomaly users by tracking varying θ. Fig.3 (b) shows that oIFRM can tolerate the inserted noises. On the contrary, the performance of RMFX can be significantly affected under attacks. The reason is that adopting a fixed learning step, RMFX considers every observed feedback as positive sample and fails to "filter out" those low-quality samples.

Throughput. Another important indicators of online system is throughput. If the update process cant catch up with data stream, the online system would fail. As a result, we examine the average time cost for one update operation. The experiments are conducted on the same computer with double Core i3 CPU and 2G RAM. All the comparative algorithms are implemented in Python 2.7. The statics are concluded in Table 2.

The batch models are very time-consuming compared with the online models. It is meaningless to compare these two clusters of models directly. Consequently, we compare the three online methods TP, oIFRM and RMFX. Since the dimension of latent features can affect the running time, we conduct experiments repeatedly on different dimensions $D = 5, 50, 200$. We run the program 10 times and obtain the average statical results. Surprisingly, the results show that the time cost of our model is insensitive to D. The reason is that higher dimension of latent features leads to more personality. oIFRM dynamically adjusts confidence θ which is also the controller of iteration times. Thus, more personalized predictions at the beginning will decrease θ and then reduce average iteration times. We implement TP by using a queue and an index, which makes TP the most efficient online model (time complexity $O(1)$). oIFRM achieves lower time cost compared with RMFX. Because RMFX spend some time to sample negative items from reservoir to construct a pair before updating. oIFRM can directly calculate gradient without sampling. The update process can be more efficient if we cache a few temporary variables.

In our configuration, the throughput of our recommender system is around 429 updates per second. It can be certainly improved by implementing in C, sampling from data stream, or parallel computing [12].

Table 2. Average Time Cost (milliseconds) per Update

	TP	$D = 5$		$D = 50$		$D = 200$	
		oIFRM	RMFX	oIFRM	RMFX	oIFRM	RMFX
LastFM	0.0078	2.27	5.22	2.33	5.29	2.22	5.89
Douban	0.0076	2.96	5.14	2.88	5.35	2.68	5.91

5   Conclusion

This paper addresses two problems in recommender systems, namely 1) recommendation based on implicit user feedback and 2) real-time recommendation in a stream setting. To overcome One-class problem, we first propose IFRM which is tailor-made for implicit feedback recommendation. Moreover, IFRM is flexible to incorporate side information such as item features or item temporal popularity. To increase training efficiency and capture user dynamic preference in time, we extend IFRM to an online version, oIFRM. By assuming the value of user feedback is different, oIFRM automatically adopts different learning step for each feedback. By comparing user adoption probability and user behavior confidence, oIFRM reinforces to learn the new trend while weakens to learn the habitual feedback and noise. Finally, comprehensive experiments conducted on two real-world datasets have showed that oIFRM has an obvious superior performance over the state-of-the-art approach from various perspectives, including accuracy, efficiency, diversity, interpretability, and robustness.

References

1. Abernethy, J., Canini, K., Langford, J., Simma, A.: Online collaborative filtering. Tech. rep., UC Berkeley (2011)
2. Amatriain, X.: Mining large streams of user data for personalized recommendations. ACM SIGKDD Explorations Newsletter 14(2), 37–48 (2013)
3. Chen, K., Chen, T., Zheng, G., Jin, O., Yao, E., Yu, Y.: Collaborative personalized tweet recommendation. In: Proceedings of the 35th International ACM SIGIR Conference on Research and Development in Information Retrieval, pp. 661–670. ACM (2012)
4. Chua, F.C.T., Oentaryo, R.J., Lim, E.P.: Modeling temporal adoptions using dynamic matrix factorization. In: 2013 IEEE 13th International Conference on Data Mining (ICDM), pp. 91–100. IEEE (2013)
5. Cremonesi, P., Koren, Y., Turrin, R.: Performance of recommender algorithms on top-n recommendation tasks. In: Proceedings of the Fourth ACM Conference on Recommender Systems, pp. 39–46. ACM (2010)
6. Diaz-Aviles, E., Drumond, L., Schmidt-Thieme, L., Nejdl, W.: Real-time top-n recommendation in social streams. In: Proceedings of the Sixth ACM Conference on Recommender Systems, pp. 59–66. ACM (2012)
7. Hu, Y., Koren, Y., Volinsky, C.: Collaborative filtering for implicit feedback datasets. In: Eighth IEEE International Conference on Data Mining, ICDM 2008, pp. 263–272. IEEE (2008)
8. Koren, Y.: Collaborative filtering with temporal dynamics. Communications of the ACM 53(4), 89–97 (2010)
9. Kurucz, M., Benczur, A.A., Kiss, T., Nagy, I., Szabó, A., Torma, B.: Who rated what: a combination of svd, correlation and frequent sequence mining. In: Proc. KDD Cup and Workshop, vol. 23, pp. 720–727. Citeseer (2007)
10. Pan, R., Zhou, Y., Cao, B., Liu, N.N., Lukose, R., Scholz, M., Yang, Q.: One-class collaborative filtering. In: Eighth IEEE International Conference on Data Mining, ICDM 2008, pp. 502–511. IEEE (2008)
11. Rendle, S., Freudenthaler, C., Gantner, Z., Schmidt-Thieme, L.: Bpr: Bayesian personalized ranking from implicit feedback. In: Proceedings of the Twenty-Fifth Conference on Uncertainty in Artificial Intelligence, pp. 452–461. AUAI Press (2009)
12. Zhuang, Y., Chin, W.S., Juan, Y.C., Lin, C.J.: A fast parallel sgd for matrix factorization in shared memory systems. In: Proceedings of the 7th ACM Conference on Recommender Systems, pp. 249–256. ACM (2013)

A Self-learning Rule-Based Approach for Sci-tech Compound Phrase Entity Recognition

Tingwen Liu, Yang Zhang, Yang Yan, Jinqiao Shi, and Li Guo

Institute of Information Engineering, Chinese Academy of Sciences, Beijing, China
{liutingwen,zhangyang,yanyang9021,shijinqiao,guoli}@iie.ac.cn

Abstract. Sci-tech compound phrase entity (*e.g.*, the names of projects, books and patents) recognition is a fundamental task of science and technology data processing and discovery. However, much less work have been done on the problem. In this paper, we first give the characteristics of sci-tech entities that are different from personal name and other traditional entities. Then we introduce a self-learning rule-based approach to address the problem. The approach consists of three stages, namely rule-based text truncation, BlackPOS-based text split and WhiteKey-based confirmation. Constructing the best WhiteKey list is a NP-hard problem. We further design dynamic programming and greedy algorithms to address the problem. Experimental results show that our approach achieves 94.78% precision rate, 89.19% recall rate and 91.9% F_1 measure in average. Moreover, our approach is universal and orthogonal to prior named entity recognition work.

Keywords: sci-tech compound phrase entity, self-learning, scholar data discovery

1 Introduction

Named entity recognition (NER), as a fundamental task of natural language processing, seeks to locate and classify elements in texts into pre-defined categories, such as the names of persons, organizations, locations *etc.* However, much less work have been done on the recognition of a special type of named entity, namely sci-tech compound phrase entity recognition (STCPER). Sci-tech compound phrase entities (STCPE) are phrases that contain one or more modifiers and known entities in the fields of science and technology. Project names, book names and patent names are typical examples in the type. STCPER is an important problem in the field of science and technology data processing and discovery, which needs to be addressed urgently in many applications, such as scholar data discovery, open-source intelligence [18], intellectual property protection [17].

In this paper, we describe the STCPER problem as follows: for a given text t that contains a STCPE named p, we want to extract a substring of t, denoted as $\hat{p}$ hereafter, that is similar with p as much as possible. Obviously, the best result we want to get is that $\hat{p}$ and p are the same. As text t has too many substrings, the biggest technique challenge we need to address is how to find the correct answer from so many substrings.

We first observe the structures of STCPE, and conclude three unique characteristics. Based on our observations we propose a rule-based approach, which is the first work to address the problem to our best knowledge. The main idea is to exclude noise contents in t gradually, which are located by the characters when describing STCPE. Our approach consists of three stages, namely rule-based text truncation, BlackPOS-based text split and WhiteKey-based confirmation. Our approach is self-learning, because all the rules used in our approach are automatically learning from a training set, without any human involvement. We prove that constructing the best WhiteKey list is NP-hard. We design two algorithms, namely the dynamic programming algorithm and the greedy algorithm, to address the problem.

We make three key contributions in this paper. First, we introduce the problem of locating the names of STCPE from web texts, which is the first of its kind to the best of our knowledge. Second, we give an efficient approach to address the problem. And we argue that our approach is universal and orthogonal to prior NER work, such as HMM-based (Hidden Markov Models) and CRF-based (Conditional Random Fields) approaches. That means prior work can be integrated into our approach to improve the identification of STCPE names. Third, we implemented our algorithm in JAVA and conducted experiments on real-world project name sets and web text sets. Our experimental results show that our approach achieves 94.78% precision rate, 89.19% recall rate and 91.9% F_1 measure in average.

2   Related Work

Research on NER starts at 1990s. In 1991, Rau first described the concept of information extraction and recognition, which implements heuristic algorithms and manually managed rules to extract targeted information. In 1996, the NER problem was import as track in MUC-6. In the following competition such as MUC-7, CoNLL-2002 and CoNLL-2003, Chinese NER was introduced. English NER does not need word segmentation, which makes it easy to achieve better result. According to the MUC and ACE tracks record, English NER can achieve as high as 90% precision rate, recall rate and F-measure.

Although much NER work have been proposed, little of them try to address the STCPER problem. Prior NER work mainly focus on identifying the names of persons, organizations, locations *etc.*, which can be classified into two categories: rule-based approaches and stochastic-based approaches.

Rule-Based Approaches. [3,8,11,5] utilize pattern-matching techniques as well as heuristics derived either from the morphology or the semantics of the input sequence. The conceit of handling NER is to look for clues within the structure and grammars of the text which is the prompt of some named entities. These insights can be used as classifiers in machine-learning approaches, or as candidate taggers words in gazetteers. Some applications can also make effective use of stand-alone rule-based systems, but they are prone to be both overreach and skipping over named entities. Nadeau *et al.* provide a table of word-level features to recognize a person's name or *etc.* in [6]. For example, the word is capitalized or all capital letters (or indeed, includes a mixed case), punctuation, morphology. Features like these, especially capitalization, form the basis for most heuristics. Word-level features can then be combined with contextual features to further identify names, called sure-fire rules as described in [12], which rely on contextual trigger words either prepended or appended to a series of capitalized tokens, whose Part Of Speech (POS) is unknown or are generically referenced as nouns.

Stochastic-based approach is the current dominant technique for addressing the NER problem. Stochastic-based techniques include Hidden Markov Models (HMM) [1], Decision Trees [13], Maximum Entropy Models (ME) [2], Support Vector Machines (SVM) [15], and Conditional Random Fields (CRF) [10]. These work are all variants of the stochastic-based approaches that typically consist of a system that reads a large annotated corpus, memorizes lists of entities, and creates disambiguation rules based on discriminative features.

Besides the above work, Cucerzan *et al.* [4] presented a large-scale system for the recognition and semantic disambiguation of named entities based on information extracted from Wikipedia and Web search results. Mann and Yarowsky [9] developed an unsupervised method for personal name disambiguation that exploits a rich set of biographic features, such as birth date, birth place and others.

Our approach is universal and orthogonal to prior NER work. Thus we can integrate prior work into our approach to improve the precision rate and the recall rate of the STCPER problem.

3 Our Approach

The approach we explore in this paper is illustrated in Figure 1. Our STCPER process consists of three stages. First, we use linguistic cues to truncate the input text t; then, we split the substring outputs of the first stage by BlackPOSes; in the final stage, we use a white list of STCPE keywords to confirm whether the output substrings of the second stage are STCPE or not.

In this section, we first show our observations on the characteristics of STCPE, which are the starting points of our apporach. Then we describe the above three stages in detail in the following three subsections. At last, we introduce how to automatically generate the black list of STCPE POSes and the white list of STCPE keywords.

3.1 Characteristics of STCPE

We observe that STCPE have three novel characteristics that are significantly different from traditional named entities, such as personal names.

Multi-entities Compound: each STCPE consists of multiple noun entities in many different fields, such as entities in the chemical or computer field.

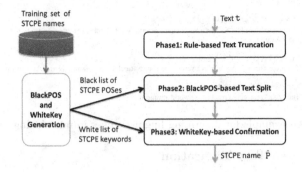

Fig. 1. System architecture of our approach.

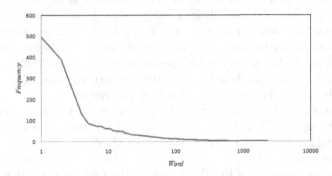

Fig. 2. Distribution of the appearance times for each word.

While the structures of personal names and other traditional named entities are relatively simple.

Limited Missing POSes: We observe that the number of the types of Part-Of-Speeches (POSes) in STCPE names does not increase after the number of STCPE names exceeds a certain size. As the total number of the types of POSes is fixed, some POSes are highly unlikely to appear in STCPE names, such as be verbs (be/is/are/am/was/were etc.) and demonstrative pronouns (you/me/I/my/him/her etc.).

Multiple Indicative Keywords: We note that people usually follow strong linguistic laws when they give one STCPE a name. We collect 813 project names from the Internet and obtain 2297 different words from them. The distribution of the appearance times for each word in these project names is extremely skewed, as shown in Figure 2. For example, project names usually have their own indicative keywords, such as "research", "system", "application" *etc.* Moreover, most of STCPE have more than one indicative keyword. And a sentence without any indicative keyword is very unlikely to contain any STCPE. Note that personal names and other traditional named entities do not have this characteristic.

— "XXX"项目获得了国家科学技术进步奖一等奖
The project named "XXX" won the National Science and Technology Progress Award First
Class Award

— "YYY"项目被授予国家自然科学奖二等奖
The project named "XXX" was awarded the State Natural Science Award Second Class Award

— "ZZZ"项目达到了世界先进水平
The "ZZZ" project has reached the world advanced level

Fig. 3. An example for describing linguistic cues in introducing projects.

3.2   Rule-Based Text Truncation

From a linguistic view, the names of STCPE in the texts of natural language are expressed either implicitly or explicitly. The difference is that implicit STCPE names are expressed without obvious cue phrases. In order to reduce the influence of unrelated text segments in classifying STCPE names, we use explicit connective markers as linguistic cues to truncate the original input text. Figure 3 shows an example that describes the specific details for three different projects respectively. For the example in Figure 3, the linguistic cues are "won [*] award", "was awarded" and "has reached [*] level" respectively.

According to prior research experiences, we focus on the most frequent and least ambiguous cues. The rules used in our work to truncate texts are written by Perl Compatible Regular Expressions (PCRE). We use these linguistic cues to find texts that we are interested in, and use the capturing parentheses and back references functions of PCRE to capture the text segments that are likely to contain STCPE names. In this way, we can use a few PCRE rules to achieve our goal by the help of the powerful expressive, efficient, and flexible capacity of PCRE. And we do not need to enumerate each linguistic cue in a sperate rule.

Obviously, the recall of texts that contain STCPE names can be improved if we use more comprehensive and more relaxed cues. However, one thing should be noted that using more comprehensive and more relaxed cues will retrieve more texts that do not contain any STCPE name mistakenly at the same time.

3.3   BlackPOS-Based Text Split

This stage is motivated by the limited missing POSes characteristic. We call these missing POSes BlackPOSes which means that they are "prohibited" to appear in STCPE names. All the BlackPOSes constitute a black list of STCPE POSes. However, BlackPOSes may frequently appear in web texts. Thus, we can shorten the substring outputs of the rule-based text truncation step further to make $\hat{P}$ more similar with P in the following method: we can split the substring outputs of the rule-based text truncation step by BlackPOSes, and use the split results as the substring outputs of the BlackPOS-based text split step, which are the inputs of the next and last step, namely WhiteKey-based confirmation.

Note that we do not use keywords to split texts here, because the keywords that appear and not appear in STCPE names are both huge. It is impossible to enumerate them.

The detail of choosing which POSes as BlackPOSes will be described in Section 3.5.

3.4   WhiteKey-Based Confirmation

The main idea of our WhiteKey-based confirmation step is based on the multiple indicative keywords characteristic. Motivated by the characteristic, we can choose some keywords as Whitekeys and use them to constitute a white list of STCPE keywords. If a substring output of the BlackPOS-based text split step contains more than one word in the white list, we regard the substring as a STCPE name.

Note that we need to exclude the existence of noise words to avoid making a mistake. For example, if the substring is "Senior researcher Alex will give us a technical report tomorrow", and word "research" is exactly a WhiteKey, then we will mistakenly regard the substring as a STCPE name. We can use the negative lookahead assertion function of PCRE to achieve the goal. The PCRE for the above example is "research(?!er)".

The detail that which words are considered as WhiteKeys will be described in the next subsection (Section 3.5).

3.5   BlackPOS and WhiteKey Generation

There are still two open problems to be addressed in our approach: how to generate the black list of STCPE POSes and the white list of STCPE keywords. The input of the two problems are the same: a training set that consists of the names of n STCPE. We denote the training set as $P = \{p_1, p_2, \cdots, p_n\}$, where p_i is the name of the i-th STCPE.

The the first problem can be addressed in a simple and intuitive way: we conduct word segmentation and POS tagging on each STCPE name and get the black list which is composed of all POSes that do not appear in any tagging result. We admit that the black list may contain POSes that appear in the tagging results of STCPE names that are not in the training set P, especially when P is small. However, we argue that the black list may be accurate enough as long as P is large enough. Moreover, we can seek the help of linguists to adjust the black list.

To address the second problem, we also need to conduct word segmentation on each STCPE name first. Note that each keyword is a segmented word. As there are lots of segmented words, how to select some appropriate members from them to constitute the white list are the biggest challenge to be addressed. On one hand, we want the white list has as few words as possible, as more words mean much more substrings of input texts will be identified as STCPE names mistakenly. On the other hand, we want these selected keywords cover as many STCPE as possible, where a STCPE name is covered if at least one member in the white list appears in the STCPE name. Because more STCPE are covered means much more substrings that are really STCPE names will be correctly identified. These two goals are unfortunately conflicting. With the white list consisting of

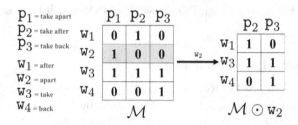

Fig. 4. Description of $\odot$ with $\mathcal{M} \odot w_2$ as an example.

all segmented words, all STCPE names in the training set are covered, but the white list is the biggest. With the white list being empty, the white list is the smallest, but it cannot out identify any STCPE name.

For the given training set $P = \{p_1, p_2, \cdots, p_n\}$, we can get m words after word segmentation, which constitute a set $W = \{w_1, w_2, \cdots, w_m\}$, where w_i is the i-th word. The problem of selecting Words with the LeAst Number (denoted as WLAN problem hereinafter) can be defined as follows: we want to find a minimum subset $S \subseteq W$ that satisfies the following two limitations at the same time. First, S covers P, which means that each STCPE name in P is covered by at least one word in S. Second, $|S|$ is the least where $|X|$ represents the number of members in set X. The following Theorem 1 shows the computing complexity of WLAN problem.

Theorem 1. *The WLAN problem is NP-hard.*

Proof. For each word $w \in W$, we can construct a set $A(w) \subseteq P$ which consists of all STCPE names that are covered by w. Then we can infer that $A(W) = \{A(w_1), A(w_2), \cdots, A(w_m)\}$, which consists of m subsets of the finite set P. Then the WLAN problem becomes the classical minimum set cover problem: find a subset $AS \subseteq A(W)$ such that every element in P belongs to at least one member of AS and minimize $|AS|$, as shown in [16].

In this paper, we use the dynamic programming strategy to obtain the optimal result and use the greedy strategy to obtain a near-optimal result. Before describing the two algorithms, we first introduce some symbols. We can construct a $m \times n$ boolean matrix $\mathcal{M}$, where each row represents an element in W and each column represents an element in P. If the i-th word w_i appears in the j-th STCPE name p_j, then $\mathcal{M}_{ij} = 1$; otherwise $\mathcal{M}_{ij} = 0$. We define a new binary operator '$\odot$' with a boolean matrix as the first argument and a word as the second argument. The operational result is another binary matrix got by removing the corresponding rows of the second argument from the first argument, and only retaining these columns that are not covered by the second argument. For example, assuming $P = \{p_1, p_2, p_3\}$, four words will be obtained after word segmentation: $W = \{w_1, w_2, w_3, w_4\}$, and $\mathcal{M}$ is a 4×3 boolean matrix as shown in Figure 4. The output of $\mathcal{M} \odot w_2$ is a 3×2 boolean matrix: the w_2 row and the p_1 column are removed from $\mathcal{M}$.

We use the function F with a boolean matrix as the input to get the optimal result of the WLAN problem. Then we can infer that

$$F(\mathcal{M}) = \min\{F(\mathcal{M} \odot \mathbf{w}_i) + 1 \mid 1 \leq i \leq m\} \tag{1}$$

Based on Equation 1, we can design a dynamic programming algorithm to get the optimal result of WLAN problem.

We can also design a greedy algorithm to get a near-optimal result. Initially, we set $\mathcal{M}' = \mathcal{M}$. For the current boolean matrix $\mathcal{M}'$, we choose the word $\mathbf{w}$ that has the biggest number of value '1' in the corresponding row of $\mathcal{M}'$, and set $\mathcal{M}' = \mathcal{M}' \odot \mathbf{w}$. The above step repeats until $\mathcal{M}'$ becomes a 0×0 boolean matrix. For the example in Figure 4, the output of our greedy algorithm is $\{\mathbf{w}_3\}$, which is also the optimal result of the corresponding WLAN problem.

Note that selecting words of the least number directly may not be the best strategy for different training sets. Next, we try to define the problem of selecting Words with a given Fixed Number k (denoted as k-WFN problem hereinafter) and discuss how to choose an appropriate value k.

The k-WFN problem can be defined as follows: we want to find a minimum subset $S' \subseteq W$ that satisfies the following three limitations at the same time. First, S' covers P, which means that each STCPE in P is covered by at least one word in S'. Second, $|S'| \leq k$. Third, $\sum_{\mathbf{w}_i \in S'} |A(\mathbf{w}_i)|$ should be maximized, namely the goal is to select no more than k indicative words that appear in as many STCPE names as possible.

In this paper, we also use the dynamic programming strategy to obtain the optimal result and use the greedy strategy to obtain a near-optimal result. We use the function H with a boolean matrix and an integer as the inputs to get the optimal result of the k-WFN problem. We can infer that:

$$H(\mathcal{M}, k) = \max\{H(\mathcal{M}, k-1), H(\mathcal{M}, k-1 \mid \mathbf{w}_i)\} \tag{2}$$

where $H(\mathcal{M}, k-1 \mid \mathbf{w}_i) = \max\{|A(\mathbf{w}_i)| + H(\mathcal{M} \odot \mathbf{w}_i, k-1) \mid 1 \leq i \leq m\}$, means that the word $\mathbf{w}_i$ must be selected. Note that $H(\mathcal{M}, k)$ must be more than $H(\mathcal{M}, k-1)$ for k-WFN problem when k is no more than the number for rows in $\mathcal{M}$ and no less than $F(\mathcal{M})$. Thus, we know that $\max\{|A(\mathbf{w}_i)| + H(\mathcal{M} \odot \mathbf{w}_i, k-1) \mid 1 \leq i \leq m\}$ must be more than $H(\mathcal{M}, k-1)$, otherwise $H(\mathcal{M}, k)$ will be equal to $H(\mathcal{M}, k-1)$. Then Equation 2 becomes as follows:

$$\begin{aligned} H(\mathcal{M}, k) &= H(\mathcal{M}, k-1 \mid \mathbf{w}_i) \\ &= \max\{|A(\mathbf{w}_i)| + H(\mathcal{M} \odot \mathbf{w}_i, k-1) \mid 1 \leq i \leq m\} \end{aligned} \tag{3}$$

Based on Equation 3, we can design a dynamic programming algorithm to get the optimal result of k-WFN problem.

We can also design a greedy algorithm to get a near-optimal result of k-WFN problem. Note that using the same greedy strategy (as shown in the above greedy algorithm of WLAN problem) directly does not generate to give a result that satisfies the above three limitations at the same time. The result may have bigger than k members even if the greedy algorithm terminates when it finds that the first limitation is satisfied. Thus we design a new greedy algorithm for k-WFN problem. First, we try to obtain less than k words to constitute S' which can satisfy the first two limitations at the same time. Then, we add some other words into S' to maximize $\sum_{\mathbf{w}_i \in S'} |A(\mathbf{w}_i)|$. Note that the second limitation must be satisfied, namely $|S'| \leq k$.

Next, we discuss how to choose an appropriate value for k. We can think $H(\mathcal{M}, k+1) - H(\mathcal{M}, k)$ as the marginal benefit of adding a new word into set S' of k elements. Theoretically, the marginal benefit $H(\mathcal{M}, k+1) - H(\mathcal{M}, k)$ should decrease with the increase of k. Thus, we introduce a marginal benefit threshold α, and we want to choose value k such that $H(\mathcal{M}, k+1) - H(\mathcal{M}, k) < \alpha$ and $H(\mathcal{M}, k) - H(\mathcal{M}, k-1) \geq \alpha$.

4   Experimental Resutls

4.1   Experimental Setup

Project names are typical STCPE names. In our experiments, we evaluate our approach on real-world project name sets and web text sets that are collected form the Internet.

The training set used to generate the white list of project keywords and the black list of project POSes consists of 1119 project names. Each element in the training set is the name of a real project supported by the National Natural Science Foundation of China in 2014. All the 1119 projects are undertaken by 8 universities.

We collect 1522 project names that each project won the National Science and Technology Progress Award between 2005 and 2014. We take advantage of search engines to retrieve web texts. Each search uses a different project name as the search keyword. Finally, we get 734 web texts that each of them contains the project name used to search it. We also collect 766 web texts that each of them do not contain any project name from Sogou Lab Data [14]. All the 1500 (734+766) web texts consist of text set T.

We evaluate our approach on the metrics of precision rate, recall rate and F-measure. We denote the i-th text in T as $\mathbf{t}_i$, and denote the corresponding project name as $\mathbf{p}_i$, and denote the project name that our approach gives for $\mathbf{t}_i$ as $\hat{\mathbf{p}}_i$. For text $\mathbf{t}_i$, its precision rate is defined as the ratio of the length of the longest common substring between $\mathbf{p}_i$ and $\hat{\mathbf{p}}_i$ and the length of $\hat{\mathbf{p}}_i$, denoted as $Pr(\mathbf{t}_i) = \frac{|L(\mathbf{p}_i, \hat{\mathbf{p}}_i)|}{|\hat{\mathbf{p}}_i|}$, where $L(\mathbf{p}_i, \hat{\mathbf{p}}_i)$ is the longest common substring between $\mathbf{p}_i$ and $\hat{\mathbf{p}}_i$; its recall rate is defined as the ratio of the length of the longest common substring between $\mathbf{p}_i$ and $\hat{\mathbf{p}}_i$ and the length of $\mathbf{p}_i$, denoted as $Re(\mathbf{t}_i) = \frac{|L(\mathbf{p}_i, \hat{\mathbf{p}}_i)|}{|\mathbf{p}_i|}$. If both $\mathbf{p}_i$ and $\hat{\mathbf{p}}_i$ are null, $Pr(\mathbf{t}_i) = Re(\mathbf{t}_i) = 1$; if $\mathbf{p}_i$ is null and $\hat{\mathbf{p}}_i$ is not null, then $Pr(\mathbf{t}_i) = 0$ and $Re(\mathbf{t}_i) = 1$; if $\mathbf{p}_i$ is not null and $\hat{\mathbf{p}}_i$ is null, then $Pr(\mathbf{t}_i) = 1$

and $Re(\mathbf{t}_i) = 0$. Then for text set T, its precision rate is defined as the ratio of the total precision rates of its elements and the size of T, namely

$$Pr(T) = \frac{\sum\limits_{\mathbf{t}_i \in T} Pr(\mathbf{t}_i)}{|T|}$$

and its recall rate is defined as the ratio of the total recall rates of its elements and the size of T, namely

$$Re(T) = \frac{\sum\limits_{\mathbf{t}_i \in T} Re(\mathbf{t}_i)}{|T|}$$

F-measure is the harmonic mean of precision and recall. The general formula of F-measure for positive real β is: $F_\beta = (1 + \beta^2) \times \frac{Pr \times Re}{\beta^2 \times Pr + Re}$. In this paper, we use $F_1 = 2 \times \frac{Pr \times Re}{Pr + Re}$ to evaluate our approach, as the way used in many papers.

We implement our approach in Java. All the word segmentation and POS tagging operations are conducted with the widely-used ICTCLAS [7] tool.

4.2   Effect of Our Approach

For the training set of 1119 project names, there are totally 61 different kinds of POSes appear in the elements of the training set. We know that there are 96 different kinds of POSes in ICTCLASS. Thus 35 POSes do not appear in any of the 1119 project names. In our experiments, we use these 35 POSes to constitute the black list.

After word segmentation, the 1119 project names are divided into 3397 words. In our experiments, we remove the stop words form the 3397 words. We also remove the words of length 1, in order to avoid the white list being over-matched. Our experimental result shows the least number of words required to cover the training set is 72. Figure 5 shows the change of $H(\mathcal{M}, k+1) - H(\mathcal{M}, k)$ with the increase of k where $k \geq 72$. The values of $H(\mathcal{M}, k+1) - H(\mathcal{M}, k)$ for different k are got by our greedy algorithm. From Figure 5 we can find that: 1) $H(\mathcal{M}, k+1) - H(\mathcal{M}, k)$ is always bigger that 0, namely $H(\mathcal{M}, k+1)$ is bigger than $H(\mathcal{M}, k)$ for any k that is less than the number of rows in $\mathcal{M}$; 2) the bigger the value of k, the smaller the marginal benefit we can get. If the marginal benefit threshold α is set to be 30, then the value of k is 102.

We use the 35 BlackPOSes and 102 WhiteKeys to perform our project name recognition. Our experimental result shows that our approach can achieve 90.97% precision rate, 77.9% recall rate and 83.93% F_1 measure over the 734 web texts that each text contains a predefined project name. Note that the recall rate is a little low, because of the following two reasons. First, the white list of project keywords is incomplete, as the training set does not contain all kinds of project names, especially when the training set and the testing set come from different sources. Second, the results of POS tagging are not entirely accurate, thus some project names are fortunately cut off by the black list of project POSes.

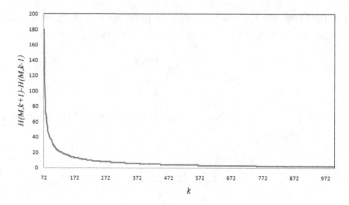

Fig. 5. Change of $H(\mathcal{M}, k+1) - H(\mathcal{M}, k)$ with the increase of k.

Experimental results show that our approach can achieve 98.43% precision rate, 100% recall rate and 99.21% F_1 measure over the 766 web texts that do not contain any project name. The primary reason that the precision rate is close to but not 100% is that some substrings contain the elements in the white list of project keywords, and are identified as project names.

Overall, our approach achieves 94.78% precision rate, 89.19% recall rate and 91.9% F_1 measure over the set T of 1500 web texts.

5   Conclusion and Future Work

In this paper, we propose the first work to address the problem of recognizing sci-tech compound phrase entities. The process of our STCPER approach consists of three phrases. First, we use linguistic cues to truncate the input text; then, we split the substring outputs of the first stage by BlackPOSes; in the final stage, we use a white list of STCPE keywords to confirm whether the substring outputs of the second stage are STCPE names or not. We reduce the problem of generating the white list of STCPE keywords into the classical minimum set cover problem. And the we design two algorithms, namely the dynamic programming algorithm and the greedy algorithm, to address the problem. All the rules in our approach are automatically learning from the training set without any human involvement. Experimental results on real-world project names and web texts shows that our approach achieves quite high precision rate and recall rate on this complex problem.

Our approach is universal and orthogonal to prior NER work. In the future, we will integrate HMM-based NER approach and CRF-based NER approach into our work to improve the precision rate and recall rate of identifying STCPE names. Mover, we will extend our work from multiple views, such as the dependency trees of STCPE names.

Acknowledgement. Supported in part by the Strategic Priority Research Program of the Chinese Academy of Sciences under Grant No. XDA06030200.

References

1. Bikel, D.M., Miller, S., Schwartz, R., Weischedel, R.: Nymble: A High-performance Learning Name-finder. In: Proc. of ANLC, pp. 194–201 (1997)
2. Borthwick, A., Sterling, J., Agichtein, E., Grishman, R.: NYU: Description of the MENE Named Entity System as Used in MUC-7. In: Proc. of MUC (1998)
3. Chiticariu, L., Krishnamurthy, R., Li, Y., Reiss, F., Vaithyanathan, S.: Domain Adaptation of Rule-based Annotators for Named-Entity Recognition Tasks. In: Proc. of EMNLP, pp. 1002–1012 (2010)
4. Cucerzan, S.: Large-Scale Named Entity Disambiguation Based on Wikipedia Data. In: Proc. of EMNLP-CoNLL 2007, pp. 708–716 (2007)
5. Cucerzan, S., Yarowsky, D.: Language independent named entity recognition combining morphological and contextual evidence. In: Proc. of EMNLP and VLC, pp. 90–99 (1999)
6. Farmakiotou, D., Karkaletsis, V., Koutsias, J., Sigletos, G., Spyropoulos, C.D., Stamatopoulos, P.: Rule-Based Named Entity Recognition For Greek Financial Texts. In: Proc. of COMLEX, pp. 75–78 (2000)
7. Zhang, H.: NLPIR/ICTCLAS (2012), http://ictclas.nlpir.org/
8. Krishnan, V., Manning, C.D.: An Effective Two-stage Model for Exploiting Non-local Dependencies in Named Entity Recognition. In: Proc. of ACL, pp. 1121–1128 (2006)
9. Mann, G.S., Yarowsky, D.: Unsupervised Personal Name Disambiguation. In: Proc. of CONLL at HLT-NAACL 2003, vol. 4, pp. 33–40 (2003)
10. McCallum, A., Li, W.: Early Results for Named Entity Recognition with Conditional Random Fields, Feature Induction and Web-enhanced Lexicons. In: Proc. of CONLL at HLT-NAACL 2003, vol. 4, pp. 188–191 (2003)
11. Mikheev, A., Moens, M., Grover, C.: Named Entity Recognition without Gazetteers. In: Proc. of EACL, pp. 1–8 (1999)
12. Nadeau, D., Sekine, S.: A Survey of Named Entity Recognition and Classification (2007), http://brown.cl.uni-heidelberg.de/ sourjiko/NER_Literatur/ survey.pdf
13. Sekine, S.: NYU: Description of the Japanese NE system used for MET-2. In: Proc. of MUC (1998)
14. Sogou Labs: Sogou Text Classification Corpus (2008), http://www.sogou.com/labs/dl/c.html/
15. Takeuchi, K., Collier, N.: Use of Support Vector Machines in Extended Named Entity Recognition. In: Proc. of COLING, vol. 20, pp. 1–7 (2002)
16. Viggo Kann: Minimum Set Cover (2000), http://perso.ensta-paristech.fr/ ~diam/ro/online/viggo_wwwcompendium/node146.html#6062
17. Wikipedia: Intellectual Property Protection, en.wikipedia.org/wiki/Intellectual_property
18. Wikipedia: Open-Source Intelligence, en.wikipedia.org/wiki/Open-source_intelligence

Batch Mode Active Learning for Geographical Image Classification

Zengmao Wang[1], Bo Du[1,*], Lefei Zhang[1],
Wenbin Hu[1], Dacheng Tao[2], and Liangpei Zhang[3]

[1] School of Computer, Wuhan University, Wuhan, China
[2] Center for Quantum Computation and Intelligence System,
University of Technology, Sydney, Sydney, Australia
[3] State Key Laboratory of Information Engineering in Surveying,
Mapping and Remote Sensing, Wuhan University, Wuhan, China
gunspace@163.com

Abstract. In this paper, an innovative batch mode active learning by combining discriminative and representative information for hyperspectral image classification with support vector machine is proposed. In the past years, the batch mode active learning mainly exploits different query functions, which are based on two criteria: uncertainty criterion and diversity criterion. Generally, the uncertainty criterion and diversity criterion are independent of each other, and they also could not make sure the queried samples identical and independent distribution. In the proposed method, the diversity criterion is focused. In the innovative diversity criterion, firstly, we derive a novel form of upper bound for true risk in the active learning setting, by minimizing this upper bound to measure the discriminative information, which is connected with the uncertainty. Secondly, for the representative information, the maximum mean discrepancy(MMD) which captures the representative information of the data structure is adopt to match the distribution of the labeled samples and query samples, to make sure the queried samples have a similar distribution to the labeled samples and guarantee the queried samples are diversified. Meanwhile, the number of new queried samples is adaptive, which depends on the distribution of the labeled samples. In the experiment, we employ two benchmark remote sensing images, Indian Pines and Washington DC. The experimental results demonstrate the effective of our proposed method compared with the state-of-the-art AL methods

Keywords: Active learning, Discriminative and Representative, Empirical Risk Minimization, Maximum Mean Discrepancy, Classification.

1 Introduction

Data classification has been an active research topic in the machine learning community for many years. To automatically assign data samples to a set of

---

* Corresponding author.

© Springer International Publishing Switzerland 2015
R. Cheng et al. (Eds.): APWeb 2015, LNCS 9313, pp. 744–755, 2015.
DOI: 10.1007/978-3-319-25255-1_61

predefined classes is the goal of data classification tasks. One prerequisite for data classification is to obtain the labeled samples beforehand. So data collection and annotation is a fundamental problem, but actually it is very time and labor consuming to label the samples. In order to reduce the effort in acquiring labeled samples, active learning is a very useful tool in such situations when unlabeled data is cheap to collect but labeling them is expensive. Now a number of active learning methods [5] have been developed for data classification [22]. The key idea of active learning is to identify the samples that are most informative with respect to the current classification model. In current research, two points should be taken into account in the query function adopted for selecting the batch of the most informative samples to design the practical active learning algorithms: the uncertainty and the diversity [3,4,16]. The uncertainty aims to choose the samples the current classifier is most uncertain about [1,2]. For multi-class classification problems, the margin based uncertainty sampling is a popular method which aims to select the samples which are closest to the separating hyperplane[11]. In such methods, the queried samples may have redundancy. The diversity of active learning aims to alleviate this problem by querying the most representative samples for the overall patterns of the unlabeled data[6]. Since using either kind of criterion alone is not sufficient to get the optimal result, there are several works trying to query the unlabeled samples with both high information and high representation[3,16]. Usually they are either heuristic in designing the specific query criterion or ad hoc in measuring the informativeness and representativeness of the samples. In this paper, the proposed method combines the discriminative information and representative information into one optimal formula as a diversity criterion to remove the redundancy samples in the uncertainty criterion. For the discriminative information, we derive a novel form of upper bound for true risk in the active learning setting, by minimizing this upper bound to measure the discriminative information. For the representative information, we adopt the maximum mean discrepancy (MMD) to match the distribution between the labeled samples and queried samples. We test the proposed method on two benchmark hyperspectral images. The rest of this paper is organized as follows: Section 2 briefly reviews the related work. Section 3 introduces the general framework of the proposed active learning method in detail. Section 4 presents the experiments and interprets the results. Finally, we conclude the paper and propose in Section 5.

2 Related Work

In recent research, a number of criterions for active learning have been proposed for classification problems. One common criterion is called uncertainty sampling[1,2], which aims to query the samples with the least certainty to the current model. The other common criterion is called diversity sampling[3,4,16], whose goal is to reduce the redundancy of the uncertain samples. For the uncertainty sampling, there has been a large amount of research into the study of the uncertainty criterion, which can be grouped into three main areas: 1) query by committee [5,6], where the uncertainty is evaluated by measuring the disagreement of a committee of classifiers; 2) the posterior probability[8,9], in which the posterior probability is used to measure the uncertainty of the candidates; and 3) the large margin heuristic, where the distance to the margin is calculated to measure the uncertainty, which is well-suited for the margin-based classifiers such as support vector machine (SVM)[10,11]. However, in the current research, less attention is being paid to the diversity criterion. In general, cluster algorithms, such as k-means[12] and its kernel version[13,14], can be used to select the most uncorrelated samples by extracting from each cluster the representative one that is closest to the corresponding cluster center for labeling. However, clustering algorithms depend on the correctness of the convergence and are usually influenced by the adequacy of initialization. Meanwhile, in such methods, the queried data are not guaranteed to be i.i.d (identical and independent distribution), and the sampled data could not stay the same with the original data distribution, as they are selectively sampled based on the active learning criterion[15].

3   The Innovative Batch Mode Active Learning

This part mainly describes the active learning method proposed in this paper. Since using either kind of criterion alone is not sufficient to get the optimal result and less attention is paid on the diversity criterion, there are several works trying to query the unlabeled samples with both high information and high representation[16,4], so the proposed method is mainly focused on the diversity criterion and combines the discriminative information and representative information into one formulation as the diversity criterion. For uncertainty criterion, the proposed method adopts the two advanced the large margin heuristic method, namely MS, MCLU[17,18]. Then the mechanisms of the two uncertainty criterion is briefly summarized.

3.1   Uncertainty Criterion

The MS Active Learning Algorithm. The margin sampling (MS) heuristic takes the advantage of the distance to the hyperplane in SVM which builds a linear decision function in the high dimensional feature space H where the samples are more likely to be linearly separable. Consider the decision function of the two-class SVM:

$$f(q_i) = sign(\sum_{j=1}^{n} y_i \alpha_i K(x_j, q_i)) \tag{1}$$

Therefore, the candidate included in the training set is the one that respects the condition.

$$\hat{x} = \arg \min_{q_i \in Q} |f(q_i)| \tag{2}$$

The MCLU Active Learning Algorithm. The adopted MCLU technique selects the most uncertainty samples according to a confidence value, $c(x), x \in U$, which is defined on the basis of their functional distance $f_i(x), i = 1, 2, \ldots, n$ to the n decision boundaries in the binary SVM classifiers included in the OAA architecture[16]. In this technique, the distance of each sample $x \in U$ to each hyperplane is calculated, and a set of n distance values is obtained. Then, the confidence value $c(x)$ can be calculated using different strategies[18]. In this paper, the difference function $c_{diff}(x)$ strategy is used, which considers the difference between the first and second largest distance values to the hyperplanes.

$$
\begin{aligned}
r_{1\,max} &= \arg \max_{i=1,2,\ldots,n} \{f_i(x)\} \\
r_{2\,max} &= \arg \max_{j=1,2,\ldots,n \neq r_{1\,max}} \{f_j(x)\} \\
c_{diff}(x) &= r_{1\,max} - r_{2\,max}
\end{aligned}
\tag{3}
$$

3.2 The Proposed Diversity Criterion

Suppose we are given a dataset with n samples $D = \{x_1, x_2, \ldots, x_n\}$ of d dimensions. Firstly, select l samples as the initial training set $L = \{(x_1, y_1), (x_2, y_2), \ldots, (x_n, y_n)\}$, $y_i \in \{1, 2, \ldots, N_c\}$, N_c is the number of classes. The remaining $u = n - l$ samples form the unlabeled set $U = \{x_{l+1}, x_{l+2}, \ldots, x_{l+u}\}$, which is candidate dataset for the active learning. In the following, we will discuss the discriminative and informative criterion in detail. As mentioned above, the proposed diversity criterion combines the discriminative information and representative information into one optimal formula. Now we represent discriminative information and representative information, respectively.

Discriminative Information Determined by the Minimum Margin. In this paper, we derive a novel form of upper bound for true risk in the active learning setting, by minimizing this upper bound to measure the discriminative information.

$$
\max_{Y_i : \forall x_j \in Q} \min_{Q,f} \sum_{x_i \in L} (Y_i - f(x_i))^2 + \sum_{x_j \in Q} (Y_j - \hat{f}(x_j))^2 + \lambda \|f\|_F^2
\tag{4}
$$

In formula (4), $Y_i \in \{1, -1\}f$ is the classifier, $\|f\|_F^2$ is used to constrain the complexity of the classifier class. It is known that, generally speaking, the data classification is a multi-class problem. Now we develop it to a multi-class formula, and assume there are classes in a dataset. According to formula (4), binary class problem could be built, so the multi-class optimal formula could be represented as:

$$
\min \sum_{c=1}^{N_c} \left\{ \max_{\hat{Y}} \min_{Q_c, f_c} \sum_{x_i \in L} (Y_i - f(x_i))^2 + \sum_{x_j \in Q_c} (\hat{Y}_j - \overset{\wedge}{f_c}(x_j))^2 + \lambda \|f_c\|_F^2 \right\}
\tag{5}
$$

where $x_i \in L$, if $y_i = c, Y_i = 1$ else $Y_i = -1. \hat{Y}_j = -sign(f(x_j))$,which is a pseudo label;Q_c is query samples set and is get under the binary problem when the class c label is 1 and the other classes label is -1. Since the multi-class optimal formula is composed by N_c binary classes, and it is independent for each binary class, so we can solve (5) by solve each binary class independently. Now we use the linear regression model in the kernel space as the classifier, which is in the form of $f(x) = w^T \phi(x)$, with the feature mapping $\phi(x)$. By minimizing the worst case, the objective function (5) becomes:

$$\min \sum_{c=1}^{N_c} \min_{Q_c, w_c} \left\{ \sum_{x_i \in L} (Y_i - w_c^T \phi(x_i))^2 + \left[(w_c^T \phi(x_j))^2 + 2 \left| w_c^T \phi(x_j) \right| \right] + \lambda \|w\|^2 \right\} \tag{6}$$

Representative Information Determined by MMD. The representative part in proposed method is the MMD term. MMD is used to constrain the distribution between the labeled and queried samples, and makes it as similar to the overall sample distribution as possible. It captures the representative information of the data structure in the proposed method. According to[25], the MMD can be expressed as

$$\min_{\alpha^T 1_u = b} \frac{1}{2} \alpha^T K_{UU} \alpha + \frac{u-b}{n} 1_l K_{LU} \alpha - \frac{l+b}{n} 1_u K_{UU} \alpha \tag{7}$$

In formula (7), $1_l, 1_u$ are two vectors of length l and u, respectively, with all entries 1; α is the indicator vector with u elements and each element $\alpha_i \in \{0,1\}$; $\alpha^T 1_u = b$; b is the query number for a binary class problem. K is the kernel matrix with its elements as $K_{ij} = k(x_i, x_j) = \phi(x_i)^T \phi(x_j)$, and K_{AB} denotes its sub-matrix between the samples from set A and B. The objective function (6) can be further simplified as

$$\min_{\alpha^T 1_u = b} \alpha^T K_1 \alpha + k\alpha \tag{8}$$

where $K_1 = \frac{1}{2} K_{UU}, k = k_3 - k_2, \forall x_i \in U, k_2(i) = \frac{l+b}{n} \sum_{x_j \in U} K(i,j), k_3(i) = \frac{u-b}{n} \sum_{x_j \in L} K(i,j)$

The Proposed Discriminative and Representative Diversity Criterion. We combine the discriminative and representative information into one optimal formula, which use a weight to balance the discriminative information term and representative information term. Firstly, build the optimal formula for the binary problem is built and developed it to multiclass problem. So we can put (4) and (8) together, and obtain the binary class optimal formula:

$$\min_{w, \alpha^T 1_u = b} \sum_{x_i \in L} (Y_i - w^T \phi(x_i))^2 + \sum_{j=1}^{u} \alpha_j \left[(w^T \phi(x_j))^2 + 2 \left| w^T \phi(x_j) \right| \right] + \lambda \|w\|^2$$
$$+ \beta (\alpha^T K_1 \alpha + k\alpha) \tag{9}$$

Like the formula (5), we develop it to multi-class problem.

$$\min \sum_{c=1}^{N_c} \{ \min_{w_c, \alpha_c^T 1_u = b} \sum_{x_i \in L} (Y_i - w_c^T \phi(x_i))^2 + \sum_{j=1}^{u} \alpha_{cj} \left[(w_c^T \phi(x_j))^2 + 2 \left| w_c^T \phi(x_j) \right| \right]$$
$$+ \lambda \|w_c\|^2 + \beta_c (\alpha_c^T K_1 \alpha_c + k_c \alpha_c) \}$$
(10)

The formula (9) is the main part in the proposed method. The alternating optimization strategy is employed to solve it[3]. Since the multiclass is formed by many binary problems, (10) can be solved by a binary model. For a binary problem, if the query index α is fixed, the objective is to find the best classifier based on the current labeled and query samples:

$$\min_w \sum_{i=1}^{l} (y_i - w^T \phi(x_i))^2 + \lambda \|w\|^2 + \sum_{j=1}^{b} \left[\left\| w^T \phi(x_j) \right\|_2^2 + 2 \left| w^T \phi(x_j) \right| \right] \quad (11)$$

The equation (11) can be solved by the alternating direction method of multipliers (ADMM)[20]. If w is fixed, the objective becomes

$$\min \sum_{j=1}^{u} \alpha_j \left[(w^T \phi(x_j))^2 + 2 \left| w^T \phi(x_j) \right| \right] + \beta(\alpha^T K_1 \alpha + k\alpha) \quad (12)$$

which can be rewritten as

$$\min_{\alpha^T 1_u = b} \beta \alpha^T K_1 \alpha + (\beta k + a) \alpha \quad (13)$$

where $a_j = (w^T \phi(x_j))^2 + 2 \left| w^T \phi(x_j) \right|$, (13) is a standard quadratic programming. If we relax α to a continuous value range $[0,1]^u$, the b samples in the pool U corresponding to the largest b elements are the sample with discriminative and representative information.

3.3 The Proposed Batch Mode Active Learning

In the novel discriminative and representative criterion, a QP problem should be solved, so if the unlabeled data set is big, the solution will be slow. In order to overcome this problem, for each iteration, the proposed method firstly uses the uncertainty criterion to select the N most uncertain samples from the unlabeled samples pool U as the new unlabeled samples U_n. The optimal methods include MS, MCLU. For each class, the training set L and the unlabeled samples U_n are used to establish a binary class problem, so when there is N_c classes, N_c binary class problem can be built. For each binary class the discriminative and representative criterion are employed to select b batch size samples, so we can get $b(N_c)$ query samples in the query set G. Since we apply the discriminative and representative criterion on the same unlabeled samples U_n, the same samples may be queried in different binary class, so the $b(N_c)$ query samples may contain the repetitive samples. Firstly, the repetitive samples in G should be merged to obtain the new dataset S in which the samples are different. In this step the size of S is adaptive. In order to control the size of the querying samples added into the training set in each iteration and reduce the spatial redundancy, the maximum number of the new queried samples is set as h. The last samples we queried according to the size of S. If the size of S is greater than the initial query batch size h, the h samples are chosen corresponding to the h smallest values of formula (9) from the set S. In this way, the size of query samples is always less than or equal to the initial h. The progress of proposed method is shown in Algorithm 1.

Algorithm 1. Algorithm 2 The Progress of Proposed Method

Input:
 Initial training data set $x^{iter} = \{x_i, y_i\}_{i=1}^{l}, iter = 1, y_i \in \{1, 2, \ldots, N_c\}$
 The number of the most Uncertain Unlabeled Samples N
 The number of selected discriminative and representative unlabeled samples in each one to all Discriminative and Representative Criterion b
 The number that adds to training set at one iteration h
 The number of iteration it

Output:
 for $iter = 1 : it$
 1: Train the current Labeled data set X^{iter} with OAA SVM;
 2: Train the Unlabeled data set $U^{iter} = \{x_i\}_{i=l+1}^{l+u}$ and select the N most Uncertainty Unlabeled Samples
 3: Initial discriminative and representative set $G = \Phi$
 for $c = 1 : N_c$
 4: Set the one to all labeled set $X = \{x_i, y_i\}_{i=1}^{l}, Y_i = \{1, -1\}$(if y==c, Y=1 else Y=-1 end)
 5: Select the b discriminative and representative samples add to set G from the N most uncertainty unlabeled samples using the discriminative and representative criterion (Algorithm 1) with the best weight β
 6: Merge the same samples in the discriminative and representative set G
 end
 7: Update $X^{iter+1} = X^{iter} \cup S$ and $U^{iter+1} = U^{iter} \backslash S$
 end

---

4 Experiments

In this section, two benchmark hyperspectral images have been used for experiments. The proposed method is compared with the state-the-art batch mode active learning methods and at last the experimental results are analyzed. We list all the methods we compared in the experiments in Table 1.

Two benchmark hyperspectral image datasets are Indian Pines and Washington DC[23]. Indian Pines was acquired by NASAs Airborne Visible Infrared Imaging Spectrometer (AVIRIS). A total of 10249 samples from 16 classes are used. Meanwhile, Washington DC Mall data set is a Hyperspectral Digital Imagery Collection Experiment airborne Hyperspectral Image. A total of 8424 pixels from seven classes were used in our experiments on it. We summarize the characteristics of the data sets in Table 2.

Table 1. Different combinations of the uncertainty criterion and diversity criterion

Uncertainty Criterion	Diversity Criterion	Method
Marginal Sampling	Closest-Support Vector	MS-cSV
Marginal Sampling	Kernel K-means	MS-Kkmeans
Marginal Sampling	The Proposed DR Criterion	MS-DR
Multiclss-Level Uncertainty	/	MCLU
Multiclss-Level Uncertainty	Kernel K-means	MCLU-ECBD
Multiclss-Level Uncertainty	The Proposed DR Criterion	MCLU-DR
Query by Committee	/	EQB

Table 2. Characteristics of the data sets, including the numbers of the corresponding classes and samples.

Indian Pines				Washington DC	
Class	#Samples	Class	#Samples	Class	#Samples
Alfalfa	46	Oats	20	Water	1433
Corn-notill	1428	Soybean-notill	972	Road	1463
Corn	237	Soybean-clean	593	Roof	1342
Grass-pasture	483	Grass-pasture-mowed	28	Trail	1033
Grass-trees	730	Hay-windrowed	478	Shadow	1059
Wheat	205	Buildings-Grass-Trees-Drivers	386	Grass	1026
Woods	1265	Stone-Steel-Towers	93	Tree	1068
Corn-mintill	830	Soybean-mintill	2455		

In the experiments, for each data set, 15 samples are randomly selected for each class as the initial training set, and the rest are regarded as the unlabeled samples. the total samples are took as the testing set. In the uncertainty criterion, 400 uncertain samples are chose to input into our proposed diversity criterion. Because the queried samples of the proposed method is adaptive, so here the maximum value of h which is the number of samples add to the training set should be set in each iteration. For Indian Pines data set which includes 16 classes, in the experiment, we set h=20, and set the batch size b=2 which is the number of samples queried from a binary class. For the Washington DC data set which includes 7 classes, we set h=15 and b=3. But for the other compared method, the number of queried samples is manually fixed, so for Indian Pines the number of queried samples is 20 and for Washington DC the number is 15. There are some parameters in the compared methods, we adopt the values in the original papers. In our proposed method, we set the regularization weight , and the trade-off parameter is chosen from a candidate set by a cross validation. For each data set, we use the same kernel for all methods, which is properly chosen from the linear kernel or RBF kernel with the optimal kernel width. For the fairness, the same SVM classifier is used for all methods to evaluate the informativeness of the selected samples and set the number of iterations as 60 . The accuracy curve of the SVM classifier are reported after each query. The SVM provided by LIBSVM[21].

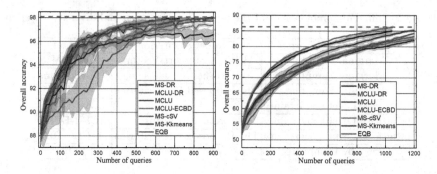

Fig. 1. Comparison of different active learning methods on two datasets with the average result of 10 runs. Left: The curves of DC data. Right: The curves of Indian Pines data.

4.1   Results and Analysis

For each data set, we run the experiment independently for several runs, and present the results in Fig. 1. Intuitively, for the two benchmark hyperspectral image datasets experimental results, we can observe that our method outperforms the competitors in five aspects. Firstly, the proposed methods MS-DR and MCLU-DR show the curves higher learning rate than the other methods. Secondly, when the overall accuracy is same, the queried samples of the proposed method are less than the state-the-art methods. Thirdly, when the iteration time is same, the proposed method can obtain a higher overall accuracy with less queried samples than the other methods. And the standard deviation of the proposed method is smaller than the compared methods. Finally, the curves of the proposed method are much more smooth than the state-the-art methods, especially for the Washington DC experiments. These results demonstrate that both discriminative and representative information are critical for active learning, and a proper balance of these two sources of information will boost the active learning performance.

In the proposed method, the important parameter is the balance weight β between discriminative information and representative information. In the experiment, for each iteration, we run our algorithm from a candidate set $\{10^{-5}, 10^{-4}, 10^{-3}..., 10^{5}\}$, and choose the best value which make the overall accuracy highest. We report the results on the two hyperspectral image data set. Because for each binary class there is a best value, so in each iteration the value should choose N_c times, N_c is the number of classes. Now we show the value statistics of each class in 60 iterations in figure 2. From fig.2, we can observe the distribution of trade-off parameter β on two data sets are similar in the experiments. We can see that when the β is greater than 1, these values occupation is less than 30%, this part values is mainly to adjust the training set when the uncertain information is not enough to select the most informative samples. But for the β is less than 1, this part values take an occupation more than 70%. For the proposed method, when β is greater than 1, the representative information is predominantly, or the discriminative information is major. According to that, all the experiments demonstrates that the discriminative information is predominant. So if the user does not require selecting the best value, he can choose a compromised value which not only pays attention on the discriminative information but also considers the representative information. $\beta = 10^{-3}$ can be used in all the iterations.

4.2 Sensitivity Analysis

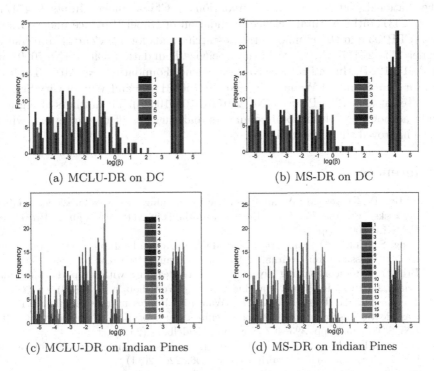

(a) MCLU-DR on DC   (b) MS-DR on DC

(c) MCLU-DR on Indian Pines   (d) MS-DR on Indian Pines

Fig. 2. The trade-off parameters statistics on two benchmark data sets of each class in 60 iterations.

5 Conclusion

This paper developes a dicriminative and representative information based batch-mode active learning, and applys it to Hyperspectral image classification. According to the experients on two benchmark remote sensing image data sets, the proposed AL method shows a superior performance when compared with the other state-of-the-art AL methods. Firstly, the proposed diversity can effectively reduce the number of querying samples with the adaptively number of the new queried samples. Secondly, by considering the uncertainty information in the diversity, the proposed diversity criterion measures the uncertainty information by the disiciminative information, and combines the representative information into one formula. Finally, maximum mean discrepancy (MMD) is used to measure the representative information, and makes sure the new queried samples are i.i.d between each other. That also makes the new queried samples have consistency of distribution in the original space.

Acknowledgments. This work was supported in part by the National Basic Research Program of China (973 Program) under Grant 2012CB719905, the National Natural Science Foundation of China under Grants 61471274 and 41431175the Natural Science Foundation of Hubei Province under Grants 2014CFB193 and the Fundamental Research Funds for the Central Universities under 211−274175. National Natural Science Foundation, China (No.70901060 and 61471274), Hubei Province Natural Science Foundation (No. 2011CDB461), Youth Plan Found of Wuhan City (No.201150431101) and Wuhan City Science and Technology Plan Project (No. 2015010101010023). The authors also gratefully acknowledge the helpful comments and suggests of the reviewers, which have improved the presentation.

References

1. Settles, B., Craven, M.: An analysis of active learning strategies for sequence labeling tasks. In: Proc. Int. Conf. Emprical Methods in Natural Language Processing (EMNLP 2008), pp. 1070–1079 (2008)
2. Tong, S., Koller, D.: Support Vector Machine Active Learning with Applications to Text Classification. Journal of Machine Learning Research 2, 45–66 (2001)
3. Brinker, K.: Incorporating diversity in active learning with support vector machines. In: Proc. Int. Conf. Machine Learning Workshop, p. 59 (2003)
4. Xu, Z., Yu, K., Tresp, V., Xu, X., Wang, J.: Representative sampling for text classification using support vector machines. In: Sebastiani, F. (ed.) ECIR 2003. LNCS, vol. 2633, pp. 393–407. Springer, Heidelberg (2003)
5. Argamon-Engelson, S., Dagan, I.: Committee-based sample selection for probabilistic classifiers. arXiv preprint arXiv:1106.0220 (2011)
6. Melville, P., Mooney, R.J.: Diverse ensembles for active learning. In: Proc. Int. Conf. Machine Learning Workshop, p. 584 (2004)
7. Nigam, K., McCallum, A.K., Thrun, S., Mitchell, T.: Text classification from labeled and unlabeled documents using EM. Mach. Learn. 39, 103–134 (2000)
8. Luo, T., Kramer, K., Goldgof, D.B., Hall, L.O.: Active learning to recognize multiple types of plankton. In: Proc. 17th Int. Conf. Pattern Recognition, ICPR 2004, pp. 478–481 (2004)
9. Roy, N., McCallum, A.: Toward optimal active learning through monte carlo estimation of error reduction. In: ICML, Williamstown (2001)
10. Campbell, C., Cristianini, N., Smola, A.: Query learning with large margin classifiers. In: Proc. Int. Conf. Machine Learning Workshop, pp. 111–118 (2000)
11. Schohn, G., Cohn, D.: Less is more: Active learning with support vector machines. In: Proc. Int. Conf. Machine Learning Workshop, pp. 839–846 (2000)
12. Kanungo, T., Mount, D.M., Netanyahu, N.S.: An efficient k-means clustering algorithm: Analysis and implementation. IEEE Trans. Pattern Anal. Mach. Intell. 24, 881–892 (2002)

13. Shawe-Taylor, J., Cristianini, N.: Kernel methods for pattern analysis. Cambridge University Press (2004)
14. Dhillon, I.S., Guan, Y., Kulis, B.: Kernel k-means: spectral clustering and normalized cuts. In: Proc. 10th ACM SIGKDD Int. Conf. Knowledge Discovery and Data Mining, pp. 551–556 (2004)
15. Beygelzimer, A., Dasgupta, S., Langford, J.: Importance weighted active learning. In: Proceedings of the 26th International Conference on Machine Learning (ICML), pp. 49–56 (2009)
16. Nguyen, H.T., Smeulders, A.: Active learning using pre-clustering. In: Proceedings of the 21st International Conference on Machine Learning (ICML), New York, NY, USA, pp. 79–86 (2004)
17. Tuia, D., Ratle, F., Pacifici, F., Kanevski, M., Emery, W.J.: Active learning methods for remote sensing image classification. IEEE Trans. Geosci. Remote Sens. 47(7), 2218–2232 (2009)
18. Demir, B., Persello, C., Bruzzone, L.: Batch-mode active-learning methods for the interactive classification of remote sensing images. IEEE Trans. Geosci. Remote Sens. 49, 1014–1031 (2011)
19. Bezdek, J.C., Hathaway, R.J.: Convergence of alternating optimization. Neural, Parallel and Scientific Computations 11(4), 351–368 (2003)
20. Dasgupta, S.: Two faces of active learning. Theoretical Computer Science 41(19), 1767–1781 (2011)
21. Chang, C.C., Lin, C.J.: LIBSVM: a library for support vector machines. ACM T. Intell. Syst. Technol. (TIST) 2, 27 (2011)
22. Wenbin, C., Ya, Z., Jun, Z.: Maximizing expected model change for active learning in regression. In: Proc. 13th Intl. Conf. Data Mining (2013)
23. Chakrabarti, A., Zickler, T.: Statistics of real-world hyperspectral images. In: IEEE Conference on Computer Vision and Pattern Recognition (CVPR), pp. 193–200 (2011)
24. Settles, B.: Active Learning. Morgan and Claypool (2012)
25. Chattopadhyay, R., Wang, Z., Fan, W., Davidson, I., Panchanathan, S., Ye, J.: Batch mode active sampling based on marginal probability distribution matching. In: Proc. 10th ACM SIGKDD Int. Conf. Knowledge Discovery and Data Mining (KDD), pp. 741–749 (2012)

A Multi-view Retweeting Behaviors Prediction in Social Networks

Bo Jiang, Ying Sha, and Lihong Wang

Institute of Information Engineering, Chinese Academy of Sciences,
Beijing, 10093, China
{jiangbo,shaying}@iie.ac.cn

Abstract. Retweeting is the most prominent feature in online social networks. It allows users to reshare another user's tweets for her followers and bring about second information diffusion. Predicting retweeting behaviors is an important and essential task for advertising product launch, hot event detection and analysis of human behavior. However, most of the methods and systems have been developed for modeling the retweeting behaviors, it has not been fully explored for this problem. In this paper, we first cast the problem of retweeting behaviors prediction as a classification task and propose a formally definition. We then systematically summarize and extract a lot of features, namely user status, content, temporal, and social tie information, for predicting users' retweeting behaviors. We incorporate these features into Support Vector Machine (SVM) model for our prediction problem. Finally, we conduct extensive experiments on a real world dataset collected from Twitter to validate our proposed approach. Our experimental results demonstrate that our proposed model can improve prediction effectiveness by combining the extracted features compared to the baselines that do not.

Keywords: retweeting behaviors, online social networks, SVM, extract feature, classification.

1 Introduction

Online social networks such as Sina Weibo, Twitter and Facebook have become an important information service platform for all walks of life. People not only share interesting information each other but also express their views on hot topic occurred in the real life. According to the study in [16], users post more than 500 million tweets every day in Twitter. These information is widely spread with thousands of millions of users participating over Twitter through retweeting mechanism. Retweeting is a social service function for users that one can reshare any users' tweets to his/her timeline. Through the feature, tweets can quickly reach all of their followers. This may cause information cascade where a tweet is reposted from one user to another or from one community to another. Moreover, retweeting has proven to be a significant factor for the form of large information cascades in social network [5]. Therefore, understanding the mechanisms of information diffusion and predicting users' retweeting behaviors are an

© Springer International Publishing Switzerland 2015
R. Cheng et al. (Eds.): APWeb 2015, LNCS 9313, pp. 756–767, 2015.
DOI: 10.1007/978-3-319-25255-1_62

important tasks for effective monitoring the trend of information diffusion and maximizing the popularity of new product. Specifically, in this paper, our goal is to choose a focal user and then try to predict who will retweet an incoming tweet published by the focal user in the near future. We propose a prediction model combining user status, content, temporal, and social tie information to model users' retweeting behaviors.

The main contributions of this work can be summarized as follows:

- We formulate the problem of retweeting behavior prediction as a classification task. Specifically, given an incoming tweet posted by a publisher, our goal is to predict who will retweet.
- We systematically summarize and extract a lot of features which are closely related with retweeting behavior prediction. We then propose a prediction model to incorporate user status, content, temporal, and social tie information for predicting users' retweeting behaviors. Meanwhile, we introduce semantic enrichment technologies to measure interests relevance between publishers, followers and transmissible tweets.
- We collect a large number of tweets from Twitter service. Experimental results on the constructed dataset demonstrate that our proposed method outperforms other baselines with a significant margin.

The rest of this paper is organized as follows. We review the related work in Section 2. In Section 3, we introduce a clear definition of retweeting prediction task. Section 4 discusses how to extract effective features for the problem. We present experiments and empirical analysis of our models in Section 5. Finally, conclusions are given in Section 6.

2   Related Work

The studies of retweeting behavior in social network have exist an explosion of research. We can roughly divide these works into two categories of models in this scope: (i) explanatory models and (ii) predictive models. In the following, we will summarize and review some representative efforts in both of them.

The goal of explanatory models is to understand why people retweet and analysis which factors impact retweet. These models are very useful to understand how human make decision and how information spread. For example, Boyd et al. [2] conduct a user survey to analysis the reasons on how people retweet, why people retweet, and what people retweet. Suh et al. [15] firstly collect a large number of Twitter data and extract a number of features to identify factors that might affect retweetability of tweets. Yang et al. [3] analyze how the retweeting behaviors is influenced based on user, message and time factors. They find that almost 25.5% of the tweets posted by users are actually retweeted from their friends. Macskassy et al. [11] make a better understanding of what makes people spread information in Twitter through the use of retweeting. Abdullah et al. [1] conduct a user survey to investigate what is the user's action towards spread message and why user decide to perform on the spread message. Their results reveal that users retweet a message due to the important and interesting of content and author's influence.

The aim of predictive models is to predict who will repost a tweet and the scale and depth of retweeting in a given network based on user, content, social and/or temporal features. For instance, Liu et al. [9] propose a probabilistic graph model to measure topic-level influence between users in order to predict user behaviors. Zhang et al. [18] propose the notion of social influence locality and construct a large ego network to study users' retweeting behaviors. Naveed et al. [12] only employ the content-based features of retweets to predict the probability of a tweet to be retweeted. Can et al. [4] exploit content- and structure-based features as well as image-based features to predict the retweet count of the tweets that contain links to images. Zaman et al. [17] present a collaborative filtering approach using user feature and tweet feature to predict individual retweets in Twitter. Luo et al. [10] use a wide range of followers' features, such as retweet history, social status and interests to predict retweet occurrences. Feng et al. [7] develop a feature-aware factorization model to recommend the tweets based on the their probability of being retweeted. Peng et al. [13] propose using conditional random fields (CRFs) to model and predict the retweet patterns with three types of user-tweet features. Petrovic et al. [14] propose a time-sensitive model combining social features and tweets features to predict retweets. Zhang et al. [19] develop using non-parametric Bayesian model adapted from the hierarchical Dirichlet process for predicting retweeting behaviors.

However, prior research work ignore some critical factors for users' retweeting behaviors. First, the more interactions between users show that they have a strong relationship, and retweeting behaviors are more likely to happen. Second, user's topics of interest play an important role when retweeting, previous work don't precisely identify the users interests due to only use Bag Of Words approach. We focus on how to tackle these problems in this paper.

3   Problem Statement

Retweeting is an important social function for information diffusion in social networks. It allows users to directly repost a tweet using the form of RT @username.

For convenience, we name RBP (**R**etweeting **B**ehavior **P**rediction) for our proposed model. Meanwhile, we formally define RT @username as a three tuple representation of retweeting as follows.

Definition 1. (RT @username) Suppose given a tweet t posted by a user u and a user v repost the tweet t to her timeline. An RT @username is a retweeting relation triple (u, t, v). Specifically, we denote u as retweetee, t as transmission tweet and v as retweeter, respectively.

We formally define the problem of user retweeting prediction as a classification problem. More specifically, given a tweet t published by a user u, and the candidate set C that is consist of followers and the interactive users, the goal of the work is to find who will retweet the tweet t in C where every candidate

$c_i(\in C)$ is tagged whether she retweeted tweet t in training data or not. Therefore, the problem can be solved by employing effective features in a supervised learning framework.

4   Features for Retweeting Prediction

In this section, we define a lot of features for modeling the retweeting behaviors. These features are roughly divided into four main categories, namely user status, content, temporal and social tie information. Next, we will introduce the extraction methods of these features.

4.1   User Status Features

User's personal attributes are an important feature for effecting one's retweeting or retweeted. Specifically, we also consider the personal profile of publisher and follower, including *the number of followers, followees, tweets, favorites and listed, verification status, the age of the account, location, whether the user profile has a self-description, and whether the user profile has a URL.* Intuitively, the richer profile user has, the stronger the social credibility. Hence, their tweets are more likely to be retweeted than those who have recently just create a new account or have a mall number of content in their profile.

4.2   Content Features

The intuition behind is that whether a user retweet an incoming tweet or not depends on content's self-feature and user's interest to a certain extent. Hence, in the subsection, we extract three categories of features for retweeting behavior prediction, namely *self-feature, the interest similarity between publishers and followers, the topic similarity between an incoming tweet and followers.*

For tweet's self-feature, we extract a set of feature, such as the number of hashtags, URLs, media (containing images and videos) and mentions (referencing other users in tweet text). Previous study [15] has been found that these features have strong relationships with retweetability. Moreover, a lot of tweets are an express of personal sentiments in social network. [8] has shown that whether a tweet contains sentiment words or not may effect the retweetability of the tweet. To examine the influence of emotion in the retweeting, we use Stanford CoreNLP[1] that is a natural language analysis library including sentiment analysis tools to identify tweet emotional content. We classify sentiment into three classes, such as positive, neutral and negative. 1 represents the positive emotion of a tweet, 0 represents neutral, and -1 represents negative.

For the second feature, previous work [1,10] has been studied that the match of topics of interest between publishers and followers is an key important feature for retweeting behavior prediction. When facing a lot of incoming tweets, a follower is more likely to retweet these tweets that he/she is interested in.

---

[1] http://nlp.stanford.edu/software/corenlp.shtml

To calculate the match, an intuitive way is to leverage topic modeling methods like Latent Dirichlet Allocation (LDA) for extracting user topic distribution. However, these topic modeling methods may not fit for tweets due to their short text, noisy, ambiguous and dynamic nature features [6]. Consequently, we use semantic enrichment methods to generate user's topics of interest. The semantic enrichment methods explicitly attach entities in their tweets to semantic annotations by applying Linked Data, and therewith allows for making explicit topic tags. Since the focus of our work is on user's topics of interest identification and not semantic annotations technologies, we employee an existing solution.

A plethora of semantic annotation with linked data techniques and systems is available in general [6]. We opt to use OpenCalais[2] for our work because of the following reasons: (1) OpenCalais maps the entities identified in tweets to their corresponding topic label; (2) OpenCalais incorporates state-of-the-art semantic function with content; (3) OpenCalais provides a relatively high rate limit to other services, the API default usage quotas are 50,000 transactions per day, 4 transactions per second[3]. OpenCalais now offers 18 topics categorization, such as Sports, Education, Environment and Politics. Therefore, in this paper, we construct a 18 dimensions vector representation to profile user's topics of interest. We can then compute a cosine similarity of user's interest between publishers and followers. Specifically, give the publisher u and his/her follower v, u's topics vector is denoted as $P(u)$, and v's topics vector is denoted as $P(v)$. We denote *interest similarity* between u and v as follows:

$$InterSim(u, v) = \frac{P_u \cdot P_v}{\parallel P_u \parallel \parallel P_v \parallel} \tag{1}$$

Last but not least, the interesting tweet content also is a key factor in pushing the retweetablity of the users. Therefore, in this paper, we also propose a feature to measure the retweetable of the followers for an incoming tweet. Analogously, given a tweet t and the user v, we denote the topic vector of tweet t as $P(t)$, and then *topic similarity* between tweet t and user v can be estimated as below:

$$TopicSim(t, v) = \frac{P_t \cdot P_v}{\parallel P_t \parallel \parallel P_v \parallel} \tag{2}$$

4.3   Temporal Features

Intuitively, Twitter users have the same or similar activity time, the retweet action is more likely happen each other. This is because tweets published by one user in the morning are often overwhelmed by other users in the afternoon because of them being replaced by the more recent tweets. In this study, we consider four time based features as the recency of retweeting a tweet. The first feature can be extracted from user profile that whether two users are in the same

---

[2] http://www.opencalais.com/
[3] http://www.opencalais.com/documentation/calais-web-service-api/
usage-quotas

timezone or not. We use *timezone*(1 indicates being lived in the same area and 0 indicates not being lived) to represent the time difference of user living city.

We also propose another feature *activity time overlap*, which is defined as the overlapping degree of posting time between publisher and follower. Specifically, we divide one day into 24 units per one hour and map user's activity time span to discrete time slice. Assume given publisher u and follower v, we denote u's activity time span as $t_a(u)$, and v's activity time span as $t_a(v)$. Similar to Jaccard Coefficient, we define *activity time overlap* between publisher u and follower v as follows:

$$ATO(u,v) = \frac{|t_a(u) \bigcap t_a(v)|}{|t_a(u) \bigcup t_a(v)|} \tag{3}$$

This feature could measure the consistency of publisher u's posting time habit and follower v's posting time.

Moreover, the more recent the interaction happen, the more likely the behavior to be occurred in the near future. Specifically, we denote the timestamp of v first interact with u as $I_s(v,u)$, and the timestamp of their last interaction as $I_e(v,u)$. We define *interaction span* from v to u is defined as:

$$IS(v,u) = I_e(v,u) - I_s(v,u) \tag{4}$$

Correspondingly, *interaction frequency* is the average interacting interval from v to u:

$$IF(v,u) = \frac{freq(v,u)}{IS(v,u)} \tag{5}$$

where $freq(v,u)$ is the number of times v interact with u in the above given time interval.

Furthermore, the more recent the retweeting happen, the more likely the behavior to be occurred in the near future. Specifically, we denote the timestamp of follower v last retweet tweets posted by publisher u as $R_e(v,u)$. We then calculate *recent retweeting interval*, which is defined as the interval between $R_e(v,u)$ and the timestamp $T_t(u)$ of an incoming tweet t posted by user u:

$$RRI(u,t,v) = T_t(u) - R_e(v,u) \tag{6}$$

4.4   Social Tie Features

The intuition behind is, the more strong social tie between users has each other, the more likely the retweeting happen. We measure the strength of social tie combining structural, relationship, and interaction information together to predict users' retweeting behaviors.

To extract structural feature, we first construct an explicit network G_e by utilizing the followees and followers relationships in the data collection. In G_e, the nodes represent users and the directed edges represent following or followed relationship between user u and user v. We extract two features between two users by the number of *mutual followees* and *mutual followers* as prediction indicator.

In addition to the explicit network, we also construct an implicit network G_i that is consist of retweet network, reply network and mention network. We call G_i as interaction network. Therefore, we extract *interaction num* that sum of the number of retwteet, reply and mention from follower to publisher as one of the prediction features.

Moreover, we observe that users have four types of relationship each other in social network such as stranger, followee, follower, friend where stranger denotes no link between users, followee and follower denote a unidirectional follow relationship, friend denotes a bidirectional follow relationship. We define *relationship type* to measure the familiarity between users. Specifically, 3 represents friend relationship, 2 represents following relationship, 1 represents follower relationship and 0 represents stranger relationship.

To sum up, Table 1 gives a complete list of features described in this section for retweeting prediction task, where u denotes retweetee, t is transmission tweet and v denotes retweeter.

Table 1. The summary of features for retweet prediction model

Category	Feature	Description
User Status Feature	num_follower	number of who one is followed
	num_friend	number of whom one is following
	num_tweet	number of tweets that one post
	num_favourites	number of tweets that one like
	num_listed	number of group that one list
	is_verification	whether account is verified or not
	age_account	the account create time
	location	whether user's location enable or not
	self-description	whether profile has a self-description or not
	URL	whether profile has a URL or not.
Content Feature	num_hashtag	number of hashtag that a tweet contain
	num_URL	number of URL that a tweet contain
	num_media	number of media that a tweet contain
	num_mention	number of mention refer to another user
	sentiment	tweet emotional class
	u2u_interest	topic based user similarity between users
	t2u_interest	topic based preference of user v for t
Temporal Feature	timezone	whether u and v are lie the same area or not
	ATO	activity time overlap between u and v
	IF	interaction frequency between u and v
	RRI	recent retweeting interval between u and v
Social Tie Feature	mutual_followee	number of mutual followee between u and v
	mutual_follower	number of mutual follower between u and v
	num_interaction	interaction number between u and v
	relationship	the type of relationship between u and v

5 Experiments

In this section, we first describe the approach of data collection from Twitter service. Then the baseline methods and evaluation metrics are proposed. Finally, we compare the effectiveness of our approach with these baseline methods and analysis the results.

5.1 Data Collection

For the work executed for this paper, we collect the microblog data from Twitter service. More specifically, we use Twitter API[4] to collect our experiment dataset from 15th September to 20th December 2014. The dataset was collected in the following ways. First, we randomly select 50 users as seed users. For each seed, we crawl four types of information including personal profile, follower and followee list, and all tweets. In the same way, we also collect all these information of their followers and followees. Finally, the dataset contains 14325 users, 63 millon edges relationships among them and 22 millon tweets.

We focus on retweeting behaviors in social networks. Thus, we preprocess the dataset by extracting popular tweets which are retweeted larger than 30 times from the data set. Each diffusion process contains the original retweetee and all its retweeters. After preparation, we have 2615 publishers, 31359 original tweets which give rise to 1,849,596 retweet instances. Table 2 lists statistics of the retweeted data.

Table 2. Retweeters data statistics

Dataset	#Users	#Relationships	#OriginalTweets	#Retweets
Twitter	2615	4,852,240	31,359	1,849,596

In order to limit problems like overfitting, we first utilize a ten-fold cross validation and split the data into training and testing data. We then report the average performance in ten rounds of tests.

5.2 Comparison Methods

To evaluate the performance of our prediction models, we compare our prediction results with four baseline prediction models as follows:

- **Random Guess(RG):** We randomly selects users and randomly assign the class label to each user with the prior probability.
- **Majority Vote(MV):** We observe that a lot of users ever retweet the same user's tweets many times. Therefore, we first rank candidates in our dataset by the number of times they ever retweeted the publisher's previous tweets before. Then, we choose top ranked users to retweet. This simple but powerful baseline has been used in most existing studies.

---

[4] https://dev.twitter.com/docs/api/1.1

- **Who Will Retweet Me(WRM):** Our work is similar to previous study [10] focusing on a wide range of features, such as retweet history, followers status, followers active time and followers interests to find retweeters in Twitter. We implement the method as the baseline.
- **LRC-BQ:** [18] formally define the feature of social influence locality, and also combine additional features(personal attributes, instantaneity and topic propensity). The authors use these features to train a logistic regression classifier for predicting users' retweeting behaviors. Moreover, the method and dataset mentioned in the paper has been released[5]. We employee the method as the baseline.

5.3   Evaluation Metrics

We use four common metrics to evaluate the performance of our prediction model, namely Precision, Recall, F1-measure, Accuracy.

Specifically, we assume Γ is the set of testing samples and $N = |\Gamma|$ is the size of testing samples. The ground truth of retweeted tweets is notated as Θ and $M = |\Theta|$ is the number of true retweeted tweet. We then denote an indicator function θ to indicate whether a tweet t is retweeted ($\theta = 1$) or not ($\theta = 0$). Let $\hat{y} = \{\hat{y_1}, \hat{y_2}, \cdots, \hat{y_n}\}$ be our prediction result vector and $y = \{y_1, y_2, \cdots, y_n\}$ be the ground truth vector. Therefore, the Precision, Recall, F1-measure, and Accuracy can be computed as follows:

$$Precision = \frac{\sum_{i=1}^{N} \theta_i}{N} \qquad Recall = \frac{\sum_{i=1}^{N} \theta_i}{M} \qquad (7)$$

$$F1 = \frac{2 \times Precision \times Recall}{Precision + Recall} \qquad Accuracy = \frac{\sum_{i=1}^{N} \{\hat{y_i} = y_i\}}{N} \qquad (8)$$

5.4   Results and Analysis

Overall Results and Analysis. In Table 3, we compare the baseline methods and our proposed approach (named as RBP) for different metrics. From the table, we can observe that our proposed prediction method significantly outperforms other baseline methods. Through comparing F1-score than other the methods, we can see that extracted these features which we proposed in the previous section is a better indicator for predicting users' behaviors. Meanwhile, we have found that the performance of WRM in the experiments dataset is lower than that of reported in [10]. The most likely cause of the low performance is the noisy and tweets' diverse of content. In addition, the performance of LRC-BQ in our experiment results is basically the same comparable with that of reported in [18]. This shows the robustness of influence locality method, and social network platforms between Twitter and Weibo also exist great similarities. In addition, we also compare other popular models: Naive Bayes, Logistic Regression, Random Forest. The experimental results show that SVM classifer outperforms other methods for the task. We omit the details due to space restrictions.

---

[5] http://arnetminer.org/billboard/Influencelocality

Table 3. Performance of retweeting behaviors prediction

Model	Precision	Recall	F1	Accuracy
RG	0.342	0.338	0.340	0.339
MV	0.475	0.478	0.477	0.476
WRM	0.514	0.505	0.509	0.508
LRC-BQ	0.708	0.643	0.674	0.673
RBP	**0.876**	**0.868**	**0.872**	**0.871**

Feature Evaluation. As discussed in Section 4, we extract a lot of features and then roughly divide them into four categories: user status, content, temporal and social tie features. In order to explore the importance of the four features to the prediction performance, here we eliminate one feature at a time from our proposed model. Specifically, we denote each comparison method and a simple explanation as follows: NU-RBP represents the model without user based features into consideration, NC-RBP represents the model without content based features into consideration, NT-RBP represents the model without temporal based features into consideration, NS-RBP represents the model without social based features into consideration.

Table 4. The performance of RBP with deleting the kth feature.

Model	Precision	Recall	F1	Accuracy
NU-RBP	0.770	0.603	0.676	0.683
NC-RBP	0.763	0.604	0.674	0.674
NT-RBP	0.767	0.588	0.666	0.688
NS-RBP	0.759	0.623	0.684	0.723
RBP	0.876	0.868	0.872	0.871

The measurement results of above mentioned features are shown in Table 4, where the larger the value is, the less important the feature is. From the table, we clearly observe that basically the descending order of importance for all features is S > C > T > U. More specifically, we first can see that social based features are the most important factors, which means that the stronger social ties are individually more influential, thus retweeting behaviors are more likely to happen each other. This conclusion agrees with the views reported in [18]. Second, an interesting content is the significant indictor to trigger more retweeting. Third, time based feature is relatively important. This is because a large number of tweets are generated in all the time on social network platforms, if a follower has no the same habit of posting time with her followee, she is more likely to not see the tweets posted by the followee. Thus, time period especially most recent interaction affects future retweeting behaviors. Lastly, the performance for user based feature works the worst than other. On the one hand, one possible reason is that the type of selected users is lack in our constructed dataset, this causes less obvious differentiate with different users. On the other hand, it also indicates that social authority is also a significant factor for predicting users' retweeting behaviors.

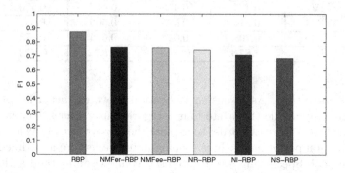

Fig. 1. Comparison of importance of each feature in the social tie feature

As we see that the best features are from the group of social tie feature, we further explore the effectiveness of each feature in the group by removing each feature and examining how the prediction performance is affected in terms of F1. Similarly, we denote RBP without mutual_follower based feature as NMFer-RBP, RBP without mutual_followee based feature as NMFee-RBP, RBP without num_interaction based feature as NI-RBP, and RBP without relationship based feature as NR-RBP. The performance is shown in Figure 1. From the figure, we can clearly conclude that past frequent interactions are more likely to retweet in the future.

6   Conclusion

In this paper, we focus on retweeting behaviors prediction in social network. Specifically, we cast the retweet prediction problem as a classification task. We extract four categories of features including user status, content, temporal, and social tie features from the observed retweets for the prediction task. Furthermore, these features are incorporated into Support Vector Model to predict the class label of candidate. Finally, we collect a large number of data from Twitter, and validate the effectiveness and efficiency of our approach on the constructed dataset. The experimental results clearly show that our approach outperforms other baselines with a significant margin.

Acknowledgments. This work was supported by National Key Technology R&D Program(No.2012BAH46B03), and the Strategic Leading Science and Technology Projects of Chinese Academy of Sciences(No.XDA06030200).

References

1. Abdullah, N.A., Nishioka, D., Tanaka, Y., Murayama, Y.: User's Action and Decision Making of Retweet Messages towards Reducing Misinformation Spread during Disaster. Journal of Information Processing 23(1), 31–40 (2015)
2. Boyd, D., Golder, S., Lotan, G.: Tweet, tweet, retweet: Conversational aspects of retweeting on twitter. In: HICSS, pp. 1–10 (2010)
3. Yang, Z., Guo, J., Cai, K., Tang, J., Li, J., Zhang, L., Su, Z.: Understanding retweeting behaviors in social networks. In: CIKM, pp. 1633–1636 (2010)
4. Can, E.F., Oktay, H., Manmatha, R.: Predicting retweet count using visual cues. In: CIKM, pp. 1481–1484 (2013)
5. Cheng, J., Adamic, L., Dow, P.A., Kleinberg, J.M., Leskovec, J.: Can cascades be predicted? In: WWW, pp. 925–936 (2014)
6. Derczynski, L., Maynard, D., Aswani, N., Bontcheva, K.: Microblog-genre noise and impact on semantic annotation accuracy. In: HT, pp. 21–30 (2013)
7. Feng, W., Wang, J.: Retweet or not?: personalized tweet re-ranking. In: WSDM, pp. 577–586 (2013)
8. Kanavos, A., Perikos, I., Vikatos, P., Hatzilygeroudis, I., Makris, C., Tsakalidis, A.: Modeling ReTweet diffusion using emotional content. In: Iliadis, L. (ed.) AIAI 2014. IFIP AICT, vol. 436, pp. 101–110. Springer, Heidelberg (2014)
9. Liu, L., Tang, J., Han, J., Jiang, M., Yang, S.: Mining topic-level influence in heterogeneous networks. In: CIKM, pp. 199–208 (2010)
10. Luo, Z., Osborne, M., Tang, J., Wang, T.: Who will retweet me?: finding retweeters in Twitter. In: SIGIR, pp. 869–872 (2013)
11. Macskassy, S.A., Michelson, M.: Why do people retweet? anti-homophily wins the day! In: ICWSM (2011)
12. Naveed, N., Gottron, T., Kunegis, J., Alhadi, A.C.: Bad news travel fast: A content-based analysis of interestingness on twitter. In: WebSci, p. 8 (2011)
13. Peng, H.K., Zhu, J., Piao, D., Yan, R., Zhang, Y.: Retweet modeling using conditional random fields. In: ICDMW, pp. 336–343 (2011)
14. Petrovic, S., Osborne, M., Lavrenko, V.: RT to win! predicting message propagation in twitter. In: ICWSM (2011)
15. Suh, B., Hong, L., Pirolli, P., Chi, E.H.: Want to be retweeted? large scale analytics on factors impacting retweet in twitter network. In: SOCIALCOM, pp. 177–184 (2010)
16. Yang, S.H., Kolcz, A., Schlaikjer, A., Gupta, P.: Large-scale high-precision topic modeling on twitter. In: SIGKDD, pp. 1907–1916 (2014)
17. Zaman, T.R., Herbrich, R., Van Gael, J., Stern, D.: Predicting information spreading in twitter. In: NIPS, pp. 17599–17601 (2010)
18. Zhang, J., Liu, B., Tang, J., Chen, T., Li, J.: Social influence locality for modeling retweeting behaviors. In: IJCAI, pp. 2761–2767 (2013)
19. Zhang, Q., Gong, Y., Guo, Y., Huang, X.: Retweet behavior prediction using hierarchical dirichlet process. In: AAAI (2015)

Probabilistic Frequent Pattern Mining by PUH-Mine

Wenzhu Tong[1,2,3], Carson K. Leung[1,*], Dacheng Liu[1,2,4], and Jialiang Yu[1]

[1] University of Manitoba, Winnipeg, MB, Canada
[2] Wuhan University, Wuhan, Hubei, China
[3] University of Illinois at Urbana-Champaign, Urbana, IL, USA
[4] Carnegie Mellon University, Pittsburgh, PA, USA
kleung@cs.umanitoba.ca

Abstract. To mine frequent itemsets from uncertain data, many existing algorithms rely on expected support based mining. An alternative approach relies on *probabilistic based mining*, which captures the frequentness probability. While the possible world semantics are widely used, the exponential growth of possible worlds makes the probabilistic based mining computationally challenging when compared to the expected support based mining. In this paper, we propose two efficient approximate hyperlinked structure based algorithms, which generate a collection of all potentially probabilistic frequent itemsets with a novel upper bound and verify if they are truly probabilistic frequent. Experimental results show the efficiency of our algorithms in mining probabilistic frequent itemsets from uncertain data.

1 Introduction and Related Works

Data mining aims to discover implicit, previously unknown, and potentially useful knowledge from data (e.g., shopper market, social network data). Common data mining tasks include classification [11], social network mining [10,23,25], and *frequent itemset mining* [2]. The latter finds those frequently co-occurring items, objects, or events. Over the past two decades, numerous frequent itemset mining algorithms [8] have been proposed. While many of them are designed to mine *precise* data in which the existence of data items is certainly known, there is also demand for mining *uncertain* data (e.g., sensor network data, clinic reports) [9,14,15,19,30] in many other real-life applications. In these applications, each item is associated with an existential probability expressing the likelihood of the presence of that item. See Table 1 for a sample database consisting of 5 transactions of uncertain data. In comparison, the traditional precise data can be seen as being present with a 100% possibility.

To mine uncertain data, many existing algorithms use *expected support based* mining [7] together with the "possible worlds" semantics. An itemset X is **frequent** if its expected support meets or exceeds the user-specified *minsup*

---

* Corresponding author.

© Springer International Publishing Switzerland 2015
R. Cheng et al. (Eds.): APWeb 2015, LNCS 9313, pp. 768–780, 2015.
DOI: 10.1007/978-3-319-25255-1_63

threshold; i.e., $expSup(X) \geq minsup$. Over the past few years, researchers have extended precise frequent pattern mining algorithms to accommodate uncertainty [13,16,17,18,22].

In addition to expected support based mining, some other existing algorithms use *probabilistic based* mining [3,4,21]. An itemset is **probabilistic frequent** if the probability of its existence in at least *minsup* transactions (i.e., frequentness probability) meets or exceeds the user-specified *minprob* threshold; i.e., $P(sup(X) \geq minsup) \geq minprob$. The *frequentness probability* $P(X)$, which can be derived from a *support probability distribution* (*spd*), computes the probability that X occurs in at least *minsup* transactions. Existing probabilistic based mining algorithms mostly apply a candidate generate-and-test approach (i.e., Apriori-based). To prune candidates early, Sun et al. [24] used Chernoff bound [29]. However, their algorithm is designed to handle tuple-level uncertainty (instead of attribute-level uncertainty). Besides these exact algorithms, others adopt the normal distribution [26,28] or Poisson distribution [6] to approximate the spd for a shorter runtime. However, these approximate algorithms only discover *very likely* probabilistic frequent itemsets, which means they may miss some probabilistic frequent itemsets and may find some false positives. To avoid generating-and-testing candidates, a tree-based algorithm called ProFP-Growth was proposed [5]. However, it suffers from another problem—namely, large memory consumption. To reduce computation, some algorithms [20,27] find probabilistic *closed* frequent itemsets.

Tong et al. [26] surveyed and reported that, when mining expected support based frequent itemsets, *hyperlinked structure-based* algorithms outperformed both Apriori-based and tree-based algorithms. In this paper, we propose two algorithms for probabilistic uncertain hyperlinked structure-based mining to find all and only those probabilistic frequent itemsets. Our key contributions include:

1. a hyperlinked structure-based algorithm called B-PUH, which finds exactly those probabilistic frequent itemsets without expensive candidate generation;
2. a novel upper bound of frequentness probability to prune infrequent itemsets early; and
3. a variation of basic B-PUH called VI-PUH, which pushes the pruning step early.

The remainder of this paper is organized as follows. The next section provides some background. In Section 3, we present the calculation of frequentness probability and its upper bound. We then describe our proposed B-PUH and VI-PUH algorithms in Section 4. Finally, experimental results and conclusions are given in Sections 5 and 6, respectively.

2   Background

Let (i) I be the set of all possible items and (ii) T represent the uncertain database, where a transaction $t_j \in T$ is a set of uncertain items, i.e., $t_j \subseteq I$. Unlike the traditional precise database, each item $x_i \in t_j$ is associated with an *existential probability* $P(x_i, t_j) \in (0,1]$, which denotes the likelihood that x_i is truly present in t_j. For an itemset $X \subseteq t_j$, based on the common assumption that items in X are independent [1,7,12], the existential probability $P(X \subseteq t_j) = \prod_{x \in X} P(x, t_j)$. The *expected support* of X is then the sum of existential probability over all transactions $expSup(X) = \sum_{t_j \in T} P(X \subseteq t_j)$.

The frequentness probability $P(X)$ [4] of X can be computed by summing $P_i(X)$ for all $i \geq minsup$:

$$P(X) = \sum_{i=minsup}^{|T|} P_i(X), \tag{1}$$

where $P_i(X)$ records the probability that X occurs in exactly i transactions:

$$P_i(X) = \sum_{\substack{S \subseteq T \\ |S|=i}} \left(\prod_{t \in S} P(X \subseteq t) \prod_{t \in T-S} (1 - P(X \subseteq t)) \right). \tag{2}$$

If $P(X) \geq minprob$, then X is a *probabilistic frequent itemset*.

Given (i) an uncertain transactional database T, (ii) a minimum support threshold *minsup* and (iii) a minimum frequentness probability threshold *minprob*, the research problem of **probabilistic frequent itemset mining** is to find all and only those probabilistic frequent itemsets; i.e., find every X such that $P(X) \geq minprob$.

Recall that frequent itemsets can be mined from uncertain data using Apriori-based, tree-based, and hyperlinked structure-based approaches. As an example of the latter, the UH-Mine algorithm [1] adopts a pattern-growth manner with a hyperlink data structure called UH-struct, which is a two-dimensional array. Every row in the structure represents a transaction. Every element in the row corresponds to an item, which contains (i) an item ID, (ii) its existential probability and (iii) a pointer linked to the next transaction within the same partition. UH-Mine first scans the database once to construct the UH-struct. Consequently, all the singleton frequent itemsets are discovered, and infrequent ones are removed. Then, a header table keeps the information of the items that could be extended in the next step. Every item i in the header table consists of (i) an item ID, (ii) its expected support and (iii) a pointer to the head of the corresponding projected list. The support of the itemset $P \cup \{i\}$ can be computed by following the links from its header. If the new itemset is frequent (with the expected support $\geq minsup$), further extensions can be recursively mined within the projected database of $P \cup \{i\}$.

3   Computation of Frequentness Probability and Its Upper Bounds

To compute the frequentness probability with the spd, we adopt the approach of a generating function [5] with some modifications as follows. Consider a generating function $F^j(x) = \prod_{n=1}^{j} (a_n x + b_n) = \sum_{i=0}^{j} c_i^j x^i$, where (i) $a_n = P(X \subseteq t_n)$,

Table 1. A transaction database T of uncertain data

TID	Transaction	TID	Transaction
t_1	{A:0.4, D:0.5, E:1.0}	t_4	{A:0.5, C:0.8, E:0.8}
t_2	{A:1.0, B:0.2, C:0.1, D:0.4, E:1.0}	t_5	{A:0.2, B:0.1, C: 0.3, D:1.0}
t_3	{B:0.5, C:0.3}		

(ii) $b_n = 1 - P(X \subseteq t_n)$ and (iii) c_i^j is the probability that X occurs in i transactions among the first j transactions (for $i \leq j$). When $j=|T|$, $c_i^j=P_i(X)$ in spd (i.e., the probability that X is present in exactly i transactions in the whole database T).

Theorem 1. *Given an uncertain transactional database T, the probability that an itemset X is frequent among the first j transactions, denoted by $P^j(X)$, can be recursively computed as follows:*

$$P^j(X) = P^{j-1}(X) + c_{minsup-1}^{j-1}P(X \subseteq t_j), \qquad (3)$$

where

$$c_i^j = c_{i-1}^{j-1}P(X \subseteq t_j) + c_i^{j-1}(1 - P(X \subseteq t_j)). \qquad (4)$$

The frequentness probability of X is $P^{|T|}(X)$. □

Example 1. Consider the uncertain database shown in Table 1. Let $minsup=2$ and $minprob=0.5$. So, we keep two coefficients: c_1 and c_0. The singleton itemset {A} occurs in transactions t_1, t_2, t_4 & t_5 with existential probabilities 0.4, 1.0, 0.5 & 0.2, respectively. So, initially, $P^0(\{A\})=0$, $c_1^0=0$ and $c_0^0=1$. After reading t_1, $P^1(\{A\}) = 0 + 0 \times 0.4 = 0$ using Eq. (3). According to Eq. (4), $c_1^1 = 1 \times 0.4 + 0 \times 0.6 = 0.4$ and $c_0^1 = 1 \times 0.6 = 0.6$. Next, $P^2(\{A\}) = 0 + 0.4 \times 1.0 = 0.4$, $c_1^2 = 0.6 \times 1.0 + 0.4 \times 0 = 0.6$ and $c_0^2=0.6 \times 0 = 0$. We skip t_3 (from which A is absent). Then, after reading t_4, $P^3(\{A\}) = 0.4 + 0.6 \times 0.5 = 0.4$, $c_1^3 = 0.6 \times 1.0 + 0.4 \times 0 = 0.6$ and $c_0^3 = 0 \times 0.5 = 0$. Finally, the frequentness probability $P(\{A\}) = P^4(\{A\}) = 0.7 + 0.3 \times 0.2 = 0.76$. Hence, {A} is probabilistic frequent. □

Observe from Example 1 that we only need one multiplication and addition to compute $P^j(X)$ (cf. $minsup$ additions for each transaction in many existing approaches). However, in order to update $c_{minsup-1}^{j-1} = c_{minsup-2}^{j-2}P(X \subseteq t_{j-2}) + c_{minsup-1}^{j-2}(1-P(X \subseteq t_{j-2}))$ in each iteration, we have to keep track of $c_{minsup-2}^{j-2}$, etc., for to all the coefficients c_i for $i < minsup$. This results in a complexity of $O(|T| \cdot minsup)$. However, the frequentness probability computed by Eq. (3) is monotonically increasing. So, once $P^j(X) \geq minprob$, X is guaranteed to be probabilistic frequent. For instance, in Example 1, we can stop the calculation after t_4 because $P^3(\{A\})=0.7 > minprob$.

3.1 Our Enhancement #1: The 1-Upper Bound

Here, we introduce our first upper bound to prune infrequent itemsets early without the expensive computation of *exact* frequentness probability. The upper bound $UB(P^j(X))$ to $P^j(X)$ shown in Eq. (3) can be computed by

$$UB(P^j(X)) = UB(P^{j-1}(X)) + UB(c_{minsup-1}^{j-1})P(X \subseteq t_j), \qquad (5)$$

where $UB(c_{minsup-1}^{j-1})$ is an upper bound to $c_{minsup-1}^{j-1}$.

To avoid updating the entire coefficient array (from c_0 all the way up to $c_{minsup-1}$) in each iteration during the computation of $c_{minsup-1}^{j-1}$, we compute its upper bound $UB(c_{minsup-1}^{j-1})$ without the information of the lower coefficients and thus reduce the computation overhead. Recall that $c_i^j = c_{i-1}^{j-1}P(X \subseteq t_j) + c_i^{j-1}(1 - P(X \subseteq t_j))$, which represents the probability that X occurs i times in the first j transactions. Note that $\sum_{i=0}^{j-1} c_i^{j-1} = 1$ and

$$\begin{aligned}
c_{minsup-1}^j &= c_{minsup-2}^{j-1}P(X \subseteq t_j) + c_{minsup-1}^{j-1}(1 - P(X \subseteq t_j)) \\
&\leq (1 - c_{minsup-1}^{j-1})P(X \subseteq t_j) + c_{minsup-1}^{j-1}(1 - P(X \subseteq t_j)) \\
&= (1 - 2P(X \subseteq t_j))c_{minsup-1}^{j-1} + P(X \subseteq t_j) \qquad (6)
\end{aligned}$$

In this equation, if $P(X \subseteq t_j) < 0.5$, then $c_{minsup-1}^j$ is increasing w.r.t. $c_{minsup-1}^{j-1}$. We can recursively compute this upper bound. Otherwise, $(1 - 2P(X \subseteq t_j))$ becomes negative. This leads to the following definition. We call it **1-upper bound** (or **1-UB** in short) because the sum of c_i^j equals 1 when j is fixed.

Definition 1. *The **1-upper bound** of $c_{minsup-1}^j$ can be computed as*

$$UB(c_{minsup-1}^j)$$
$$= \begin{cases} (1 - 2P(X \subseteq t_j))UB(c_{minsup-1}^{j-1}) + P(X \subseteq t_j) & \text{if } P(X \subseteq t_j) < 0.5 \\ P(X \subseteq t_j) & \text{otherwise} \end{cases} \qquad (7)$$

Example 2. Revisit Example 1 by substituting Eq. (7) into Eq. (5). Initially, $UB(P^0(\{A\}))=0$ and $UB(c_1^0)=0$. After reading t_1, as $P(\{A\} \subseteq t_1)=0.4 < 0.5$, $UB(P^1(\{A\})) = 0 + 0 \times 0.4 = 0$ and $UB(c_1^1) = (1-2\times0.4)\times0 + 0.4 = 0.4$. Next, as $P(\{A\} \subseteq t_2)=1.0 \geq 0.5$, $UB(P^2(\{A\})) = 0 + 0.4\times1.0 = 0.4$ and $UB(c_1^2)=1.0$. We skip t_3 (from which A is absent). Afterwards, as $P(\{A\} \subseteq t_4)=0.5 \geq 0.5$, $UB(P^3(\{A\})) = 0.4 + 1\times0.5 = 0.9$ and $UB(c_1^3) = P(\{A\} \subseteq t_4)=0.5$. Finally, as $P(\{A\} \subseteq t_5)=0.2 < 0.5$, $UB(P^4(\{A\})) = 0.9 + 0.5\times0.2 = 1.0$. Hence, $\{A\}$ is potentially probabilistic frequent. Again, we can the stop the calculation after t_4 because $UB(P^3(\{A\}))=0.9 > minprob$. ☐

3.2 Our Enhancement #2: The Subset Upper Bound

Although 1-UB simplifies the calculation of $c_{minsup-1}^{j-1}$ in each iteration, it suffers from the problem that the upper bound to $c_{minsup-1}^{minsup}$ could jump to at least $P(X \subseteq t_{minsup})$. In order to reduce this gap, we propose another definition. We call it **subset upper bound** (or **SUB** for short) because of its subset-selection nature. See Lemma 1 and Definition 2.

Lemma 1. $c_{minsup-2}^j \leq \frac{1}{j-minsup+2} c_{minsup-1}^j \sum_{i=1}^j \frac{1-P(X \subseteq t_i)}{P(X \subseteq t_i)}.$ ☐

Definition 2. *The* **subset upper bound** *of* $c_{minsup-1}^j$ *can be computed as*

$$UB(c_{minsup-1}^j)$$

$$= UB(c_{minsup-1}^{j-1})\left(1 - P(X \subseteq t_j) + \frac{P(X \subseteq t_j)}{j - minsup + 1}\sum_{i=1}^{j-1}\frac{1 - P(X \subseteq t_i)}{P(X \subseteq t_i)}\right). \quad (8)$$

Example 3. Revisit Examples 1 and 2 by substituting Eq. (8) into Eq. (5). After reading t_1, $UB(P^1(\{A\}))=0$, $UB(c_1^1)=0.4$ and $\sum_{i=1}^{1}\frac{1-P(\{A\}\subseteq t_i)}{P(\{A\}\subseteq t_i)} = \frac{1-0.4}{0.4} = 1.5$. Next, $UB(P^2(\{A\})) = 0 + 0.4\times1.0 = 0.4$ and $UB(c_1^2) = 0.4(1 - 1.0 + \frac{1.0}{2-2+1}\times1.5) = 0.6$ (which is much smaller than its counterpart value of 1 in Example 2, and this difference leads to a tighter upper bound to frequentness probability in the following computation). $\sum_{i=1}^{2}\frac{1-P(\{A\}\subseteq t_i)}{P(\{A\}\subseteq t_i)}$ is updated by adding $\frac{1-P(\{A\}\subseteq t_2)}{P(\{A\}\subseteq t_2)}=0$ to the previous value of 1.5, which results in 1.5. Again, we skip t_3 (from which A is absent). Afterwards, $UB(P^3(\{A\})) = 0.4 + 0.6\times0.5 = 0.7$ and $UB(c_1^3) = 0.6(1 - 0.5 + \frac{0.5}{3-2+1}\times1.5) = 0.525$. $\sum_{i=1}^{3}\frac{1-P(\{A\}\subseteq t_i)}{P(\{A\}\subseteq t_i)}$ is updated by adding $\frac{1-P(\{A\}\subseteq t_4)}{P(\{A\}\subseteq t_4)}=1$ to the previous value of 1.5, which results in 2.5. Finally, $UB(P^4(\{A\})) = 0.7 + 0.525\times0.2 = 0.805$. Hence, $\{A\}$ is potentially probabilistic frequent. □

3.3 Our Enhancement #3: The Combined Upper Bound

Note that, the 1-UB sometimes provides a tighter upper bound to $c_{minsup-1}^{j-1}$ than the SUB, and the SUB sometime provides a tighter bound than the 1-UB. Hence, a logical solution is to incorporate these two bounds by picking their minimum in each iteration. This leads to the **combined upper bound** (or **CUB** for short).

Definition 3. *The* **combined upper bound** *of* $c_{minsup-1}^j$ *can be computed as*

$$UB(c_{minsup-1}^j) = \min\{1\text{-}UB(c_{minsup-1}^j), SUB(c_{minsup-1}^j)\}, \quad (9)$$

where (i) $1\text{-}UB(c_{minsup-1}^j)$ is the 1-upper bound to $c_{minsup-1}^j$ given in Eq. (7) and (ii) $SUB(c_{minsup-1}^j)$ is the subset upper bound given in Eq. (8). Substituting $UB(c_{minsup-1}^j)$ in Eq. (9) into Eq. (5) leads to the following:

$$UB(P^j(X)) = UB(P^{j-1}(X)) + UB(c_{minsup-1}^{j-1})P(X \subseteq t_j). \quad □$$

To compute the CUB, we need to update (i) $UB(P^j(X))$ with Eq. (5), (ii) $UB(c_{minsup-1}^j)$ with Eq. (9) and Definitions 1 & 2, as well as (iii) the cumulative sum $\sum_{i=1}^{j}\frac{1-P(X\subseteq t_i)}{P(X\subseteq t_i)}$ in Eq. (8) for each iteration, so the time complexity would be $O(|T|)$, which is much better than the complexity $O(|T| \cdot minsup)$ of computing the spd and exact frequentness probability.

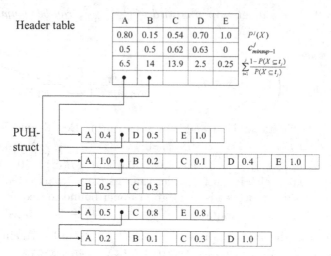

Fig. 1. PUH-struct

4 Our PUH Mining Algorithms for Discovering Probabilistic Frequent Patterns from Uncertain Data

In this section, we propose two algorithms for mining probabilistic frequent itemsets from uncertain data using the hyperlinked structure-based mining model. The first algorithm—called **Basic PUH (B-PUH)**—first finds all the potential probabilistic frequent itemsets using the upper bound proved in Section 3, and then verifies the candidates with a database scan at the end of the mining process. The second algorithm—called **Verified Immediately PUH (VI-PUH)**—verifies the candidate as soon as it is generated.

4.1 Our Basic PUH (B-PUH) Mining Algorithm

On the one hand, like UH-Mine described in Section 2, our B-PUH algorithm also uses a hyperlinked structure. On the other hand, unlike UH-Mine, our proposed B-PUH algorithm computes the Combined Upper Bound (CUB) of frequentness probability $UB(P(X))$ using Definition 3. In addition, we also capture $UB(c_{minsup-1}^{j-1})$ and $\sum \frac{1-P(X \subseteq t_i)}{P(X \subseteq t_i)}$ in the header table. This new structure is called the *PUH-struct*. Fig. 1 shows the global PUH-struct built for the uncertain database in Table 1. As we only compute an upper bound (instead of the exact frequentness probability), what we discover from the mining process is a superset of the truly probabilistic frequent itemsets. To remove those false positives, we quickly scan the database once more to compute the exact frequentness probability with the method described in Section 3.

4.2 Our Verified Immediately PUH (VI-PUH) Mining Algorithm

Recall from Section 3 that the computation of exact frequentness probability for a single itemset requires $O(|T| \cdot minsup)$ runtime. However, we also observed that the use of upper bounds such as CUB (instead of exact frequentness probability) may increase the number of false positives, especially when mining dense datasets. Mining the extensions of these false positives not only prolongs the mining process, it also plagues the verification process due to a larger number of candidates. For instance, $UB(P(\{C\}))$=0.542 $\geq$ $minprob$=0.5 for the uncertain database shown in Table 1 and Fig. 1. As this upper bound exceeds the $minprob$ threshold, our B-PUH algorithm recursively mines its extensions. On the other hand, as its exact frequentness probability $P(\{C\})$=0.4736 < $minprob$=0.5, $\{C\}$ is *not* probabilistic frequent and thus we do not need to consider all its supersets (due to the *downward closure* property for probabilistic frequent itemsets: All subset of a probabilistic frequent itemset must also be probabilistic frequent).

In order to avoid redundant mining, we propose the VI-PUH algorithm. After getting all the information in the header table with respect to the prefix itemset P, the VI-PUH algorithm prunes those infrequent extensions according to their small CUB. Then, VI-PUH immediately verifies those remaining potentially frequent itemsets $P \cup \{i\}$ in the P-projected list. If $P \cup \{i\}$ is truly (probabilistic) frequent, we add it to the result, and recursively discover its extensions. Otherwise, we will not consider it anymore. Compared with B-PUH, this VI-PUH algorithm verifies candidates early and prunes those false positives early, thus reducing the potential overhead of extending and verifying infrequent itemsets. This generate-and-verify-immediately framework is no confined to probabilistic frequent itemset mining; it is also applicable to other itemset mining algorithms that exploit the upper bounds.

Example 4. Consider the uncertain database in Table 1. When building the global header table with CUB, we find that $\{A\}$, $\{C\}$, $\{D\}$ & $\{E\}$ are potentially frequent, but $\{B\}$ is not and thus removed. Then, we compute the exact frequentness probability for these four candidates immediately, and confirm that $\{A\}$, $\{D\}$ & $\{E\}$ are truly probabilistic frequent itemsets, but $\{C\}$ is not. Hence, C will not appear in $\{A\}$'s header table. Moreover, we also skip C after mining the $\{A\}$-projected database.   □

5   Experimental Results

We evaluated the algorithms using both synthetic and real datasets. The (sparse) synthetic dataset is generated by the IBM data generator [2], which contains 1000 items and 1M transactions with an average length of 25. Real datasets include mushroom, retail, and connect4 from the UCI Machine Learning Repository and Frequent Itemset Mining Implementation Repository. To avoid repetition, we use the (dense) mushroom dataset as a representative in our experimental results. We assigned a random existential probability to each item in these datasets. The default $minsup$ values are 0.5% and 1% for the synthetic and mushroom datasets, respectively; the default $minprob$ is 0.8 for both datasets. The longest frequent itemsets contain 2 items for synthetic dataset and 6 items for mushroom. All programs were written in C++, compiled with Visual Studio

2010, and run on Windows XP on a 1.8 GHz dual-core AMD processor with 2GB RAM.

5.1   Scalability

We first evaluated the scalability of both B-PUH and VI-PUH algorithms with existing algorithms such as Apriori-based PFIM DynamicOpt+P algorithm [4] and tree-based ProFP-Growth [5] on different numbers of transactions to demonstrate their scalability. The database size ranges from 5K up to 1M. Fig. 2 shows that our PUH-mining algorithms (i.e., B-PUH and VI-PUH) scale the best compared with the two existing algorithms. The reason is that, with increasing database size, the computation of exact frequentness probability becomes much more costly. Thus, the benefits of pruning those infrequent itemsets by our upper bounds become more significant.

Between our two PUH-mining algorithms, VI-PUH is observed to take shorter runtime than B-PUH. This observation is expected due to the benefits of pushing the pruning step early. Due to its shorter runtimes, only VI-PUH will be shown in the following experimental results.

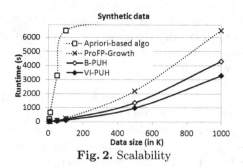

Fig. 2. Scalability

5.2   Runtime

We then evaluated the runtime of different algorithms on different datasets with respect to *minsup* and *minprob*. We compared our VI-PUH algorithm with existing algorithms such as Apriori-based PFIM DynamicOpt+P algorithm [4] and tree-based ProFP-Growth [5]. Figs. 3(a) & 3(b) show the results on the first 10K transactions of the synthetic dataset with respect to a range of *minsup* and *minprob*. The Apriori-based algorithm took a significantly longer runtime than the others. Figs. 3(c) & 3(d) show the runtime on the mushroom dataset. Our VI-PUH mining algorithm took much shorter runtimes on the sparse synthetic dataset, while performing consistently well on the dense mushroom dataset. The primary reason is that the usage of upper bounds to prune infrequent itemsets saves a lot of runtime for the expensive spd computation (recall from Section 3 that the complexity is $O(|T| \cdot minsup)$), especially for big databases. Another reason lies in the pattern-growth framework, which avoids the candidate generation.

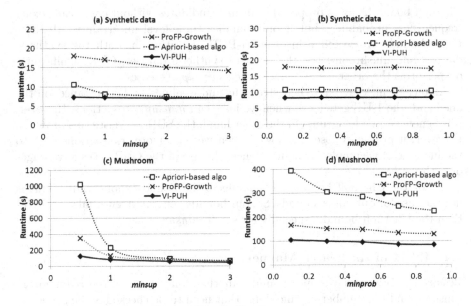

Fig. 3. Runtime

Regarding effects of varying *minprob*, the performances of all the algorithms generally remain stable for different *minprob*. Hence, we will only consider the effect of *minsup* in the following experiments.

5.3   Pruning Effect: The Number of Candidates

Since the computation of the spd is the most time-consuming part, the number of candidates (i.e., itemsets whose spd are computed) makes a big difference to the overall performance. Here, we evaluated the pruning effect of our CUB as well as the generate-and-verify-immediately search manner, which helps reduce the number of candidates and the runtime.

Fig. 4 shows the effect of our three upper bounds (1-UB, SUB, and CUB) when compared with the Chernoff Bound [24]. Recall from Section 1 that the

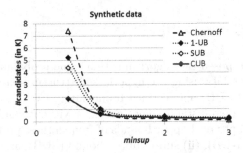

Fig. 4. Pruning Effect: The Number of Candidates

Chernoff bound [24] was proposed to prune candidates without computing exact frequentness probability, but it only handles tuple-level uncertainty. All these upper bounds are tested using the VI-PUH algorithm on both the synthetic dataset and mushroom. For the (sparse) synthetic dataset, CUB significantly outperforms the other three bounds when *minsup* decreases.

Among them, they have their strengths and weaknesses. For instance, on the one hand, the 1-UB gets loose when we just scan *minsup* transactions. However, it is beneficial in a long run. On the other hand, the SUB alleviates this problem. However, it gradually loses its power when we scan more transactions (due to the increase of accumulative sum). Hence, the use of the CUB takes advantages of both worlds.

As for the Chernoff bound, *minsup* affects the bound exponentially. This is not the case for our upper bounds. So, our bounds generated fewer false positives than the Chernoff bound, especially for small *minsup*.

5.4   Effect of the Search Manner

Besides the tightness of upper bounds, another contributor in the reduction in runtimes and the number of candidates that need to be checked is the generate-and-verify-immediately search manner. So, we evaluated this aspect of our PUH algorithms by comparing them with the Apriori-based PFIM DynamicOpt+P algorithm [4]. Between the two PUH algorithms, they produce almost the same number of candidates. For readability, we only pick one (say, VI-PUH) in Fig. 5, which shows that VI-PUH provides fewer candidates for the synthetic dataset.

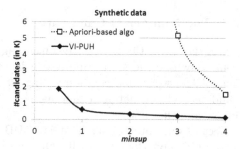

Fig. 5. Effect of the Search Manner

6   Conclusions

In this paper, we proposed two algorithms that mine probabilistic frequent patterns from uncertain data using the hyperlinked structure-based mining model. Both Basic PUH (B-PUH) and Verified Immediately PUH (VI-PUH) efficiently mine all and only probabilistic frequent patterns without candidate generation. Both algorithms adopt novel upper bounds of frequentness probability—namely, (i) 1-upper bound (1-UB), (ii) subset upper bound (SUB), and (iii) their combination called the combined upper bound (CUB)—to prune infrequent patterns early to reduce the expensive computation of the spd and exact frequentness probability. Experimental results showed the efficiency of our algorithms, especially the VI-PUH that performs consistently well on both sparse synthetic dataset and dense real datasets in mining probabilistic frequent patterns.

Acknowledgement. This project is partially supported by (i) CSC (China), (ii) Mitacs (Canada), (iii) NSERC (Canada), (iv) U. Manitoba & (v) Wuhan U.

References

1. Aggarwal, C.C., Li, Y., Wang, J., Wang, J.: Frequent pattern mining with uncertain data. In: ACM KDD 2009, pp. 29–38 (2009)
2. Agrawal, R., Srikant, R.: Fast algorithms for mining association rules. In: VLDB 1994, pp. 487–499 (1994)
3. Bernecker, T., Cheng, R., Cheung, D.W., Kriegel, H.-P., Lee, S.D., Renz, M., Verhein, F., Wang, L., Züfle, A.: Model-based probabilistic frequent itemset mining. KAIS 37(1), 181–217 (2013)
4. Bernecker, T., Kriegel, H.-P., Renz, M., Verhein, F., Züfle, A.: Probabilistic frequent itemset mining in uncertain databases. In: ACM KDD 2009, pp. 119–128 (2009)
5. Bernecker, T., Kriegel, H.-P., Renz, M., Verhein, F., Züfle, A.: Probabilistic frequent pattern growth for itemset mining in uncertain databases. In: Ailamaki, A., Bowers, S. (eds.) SSDBM 2012. LNCS, vol. 7338, pp. 38–55. Springer, Heidelberg (2012)
6. Calders, T., Garboni, C., Goethals, B.: Approximation of frequentness probability of itemsets in uncertain data. In: IEEE ICDM 2010, pp. 749–754 (2010)
7. Chui, C.-K., Kao, B., Hung, E.: Mining frequent itemsets from uncertain data. In: Zhou, Z.-H., Li, H., Yang, Q. (eds.) PAKDD 2007. LNCS (LNAI), vol. 4426, pp. 47–58. Springer, Heidelberg (2007)
8. Cuzzocrea, A., Jiang, F., Lee, W., Leung, C.K.: Efficient frequent itemset mining from dense data streams. In: Chen, L., Jia, Y., Sellis, T., Liu, G. (eds.) APWeb 2014. LNCS, vol. 8709, pp. 593–601. Springer, Heidelberg (2014)
9. Cuzzocrea, A., Leung, C.K., MacKinnon, R.K.: Mining constrained frequent itemsets from distributed uncertain data. Future Generation Computer Systems 37, 117–126 (2014)
10. Jiang, J., Lu, H., Yang, B., Cui, B.: Finding top-k local users in geo-tagged social media data. In: IEEE ICDE 2015, pp. 267–278 (2015)
11. Lee, W., Song, J.J., Leung, C.K.-S.: Categorical data skyline using classification tree. In: Du, X., Fan, W., Wang, J., Peng, Z., Sharaf, M.A. (eds.) APWeb 2011. LNCS, vol. 6612, pp. 181–187. Springer, Heidelberg (2011)
12. Leung, C.K.-S.: Uncertain frequent pattern mining. In: Aggarwal, C.C., Han, J. (eds.) Frequent Pattern Mining, pp. 417–453. Springer, Switzerland (2014)
13. Leung, C.K., Jiang, F.: A data science solution for mining interesting patterns from uncertain big data. In: IEEE BDCloud 2014, pp. 235–242 (2014)
14. Leung, C.K.-S., MacKinnon, R.K.: BLIMP: A compact tree structure for uncertain frequent pattern mining. In: Bellatreche, L., Mohania, M.K. (eds.) DaWaK 2014. LNCS, vol. 8646, pp. 115–123. Springer, Heidelberg (2014)

15. Leung, C.K., MacKinnon, R.K., Jiang, F.: Reducing the search space for big data mining for interesting patterns from uncertain data. In: IEEE BigData Congress 2014, pp. 315–322 (2014)
16. Leung, C.K., MacKinnon, R.K., Tanbeer, S.K.: Fast algorithms for frequent itemset mining from uncertain data. In: IEEE ICDM 2014, pp. 893–898 (2014)
17. Leung, C.K., Mateo, M.A.F., Brajczuk, D.A.: A tree-based approach for frequent pattern mining from uncertain data. In: Washio, T., Suzuki, E., Ting, K.M., Inokuchi, A. (eds.) PAKDD 2008. LNCS (LNAI), vol. 5012, pp. 653–661. Springer, Heidelberg (2008)
18. Leung, C.K., Tanbeer, S.K.: Fast tree-based mining of frequent itemsets from uncertain data. In: Lee, S.-g., Peng, Z., Zhou, X., Moon, Y.-S., Unland, R., Yoo, J. (eds.) DASFAA 2012, Part I. LNCS, vol. 7238, pp. 272–287. Springer, Heidelberg (2012)
19. Leung, C.K., Tanbeer, S.K.: PUF-tree: a compact tree structure for frequent pattern mining of uncertain data. In: Pei, J., Tseng, V.S., Cao, L., Motoda, H., Xu, G. (eds.) PAKDD 2013, Part I. LNCS, vol. 7818, pp. 13–25. Springer, Heidelberg (2013)
20. Liu, C., Chen, L., Zhang, C.: Mining probabilistic representative frequent patterns from uncertain data. In: SDM 2013, pp. 73–81 (2013)
21. Lv, Y., Chen, X., Sun, G., Cui, B.: A probabilistic data replacement strategy for flash-based hybrid storage system. In: Ishikawa, Y., Li, J., Wang, W., Zhang, R., Zhang, W. (eds.) APWeb 2013. LNCS, vol. 7808, pp. 360–371. Springer, Heidelberg (2013)
22. MacKinnon, R.K., Strauss, T.D., Leung, C.K.: DISC: efficient uncertain frequent pattern mining with tightened upper bounds. In: IEEE ICDM Workshops 2014, pp. 1038–1045 (2014)
23. Pham, T.-A.N., Li, X., Cong, G., Zhang, Z.: A general graph-based model for recommendation in event-based social networks. In: IEEE ICDE 2015, pp. 567–578 (2015)
24. Sun, L., Cheng, R., Cheung, D.W., Cheng, J.: Mining uncertain data with probabilistic guarantees. In: ACM KDD 2010, pp. 273–282 (2010)
25. Tanbeer, S.K., Leung, C.K.: Finding diverse friends in social networks. In: Ishikawa, Y., Li, J., Wang, W., Zhang, R., Zhang, W. (eds.) APWeb 2013. LNCS, vol. 7808, pp. 301–309. Springer, Heidelberg (2013)
26. Tong, Y., Chen, L., Cheng, Y., Yu, P.S.: Mining frequent itemsets over uncertain databases. PVLDB 5(11), 1650–1661 (2012)
27. Tong, Y., Chen, L., Ding, B.: Discovering threshold-based frequent closed itemsets over probabilistic data. In: IEEE ICDE 2012, pp. 270–281 (2012)
28. Wang, L., Cheng, R., Lee, S.D., Cheung, D.: Accelerating probabilistic frequent itemset mining: a model-based approach. In: ACM CIKM 2010, pp. 429–438 (2010)
29. Xia, Y.: Two refinements of the Chernoff bound for the sum of nonidentical Bernoulli random variables. Statistics & Probability Letters 78(12), 1557–1559 (2008)
30. Zhang, M., Chen, S., Jensen, C.S., Ooi, B.C., Zhang, Z.: Effectively indexing uncertain moving objects for predictive queries. PVLDB 2(1), 1198–1209 (2009)

A Secure and Efficient Framework for Privacy Preserving Social Recommendation

Shushu Liu[1,2], An Liu[1,2], Guanfeng Liu[1,2], Zhixu Li[1,2], Jiajie Xu[1,2], Pengpeng Zhao[1,2], and Lei Zhao[1,2]

[1] School of Computer Science and Technology, Soochow University, China
[2] Collaborative Innovation Center of Novel Software Technology and Industrialization, Jiangsu, China
shushu.suda@gmail.com,
{anliu,gfliu,zhixuli,xujj,ppzhao,zhaol}@suda.edu.cn

Abstract. The well-known cold start problem in traditional collaborative filtering based recommender systems can be effectively addressed by social recommendation, which has been witnessed by a number of researches recently in many application domains. The social graph used in social recommendation is typically owned by a third party such as Facebook and Twitter, and should be hidden from recommender systems for obvious reasons of commercial benefits, as well as due to privacy legislation. In this paper, we present a secure and efficient framework for privacy preserving social recommendation. Our framework is built on mature cryptographic building blocks, including Paillier cryptosystem and Yao' protocol, which lays a solid foundation for the security of our framework. Using our framework, the owner of sales data and the owner of social graph can cooperatively compute social recommendation, without revealing their private data to each other. We theoretically prove the security and analyze the complexity of our framework. Empirical study shows our framework has a linear complexity with respect to the number of users and items in recommender systems and is practical in real applications.

Keywords: recommender system, social recommendation, privacy preserving, secure two-party computation.

1 Introduction

A Recommender system automatically generates meaningful recommendations to a collection of users for items such as books and movies that might interest them. Collaborative filtering (CF) is a mature technique adopted by most modern recommender systems such as Amazon and Netflix. The basic idea of CF is that similar users will like the same items (user-based CF), and that a user will prefer items that are similar to other items he or she has purchased in the past (item-based CF). One typical challenge of CF in practice is the well-known *cold start* problem: new users should rate sufficient number of items to get their preferences captured by the recommender system, and new items should also

© Springer International Publishing Switzerland 2015
R. Cheng et al. (Eds.): APWeb 2015, LNCS 9313, pp. 781–792, 2015.
DOI: 10.1007/978-3-319-25255-1_64

be rated by sufficient number of users before they could be recommended to users. To address the cold start problem, a wide variety of social recommendation approaches [7, 17] have appeared in the literature. The motivation of social recommendation is that the explicit social relations among users provided by a social graph could be used as a data source to calculate user similarities required in CF. Recent studies have shown that social recommendation outperforms traditional, non-social approaches in many application domains.

Though the benefits brought by recommender systems is significant, it also poses a serious threat to personal privacy. In non-social recommendation, the private user transaction data, such as which item were bought by a user and how the user rates these items, is required to report to the recommender system. Personal information like gender, age, health condition can be easily inferred from these data. In addition, the recommender system must have full access to user social relations to enable social recommendation, which clearly violates against personal privacy.

The public concern over privacy stimulated lots of research efforts in privacy preserving recommender systems recently [15, 13, 12, 3, 11, 5, 8]. While most studies have focused on non-social recommendation, little attention has been paid to privacy preserving social recommendation. In [5], the authors study this problem under the differential privacy paradigm. This work prevents the inference of any single user's data from the recommendation results, but has a strong assumption that the underlying data, including user-item matrix and social graph, reside with a trusted and secured central party, and that the recommender has full, unfettered access to the data. In practice, however, social graph is owned by a third party such as Facebook or Twitter (we refer to the social networking service provider as Bob). The proprietary social graph is an important asset with inestimable value, thus Bob must keep it secret for obvious reasons of commercial benefits, as well as due to privacy legislation. For the same reason, the recommender system (typically established by an electronic commerce platform like Amazon and we refer to it as Alice) must keep his historic sales data (e.g., the user-item matrix) secret. Therefore, we face a scenario in which Alice holds only historic sales data and Bob holds only the social graph. It is therefore necessary to design a method that enables Alice and Bob to cooperatively perform social recommendation without any information leakage about their individual data.

In this paper, we present a secure and efficient framework for privacy preserving social recommendation. The main contributions of our work can be summarized as follows:

1. We combine homomorphic cryptosystem and Yao's garbled circuit to design secure and efficient protocols that enable two distinct parties to cooperatively compute social recommendation, without revealing their private data to each other. To our best knowledge, this is the first work to keep individual data secret in the course of social recommendation.
2. We theoretically analyze the complexity of the proposed framework and present the performance of our framework on four real-world datasets. Both

theoretical and empirical results show our approach has a linear complexity with respect to the number of users and items in recommender systems and thus is practical in real applications.

The rest of paper is organized as follows. Section 2 formalizes our model and definitions, while Section 3 reviews some background knowledge and related work. In Section 4, we present our framework for privacy preserving social recommendation. We report the experimental results on real-world datasets in Section 5 and conclude our work in Section 6.

2 Problem Definition

2.1 Data Model

Following the model presented in [5], we have two parties Alice and Bob. Alice provides an electronic commerce platform and has the historic transaction data. Bob provides a social networking platform and has the social relation data. The historic transaction data and the social relation data are modeled as a *transaction graph* and a *social graph*, respectively.

Definition 1 *(Transaction Graph). A transaction graph, $G_t = (U, I, E_t)$, is a bipartite graph where U is a set of users, I is a set of items, and $E_t \subseteq U \times I$ is a set of directed edges. Every edge (u, i) is associated with a weight $w(u, i) \geq 0$, indicating user u has not purchased item i if $w(u, i) = 0$ and otherwise the rating user u gives to item i.*

Definition 2 *(Social Graph). A social graph, $G_s = (U, E_s)$, consists of a set of users, U, and a set of edges, $E_s \subseteq U \times U$, where an edge $(u, v) \in E_s$ represents a social relation between two user $u, v \in U$.*

Note that the transaction graph could be easily extended to model data in other application domains. For example, an edge (u, i) might indicate that a user u of Last.fm has listened to a song by artist i, or that a user u of Brightkite.com has visited location i. In these cases, the weight $w_{u,i}$ might be the number of times that u listened to a song by artist i, or that u visited location i. Also note that many real-world online social networks can be modeled with the social graph. For example, an edge (u, v) in the social graph might indicate that the "friendship" relation between two users u and v in Facebook, or the "following" relation between two users u and v in Twitter.

2.2 Privacy Preserving Social Recommendation

For each user u of the electronic commerce platform, Alice recommends a size K item set $R_u \subset I$ with the highest scores. Here, the score of recommending item i to user u (or just the score of item i where the context is clear), is denoted as $s(u, i)$, and is computed as follows:

$$s(u, i) = \sum_{v \in U, v \neq u} sim(u, v) \times w(v, i) \tag{1}$$

where $sim(u, v)$ is the similarity between two users $u, v \in U$. In non-social recommendation, $sim(u, v)$ is typically evaluated based on the purchase profiles $\mathbf{p}$ of u and v, such as the cosine similarity $\cos(\mathbf{p}_u, \mathbf{p}_v)$, where $\mathbf{p}_u = (w_{u,i_1}, \cdots, w_{u,i_{|I|}})$. In social recommendation, however, $sim(u, v)$ is usually evaluated based on social similarity measures[10], such as common neighbors (CN), Katz, and random walk with restart (RWR), just to name a few.

An important observation here is that the calculation of $sim(u, v)$ in non-social recommendation can be conducted solely on the transaction graph G_t, while in social recommendation the social graph G_s is indispensable for this calculation. Clearly, to enable social recommendation, Alice must have full, unfettered access to G_s that is owned by Bob. As stated before, however, Bob wants to keep it secret for both commercial benefits and privacy legislation. This challenging situation inspires us to give the definition of privacy preserving social recommendation as follows:

Definition 3 *(Privacy preserving social recommendation). Given a private transaction graph $G_t = (U, I, E_t)$ hold by Alice, a private social graph $G_s = (U, E_s)$ hold by Bob, and a social similarity measure sim, Alice and Bob cooperatively compute, for each user $u \in U$, a size K item set $R_u \subset I$ with the highest scores, without revealing their private data (i.e., G_t and G_s) to each other.*

2.3   Assumptions

Static Recommendation. For simplicity, we only consider in this paper static social recommendation, that is, we take a snapshot of G_t and G_s at some time and all the recommendations are generated based on these fixed graphs at that time. Dynamic behaviors on G_t and G_s, such as the insertion/deletion of new/old nodes and edges, will cause Alice and Bob to run the proposed protocols from the beginning. We leave extending our protocols to dynamic recommendation as a subject for future work.

Adversarial Model. We assume that Alice and Bob are *semi-honest*, also known as "honest but curious". They run the protocol exactly as specified (no deviations, malicious or otherwise), but may try to learn as much as possible about the input of the other party from their views of the protocol. It should be noted that though secure protocols against malicious adversaries exist, they are far too inefficient to implement and be used in practice. Secure protocols against semi-honest adversaries, however, are not only useful in practice but also the foundation of designing secure protocols against malicious adversaries.

3   Background and Related Work

3.1   Paillier Cryptosystem

We use the Paillier cryptosystem [14] to encrypt the private data of Alice and Bob. The encryption function E is defined as $E_{pk}(m, r) = g^m \times r^N \mod N^2$

where $m \in \mathbb{Z}_N^*$ is a message for encryption, N is a product of two large prime numbers p and q, g generates a subgroup of order N, and r is a random number in $\mathbb{Z}_{N^2}$. The public key for encryption is (N, g) and the secret key for decryption is (p, q). The details of decryption function D with secret key sk can be found in [14]. The properties of the Paillier cryptosystem include homomorphic addition and semantic security. *Homomorphic addition*: The product of two ciphertexts will be decrypted to the sum of their corresponding plaintexts, and the kth power of a ciphertext will be decrypted to the product of k and its corresponding plaintext. *Semantic security*: Given a set of ciphertexts, an adversary cannot deduce any information about the plaintexts.

3.2   Yao's Protocol

Yao's protocol [16, 9] (a.k.a. garbled circuits) allows two semi-honest parties holding inputs x and y, respectively, to evaluate an arbitrary function $f(x, y)$ without leaking any information about the inputs beyond what can be deduced by the function output. The basic idea is that one party (the garbled-circuit *constructor*) constructs a garbled version of a circuit to compute f, while the other party (the garbled-circuit *evaluator*) obliviously computes the output of the circuit without learning any intermediate values. Two simple circuits will be used in this paper to realize secure integer comparison required in the top-K selection. An ADD circuit takes two σ-bit integers x and y as input, and outputs a $(\sigma+1)$-bit integer z, such that $z = x+y$. A CMP circuit takes two σ-bit integers x and y as input, and outputs 1 if $x > y$ and 0 otherwise. The details of ADD and CMP circuits can be found in [6].

3.3   Privacy Preserving Recommender Systems

The need for privacy preserving in recommender systems triggered lots of research efforts in the past years. Shokri *et al.* present a recommender system built on distributed aggregation of user profiles, which suffers from the trade-off between privacy and accuracy [15]. In [13, 12], Nikolaenko *et al.* consider two basic problems in model-based recommendation algorithms: matrix factorization and ridge regression. For the first problem, they propose a solution based on Yao's protocol. For the second problem, they design a hybrid method by combing Yao's protocol and homomorphic encryption. In [3], the authors present a solution for privacy preserving recommendation via homomorphic encryption and data packing. These efforts focus on non-social recommendation and what has being protected is users' individual data, for example, which items were bought by a user and how the user rate these items. That is, Alice does not have full access to the transaction graph G_t. In this paper, however, we aim at privacy preserving in social recommendation. Following the model defined in [5], we assume Alice has full access to G_t and Bob has full access to G_s. In the course of recommendation, Alice is not willing to reveal her private data G_t to Bob, and vice versa. Privacy preserving social recommendation is also studied in [8] where the authors present a framework for secure social recommendations in geosocial networks. They introduce a mix network for message transmitting to prevent the service provider (i.e., Alice in our model) from learning users' social relations. However, this work is not secure as the service provider can easily learn a lot of social relations by statistical analysis.

There is also a number of achievements that aim at privacy preserving recommender systems under the differential privacy paradigm. McSherry and Mironov [11] integrates differential privacy into non-social recommender systems. However, their work will lead to an unacceptable loss of utility when applied to the social recommendation. To overcome this weakness, Jorgensen and Yu [5] incorporates a clustering procedure that groups users according to the natural community structure of the social network and significantly reduces the amount of noise. This kind of work is orthogonal to ours. In differential privacy, the recommender system has full access to the data (G_t in the non-social recommendation, while G_t and G_s in the social recommendation). The privacy threat arises from releasing a function (e.g., get recommendation for a given user) over the data to a third party, who may use it to infer data values of users in the database. Whether the final results expose personal privacy is not the concern of our work. Instead, we are interested in keeping data secret during computation.

4   Privacy Preserving Social Recommendation

In this section, we present our framework for privacy preserving social recommendation. In Section 4.1, we present a protocol that computes the scores of items while keeping both Alice's and Bob's data secret. As these scores are in the encrypted form, it is impossible to compare them directly. Instead, we combine Yao's garbled circuits with homomorphic encryption to realize secure comparison between two encrypted values, based on which an efficient protocol for secure top-K selection is proposed, as discussed in Section 4.2. Section 4.3 analyzes the security and complexity of the proposed protocols.

4.1   Privacy Preserving Recommendation Score Computation

For a target user u and every user $v \in U \setminus \{u\}$, the similarity between them, $sim(u, v)$, should be available before computing the score of recommending an item i to user u, as seen in Formula 1. Once the similarity measure sim is fixed, the computation can be carried out directly for a simple similarity measure such as *common neighbors*. For a complex similarity measure, some mature algorithms can be adopted, for example, the *bookmark-coloring* algorithm [1] can be used to efficiently calculate the similarity measure *random walk with restart*. Social similarity measures and their efficient calculations is an interesting topic but not the focus of this paper, so we simply assume that Bob is able to compute $sim(u, v)$ based on his private social graph G_s and the given similarity measure sim.

These similarities, however, cannot be sent directly to Alice, as this might lead to serious privacy leakage. Specifically, Alice might learn part even the whole social graph G_s, the private data of Bob. Suppose *common neighbors* is used for similarity computation and Alice has the background knowledge that there are totally 3 users u, v, w in the social network. If Alice learns $sim(u, v) = 1$ and $sim(u, w) = 1$, she can infer that any two of them are friends, that is, she knows the whole social graph. To keep his private data secret, Bob adopts the Paillier cryptosystem to encrypt these similarities. More specifically, he first chooses a key pair (pk, sk) in the Paillier cryptosystem and encrypts $sim(u, v)$ for all $v \in U \setminus \{v\}$ using the public key pk. These encrypted similarities as well as the public key pk can then be safely sent to Alice. Provided Alice does not know Bob's secret key sk, she cannot deduce any information from these encrypted data, which is guaranteed by Paillier's *semantic security*.

After receiving the encrypted similarities, Alice needs to compute the score of recommending item i to the target user u for every i that u has not purchased yet, as defined in Formula 1. This computation is feasible on ciphertexts as Paillier supports *homomorphic addition*. In particular, $s(u, i)$ can be calculated by Alice as follows:

$$E_{pk}(s(u,i)) = \prod_{v \in U \setminus \{u\}} E_{pk}(sim(u,v))^{w(v,i)} \tag{2}$$

Note that, $w(v, i)$ is a plaintext as it is the private data of Alice, but the final result $E_{pk}(s(u, i))$ is still an encrypted value as Alice only knows the public key pk of Bob.

The above procedure is summarized in Algorithm 1. Alice and Bob cooperatively run this protocol for privacy preserving recommendation score computation. In the end, Alice obtains the scores of recommending new items to the target user u in the encrypted form.

Recall that in social recommendation Alice recommends a size K item set $R_u \subset I_u^n$ with the highest scores. This is a trivial task on scores in the plaintext. On ciphertext, however, this becomes a much harder task as comparison is not supported by the Paillier cryptosystem. Considering only Bob has the secret key, a simple solution is that these encrypted scores are sent to Bob for decryption. However, this is not secure, as for each decrypted score, Bob can create a system of linear equations based on Formula 1. Clearly, there are $|U| - 1$ equations and $|U| - 1$ variables. Therefore, Bob can know Alice's private data $w(v, i)$ for all $v \in U \setminus \{u\}$ by solving these equations.

Next, we will present an efficient protocol for top-K selection on encrypted scores. Based on these protocols, we can build a secure and efficient framework for privacy preserving social recommendation.

4.2 Secure Top-K Selection

Recall that the objective of social recommendation is to recommend a set of K items with the highest scores. As we are not interested in the order among

---

Algorithm 1. Privacy Preserving Recommendation Score Computation

Input: transaction graph $G_t = (U, I, E_t)$ hold by Alice;
social graph $G_s = (U, E_s)$ hold by Bob;
social similarity measure sim;
$u \in U$ is the target user
Output: $E_{pk}(s(u, i))$ for every $i \in I_u^n$, hold by Alice

1: Bob computes $sim(u, v)$ for all $v \in U \setminus \{u\}$ based on G_s and sim
2: Bob chooses a key pair (pk, sk) in the Paillier cryptosystem
3: Bob encrypts $sim(u, v)$ for all $v \in U \setminus \{u\}$ using the public key pk
4: Bob sends $E_{pk}(sim(u, v))$ for all $v \in U \setminus \{u\}$ and pk to Alice

5: Alice receives the encrypted similarities and the public key from Bob
6: **for** each item $i \in I_u^n$ **do**
7:     Alice computes $E_{pk}(s(u, i)) = \prod_{v \in U \setminus \{u\}} E_{pk}(sim(u, v))^{w(v, i)}$
8: **end for**

---

these K items, we build our protocol for secure top-K selection based on the well-known randomized-selection algorithm with expected linear time [2]. At the beginning of Algorithm 2, an array A is used to store the indices of items that the target user u has not purchased yet (line 1). Then, a value k is selected randomly from the range $[l, h]$ for the index of the pivot (line 3), followed by a partition on the subarray $A[l..h]$ that returns the pivot's new index (line 4). This partition has the property that for any $k' < k$ and $k'' > k$, $sim(u, i_{A[k']} > sim(u, i_{A[k'']}))$ holds. In other words, the subarray $A[1..k]$ contains the indices of all items with the highest scores. Therefore, if k equals to K, the protocol terminates by returning $(i_{A[1]}, \cdots, i_{A[k]})$ as the recommendation result (line 6). Otherwise, in lines 7-8, a new partition is performed on a new subarray ($A[k+1..h]$ if $k < K$ or $A[l..k-1]$ if $k > K$) and this process continues until the returned pivot's index equals exactly to K.

The key operation in the partition procedure is the comparison of two scores encrypted by the Paillier cryptosystem (lines 13 and 16). As Paillier does not support comparison over ciphertexts, we first make these scores additively secret shared between Alice and Bob. More specifically, a score s is additively secret shared between two parties Alice and Bob if Alice holds a uniformly distributed random number r sampled from a sufficiently large domain, Bob holds s', and $r + s' = s$. Clearly, neither Alice nor Bob knows the actual value of s. To achieve secret sharing of the list of scores $L = \{s(u, i_1), \cdots, s(u, i_N)\}$, Alice first picks $N = |I_u^n|$ random integers and computes the product of $E_{pk}(s(u, i_j))$ and $E_{pk}(-r_j)$ for each random integer r_j based on Bob's public key pk. Clearly, Alice now has two lists L_r and $L_{E(s-r)}$ in hand where $L_r = \{r_1, \cdots, r_N\}$ and $L_{E(s-r)} = \{E_{pk}(s(u, i_1) - r_1), \cdots, E_{pk}(s(u, i_N) - r_N)\}$. She then sends $L_{E(s-r)}$ to Bob. As Bob has the secret key sk, he can decrypt all elements in $L_{E(s-r)}$ and obtain a list of randomized scores ($L_{s-r} = \{s(u, i_1) - r_1, \cdots, s(u, i_N) - r_N\}$. Clearly, every score $s(u, i_j)$ in L is now additively secret shared between Alice and Bob, because Alice holds r_j and Bob holds $s(u, i_j) - r_j$. Then, Alice prepares a garbled circuit to compare two additively secret shared values and makes it available to Bob. This circuit is quite simple as it only consists of two ADD circuits and a CMP circuit. One ADD circuit takes for example r_1 and $s(u, i_1) - r_1$ as input and the other takes for example r_2 and $s(u, i_2) - r_2$ as input, while their outputs (i.e., $s(u, i_1)$ and $s(u, i_2)$ (or s_1 and s_2 for short) in the form of internal encoding of garbled circuits) serve as the inputs of the CMP circuit.

Algorithm 2. Secure Top-K Selection

Input: $L_{E(s)} = \{E_{pk}(s_1), \cdots, E_{pk}(s_N)\}$ hold by Alice;
Output: A size K item set R with the highest scores
1: $l \leftarrow 1, h \leftarrow N, A[i] \leftarrow i$ for $1 \le i \le K$
2: **loop**
3:      $k \leftarrow$ RANDOM(l, h)
4:      $k \leftarrow$ PARTITION(l, h, k)
5:      **case**
6:          $k = K$: **return** $(i_{A[1]}, \cdots, i_{A[k]})$
7:          $k < K$: $l \leftarrow k + 1$
8:          $k > K$: $h \leftarrow k - 1$
9:      **endcase**
10: **end loop**

    procedure PARTITION(l, h, k)
11: $p \leftarrow l, q \leftarrow h, m \leftarrow k$
12: **loop**
13:     **while** COMPARE$(L_{E(r)}[p], L_{E(r)}[m]) = 1$ **do**
14:         $p \leftarrow p + 1$
15:     **end while**
16:     **while** COMPARE$(L_{E(s)}[q], L_{E(s)}[m]) = 0$ **do**
17:         $q \leftarrow q - 1$
18:     **end while**
19:     **if** $p < q$ **then**
20:         exchange $L_{E(s)}[p]$ with $L_{E(s)}[q]$ and $A[p]$ with $A[q]$
21:     **else**
22:         **return** p
23:     **end if**
24: **end loop**

4.3   Theoretical Analysis

Security Analysis. In Algorithm 1, Bob sends $E_{pk}(sim(u, v))$ and pk to Alice. These data do not reveal any private information of Bob due to the semantic security of the Paillier cryptosystem. In Algorithm 2, Alice sends a list of encrypted values $L_{E(s-r)}$ to Bob for secret sharing. As Bob can decrypt them using his private key sk, we build a simulator S to simulate this kind of information. As all the values obtained by Bob are masked by random numbers chosen from a sufficiently large domain, they are independent from Alice's input. S can also randomly choose some elements in the same domain. It is clear that there does not exist an adversary that is able to distinguish between interaction with Alice verses interaction with S, which means revealing this kind of information will not harm the private information of Alice. After secret sharing, Alice and Bob cooperatively execute a garbled circuit, so Algorithm 2 is secure as long as Yao's garbled circuit is secure, which has been proved in [9].

Complexity Analysis. In Algorithm 1, Bob needs to perform $|U|$ encryptions and Alice needs to do $|I|$ exponentiations. In Algorithm 2, $|I|$ encryptions and exponentiations are required to be done by Alice and $|I|$ decryptions by Bob for secret sharing. The garbled circuit prepared by Alice consists of two ADD circuits and one CMP circuit, thus containing $3\sigma + 1$ non-free gates given the input of ADD circuits are σ-bits long. As this garbled circuit will be executed $|I|$ times in average, the total cost is $(3\sigma + 1)|I|$ non-free gates.

Table 1. Summary of datasets

	Last.fm	Flixster	Brightkite	Gowalla
$|U|$	1892	786,936	58,228	196,591
$|E_s|$	12,717	7,058,819	214,078	950,327
$|I|$	17,632	48,796	314,417	1,280,969
$|E_t|$	92,198	8,196,077	4,491,143	6,442,890

Table 2. Running time of our protocols on four real-world datasets

dataset	Algorithm 1 (min)	Algorithm 2 (min)	Total (min)	Avg. 1 (ms)	Avg. 2 (ms)
Last.fm	0.88	2.55	3.40	2.60	8.69
Flixster	38.87	6.66	45.53	2.79	8.20
Brightkite	13.05	44.33	57.38	2.10	8.46
Gowalla	56.11	185.05	241.17	2.27	8.66

5   Experiments

Four public available datasets are used to evaluate the performance of our framework. Last.fm[1] is a relatively small dataset containing social networking and music artist listening information. Flixter.com[2] is a social movie site allowing users to share movie ratings, discover new movies and meet others with similar movie taste. Brightkite.com[3] and Gowalla.com[4] are location-based social networking

---

[1] http://www.lastfm.com
[2] http://www.cs.ubc.ca/~jamalim/datasets/
[3] http://snap.stanford.edu/data/index.html
[4] http://snap.stanford.edu/data/index.html

websites where users share their locations by checking-in. The statistics of these four datasets are summarised in Table 1.

We adopt *random walk with restart* [10] as the social similarity measure and use the *bookmark-coloring* algorithm [1] to calculate item scores. For Paillier cryptosystem, we also use a public available Java implementation[5]. The Paillier encryption key size is set to 1024. Note that this size will make encryption and decryption much more time-consuming, but will increase the level of security at the same time. Besides, we implement Yao's garbled circuits based on FasterGC [4]. The size K of recommended item set is set to 10. All experiments all performed on a PC with 3.4GHz CPU, 16GB RAM, JDK 7, and OS X Yosemite.

As shown in Table 2, our framework takes less than 4 minutes to generate recommendations on Last.fm dataset that contains 1892 users and 17632 items. Even for the biggest dataset Gowalla that contains 196,591 users and 1,280,969 items, our framework only requires about 4 hours to get the final result. We believe this running time is acceptable in practice considering the computation could be done offline and high performance servers and clusters could be used.

It is also worth noting that the running time of Algorithm 1 increases linearly with respect to the number of users and items. As seen from the fifth column of Table 2, the average time is about 2ms. Similarly, the running time of Algorithm 2 increases linearly with respect to the number of items, and the average time is roughly 8ms (see the 6th column of Table 2) . All these experimental results coincide with our theoretical analysis. Therefore, our framework has a good scalability and is practical in real applications.

6   Conclusion

In this paper, we have presented a secure and efficient framework for privacy preserving social recommendation. We have designed a protocol that computes the scores of recommending items to users securely in the sense that the two parties involved in the computation do not know the data of the other party. We have also presented an efficient protocol for secure top-k selection with the linear time. We have theoretically and empirically shown the performance of the proposed protocols. The results on four real-world datasets show our framework has a good scalability and is practical in real applications.

Acknowledgments. This work was partially supported by Natural Science Foundation of China (Grant Nos. 61232006, 61303019, 61402313) and Collaborative Innovation Center of Novel Software Technology and Industrialization, Jiangsu, China.

---

[5] http://www.csee.umbc.edu/~kunliu1/research/Paillier.html

References

[1] Berkhin, P.: Bookmark-coloring algorithm for personalized PageRank. Internet Mathematics 3(1), 41–62 (2006)

[2] Cormen, T.H., Leiserson, C.E., Rivest, R.L., Stein, C.: Introduction to Algorithms, 4th edn. MIT Press (2009)

[3] Erkin, Z., Veugen, T., Toft, T., Lagendijk, R.L.: Generating Private Recommendations Efficiently Using Homomorphic Encryption and Data Packing. IEEE Trans. Information Forensics and Security 7(3), 1053–1066 (2012)

[4] Huang, Y., Evans, D., Katz, J., Malka, L.: Faster secure two-party computation using garbled circuits. USENIX Security (2011)

[5] Jorgensen, Z., Yu, T.: A privacy-preserving framework for personalized, social recommendations. In: EDBT 2014, pp. 571–582 (2014)

[6] Kolesnikov, V., Sadeghi, A.-R., Schneider, T.: Improved garbled circuit building blocks and applications to auctions and computing minima. In: Garay, J.A., Miyaji, A., Otsuka, A. (eds.) CANS 2009. LNCS, vol. 5888, pp. 1–20. Springer, Heidelberg (2009)

[7] Konstas, I., Stathopoulos, V., Jose, J.M.: On social networks and collaborative recommendation. In: ACM, pp. 195–202 (2009)

[8] Liu, B., Hengartner, U.: Privacy-preserving social recommendations in geosocial networks. In: IEEE, pp. 69–76. (2013)

[9] Lindell, Y., Pinkas, B.: A Proof of Security of Yao's Protocol for Two-Party Computation. J. Cryptology 22(2), 161–188 (2009)

[10] Lü, L., Zhou, T.: Link prediction in complex networks: A survey. Physica A 390(6), 1150–1170 (2011)

[11] McSherry, F., Mironov, I.: Differentially private recommender systems: Building privacy into the Netflix prize contenders. In: KDD 2009, pp. 627–636 (2009)

[12] Nikolaenko, V., Weinsberg, U., Ioannidis, S., Joye, M., Boneh, D., Taft, N.: Privacy-preserving ridge regression on hundreds of millions of records. In: S&P 2013, pp. 334–348 (2013)

[13] Nikolaenko, V., Weinsberg, U., Ioannidis, S., Joye, M., Boneh, D., Taft, N.: Privacy-Preserving matrix factorization. In: CCS 2013, pp. 801–812 (2013)

[14] Paillier, P.: Public-key cryptosystems based on composite degree residuosity classes. In: Stern, J. (ed.) EUROCRYPT 1999. LNCS, vol. 1592, pp. 223–238. Springer, Heidelberg (1999)

[15] Shokri, R., Pedarsani, P., Theodorakopoulos, G., Hubaux, J.: Preserving privacy in collaborative filtering through distributed aggregation of offline profiles. In: RecSys 2009, pp. 157–164 (2009)

[16] Yao, A.C.-C.: How to generate and exchange secrets. In: FOCS 1986, pp. 162–167 (1986)

[17] Yuan, Q., Zhao, S., Chen, L., et al.: Augmenting collaborative recommender by fusing explicit social relationships. In: Recsys (2009)

DistDL: A Distributed Deep Learning Service Schema with GPU Accelerating

Jianzong Wang and Lianglun Cheng*

Faculty of Computer Science, Guangdong University of Technology, Guangzhou,
P.R. China 510006
Faculty of Automation, Guangdong University of Technology, Guangzhou,
P.R. China 510006
{jzwang,llcheng}@gdut.edu.cn

Abstract. Deep Learning is a hot topic developed by the industry and academia which integrates the broad field of artificial intelligence with the deployment of deep neural networks in the big data era. Recently, the capability to train neural networks has resulted in state-of-the-art performance in many domains such as computer vision, speech recognition, recommended system, natural language processing, drug discovery and behavioural analysis etc. However, existing deep learning systems are inadequate for scaling, especially in current cloud infrastructures where the nodes are distributed across multiple geographical location. The tendency is evolving that deep learning must be collaborative optimized by the fields both machine learning and systems.

In this paper, we have presented *DistDL*, a novel distributed deep learning service schema in order to reduce the training time consumption along with communication overhead and achieve extremely parallelism with data and model. Additionally, we also took into consideration GPUs inside *DistDL* by leveraging the remarkable competency of heterogeneous computation. The results of our experiments in the benchmarks suggest that *DistDL* is adaptive to various scaling patterns with the same accuracy while minimising training time by adopting the GPU, up to 80% speed up.

Keywords: Deep Learning, GPU, Spark, Data Partition, Model Parallelism.

1 Introduction

Along with the development of data explosion and the coming of big data age, it is possible to distill more informative knowledge from the huge size of data. Deep Learning (DL) is a powerful artificial intelligence tool that enables us to make sense of the story behind data. And it's providing the solution on previously unaddressed problems. In essence, neural networks are revitalised by DL in the big data era, where made possible by massive amounts of information to

---

* Corresponding author.

© Springer International Publishing Switzerland 2015
R. Cheng et al. (Eds.): APWeb 2015, LNCS 9313, pp. 793–804, 2015.
DOI: 10.1007/978-3-319-25255-1_65

train them, so that they approach behavioural pattern of human cognition when dealing with various problems.

In order to handle challenging applications with accurate result, the training data sizes are increasing exponentially along with the complexity increase of topology in neural networks. However, the computation overhead of using a large scale deep learning network could be really time-consuming under current single node implementation. A single node is often inadequate since the model and data may be too large to fit in memory. Therefore, it is impossible to satisfy the requirements of real-time computing by single machine in that DL systems are scaling to cluster environments for purpose of harnessing additional memory, CPU/GPGPU processors and storage resources.

These scalability needs arise from at least two aspects: massive data volume, such as the size of Facebook social photos posted per day with up 10TB [1]; and abundant model size, such as the Google Brain deep neural network [2] containing billions of parameters. Many important DL algorithms iteratively update a large set of parameters with 9 to 12 orders of magnitude [3]. Hence, addressing the DL questions is transferring from machine learning field to system problems result in the need to design a scalable application framework.

1.1   Motivation Example

In order to accelerate the convergence of deep neural networking training, the iterative DL algorithm must run in parallel. The traditional way in the single node is to carry out in one machine by leveraging customized heterogeneous hardware, for instance GPGPU, FPGA etc whereas the pure hardware solution makes the expenditure increasingly. Adopting the parallel version of multi-threads is another option, but which is of no code reusing and less readable. It urges the scholars to look up efficient distributed frameworks with the large-scale adoption of clusters for DL applications. A naturally desirable goal for distributed deep learning systems is to pursue a system with maximally unleashing the combined computational power in a cluster of any given size .

There are a good deal of open questions remain for the design of distributed DL training frameworks. Integrated with the MapReduce-compatibled frameworks is an easy thing to think about for our purpose. How to achieve the model synchronization between nodes for the next iteration training need be taken into consideration primarily. Otherwise, MapReduce has no communication API and heavy model transformation cost. Parallel implementations of iterative convergent algorithms are naturally expressed using the Bulk Synchronization Parallel [4] model of computation. For example, a typical distributed computing architecture of BSP, HAMA [5] based on Hadoop [6] needs vast synchronous overhead. As the Figure 1 shown based on the expansibility test, we could always reduce computation time in a near-linear fashion according by the increasing of the number of machines. But, the communication time increased as number of machines increased. Furthermore, the stragglers' problem where every transient slowdown of any node can delay all other nodes should be avoided.

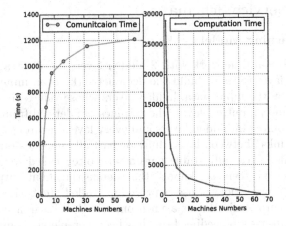

Fig. 1. The Relationship between Scalability and Cost

Therefore, building a scalable deep learning training system is the key to unlocking the full potential of the age of big data. It is urgent to explore a novel deep learning service schema to satisfy the requirements of scaling and communication spending. The customized acceleration techniques for DL such as *cuDNN* [7] also enable the opportunity to further improve the performance of deep neural network.

1.2 Main Contributions

Motivated by above-mentioned need, we propose in this paper our work *DistDL*, a scalable and robust framework for deep learning services. *DistDL* aims to provide Deep Learning as a Schema(DLaaS) for both research and application users. The schemes of our presented distributed deep learning Service *DistDL* are listed below:

- *DistDL* proposed the hierarchical architecture of data partition with the seamless channel to parameter nodes result in achieving the extremely data & model parallelism.
- *DistDL* leverages Bloom Filter to build a Bitmap through encoding the data & computing information in every work node for the purpose of reduce the computation cost and data movement.
- Since GPUs have been tremendously successful in adding compute horsepower to neural networks, *DistDL* can speed up the training rate of convergence by utilizing the GPU heterogenous resources.

Our prototype is rapidly developing based on Spark [8], cuDNN [7], and Cassandra [9]. The performance evaluation proved that *DistDL* is of scalability and efficiency.

2   *DistDL* Service Schema

2.1   Deep Learning Model

A full-connected neural network with 2 hidden layers are illustrated in Fig. 2. There are three layer types to build the deep network, as an input layer whose values are fixed by the input data, hidden layers whose values are derived from previous layers and an output layer whose values are derived from the last hidden layer. The number of hidden layers and every layers' neurons can be configured with the "links" between layers fully or partially depends on the training algorithms details. These adjustable parameters, often called weights, are real numbers that can be seen as "knobs" that define the input-output function. In practice, most practitioners use a procedure called stochastic gradient descent (SGD) synchronously [10] or asynchronously [11]. Each layer can only be fully computed after its preceding layer has been completely computed by the feed forward algorithm. The back propagation uses for getting the errors, computing the average gradient, and adjusting the weight preparing for the next iterations [12].

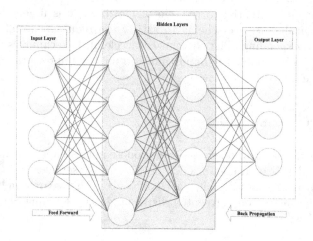

Fig. 2. The Standard Deep Learning Model

2.2   Proposed System Architecture

The proposed frameworks and data flows of *DistDL* are shown in Fig. 3 (a) and (b) separately with the features of data partition, model parallelism, model updating, consistence control, and fault recovery etc. We will give more details later. In the *DistDL* architecture, these nodes of clusters are split into three sets of nodes: worker, name and parameter. The DL instances with executable codes are deployed into the worker nodes to train the partitioned data by leveraging GPU resources. The name nodes are responsible for the data metadata managements, job scheduling, parameters exchanging and system load balancing. The model partition is handled by the parameter nodes with push and pull operations to carry out the model updating. The choice between synchronization and asynchronism ways to update the weights is a big concerning for performance tradeoff consideration. The entire design of *DistDL* aims to provide flexible consistency, elastic scalability, and continuous fault tolerance.

2.3   *DistDL* Data Flowing

The training data is divided into several parts by name nodes, then move to the worker nodes as described in Fig. 3 (b). Each work node only handles its own part, sends the results to other workers in every iteration, and obtains updates from other worker nodes through parameter nodes. In the model updating processing, there are four elements: Weights(w), Bias(b), Input (x) and Output (y). Weights were possessed by each units their own and we can increase or decrease the weight by whether they are active or inactive. The learning algorithm will pick a random training case (x,y), then run the neural network on input y, finally modify connection weights (w) to make prediction closer to y. This calculation can be mathematically represented as equation 1. The novel design for computing in parallel for data $(x, (y))$ and model(w,b) can accelerate the training rate significantly.

$$y = b + \sum_{i}^{n} x(i) * w(i) \tag{1}$$

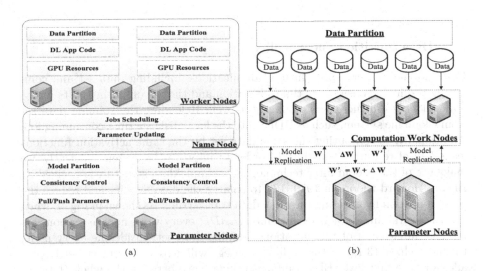

Fig. 3. (a) The *DistDL* Frameworks; (b) The *DistDL* Data Flows

3 The Design of Components in *DistDL*

Our design has brought out some tricks for the components in our frameworks. The mechanisms of Data Partition, Model Updating, Parameter Servers, and Fault Recovery will be addressed in the following subsections.

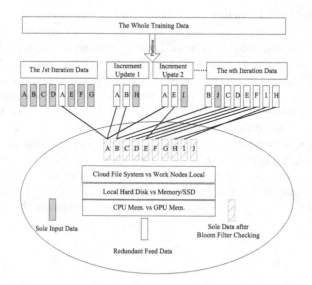

Fig. 4. Hierarchical Data Partition Architecture in *DistDL*

3.1 Data Partition and Locality

Conventionally, there exists huge of duplications, including data set and computing results between every time's iteration and the re-computing units whose data set only did increment updating as well. Reloading the training data set from Cloud file system, such as HDFS into the local disk is a time-consuming job. Additionally, re-computing the same data set by the same activation functions leads lots of resources' wasting. In order to improve the system efficiently, we make use of the Bloom filter technologies, a simple data structure conceived by Burton Howard Bloom in 1970 [13], to enlarge the data locality and reduce the communication by avoiding the same data transferring. As shown in Fig. 4, when the whole training data is feeding into *DistDL*, every work node is responsible for $1/N$ mini-batch of this iterations training data set. Inside the every node, the hierarchical (3-layer) Bloom filter strategy will help us to achieve better parallelism. The first layer Bloom filter is taking place between the whole data set in cloud file system and the work node. The second layer comes up between the CPU memory/SSD and hard disk inside the work node. The third one exists between CPU and GPU memory locally.

3.2   Model Updating and Caching

As demonstrated in Fig. 5, we will build a Bitmap generated by Bloom filtering algorithm through encoding the data/computing information in every work node. Each work node could share the Bloom filters with the central coordinator - *Akka*, an *Actor* model [14] implementation in *Scala* which the *DistDL* is using. Every training data loading/re-computing does essentially the same checking: look at data in "chunks" (all data is operating in block level) and store only a single copy of each unique chunk locally. Finally, the sole data sets will be saving in the work node persistently in order to reduce the frequent communication between the data nodes and work nodes.

In other words, every work node will build the bitmap based on the initial training data set. This bitmap includes m bits, all set to 0. There must also be k different hash functions defined, each of which maps or hashes the set element to one of the m array positions with a uniform random distribution. This bitmap is to discern the possible replicated data in order to reduce redundant computing/data storing.

When the data-training request is arrived, the data analyzer will calculate the data set by k hash functions, then lookup in the built bitmap to check it is in the local disk. If the data is already stored in this work node, the data loading could be avoided. If not, the data is placed into the local disk and the bitmap updated. The similar checking and bitmap updating operation will happen in the other layers of bloom filters.

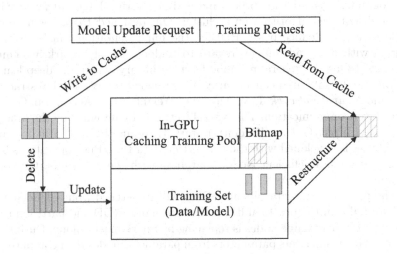

Fig. 5. Mode Updating and and Caching Procedure

3.3   Parameter Servers and Fault Recovery

The parameter server is the key to achieve the model/data parallelism and reduce the communication cost dramatically. A model consists of two parts, the architecture parts (network topology and related functions) and the parameters. How to positively alternate the model towards an effective representation is the main mission of deep learning. In the iterative learning procedure, the parameters are accessed and updated at different time and locations. Large models are also partitioned and parameters are maintained by the parameter nodes. This component is responsible for user-facing model description, and low latency access and policy orientated update for parameters.

Currently all the parameters are stored on the group of parameter nodes. Parameters should be partitioned across multiple servers when the parameter size get large. Parameter server client should route requests to different servers accordingly. To get better performance, client can cache the parameter and store updates through Bloom filters Bitmaps. For the purpose of system reliability, it should be restarted on another node when a parameter node crashes. In order to achieve fast fault recovery, Data of a parameter server should be periodically check-pointed so it can be transferred when a server is restarted. When a task is restarted, it should not rerun finished iterations.

4   *DistDL* Implementation

Big data processing frameworks (BDPF) are of unexpectedly excellent performance for massive data processing, which are the key for knowledge mining in the big data era. Most BDPFs are based on MapReduce [15] paradigms, which is a programming model for processing and generating large data sets with a parallel, distributed algorithm on a cluster. Our designed *DistDL* will be compatible with MapReduce based BDPFs, the implementation will be based on Spark [8] with many advantages regard to Hadoop. Firstly, Spark has embeds the MLlib [18] as the general machine learning library, where the deep learning functions can be inherited conveniently. We provided the two level abstractions: *Layer* and *NeuralNetwork*. Layers includes Weights, Bias, Activation, Gradient & Inter-layer communication. The Neural Network are designed by the functions of Topology (Input/Output/Hidden), Iterate, Error and Predict. The parameter nodes cluster is developed with the supporting data Pull/Push operations based on Cassandra [9], an efficient NoSQL database with the well support by Spark community.

In the procedure of data pipeline, Data will be fetched from Cloud file system, like HDFS in the practical flows, then store into RDD and partition across workers. The role of name nodes is replacing by Spark master node. During each iteration, each worker gets parameters from parameter nodes then computes new parameters based on old parameters and training data in this partition. Finally each worker updates parameters to parameter nodes. Worker communicates with parameter nodes through a parameter daemon, which is initialized in "TaskContext" of this worker. It enables in memory data sharing to decrease the overhead of disk I/O, thus less data movement. Additionally, It provides interactive shell for easy interface and has built in fault tolerance to automatically rebuild on failures.

DistDL integrated GPU accelerators into current Spark platform to achieve further calculation acceleration. The users can obtain Plugin style design C current Spark applications can choose to enable/disable GPU acceleration by changing in configuration file. Existing system codes can be easily imported to the heterogeneous platform. We adopted the NVIDIA cuDNN [7] as our GPU accelerating backend. In the point of view users, their code will be reused after the GPU backend scheduler called the corresponding API to carry out the matrix calculation according to the configuration.

5   Quantitive Experimental Analysis

5.1   Testbed Configuration

We employed the MNIST [16], a benchmark of handwritten digits, with 60,000 samples from 250 writers in 28x28 gray scale pixels. The total dataset has a training set of 60,000 examples, and a test set of 10,000 examples. And all of them are labeled. Our cluster is group by the EC2 instances in AWS cloud platform. We leased 12 *t2.medium* for the scaling test. In the GPU parts, the *G*2 instances with High-performance NVIDIA GPUs, each with 1,536 CUDA cores and 4GB of video memory are employed for the GPU accelerating testing.

Table 1. The Training Time Comparison between GPU and CPU

Iterations	2500 Samples CPU	2500 Samples GPU	5000 Samples CPU	5000 Samples GPU	Speed Up
1000	63	40	98	56	37%-43%
2000	102	69	182	101	32%-45%
3000	160	99	279	158	38%-43%

5.2   Scalability Evaluation

We conduct the experiments on speed-up gained from Spark. The efficiency and scalability of *DistDL* is tested on cloud clusters with 2,4,6,8,10, and 12 nodes with 60000 samples. The test input is the whole MNIST data set. From our results, we can clearly conclude that the running time of tasks is linearly dependent on the aspects of nodes numbers in the cluster. The slightly mismatch between result and our anticipation is because of the overhead of system architecture. Our result proves *DistDL* have an excellent speed-up performance. Fig. 6 (a) shows the running time has an inverse ratio relationship with the nodes number approximately. Our result proves our schema has an expected scalability.

5.3   GPU Evaluation

Due to the impressive price/performance and performance/watt curves, GPU has become the wisest option for deep neural network training. The preliminary

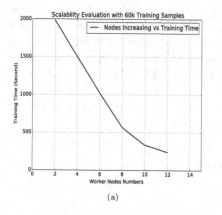 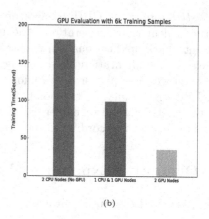

(a)   (b)

Fig. 6. (a) Scalability Evaluation Performance Result; (b) The Preliminary Results by GPU Accelerating

result to compare the performance of CPU and GPU with various iteration and samples is introduced in table 1, get 40% decrease in average compared with CPU training. What's more, we have trained in the *DistDL* with 1500 iterations with 6000 samples based on Two CPU Nodes, One CPU & One GPU Node, and Two GPU Nodes. The result is shown in Fig. 6 (b). We noticed that the time consumption can reduce up to 80% compared the no-GPU manner with GPU accelerating.

6   Related Works

A lot of efforts has been made to build distributed machine learning systems. Mahout [17], based on Hadoop [6] and MLI [18], based on Spark [8], adopt the iterative MapReduce [15] frameworks. [2,3,19] have successfully trained deep neural networks. The Google Brain deep neural network [2] containing billions of parameters, where described two complementary (synchronous) variants: they decompose the set of variables over several different and compute different parts of the objective function respectively. Many important DL algorithms iteratively update a large set of parameters with 9 to 12 orders of magnitude. Mu Li et al [3] present their design and implementation of a distributed parameter server for general machine learning cases. Their work presented the Parameter Servers to partition the workloads and data into cluster computing environments in order to address the distributed deep learning problems and get a quick result on a very large dataset. Microsoft [19] brought out the Project Adam, similar to Parameter Servers [3], an efficient and scalable deep learning, which can enhance 2 times accuracy compared with the previous benchmarking recorders on ImageNet [20] object classification benchmark. Petuum [21]is a framework for iterative-convergent distributed ML, by treating data, parameter and variable blocks as computing units to be dynamically scheduled and updated in an error-bounded manner.

Since GPU's computing model makes it 1~10X times faster than CPU, using GPU to speed-up training is a popular approach in the machine learning community. Baidu Research built a large supercomputer, named Minwa [22] dedicated to training deep neural networks with 36 server nodes & 4 Nvidia Tesla K40m GPUs in every node. By training a large model using high-resolution images and seeing a large number of examples, Minwa has achieved excellent results on multiple benchmarks. The resulting model achieved a top-5 error rate of 4.58% on ImageNet [20]. NVIDIA released the cuDNN [7] library, a deep learning and GPU-accelerated convolutional net library that runs efficiently on their GPUs, also as our GPU backend. It emphasizes performance, ease-of-use, and low memory overhead. A cluster of GPU servers with InfiniBand interconnects and MPI is proposed to training the network in DistBelief [2] by leveraging inexpensive computing power in the form of GPUs [23]. By assimilating the advantages of the above systems' design and trade off the various elements should be considered, *DistDL*, aims at providing a deep learning service schema where cluster's nodes can share data and model by harnessing GPU capability of heterogenous.

7   Conclusions and Future Works

Briefly, *DistDL* not only serves large scale models, but also speed up the training time by leverage the GPU heterogeneous resource. Currently, we have developed the prototype based on Spark, Cassandra and NVIDIA cuDNN as the GPU backend. In general, we successfully designed and implemented a deep learning system to valid in a typical benchmarking on large scale of data which running on top of Amazon Web Service platform. We are convinced that with the application of GPUs, *DistDL* is capable of successfully enhancing the recognizing handwriting digits with great efficiency and accuracy. Improvements on efficiency and accuracy have been achieved by our system. We also observe that the running time of processing data is decreasing as the number of nodes increasing, almost in an inverse linear mode. Training time by adopting the GPUs can be reduce up to 80% off. The slightly mismatch between experiments and theory is because of the system overhead when larger number of nodes is introduced into the system. In the future we are planning to apply our algorithm on more complex deep learning applications including face recognition, speech recognition and human language processing. Currently we have just implemented BSP protocol, in that Stale Synchronous Parallel (SSP) consistency model will be supported.

Acknowledgement. We thank the anonymous reviewers of Asia Pacific Web Conference (APWeb) for their feedback on previous versions of this paper. This Project supported by the National Natural Science Foundation of China (No. U2012A002D01).

References

1. Doug, B., Kumar, S., Li, H., Sobel, J., Vajgel, P.: Finding a needle in haystack: facebook's photo storage. In: OSDI (2010)

2. Dean, J., Corrado, G., Monga, R., Chen, K., Devin, M., Mao, M., Senior, A., Tucker, P., Yang, K., Le, Q., et al.: Large scale distributed deep networks. In: NIPS (2012)
3. Li, M., Andersen, D., Park, J.W., Smola, A., Ahmed, A., Josifovsiki, V., Long, J., Eugene, S., Su, B.: Scaling distributed machine learning with the parameter server. In: Proceedings of OSDI (2014)
4. Leslie, V.G.: A Bridging Model for Parallel Computation. Communications of the ACM 33(8), 103–111 (1990)
5. Seo, S., Yoon, E.J., Kim, J., Jin, S., Kim, J., Maeng, S.: Hama: An efficient matrix computation with the mapreduce framework. In: CloudCom (2010)
6. Hadoop. http://hadoop.apache.org
7. Chetlur, S., Woolley, C., Vandermersch, P., Cohen, J., Tran, J., Catanzaro, B., Shelhamer, E.: cuDNN: Efficient primitives for deep learning. In: Proc. Deep Learning and Representation Learning Workshop NIPS (2014)
8. Zaharia, M., Chowdhury, M., Das, T., Dave, A., Ma, J., McCauley, M., Franklin, M., Shenker, S., Stoica, I.: Resilient distributed datasets: A fault-tolerant abstraction for In-memory cluster computing. In: Proceedings of NSDI (2012)
9. Apache Cassandra. http://cassandra.apache.org/
10. Zinkevich, M., Wermer, M., Li, L., Smola, A.: Parallelized stochastic gradient descent. In: Advamces in Neural Information Processing Systems, pp. 2595–2603 (2010)
11. Niu, F., Retcht, B., Re, C., Wright, S.J.: Hogwild! a lock-free approach to parallelizing stochastic gradient descent. In: Proc. NIPS (2011)
12. LeCun, Y., Bengio, Y., Kim, G., Jin, S., Kim, J., Maeng, S.: Deep Learning. Nature 521, 436–444 (2015)
13. Bloom, B.H.: Space/time trade-off in Hash Coding with Allowable Errors. Communications of the ACM 13(7), 422–426 (1970)
14. Agha, G.: An Overview of Actor Languages. ACM 21(10), 58–67 (1986)
15. Dean, J., Ghemawat, S.: Mapreduce: Simplified Data Processing on Large Clusters. Communications of the ACM 51(1), 107–113 (2008)
16. Cortes, C., LeCun, Y., Burges, C.J.: The MNIST Database of Handwritten digits (2013). http://yann.lecun.com/exdb/mnist/
17. Mahout. http://mahout.apache.org
18. Sparks, E., Talwalkar, A., Smith, V., et al.: Mli: An API for distributed machine learning. In: 2013 IEEE 13th International Conference on Data Mining (ICDM), pp. 1187–1192 (2013)
19. Chilimbi, T., Suzue, Y., Apacible, J., Kalyanaraman, K.: Project adam: building an efficient and scalable deep learning training system. In: Proceedings of OSDI (2014)
20. ImageNet. http://www.image-net.org/challenges/LSVRC/2014/
21. Dai, W., Wei, J., Zheng, X., Kim, J., Lee, S., Yin, J., Ho, Q., Xing, E.: Petuum: A New Platform for Distributed Machine Learning on Big Data. arXiv preprint, arXiv:1312.7651 (2013)
22. Wu, R., Yan, S., Shan, Y., Dang, Q., Sun, G.: Deep Image: Scaling up Image Recognition. arXiv preprint, arXiv:1501.02876 (2015)
23. Coates, A., Huval, B., Wang, T., Wu, D., Andrew, N., Catanzaro, B.: Deep learning with COTS HPC systems. In: Proceedings of ICML, pp. 1337–1345 (2013)

A Semi-supervised Solution for Cold Start Issue on Recommender Systems

Zhifeng Hao, Yingchao Cheng, Ruichu Cai, Wen Wen, and Lijuan Wang

Faculty of Computer, Guangdong University of Technology, GuangZhou, China
zfhao@gdut.edu.cn, 981173185@qq.com, cairuichu@gmail.com

Abstract. The recommender system is the most competitive solution to solve information overload problem, and has been extensively applied. The current collaborative filtering based recommender systems explore users' latent interest with their historical online behavior records. They are all facing the cold start issue. In this work, we proposed a background-based semi-supervised tri-training method named BSTM to tackle this problem. In detail, we capture fine-grained users' background information by using a factorization model. By exploring these information, the performance of our recommendation can be significantly promoted. Besides, we proposed a semi-supervised ensemble algorithm, which got both labeled and unlabeled data involved. This algorithm assembled diverse weak prediction models which are generated by exploring samples with diverse background information and by tri-training tactic. The experimental results show that, with this solution, the accuracy of recommendation is significantly improved, and the cold-start issue is alleviated.

Keywords: cold-start, recommendation, semi-supervised learning.

1 Introduction

Nowadays, Internet has deeply integrated into people's daily lives. It makes the circumscription of life online and offline faded, and uploads nearly all the information to the Internet. Thereby, it aggravated the problem of so-called "information overload" [1]. To solve this problem, numerous methods have been proposed: from information indexing tools, portals, search engine to recommender system which is the most potential one. Recommender systems can do the statistical analysis on users' historical activity records, tap the users' preferences, then automatically find the matching content from the mass of information and make right recommendation [2,3,4].

Nevertheless, the mainstream of recommender systems are troubled by cold start issue. There is rare information of the fresh users; meanwhile, the new items have little chance to be selected [5]. So, it is difficult to give users precise recommendation. This defines the so-called cold start issue. If this issue could be resolved satisfactorily, users' experience on recommendation will be impressively improved, so as to enjoy much more personalized Web services.

© Springer International Publishing Switzerland 2015
R. Cheng et al. (Eds.): APWeb 2015, LNCS 9313, pp. 805–817, 2015.
DOI: 10.1007/978-3-319-25255-1_66

Lots of remarkable related works, which inspired us, try to solve this issue in different field, but the works have their limitations:

- It has been proved that new users are more likely to choose the popular items [6], which indicates that the hot-selling list can do something for cold start issue. This method could meet the basic need, but it's helpless for further requirements.
- Some researchers believe the extensive application of labelling systems could solve the cold start issue [7]. Because labels indicate the content of items and preference of users, which could be used to calculate the similarity among items. So as to solve the issue properly. But it is also helpless for sheer cold start, these users have labeled nothing.
- There are many other constructive solutions, including mining the multidimensional data on overlapped social networks [8,9] and transfer learning algorithm [10]. R. Sinha et al. point out that users prefer recommendation from friends rather than the recommender systems [11]. Social impact is considered more important than the similarity of historical behavior [12]. Social recommendation from a friend can increase the popularity of items and their reviews [13], which could compensate the lack of ratings. However, these solutions' performance on cold start issue is usually unstable [14,15].

Our motivation of solving cold start issue is derive from life. The sophisticated men have kind of personality-recognition ability. They are so popular in the crowd that nearly everyone is interested in them. When they meet a gay, they exchange greetings: "what's your name?", "your job?" and so on. This is their first step: to collect person's basic background information. Then they use their rich life experience to do the personality-recognition work. At last, they start a conversation you are interested in. They feel like old friends. This phenomenon inspired us. The recommender systems can do the same thing. Input users' basic background information and vast historical online activity records, output user's interest and requirement. Finally, do the recommendation job precisely.

In this work, we get attributes similarity between users' background information. These information is analyzed to get users' social pictures. The detailed description will be given later. At last, the proposed method is verified on Movie Lens dataset.

Organization. Section 1 formulates the problem and discusses related work; In section 2: 2.1 presents a fine-grained users' background information model, 2.2 describes the proposed background-based semi-supervised tri-training method (BSTM); Section 3 presents the experimental results and Section 4 is conclusion and future work.

2   Methods

2.1   BBPM

On Internet, collaborative filtering methods have been applied to many recommender systems [16]. Take equation (1) for example [16], it established a correlation model which reflected the relationship between users' rating and their rating time\rating frequency. The predicted ratings composed by four parts: the average score μ of all available movies, the users' bias b_u on rating, the society's common bias b_i on the movie i, a certain user's personal preference $q_i^T p_u$ on a certain movie i. All these unknown parameters $[\mu, b_u, b_i, q_i^T p_u]$ can be estimated by minimizing the deviation between the predicted ratings and real ratings in the labeled training set.

$$\hat{r}_{ui} = \mu + b_u + b_i + q_i^T p_u \tag{1}$$

While, the collaborative filtering methods suffer from cold start issue. Researchers have to not only take advantage of users' historical behavior data, but also take the movie information and user attributes into consideration [3].

$$b_{bg} = \frac{\sum_{i \in genres(g)} b_{ui}}{|genres(g)|} + b_{ia} + b_{io} + b_{is} \tag{2}$$

In equation (2), $\frac{\sum_{i \in genres} b_{ui}}{|genres(g)|}$ indicates movies' information and b_{ia}, b_{io}, b_{is} indicates attributes of users' background, respectively. Generally speaking, when the users rate for a movie, it is possible to give a pretty high score to a certain genre they loved, and vice versa. Therefore, we can quantify the impact of the movie genre by $\frac{\sum_{i \in genres} b_{ui}}{|genres(g)|}$. b_{ui} indicates the users' bias on movie i which belong to genre g; $b_{ia}/b_{io}/b_{is}$ indicates the bias of certain age/occupation/sex of users on the movie i, respectively. The parameters can be obtained by stochastic gradient descent algorithm.

In order to achieve continuous improvement on resolving cold start issue, we need to consider more factors. The more factors been taken into consideration, the better of the recommendation.

Apart from calculating the impact (as shown in equation (1) & (2)) of movie genre and users' background information, we focus on accurately characterizing the mutual impact of different user background attributes (as shown in equation (3) & (4) & (5)). Adding these factors to model construction makes the model much more comprehensive and integrated. We built such a model called BBPM (short for background-based prediction model). It could compensate the insufficient of rating information on the websites and make more precise recommendation.

Life always inspires us: the elderly with abundant experience tend to have a good knowledge of youths' thoughts; old friends are able to interest each other for knowing each other's hobbies; and so on. They can make it, because they know each other very well. If we are well versed about users, it is possible to make recommender systems much more precise.

Therefore, we add more users' background factors into BBPM. For example, users' age which can be classified into seven intervals. Users who are in different age interval demonstrated certain age-related disposition and aesthetic taste. And users' occupation/sex have the similar impact. Now we'll explore the mutual impact of users' age, occupation and sex. These factors really matter for discovering users' tastes.

Firstly, the mutual impact of users' age and sex, is a significant factor for predicting users' interest. E.g. a young girl may prefer romantic movies, the middle-aged males usually like Westerns. Thus, the model can be written as follow:

$$\hat{r}_{ui} = \mu + b_u + b_i + q_i^T p_u + b_{bg} + b_{ias} \tag{3}$$

Here, b_{ias} indicates the bias of users on movie i, who are in age a and whose sex is s.

Secondly, users' occupation and sex also can produce a chemical reaction. Such as: working in a same hospital, male doctors probably like serious documentary "Inside the Living Body"; while female doctors may prefer the interesting drama "Grey's Anatomy". Therefore, the model evolves:

$$\hat{r}_{ui} = \mu + b_u + b_i + q_i^T p_u + b_{bg} + b_{ias} + b_{ios} \tag{4}$$

Here, b_{ios} indicates the bias of users on movie i, who engaged in occupation o and whose sex is s.

Finally, the mutual impact between age and occupation may manifest like this: a prentice young civil servant could be attracted by Machiavellian stories in movie "The Godfather"; but the elder civil servants, who are approaching retirement, generally like "Forrest Gump". Thus, we finish the comprehensive model:

$$\hat{r}_{ui} = \mu + b_u + b_i + q_i^T p_u + b_{bg} + b_{ias} + b_{ios} + b_{iao} \tag{5}$$

Here, b_{iao} indicates the bias of users on movie i, who are in age a and engaged in occupation o.

We can adopt stochastic gradient descent (SGD) to train these parameters. Since $[b_{ias}, b_{ios}, b_{iao}]$ are certain kind of users' bias for a particular movie i, only the ratings of movie i can impact the training process of these certain group of users. And we can obtain the model BBPM by minimizing the recommendation errors (equation (6)) on the entire training set.

$$\min_{p*, q*, b*} \Sigma_{(u,i) \in T} (r_{ui} - \hat{r}_{ui})^2 + \lambda(\|p_u\|^2 + \|q_i\|^2 + \Sigma b_*^2) \tag{6}$$

In the above equation, $\hat{r}_{ui}$ represents the movie ratings predicted by BBPM, $\|\cdot\|$ represents the Euclidean distance of corresponding parameters, and λ is the regularization rate. Equation (6) can be resolved by stochastic gradient descent (SGD). SGD ultimately acquire various parameters of BBPM via training samples one by one, and gradually update the parameters.

2.2 BSTM

We proposed a background-based semi-supervised tri-training method (referred to BSTM) to solve cold start issue. BSTM is intended to establish a semi-supervised learning process. Try to exploit both the labeled data and unlabeled data to optimize the generalization error of a supervised learning algorithm [13]. In this work, according to Bagging method and users' background information, we formulate diversified models. Then we integrate the multiple models to improve the accuracy of recommendation. More specifically, BSTM involves three main steps:

Firstly, formulate three regressors. Recommendation is a regression issue, so we formulate three regressors (r_1 , r_2 and r_3) from training set T; then use these regressors separately to label the unlabeled training samples; at the last of first step, these regressor-labeled samples are used to optimized other regressors.

Secondly, tri-training. During this process, three regressors learn from each other: samples with high confidence is chosen to be labeled by one of the regressors (give the samples predicted ratings), and then use those samples to train other regressors.

Finally, integration of the results. The results obtained by each regressor are integrated according to certain principle, thus receive the ultimate recommendation.

To be more intuitive, we give the corresponding algorithm flow chart as follow:

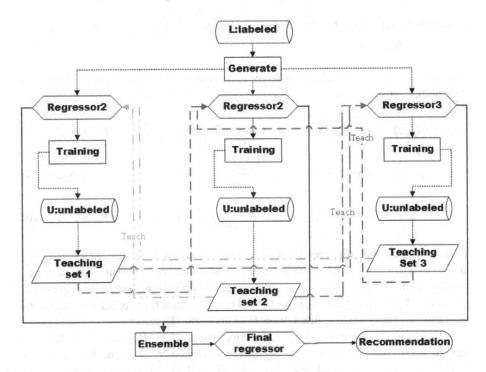

Fig. 1. Algorithm flow chart of BSTM

This is the specific technical route of our work. We construct BBPM (background-based prediction model), and then use it to extract different regressors, eventually integrate the recommendation results of all regressors.

In the cause of establishing a plurality of regressors, we operate by two schemes: one is that the BSTM be trained with diverse samples by operating the training data set, this scheme is represented by O_s; the other is to take different factor into consideration by operating background attributes, this scheme is represented by O_a.

A common method to build multiple training subset is Bagging, which randomly select n samples from the original training set to form new training subsets in each run of training process. We use Bagging to construct three training subsets for regressor learning. The favorable diversity among regressors is conducive for the performance of joint training [17]. The diversity of regressors depends on the deviation of training samples. In addition, there should be enough samples in the subsets to ensure considerable accuracy of the regressors. But too many samples can damage the diversity of the training subsets. These two factors are contradictory. So it must be done in the experiments to find proper capacity of subsets. When this work has been done, we take equation (7) as fundamental learner for formulating the regressors. Then we train the learner on three diverse training subsets and get different regressors. They cooperate during tri-training process. This is the O_s scheme.

$$r_1(u,i) = r_2(u,i) = r_3(u,i) = \mu + b_u + b_i + q_i^T p_u + \frac{\sum_{i \in genres(g)} b_{ui}}{|genres(g)|} + b_{ia} + b_{io} +$$
$$b_{is} + b_{ias} + b_{ios} + b_{iao} \tag{7}$$

The second scheme to establish different regressors is to split background attributes into multiple points.

$$A: \quad r_1(u,i) = \mu + b_u + b_i + q_i^T p_u + \frac{\sum_{i \in genres(g)} b_{ui}}{|genres(g)|} + b_{ia} + b_{io} + b_{is} + b_{ias} \tag{8}$$

$$B: \quad r_2(u,i) = \mu + b_u + b_i + q_i^T p_u + \frac{\sum_{i \in genres(g)} b_{ui}}{|genres(g)|} + b_{ia} + b_{io} + b_{is} + b_{ios} \tag{9}$$

$$C: \quad r_3(u,i) = \mu + b_u + b_i + q_i^T p_u + \frac{\sum_{i \in genres(g)} b_{ui}}{|genres(g)|} + b_{ia} + b_{io} + b_{is} + b_{iao} \tag{10}$$

Intuitively, users with different pairwise attributes do show differentiated interest. The above regressors A/B/C characterize the mutual impact of different pairwise background attributes (age/sex/occupation), respectively. This is source of regressors' diversity. These regressors would be trained on the overall training set to ensure considerable accuracy, which is differ from O_s. This is the O_a scheme.

As we can see, the diversity of regressors generated by O_s came from different training subset and the diversity of regressors generated by O_a came from different models. At last, we take advantage of both O_s and O_a to generate more diverse regressors, which means getting regressors by training models (A/B/C) on training subsets. We call this hybrid scheme as O_{sa}.

We've got multiple regressors, a following work is tri-training different regressors. Firstly, we need to build a teaching dataset D_j for each regressor. Then regressors would be iteratively optimized by their peer regressors. Here comes the problem, how to select samples from the un-rating dataset U' to found the teaching dataset? We do this work under the criterion of data confidence. Confidence indicate samples contribution to the accuracy of the prediction. How to describe the confidence? The accurcy of recommendation increases with the rising number of rating data. So, the confidence of regressors' predicted rating is defined as follows:

$$C_r(x_{ui}) = \frac{d_u^{(r)} \times d_i^{(r)}}{N} \quad (11)$$

Equation (11) expresses the confidence on a recommendation made by regressor r for sample x_{ui}. While N is the regularization coefficient; $d_u^{(r)}$ expresses the activeness of user u, and its value associated with users' preference parameters b_u and p_u; $d_i^{(r)}$ expresses the watching frequency of movie i, its value associated with movies' attribute parameters b_i and p_i. That is to say, if the users are more active and the movies are watched more frequently, the confidence of the data is higher. The accuracy of recommendation derived from these data could be better.

Furthermore, confidence (11) can be improved as the following form:

$$C_r(x_{ui}) = \frac{d_u^{(r)} \times d_i^{(r)} \times \prod_{c \in G,O,A,S} d_c^{(r)}}{N} \quad (12)$$

In the above equation (12), c expresses the various movie information and users' background attributes: genre, occupation, age, sex. d_c expresses the popularity of particular type of movie and activeness of specific users. For example, d_{g7} expresses the popularity of "Documentary" movies, d_{g18} expresses the popularity of "Western" movies, and so on. In total, there are $G=\{g_0...g_{18}\}$ 19 movie genres. Similarly, d_a, d_o, d_s expresses A= $\{a_0...a_6\}$ 7 age intervals, S=\{male, female\} 2 kind of sex and O=$\{o_0...o_{20}\}$ 21 occupations, respectively.

According to this definition, the value of d_u, d_i and d_c associated with the three regressors (O_s) is different. That is to say, in terms of the same user u: $d_u^{(1)}/d_u^{(2)}/d_u^{(3)}$ (associated with the confidence of regressor $r_1/r_2/r_3$) expresses the rating number of user u (assigned to the first/second/third training subset, respectively). And so is d_i. Therefore, the confidence of the three regressors can be very different. But the confidence $C_j(x_{ui})$ is just come up to select samples for a certain regressor during the semi-supervised learning process. So it is not comparable.

For each regressor r (u, i), we must eliminate the situation that a teaching dataset D_j is full of popular movies and active users. We abandon to directly take confidence value as selection criteria. Instead, we use a confidence-based Roulette algorithm to pick samples for each teaching dataset. The probability of selecting a sample is:

$$Pr(x_{ui}, j) = \frac{c_j(x_{ui})}{\sum_{x_k \in U'} c_j(x_k)} \tag{13}$$

The final step of the solution is to integrate the prediction of three regressors. We take confidence as the integration weight. And the final regressor is:

$$r(u, i) = \sum_{j=1...l} \frac{c_j(x_{ui})}{\sum_{k=1...l} c_k(x_{ui})} r_j(u, i) \tag{14}$$

It should be noted that integration step should be applied after the semi-supervised training, which can ensure that both labeled data (with ratings) and unlabeled data is used.

3   Results

As mentioned above, we use a public significant dataset (Movie Lens) to validate our solution. In order to verify the robustness and efficiency, the solution is accomplished at four datasets with different dimension in the experiments. The first one contains 50,000 ratings of 1,466 movies awarded by 489 users; there are 100,000 ratings from 943 users on 1682 movies in the second one; The third one contains 200,000 ratings of 3,266 movies awarded by 1429 users; and 1,000,209 ratings of approximately 3,900 movies made by 6,040 users are in the fourth one. These are respectively referred to M_1, M_2, M_3, M_4.

The data states the ratings with users' background information and movie genre. This characteristic satisfies BBPM's requirement for data and make it possible to implement BSTM via operating the background attributes. Specifically, Movie Lens datasets depict users' background information, involving age, occupation and sex. In the experiments, the significant ratings were partitioned into three subsets: (1) validation set contain each users' 10% ratings, which is utilized for inchoate termination and meta-parameters adjustment; (2) training set possess 80% of ratings, which cooperate with validation set to tune the meta-parameters and restructure our ultimate model; (3) test set encompass the rest, and we give the solution's results (RMSE) on it, as shown in table 1.

In our experiments, the accuracy of the recommendation results is indicated by the root mean square error:

$$RMSE = \sqrt{\sum_{(u,i \in TestSet)} \frac{(r_{ui} - \hat{r}_{ui})^2}{|TestSet|}} \tag{25}$$

r_{ui} represents the true rating of users and $\hat{r}_{ui}$ represents the predicated rating. If the RMSE is lower, the result will be better. Low RMSE indicates high accuracy of the result. We compared different solutions: firstly, we validate the state-of-the-art algorithms; than we validate the solutions proposed recently; at last, we validate our work on Movie Lens data.

Table 1. The overall RMSE performance of different solutions on the four datasets

Models	M_1	M_2	M_3	M_4
UB k–NN	1. 0250	1. 0267	1. 0224	0. 9522
IB k–NN	1. 0733	1. 0756	1. 0709	1. 0143
CF	0. 9310	0. 9300	0. 9260	0. 8590
fMF	0. 9508	0. 9439	0. 9432	0. 9413
LFfMG	0. 9764	0. 9617	0. 9515	0. 9520
CSEL	0. 8987	0. 8966	0. 8963	0. 8334
BSTM	0. 8832	0. 8831	0. 8824	0. 8305

Specifically, the solutions are:

- the classic and common used user-based k-nearest neighbor collaborative filtering algorithm, short for UB k-NN;
- item-based k-nearest neighbor collaborative filtering algorithm (IB k-NN);
- the factorization-based collaborative filtering (CF) method in [16];
- a functional matrix factorization (fMF), which extract new users' preference with a decision tree by progressively querying user responses through an initial interview process [18];
- in the work [19], discovering latent factors from movies genres (LFfMG) based on a factorized matrix;
- CSEL, the main compare method in the work [3].
- BSTM, our background-based semi-supervised tri-training method.

Experiment results show that the overall recommendation performance of our solution is better than others. Besides the solution's overall recommendation accuracy, we also investigated its performance on cold start issue. BSTM take advantage of the mutual impact of each pairwise background attribute factors. We evaluate activity of each user by the times of watching. And then divide users into 10 groups by users' activity. At last, we measure RMSE separately on each user group. And the RMSE show us the performance of the solutions on cold start issue as the following Table 2:

In Table 2, *Bin(u)1* is the group of most active users; *Bin(u)10* is the group of most inactive users. The property of these algorithms (k-NN, CF ···), additionally, make it clear that most common recommender systems suffer from the cold-start issue. It is because the performance of these solutions relays on the similarity of users or items. While, the result reveals the good efficiency of our solution on cold-start issue.

Table 2. The cold start RMSE performance of different solutions on dataset M_4.

M_4	Total RMSE	Bin(u)1	Bin(u)2	Bin(u)3	Bin(u)4
Rating #	106 (AVG#)	340	207	148	110
UB k-NN	1.0267	0.9816	0.9939	1.0191	1.0292
IB k-NN	1.0756	1.0156	1.0470	1.0536	1.0729
CF	0.9300	0.8929	0.9019	0.9100	0.9432
fMF	0.9051	0.8749	0.8895	0.8978	0.9176
LFfMG	0.9083	0.8723	0.8952	0.9011	0.9052
CSEL	0.8966	0.8694	0.8752	0.8832	0.9078
BSTM	**0.8753**	**0.8608**	**0.8652**	**0.8684**	**0.8748**

M_4	Bin(u)5	Bin(u)6	Bin(u)7	Bin(u)8	Bin(u)9	Bin(u)10
Rating #	77	57	44	33	26	21
UB k-NN	1.0696	1.0739	1.1043	1.1145	1.1381	1.1694
IB k-NN	1.1041	1.1245	1.1396	1.1446	1.1852	1.2768
CF	0.9789	0.9933	1.0291	1.0435	1.0786	1.0923
fMF	0.9292	0.9365	0.9704	0.9937	1.0058	1.0313
LFfMG	0.9326	0.9931	0.9722	0.9911	0.9987	1.1926
CSEL	0.9215	0.9223	0.9687	0.9835	0.9918	1.0023
BSTM	**0.8911**	**0.8909**	**0.9334**	**0.9695**	**0.9817**	**0.9936**

The regressors' collaboration is one of the key in our solution. In the experiment, we validate the effect of different numbers of regressors' co-work. The main idea is expressed in section 2.2. We give each regressor a subset of the training set. The subset contains unlabeled samples, and the corresponding regressor will give all the samples, in the subset, a rating, which means the subset has been labeled. Then all the subsets are used to train one regressor, except the subset that labeled by this regressor. This is the process of multiple regressors' collaboration.

We managed to study the effect of different numbers of regressors' collaboration, and have showed the result in Fig. 2. X axis represents the numbers of regressors and y axis represents RMSE. The bars in this diagram show us the different accuracy (RMSE) of different collaboration. And the line show us the cost and complexity of different collaboration. The experiments reveal that three is the appropriate number of regressors for multi-training. If the regressors less than three, the solution is easier to

be implemented, but the accuracy is dissatisfactory. While more than three regressors' collaboration could bring in the noise of data and face the over fitting problem. What's more, too many regressors do nothing to improve the accuracy but increase the complexity and cost of the solution. Actually, when the number of regressor is more than three, the accuracy is lower compared to the accuracy of three regressors' collaboration.

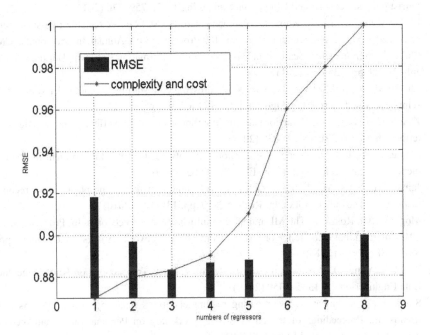

Fig. 2. The performances of different numbers of regressors' collaboration

4   Conclusion and Future Work

In this work, we use the semi-supervised learning methods to solve the cold-start issue in Recommender Systems. Firstly, we proposed the BBPM to address the lack of ratings problem by bringing the mutual impact among the attributes of users' background information into account. Then, we explore a semi-supervised tri-training method to involve the unlabeled samples. The experimental results show that our solution improves the overall system performance. This solution makes a remarkable progress in solving the cold start issue. However, the instantaneity requirement is still great challenge of our solution, which can be considered as the future work. The possible solution includes, selecting the most influential users to reduce the sample size, using parallel training method to accelerate the training procedure, and so on.

References

1. Dean, D., Webb, C.: Recovering from information overload. McKinsey Quarterly (2011)
2. Wang, S.L., Wu, C.Y.: Application of context-aware and personalized recommendation to implement an adaptive ubiquitous learning system. Expert Systems with Applications 38, 10831–10838 (2011)
3. Gavalas, D., Kenteris, M.: A web-based pervasive recommendation system for mobile tourist guides. Personal and Ubiquitous Computing 15(7), 759–770 (2011)
4. Li, L., Wang, D., Li, T., Knox, D., Padmanabhan, B.: SCENE: A scalable two-stage personalized news recommendation system. In: Proc. the 34th Annual International ACM SIGIR Conference on Research and Development in Information Retrieval, Beijing, China, July 25–29, pp. 124–134 (2011)
5. Zhang, M., Tang, J., Zhang, X., Xue, X.: Addressing cold start in recommender systems: A semi-supervised co-training Algorithm. In: SIGIR 2014, pp. 73–82 (2014)
6. Zhang, C.-J., Zeng, A.: Behavior patterns of online users and the effect on information filtering. Physica A 391, 1822–1830 (2012)
7. Zhang, Z.-K., Liu, C., Zhang, Y.-C., Zhou, T.: Solving the cold-start problem in recommender systems with social labels. EPL 92, 28002 (2010)
8. Jamali, M., Ester, M.: A matrix factorization technique with trust propagation for recommendation in social networks. In: RecSys 2010, pp. 135–142 (2010)
9. Mognani, M., Rossi, L.: The ML-model for multi-layer social networks. In: Proceedings of 2011 International Conference on Advances in Social Networks Analysis and Mining, pp. 5–12. IEEE Press (2011)
10. Pan, S.J., Yang, Q.: A survey on transfer learning. IEEE Transactions on Knowledge and Data Engineering 22, 1345–1359 (2010)
11. Sinha, R., Swearingen, K.: Comparing recommendations made by online systems and friends. In: Proceedings of the DELOS-NSF Workshop on Personalization and recommender systems in Digital Libraries (2001)
12. Bonhard, P., Sasse, M.A.: Knowingme knowing you—using profiles and social networking to improve recommender systems. BT Technology Journal 24, 84–98 (2006)
13. Huang, J., Cheng, X.-Q., Shen, H.-W., Zhou, T., Jin, X.: Exploring social impact via posterior effect of word-of-mouth recommendations. In: Proceedings of the 5th ACM International Conference on Web Search and Data Mining, pp. 573–582. ACM Press, New York (2012)
14. Xu, B., Bu, J., Chen, C., Cai, D.: An exploration of improving collaborative recommender systems via user-item subgraphs. In: Proceedings of the 21st International Conference on World Wide Web, pp. 21–30. ACM Press, New York (2012)

15. Yuan, Q., Chen, L., Zhao, S.: Factorization vs. regularization: fusing heterogeneous social relationships in Top-N recommendation. In: Proceedings of the 5th ACM Conference on Recommender Systems, pp. 245–252. ACM Press, New York (2011)
16. Koren, Y.: The BellKor solution to the Netflix Grand Prize (2009).
    `http://www.netflixprize.com/assets/`
    `GrandPrize2009_BPC_BellKor.pdf`
17. Zhou, Z.-H., Li, M.: Semi-supervised regression with co-training style algorithms. IEEE Transactions on Knowledge and Data Engineering 19(11), 1479–1493 (2007)
18. Zhou, K., Yang, S.-H., Zha, H.: Functional matrix factorizations for cold-start recommendation. In: Proceedings of the 34th International ACM SIGIR Conference on Research and Development in Information Retrieval, Beijing, China, July 24-28 (2011)
19. Manzato, M.G.: Discovering latent factors from movies genres for enhanced recommendation. In: Proceedings of the Sixth ACM Conference on Recommender Systems, Dublin, Ireland, September 09-13 (2012)

Industry Track

Hybrid Cloud Deployment
of an ERP-Based Student Administration System

Simon K.S. Cheung

The Open University of Hong Kong
Good Shepherd Street, Homantin, Kowloon, Hong Kong
kscheung@ouhk.edu.hk

Abstract. Conceptually, a hybrid cloud approach to deploying an ERP system can effectively achieve a cost-performance balance, especially during seasonal peaks of concurrent accesses. This paper describes an actual implementation of an ERP-based student administration system, where the web tier and application tier are deployed on hyrbid cloud with the databases resided on premises. The key challenge of this implementation is to set up the web tier and application tier for the right cost-performance balance while attaining an appropriate level of resilience. In this paper, technical evaluation is conducted to demonstrate the cost efficiency of the hybrid cloud deployment, and the benefits are discussed. These provide a useful reference on hybrid-cloud deployment of ERP systems.

Keywords: hybrid cloud, ERP system, student administration system.

1 Introduction

In many comprehensive universities, a student administration system used to support administrative workflows for over ten thousand students. Throughout an academic year, there are some occasions, such as the enrollment of courses and the release of examination results, where a large number of concurrent accesses are recorded within a short period of time. These accesses are made both within and outside the campus. In order to accommodate a large volume of accesses, especially concurrent accesses, the capacity planning on system infrastructure must be done accurately. One difficult challenge in this capacity planning is to achieve the right cost-performance balance, especially during some seasonal peaks of concurrent accesses. Cloud computing is a feasible solution to meet the challenge.

Cloud computing is characterized by the ubiquitous, convenient, on-demand access to a shared pool of configurable resources, such as networks, computer servers, operating systems, storage, applications and processing services, that can be rapidly provisioned and released with minimal management effort, control or service provider interaction [1]. Typically, cloud computing providers offer three models of services, namely, infrastructure as a service, platform as a service, and software as a service, and four models of deployment, namely, private cloud, community cloud, public cloud and hybrid cloud [1, 2].

© Springer International Publishing Switzerland 2015
R. Cheng et al. (Eds.): APWeb 2015, LNCS 9313, pp. 821–831, 2015.
DOI: 10.1007/978-3-319-25255-1_67

This paper describes a successful case of hybrid-cloud deployment of an Enterprise Resource Planning (ERP) based student administration system at the Open University of Hong Kong [3]. The University is currently adopting an ERP solution, Peoplesoft Campus Solution which is a very popular in higher education institutions in the North America, for its student administration [4]. The University offer programmes in both full-time and distance-learning modes, and therefore, substantial customizations had to be made on the student administration system in order to support both the full-time and distance-learning student administration. Although a cloud option of Peoplesoft Campus Solution is available in the form of Software as a Service, it does not support customizations of the ERP functional modules [5].

Typically, a web-based ERP system adopts a three-tier architecture, namely, the web tier, the application tier and the database tier. In our implementation, Microsoft Azure is subscribed for platform as a service [6]. The web tier and application tier are deployed to hybrid cloud while the database tier resides on premises (i.e. on campus). For this implementation, two technical issues need to be addressed. First, the hybrid cloud is configured to support customization of functional modules of the system. Second, the databases are resided in the campus, where accesses are controlled by both external firewalls and internal firewalls. This hybrid-cloud deployment was successfully implemented in 2014. Evaluation with respect to cost and performance was conducted, and promising results are obtained.

In this paper, our experience in the implementation is shared, and the benefits are discussed. The rest of this paper is structured as follows. Section 2 elaborates the motivation of hybrid-cloud deployment for the student administration system at the Open University of Hong Kong. Section 3 then describes the implementation of this hybrid-cloud deployment. Section 4 shows a technical evaluation of the hybrid-cloud deployment. Section 5 concludes this paper.

2   Motivation of a Hybrid-Cloud Deployment

Like many universities in the North America, the Open University of Hong Kong is using Peoplesoft Campus Solution, which is a major ERP solution in the market, to support its student administration workflows [4]. At present, the University has about 20,000 active students, 8,000 being full-time students and 12,000 being part-time and distance-learning students. The system is a web-based application, where all accesses are made through web browsers. There are on average thousands of accesses per day. About 80% of the accesses are made outside the campus while the rest 20% of the accesses are made within the campus.

Throughout an academic year, there are several occasions where excessive volume of concurrent accesses is made within a short period of time. Typical examples are the enrolment of courses before the start of a semester, and the release of examination results at the end of a semester. In order to cater for a large volume of concurrent accesses, the traditional way is to expand the capacity of web tier and application tier. The Internet bandwidth is also increased accordingly. However, the investment is only for several seasonal peaks of concurrent accesses, and the total duration of these seasonal peaks amounts to less than 20 days per year. Most of the time, the expanded capacity is not fully used.

A definite advantage of cloud computing services is to pay as you go. With a high degree of leverage, this relieves the occasional high processing demands on the web tier and application tier in a cost-effective way. In 2013, the University started to investigate the possibility of subscribing cloud computing services and deploying its ERP-based student administration system on cloud, with an aim to devise a feasible implementation that achieves the cost-performance balance.

According to the National Institute of Standards and Technology, cloud computing model exhibits five essential characteristics. They include on-demand self-services, broad network access, resource pooling, rapid elasticity and measured services [1]. There are three service models of cloud computing, as listed below.

- *Software as a Service.* This refers to the services provided for the customers to use the service provider's applications running on a cloud environment. The customers need not manage or control the underlying infrastructure, including network, servers, operating systems, storage servers and application software packages or solutions, with possible exception of limited specific application configuration setting.

- *Platform as a Service.* This refers to the services provided for the customers to deploy on the cloud environment customer-created or acquired applications. The customers need not manage or control the underlying infrastructure, including network, servers, operating systems and storage servers, but has control over the deployed applications and possibly configuration settings and even customizations.

- *Infrastructure as a Service.* This refers to the services provided for the customers on network, storages and other fundamental computing resources where the customers are able to deploy and run arbitrary software, which include operating systems and applications. The customers need not manage or control the underlying infrastructure but has control over the operating system, storage and deployed applications.

In general, there are four options of cloud deployment to be offered by the cloud computing services providers.

- *Private Cloud.* This refers to a cloud infrastructure solely for an organization. It may be owned, managed and operated by the organization, an external party, or a combination of both.

- *Community Cloud.* This refers to a cloud infrastructure solely for a specific community or a group of organizations that have common or shared concerns, such as security requirements, policies and compliance considerations. It may be owned, managed or operated by the organizations within the community or group, an external party, or a combination of both.

- *Public Cloud.* This refers to a cloud infrastructure that is open for the general public. It may be owned, managed or operated by a business, academic or government organization.

- *Hybrid Cloud.* This refers a combination of two distinct cloud infrastructures, typically a combination of private cloud and public cloud, that attend benefits from both infrastructures.

The Open University of Hong Kong is now using the Peoplesoft Campus Solution, version 9.0, with functional modules including academic advisement, campus self-service, contributor relations, financial aid, gradebook, recruiting and admission, student administration, student financials, and student records [4]. The Peoplesoft PeopleTools, version 8.53, is being used [7]. Besides, the back-end database system is running Oracle Database, version 12c, with the In-Memory option enabled [8]. Before moving to the cloud environment, the student administration system was deployed on premises. An in-house ERP infrastructure, comprising web tier, application tier and database tier, is used.

The student administration system used to support two different sets of student administration workflows, one for full-time students and the other for part-time or distance-learning students. These workflows share common functionalities, such as on enrollment, tuition payment, examination arrangement and graduation processing, whilst having their own specific requirements. For example, the tuition fees for full-time programmes are calculated and collected on programme-base while the tuition fees for distance-learning programmes are calculated and collected on course-base. Full-time programmes and distance-learning programmes have different programme structures and graduation requirements. In order to fulfill the requirements of both full-time student administration and distance-learning student administration, some customizations were made on system.

In exploring the possibility of subscribing cloud computing services and deploying the student administration system on cloud, the first consideration is to identify which service model is to be adopted. The ERP solution provider, Peoplesoft, offers a cloud option for its Campus Solution version 9.0, which is essentially Software as a Service, where the application can be accessible through a thin client interface such as a web browser [5]. Although it allows configuration the application and functional modules, there are many limitations. More important, customizations are not allowed. Hence, the Software as a Service model is not feasible.

Among other service models of cloud computing, the Platform as a Service model is apparently suitable to the University, as it offers an infrastructure that support customer-created applications. On this service model, the University can retain the flexibility of deploying the configured and customized functional modules of PeopleSoft Campus Solution on the cloud. The service provisions of a number of cloud providers, such as Microsoft [6], Amazon [9] and Citric [10], were reviewed. Microsoft Azure was finally selected as the cloud computing services, operating on the model of Platform as a Service.

At present, there are thousands of accesses to the student administration system every day. About 20% of accesses come from the campus while the majority (80% of accesses) are off-campus accesses. The hybrid cloud deployment essentially aims for these off-campus accesses. Among different options of cloud deployment, the hybrid cloud deployment option is selected because it can effectively balance the processing loads. Better system availability and data security can be attained. On the other hand, it supports customization of functional modules and offers good user experience. Table 1 shows a comparison of different cloud deployment models. It should be noted that, although the implementation of hybrid cloud is more complicated than private cloud and public cloud, the hybrid cloud deployment model yields maximal benefits in both the operation and technical aspects.

Table 1. Comparison of different options of cloud deployment

Items	Private Cloud	Community Cloud	Public Cloud	Hybrid Cloud
Set-up Costs	High	High	Low	Medium
Operation Costs	High	High	Low	Low
Scalability	High	High	High	High
Customization	Supported	Supported	Not supported	Supported
Time for implementation	Long	Long	Short	Long
User Experience	Good	Good	Acceptable	Good
Availability	Good	Good	Best	Best
Data Security	High	High	Medium	High

3   The Hybrid-Cloud Deployment Model

This section shows the hybrid-cloud deployment of the student administration system at the Open University of Hong Kong.

In general, an ERP system architectural framework comprises three tiers, namely, web tier, application tier and database tier. The web tier covers web services and load balancing services. The application tier hosts the applications and functional modules while the database tier refers to the database management system resided in storage servers, virtual machines or SAN. Precisely, in our implementation, the web tier and application tier are deployed on hybrid cloud. This hybrid cloud essentially covers a private cloud portion and a public cloud portion. The private cloud portion includes an internal database management system, storage servers and an internal network. The public cloud portion includes VPN gateways, cloud-based storage, virtual machines, virtual networks, etc. A fast Internet link is also provided. On the other hand, the database tier, which resides in campus, is protected by internal network firewall. Figure 1 shows the 3-tier architecture.

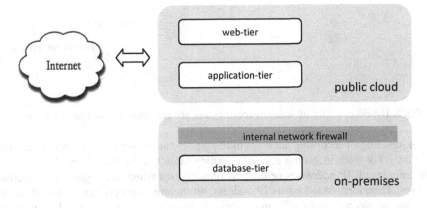

Fig. 1. The 3-tier architecture under a hybrid cloud deployment model.

This hybrid cloud deployment aims for off-campus accesses which account for the majority (80%) of accesses. For in-campus accesses, it may not be effective to route the accesses through this hybrid cloud deployment. Internal private cloud service is therefore implemented to cater for those in-campus accesses (20% of accesses). In essence, a dual-cloud environment is essentially implemented – a hybrid cloud for off-campus accesses, and a private cloud for in-campus access. For the hybrid cloud environment, Microsoft Azure is used for hosting the web tier and application tier whilst the database tier resides on premises. For the private cloud environment, both the web tier and application tier are hosted inside campus. Figure 2 shows the detailed configuration of dual-cloud environment.

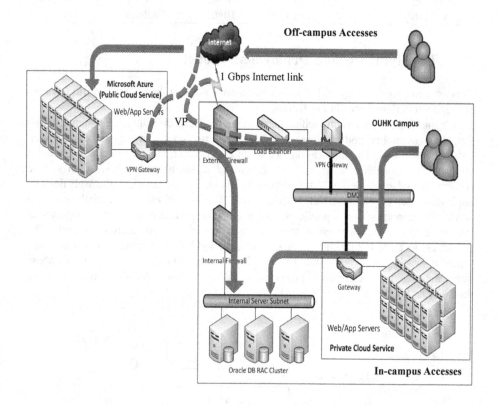

Fig. 2. Hybrid cloud for off-campus accesses and private cloud for in-campus accesses.

The handling of incoming accesses to the student administration system is outlined as follows. For each incoming access, a routing table will be looked up. This routing table contains all IP addresses or domain IP addresses which are regarded as internal accesses. Therefore, based on its IP address, an incoming access can be identified as either in-campus access or off-campus access. In-campus accesses will be routed to the private cloud environment while off-campus accesses will be routed to the hybrid cloud environment.

Figure 3 shows the logic for processing every incoming access to the Student Administration System.

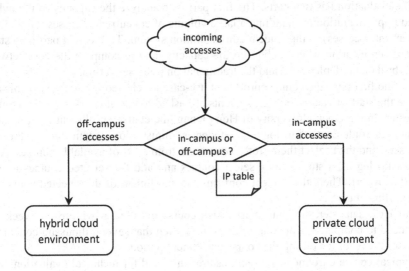

Fig. 3. Processing of incoming accesses to the student administration system.

For the hybrid cloud deployment, Microsoft Azure is subscribed [6]. It basically offers a Platform as a Service. The steps of the implementation are briefly described as follows.

- *Set up the virtual network.* Multiple virtual networks are set up for load balancing of incoming accesses.

- *Set up the virtual machines.* Virtual machines with 2-core to 32-core CPUs are available. Decisions on the number of cores and amount of memory to be used should be based on stress test results. In the student administration system, virtual machines with 2-core CPUs and 14 GB memory are selected. It should be noted that more cores for virtual machines would incur higher license costs of the underlying software.

- *Choose an image.* This involves the detailed set-up of virtual machines, based on a set of standard templates. The standard templates are usually provided by the cloud service providers.

- *Set up the VPN gateway.* On Microsoft Windows Server, routing services and remote access services are used to serve as the VPN gateway, with one-to-one mappings between the VPN gateway and the virtual network.

- *Set up the network firewall and load balancer.* VPN tunnels are established between the in-campus site and the remote public cloud. Firewall rules are set out for security control. Also, load balancers are configured to route accesses to public cloud portion and private cloud portion, based on a table identifying those in-campus IP addresses and domain IP addresses.

4   Technical Evaluation

The University conducted an evaluation of the hybrid cloud deployment with respect to cost and performance. This section reports the evaluation results.

The evaluation has two parts. The first part is to analyze the capacity of the hybrid cloud implementation for handling a large volume of concurrent accesses (up to 3,000 concurrent accesses) to the student administration system. The second part is to study the cost implication when scaling up the capacity, and to compare the cost between the hybrid cloud deployment and the traditional on premises set-up.

For the first part, the course enrollment use-case is selected for analysis. Typically, before the start of each semester, students are asked to enroll courses for the coming semester. In the Open University of Hong Kong, the course enrollment must be done online. All students need to login the student administration system through the web browsers, and then select their preferred courses from a list of available courses. After successful log-in, a student may select courses and add the selected courses into an enrollment cart. Then, the student confirms the enrollment of the selected courses in the enrollment cart.

For each semester, students must finish course enrollment within one week. As they need to compete for popular courses, it is often that several thousand concurrent accesses are recorded during the course enrollment period.

Typical course enrolment transactions are simulated for technical evaluation. Two separate stress tests are conducted for 1,000, 1,500, 2,000, 2,500 and 3,000 concurrent accesses. One is conducted on the hybrid cloud deployment whilst the other is on a similar on-premises set-up. For each transaction, the average request response time and average page response time are recorded. Neoload tool is used in this stress test [11].

The technical configurations of the two environment are summarized as follows. The hybrid cloud environment comprised 9 virtual machines for the web tier and 9 virtual machines for the application tier on Microsoft Azure. Each virtual machine is equipped with 2-core CPU at 2.2GHz, 14G memory and 100G storage. A similar configuration is set up for the on-premises environment, where a total of 9 physical servers are used for both the web tier and application tier. Each physical server is equipped with 8-core CPU at 2.9GHz, 24G memory and 150G storage. These servers are identical blade servers [12].

Table 2 shows the average request response time and average page response time (for each course enrolment transaction) recorded in the hybrid cloud deployment from 1,000 concurrent accesses to 3,000 concurrent accesses. Table 3 shows the figures in the on-premises environment.

For the hybrid cloud environment, the average request response time is maintained at around 3 second and the average page request response time at around 15 second for 2,000, 2,500 and 3,000 concurrent accesses. For the on-premises environment, the average request response time and average page request response time steadily grow with the number of concurrent accesses. When there are 3,000 concurrent accesses, the average request response time and average page response time reach 6.31 second and 24.40 second, respectively.

Table 2. Request response time and page response time (hybrid cloud).

No. of concurrent logins / accesses	No. of SA login	No. of requests	Average request response time	Average page response time
1,000	987	987	1.67 s	6.86 s
1,500	1498	1498	2.88 s	12.96 s
2,000	1,999	1,999	3,16 s	15.10 s
2,500	2,499	2,465	2.91 s	14.90 s
3,000	2,940	2,647	3.03 s	15.50 s

Table 3. Request response time and page response time (on premises).

No. of concurrent logins / accesses	No. of SA login	No. of requests	Average request response time	Average page response time
1,000	996	911	1.87 s	8.18 s
1,500	1,480	1,229	2.96 s	12.10 s
2,000	1,998	1,570	3.04 s	16.00 s
2,500	2,482	1,797	3.86 s	17.80 s
3,000	2,831	2,254	6.31 s	24.40 s

The second part of the evaluation is a cost comparison between the hybrid cloud deployment and premise set-up. Suppose the performance benchmarks are set at no more than 10 second per request response time and no more than 30 second per page response time for each course enrolment transaction. If fulfilling these performance benchmarks, the costs of the set-up for handling 3,000, 4,000, ..., 12,000 concurrent accesses are estimated as follows.

For the hybrid cloud deployment, in order to support 3,000 concurrent accesses, a total of 18 virtual machines (9 for the web tier and 9 for the application tier, each equipped with 2-core CPU at 2.2GHz, 14G memory and 100G storage) on Microsoft Azure are required. The cost (including the subscription of virtual servers, network bandwidth, maintenance of load balancer and network firewall servers) is estimated to HK$200,000 per year. More virtual machines (same capacity) are required for more concurrent accesses. The cost will grow at a steady rate of around HK$10,000 per 1,000 additional concurrent accesses.

For the on-premises environment, to support 3,000 concurrent accesses within the performance benchmarks, a total of 9 physical servers (for both the web tier and the application tier, each equipped with 8-core CPU at 2.9GHz, 24G memory and 150G storage) are required. The cost (including the server hardware, network bandwidth, maintenance of load balancer and network firewall servers, and assuming server hardware with a usage-life of 3 years) is estimated to around HK$270,000 per year for 3,000 and up to 6,000 concurrent accesses. More physical servers (same capacity) are required for more concurrent accesses. The cost will grow to HK$390,000 per year for 6,000 and up to 9,000 concurrent access, and HK$510,000 per year for 9,000 and up to 12,000 concurrent accesses.

Figure 4 shows a graph projecting the annual cost of the hybrid cloud deployment and on-premises set-up from 3,000 to 12,000 concurrent accesses, based on the current price level. It is clearly shown that the hybrid-cloud deployment yields more cost-performance benefits than the on-premises set-up, and that the cost difference becomes more and more significant as the number of concurrent accesses grows. At 10,000 concurrent accesses, the cost of on-premises set-up is almost double of the cost of hybrid cloud deployment.

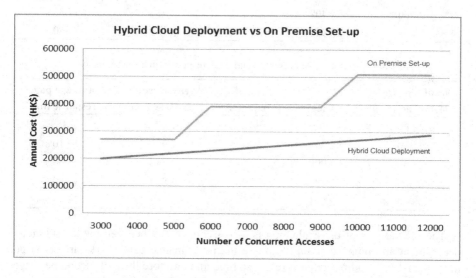

Fig. 4. Cost comparison between the hybrid cloud deployment and on premise set-up.

5   Conclusion

This paper shows the implementation of an ERP-based student administration system, where the web and application tier is deployed as a hybrid-cloud model. This hybrid-cloud deployment effectively achieves cost-performance balance, especially during occasional peak periods of concurrent accesses. Whilst this hybrid cloud deployment is adopted for the web and application tier, the database tier is still maintained within the campus and protected by appropriate firewalls. This protects the student data from exposing the database server to the public cloud.

Although the implementation of hybrid cloud is much more complicated than that of private cloud or public cloud, there are many benefits, such as achieving the right balance between on-campus and off-campus accesses and allowing customization of functional modules of the system for better user experience. Maximal benefits in both operation and technical aspects are yielded. All these are discussed, and the key steps of the implementation are described. Technical evaluation is reported to demonstrate the cost benefits of a hybrid cloud deployment. Our findings and experience would provide a useful reference for industrial practitioners in planning and designing a hybrid-cloud deployment for ERP-based systems.

References

1. Mell, P., Grance, T.: The NIST Definition of Cloud Computing, Report No. 800-145, Information Technology Laboratory, National Institute of Standards and Technology, Department of Commerce, The United States of America (2011)
2. Buyya, R., Broberg, J., Goscinski, A.: Cloud Computing: Principles and Paradigms. John Wiley (2011)
3. Website of Open University of Hong Kong, http://www.ouhk.edu.hk (retrieved) 15 April 2015
4. Oracle, Campus Solutions for Higher Education, Oracle Data Sheet, Oracle Document Library, Oracle (2013)
5. Oracle, Campus Solutions in the Cloud, Oracle Data Sheet, Oracle Document Library, Oracle (2013)
6. Website of Microsoft Azure, https://azure.microsoft.com (retrieved) 15 April 2015
7. Oracle, PeopleTools version 8.53: Getting Started with PeopleTools, Oracle Document Library, Oracle (2013)
8. Oracle, Oracle Database version 12c: Product Family, Oracle White Paper, Oracle Document Library, Oracle (2015)
9. Website of Amazon Web Services (AWS), http://aws.amazon.com (retrieved) 15 April 2015
10. Website of Citrix Cloud Services, http://www.citrix.com/solutions/cloud-services/overview.html (retrieved) 15 April 2015
11. Website of Neoload, http://www.neotys.com/product/overview-neoload.html (retrieve) 15 April 2015
12. Mani, K., Jee, B.: On the Edge: A Comprehensive Guide to Blade Server Technology. Wiley (2007)

A Benchmark Evaluation
of Enterprise Cloud Infrastructure

Yong Chen, Xuanzhong Xie, and Jiayin Wu

China Telecom Corporation Ltd. Guangzhou Research Institute
Guangzhou, China
{cheny,xxz,wujiayin}@gsta.com

Abstract. As cloud computing matures in enterprise IT system, agile and scalable evaluation of cloud infrastructure becomes critical. A lightweight benchmark approach is introduced, to meet the requirement of low cost, scalability and relevance of typical enterprise environment. Benchmark results are analyzed and compared to industry benchmark data, conclusion is given on implementing a scientific evaluation of enterprise cloud infrastructure.

Keywords: cloud computing, enterprise IT infrastructure, benchmark.

1 Introduction

Cloud computing becomes popular in enterprise IT infrastructure as result of matured technology and product, such as hypervisor and IaaS / PaaS platform. But the fast development of cloud technology and product, the distinct characteristic of rapid elasticity [1] comparing to traditional infrastructure, bringing challenges to enterprise to adopt cloud computing. Agile and scalable evaluation of the cloud infrastructure become critical for enterprise to make technical decision, while the service level agreement (SLA) of enterprise system remains valid, such as performance, reliability and energy efficiency.

Benchmarking is an effective method to evaluate technology and product, many industry benchmark and results are available and of great value for enterprise to make technical decision [2, 3]. In cloud computing era, changes are required in following area: a) agility, a fast and low cost evaluation on the mainstream cloud platform is essential, for a better emulation of cloud environment; b) scalability and portability, the horizontal scale-out and heterogeneous platform are key features of cloud computing.

Two years ago we wrote a paper of the design and implementation of application benchmark for enterprise server evaluation [4], summarized the practice of tailoring a real production IT system for benchmark usage. In cloud computing era, we change our methodology to meet the challenges, whereas the Online Transaction Processing (OLTP) scenario remains unchanged, since it is still the core of enterprise IT system, and is capable to evaluate different metrics of cloud infrastructure like comprehensive performance, reliability and energy efficiency.

© Springer International Publishing Switzerland 2015
R. Cheng et al. (Eds.): APWeb 2015, LNCS 9313, pp. 832–840, 2015.
DOI: 10.1007/978-3-319-25255-1_68

We switch from real application to the customized open-source tools, HammerDB [5], for load generation, to utilize the features of scalability and portability to different database and operation system. The customization includes changing the source code to meet specific requirement, setting up the testing rules for different system under test (SUT), and optimizing works to fully utilize the SUT capacity.

In this paper, we introduce the practice of agile and scalable benchmark test on systems designed for cloud computing. Analysis is made on acknowledged results, and compared with industry benchmark. Finally we give a conclusion of benchmark methodology in the cloud computing era.

2   Tool Selection and Customization

A feasible benchmark has three components: load generator system, specification document and test result. In this chapter we introduce the third-party tool selection as load generator system, and how to customize it and setup a cloud infrastructure benchmark specification.

2.1   Principles for Tool Selection

In cloud computing era open source software becomes mainstream as the requirement of open architecture and the capability of customization. We choose HammerDB as the load generator system for cloud infrastructure benchmark, which has following features for the evaluation:

- Agile and customizable: HammerDB is a ready-to-use tool emulating a simplified TPC-C scenario, a famous industry OLTP benchmark [3], making it understandable for most stakeholders. HammerDB greatly reduces testing cost by simplifying the TPC-C rules, such as the think time restriction of virtual user, significantly decrease the terminal demand. As an open source project, HammerDB can be modified and run on customized environment, such as the PCIE SSD we used as main storage media instead of traditional HDD storage, saving space and electricity while IO performance is better. The factors above lead to a low cost, fast deploy and customizable solution for benchmark test.
- Scalable and portable: TPC-C benchmark is scalable by varying the warehouse number in the database, the SUT of official results range from high-end clustered server to entry level one socket server [6], we customize the HammerDB source code to enhance the scaling ability and setup a data scale rule fitting the SUT. Currently HammerDB support multiple databases like Oracle, MySQL and Redis, running on a wide range of operation system, meeting the requirement of the elastic and heterogeneous.
- Multipurpose and comprehensive: OLTP application is a typical enterprise IT scenario, running OLTP benchmark can effectively verify the comprehensive performance and multiple metrics under an environment relevant to real world, like the computing performance of server and hypervisor, the IO performance of storage product, the reliability and energy efficiency of infrastructure.

2.2   Specification Setup

Apart from the fundamental characteristics of repeatable, scalable and comparable, a feasible cloud infrastructure benchmark specification shall be capable to evaluate multiple cloud scenarios, running workload on different platform like physical server, virtual machine over hypervisor, traditional database and storage, PaaS database, software defined storage, etc., under a complete set of benchmark test rules to ensure the correctness and fairness of the evaluation.

We setup a basic specification to meet the urgent requirement of evaluating the high-end 8 and 16 socket x86 server from different vendors, and later we extend the specification for other enterprise cloud scenario. We name this benchmark specification as CTB-H (China Telecom Benchmark, HammerDB) [7]. Fig. 1 shows the CTB-H system deployment diagram.

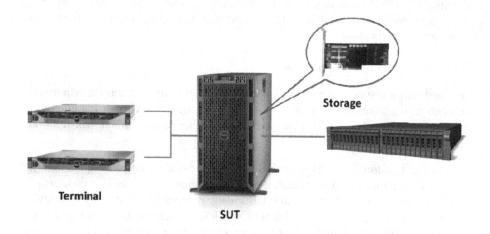

Fig. 1. CTB-H system deployment diagram

The CTB-H benchmark specification is consisted from three parts:

2.3   The Scenario Definition and Scaling Rule

Currently CTB-H specification defines a TPC-C load generating method to evaluate maximum performance, scalability, reliability and energy efficiency metrics of multiple cloud infrastructures, such as server, storage, database and hypervisor. Multiple scenarios are defined for different group of metrics, such as comprehensive performance evaluation scenario, energy efficiency evaluation scenario, etc. The CTB-H scenarios can be expanded to meet the enterprise requirement.

Scaling rules and system configuration rules are the essence of scenario definition. A scalable dataset can verify the elasticity of computing, scheduling and IO handling, which is the key of cloud infrastructure. The official TPC-C specification defines a rigid data scaling rule [3], although scientifically designed but leads to a high testing cost because of the storage and terminal demand and long testing period to achieve the best result, impacting the feasibility. For enterprise cloud planning, the scenario and system configuration can be specified in advance, the precision can be reduced, thus a simplified solution can be achieved.

We defined a static data scaling rule considering the capability of load generator system and the configuration of server running workload, so as to evaluate the SUT under an acceptable cost and precision, the rule would be changed according to the fast development of cloud infrastructure. The data scaling rule is listed below:

Table 1. CTB-H Data scaling rule

Warehouse number	Server/VM configuration
1,000	2 * E5 CPU
2,000	4 * E5 / E7 CPU
4,000	8 * E7 CPU
8,000 / 2* 4,000 in two logical instance	16 * E7 CPU

HammerDB is capable to generate a scalable OLTP workload by scientifically designed codes in terminal side and database side, we change the database schema definition code and terminal side TCL script, to extend the data scaling range, making CTB-H test feasible to evaluate different SUT, even a customized scenario for evaluating the pluggable database (PDB) feature of Oracle 12c database [8], which is a typical PaaS cloud environment.

2.4 System Configuration Rule

Cloud technology is continuously advancing, both hardware and software platform are fast developing, the system configuration rule is to matching the technology and product to the enterprise requirement. For the scenario of running Oracle database on physical server, we specify the hardware and software configuration that are up-to-date, cost-effective and complying with the enterprise SLA, thus we simplify the benchmark test execution and make the test result relevant to the actual enterprise environment.

Besides the static hardware and software configuration, the runtime configuration rules are also included in the specification, to ensure a reasonable testing environment that can be apply to production system. These rules can help to avoid radical method from just achieving best result without being able to be used in real world, which can be seen in some third party benchmark and is of no value for enterprise.

We summarize the system configuration rules below for the scenario of Oracle database running on physical server:

Table 2. CTB-H system configuration rules for Oracle database running on physical server

Item	Rule
CPU	XEON E5v2, E7v2 mid-range model
Operation System	RHEL 6.5 or above
Database version	Oracle 11gR2, or 12cR1 for multi-tenant scenario
Memory chip allocation	Every memory channel has one 8G DDR3 memory chip
HammerDB version	2.16
Storage	2 * Local PCIE SSD
Oracle runtime setting	7 explicit rules of runtime setting, like archive mode setting, parameter restriction; Manual checking of test environment to exclude unrealistic configuration.

2.5   Definition of Result Metrics and Measuring Rules

Scientific and quantitative evaluation is the goal of benchmarking, the definition of result metrics and measuring rules are key elements for this goal.

TPC-C benchmark defines a tpmC result metric, representing the overall transaction throughput and the comprehensive performance of SUT under the strict specification, HammerDB transform tpmC to a similar metric NOPM which we use in CTB-H. Based on NOPM and the external power meter, we add an energy efficiency metric, NOPM per watt, referencing the industry benchmark SPEC Power 2008 [9], measured by getting the performance and power consumption at 4 intermediate point between maximum transaction throughput to active-idle and calculate the final average as result metric, giving a comprehensive indication of the energy efficiency of the SUT.

The measuring rule includes the duration of the measuring period and the data source, a proper measuring duration is to overcome the randomness of the SUT behavior and take the backend workload into account, which may be omitted if the measuring duration is not long enough, such as the log file switch of the database. For the scenario of evaluating maximum performance of a physical server, the duration of measurement is 30 minutes, for reliability evaluating scenario, it is 24 hours.

3   Benchmark Results Analysis

In this chapter, we analyze some acknowledged CTB-H benchmark results and compare with the industry benchmark results, to illustrate the feasibility of evaluating enterprise cloud infrastructure and verify the correctness.

3.1   Scalability Analysis

We choose IBM x3850 X6, x3950 X6 and HP Superdome-X server models for analyses, which are 4-way, 8-way and 16-way server respectively, as shown in Table 3, comparison between servers of same configuration is not illustrated here to avoid possible conflict. These servers are up-to-date and mainstream servers for enterprise mission critical system, capable to support cloud platforms like Linux, VMWare and Oracle 12c. These servers are designed for complicated and intensive business workload, their performance, reliability, energy efficiency and compatibility to cloud platform are the main concern of enterprise, a comprehensive evaluation is executed through our CTB-H performance and energy efficiency scenario.

Table 3. Server system invoked

Vendor	System	CPU	CPU socket	Total server cores
IBM	x3850 X6	XEON E7-4870v2	4	60
IBM	x3950 X6	XEON E7-8870v2	8	120
HP	Superdome-X	XEON E7-2890v2	16	240

The performance of these three servers is measured via CTB-H and compared with industry SPEC CInt2006 rates benchmark results [10]. For extruding performance variation with core scaling, values of results are transformed into ratios by setting x3850 X6 as baseline. The CPU of Superdome-X is more powerful in computing because of the frequency and cache configuration, making the SPEC CInt2006 rates metric ratio is bigger than CPU core ratio.

As Figure 2 shows, a good agreement can be seen, verifying the scalability and feasibility of CTB-H.

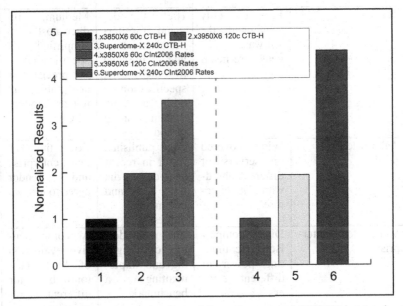

Fig. 2. Normalized results with core scaling of enterprise servers (x3850 X6 is treated as baseline for normalization)

3.2 Comparison with Industry Benchmark

SPEC CInt2006 benchmark is a major tool for enterprise to evaluate server computing performance, the results in Fig.2 are practically linear with the CPU core scaling because of the independent batch workload. But in real enterprise application the workload are inter-related with latency on scheduling and IO handling, thus SPEC CInt2006 does not satisfy the enterprise requirement of predicting the system behavior in a typical production environment.

Whereas CTB-H benchmark provides a comprehensive evaluation of the SUT, takes the overhead of the platform scheduling and IO handling into account, gives a better verification and a scalable basis of comparison, which are of great value for enterprise to make technical decision.

Table 4 gives a brief comparison of two commonly used industry benchmarks and CTB-H, shows the pro and con of different approaches:

Table 4. Benchmark comparison

	SPEC CInt2006	TPC-C	CTB-H
Workload type	Independent batch workload, only computing performance of SUT is measured.	An OLTP application of wholesale management, generate a comprehensive workload on SUT.	The same as TPC-C.
Test cost	Lowest, only SUT and the software license are needed.	High, the demand for terminal and storage is big due to the strict specification, with a long system tuning period	Medium, reduce cost by a simplified specification, the precision is acceptable but not as good as TPC-C.
Acknowledged result	Widely covered on servers of different hardware architecture.	Few published result in recent year due to the test cost and commercial interest.	Cover the SUT that enterprise and vendor agree to test.
Value for enterprise	Direct comparison of computing power of different servers	The methodology of designing and implementing a benchmark.	A comprehensive evaluation of SUT, customizable for different requirement.

3.3 Approaches of System Optimization

Besides the evaluation of SUT, CTB-H tests provide approaches for eliminating the bottleneck in the real workload. For a compute and IO intensive database system, the IO contention and synchronize control are common bottleneck. The principles for optimizing the system to better carry out workload are listed below:

- Make full use of the platform IO capacity, like Oracle 12c multiple LGWR processes feature, optimized storage driver for specific platform like RHEL operation system, to minimize the IO latency;
- Find ways to scale-out the frequently accessed data objects, like sharding to different database user or instance, allocate to dispersed storage, to minimize the IO contention and synchronization control latency;
- Make use of the cache control feature of specific database platform on extremely contentious data object, like Oracle database DBMS_SHARED_POOL.MARKHOT procedure, to minimize the synchronization control latency;

4 Conclusion

Cloud computing infrastructure is fast developing nowadays, different technology and product are continuously launch to market, technical decision become more and more challenging for enterprise while the SLA remain valid. Benchmarking is an important and effective method to quantitative evaluation, helping to scientific comparison and pre-production verification. Agility, scalability and portability are key factors for implementing cloud computing infrastructure benchmark, relevance of production scenario is also important. A careful designed benchmark is capable in different evaluation scenario like performance, reliability, energy efficiency, thus maximize the outcome of effort to execute the benchmark.

Acknowledgment. This work was supported by the Corporate Informatization and Procurement Department of China Telecom Corp., Intel Shanghai R&D center.

References

1. Mell, P., Grance, T.: The NIST Definition of Cloud Computing, NIST Special Publication 800-145 (2011)
2. SPEC CPU (2006), http://www.spec.org/cpu2006/
3. Transaction Processing Performance Council (TPC), TPC Benchmark C Standard Specification Revision 5.11, http://www.tpc.org/tpcc/spec/tpcc_current.pdf
4. Chen, Y., Wu, J., Xie, X., Li, X.: Design and Analysis of Application Benchmark for Enterprise Server Evaluation. In: 2013 IEEE International Conference of Region 10, Conference Proceedings, Xi'an, pp. 1–4 (2013)
5. Introduction to Transactional (OLTP) Load Testing for all Databases, http://hammerora.sourceforge.net/ hammerdb_transactionintro.pdf

6. TPC-C – All Results,
   http://www.tpc.org/tpcc/results/tpcc_results.asp
7. China Telecom Corp., The server benchmark specification of China Telecom Corporation V3.0 (2014)
8. Oracle Corporation: Oracle Database 12c Feature: Multitenant Database,
   http://www.oracle.com/technetwork/articles/database/
   multitenant-part1-pdbs-2193987.html
9. SPECpower_ssj® (2008), http://www.spec.org/power_ssj2008/
10. SPEC CPU2006 Results, http://www.spec.org/cpu2006/results/

A Fast Data Ingestion and Indexing Scheme for Real-Time Log Analytics*

Haoqiong Bian, Yueguo Chen**, Xiongpai Qin, and Xiaoyong Du

Key Laboratory of Data Engineering and Knowledge Engineering (MOE),
Renmin University of China, Beijing 100872, China
dingxiaoou_hit@163.com

Abstract. Structured log data is a kind of append-only time-series data which grows rapidly as new entries are continuously generated and captured. It has become very popular in application domains such as Internet, sensor networks and telecommunications. In recent years, many systems have been developed to support batch analysis of such structured log data. But they often fail to meet the high throughput requirements of real-time log data ingestion and analytics. An efficient index is very important to accelerate log data analytics, and at the meanwhile to support high throughput data loading. This paper focuses on designing a specialized indexing scheme for real-time log data analytics. The solution adopts a dynamic *global hash index* to partition the tuples into hash buckets. Then the tuples in the hash buckets are sorted and buffered in the *sort buffer queue*. When the amount of data in the queue reaches a threshold, the data is packed into *segments* before spilling to the disks. Moreover, an *intra-segment index* is maintained by *meta database*. With such an indexing scheme, the database system achieves high throughput and real-time data loading and query performance. As shown in the experiments, the data loading throughput reaches 5 million tuples per second per node. The delay of data loading does not exceed 10 seconds, and a sub-second query performance is achieved for the given queries.

Keywords: log data analytics, index, real-time, high throughput.

1 Introduction

Log data has become a major type of big data. Some typical examples of log data include user click logs of Web pages, transaction records of sensor networks, or machine-generated records in a telecommunication network. Real-time analysis of log data is critical for many applications such as anomaly detection and event tracking. In enterprises, although SQL-on-Hadoop systems such as Pig [21] and Hive [27] have been widely used for batch analysis of big log data, MPP (Massively Parallel Processing) databases are however more efficient in handling real

---

* This work is supported by the Outstanding Innovative Talents Cultivation Funded Programs 2014 of Renmin University of China.
** Corresponding author.

© Springer International Publishing Switzerland 2015
R. Cheng et al. (Eds.): APWeb 2015, LNCS 9313, pp. 841–852, 2015.
DOI: 10.1007/978-3-319-25255-1_69

time analytical workloads [23,25]. In MPP databases, various indexes can be easily applied to further improve the query performance.

Real-time log data analytics however brings some challenges to the existing indexing solutions of MPP databases. In the last second Chinese national contest of big data technical innovation[1], as a leading database manufacturer in China, GBase proposed the problem of designing an efficient indexing solution for GBase MPP cluster products [9] to support real-time log data analytics. The contest provides a typical real-world log data analytical workload. The queries in the workload aim to track the crucial Internet usage information from specified IP addresses. The data grows in a high speed and the queries must be capable of analyzing the fresh data in real time. Current index solutions in RDBMSs can hardly handle such real-time analytical workload of high volume log data.

To address the problem, we propose an innovative indexing scheme for real-time log data analytics: (1) The scheme, from the perspective of system architecture, is a multi-layer solution. Each layer is scalable so that the solution can adapt to the dynamics of data loading throughputs. (2) The layers in the scheme are pipelined, and each layer can be divided and conquered so that the schema can take full advantage of the hardware resources. (3) The in-memory buffers are separated to the mutable and immutable parts. As shown in Section 3.1 and 3.2, the *hash buckets* are the mutable part which is very small and the *sorted buffer queues* are the immutable part which is relative larger and available to queries. The data in hash buckets are continuously flushed into the sorted buffer queues so that the queries can access the fresh data in memory. (4) The scheme combines the features of hash index, in-cluster index and statistical information to support high performance indexing on the time-series data, and fast query processing for queries on attributes of high selectivity. (5) The scheme merges in-memory data into large segments. The segments are spilled to disks as sequential files. Within the segment, rows are split to row groups. *Intra-segment index* is built to address the right row groups in the segment. Within each row group, other database techniques such as column store [26], bloom filter [12] and other indexes can also be applied.

In this paper, we first give details of the real-world workloads used in the contest. Then we take the workloads as an example and summarize the main challenges and features of typical real-time log data analytics. After that, We introduce the details of our scheme and show some implementation details. Finally, we perform experimental evaluations of our solution, give the related work study, and conclude the paper.

2   Preliminaries

Real-time log data analytical workloads are different from the workloads in traditional decision-support systems. We introduce a typical workload in this section and summarize the challenges and features of real-time log data analytics.

---

[1] http://bigdatacontest.ccf.org.cn/gindex.html

2.1   Typical Data and Workload

The log data used in this paper is the records of Internet accesses. The log records are stored in a large table. The schema of the table is shown in Table 1. The name of the table is *WebInfo*.

Table 1. Table schema of *WebInfo*

Field name	Data type	Description	Cardinality
SrcIP	integer	source IP	~30M
DestIP	integer	destination IP	~30M
SrcPort	integer	source port	~300
DestPort	integer	destination port	~300
CaptureTime	integer	capture time (seconds)	
Flag	integer	flag bits	~200
Protocol	integer	protocol code	<10
ISP	string	Internet Service Provider	<10
Area	string	area of location	~2000
QueryType	integer	code of request type	<20

The *WebInfo* table is stored in Gbase MPP cluster. The data is loaded into the table in a streaming manner. The coming data is partitioned to the *GNodes* in the cluster. On each *GNode*, 30 billion tuples (log records) of data must be loaded into its local RDBMS engine per day. In peak hours, about 3 billion tuples of data should be loaded within one hour. The data must be loaded into the table in real time, i.e. the received data needed to be loaded into the table in seconds and be available to analytical queries immediately.

The queries in the workloads select the data by a range condition on the *CaptureTime* column and two equi-conditions on the *SrcIP* and *DestIP* columns. The range of the capture time may be relatively wide, for example several hours, one day or even wider. Generally, the data in the past one month is considered as warm data and are analyzed by queries in the MPP database cluster. More earlier historical data are archived in other low-cost batch analytical systems such as Hive. The equi-conditions on *SrcIP* and *DestIP* specify the target Internet usage information we want to track. Based on the selected data with these three conditions, the query may further conduct aggregate and order-by operations. An example of the basic form of the query is shown as:

```
SELECT ISP, Area FROM WebInfo
WHERE CaptureTime BETWEEN 1377221057 AND 1377303456
AND SrcIP=<IP 1>
AND DestIP=<IP 2>
```

The selectivity of *SrcIP* and *DestIP* is relatively high. With the equi-conditions on these two columns, only a small amount (generally several hundreds or thousands) of tuples are selected from the billions of total records. But the workload is time critical, so that each query in the workload are expected to be executed and finished in sub-second. To support the real-time analytical workloads on the log data, we need to build appropriate indexes for a large volume of data stored in the table.

2.2   Challenges and Features

We can summarize the challenges of building indexes for real-time log data analysis as following. The index solution must give a good tradeoff between the following two major challenges.

1. **Real-time ETL.** Data comes to database systems in a high throughput, and ETL (extract, transform and load) must be done in real time. Log data are usually generated by machines or user activities. So the data are growing rapidly and the amount of accumulated data is very large. In the example in Section 2.1, each node in the cluster has to load 830,000 tuples into the table per second on average during the peak hours. Such real-time ETL brings huge challenges to the storage module of database systems.
2. **Real-time query evaluation.** Queries in real-time log data analysis need return the results in sub-second. The queries in the workload are relative simple, but the queries may be submitted by many external clients such as web site users, unlike the report and decision support queries which are generally submitted by a few data analysts. So that the throughput of the queries may be relatively high and the clients can not wait for couples of seconds or even minutes.

The above two challenges are the two ends of a teeterboard: the first one is the challenge on write performance and the second one is the challenge on read performance. Tradeoff between these two challenges must be well considered in the indexing solution. Besides the above challenges, real-time log data analysis also has some other important features, from which we are able to develop dedicated indexing solutions to address the above challenges.

1. **Easy to scale out.** The queries in real-time log data analytical workloads are relatively simple and easy to scale out. Unlike typical report or decision support queries, there are neither joins between two large tables, nor complex aggregations on massive intermediate results. The main operations of the query in such log data analytical workloads is selection and projection, which are easy to scale out for a share nothing architecture.
2. **High selectivity columns.** The table has several columns with high selectivity. Most of the queries have predicates on these columns. Real-time queries should not touch a major part of the large data set. Otherwise it will be impossible for databases to support real-time and high throughput queries.
3. **Time series.** Log is a type of time series data. Generally, there is a timestamp field in the table schema of the log data. The field records the occurrence time of events or the generating time of the log entries. Besides,

the value of the log data decreases as time goes by. The latest data is usually most important and valuable.

3   Indexing Solution

In this section, we introduce the details of our indexing solution. The framework of the solution, which has four layers, is shown in Figure 1. From top to bottom, in the first layer, we use global hash index to shuffle the log data based on the hash key. In the second layer, the data from each hash bucket is sorted and buffered in the corresponding sorted buffer queue. In the third layer, when the number of splits in a certain sorted buffer queue reaches a specified threshold, the splits are merged into a segment. An intra-segment index is built for each segment. In the bottom layer, the intra-segment index is stored in the meta database, the segment is stored into the file system. The technical details of the framework are discussed in the following subsections.

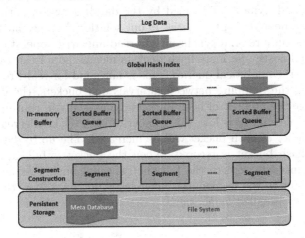

Fig. 1. Framework of the indexing solution

3.1   Global Hash Index

We build global index on the column(s) with high selectivity. If there are more than one column that always co-occur in query predicates, they can be considered as a combined key. This key is called *hash key*. For example, in the typical workloads shown in Section 2.1, $SrcIP$ and $DestIP$ are used by all the queries so that can be combined as a 64-bit integer ($SrcIP$ stored in the first 32 bits and $DestIP$ stored in the last 32 bits).

For the global index, we apply a strategy like consistent hashing [18] to supply a scalable hash index. We have a hash function $h = H(key)$ and calculate the hash value of the hash key. The range of the hash value is split to k intervals. The intervals are mapped to a set of hash buckets. An example is shown in Figure 2. The hash value ranges from 0 to $(2^{32} - 1)$. The range is split into 8 intervals

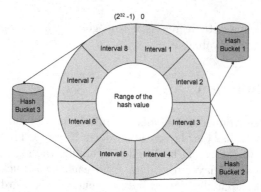

Fig. 2. An illustration of the global hash index

which are mapped to 3 hashing buckets. The hash bucket is a small container in main memory (2MB by default) in which the tuples are temporarily buffered.

The global hash index is scalable so that the number of hash buckets are automatically and dynamically tuned by the database system. Tuples in hash buckets can not be accessed by queries in our solution, because hash bucket is a mutable data structure, and synchronizing read and write operations on it will be complex and costly. So more hash buckets means more fresh tuples are not addressable by the analytical queries. This is not good for real-time ETL.

By changing the mapping between intervals and buckets, the database system can change the number of hash buckets dynamically in runtime. When the throughput (of new data) is low, the number of hash bucket decreases. At peak hours, the number of hash bucket will be increased. The up to date hash value range of the hash bucket is recored. When the bucket is full, an empty hash bucket is instanced to replace the full one. The full bucket is then set to immutable and sent to the in-memory buffer layer. In this paper, the immutable full hash buckets is called *buffer split*.

3.2 Sorted Buffer Queue

In the in-memory buffer layer, there are a set of *sorted buffer queues*. One hash bucket in the global hash index corresponds to one sorted buffer queue. The sorted buffer queue is a queue for the *buffer splits*. It is stored in main memory and can be accessed by the analytical queries.

After the buffer split is received by the in-memory buffer layer, it is sorted on the *sort key* and put into the corresponding sorted buffer queue. The *sort key* can be the same column used by *hash key* or another frequently used column in the table schema. The sorted buffer queue has a threshold of maximal size. If the number of buffer splits in the queue reaches the threshold or the flush buffer signal is emitted by the database system, an empty sorted buffer queue is instanced to replace the current sorted buffer queue. The replaced sorted buffer queue is then send to the *intra-segment index* layer to be merged into a segment and to build the intra-segment index.

The sorted buffer queues in the in-memory buffer layer must be flushed before the global hash index is scaled in/out. When the database system decides to scale the global hash index by changing the number of hash buckets and the mapping between the hash value intervals and the hash buckets, it must emit a flush buffer signal to the in-memory buffer layer. After the sorted buffer queues are flushed, the global hash index and the in-memory buffer can be scaled.

3.3   Build Segments and Intra-Segment Indexes

The *segment construction layer* is responsible for constructing the *segment* and *intra-segment index* from the *sorted buffer queue*. Intra-segment index is compatible with other database techniques such as column store [26] and other indexes such B$^+$-tree and bitmap. The diagram of building segment and intra-segment index is shown in Figure 3.

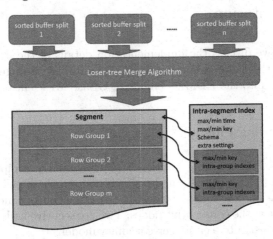

Fig. 3. Diagram of building segment and the intra-segment index

The sorted buffer queue contains a set of buffer splits which are sorted on the sort key. The sorted buffer queue also has attributes to record the min and max value of the hash key, the sort key and the capture time. The sorted buffer splits in the sorted buffer queue is merged by loser-tree merge algorithm [5] of which the computation complexity is $O(tlog^n)$, where t is the number of tuples in all the buffer splits and n is the number of buffer splits.

After merged by the loser-tree merge algorithm, the tuples form a *segment* which is split into m row groups. Then the *intra-segment index* is built for the segment. The intra-segment index contains the meta information of the segment and the meta information for each row group. The meta information of a segment includes (1) min and max values of the capture time, hash key and sort key, (2) the schema of the tuples and (3) the extra settings of the segment such as the compression codec of the row groups. The meta information of row group includes (1) the offset of the row group in the segment, (2) the min and max values of the sort key, and (3) the intra-group index.

The intra-group index is optional, it is the index on other columns except the capture time, hash key and sort key columns. It can be any type of index such as bloom filter[12], bitmap index[14] and B$^+$-tree. Intra-group index can be applied to accelerate the queries which have predicates on other columns.

It can be seen that within a row group, database techniques such as column store [26], B$^+$-tree, R-tree [16] and bitmap can be applied. Default in this paper, the size of the segment is 64MiB and the size of the row group is 4MiB which is a suitable size for column store of big Internet data [17]. The constructed segment and intra-segment index are then stored in the file system and the meta database. The file system can be local file system or distributed file system such as HDFS or GlusterFS [10]. The meta database can be stored in transactional databases such as MySQL or PostgreSQL.

4   Implementation

In this section, we discuss the implementation details of the indexing solution. The solution in this paper is implemented as a database storage engine in Java, and it is moving into the RDMDS engine of GBase MPP cluster.

4.1   Data Loading Pipeline

The framework of the indexing solution is implemented as a pipeline. As shown in Figure 4, the pipeline has six nodes. Each node in the pipeline is a group of parallel threads. The threads of a node do the same things concurrently. There are buffers between the consecutive nodes. Two consecutive nodes and the buffer between them constitute a producer-consumer model.

The number of threads in each node of the pipeline is automatically tuned by the database system. The indexing solution provides APIs to monitor the number of items in the buffers of the pipeline. So that the database system can tune the number of threads if any buffer in the pipeline is always full or empty.

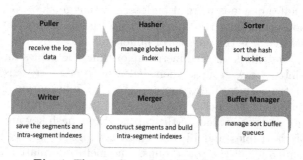

Fig. 4. The pipeline of data loading threads.

4.2 High Availability of Sorted Buffer Queue

The sorted buffer queues in the framework (Figure 1) is managed by *buffer manager* in the pipeline. To avoid losing data when the database system is crashed, the buffer manager stores the buffer splits in a local Redis[4] instance. The buffer split is stored as the binary value in Redis. Redis is run in another process and has data dumps and HA solutions. We define the abstract interface between our indexing solution and Redis, so Redis also can be replaced by other memory caches.

4.3 Lock on the Segment Construction Layer

During the time that the sorted buffer queue is merged and the intra-segment index is constructed, the system is in an uncertain condition. The data in the constructing segment is not available to queries. To solve this problem, we use a file lock on the constructing segment, so that the segment construction thread and the query execution threads are synchronized. The time of constructing or reading a segment is generally no longer than 200 milliseconds on PC servers. So this lock is a lightweight lock.

4.4 Avoiding Unnecessary Segment Accesses

The chance of synchronizing the queries and the segment construction threads is not very large because they are more likely to handle different segments. But to further improve the performance, we try to avoid unnecessary segment accesses of the queries by recording the earliest capture time of the tuples in all the buffer splits in memory. If a query is willing to access the data with earlier capture time than that, the query only need to access the segments in the file system. This helps to reduce the chance of lock competition.

4.5 Metadata Management

The meta database is responsible for storing the in intra-segment indexes. It can be an RDBMS such as MySQL. The queries firstly query the meta database to find the right segment for themselves. The intra-segment indexes are also save as the header of the corresponding segments. The intra-segment index includes all the metadata which is needed when reading the segment. So, if the segments are copied to another cluster node, the meta database can be easily upgraded.

5 Experimental Evaluations

5.1 Experiment Environment

We are focusing on the indexing solution in the storage engine on nodes of a MPP database cluster. We care about the performance of the indexing solution on one node. So in our experimental evaluations, we only use one PC Server. The environment of the server is shown in Table 2.

Table 2. Settings of the experiment environment

Item	Setting
CPU	Intel Xeon E5-2640 2.50GHz × 2
Memory	64 GB
Disk	1TB 7200RPM SAS × 4, RAID5
Operating System	CentOS release 6.4 x86-64 (Final)

5.2 Data Loading Performance

We use the data in Section 2.1. We use 3 billion tuples of data to test the data loading performance. MonetDB and TokuDB support insert while Infobright does not. But the insert is an transactional operation of which the throughput is generally lower than 10,000 ops/s. So we split the data into mini batches, each mini batch contains 0.1 million tuples. The mini batches are loaded with the copy/load command in each system. The data loading throughput is shown in Figure 5. It can be seen that the throughput of our solution is one order of magnitude higher than those of the three systems. Infobright does not supprt index. Indexes are built on *SrcIP*, *DestIP* and *Capture Time* on the other three. The time between the moment at which data is passed to the system and the moment at which data is available to the queries of all the four are no longer than 10 seconds.

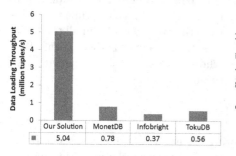

Fig. 5. Data loading throughput   **Fig. 6.** Query elapsing time

5.3 Query Performance

After the data is loaded, we execute the query in Section 2.1 to evaluate the query performance. The result is shown in Figure 6. It can be seen that the query performance of our solution is two order of magnitudes higher than Infobright and MonetDB and one order of magnitude higher than ToluDB. The reason why Infobright and MonetDB is so slow is that they do not support the real index. Index is just an advice to MonetDB and may be ignored. More important is that, when loading data, queries on the other three systems are almost blocked while in our solution the queries are not significantly affected.

6   Related Work

Traditional index such as B-Tree is widely used in databases and data warehouses[1]. It supports range condition and equi-condition on columns. But it is a kind of heavy wait index. When log records arrive rapidly, the b-tree will keep updating the tree nodes and hierarchical structure frequently[20], which limits the throughput and performance of data insertion. Write times of b-tree grow considerably as the data becomes scattered in different disk sectors over time. More importantly, when the index is updated frequently, the select tasks and the data insertion tasks must be frequently synchronized and the performance of both select and insertion will be restricted. Hash index used in most RDBMS does not support range queries so that it is not suitable for our workload although it is faster than b-tree[6]. Bitmap index is also a type of common used index in RDBMSs, but it is only suitable for low cardinality columns[11].

Besides the traditional b-tree, hash and bitmap index, there are some other research prototypes and system implementations of novel indexing techniques. log-structured merge-tree (LSM-tree) is a disk-based data structure designed to provide low-cost indexing for high-rate record inserts over an extended period[22]. But LSM-tree sorts the data on the index key so that it can only index one column in the table. Systems such as LevelDB[7], Bigtable[15], Hbase[3] and Cassandra[19] implement LSM-tree. They are key-value storage systems do not support both equi-conditions on the high selectivity column and the range conditions on the time series column. But another index called *fractal tree* is a tree data structure supplies the same queries performance as a B-tree but with insertions and deletions that are asymptotically faster than a B-tree[2]. It is currently used in TokuDB[8].

In recent years, column store is a hot topic in analytical database area. Techniques to improve the query and insertion performance of DBMS are well researched. In columnar databases such as C-store[26], infobright[24] and monetdb [13], column statistics and just-in-time indexes are built to speed up analytical queries[24,13]. These can be considered as lightweight indexes and are low-cost and efficient for data insertion. Infobright is a disk-based columnar storage engine while MonetDB uses memory aggressively. In this paper, we compare our solution with Infobright, MonetDB and TokuDB.

7   Conclusion

Index is very important to real-time log data analysis. It can be seen from Section 5.3 that the two state-of-the-art open-source analytical databases, MonetDB and Infobright are failed to provide good query performance due to the lack of indexes. To support high throughput and real-time data loading and query evaluation, we designed an efficient indexing scheme of which the loading throughput and query performance are 1-2 order of magnitudes higher than the existing baselines.

References

1. http://docs.oracle.com/cd/b28359_01/server.111/b28313/indexes.htm#i1006549
2. http://en.wikipedia.org/wiki/fractal_tree_index
3. http://hbase.apache.org/
4. http://redis.io
5. http://sandbox.mc.edu/~bennet/cs402/lec/losedex.html
6. https://dev.mysql.com/doc/refman/5.5/en/index-btree-hash.html
7. https://github.com/google/leveldb
8. https://github.com/tokutek/tokudb-engine
9. http://www.gbase.cn/comcontent_detail1/&i=30&comcontentid=30.html
10. http://www.gluster.org
11. http://www.oracle.com/technetwork/articles/sharma-indexes-093638.html
12. Bloom, B.H.: Space/time trade-offs in hash coding with allowable errors. CACM 13(7), 422–426 (1970)
13. Boncz, P.A., Zukowski, M., Nes, N.: Monetdb/x100: Hyper-pipelining query execution. In: CIDR, vol. 5, pp. 225–237 (2005)
14. Chan, C.-Y., Ioannidis, Y.E.: Bitmap index design and evaluation. In: SIGMOD, vol. 27, pp. 355–366 (1998)
15. Chang, F., Dean, J., Ghemawat, S., Hsieh, W.C., Wallach, D.A., Burrows, M., Chandra, T., Fikes, A., Gruber, R.E.: Bigtable: A distributed storage system for structured data. In: OSDI (2006)
16. Guttman, A.: R-trees: A dynamic index structure for spatial searching. In: SIGMOD, pp. 47–57 (1984)
17. He, Y., Lee, R., Huai, Y., Shao, Z., Jain, N., Zhang, X., Xu, Z.: Rcfile: A fast and space-efficient data placement structure in mapreduce-based warehouse systems. In: ICDE (2011)
18. Karger, D., Lehman, E., Leighton, T., Panigrahy, R., Levine, M., Lewin, D.: Consistent hashing and random trees: Distributed caching protocols for relieving hot spots on the world wide web. In: STOC, pp. 654–663 (1997)
19. Lakshman, A., Malik, P.: Cassandra: a decentralized structured storage system. SIGOPS 44(2), 35–40 (2010)
20. Lehman, P.L., et al.: Efficient locking for concurrent operations on b-trees. TODS 6(4), 650–670 (1981)
21. Olston, C., Reed, B., Srivastava, U., Kumar, R., Tomkins, A.: Pig latin: a not-so-foreign language for data processing. In: SIGMOD (2008)
22. Neil, P.O., Cheng, E., Gawlick, D., ONeil, E.: The log-structured merge-tree (lsm-tree). Acta Informatica 33(4), 351–385 (1996)
23. Pavlo, A., Paulson, E., Rasin, A., Abadi, D.J., DeWitt, D.J., Madden, S., Stonebraker, M.: A comparison of approaches to large-scale data analysis. In: SIGMOD (2009)
24. Ślȩzak, D., Wróblewski, J., Eastwood, V., Synak, P.: Brighthouse: an analytic data warehouse for ad-hoc queries. PVLDB 1(2), 1337–1345 (2008)
25. Stonebraker, M., Abadi, D., DeWitt, D.J., Madden, S., Paulson, E., Pavlo, A., Rasin, A.: Mapreduce and parallel dbmss: Friends or foes? CACM, 53(1), January 2010
26. Stonebraker, M., Abadi, D.J., Batkin, A., Chen, X., Cherniack, M., Ferreira, M., Lau, E., Lin, A., Madden, S., O'Neil, E.J., O'Neil, P.E., Rasin, A., Tran, N., Zdonik, S.B.: C-store: A column-oriented DBMS. In: VLDB, pp. 553–564 (2005)
27. Thusoo, A., Sarma, J.S., Jain, N., Shao, Z., Chakka, P., Zhang, N., Anthony, S., Liu, H., Murthy, R.: Hive - a petabyte scale data warehouse using hadoop. In: ICDE (2010)

Demonstration Track

A Fair Data Market System with Data Quality Evaluation and Repairing Recommendation

Xiaoou Ding, Hongzhi Wang, Dan Zhang, Jianzhong Li, and Hong Gao

Harbin Institute of Technology, Harbin, China
{dingxiaoou_hit}@163.com,
{wangzh}@hit.edu.cn

Abstract. With the development of data market, data resources play a key role as the part of business resources. However, existing data markets are too simple to reveal the real data values in practical application. Motivated by the effectiveness and fairness of the data market, we develop a fair data market system that takes data quality into consideration. In our system, we design a fair data price evaluation mechanism, which aims at meeting the needs of both supply and demand. For the data quality issues in the data market, several critical factors, including accuracy, completeness, consistency, and currency, are integrated in order to show comprehensive assessment of the data. Moreover, our system can also provide data repairing recommendation based on data quality evaluation.

1   Introduction

With the widespread of business needs of the big data, its unique value arrests more attention [1]. Generally, the content and pricing of data in the traditional offline data market have no uniform standard. With the rapid development of big data, several drawbacks of the existing data market have come out.

(1) **Costly.** When buyers and sellers attempt to start a transaction, both of them need to invest considerable efforts for consultation and trading.

(2) **Neglecting of Data Quality Issues.** A wide range of techniques have been proposed to assess and improve the data quality in the literature [1]. However, existing data markets have not involved data quality issues in data pricing models.

(3) **Poor Fairness.** There is a lack of standard mechanism for pricing the data commodities. It is difficult for buyers to obtain a reliable assessment.

To overcome these drawbacks, we suggest a correlation between data quality and data pricing. We designed and implemented a data market system, in which multiple data quality dimensions are taken account together affecting data pricing [1][3]. The fairness and security mechanisms are developed to ensure an organized system.

The major features of the system are shown as follows.

(1) **Data Quality Sensitive.** Our data market provides complete assessment oriented to multiple data quality dimensions, which have potent effects on data usability.

(2) **Fair Trading.** The system develops uniform standards on evaluation, pricing and trading course.

© Springer International Publishing Switzerland 2015
R. Cheng et al. (Eds.): APWeb 2015, LNCS 9313, pp. 855–858, 2015.
DOI: 10.1007/978-3-319-25255-1_70

(3) **Effectiveness.** The system involves an efficient data pricing mechanism based on data quality evaluation. In addition, it will repair the data set using reliable and efficient mechanism. As a result, the effectiveness of the system is ensured.

This paper is organized as follows. In Section 2, we overview the system to assess and set a price on data in brief. Then, we introduce key techniques in Section 3, after which we discuss the plan of demonstrations in Section 4.

2   System Overview

In this section, we introduce the architecture of our system. The organization of the system is shown in Fig. 1. The whole system is divided into 4 modules.

Fig. 1. System Overview   **Fig. 2.** The IDRS function progress

In Data Quality Evaluation Module, the quality of trading data is evaluated. In Fairness Pricing Module, the system sets a price grade on data according to the evaluation result. In Repairing Recommendation Module, the system shows possible repairs for data with poor quality and the reliability level of the recommended repairs. The Procedure Security Module provides security assurance in key process management.

3   Key Technologies

In this section, we introduce key technologies in our system including data quality evaluation techniques and data repairing recommendation strategies.

3.1   Data Quality Evaluation

It is acknowledged in the research community that accuracy, completeness, consistency, currency are major dimensions of the data quality [4]. On account of this, we carry out the data quality evaluation mainly based on these four dimensions in our system. Here we only discuss the evaluation strategy of other two dimensions for the interest of space. We perform accuracy and consistency evaluation according to the methods introduced in [3] and [5], respectively.

Completeness Evaluation. Completeness reflects the degree to which values are present in a dataset [4]. Appropriate amount of data, adequate attributes, few missing values in the table are important component when analyzing completeness dimensions. Thus, completeness is calculated according to (1), where n_{min} and n_{nec} represent the minimum tuples required to guarantee the usability of dataset and the minimum attributes to describe sufficient features of an object, respectively.

$$\eta_{com} = -\log \left(w_{com1} \cdot \frac{n_{min}}{n} + w_{com2} \cdot \frac{n-n_{nec}}{n} + w_{com3} \cdot \frac{n_{mis}}{m \cdot n} \right) \quad (1)$$

Currency Evaluation. Currency is the degree to which a dataset is up-to-date. With the redundant records and currency constraint recovery, we approach data currency according to [1] [6]. We measure the volatility ratio of the data set. T_{vol} measures the period of validity of data records. T_{IS} represents the time in which data are stored in the system, and $T_{RW_{update}}$ represents the time in which data are updated in the real world. When η_{curr1} of the data record is low, it will be labelled as $n_{outdated}$.

$$\eta_{curr1} = (\max \left(0, \frac{T_{IS} - T_{RW_{update}}}{T_{vol}} \right))^n \quad (2)$$

We measure currency of a certain dataset in the following formula.

$$\eta_{cur} = -\log \left(\frac{n_{outdated}}{n} \right) \quad (3)$$

3.2   Repairing Recommendation

In the repairing recommendation module, the part of the poor-quality data can be corrected by our system automatically.

Inaccuracy Repairing. According to the specific rules defined in the rules base, the field value illegal, out of range, or violating the attribute pattern will be corrected are replaced with more accurate value.

Inconsistency Repairing. In this part, the inconsistent data repairing system (IDRS) is implemented, and Fig.2 shows how it functions. After detecting the data which violates CFDs [5], it provides appropriate repairing solutions.

4   Demonstration

In this section, we introduce demonstration plan of our system. The features of our system will be shown in two aspects.

From the aspect of data suppliers, the system first evaluates the quality of the dataset and returns the result (as shown in Fig.3), which shows the data-quality-sensitive feature of the system. Secondly, suppliers could interact with the user-friendly interface to obtain the repaired dataset in which the wrong data will be marked with eye-catching colors (Fig.4). Suppliers have the option to upload the data commodity after assessment with an authorized and fair price which is beyond artificial revision. From the aspect of data consumers, they can find the resource by classification search and keyword search. Furthermore, popular recommendation also assists in making purchase decisions (Fig.5). Consumers can make fair trades with authorized data suppliers.

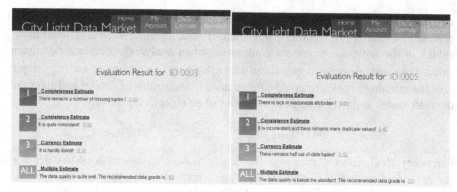

Fig. 3. Data quality evaluation interfaces

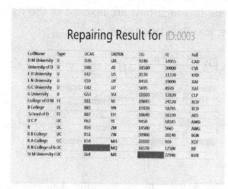

Fig. 4. Part of the repairing result interface   **Fig. 5.** Popular recommendation display interface

Acknowledgement. This paper was partially supported by NGFR 973 grant 2012CB316200, NSFC grant 61472099, 61133002 and National Sci-Tech Support Plan 2015BAH10F00.

References

1. Li, M., Li, J., Gao, H.: Evaluation of Data Currency. Chinese Journal of Computers 35(11) (2012)
2. Batini, C., Cappiello, C., Francalanci, C., Maurino, A.: Methodologies for data quality assessment and improvement. ACM Computing Surveys (CSUR) 41, 16 (2009)
3. Zhang, Y., Wang, H.: Accuracy Evaluation for Sensed Data. In: Cai, Z., Wang, C., Cheng, S., Wang, H., Gao, H. (eds.) WASA 2014. LNCS, vol. 8491, pp. 205–214. Springer, Heidelberg (2014)
4. Sidi, F., Shariat, P., Payam, H., et al.: Data Quality: A Survey of Data Quality Dimensions. In: Information Retrieval & Knowledge Management, pp. 300–304. IEEE (2012)
5. Bohannon, P., Fan, W., Geerts, F., Jia, X., Kementsietsidis, A.: Conditional functional dependencies for data cleaning. In: ICDE 2007 (2007)
6. Li, M., Li, J.: Algorithms for Improving Data Currency. In: NDBC 2014 (2014)

HouseIn: A Housing Rental Platform with Non-redundant Information Integrated from Multiple Sources

Jian Zhou[1], Zhixu Li[1,*], Qiang Yang[1], Jun Jiang[1], Jia Zhu[2], An Liu[1], Guanfeng Liu[1], and Lei Zhao[1]

[1] School of Computer Science and Technology, Soochow University, China
[2] School of Computer Science, South China Normal University, China
{jzhou_jz,qiangyanghm,jiangjun_void}@hotmail.com,
{zhixuli,anliu,gfliu,zhaol}@suda.edu.cn, jzhu@m.scnu.edu.cn

Abstract. Housing Rental Platforms (HRPs) such as *rentalhouses.com, 58.com, ganji.com* provide convenient ways to find accommodations, but the redundancy of the rental advertising information within and across platforms brings unpleasant user experience. Besides, rental advertisements are usually presented in the form of a list, which can hardly give users a straightforward big picture about the housing rental market of a particular interested area. In this demonstration, we introduce HouseIn, a novel HRP that expect to: 1) provide users a clear big picture in several aspects about the housing rental market of a particular interested area; and 2) detect those advertisements referring to the same property, such that to give users a price comparison between various platforms for the same apartment for rental. The core challenge in implementing HouseIn lies on the Record Matching problem between these advertisements, given that there is no exact Unit/Building Number of the apartment for the sake of privacy issue, and the detailed information about the house can be various given by different agencies. To handle the Record Matching (RM) problem between these advertisements, we employ several state-of-the-art RM methods plus Information Extraction (IE) techniques to use both structured and unstructured information in the advertisements. We show the advantages of HouseIn with several demonstration scenarios.

Keywords: Housing Rental, Data Integration, Record Matching.

1 Introduction

In the modern society, Housing Rental Platforms (HRPs) such as *rentalhouses.com, 58.com, ganji.com* provide people convenient ways to find their accommodations. Along with the development of those platforms, however, the redundancy of the rental advertising information within and across platforms brings unpleasant user experience. Besides, rental advertisements are usually presented in the form of a list, which can hardly give users a straightforward big picture about the housing rental market of a particular interested area.

---

* Corresponding author.

© Springer International Publishing Switzerland 2015
R. Cheng et al. (Eds.): APWeb 2015, LNCS 9313, pp. 859–862, 2015.
DOI: 10.1007/978-3-319-25255-1_71

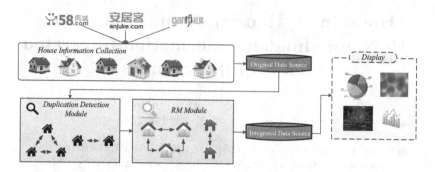

Fig. 1. The System Overview of HouseIn

We develop HouseIn, a novel HRP that expects to: 1) provide users a clear big picture from several aspects about the housing rental market of a particular interested area; and 2) detect those advertisements referring to the same property, such that to give users a price comparison between various platforms for the same apartment for rental.

The core challenge in implementing HouseIn lies on how we identify duplicate advertising records referring to the same apartment within and across sources, given that there is usually no exact Unit/Building Number of the apartment for the sake of privacy issue. Besides, the detailed information (either structured or unstructured description information) about the apartment can be various or missing due to different accommodation agencies in different HRPs. For instance, the four posts (G2, G4, T2, and A3) highlighted in Fig. 3 actually refer to the same apartment, although they do not look the same. Plenty of work [1–3] has been done on RM, but they can hardly handle the RM problem between these housing rental posts in our case, given that the inconsistency of attribute values within and across data sets and the existence of missing attribute values. To tackle the problem, we employ both structured and unstructured information in these posts in two ways: (1) we employ a decision tree based model [4] to leverage the structured attribute values such as "community, location, house type, area and floor" to find possible duplicate posts in an effective and efficient way; and (2) we make use of the detailed description information in each post (unstructured textual information) to further detect duplicate posts that can not identified based on structured information only. In this demonstration, we will show the advantages of HouseIn with several demonstration scenarios.

2 System Overview

The system overview of HouseIn is depicted in Fig. 1. There are basically three modules in HouseIn: Duplicates Detection (DD) Module, the Record Matching (RM) module and the Display module.

(1) DD Module: This module detects duplicate posts referring to the same apartment within one data source. The posts sharing the same URL are firstly merged, since they are the updates of the same post. The challenge lies on how to

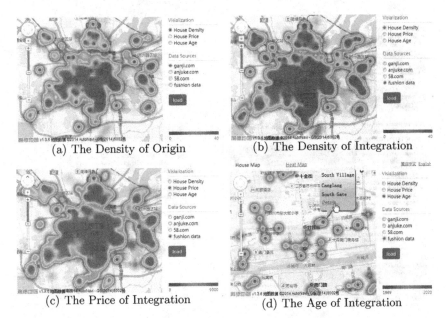

(a) The Density of Origin (b) The Density of Integration

(c) The Price of Integration (d) The Age of Integration

Fig. 2. The House Metrics Demonstration

detect the duplicate posts having different URLs, given that these records may have inconsistent or missing attribute values. Most existing methods [1–3] work on measuring the similarity between the key attribute values of record pairs, but can hardly handle the problem well when the key attribute values look dissimilar. As an alternative, we employ a novel method called NokeaRM [4], which uses both key and non-key attribute values in doing RM. It is a rule-based algorithm based on a tree-like structure, which can not only deal with noisy and missing values, but also greatly improve the efficiency of the method by finding out matched instances or filtering unmatched instances as early as possible [4].

(2) RM Module: The RM module is responsible for matching records between multiple sources. The same techniques used in the DD module can be applied here. But for those matched records that can not be identified, we use the unstructured descriptive information to assist the process of matching. We first find out important terms that can help to identify the similarity between these records from all these records' descriptions, and then calculate the similarity between these records based on the TF/IDF scores of these terms.

(3) Display Module: The display module of HouseIn is especially designed to provide a user-friendly interface with great user experience. In particular, based on the record matching results, we do real-time statistic analysis to get some general information such as the average price, density, and age of the houses in different areas, and then map the data into a city map to show a big picture about the rental market in their interested area. Meanwhile, we provide users a price comparison between various platforms/posts for the same apartment for rental.

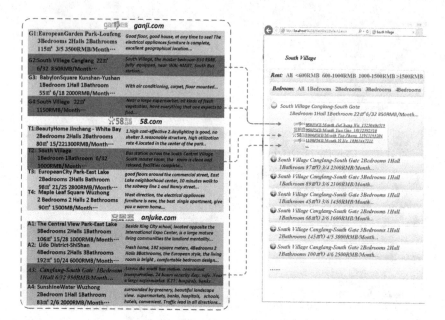

Fig. 3. The Demonstration of Integrated Data

3   Demonstration Scenarios

We demonstrate three scenarios to show the advantages of HouseIn, where the datum are from three popular HRPs in China including GanJi, 58 and Anjuke.

1) Map-based Market Overview. HouseIn provides an overview of the rental market in a user-selected area from the following several aspects: (1) the distribution of the house supply density (Fig. 2(a)(b)); (2) the distribution of the rental price (Fig. 2(c)); and (3) the distribution of the house age (Fig. 2(d)).

2) Ad. List with Advanced Search. Similar with popular HRPs, HouseIn also let users to input the price range, interested location etc. to quickly get short-listed advertisements, where all duplicate advertisements are removed. Besides, users can also direct click on a point in the city map to get the advertisement list of the area as shown in Fig. 2(d).

3) Price Comparison. For every distinct apartment for rental, we list different advertisements (with different rental prices) referring to the apartment.

References

1. Bilenko, M., Mooney, R.J.: Adaptive duplicate detection using learnable string similarity measures. In: KDD, pp. 39–48 (2003)
2. Elmagarmid, A.K., Ipeirotis, P.G., Verykios, V.S.: Duplicate record detection: A survey. TKDE 19(1), 1–16 (2007)
3. Han, J., Kamber, M.: Data Mining, Southeast Asia Edition: Concepts and Techniques. Morgan kaufmann (2006)
4. Yang, Q., Li, Z., Jiang, J., Zhao, P., Liu, G., Zhu, J.: NokeaRM: Employing non-key attributes in record matching. In: Li, J., Sun, Y. (eds.) WAIM 2015. LNCS, vol. 9098, pp. 438–442. Springer, Heidelberg (2015)

ONCAPS: An Ontology-Based Car Purchase Guiding System

Jianfeng Du[1], Jun Zhao[2], Jiayi Cheng[1], Qingchao Su[1], and Jiacheng Liang[1]

[1] Guangdong University of Foreign Studies, Guangzhou 510006, China
jfdu@gdufs.edu.cn
[2] Lancaster University, Lancaster LA1 4WA, United Kingdom
j.zhao5@lancaster.ac.uk

Abstract. In this paper, an ontology-based car purchase guiding system, ONCAPS, is presented. The system integrates Semantic Web technologies to provide car information to people who want to buy cars. It employs ontology-based data integration technology to gather data from multiple car selling websites in China. It captures personalized weights on car features by an analytic hierarchy process based approach. It computes the personalized distance between a car and a given requirement by a weighted sum model where ontology reasoning is involved. For purchase guiding, ONCAPS suggests the user all cars that have the minimum personalized distance to the given requirement. The suggestions are shown with explanations on why these cars are recommended.

1 Motivation and Novelty

The car market in China is growing rapidly as more people can afford private cars. Nevertheless, cars are still expensive, and thus most people need to consider carefully before making their final purchase decisions. A number of car selling websites have emerged in China to help customers purchase cars. Some car selling websites such as Bit Auto (http://www.bitauto.com/) provide only simple search interfaces to the users. Users can only use keywords about brands or car series to express their requirements. Other websites such as Auto Home (http://www.autohome.com.cn/) provide advanced search interfaces to the users. Although users can express more detailed requirements in such a website, the website cannot suggest alternative cars to the users when no exact matching can be found. Existing car selling websites are hugely insufficient in helping potential buyers make their decisions. A car selling website that can guide users to express their detailed requirements and suggest relevant cars when no exact matching can be found is therefore highly desirable.

In this paper we present an ontology-based car purchase guiding system, ONCAPS, which can achieve the above desirable features. It has the following novelties compared to existing car selling websites. Firstly, it integrates data from multiple car selling websites and provides more complete information on the market. Secondly, it captures user preferences on car features by an analytic hierarchy process (AHP) based approach, i.e., by organizing car features in a

R. Cheng et al. (Eds.): APWeb 2015, LNCS 9313, pp. 863–866, 2015.
DOI: 10.1007/978-3-319-25255-1_72

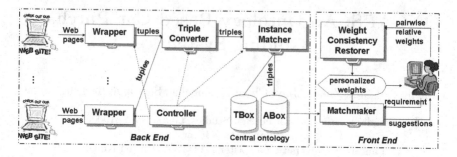

Fig. 1. The architecture of ONCAPS

hierarchy and capturing pairwise relative weights for sibling features. Thirdly, it computes personalized distances between cars and requirements by a weighted sum model, where ontology reasoning is involved. Finally, it tackles the problem of no exact matching by showing alternative options that have the minimum personalized distance to the given requirement. All options are shown with explanations to tell e.g. the differences between a recommended car and the requirement as well as how the personalized distances are computed.

2   System Architecture and Key Methods

ONCAPS has two parts, the back end and the front end, as shown in Fig. 1. The back end provides facilities for integrating data from multiple car selling websites to a central ontology, which is expressed in the tractable RL profile of OWL 2 [3], the newest W3C recommended language for modeling ontologies. The central ontology consists a TBox declaring logical relations between classes and properties and an ABox asserting instances of classes or properties, where the TBox is constructed manually and the ABox is generated automatically through a sequence of components in the back end. Every wrapper component extracts tuples from a certain car selling website, where the field names of extracted tuples are mapped to class or property names in the central ontology. The triple converter component directly translates tuples to RDF triples and feed them to the ABox. The instance matcher component identifies the same cars from different websites. It selects a representative identifier for every group of cars that are recognized to be the same and then replaces car identifiers in triples with their corresponding representative identifiers. It also adds to the ABox all triples that can be entailed from the ontology by applying the standard forward inference method in OWL 2 RL systems [3]. The controller component periodically detects data updates in the crawled car selling websites and schedules the incremental execution of all wrappers, the triple converter and the instance matcher.

The frond end provides facilities for capturing user preferences on car features and for computing recommended cars to a given requirement. The weight consistency restorer component resolves inconsistency for the pairwise relative weights captured by the AHP based approach. The matchmaker component computes the personalized distances of all cars to a given requirement by exploiting a weighted sum model, where the weights are the personalized ones stored previously for the user who raises the requirement. It presents users all the cars that have the minimum personalized distance.

The key methods used in ONCAPS are described below.

Instance Matching. For integrating data from multiple websites, we need to firstly match field names in every website with class or property names in the TBox, and secondly, match individuals such as car identifiers between different websites. While the former task can be done by manually defined rules and is relatively easy, the latter task is hard because there are many individuals in a website and it is infeasible to manually retrieve all individuals in advance. Hence we employ the instance matching method proposed in [4] to match individuals. The method builds the name vector and the virtual document vector for each individual and generates two inverted indexes for these two groups of vectors. Then all matches are selected by several heuristic rules defined on the indexes.

Personalized Weight Capturing. To capture personalized weights on car features, it is impractical to ask a user to directly specify the weight of every feature. This is because the user can easily confuse with a mass of car features and is hard to balance the weights. To alleviate the effort for capturing weights, AHP [1] can be employed. It organizes all car features in a hierarchy and requires a user to input pairwise relative weights for sibling features only. By following [1], we construct a two-level hierarchy on car features and establish a many-to-one correspondence between class or property names in the TBox and the car features. AHP comes with efficient algorithms for computing absolute weights from the matrix of pairwise relative weights (called the *pairwise comparison matrix* or simply *PCM*) under the constraint that the sum of all weights is one. However, these algorithms cannot work out reasonable weights unless the given PCM is *sufficiently consistent*, i.e., the consistency measure satisfies a threshold condition, where a PCM $\{a_{ij}\}_{1 \leq i \leq n, 1 \leq j \leq n}$ is said to be *consistent* if $a_{ij}a_{jk} = a_{ik}$ for all $1 \leq i, j, k \leq n$, and where a_{ij} is the relative weight of the ith feature over the jth feature. To resolve inconsistency for a given PCM, we employ an efficient method proposed in [2] to compute a new PCM which is sufficiently consistent and is as close as possible to the given PCM.

Personalized Distance Computation. In ONCAPS, a requirement on cars to be purchased is represented by a conjunctive query of the form

$$\bigwedge_{i=1}^{k} A_i(x) \wedge \bigwedge_{i=1}^{m} r_i(x, u_i) \wedge \bigwedge_{i=1}^{n} (s_i(x, y_i) \wedge y_i \geq v_{i1} \wedge y_i \leq v_{i2}) \text{ for } k, m, n \geq 0, \quad (1)$$

where x and y_i are variables, u_i is an individual or a string, v_{ij} is a number, A_i is a class name, r_i and s_i are property names, and the atom $y_i \geq v_{i1}$ or $y_i \leq v_{i2}$ can be absent. This conjunctive query asks for all car identifiers whose substitution for x makes the query entailed by the central ontology. To avoid reasoning over the TBox on the fly, the instance matcher component computes all entailments of the ontology and adds them to the ABox. In this way all answers to a conjunctive query can directly be retrieved from the ABox without considering the TBox. The matchmaker component computes the personalized distances between cars and the given conjunctive query and presents the user all cars that have the minimum distance. The personalized distance is defined as the weighted sum of distances between a car and all atoms in the query by treating the predicates of atoms as car features. More precisely, a distance between a car whose identifier is c and a conjunctive query of the form (1) is defined as

$$\sum_{i=1}^{k} w(A_i) \cdot d(A_i(c), \mathcal{A}) + \sum_{i=1}^{m} w(r_i) \cdot d(r_i(c, u_i), \mathcal{A})$$

$$+ \sum_{i=1}^{n} w(s_i) \cdot d(s_i(c, y) \wedge y \geq v_{i1}, \mathcal{A}) + \sum_{i=1}^{n} w(s_i) \cdot d(s_i(c, y) \wedge y \leq v_{i2}, \mathcal{A}),$$

where $\mathcal{A}$ is the ABox with all entailments added and $w(f)$ is the personalized weight of the car feature f. The following shows how $d(\phi, \mathcal{A})$ is defined, where $\bowtie$ is $\geq$ or $\leq$, $P(r_i = u_i)$ is the probability that the value of r_i is u_i, $P(s_i \geq v_i)$ (resp. $P(s_i \leq v_i)$) is the probability that the value of s_i is not less (resp. greater) than v_i, and where all the probabilities are estimated from $\mathcal{A}$.

$$d(A_i(c), \mathcal{A}) = \begin{cases} 0 & A_i(c) \in \mathcal{A} \\ 1 & A_i(c) \notin \mathcal{A} \end{cases}$$

$$d(r_i(c, u_i), \mathcal{A}) = \begin{cases} 0 & r_i(c, u_i) \in \mathcal{A} \\ 1 - P(r_i = u_i) & \nexists u : r_i(c, u) \in \mathcal{A} \\ 1 & \exists u \neq u_i : r_i(c, u) \in \mathcal{A} \end{cases}$$

$$d(s_i(c, y) \wedge y \bowtie v_i, \mathcal{A}) = \begin{cases} 0 & \exists v : s_i(c, v) \in \mathcal{A}, v \bowtie v_i \\ 1 - P(s_i \bowtie v_i) & \nexists v : s_i(c, v) \in \mathcal{A} \\ \min(\{\frac{|v - v_i|}{|v_i|} \mid s_i(c, v) \in \mathcal{A}\}) & \forall v : s_i(c, v) \in \mathcal{A}, v \not\bowtie v_i \end{cases}$$

3   What Will Be Demonstrated?

We will demonstrate two scenarios about how the front end of ONCAPS works. One is on capturing personalized weights through AHP, while the other is on suggesting cars with explanations for a requirement on cars. More information can be found at http://www.dataminingcenter.net/oncaps/.

Acknowledgements. This work is partly supported by NSFC (61375056), Guangdong Natural Science Foundation (S2013010012928), and the Undergraduate Innovative Experiment Project in Guangdong University of Foreign Studies.

References

1. Byun, D.: The AHP approach for selecting an automobile purchase model. Information & Management 38(5), 289–297 (2001)
2. Du, J., Jiang, R., Hu, Y.: Multi-criteria axiom ranking based on analytic hierarchy process. In: Qi, G., Tang, J., Du, J., Pan, J.Z., Yu, Y. (eds.) CSWS 2013. CCIS, vol. 406, pp. 118–131. Springer, Heidelberg (2013)
3. Grau, B.C., Horrocks, I., Motik, B., Parsia, B., Patel-Schneider, P.F., Sattler, U.: OWL 2: The next step for OWL. Journal of Web Semantics 6(4), 309–322 (2008)
4. Li, J., Wang, Z., Zhang, X., Tang, J.: Large scale instance matching via multiple indexes and candidate selection. Knowledge Based Systems 50, 112–120 (2013)

A Multiple Trust Paths Selection Tool in Contextual Online Social Networks

Linlin Ma[1], Guanfeng Liu[1], Guohao Sun[1], Lei Li[2], Zhixu Li[1], An Liu[1], and Lei Zhao[1]

[1] School of Computer Science and Technology, Soochow University, China
[2] School of Computer and Information, Heifei University of Technology, China 230009
{gfliu,lizhixu,anliu,zhaol}@suda.edu.cn,
lilei@hfut.edu.cn

Abstract. Online Social Network (OSN) is becoming popular, where trust is one of the most important factors for participants' decision making. This requires to efficiently and effectively select those social trust paths that can yield the most trustworthy trust evaluation results between two unknown participants to establish reasonable trust relationships between them. Thus, we develop a Multiple Trust Paths (MTP) selection tool based on the state-of-the-art trust paths selection method proposed by us, which considers the social contexts, like the social trust and social intimacy degree between participants, and the role impact factor of participants in trust paths selection. Our tool could help users evaluate the trustworthiness of unknown participants in a variety of OSN based applications, for instance, to help a retailer identify new trustworthy customers and introduce products to them, or help users select the trustworthy workers in OSN based crowdsourcing platforms.

Keywords: online social network, trust, paths selection.

1 Introduction

Online Social Networks (OSNs) have been used as a means for a variety of activities, like employment, movie recommendation and e-commerce, etc. A Contextual OSN (COSN) [1] can be modeled as a graph depicted in Fig. 1, where each node represents a participant and each link corresponds to the real-world interactions or online interactions between them. In addition to the social network structure, the COSN also contains social contexts including social trust, social intimacy degree between participants and role impact factor of participants denoted as T, r and ρ, as shown in Fig. 1. Selecting the trust paths that can yield the most trustworthy trust evaluation results based on the preferences of users can be modelled as an classical NP-Complete multi-constrained optimal path selection problem [2].

This tool provides three types of social paths for supporting users decision making, i.e., (1) the shortest path between a pair of source and target, (2) the path with the maximum T value based on the method proposed in [3], and (3) the multiple constrained K optimal social paths based on our efficient and effective approximation algorithm, D-MCBA proposed in [1], which considers user's preference and social context in trust paths selection. Our MTP selection tool can help establish reasonable trust relationships between unknown participants, which can be a backbone in many OSN based

R. Cheng et al. (Eds.): APWeb 2015, LNCS 9313, pp. 867–870, 2015.
DOI: 10.1007/978-3-319-25255-1_73

applications.For example, it can help an employer find potential trustworthy employees in OSNs, help a retailer find trustworthy loyal customers in an OSN based CRM system, and help find reliable workers in OSN based crowdsourcing platforms.

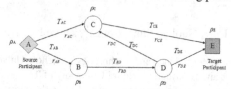

Fig. 1. A contextual OSN

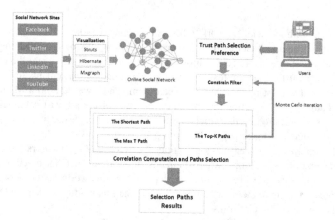

Fig. 2. The architecture of MTP selection tool

2   System Overview

Our tool is based on Struts and Hibernate, the server runtime environment is Tomcat 7 and JDK 7. The social network is stored by MySQL database and the social network display is developed by using mxGraph[1], a Javascript library for graph drawing. Fig. 2 shows the architecture of our tool. The MTP selection tool can divided into two components: (1) *Source-Target selection component* (Fig. 3), and (2) *paths selection component* (Fig. 4 and Fig. 5). Our MTP selection tool can be accessed via the web link http://mtp.cnzhujie.cn/ (*compatible with Google Chrome*).

2.1   Source-Target Selection Component

The Source-Target selection component displays the social network when loading the tool as shown in Fig. 3. The red and big nodes have larger indegree than that of the purple and small ones. We name the red nodes as *Popular Nodes* and the purple nodes as *General Nodes*. When an user move the mouse on a node in the social network, he/she can view the brief information of the participant. When an user selects a node from the displayed social network structure, the corresponding participant's head image will be added into the node selection component on the left.

---

[1] http://jgraph.github.io/mxgraph/

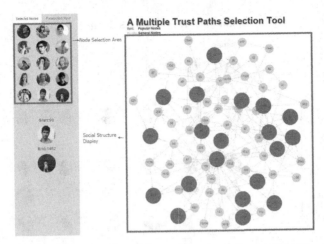

Fig. 3. The interface of Source-Target selection component

2.2   Paths Selection Component

The paths selection component is divided into two areas, *the parameter input area* on the left of MTP selection tool (see Fig. 4) and *the result display area* on the right (see Figure 5). The top of the parameter input area is for users to set the simulation times of the Monte Carlo interaction [4] and the end-to-end constraints for T , r and ρ in the scope of [0,1] as the requirements for the K optimal trust paths selection. Then, users can set T , r and ρ weights in the scope of [0,1], and the sum of them equals to 1. In addition, the middle of the parameter input area is to (1) help filter those nodes with small indegree and outdegree, and (2) control the maximal length of the selected social trust paths. Furthermore, at the bottom of the parameter input area, there is a checkbox list for users to determine which type of paths they would like to select and display. The result of paths selection can be viewed in paths selection result display area as shown in Fig. 6, where three types of paths between a source (ID=14) and the target (ID=34) are displayed, and the T and r values between participants are displayed on edges, and the ρ values of participants are displayed on the top of the head images.

3   Demo

After opening our MTP selection tool, the social network structure can be displayed in the social network display area. Then an user could select a source and a target by clicking the nodes from the structure. The corresponding head images of the selected nodes will be displayed in the node selection area. Then the user can input parameter values to select social trust paths based on his/her preferences. After clicking the button to submit the parameters, the trust path selection results will be displayed in the path display area. Fig. 5 shows the three types of paths between a source (ID=14) and the target (ID=34).

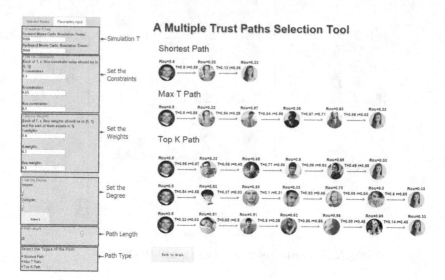

Fig. 4. The parameter input area

Fig. 5. Three types of trust paths between ID=14 and ID=34

4 Conclusion

In this paper, we introduce a tool to select multiple trust paths in Contextual Online Social Networks (COSNs). By using our MTP selection tool, an user can evaluate the trustworthiness of other unknown participants, which can be used in many OSN based applications, like finding potential trustworthy employees in OSNs, finding trustworthy loyal customers for a retailer in CRM systems, and finding reliable workers in OSN based crowdsourcing platforms.

Acknowledgement. This work was partially supported by Natural Science Foundation of China (Grant Nos. 61303019, 61402312, 61402313, 61232006, 61003044, 61440053), Doctoral Fund of Ministry of Education of China (20133201120012), and Collaborative Innovation Center of Novel Software Technol- ogy and Industrialization, Jiangsu, China.

References

1. Liu, G., Liu, A., Wang, Y., Li, L.: An effcient multiple trust paths finding algorithm for trustworthy service provider selection in real-time online social network environments. In: IEEE ICWS, pp. 121–128 (2014)
2. Liu, G., Wang, Y., Orgun, M.A.: Optimal social trust path selection in complex social networks. In: AAAI, pp. 1391–1398 (2010)
3. Hang, C.-W., Wang, Y., Singh, M.P.: Operators for propagating trust and their evaluation in social networks. In: AAMAS, pp. 1025–1032 (2009)
4. Morton, D.P., Popova, E.: Monte-carlo simulations for stochastic optimization monte-carlo simulations for stochastic optimization. In: Encyclopedia of Optimization, pp. 2337–2345. Springer (2009)

EPEMS: An Entity Matching System for E-Commerce Products

Lei Gao[1], Pengpeng Zhao[1], Victor S. Sheng[2], Zhixu Li[1], An Liu[1], Jian Wu[1], and Zhiming Cui[1]

[1] School of Computer Science and Technology, Soochow University, Suzhou, 215006, P.R. China
[2] Computer Science Department, University of Central Arkansas, USA
ppzhao@suda.edu.cn

Abstract. Entity Matching is used to identify records representing the same entities in the real world. As e-commerce is developing rapidly, online products grow explosively in both amount and variety. Applying entity matching to e-commerce data and finding records representing the same products make customers convenient to compare prices. This paper proposes an entity matching system for e-commerce data, called EPEMS. Compared with existing systems, we improve an existing sorted neighborhood blocking method, which is used to reduce the number of comparisons. At the same time the similarity of product pictures is used to improve matching results.

Keywords: Entity Matching, E-commerce Data, Blocking, Picture Similarity.

1 Introduction

Nowadays with the rapid development of e-commerce, both the amount and the variety of online products are increasing constantly. Same products are offered by different websites and different merchants with disorganized descriptions. Applying Entity Matching (EM for short) to e-commerce data, means finding records representing the same products from multiple sources, makes it convenient for customers to compare prices from different sellers.

So far, plenty of research work has been done on EM. To avoid expensive comparisons of all records, some blocking methods are proposed [1], which partition candidate records into smaller subsets and reduce the number of required comparisons actually. Classifiers are trained to identify record pairs as match or non-match. However, we notice that in the field of e-commerce, rare work has been done. Due to inconsistent description, loss of attribute values, cacography and some other reasons, EM for e-commerce data can hardly achieve good results. In fact, a recent study shows, current solutions can achieve only about 30% to 70% F-measure for product matching.

In this paper we present an EM system called EPEMS especially for e-commerce data. Compared to existing systems, EPEMS has following advantages: (1) a more appropriate sorting-based blocking method is developed for

© Springer International Publishing Switzerland 2015
R. Cheng et al. (Eds.): APWeb 2015, LNCS 9313, pp. 871–874, 2015.
DOI: 10.1007/978-3-319-25255-1_74

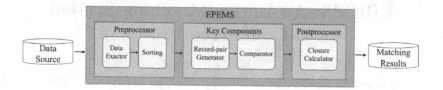

Fig. 1. The Architecture of EPEMS

e-commerce data to minimize the number of comparisons, rather than using a traditional SNM [2] (Sorted Neighborhood Method); (2) the similarity of product pictures is used to improve matching results.

2   System Overview

The architecture of EPEMS is shown in Fig. 1. The most important components of our system include the record-pair generator and the comparator. The record-pair generator uses a blocking method to reduce the number of comparisons. It generates record pairs in which records are likely to match each other. The comparator is used to compare two records in a pair and decides if they match or not.

We first extract data from a data source and convert the extracted data into the internal format. Then, a sorting key is defined and all records are sorted by this key. The key is not necessarily unique and can be generated by concatenating attribute values (or substrings of values) in the hope that matching records are assigned similar sorting keys and are thus close. The functionalities of the two main components record-pair generator and comparator will be introduced in detail in following paragraphs. At the end we use Warshall's algorithm [3] to calculate the transitive closure for all identified matches.

2.1   Record-Pair Generator

The record-pair generator is used to select pairs of records that should be compared. Existing sorting-based blocking methods always slide a window over the sorted data. Only records in the same window will be compared with each other. Among these strategies that adapt the window size, DCS++ [4] (Multiple Duplicate Count Strategy) has the best performance.

DCS++ only compares the first record in a window with the successors, and window size is varied according to the number of identified matches. Let w be the initial window size, t_i be the i-th record in the sorted record list, $W(i, i + w - 1)$ be the initial window which starts by t_i. Whenever a match is identified, the $w - 1$ adjacent records will be added to the current window. Assuming that the pairs (t_i, t_j) and (t_j, t_k) are matches with $i < j < k$, an additional matching pair (t_i, t_k) can be returned. Hence, $W(j, j + w - 1)$ can be skipped because matches in this window can be returned by calculating the transitive closure.

As EM for e-commerce data always has a poor performance, if a non-match pair is identified as match, a window will be skipped by error. Matches in this window will be missed and calculating transitive closure for the wrong match will lead to more wrong matches. Therefore, we propose a novel strategy based on DCS++, which performs a different approach to extend windows for equivocal matches.

In $W(i, i + w - 1)$, t_i is first compared with the $w - 1$ successors. If t_j in the current window is identified as an exact matching record of t_i, then what we should do is the same as DCS++. If t_j is a likely match of t_i but we cannot make sure, the next $w - 1$ adjacent records will be added into the current window without skipping $W(j, j + w - 1)$. Thus, we don't need to calculate the transitive closure for it.

2.2 Comparator

Most traditional comparators are text-based. They calculate the similarities between attribute values of different records, using metrics such as Levenshtein distance, SoundEx, etc. A classification algorithm is used to train a model based on those similarities. This model will be used to identify record pairs.

E-commerce data always contains pictures. The same products usually have the same or similar displayed pictures. Therefore, we use the similarity of pictures to improve matching results. When the text-based comparator can't achieve an assured result, image similarity will be utilized.

We extract features from all pictures using a scale-invariant feature transform algorithm, and then use Hierarchical Clustering to partition these extracted features into clusters and build a vocabulary tree [5]. The leaf nodes of the resulting tree are defined as words, and the pictures are regarded as documents. That is, pictures can be described by words, we can calculate the similarity between two pictures according to the number of the same words they contain. We adjust the weights of textual similarity and image similarity to decide whether records match or not.

3   Demonstration

The system is developed in Java. Its deployment environment is JDK1.7 with the database system MySQL.

We have built a simple demonstration system[1]. The data set for the demonstration contains some book records collected from several online bookstores, such as ZOL, Boku, JD, etc. Both inconsistent description and loss of attribute values exist. And displayed pictures of those books are also available. The snapshot of the demonstration system is shown in Fig. 2.

The system can be demonstrated with the following steps:

---

[1] This simple demonstration system of our EPEMS is available online at: http://ada.suda.edu.cn:8080/EPEMS.

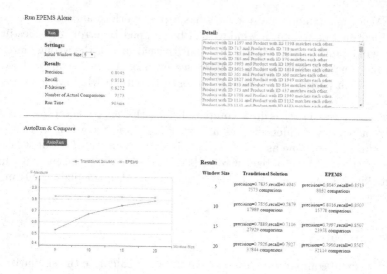

Fig. 2. A Snapshot of the Demonstration System

Step 1. A user can run EPEMS alone, some parameters have been set and the user can choose an initial window size. The system need some time to operate, then the result including precision, recall and matching details will be displayed on the screen.

Step 2. A comparison between EPEMS and a traditional solution for EM is also available. The result in Fig. 2 shows the advantage of EPEMS over the existing solution.

Acknowledgment. This work was partially supported by Chinese NSFC project (61170020, 61402311, 61440053, 61402313, 61472263) and the US National Science Foundation (IIS-1115417).

References

1. Christen, P.: A survey of indexing techniques for scalable record linkage and deduplication. IEEE Transactions on Knowledge and Data Engineering 24(9), 1537–1555 (2012)
2. Hernández, M.A., Stolfo, S.J.: The merge/purge problem for large databases. ACM SIGMOD Record 24, 127–138 (1995)
3. Warshall, S.: A theorem on boolean matrices. Journal of the ACM (JACM) 9(1), 11–12 (1962)
4. Draisbach, U., Naumann, F., Szott, S., Wonneberg, O.: Adaptive windows for duplicate detection. In: 2012 IEEE 28th International Conference on Data Engineering (ICDE), pp. 1073–1083. IEEE (2012)
5. Nister, D., Stewenius, H.: Scalable recognition with a vocabulary tree. In: 2006 IEEE Computer Society Conference on Computer Vision and Pattern Recognition, vol. 2, pp. 2161–2168. IEEE (2006)

PPS-POI-Rec: A Privacy Preserving Social Point-of-Interest Recommender System

Xiao Liu[1,2], An Liu[1,2], Guanfeng Liu[1,2], Zhixu Li[1,2], Jiajie Xu[1,2], Pengpeng Zhao[1,2], and Lei Zhao[1,2]

[1] School of Computer Science and Technology, Soochow University, China
[2] Collaborative Innovation Center of Novel Software Technology and Industrialization, Jiangsu, China
20145227012@stu.suda.edu.cn,
{anliu,gfliu,zhixuli,xujj,ppzhao,zhaol}@suda.edu.cn

Abstract. Point-of-Interest (POI) recommendation is an important task for location based service (LBS) providers. Social POI recommendation outperforms traditional, non-social approaches as social relations among users (a.k.a. social graph) could be used as a data source to calculate user similarities, which is generally hard to evaluate due to the lack of sufficient user check-in data. However, the social graph is typically owned by a social networking service (SNS) provider such as Facebook and should be hidden from LBS provider for obvious reasons of commercial benefits, as well as due to privacy legislation. In this paper, we present PPS-POI-Rec, a novel privacy preserving social POI recommender system that enables SNS provider and LBS provider to cooperatively recommend a set of POIs to a target user while keeping their private data secret. We will demonstrate step by step how a social POI recommendation can be jointly made by SNS provider and LBS provider, without revealing their private data to each other.

Keywords: POI recommendation, social recommendation, privacy preserving recommendation.

1 Introduction

Point-of-Interest (POI) recommendation is an important task for location based service (LBS) providers. Generally, an LBS provider can employ collaborative filtering (CF) to recommend a set of POIs to a target user, given that this user has shared a sufficient number of check-in experiences with the LBS provider. However, this requirement typically cannot be met in practice, which results in the notorious *cold start* problem that is common in CF based recommender systems. To tackle this problem, a wide variety of social recommendation approaches have appeared in the literature. The motive of social recommendation is the explicit social relations among users could be used as a data source to calculate user similarities. Recent studies have shown that social recommendation outperforms non-social approaches in many application domains [3].

© Springer International Publishing Switzerland 2015
R. Cheng et al. (Eds.): APWeb 2015, LNCS 9313, pp. 875–878, 2015.
DOI: 10.1007/978-3-319-25255-1_75

In practice, social relations among users (a.k.a. social graph G) is typically owned by a social networking service (SNS) provider such as Facebook or Twitter. As pointed in [2], this proprietary social graph is an important asset with inestimable value, thus should be kept secret for obvious reasons of commercial benefits, as well as due to privacy legislation. For the same reason, an LBS provider must keep user check-in data (e.g., the user-POI matrix M) secret. Therefore, we face a challenging scenario where two parties (SNS provider and LBS provider) wish to jointly conduct POI recommendation, without revealing their private data (i.e., G and M) to each other.

In this paper, we present PPS-POI-Rec, a novel privacy preserving social POI recommender system that enables SNS provider and LBS provider to cooperatively recommend a set of POIs to a target user while keeping their private data secret. Clearly, the core challenge in implementing PPS-POI-Rec is to achieve privacy preserving. Towards this goal, SNS provider adopts Paillier cryptosystem [4] to encrypt user social similarity derived from her private data M, and LBS provider computes the scores of recommending POIs to users on the encrypted social similarities. To find top-K POIs with encrypted scores, SNS provider and LBS provider follow a secure top-k selection protocol building on Yao's protocol [5] (a.k.a. Yao's Garbled Circuits, or YGC for short). In this demonstration, we will show step by step SNS provider and LBS provider are able to recommend a set of POIs to a target user, without revealing their private data to each other.

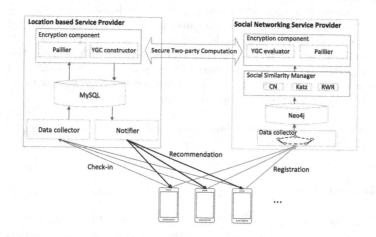

Fig. 1. System overview

2   System Overview

As shown in Fig. 1, our recommender system PPS-POI-Rec has two layers: mobile client layer and server layer. In mobile client layer, users are offered with several LBSs, including check-in sharing, trajectory recording in different temporal granularities, such as in day or in week, and POI recommendation.

The server layer consists of two parties LBS provider and SNS provider. Each party has several components and their functionalities are described as follows. *Data collector:* Both LBS provider and SNS provider have a data collector. This component is used to collect data generated by users. Specifically, LBS provider collects the check-in data submitted by users via their mobile devices, and store them in a MySQL database. SNS provider collects the social relations among users and store them in a Neo4j database.

Encryption Component: Both LBS provider and SNS provider has an encryption component, which is in charge of keeping their private data secret during POI recommendation. Paillier cryptosystem are used for data encryption. SNS provider generates a public key and a secret key, and sends the public key to LBS provider for later computation. As a YGC constructor, LBS provider prepares a Yao's garbled circuit and sends it to SNS provider who is a YGC evaluator. This garbled circuit is the key component in secure top-k selection, which is a kind of secure two-party computation.

Notifier and Social Similarity Manager: The notifier of LBS provider is responsible for pushing the recommendation result, that is, top-k POIs, to users. The social similarity manager of SNS provider calculates user similarities based on some pre-defined social similarity measures such as common neighbors (CN), Katz, and random walk with restart (RWR), just to name a few.

System Implementation: The mobile client layer is implemented as an android application based on Android SDK 4.3 and Baidu Map API. The server layer is implemented based on Hibernate and is deployed on Tomcat 7. For Paillier cryptosystem, we use a public available Java implementation[1]. Besides, we implement Yao's garbled circuits based on FasterGC [1].

Fig. 2. User check-in

Fig. 3. Recommended POIs

---

[1] http://www.csee.umbc.edu/~kunliu1/research/Paillier.html

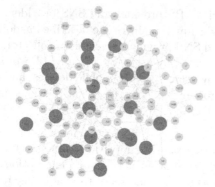

Fig. 4. Social graph stored in Neo4j   **Fig. 5.** Check-in stored in MySQL

3   Demonstration

During the demonstration, the audience could act as users and download our app in http://ada.suda.edu.cn:8080/adalab_new/apk.html. After installation, they can log in the app by creating a new account or using an existing account (e.g., username 'ada' and password '1111'). As depicted in Fig. 2, they can interact with our recommender system by checking in places of their interest (Fig. 2). Then, LBS provider and SNS provider jointly compute the recommendation result in a privacy preserving manner. Finally, LBS provider recommends top-k ($k = 5$ in the demonstration) POIs to users, as shown in Fig. 3. For demonstration purpose, SNS provider has stored a social graph containing 1,892 users in an Neo4j database (see Fig. 4). Besides, LBS provider has stored a POI dataset of Suzhou containing 193,791 POIs, and a real check-in dataset in a MySQL database (see Fig. 5).

Acknowledgments. This work was partially supported by Natural Science Foundation of China (Grant Nos. 61232006, 61303019, 61402313) and Collaborative Innovation Center of Novel Software Technology and Industrialization, Jiangsu, China.

References

[1] Huang, Y., Evans, D., Katz, J., Malka, L.: Faster secure two-party computation using garbled circuits. USENIX Security (2011)

[2] Jorgensen, Z., Yu, T.: A Privacy-Preserving Framework for Personalized, Social Recommendations. In: EDBT 2014, pp. 571–582 (2014)

[3] Konstas, I., Stathopoulos, V., Jose, J.M.: On social networks and collaborative recommendation. In: SIGIR 2009, pp. 195–202 (2009)

[4] Paillier, P.: Public-key cryptosystems based on composite degree residuosity classes. In: Stern, J. (ed.) EUROCRYPT 1999. LNCS, vol. 1592, pp. 223–238. Springer, Heidelberg (1999)

[5] Yao, A.C.-C.: How to generate and exchange secrets. In: FOCS 1986, pp. 162–167 (1986)

Incorporating Contextual Information into a Mobile Advertisement Recommender System

Ke Zhu[1], Yingyuan Xiao[1,2,*], Pengqiang Ai[1], Hongya Wang[3], and Ching-Hsien Hsu[4]

[1] Tianjin University of Technology, 300384, Tianjin, China
[2] Tianjin Key Lab of Intelligence Computing and Novel Software Tech., 300384, China
[3] Donghua University, 201620, Shanghai, China
[4] Chung Hua University, 30012 Taiwan
yyxiao@tjut.edu.cn

Abstract. The ever growing popularity of smart mobile devices and rapid advent of wireless technology have given rise to a new class of advertising system, i.e., mobile advertisement recommender system. The traditional internet advertising systems have largely ignored the fact that users interact with the system within a particular "context". In this paper, we implemented a mobile advertisement recommender prototype system called MARS. MARS captures different user's contextual information to improve recommendation results. The demonstration shows that MARS makes advertisement recommendation more effectively.

Keywords: mobile advertisement recommendation, context, probabilistic model.

1 Introduction

With the proliferation of smart mobile devices and wireless communication, mobile advertising has become a technology that allows an advertiser to promote products or services to targeted users efficiently and effectively [1, 2, 3]. The traditional internet advertising systems have largely ignored the fact that users interact with the system within a particular context [4, 5]. The concept of "context" has been studied extensively in several areas of computing. Currently, some researchers consider incorporating context into mobile advertisement recommender systems [1, 2]. However, they only take single contextual factor into account, such as location. In this paper, we build a mobile context-aware advertisement recommender prototype system called MARS, in which multiple contextual factors, such as time, location and speed, are incorporated by means of Bayesian probabilistic model.

MARS adopts a client/server model and is implemented in Java. In the mobile client end (including smart phone, PDA, laptop, etc.), MARS captures each user's current contextual information (i.e., time, location, speed, etc.) in real-time and sends them to the server through a GSM/3G/Wi-Fi service provider. In the server, MARS

---

* Corresponding author.

R. Cheng et al. (Eds.): APWeb 2015, LNCS 9313, pp. 879–882, 2015.
DOI: 10.1007/978-3-319-25255-1_76

builds the Bayesian probabilistic model based on user's profile and current contextual information to predict the recommendation list for each user.

2   System Overview

MARS is expected to provide mobile context-aware advertisement recommendation for users equipped with mobile devices. To achieve this goal, the real-time collection of user's current contextual information and the building of probabilistic model based on user's profile and current contextual information are key problems. In this section, we first describe the system framework of MARS, and then introduce how to build the Bayesian probabilistic model based on user's profile and current contextual information.

2.1   System Framework

MARS adopts a client/server model. Figure 1 illustrates the system framework of MARS. The mobile client is responsible for capturing user's current contextual information (i.e., time, location, speed, etc.) in real-time and showing the recommended advertisement list from the server. The user's characteristics (i.e., user id, age, gender, etc.) are written into profile database when a user registers into MARS. The server builds the Bayesian probabilistic model based on user's profile and current contextual information to predict the recommendation list and push it to the user.

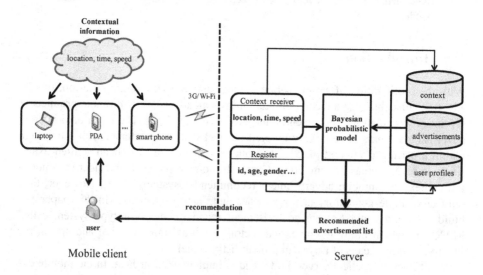

Fig. 1. System framework

2.2   Bayesian Model

Next, we use $AD = \{ad_1, ad_2, ..., ad_i\}$ to denote the set of advertisements provided by advertisers where ad_i is a triple: $<id, content, owner>$; we use $U = \{u_1, u_2, ..., u_k\}$ to denote the set of all users and $C = \{c_1, c_2, ..., c_k\}$ to denote the set of current contextual information of each user, where c_k represents the current contextual information of u_k and is expressed as a triple: $<location, time, speed>$.

Assume that the probability of pushing ad_i to u_k under c_k is $p(ad_i \mid u_k, c_k)$. According to Bayesian method, $p(ad_i \mid u_k, c_k)$ can be calculated by Equation 1.

$$p(ad_i \mid u_k, c_k) = \frac{p(u_k, c_k \mid ad_i)p(ad_i)}{p(u_k, c_k)} = \frac{p(u_k, c_k \mid ad_i)p(ad_i)}{\sum_i p(u_k, c_k \mid ad_i)p(ad_i)} \tag{1}$$

Because u_k and c_k are independent, $p(u_k, c_k \mid ad_i)$ can be calculated by Equation 2.

$$p(u_k, c_k \mid ad_i) = \prod_k p(u_k \mid ad_i) \prod_k p(c_k \mid ad_i) \tag{2}$$

Further, we can get Equation 3 having Equation 1 and Equation 2:

$$p(ad_i \mid u_k, c_k) = \frac{\prod_k p(u_k \mid ad_i) \prod_k p(c_k \mid ad_i)p(ad_i)}{\sum_i \prod_k p(u_k \mid ad_i) \prod_k p(c_k \mid ad_i)p(ad_i)} \tag{3}$$

Training the above probabilistic model with the experimental data set, MARS can calculate the probability of pushing ad_i to u_k under c_k for each user u_k. For a given user u_k within a particular context c_k, these advertisements with the higher $p(ad_i \mid u_k, c_k)$ are recommended to him/her.

3   Demonstration

In this section, we demonstrate our MARS by registering a user. After registering as a user in MARS, the user's profile is recorded into profile database. Then, we design two test scenarios: the first context is located in a commercial street at 11:20 am on Saturday and the other context is in a recreation center at 19:45 pm on Saturday.

The recommendation results of two test scenarios are shown in Figure 2 and Fingure3, respectively. In the first context, the nearby restaurants are recommended while in the second context the nearby cinemas are recommended.

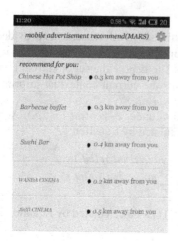
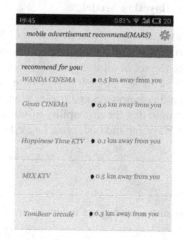

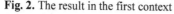

Fig. 2. The result in the first context   **Fig. 3.** The result in the second context

4   Conclusion

In this paper, we designed and implemented a mobile context-aware advertisement recommender prototype system called MARS. In the mobile client, MARS collects each user's current contextual information in real-time and sends them to the server through a GSM/3G/Wi-Fi channel. In the server, MARS builds the Bayesian probabilistic model based on user's profile and current contextual information to predict the recommendation list for each user. The demonstration shows that MARS makes advertisement recommendation more effectively.

Acknowledgment. This work is supported by the NSF of China (No. 61170174, 61370205) and Tianjin Training plan of University Innovation Team (No.TD12-5016).

References

1. Lia, K., Dub, T.C.: Building a targeted mobile advertising system for location-based services. Decision Support Systems 54(1), 1–8 (2012)
2. Dao, T.H., Jeong, S.R., Ahn, H.: A novel recommendation model of location-based advertising: Context-Aware Collaborative Filtering using GA approach. Expert Systems With Applications 39(3), 3731–3739 (2012)
3. Roh, J.-H., Jin, S.: Personalized Advertisement Recommendation System based on user profile in the Smart Phone. In: ICACT, pp. 1300–1303 (2012)
4. Xu, H., Yang, D., Gao, B., Liu, T.-Y.: Predicting advertiser bidding behaviors in sponsored search by rationality modeling. In: WWW, pp. 242–254 (2013)
5. Kim, S., Qin, T., Yu, H., Liu, T.-Y.: An Advertiser-Centric Approach to Understand User Click Behavior in Sponsored Search. In: CIKM, pp. 2121–2124 (2011)

Author Index

Ahmed, Chowdhury Farhan 389
Ai, Pengqiang 41, 879

Bai, Shuo 215
Bi, Xin 203
Bian, Haoqiong 841

Cai, Ruichu 805
Cai, Taotao 116
Cai, Xiangrui 352
Cai, Yadi 116
Cao, Yuan 304
Chang, Baobao 80
Chao, Pingfu 509
Chen, Daoxu 485
Chen, Haiming 104
Chen, Heng 340
Chen, Jie 449
Chen, Kaimeng 29
Chen, Shuang 203
Chen, Yong 832
Chen, Yueguo 841
Chen, Zhiyong 598
Chen, Zhong 129
Chen, Zhumin 178, 316, 365, 411
Cheng, Gong 695
Cheng, Jiayi 863
Cheng, Lianglun 793
Cheng, Xueqi 190
Cheng, Yingchao 805
Cheng, Yurong 227
Cheung, Simon K.S. 821
Chunxia, Zhang 401
Congli, Liu 92
Cui, Lizhen 437, 683
Cui, Wenxiang 41
Cui, Zhiming 871

Dan, Meng 401
Dequan, Wang 92
de-Rong, Shen 560
Ding, Jianwei 304
Ding, Xiaoou 855
Dong, Qiuxiang 129
Du, Bo 744

Du, Jianfeng 863
Du, Xiaoyong 841

Fang, Junhua 509
Fariha, Anna 389
Feng, Chong 141
Feng, Yansong 256
Fournier-Viger, Philippe 635
Fu, Hongping 449

Gan, Wensheng 635
Gao, Hong 424, 855
Gao, Lei 871
Gao, Ming 623
Gao, Zhu 509
Ge, Tao 80
Ge, Yu 560
Gong, Ti 497
Gu, Qing 485
Guan, Zhi 129
Guo, Li 215, 732
Guo, Shanqing 437
Guo, Weiyu 244
Guo, Xi 280
Guo, Yunchuan 610

Hao, Zhifeng 805
Hasan, Mehedi 389
He, Wei 683
He, Wenqiang 256
He, Xiaofeng 623
Hong, Tzung-Pei 635
Hong, Xiaoguang 598
Hsu, Ching-Hsien 41, 879
Hu, Wenbin 744
Hu, Yue 215
Huang, Heyan 17, 141
Huang, Tao 473

Ishikawa, Yoshiharu 280
Iwaihara, Mizuho 154

Jia, Yantao 190
Jiang, Bo 756
Jiang, Jun 859

Jiang, Zhiwei   485
Jin, Hai   340
Jin, Peiquan   29

Lee, Victor E.   497
Leung, Carson K.   768
Li, Hui   683
Li, Jianzhong   424, 855
Li, Jinyang   623
Li, Lei   867
Li, Licheng   268
Li, Qi   720
Li, Rong-Hua   116
Li, Yuming   509
Li, Zhixu   781, 859, 867, 871, 875
Liang, Jiacheng   863
Liang, Jiguang   215
Liang, Wenxin   548
Lin, Hailun   190
Lin, Jerry Chun-Wei   635
Liu, An   781, 859, 867, 871, 875
Liu, Dacheng   768
Liu, Guanfeng   781, 859, 867, 875
Liu, Jie   473
Liu, Ling   166
Liu, Peng   411
Liu, Qin   328
Liu, Qingxia   695
Liu, Shushu   781
Liu, Tingwen   732
Liu, Wei   720
Liu, Xiao   875
Liu, Ye   720
Liu, Yingbo   304
Liu, Ziyan   598
Lu, Jian   268
Lu, Wen-Yang   536
Lv, Xueqiang   658

Ma, Jun   178, 316, 365, 411
Ma, Linlin   867
Mandal, Amit   389
Mao, Rui   116
Mao, Xian-Ling   17
Mao, Xiao-Jiao   536
Miao, Lin   658

Ning, Xie   560
Niu, Zhendong   449
Nuo, Li   560

Pang, Jun   55
Pei, Wenzhe   80
Peng, Feifei   104
Peng, Peng   3
Peng, Yuwei   522
Peng, Zhaohui   598
Peng, Zhiyong   522

Qian, Junyan   610
Qin, Xiongpai   841
Qingzhong, Li   92
Qu, Guangzhi   497
Qu, Yuzhong   695

Ren, Guozhen   683
Ren, Pengjie   178, 365, 411

Sha, Ying   756
Shen, Chengguang   548
Shen, Derong   166
Sheng, Victor S.   871
Shi, Jinqiao   732
Shuo, Feng   560
Song, Tianhang   598
Song, Xiaomeng   411
Su, Lili   67
Su, Qingchao   863
Sui, Xueqin   316
Sui, Zhifang   80
Sun, Gang   485
Sun, Guohao   867
Sun, Yanwei   610

Tan, Tieniu   244
Tao, Dacheng   744
Tong, Wenzhu   768

Vasilakos, Athanasios   610

Wang, Chao   352
Wang, Chengyu   623
Wang, Guoren   203, 227
Wang, Hongya   41, 879
Wang, Hongzhi   377, 855
Wang, Jianmin   304
Wang, Jianzong   793
Wang, Jinbao   424
Wang, Kaile   328
Wang, Liang   244, 292
Wang, Lihong   756
Wang, Lijuan   805
Wang, Mengmeng   154

Wang, Weibo   598
Wang, Xiaolin   437
Wang, Yashen   141
Wang, Yuanzhuo   190
Wang, Yue   658
Wang, Zengmao   744
Wang, Zhanghui   227
Wang, Zhisheng   720
Wei, Jinmao   461
Wei, Jun   473
Wei, Yang   461
Wei, Yuhong   424
Weiping, Wang   401
Wen, Wen   805
Wen, Yanlong   352
Wu, Jian   871
Wu, Jiayin   832
Wu, Kai   178, 365
Wu, Shu   244, 292
Wu, Wenzhen   647
Wu, Xia   522
Wu, Yunfang   658
Wulamu, Aziguli   280

Xiao, Yingyuan   41, 879
Xie, Xuanzhong   832
Xie, Yonghong   280
Xiong, Jinhua   190
Xiong, Shengchao   522
Xu, Feng   268
Xu, Jiajie   781, 875
Xu, Lijie   473
Xu, Xin-Shun   437

Yan, Ming   377
Yan, Yang   732
Yan, Yang-Hui   573, 670
Yang, Donghua   424
Yang, Qiang   859
Yang, Xianxiang   141
Yang, Yao   437
Yang, Yu-Bin   536
Yang, Zhenglu   461
Yang, Zhifan   67
Yanhui, Ding   92
Yao, Yuan   268
Ye, Dan   473
Yin, Jian   720
Yin, Lihua   610
Yin, Qiyue   292
Yongxin, Zhang   92

Yu, Ge   166
Yu, Jeffrey Xu   116
Yu, Jialiang   768
Yu, Lixia   485
Yu, Man   352
Yuan, Pingpeng   340
Yuan, Xiaojie   67, 352
Yuan, Ye   424
Yue, Kou   560
Yue, Lihua   29

Zhang, Dan   855
Zhang, Feng   497
Zhang, Haiyang   707
Zhang, Lefei   744
Zhang, Li   304
Zhang, Liangpei   744
Zhang, Peng   190
Zhang, Rong   509, 623
Zhang, Xianchao   548
Zhang, Yan   377
Zhang, Yang   55, 732
Zhang, Yi-Chi   586
Zhang, Ying   67, 352
Zhang, Zhen   203
Zhao, Dongyan   3, 256
Zhao, Gansen   497
Zhao, Jun   863
Zhao, Lei   781, 859, 867, 875
Zhao, Pengpeng   781, 871, 875
Zhao, Xiangguo   203
Zhao, Xue   67
Zhao, Yuhai   227
Zheng, Hai-Tao   573, 586, 647, 670, 707
Zheng, Yiwen   328
Zhenyan, Liu   401
Zhongmin, Yan   92
Zhou, Aoying   509
Zhou, Jian   859
Zhou, Qiang   17
Zhou, Xiaofei   215
Zhou, Ying-Min   573, 670
Zhu, Feng   473
Zhu, Hang   340
Zhu, Jia   859
Zhu, Ke   879
Zhu, Lei   340
Zhu, Mingdong   166
Zhu, Qi-Hai   536
Zhu, Yuanyuan   522
Zou, Lei   3, 256

Printed in the United States
By Bookmasters